OPERATIONAL PHYSICS

OPERATIONAL PHYSICS

Modern physics increasingly replaces the object-based world-view of the old physics with an operation-based world-view. This process is supported by a similar one that is taking place in mathematics at the same time. *The International Journal of Theoretical Physics* dedicates the series "Operational Physics" to this process. Its goal is to provide scientists and researchers with textbooks, monographs, and reference works to address the growing need for information.

Prospective authors are encouraged to correspond with the Editor-in-Chief in advance of submitting a manuscript. Submission of manuscripts should be made to the Editor-in-Chief.

Editor-in-Chief

David Finkelstein
Georgia Institute of Technology
School of Physics
837 State Street
Atlanta, GA 30332-0430

Volumes Published in This Series:

Operational Quantum Theory I—Nonrelativistic Structures
by Saller, H. 2006

Operational Quantum Theory II—Relativistic Structures
by Saller, H. 2006

Heinrich Saller

Operational Quantum Theory I

Nonrelativistic Structures

Springer

Heinrich Saller
Werner-Heisenberg Institut
Max-Planck Institut für Physik
München, Germany 80805
hms@mpp.mu.mpg.de

Library of Congress Control Number: 2006920923

ISBN-10: 0-387-29199-7
ISBN-13: 978-0387-29199-4

Printed on acid-free paper.

Printed in the United States of America. (TB/MVY)

9 8 7 6 5 4 3 2 1

springer.com

I would like to thank three people from three generations without whom this book could not have been written.

First, the late Werner Heisenberg, who implanted in me the conviction that symmetries with their operations are appropriate basic concepts for understanding physical interactions and objects.

Second, David Finkelstein, who gave me the feeling, in our fruitful collaboration and work over the decades, of not being alone in giving priority to the operational approach.

Finally, I learned a lot from my first son, Christian, who is a much better mathematician than I. He taught me that many of the mathematical concepts denigrated as esoteric and academic by those physicists who have a direct pipeline to God are basic and exactly the right tools for the physical structures to be formalized. He also helped me very much by knowing and recommending the appropriate advanced mathematical literature.

Contents

INTRODUCTION

Quantum theory is connected especially with the names Planck, Bohr, Heisenberg, Pauli, and Dirac. The quantum revolution describes our deepest insight, so far, into the physical structure of nature. It is comparable only with the Copernican revolution, switching from a finally oriented anthropocentric description of physical phenomena to one using general laws with initial or boundary conditions, connected with the names Kepler, Galileo, and Newton, or with the change from tangible mass points as basic structures to Faraday's and Maxwell's field concepts and, shortly before quantum theory, with the relativization of space and time by the lonely genius Einstein.

In retrospect, the label "quantum" or, as adjective, "quantal," is too weak to characterize the extent of the revolution involved in abandoning the classical theory as a basic epistemological framework for physics. The word "quantal" – in contrast to the assumed classical "continuous" ("natura non facit saltus") – was motivated by the finite jumps and the discreteness as seen, for example, in the photoelectric effect or in the spectral lines for atoms or in the discrete split of atomic rays in Stern-Gerlach experiments.

One has to distinguish in quantum theory between two kinds of "jumps": First, the quantum structure relies on the noncommutativity of operations, e.g., of the not commuting position-momentum operator pair $[i\mathbf{p}, \mathbf{x}] = \hbar$, with a nontrivial quantum $\hbar$ (Planck's constant) or of the not anticommuting conjugate operator pair of an electron-positron field $\{\overline{\mathbf{\Psi}}(\vec{y}), \mathbf{\Psi}(\vec{x})\} = \hbar\gamma^0\delta(\vec{x} - \vec{y})$. Second, there are the jumps, characterized by integers. These jumps, as seen in the atomic spectral lines, were the starting point of quantum theory. However, after the dust has settled, they cannot be addressed as the revolutionary characteristics of quantum theory: Integers characterize compact operation groups. Take a circle, say a closed rubber string, cut it, wind it around your wrist, and glue both ends together again; the number of possible windings is always an integer. Does rubber band winding characterize quantum theory? The rubber band stands for the circle, parametrizing the compact Lie group $\mathbf{U}(1) = \exp i\mathbb{R}$ or the isomorphic group $\mathbf{SO}(2)$ with the rotations around one space axis. The irreducible representations of the circle (1-dimensional torus), as realized by the different rubber band windings and thus of all compact Lie groups involving higher-dimensional tori, come with integer winding numbers, "quantum numbers" in the narrow sense. Since bound waves in quantum mechanics are related to compact representations of the noncompact time translation group $\mathbb{R}$, they give rise to integer-related discrete (rational) quantum jumps. The same situation occurs for spin, which is related to the 3-dimensional position rotations, parametrizable by the compact volume of a sphere. However, in addition to these discrete jumps (integer winding numbers $z \in \mathbb{Z}$) continuous quantum numbers can also occur, e.g., real energies $E \in \mathbb{R}$ or momenta $\vec{q} \in \mathbb{R}^3$, or, apparently, the particle masses $m^2 \in \mathbb{R}_+$ from a continuous spectrum as eigenvalues or invariants for representations of time and space translations. Continuous numbers require operations with noncompact action groups, whereas compact groups come with rational ("quantum") numbers.

At the core of quantum theory is the relativization of the ontic structures in contrast to the absolute ontology in classical theories, e.g., of the position of mass points or of the spin direction of particles. The appropriate characterization "quantum relativity" alludes to the relativity of time and space. A quantum description starts from practic stuctures, e.g., from translations or rotations. Quantum theory describes operations with the dynamics itself an operation. Quantum theory is operation theory. A classical ontology requires a projection of the nonabelian operational framework to an abelian substructure. In a classical description, objects are primary with interactions between them as a secondary structure. In a quantum description the hierarchy is reversed: objects arise as eigenvectors of operations.

Appropriate questions in quantum theory ask for operations: What is the operational meaning of spin and mass of a particle? Invariants for rotations and spacetime translations. What is the operational meaning of a Coulomb and Yukawa potential? Representation distributions, 2-sphere spreads of position translations. What is the operational meaning of a gauge coupling constant? The relative normalization of the gauge–transformation–inducing operational Lie algebra in the Lorentz Lie algebra. What is the operational meaning of a Feynman propagator? Matrix elements of spacetime translation representations, unitary for on-shell contributions.

And one may ask even about quite specific structures: What is the operational meaning of cosines and exponentials, of Bessel and Macdonald functions, or of Laguerre polynomials, etc.? Representation coefficients of specific operations. With respect to a formulation of physics by special functions arising as solutions of "special differential equations," e.g., equations of motion in time and space, there is a unified view, initiated by Wigner and elaborated in exhaustive encyclopedic detail by Vilenkin, who writes in the introduction of his subject–related book, "a really unified view on the theory of the basic classes of special functions ... was established by employing the considerations that belong to a field of mathematics seemingly quite far from the subject under consideration, the theory of representations of Lie groups." Essentially all physically relevant special functions arise as coefficients of Lie group representations. Therefore in the following, Lie operations are of paramount importance.

Weyl was the first to connect with each other, basically and in a systematic form, "The theory of groups and quantum mechanics" in his like-named book. Wigner especially proceeded to extend the group–theoretic method in mathematical detail to relativistic quantum theory.

There was always a symmetry strain in physical theories: The Greeks started with the association of the five Platonic solids with the four basic elements: fire, water, earth, and air; the fifth polyhedron, the dodecahedron, with pentagonal sides, called quintessence, was taken as the all-encompassing cosmos. The idea was revived by Kepler in his Mysterium Cosmographicum to understand the six planets known in his time as regularly circling on the simultaneous in- and out-spheres between the five Platonic solids, nested one within the other. It is fascinating to realize how Kepler's fantastic ideas, completely wrong and without any reasonable contact with any physical dynamics, hit upon an apparently immensely important basic structure in nature: The five Platonic solids have as their sides regular triangles, squares, and pentagons. Exactly these two-dimensional symmetric Euclidean polygons characterize the symmetry operations related to simple Lie groups as classified by E. Cartan. The four main series of symmetry operations can be related, via the characterizing weight and root diagrams, to regular squares and triangles lumped together in higher and higher dimensions (details in the chapters "Simple Lie Operations" and "Rational Quantum Numbers"). All the semisimple symmetries we use in fundamental theories of particles and their interactions can be associated with those operational structures. Every particle physics student today knows the quark triangles as weight diagrams for the color operations $\mathbf{SU}(3)$. The squares as weight diagrams for orthogonal symmetries show up, for instance, in the electron occupation numbers (twice a square) of the atomic shells, $2 = 2 \times 1^2$, $8 = 2 \times 2^2$, $18 = 2 \times 3^2$, etc., originating in the nonrelativistic framework from the orthogonal group $\mathbf{SO}(4)$ desribing rotation and perihelion conservation.

The main mistake of Kepler (forgive me) was, with our knowledge today, to look for the symmetry of the objects, not for the symmetries of the dynamical law; he was no quantum theorist. The possibility in quantum field theories to have less symmetric state vectors or objects as a result of operations with a larger symmetry plays an important role in reconciling the asymetry of the world as we see it with basic symmetric operations.

The quantum concepts as a unifying picture for the basic physical laws, at least without any experimental contradiction thus far, are not "anschaulich." Particles have no positions in the naive classical sense. To call them basically "pointlike" does not make sense. All this makes our physical intuition very difficult. The classical physical concepts dissolve like Dali's clock in the desert. Let me quote from the last public talk of Heisenberg in Munich, 1975 (my translation):

"It is unavoidable that we use a language originating from classical philosophy. We ask, What does the proton consist of? Is the quantum of light elementary or composite? etc. However, all these questions are incorrectly posed since the words "divide" and "consist of" have lost almost all their meaning. Therefore it should be our task to adjust our language, our thinking, i.e., our scientific philosophy to this new situation that has been created by experiments. Unfortunately, that is very difficult. Therefore, there creep into particle physics, again and again, wrong questions and wrong conceptions....

We have to come to terms with the fact that experimental knowledge from very small and very large distances no longer provides us with an "anschauliches Bild," and we have to learn to live there without "Anschauung." In this case we realize that the antinomy of the infinitely small for the elementary particles is resolved in a very subtle way, in a way which neither Immanuel Kant nor the Greek philosophers could have imagined, the word "to divide" loses its sense.

If one wants to compare the insights of today's particle physics with any earlier philosophy, it could be only the philosophy of Plato, since the particles of today's physics are representations of symmetry groups – that is what quantum theory teaches us – and hence the particles resemble the regular Platonic polyhedra."

Physical properties are registered in experiments, i.e., they describe a relation with an observer. They are mathematically formulated as eigenvalues of operations, e.g., energy and momentum or the spin in the direction of a magnetic field. Different ontic (asymptotic) structures as projections of one practic structure (interaction) are determined by an experimental setup that distinguishes one of possibly many eigenvector bases for the operations under consideration. Behind different setups there are the characterizing invariants, e.g., the mass of a particle for the Lorentz transformation-dependent energy-momenta, as measured in different spacetime frames, or its spin as measured in one space direction, which is determined, e.g., in a Stern-Gerlach experiment by the spatial inhomogeneity of a magnetic field. An experimental setup is related, mathematically, to a diagonalization of a set of operations. Since a set of diagonalizable matrices is simultaneously diagonalizable if and only if its elements commute with each other, an ontic interpretation of a set of operations depends on the experimenter's decision, concretized in the chosen apparatus, to distinguish a subset of simultaneously diagonalizable matrices. In general, there exist many different inequivalent diagonalizable subsets. Mathematically, this is a relatively simple theorem; its physical interpretation and coordination with our daily life experience, relying on an absolute ontic description existing and remaining without an ongoing measurement, is difficult and counterintuitive. An operator is not exhaustively described by the property (eigenvalue) of one object (particle, bound state vector, eigenvector) and even more for a set with more than one operator.

A transition from operations to particle- or state-related experimental numbers has to do with a maximal diagonalization of linear transformations as introduced for the characterization of Lie groups by E. Cartan. In this sense, an experimental test of quantum operations can be maximal, but because of the basic noncommutativity, it is never complete.

A vector is not a collection (row or column) of some numbers; this is a representation of the vector in a chosen basis, physically implemented by a given experimental apparatus. Not only for a mathematician, perhaps even more for a physicist, the distinction and choice of a basis has to be justified and the imposed restrictions have to be discussed carefully.

The ontic interpretation of one operator, e.g., acting on a two-dimensional vector space and diagonalizable as the matrix $l^3 \cong \begin{pmatrix} 1 & 0 \\ 0 & -1 \end{pmatrix}$ displaying its orthonormal eigenvectors by the two columns with their property (eigenvalues) $\{\pm 1\}$, may prevent the ontic interpretation of a second operator via simultaneous eigenvalues, e.g., of $l^1 \cong \begin{pmatrix} 0 & 1 \\ 1 & 0 \end{pmatrix}$; one eigenvector basis may not be usable twice. However, that's not all. Since there exist nondiagonalizable operators, in the simplest case of an operator on a complex two-dimensional vector space with a basis representation $n \cong \begin{pmatrix} 0 & 1 \\ 0 & 0 \end{pmatrix}$ (triangular Jordan structures), quantum theory involves even operations without ontic particle interpretation at all, i.e., without eigenvector bases. Such nondiagonalizable operations really occur, e.g., in connection with the quantum gauge field describing the Coulomb interaction as one degree of freedom in the four-component electromagnetic field (potential) that has only two degrees of freedom with an ontic particle interpretation, the left and right circularly polarized photons.

In quantum theories a clear distinction has to be made between the full operational interaction language and the restricted projections to objects. Physical objects, e.g., bound state vectors or elementary particles, as seen in experiments are eigenvectors with respect to transformation groups. Particles are eigenvectors with respect to space and time translations, rotations, and electromagnetic transformations that are formalized with the real Lie groups $\mathbb{R}$, $\mathbf{SU}(2)$, and $\mathbf{U}(1)$ and give rise to the properties mass, spin, and electromagnetic charge number and, at least until now, nothing more. The bound waves of the nonrelativistic hydrogen atom are eigenvectors for the operation groups $\mathbb{R}$ and $\mathbf{SO}(4)$ with the time translations and the space rotations with perihelial transformations respectively, giving rise to the properties energy and the space rotations-related quantum numbers. Interactions are characterizable by groups that in general are larger than the asymptotic symmetry groups that determine the object's properties. Elementary interactions implement internal ("chargelike") transformation groups as used in the standard model, i.e., hypercharge $\mathbf{U}(1)$, isospin $\mathbf{SU}(2)$, and color $\mathbf{SU}(3)$, in addition to the external spacetime translations $\mathbb{R}^4$ and the orthochronous Lorentz group $\mathbf{SO}_0(1,3)$ or, more precisely, its twofold cover $\mathbf{SL}(\mathbb{C}^2)$. The projective transition from the operations characterizing the interactions to those for the objects involves a dramatic operation group reduction, e.g., in the standard model for electroweak and strong interactions

$$\text{for interactions } \Big[\underbrace{\mathbf{SL}(\mathbb{C}^2) \vec{\times} \mathbb{R}^4}_{\text{external: Poincaré}} \Big] \times \Big[\underbrace{\mathbf{U}(1) \circ \big(\mathbf{SU}(2) \times \mathbf{SU}(3)\big)}_{\text{internal: hyperisospin-color}} \Big]$$

$$\to [\mathbf{SU}(2) \times \mathbb{R}^4] \times \mathbf{U}(1) \text{ for massive particles}$$

The interaction operation groups, e.g., isospin $\mathbf{SU}(2)$ for the nuclear interactions, which vanish as symmetries for asymptotic objects, e.g., for proton and neutron with different masses, may leave their traces in multiplicities, e.g., in the two nucleons arising from an isospin doublet. Sometimes not only the symmetries may vanish, but even the related nontrivial multiplicities, as proposed for the color $\mathbf{SU}(3)$ interaction symmetry leaving asymptotically only $\mathbf{SU}(3)$-singlets (color confinement, not proved yet).

There was a development in geometry culminating in the "Erlanger Programm" (1872) of Felix Klein that can serve as an analogue for the operational point of view to characterize quantum physics. A geometry, according to Klein, can be characterized by a Lie group G acting on an analytic manifold M, in the irreducible case on the equivalence classes in the homogeneous space G/H with a subgroup $H \subseteq G$ as fixgroup ("little group") or on a vector space. An example is the spherical geometry with the rotation group $\mathbf{SO}(3)$ acting on the 2-sphere $\Omega^2 \cong \mathbf{SO}(3)/\mathbf{SO}(2)$ that parametrizes the axial rotation subgroups, or the Euclidean geometry $\mathbf{SO}(3) \vec{\times} \mathbb{R}^3$ with the rotation group acting on 3-space or the pseudo-Euclidean Poincaré geometry $\mathbf{SO}_0(1,3) \vec{\times} \mathbb{R}^4$ with the Lorentz group acting on spacetime where the Minkowski translations $\mathbb{R}^4$ can be looked on as the tangent space of the homogeneous space $\mathbf{GL}(\mathbb{C}^2)/\mathbf{U}(2)$, or the special and general linear affine geometries $\mathbf{SL}(\mathbb{R}^n) \vec{\times} \mathbb{R}^n$ and $\mathbf{GL}(\mathbb{R}^n) \vec{\times} \mathbb{R}^n$. In a Klein space $G \bullet M$ only concepts compatible with or even invariant under the operation group G make sense. For example, for general linear geometry, the invariant concepts "parallelity" and "dimension", in addition "volume" for the special linear geometry, in addition the concepts "causal order" and "length" for Poincaré geometry, and in addition "angle" and "distance" for orthogonal geometry. The decreasing group chain $G_1 \supset G_2 \supset \cdots$ is reflected in the increasing number of invariants for the space acted on: To characterize smaller subgroups one has to invoke more and more properties. In a physical interpretation of Klein's program the acting groups are the interaction governing groups like $\mathbf{SO}(4)$ for the periodic system of the atoms in nonrelativistic mechanics or the internal hypercharge-isospin-color group $\mathbf{U}(1) \circ [\mathbf{SU}(2) \times \mathbf{SU}(3)]$ for interactions in the standard model. The vector spaces with the interaction group representations, characterized by invariants, e.g., mass and spin or hyperisospin and electromagnetic charge number, contain, after symmetry reduction, the bound state vectors or the particles.

It is not the purpose of this book to teach quantum theory to the beginner; it is not an introduction, but intended for the graduate student with a good knowledge of, on the one hand, the conventional presentations of nonrelativistic quantum mechanics and canonical quantum field theory, and, on the other hand, some knowledge of groups and Lie algebras, their algebraic and topological structures and their representations. Parts of it have been used for lectures on "Algebraic Methods in Quantum Mechanics," on "Introduction to Quantum Field Theory," on the "Standard Model of Strong and Electroweak Interactions," and on "Time, Space, and Spacetime in Quantum Mechanics and Quantum Field Theory." My motivation and aim is to understand and to explain quantum physics as far as possible by operational structures: why we apply them, which structures are unavoidable, which ones are immanent already in the mathematical framework used, and which structures seem artifically complicated and should be looked at with some suspicion. I work with the prejudice that fundamental physical structures are simple, not trivial, to understand and to formulate and esthetically beautiful, in some sense definable not only by personal taste. Relevant questions, worked with, but not necessarily satisfactorily answered, are of the kind, What follows from the real Lie structure of the complex represented operations? For example, the Hilbert space formulation with probability amplitudes. Is there a connection between the causal order of time and spacetime and the probability interpretation of quantum experiments and the positivity of energy? What is the operational origin of the Yukawa and Coulomb interaction? Which transformations are represented by a Feynman propagator, by its "on-shell" and its "off-shell" contributions? Are the divergences of the canonical quantum field theories related to a misrepresentation of the operations involved? What causes the dichotomy between internal compact and external spacetime-related operations that are also noncompact? Where does the gauge structure come from?

And the deepest question is, What is the common conceptual basic root branching into the phenomenological concepts interaction, spacetime, and matter? Wigner's classification of particles as unitary representations of the Poincaré group can be taken as an indication that it is impossible to think about spacetime and matter separately. One step to further this program is to show that scattering states and interaction-bound states arise from operation group representations.

Mathematically elegant formulations in physics may leave us with an empty taste: Answers to all the questions above are physically satisfactory only if they lead also to experimentally testable numbers. Mathematics alone is not enough: The richness of mathematical forms, even esthetically appealing simple structures, seems to be inexhaustible. To paraphrase a word of Kant: Physical theories without experimental numbers are empty. The determination of one number, e.g., of a gauge coupling constant, may justify a huge theoretical building. However, also this is true: Numbers without a theoretical understanding are blind; think of numerologists. To take up the first sentence of this paragraph: Mathematically ugly formulations in physics leave us with a bad taste.

The mathematical level is not undergraduate; I have tried to use the best mathematical tools at my disposal. A. Knapp, one of the mathematical experts in the field of "Representation theory of semisimple Lie groups", writes in the preface of the like-named textbook (about 800 pages), "The subject of semisimple Lie groups is especially troublesome in this respect" (learning by logical progression). "It has a reputation for being both beautiful and difficult, and many mathematicians seem to want to know something about it. But it seems impossible to penetrate. A thorough logical-progression approach might require ten thousand pages." The application of these beautiful tools in physics would presuppose their understanding, although, I hope at least, not with the completeness and depth necessary for mathematicians. I shall try to assist this understanding by sections with mathematical tools. In the beginning, it is not necessary to master all the concepts mentioned there. The pragmatic "battle tested physical approach to mathematics" carries rather far. But in the end, a pedestrian mathematical attitude with some knowledge of the rotation group is not enough. Mathematical simplicity does not coincide with conceptual triviality. The relevant simple concepts are, in most cases, very deep.

In the historical development of physics the causal equations of motion, introduced by Newton for time development, were derived later with extremal and variational principles from Lagrangians and Hamiltonians, which, in turn, could be characterized, for important cases, by their invariance or transformation properties with respect to operation groups. In this book I will go the historical route in the opposite direction: In contrast to the familiar procedure starting with equations of motion, I start with operational structures. The equations of motion do not play the basic role. They are a Lie parameter-related formulation of the local behavior with respect to the operation group involved as expressed for a Lie group by the action of its Lie algebra (tangent space translations). Time and space for the interpretation of a physical dynamics with the conventional equations of motion are a very important, but from the operational point of view only one example of, tangent space-related structures. Therefore the time and spacetime dependence of operators or eigenvectors and equations of motion reflects properties of acting groups and Lie algebras or, to include also semigroups and symmetric spaces with their tangent translations, of acting Lie operations. Equations of motion are a powerful method to diagonalize, to find eigenvalues and invariants of the operations involved.

To illustrate this reversed procedure in the simple example of a harmonic oscillator, time operations or causality as the starting point is formalized, qualitatively and quantitatively, by the additive ordered group $\mathbb{R}$. The Lie group $\mathbb{R}$ has its irreducible complex representations in the compact group $\mathbf{U}(1)$ acting on 1-dimensional vector spaces. The represented time translations define time orbits in the representation space, especially the irreducible orbits of a dual eigenvector basis $(\mathrm{u}(t), \mathrm{u}^\star(t))$ for the two $\mathbb{C}$-isomorphic dual representation spaces with imaginary time action eigenvalues $\pm i\omega \in i\mathbb{R}$:

$$\mathbb{R} \ni t \longmapsto e^{\pm i\omega t} \in \mathbf{U}(1) \Rightarrow \left\{ \begin{array}{rcl} \mathrm{u}(t) & = & e^{i\omega t}\ \mathrm{u}(0), \\ \mathrm{u}^\star(t) & = & e^{-i\omega t}\mathrm{u}^\star(0). \end{array} \right.$$

The Lie algebra (time translation) action can be expressed by first-order differential equations for the representation orbits

$$(\tfrac{d}{dt} \mp i\omega)(\mathrm{u}, \mathrm{u}^\star)(t) = 0.$$

The Lagrangian L yields another formulation of the time translation action on dual eigenvectors

$$iL = iL_0 - iH_0 = \mathrm{u}^\star \tfrac{d}{dt}\mathrm{u} - i\omega \mathrm{u}\mathrm{u}^\star$$

with the kinetic term L_0 implementing the duality of the basic pair $(\mathrm{u}, \mathrm{u}^\star)$ and the Hamiltonian H_0 as product of the basic space identity $\mathrm{u}\mathrm{u}^\star$ and eigenvalue (frequency) ω the represented time translation (Lie algebra) basis. The dual irreducible representation characteristic invariant $|\omega|$ sets the intrinsic time unit.

The representation connected $\mathbf{U}(1)$-conjugation of the irreducible complex spaces with the time orbits implements the time reflection $t \overset{\mathrm{T}}{\leftrightarrow} -t$ and $\mathrm{u} \overset{\mathrm{T}}{\leftrightarrow} \mathrm{u}^\star$ and allows the definition of Hermitian orbits, called position-momentum $(\mathbf{x}, \mathbf{p})$. It thus becomes possible to interpret the time orbits in position and momentum space, e.g., by an oscillating spring or a pendulum. The position-momentum orbits arise from real self-dual representations of the time operations in the group $\mathbf{SO}(2)$, as Lie group isomorphic to $\mathbf{U}(1)$:

$$\begin{array}{rl} \mathbf{x} & = \ell\frac{\mathrm{u}+\mathrm{u}^\star}{\sqrt{2}} \\ i\mathbf{p} & = \frac{\mathrm{u}-\mathrm{u}^\star}{\ell\sqrt{2}} \end{array} \Rightarrow \begin{pmatrix}\mathbf{x}(t)\\ \mathbf{p}(t)\end{pmatrix} = \begin{pmatrix}\cos\omega t & \ell^2\sin\omega t\\ -\frac{\sin\omega t}{\ell^2} & \cos\omega t\end{pmatrix}\begin{pmatrix}\mathbf{x}(0)\\ \mathbf{p}(0)\end{pmatrix},$$

$$L = \mathbf{p}\frac{d}{dt}\mathbf{x} - \omega(\ell^2\frac{\mathbf{p}^2}{2} + \frac{1}{\ell^2}\frac{\mathbf{x}^2}{2}),$$

where ℓ is the characteristic length in the dual position-momentum pair, defined by the $\mathbf{SO}(2)$-metric $\begin{pmatrix}\ell^2 & 0\\ 0 & \frac{1}{\ell^2}\end{pmatrix}$ and defining together with the frequency ω two phenomenological units, the inert mass $M = \frac{1}{\omega\ell^2}$ and the spring constant $k = \frac{\omega}{\ell^2}$. The usual starting point, the classical Lagrangian $L = \mathbf{p}\frac{d}{dt}\mathbf{x} - (\frac{\mathbf{p}^2}{2M} + k\frac{\mathbf{x}^2}{2})$, encapsulating the self-dual irreducible real representations of the time operations, comes at the end of the procedure.

In quantum mechanics, much more in quantum field theory, the definition of an operator Lagrangian with explicit spacetime derivatives is in general rather difficult, if not impossible. The dual pair structure, classically encoded in the kinetic Lagrangian, e.g., in $iL_0 = \mathrm{u}^\star\frac{d}{dt}\mathrm{u}$, formulates the quantization $[\mathrm{u}^\star, \mathrm{u}] = 1$ or, for a Hermitian-anti-Hermitian pair $iL_0 = i\mathbf{p}\frac{d}{dt}\mathbf{x}$, the Born-Heisenberg relation $[i\mathbf{p}, \mathbf{x}] = 1$. The time translations are realized by the adjoint action (quantum commutator) with a Hamiltonian

$$\mathbf{H}_0 = \omega\frac{\{\mathrm{u},\mathrm{u}^\star\}}{2} = \frac{\mathbf{p}^2}{2M} + k\frac{\mathbf{x}^2}{2} \Rightarrow \left\{\begin{array}{llll} [i\mathbf{H}_0, \mathrm{u}] & = \frac{d}{dt}\mathrm{u}, & [i\mathbf{H}_0, \mathrm{u}^\star] & = \frac{d}{dt}\mathrm{u}^\star,\\ [i\mathbf{H}_0, \mathbf{x}] & = \frac{d}{dt}\mathbf{x}, & [i\mathbf{H}_0, \mathbf{p}] & = \frac{d}{dt}\mathbf{p}.\end{array}\right.$$

The time derivative $\frac{d}{dt}$ can be considered to be a shorthand notation, familiar from the classical derivative, for the adjoint-action-induced Lie algebra transformation. From this point of view the first-order time differential equations for dual pairs, e.g., for position-momentum $(\mathbf{x}, \mathbf{p})$, or the second-order equations for one Hermitian combination, e.g., for position $\mathbf{x}$, are a consequence of the quantum-implemented linear Lie algebra action, i.e., of $\frac{d}{dt} = [i\mathbf{H},\ ]$.

The conjugation group $\mathbf{U}(1)$ with the represented time operation by phase transformations $\mathbb{R} \ni t \longmapsto e^{\pm i\omega t} \in \mathbf{U}(1)$ endows the one-dimensional representation with a scalar product and a Hilbert space structure that allows Born's "probability amplitudes" for the ontological interpretation of the operations via experiments. The spectrum of the position operator $x \in \operatorname{spec}\mathbf{x}$ is used for Schrödinger wave functions $x \longmapsto \psi(x)$, which are orbits (representation coefficients) of position translations.

Also, for quantum field theory the classically oriented approach relying on differential equations of first and second orders, e.g., Dirac and Klein-Gordon equations, will not be in the foreground. Representations for external spacetime and internal unitary groups and their actions as seen, for example, in the standard model are more basic for the understanding as their projections to asymptotic particle state vectors, as used for experimental tests. An illustration of the method used in this book may be given, for instance, by a Dirac field $\boldsymbol{\Psi}$ for a massive spinor particle. Here the unitarily represented group is the Poincaré group $\mathbf{SL}(\mathbb{C}^2) \vec{\times} \mathbb{R}^4$, induced by representations of a direct product subgroup $\mathbf{SU}(2) \times \mathbb{R}^4$ involving spin $\mathbf{SU}(2)$ as double cover of position rotations $\mathbf{SO}(3)$ and spacetime translations to define the embedded particle, e.g., the electron-positron, with its spin invariant $\frac{1}{2}$ from a rational spectrum and its mass m^2 from a continuous spectrum. The Fock expectation value $\langle \ldots \rangle$ of the commutator, with Dirac matrices $\{\gamma^k\}_{k=0,1,2,3}$,

$$\langle [\overline{\mathbf{\Psi}}(y), \mathbf{\Psi}(x+y)] \rangle = \langle [\overline{\mathbf{\Psi}}, \mathbf{\Psi}] \rangle(x) = \int \frac{d^4q}{(2\pi)^3} (\gamma^k q_k + m)\delta(m^2 - q^2) e^{iqx},$$

is a matrix element of a Hilbert representation of spacetime translations. The projection to time translation representation matrix elements $e^{\pm imx_0}$ can be obtained by position integration

$$\mathbb{R} \ni x_0 \longmapsto \int d^3x \; \gamma_0 \langle [\overline{\mathbf{\Psi}}, \mathbf{\Psi}] \rangle(x) = \mathbf{1}_2 \otimes \begin{pmatrix} i \sin mx_0 & \cos mx_0 \\ \cos mx_0 & i \sin mx_0 \end{pmatrix}.$$

The corresponding position projections by time integration is trivial:

$$\int dx_0 \; \langle [\overline{\mathbf{\Psi}}, \mathbf{\Psi}] \rangle(x) = 0.$$

This is in contrast to the position projection of the time-ordered quantization anticommutator arising in the Feynman propagator. Here one obtains a Yukawa potential and force as noncompact representation coefficients $e^{-m|z|}$ of position translations, distributed with the Kepler factor $\frac{1}{r}$ on the 2-spheres in 3-dimensional position space

$$\begin{aligned} \mathbb{R}^3 \ni \vec{x} \longmapsto \int dx_0 \; \epsilon(x_0)\gamma_0 \{\overline{\mathbf{\Psi}}, \mathbf{\Psi}\}(x) &= \begin{pmatrix} \frac{\vec{\sigma}\vec{x}}{r} \frac{1+mr}{r} & m\mathbf{1}_2 \\ m\mathbf{1}_2 & -\frac{\vec{\sigma}\vec{x}}{r} \frac{1+mr}{r} \end{pmatrix} \frac{e^{-mr}}{2\pi r}, \\ \int dx dy \; \frac{e^{-mr}}{2\pi r} &= \frac{e^{-m|z|}}{m}. \end{aligned}$$

Spacetime cannot be thought of without interactions. Spacetime is perceived by its operational representations, which are given by and act on what we call quantum fields, which may or may not have particles as projections in a Hilbert space.

A customary approach to quantum structures uses ad hoc Hilbert spaces with square integrable position space functions at a very early stage. The operational approach puts the Hilbert spaces in a representational perspective. As each Lie group defines its representations, so each Lie group with real operations defines its complex Hilbert spaces on which it acts. The Hilbert spaces of nonrelativistic quantum mechanics are defined, as shown in the Stone-von Neumann theorem, by the Heisenberg Lie algebra, whose three real operations are characterized by the Lie bracket $[\mathbf{x}, \mathbf{p}] = \mathbf{I}$. Those historically first Hilbert spaces in quantum theory are not appropriate for all operation groups. They are not suited for fermionic quantum structures and not used in quantum field theory. Already quantum–mechanical scattering theory is formulated more appropriately in the Hilbert spaces defined by the Euclidean group $\mathbf{SO}(3) \vec{\times} \mathbb{R}^3$ of rotations acting on position translations. The Hilbert space for a free relativistic particle is defined, as shown by Wigner, by a representation of the Poincaré group $\mathbf{SO}_0(1,3) \vec{\times} \mathbb{R}^4$. Or there are Hilbert spaces for the Lorentz groups $\mathbf{SO}_0(1,2)$ and $\mathbf{SO}_0(1,3)$ for two or three position dimensions whose elements cannot be formulated with square integrable functions, as shown by Bargmann and Gel'fand and Naimark.

To understand the strength and appropriateness of the operational point of view it is useful to learn, to test, and to apply it in the well-established areas of nonrelativistic quantum mechanics and relativistic quantum field theory. Therefore, the first volume of the book deals essentially, after an introductory presentation of time and space translations, with the time and space-related finite-dimensional representation structures, with compact Lie operations, and, as a nonrelativistic application, with an operationally oriented formulation of the always fascinating Kepler problem.

Here arise already continuous eigenvalues and invariants for noncompact operations, which, in the context of relativistic quantum field theory with the noncompact nonabelian Lorentz group, are looked at more closely in the first part of the second volume. The representation structure of free particle fields, massive and massless, and its implementation in the familiar formalism are given. This part ends with an application of those structures to the standard model of elementary particles. Perturbation theory with its normalization-regularization procedure will not be discussed.

The second part of the second volume works with the – mathematically rather demanding – harmonic analysis of noncompact nonabelian Lie groups and their homogeneous spaces, e.g., the Lorentz and Poincaré group or the causal spacetime cone, to understand the spacetime representations in Feynman propagators and their shortcomings. One has to face the question whether the concepts of "virtual particles" ("off-shell") with the so–called energy-time uncertainty and the virtual particle-exchange in an "anschauliche" description of interactions, as suggested by Feynman diagrams, are not of the same dangerous quality as the point-particle and position-orbit concepts for electrons inside atoms to understand their spectral lines.

In the end, an attempt is made to proceed from the Wigner classification of the particles as vectors acted on with irreducible unitary Poincaré group representations, i.e., from a classification of tangent structures, to the constitution of these tangent structures. An operational spacetime model is proposed in the form of a nonlinear symmetric space whose spectrum includes as invariants particle masses and, especially, gauge coupling constants as normalization of its irreducible representations. Since this is an extremely difficult problem, such an attempt should be seen not as a solution, but as one proposal for a direction on the way to a solution.

Perhaps it is necessary to mention that essentially up to parts of the last two chapters in the second volume, the material in the following is general as concerns the results. I do not propose new theories. The aim is, on an operational basis, to understand more deeply what we are working with in quantum theory. The appropriate language and the conceptional presentation may not be so familiar.

MATHEMATICAL TOOLS

The basic mathematics used in the following is strongly influenced by the Bourbaki school. The concepts, the notation, and the names I use may be unfamiliar to many physicists. They are the usual ones in the mathematical literature and, as I found after getting used to them, also appropriate for physics. Sometimes the abstract structural concepts of mathematics are easier to probe more deeply than the ad hoc coined concepts in physics.

The structural formulation helps, as far as possible, to separate the specific problems in physics from the mathematical-logical ones. With respect to the structure of Lie groups and their representations, especially for the noncompact and nonabelian operations, I have learned much, especially from the books of Folland, Gel'fand, Helgason, Kirillov, Knapp, and Vilenkin, which are highly recommended.

In general, each chapter starts with the more physically oriented sections, which, after a summary, are followed (not always) by more mathematically oriented ones dealing with the concepts used before. Sometimes, especially in later chapters, a distinction between "mathematical" and "physical" would look too arbitrary.

Presumably, one cannot learn the mathematics only from what is given in the mathematical sections: they may already require much mathematical experience. As I know from personal experience, there is "no free mathematical lunch." The mathematical sections are intended to place the mathematical manipulations in physics in their structural context. They should define, introduce, and make familiar to some degree with or remind of the structures used, give a coarse orientation, and stimulate a deeper study of the mathematical literature, which is given with all important references, also in journals, in the books quoted above.

It is not the purpose of this book to prove mathematical theorems that can be found in mathematical textbooks. One "opens up" for the mathematical tools if one really needs them in physics. Then, many proofs become unnecessary if one dives deeply enough into the structures. The mathematical structures are treated eclectically, reflecting my personal taste and my limited abilities and avoiding cumbersome complications. Nevertheless, I am sure, that there will be mistakes I have overlooked and subtleties, even major ones, that I have not taken into account. The representation is by no means hierarchical and complete; some basic concepts are tacitly assumed as familiar and other basic concepts are briefly explained. Mathematical formulas are not always easy to read. Since, however, mathematics is the language of science, it will not be assumed to be necessary to express each formula before or after in everyday language.

The operation concept is clearly formalized in the language of *categories and functors*, which will be used only superficially, mnemotechnically, and for notational purposes. The notation $\underline{\textbf{kat}}$ denotes a category in which the objects are sets, e.g., the categories

sets:	$\underline{\mathbf{set}}$
monoids:	$\underline{\mathbf{mon}}$
groups:	$\underline{\mathbf{grp}}$
abelian groups:	$\underline{\mathbf{abgrp}}$
rings:	$\underline{\mathbf{rng}}$
modules over a ring R:	$\underline{\mathbf{mod}}_R$
vector spaces over K (abelian field):	$\underline{\mathbf{vec}}_K$
algebras over K:	$\underline{\mathbf{ag}}_K$
associative algebras over K:	$\underline{\mathbf{aag}}_K$
Lie algebras over K:	$\underline{\mathbf{lag}}_K$
differentiable manifolds over K:	$\underline{\mathbf{dif}}_K$
Lie groups over K:	$\underline{\mathbf{lgrp}}_K$
topological spaces:	$\underline{\mathbf{top}}$
measure spaces:	$\underline{\mathbf{mes}}$

Elements in the categories used are *morphisms* $(A \xrightarrow{f} B) \in \underline{\mathbf{kat}}$ (mappings, arrows). An object pair (A, B) of a category $\underline{\mathbf{kat}}$ has the set of morphisms $\mathbf{kat}(A, B) = \{A \longrightarrow B\}$, compatible with the category characterizing structure and associatively composable. Morphisms are called *endomorphisms* $\mathbf{kat}(A, A)$ for $A = B$, the *isomorphisms* $\overset{\circ}{\mathbf{kat}}(A, B)$ are called *automorphisms* $\overset{\circ}{\mathbf{kat}}(A, A)$ for $A = B$ with id_A the identity.

Isomorphies hold in a category; therefore they should be qualified, e.g., $L \overset{\mathbf{vec}_{\mathbb{R}}}{\cong} \mathbb{R}^n$ for a real vector space isomorphy of a Lie algebra. For a simpler notation, such qualifications are omitted; they should be obvious from the context.

Starting with operations as basic structures, one may use the identity operation id_A as the neutral operation in the nontrivially acting ones ("constancy in change") to define an object A. The sloppy notation $f \in \underline{\mathbf{kat}}$ and $A \in \underline{\mathbf{kat}}$ is used for morphisms and objects.

The categories above can be arranged with the inclusion order

$$\begin{array}{ccccccccccc}
\underline{\mathbf{top}} & \subset & \underline{\mathbf{mes}} & \subset & \underline{\mathbf{set}} & & & & & & \\
 & & \cup & & \cup & & & & & & \\
 & & \underline{\mathbf{dif}}_K & & \underline{\mathbf{mon}} & & & & & & \\
 & & \cup & & \cup & & & & & & \\
 & & \underline{\mathbf{lgrp}}_K & \subset & \underline{\mathbf{grp}} & \supset & \underline{\mathbf{abgrp}} & \supset & \underline{\mathbf{mod}}_R & \supset & \underline{\mathbf{vec}}_K \\
 & & & & & & & & & & \cup \\
 & & & & \cup & & & & & & \underline{\mathbf{ag}}_K \supset \underline{\mathbf{lag}}_K \\
 & & & & & & & & & & \cup \\
 & & & & \underline{\mathbf{rng}} & & & \supset & & & \underline{\mathbf{aag}}_K
\end{array}$$

Isomorphic objects of a category define classes as objects in the associated equivalence category. Categories may have additional properties. For example, they are *morphism stable* if the morphisms are objects of the same category $\mathbf{kat}(A, B) \in \underline{\mathbf{kat}}$, e.g., $\mathbf{set}(S, T) \in \underline{\mathbf{set}}$, $\mathbf{vec}_K(V, W) \in \underline{\mathbf{vec}}_K$, linear mappings constitute a vector space.

As basic operational structures, set and vector space endomorphisms (arrow monoids and arrow algebras) as well as set and vector space automorphism groups (permutation groups and linear groups) deserve special symbols

$$\begin{array}{llll} \mathbf{set}(S,S) &= \mathbf{A}(S) &\in \underline{\mathbf{mon}}, & \overset{\circ}{\mathbf{set}}\,(S,S) = \mathbf{G}(S) \quad \in \underline{\mathbf{grp}}, \\ \mathbf{vec}_K(V,V) &= \mathbf{AL}(V) &\in \underline{\mathbf{aag}_K}, & \overset{\circ}{\mathbf{vec}}_K\,(V,V) = \mathbf{GL}(V) \quad \in \underline{\mathbf{grp}}. \end{array}$$

Co- and contravariant functors are mappings for categories $\underline{\mathbf{kat}}^{1,2}$

$$\mathcal{F} : \underline{\mathbf{kat}}^1 \longrightarrow \underline{\mathbf{kat}}^2, \quad \begin{array}{c} A \\ f\big\downarrow \\ B \end{array} \longmapsto \begin{array}{c} \mathcal{F}(A) \\ \big\downarrow \mathcal{F}(f) \\ \mathcal{F}(B) \end{array} \quad \text{or} \longmapsto \begin{array}{c} \mathcal{F}(A) \\ \big\uparrow \mathcal{F}(f) \\ \mathcal{F}(B) \end{array}$$

$$\text{with } \mathrm{id}_{\mathcal{F}(A)} = \mathcal{F}(\mathrm{id}_A) \quad \begin{cases} \mathcal{F}(f \circ g) = \mathcal{F}(f) \circ \mathcal{F}(g), & \text{covariant}, \\ \mathcal{F}(f \circ g) = \mathcal{F}(g) \circ \mathcal{F}(f), & \text{contravariant}. \end{cases}$$

For example, a Lie group G has a unique Lie algebra, denoted by $\log G$, with the covariant logarithm functor

$$\log : \underline{\mathbf{lgrp}_{\mathbb{K}}} \longrightarrow \underline{\mathbf{lag}_{\mathbb{K}}}, \quad G \longmapsto \log G.$$

A functor may have additional properties, e.g., additive if direct sums of vector spaces are involved $\mathcal{F}(V_1 \oplus V_2) = \mathcal{F}(V_1) \oplus \mathcal{F}(V_2)$ or exponential $\mathcal{F}$ for (tensor) products $\mathcal{F}(V_1 \oplus V_2) = \mathcal{F}(V_1) \otimes \mathcal{F}(V_2)$.

Mappings can inherit structures of their domains, e.g., a vector space can arise from a set with mappings into a field K as expressed in the covariant *free functor* (linear extension or span functor)

$$K^{(\,)} : \underline{\mathbf{set}} \longrightarrow \underline{\mathbf{vec}_K}, \quad \begin{array}{c} S \\ f\big\downarrow \\ T \end{array} \longmapsto \begin{array}{c} K^{(S)} \\ \big\downarrow K^{(f)} \\ K^{(T)} \end{array}$$

The vector space $K^{(S)} \cong \{\sum_{\text{finite}} \alpha_s s\}$ contains the finite linear K-combinations of set elements (or the mappings $\alpha : S \longrightarrow K$ with finite support); it has S as canonical basis. For $K^{(f)}$ the set mapping is linearily extended.

Important functors arise with universal extensions (structures): Given a structure expressed with the category $\underline{\mathbf{kat}}$ there may be objects with more structure in a subcategory $\underline{\mathbf{ukat}} \subset \underline{\mathbf{kat}}$, e.g., algebras in vector spaces $\underline{\mathbf{ag}_K} \subset \underline{\mathbf{vec}_K}$ or abelian groups in abelian semigroups with cancellation rule or complete Hausdorff spaces in uniform (e.g., metric) spaces.

A *universal extension functor* $\mathcal{E}$ from a category in a more structured subcategory

$$\mathcal{E} : \underline{\mathbf{kat}} \longrightarrow \underline{\mathbf{ukat}}$$

is the solution of a universal problem if for any $A \in \underline{\mathbf{kat}}$ there exists a more structured "universal" object $\mathcal{E}(A) \in \underline{\mathbf{ukat}}$ and a natural injection ι that factorizes any kat-morphisms f to a **ukat**-object U with a unique **ukat**-morphism $\tilde{f}$ as shown in the commutative diagram[1]

$$A, \iota, f \in \underline{\mathbf{kat}}, \qquad \begin{array}{ccc} A & \xrightarrow{\iota} & \mathcal{E}(A) \\ {\scriptstyle f}\big\downarrow & & \big\downarrow{\scriptstyle \tilde{f}} \\ U & \xrightarrow[\mathrm{id}_U]{} & U \end{array}, \qquad \mathcal{E}(A), U, \tilde{f} \in \underline{\mathbf{ukat}},$$

$$f = \tilde{f} \circ \iota, \quad \mathbf{kat}(A, U) \cong \mathbf{ukat}(\mathcal{E}(A), U).$$

If $\mathcal{E}$ exists, the object $\mathcal{E}(A)$ is unique up to **ukat**-isomorphisms. The induced functor $\mathcal{E}$ is covariant: take $U = \mathcal{E}(B)$ with $B \in \underline{\mathbf{kat}}$. With a unique $\tilde{f}$ the corresponding morphism sets are set-isomorphic (equal cardinality).

An example is the linear extension functor above,

$$S, \iota, f \in \underline{\mathbf{set}}, \qquad \begin{array}{ccc} S & \xrightarrow{\iota} & K^{(S)} \\ {\scriptstyle f}\big\downarrow & & \big\downarrow{\scriptstyle \tilde{f}} \\ V & \xrightarrow[\mathrm{id}_V]{} & V \end{array}, \qquad K^{(S)}, V, \tilde{f} \in \underline{\mathbf{vec}}_K,$$

or the *tensor algebra functor* (multilinear extension functor)

$$\bigotimes : \underline{\mathbf{vec}}_K \longrightarrow \underline{\mathbf{aag}}_K, \quad V \longmapsto \bigotimes V.$$

Also, the numbers, denoted by

natural: $\mathbb{N}_k = \{k, k+1, \ldots\}$, $\mathbb{N} = \mathbb{N}_1 \supseteq \mathbb{N}_k$,
integer: $\mathbb{Z}$, rational: $\mathbb{Q}$, algebraic: $\mathbb{A}$,
number fields $\mathbb{K} \in \{\mathbb{R}, \mathbb{C}\}$ with real $\mathbb{R}$ and complex $\mathbb{C}$,
positive (negative): $\mathbb{Z}_\pm = \pm\mathbb{N}_0 = \pm|\mathbb{Z}|$, $\mathbb{R}_\pm = \pm|\mathbb{R}|$,

are examples of natural structures and basic operations. They start from an additive semigroup $\mathbb{N}$ with cancellation rule, extended to and embedded naturally into $\mathbb{Z}$, which formalizes binary operations on $\mathbb{N}$. Since $\mathbb{Z}$ forms an abelian multiplicative monoid with cancellation rule it is extendable, analogously, to $\mathbb{Q}$ formalizing binary $\mathbb{Q}$- or quartic $\mathbb{N}$-operations. $\mathbb{Q}$ allows the natural Cauchy completion to the reals $\mathbb{R}$, which formalizes approximation operations

$$\mathbb{N} \longmapsto \mathbb{Z} \longmapsto \mathbb{Q} \longmapsto \mathbb{R}.$$

Good guesses to look for universal extensions are self-relations in the set products, e.g., $\mathbb{N} \times \mathbb{N}$ for $\mathbb{Z}$ or $\mathbb{Z} \times \mathbb{Z}$ for $\mathbb{Q}$ or the countably infinite relations (Cauchy series) $\mathbb{Q}^{\aleph_0}$ for $\mathbb{R}$.

[1] If not stated otherwise, all such diagrams are commutative.

1

SPACETIME TRANSLATIONS

An immediate "naive" description and quantification of the physical phenomena in our temporal and spatial neighborhood leads to real linear spacetime structures as a first mathematical formalization: the spacetime translation vector space with its affine operations. From the Galilean-Newtonian mechanics up to Einstein's special relativity, real translations with the possibility to distinguish absolute neutral points - "now" for time and "here" for position - have played the most important role. Up to today, an interpretation of a dynamics by its experimental predictions uses decisively the concepts "mass" and "angular momentum" (spin, polarization). They characterize invariant properties with respect to spacetime translations and rotations respectively as operations in the Poincaré group.

In the interpretation of Leibniz, time and position are relational concepts, to describe and to quantify the behavior and properties of objects with respect to operations defining a dynamics and related experiments. The causal structure of spacetime operations are their most important feature. Newton's successful interpretation of time and position space as having an absolute ontology - "God-given temporal and spatial boxes for the dynamical objects" - is not favored here.

Time and position are operations. The naive interpretation of the space coordinates in classical physics as describing directly the position of a point particle gave way to the interpretation of space as a reservoir for parametrizing operations by three real numbers. Such numbers are taken from the spacetime manifold of relativity. The position coordinates in quantum-mechanical wave functions or in particle quantum fields parametrize position operations, especially translations for interaction-free structures. With the operator structure of quantum theory, the concept of a point particle has, if at all, only a significance as one experimental projection.

In this chapter, the concepts of order (causality), linear duality, and isomorphic dual linear spaces are considered for three pairs of physically relevant vector spaces: for "time translations with energies", for "position translations with momenta," and for "spacetime translations with energy-momenta." Energies and momenta are the respective eigenvalues for time and position translations. The vector spaces come with action groups, with reflections, rotations, and Lorentz transformations, as invariance groups for the metrics that define the self-duality for dual pairs. On this level time and space are, operationally, abelian Lie algebras. An operational connection of the dual position-momentum structures comes with the nonabelian Heisenberg Lie algebra $[\mathbf{x}_a, \mathbf{p}^b] = \delta_a^b \mathbf{I}$ (chapter "Quantum Algebras").

1.1 Time Translations

A linear time model $\mathbb{T}$ ("tempus") collects the *time translations* into a real 1-dimensional vector space. Time vectors should be considered as operations, not as "inert absolutely given points." "Time" in this chapter is understood always as a vector space containing the time translations. The trivial translation $0 \in \mathbb{T}$ ("now") is distinguished as the neutral element of the additive group. The spaces on which time translations act will be discussed in the chapter "Time Representations."

For time, formalized by the real numbers $\mathbb{R}$ as a 1-dimensional vector space, general vector space concepts like basis, dual space, scalar product, invariances, and topology, appear academically blown up: their importance becomes clearer if applied afterward to real 3-dimensional position translations or if time and position are embedded into a 4-dimensional vector space with spacetime translations.

Each *time translation basis*[1] gives rise to an isomorphism $x \leftrightarrow x_0$ between time and the real numbers as time coordinates

$$\mathbb{T} = \{x = x_0 \mathbf{p}^0 \mid x_0 \in \mathbb{R}\} \cong \mathbb{R}, \quad \text{basis: } \{\mathbf{p}^0\}.$$

With that, all $\mathbb{R}$-structures, called *natural*, are transportable to time or vice versa: The concept of the ordered natural numbers with their rational and real extensions may be considered to be an abstraction of the time structures. The characterization of time translations by real numbers uses the linear forms of time that constitute the dual time $\mathbb{T}^T = \{\mathbb{T} \longrightarrow \mathbb{R}\}$, the *frequency or energy space* with the eigenvalues of the time translations. A time basis $\{\mathbf{p}^0\}$ comes with a unique dual basis $\{\mathbf{x}_0\}$ of the frequency space

$$\begin{aligned}
&\mathbb{T}^T = \{p = p^0 \mathbf{x}_0 \mid p^0 \in \mathbb{R}\} \cong \mathbb{R}, \quad \text{dual bases: } \langle \mathbf{x}_0, \mathbf{p}^0 \rangle = 1,\\
&\mathbf{x}_0 : \mathbb{T} \longrightarrow \mathbb{R}, \quad \mathbf{x}_0(x) = x_0.
\end{aligned}$$

The manifold of all time bases can be obtained by operating with the general linear group on one fixed basis, i.e., by multiplying by a nontrivial scalar

$$\mathbf{GL}(\mathbb{R}) \bullet \{\mathbf{p}^0\} \cong \mathbf{GL}(\mathbb{R}) \cong \mathbb{R}^\diamond = \{\alpha \in \mathbb{R} \mid \alpha \neq 0\}.$$

With a basis, time is *totally ordered* via the real coefficients

$$x = x_0 \mathbf{p}^0 \succeq_{\mathbf{p}} 0 \iff x_0 \geq 0.$$

[1] In this chapter, basis vectors for time and position are denoted with boldface letters.

This allows the concepts *past, present, and future* with respect to each time (vector). All bases related to each other by positive multiplication $e^\lambda \in \mathbf{D}(1) = \mathbb{R}_+^\diamond$ define the same order. With respect to order there are two inequivalent classes of time bases that are isomorphic to the reflection group $\mathbb{I}(2) = \{1, -1\}$

$$\mathbf{GL}(\mathbb{R}) \cong \mathbf{D}(1) \times \mathbb{I}(2), \quad \mathbf{D}(1) = \{e^\lambda \bigm| \lambda \in \mathbb{R}\} \in \underline{\mathbf{grp}}.$$

One cannot hope for a unique order; orders come in pairs: For an order, there always exists the reflected order.

With a time basis the natural order of the real numbers defines a *time length* $\| x \|_{\mathbf{p}}$ via the modulus $|x_0| = \epsilon(x_0)x_0$:

$$\| \ \ \|_{\mathbf{p}}: \mathbb{T} \longrightarrow \mathbb{R}, \quad \| x \|_{\mathbf{p}} = |x_0| \ \| \mathbf{p}^0 \|_{\mathbf{p}} \geq 0.$$

The norm of the basis $\| \mathbf{p}^0 \|_{\mathbf{p}}$ defines a *time unit*: If x_0 is measured in seconds, the basis norm must have the dimension of a frequency, e.g., $\mu = \frac{1}{\mathrm{s}}$ (Hertz), since $\| x \|_{\mathbf{p}}$ is a pure number

$$\text{inverse time unit} \quad \| \mathbf{p}^0 \|_{\mathbf{p}} = \mu.$$

A time norm gives the natural *time topology*, the unique Hausdorff topology of $\mathbb{R}$, with, e.g., the open intervals $U_\alpha(0) = \{x \bigm| \ \| x \|_{\mathbf{p}} < \alpha\}$ as 0-basis $\{U_\alpha(0) \bigm| \alpha > 0\}$, and, furthermore, a natural *time scalar product*, which associates a real number to two time translations

$$\tau(\ ,\ ): \mathbb{T} \times \mathbb{T} \longrightarrow \mathbb{R}, \quad \begin{cases} \tau(x, y) = \frac{\|x+y\|_{\mathbf{p}}^2 - \|x-y\|_{\mathbf{p}}^2}{4} = x_0 y_0 \mu^2, \\ \mu^2 = \tau(\mathbf{p}^0, \mathbf{p}^0). \end{cases}$$

The frequency space inherits the inverse scalar product with norm and natural topology.

A time scalar product has as invariance group in all linear transformations $\mathbf{GL}(\mathbb{R})$, the 2-element *time reflection group*, identical with the orthogonal group in one real dimension $\mathbf{O}(1) = \{\pm \mathrm{id}_{\mathbb{T}}\} \cong \mathbb{I}(2)$. It is generated by the *time inversion* $\mathtt{T} = -\mathrm{id}_{\mathbb{T}}$. The time reflection group acts on the time translations in the nonabelian semidirect product group

$$\mathbf{O}(1) \vec{\times} \mathbb{R}: \quad (\mathtt{I}_1, t_1) \circ (\mathtt{I}_2, t_2) = (\mathtt{I}_1 \circ \mathtt{I}_2, t_1 + \mathtt{I}_1.t_2).$$

The general linear group as the manifold of all time bases is the direct product $\mathbf{GL}(\mathbb{R}) = \mathbf{O}(1) \times \mathbf{D}(1)$ of the discrete reflection group and the noncompact 1-parametric *scale transformations (dilations)* $\mathbf{D}(1) = \exp \mathbb{R}$, which is the connection component $\mathbf{GL}_0(\mathbb{R})$ of the unit in $\mathbf{GL}(\mathbb{R})$. Only the *orthochronous* part of $\mathbf{O}(1)$, the trivial group $\mathbf{SO}(1) = \{\mathrm{id}_{\mathbb{T}}\}$ with identity, is compatible both with a time scalar product and with a time order. The manifold of scalar products of a real 1-dimensional vector space is isomorphic to the normalizing dilations $\mathbf{GL}(\mathbb{R})/\mathbf{O}(1) = \mathbf{D}(1)$.

1.2 Position Translations

A linear position model $\mathbb{S}$ ("spatium") collects the *position translations* into a real 3-dimensional vector space with "here" $0 \in \mathbb{S}$ as the neutral element:[2]

$$\mathbb{S} = \{\vec{x} = x_a \mathbf{p}^a \bigm| x_a \in \mathbb{R}\} \cong \mathbb{R}^3, \quad \text{basis: } \{\mathbf{p}^a\}_{a=1,2,3}.$$

Position vectors should be considered as operations. "Position" in this chapter is understood always as a vector space containing the position translations. The spaces on which position translations act will be discussed in the chapter "Spin, Rotations and Position."

In many structures, the position dimension $s = 3$ is not distinguished and can be generalized to any natural number $\mathbb{S}^s \cong \mathbb{R}^s$. In contrast to time $\mathbb{T} \cong \mathbb{R}$, the transfer of the natural $\mathbb{R}$-structures to positions with $s > 1$ is not obvious and, in general, not unique. Position is related to the real numbers by a *scalar product (metric)*, a symmetric bilinear form associating a real number to two position vectors with the condition that nontriviality of a position translation is equivalent to a strictly positive length (norm):

$$\sigma(\ ,\ ) : \mathbb{S} \times \mathbb{S} \longrightarrow \mathbb{R}, \quad \begin{cases} \sigma(\vec{x}, \vec{y}) & = \sigma(\vec{y}, \vec{x}) = x_a y_b \sigma^{ab}, \\ \| \vec{x} \|_\sigma^2 & = \sigma(\vec{x}, \vec{x}) > 0 \iff \vec{x} \neq 0. \end{cases}$$

A scalar product defines the natural *position topology* as the unique Hausdorff topology with, e.g., the open spheres as 0-basis.

A scalar product explains an isomorphism to the dual position space, the *momentum space* with the eigenvalues of the position translations, naturally equipped with the inverse scalar product:

$$\begin{aligned} &\mathbb{S}^T = \{\vec{p} = p^a \mathbf{x}_a \bigm| p^a \in \mathbb{R}\} \cong \mathbb{R}^3, \quad \text{dual bases: } \langle \mathbf{x}_b, \mathbf{p}^a \rangle = \delta_b^a, \\ &\sigma = \sigma^{ab} \mathbf{x}_a \otimes \mathbf{x}_b, \quad \sigma^{-1} = \sigma_{ab} \mathbf{p}^a \otimes \mathbf{p}^b. \end{aligned}$$

The related dual isomorphism and dual bases are used in the raising and lowering of indices

$$\sigma : \mathbb{S} \longrightarrow \mathbb{S}^T, \quad \begin{cases} \sigma(\vec{x}) = \sigma(\vec{x},\ ), \\ x_a \mathbf{p}^a \longmapsto x_a \sigma^{ab} \mathbf{x}_b = x^b \mathbf{x}_b = x_a \mathbf{x}^a. \end{cases}$$

Positions and momenta are in different vector spaces. With an isomorphism between dual vector spaces one can visualize vectors and their linear forms in one vector space, e.g., position translations and momenta in position space $\vec{x}, \sigma^{-1}(\vec{p}) \in \mathbb{S}$.

The invariance group of a scalar product σ in the unital algebra $\mathbf{AL}(\mathbb{R}^3)$ of all linear position transformations with regular group $\mathbf{GL}(\mathbb{R}^3)$ is the associate *rotation group* $\mathbf{O}(3, \sigma)$, sometimes written short $\mathbf{O}(3)$ for short, a real 3-dimensional compact Lie group, defined by the following diagram involving the

[2]With Einstein's convention, there is a summation over double indices from a finite range if not stated otherwise, here $x_a \mathbf{p}^a = \sum_{a=1}^3 x_a \mathbf{p}^a$.

contragredient (inverse transposed) momentum space rotation O^{-1T}:

$$\begin{array}{ccc} \mathbb{S} & \xrightarrow{O} & \mathbb{S} \\ \sigma\downarrow & & \downarrow\sigma \\ \mathbb{S}^T & \xrightarrow[O^{-1T}]{} & \mathbb{S}^T \end{array}, \qquad \begin{array}{l} O^T \circ \sigma \circ O = \sigma, \quad \sigma(\vec{x}, O.\vec{y}) = \sigma(O^{-1}.\vec{x}, \vec{y}), \\ O = O^b_a \mathbf{p}^a \otimes \mathbf{x}_b \Rightarrow O^b_a \sigma^{ad} O^c_d = \sigma^{bc}, \end{array}$$

$$\begin{array}{l} \mathbf{O}(3) \cong \mathbf{O}(3, \sigma) = \{O \in \mathbf{GL}(\mathbb{R}^3) \bigm| O^T \circ \sigma \circ O = \sigma\} \in \underline{\mathbf{lgrp}}_{\mathbb{R}}, \\ \dim_{\mathbb{R}} \mathbf{O}(3) = 3. \end{array}$$

A rotation group is the direct product of two normal subgroups, the discrete scalar product-independent centrum $\mathbb{I}(2)$ with the *position translation reflection* $\vec{x} \overset{\mathsf{P}}{\leftrightarrow} -\vec{x}$ and its *special* rotations $\mathbf{SO}(3, \sigma)$, the connection component of the unit

$$\mathbf{O}(3) \cong \mathbb{I}(2) \times \mathbf{SO}(3), \quad \left\{ \begin{array}{l} \mathbf{SO}(3) = \{O \in \mathbf{O}(3) \bigm| \det O = 1\}, \\ \mathbb{I}(2) \cong \{\mathrm{id}_{\mathbb{S}}, -\mathrm{id}_{\mathbb{S}}\}. \end{array} \right.$$

The invariance Lie algebra of the scalar product is the real 3-dimensional *angular momentum Lie algebra* $\log \mathbf{SO}(3, \sigma)$ of the rotation group

$$\begin{array}{ccc} \mathbb{S} & \xrightarrow{\mathcal{O}} & \mathbb{S} \\ \sigma\downarrow & & \downarrow\sigma \\ \mathbb{S}^T & \xrightarrow[-\mathcal{O}^T]{} & \mathbb{S}^T \end{array}, \qquad \begin{array}{l} \sigma \circ \mathcal{O} = -\mathcal{O}^T \circ \sigma, \quad \sigma(\vec{x}, \mathcal{O}.\vec{y}) = -\sigma(\mathcal{O}.\vec{x}, \vec{y}), \\ \mathcal{O} = \mathcal{O}^b_a \mathbf{p}^a \otimes \mathbf{x}_b \Rightarrow \mathcal{O}^b_a \sigma^{ac} = \mathcal{O}^{bc} = -\mathcal{O}^{cb}, \end{array}$$

$$\begin{array}{rl} \log \mathbf{SO}(3) & = \{\mathcal{O} \in \mathbf{AL}(\mathbb{R}^3) \bigm| \sigma \circ \mathcal{O} = -\mathcal{O}^T \circ \sigma\} \\ & = \{\vec{x}_m \otimes \sigma(\vec{y}_m) - \vec{y}_m \otimes \sigma(\vec{x}_m) \bigm| \vec{x}_m, \vec{y}_m \in \mathbb{S}\} \in \underline{\mathbf{lag}}_{\mathbb{R}}. \end{array}$$

A basis of the σ-antisymmetric rotation group generators can be given in a dual position-momentum basis as follows:

$$\text{basis of } \log \mathbf{SO}(3): \quad \left\{ \begin{array}{l} \{\mathcal{O}^b_a = -\mathbf{p}^b \otimes \mathbf{x}_a + \mathbf{p}_a \otimes \mathbf{x}^b \bigm| a, b = 1, 2, 3\}, \\ \vec{x} \otimes \vec{p} - \sigma^{-1}(\vec{p}) \otimes \sigma(\vec{x}) = -x_b p^a \mathcal{O}^b_a. \end{array} \right.$$

The dual isomorphism $\mathbb{S} \cong \mathbb{S}^T$ between translations and momenta allows an antisymmetric representation of the rotation generators as power-2 tensor

$$\begin{array}{l} \mathrm{id}_{\mathbb{S}} \otimes \sigma^{-1} : \mathbb{S} \otimes \mathbb{S}^T \longrightarrow \mathbb{S} \otimes \mathbb{S}, \quad \vec{x} \otimes \vec{p} \longmapsto \vec{x} \otimes \sigma^{-1}(\vec{p}), \\ [\mathcal{O}^{ab}, \mathcal{O}^{cd}] = \sigma^{ac}\mathcal{O}^{bd} - \sigma^{ad}\mathcal{O}^{bc} - \sigma^{bc}\mathcal{O}^{ad} + \sigma^{bd}\mathcal{O}^{ac}. \end{array}$$

Via the "double trace" the angular momenta carry a rotation-invariant scalar product (the negative definite Killing form, chapter "Spin, Rotations, and Position")

$$\begin{array}{rl} & \log \mathbf{SO}(3) \times \log \mathbf{SO}(3) \longrightarrow \mathbb{R}, \quad \langle \mathcal{O} | \mathcal{O}' \rangle = \mathrm{tr}\, \mathcal{O} \circ \mathcal{O}', \\ \text{invariant:} & \mathrm{tr}\, [\mathcal{O}'', \mathcal{O}] \circ \mathcal{O}' + \mathrm{tr}\, \mathcal{O} \circ [\mathcal{O}'', \mathcal{O}'] = 0, \\ \text{definite:} & \mathrm{tr}\, \mathcal{O} \circ \mathcal{O} < 0 \iff \mathcal{O} \neq 0, \\ \text{for bases:} & \mathrm{tr}\, \mathcal{O}^{ab} \circ \mathcal{O}^{cd} = -2 \det \begin{pmatrix} \sigma^{ac} & \sigma^{ad} \\ \sigma^{bc} & \sigma^{bd} \end{pmatrix}, \text{ invariant: } \mathcal{O}_{ab} \otimes \mathcal{O}^{ab}. \end{array}$$

If one considers the special rotation group $\mathbf{SO}(s)$ for general dimension, the 1-dimensional case gives the trivial group; the 2-dimensional case gives abelian groups. Full rotation groups $\mathbf{O}(2)$ are nonabelian. Only for $s = 3$ does the dimension of the orthogonal group coincide with the dimension of the space on which it is defined

$$\begin{array}{ll} s = 1 \Rightarrow \mathbf{SO}(1) \cong \{1\}, & \log \mathbf{SO}(1) = \{0\}, \\ s = 2 \Rightarrow \mathbf{SO}(2) \text{ abelian}, & \log \mathbf{SO}(2) \cong \mathbb{R}, \\ \dim_{\mathbb{R}} \log \mathbf{SO}(s) = \binom{s}{2}, & \log \mathbf{SO}(s) \cong \mathbb{R}^s \iff s = 3. \end{array}$$

A rotation group (local, homogeneous) with its action on the position translations constitutes as semidirect product the associate *Euclidean group* $\mathbf{O}(3) \vec{\times} \mathbb{R}^3$, a real 6-parametric Lie group with the translations as abelian normal subgroup

$$\mathbf{O}(3) \vec{\times} \mathbb{R}^3: \quad (O_1, \vec{x}_1) \circ (O_2, \vec{x}_2) = (O_1 \circ O_2, \vec{x}_1 + O_1.\vec{x}_2).$$

It is represented by nonrelativistic scattering states (chapters "The Kepler Factor" and "Harmonic Analysis"). The *Lie algebra of a Euclidean group* is, as a vector space, the direct sum of two vector subspaces and, as Lie algebra, a semidirect product with the position translations an abelian ideal

$$\begin{array}{rl} \log \mathbf{SO}(3) \ \vec{\oplus}\ \mathbb{R}^3, & \log \mathbf{SO}(3) \cong \mathbb{R}^3 \cong \mathbb{S}, \\ [\log \mathbf{SO}(3), \mathbb{R}^3] = \mathbb{R}^3, & [\mathcal{O}^{ab}, \mathbf{p}^c] = \mathcal{O}^{ab}(\mathbf{p}^c) = \sigma^{ac}\mathbf{p}^b - \sigma^{bc}\mathbf{p}^a, \\ [\mathbb{R}^3, \mathbb{R}^3] = \{0\}, & [\mathbf{p}^a, \mathbf{p}^b] = 0. \end{array}$$

Compatible with the vector addition and the action of the special rotation group, there exists only the trivial order $\vec{x} \succeq \vec{y} \iff \vec{x} = \vec{y}$ as position order. The *preorder* of position via the length of the translations

$$\vec{x} \succeq \vec{y} \iff \| \vec{x} \|_\sigma \geq \| \vec{y} \|_\sigma$$

is not antisymmetric. Its totally ordered equivalence classes, the 0-centered 2-spheres, are isomorphic to the positive numbers. They determine the radial coordinate $r \in \mathbb{R}_+$ multiplying in polar coordinates the angular coordinates of the 2-sphere[3] Ω^2 (direction $\frac{\vec{x}}{r}$ of $\vec{x}$), isomorphic to the axial rotation classes

$$\begin{array}{l} \mathbb{S} \ni \vec{x} \longmapsto [\vec{x}]_\succeq = \{\vec{y} \in \mathbb{S} \mid r = \| \vec{x} \|_\sigma = \| \vec{y} \|_\sigma\} \in \mathbb{R}_+, \\ \mathbb{R}^3 \cong \mathbb{R}_+ \times \Omega^2, \quad \Omega^2 \cong \mathbf{SO}(3)/\mathbf{SO}(2). \end{array}$$

1.2.1 Axial Vectors for Rotations

Angular momenta for position $\mathbb{S} \cong \mathbb{R}^3$ determine axial vectors $\vec{\varphi}$ as their rotation axes, which, together with position translations and momenta (polar

[3]The symbols Ω^s, $\mathcal{Y}^s$, and V^s for s-dimensional spheres, hyperboloids, and light cones are chosen for their "similarity" with the 1-dimensional circle, the one branch hyperbola, and its asymptotic cone.

vectors), can be looked at in one vector space, e.g., in position space. The related isomorphism uses the scalar-product-induced position-momentum isomorphism $\mathbb{S} \overset{\delta}{\leftrightarrow} \mathbb{S}^T$ (for simplicity $\sigma \cong \delta = \mathbf{1}_3$), concatenated with a volume-element-induced isomorphism ϵ (axial vector isomorphism):

$$\begin{array}{ccccc} \mathbb{S} \otimes \mathbb{S}^T & \overset{\delta}{\longleftrightarrow} & \mathbb{S} \otimes \mathbb{S} & \overset{\epsilon}{\longleftrightarrow} & \mathbb{S}, \\ \varphi_a \mathcal{O}^a = -\varphi_a \epsilon^{abc} \mathbf{p}^b \otimes \mathbf{x}_c & \leftrightarrow & -\varphi_a \epsilon^{abc} \mathbf{p}^b \otimes \mathbf{p}^c & \leftrightarrow & -\varphi_a \mathbf{p}^a. \end{array}$$

The "infinitesimal" rotation of a translation $\vec{x}$ is expressible by the vector product

$$\varphi_a \mathcal{O}^a(\vec{x}) = -\vec{\varphi} \times \vec{x}.$$

In a box-diagonal form, the rotation axis can be chosen to define the third axis

$$\begin{array}{rcll} \exp \varphi \mathcal{O}_3 & = & \left(\begin{array}{cc|c} \cos\varphi & -\sin\varphi & 0 \\ \sin\varphi & \cos\varphi & 0 \\ \hline 0 & 0 & 1 \end{array}\right) & \in \mathbf{SO}(2) \subset \mathbf{SO}(3), \\ \begin{pmatrix} 0 \\ 0 \\ \varphi \end{pmatrix} \cong \varphi \mathcal{O}_3 & = & \left(\begin{array}{cc|c} 0 & -\varphi & 0 \\ \varphi & 0 & 0 \\ \hline 0 & 0 & 0 \end{array}\right) & \in \log \mathbf{SO}(2) \subset \log \mathbf{SO}(3). \end{array}$$

Any other axis is obtained by a rotation $O(\frac{\vec{\varphi}}{\varphi})$ whose matrix has to display an orthonormal basis in both the three columns and in the three rows:

$$\begin{pmatrix} \varphi_1 \\ \varphi_2 \\ \varphi_3 \end{pmatrix} = O(\tfrac{\vec{\varphi}}{\varphi}) \begin{pmatrix} 0 \\ 0 \\ \varphi \end{pmatrix}, \quad |\vec{\varphi}| = \varphi,$$

$$O(\tfrac{\vec{\varphi}}{\varphi}) = (\tfrac{\vec{\varphi}_\perp}{\varphi}, \tfrac{\vec{\varphi} \times \vec{\varphi}_\perp}{\varphi^2}, \tfrac{\vec{\varphi}}{\varphi}) = \frac{1}{\varphi} \left(\begin{array}{cc|c} \varphi - \frac{\varphi_1^2}{\varphi + \varphi_3} & -\frac{\varphi_1 \varphi_2}{\varphi + \varphi_3} & \varphi_1 \\ -\frac{\varphi_1 \varphi_2}{\varphi + \varphi_3} & \varphi - \frac{\varphi_2^2}{\varphi + \varphi_3} & \varphi_2 \\ \hline -\varphi_1 & -\varphi_2 & \varphi_3 \end{array}\right) \Subset \mathbf{SO}(3)/\mathbf{SO}(2).$$

One thus obtains explicitly the matrix $R(\vec{\varphi})$ for the rotation around $\frac{\vec{\varphi}}{\varphi}$ with angle φ:

$$\begin{array}{rcl} O(\tfrac{\vec{\varphi}}{\varphi}) \circ \begin{pmatrix} \varphi \mathcal{O}_3 \\ \exp \varphi \mathcal{O}_3 \end{pmatrix} \circ O(\tfrac{\vec{\varphi}}{\varphi})^{-1} & = & \begin{pmatrix} \varphi_a \mathcal{O}^a \\ \exp \varphi_a \mathcal{O}^a \end{pmatrix}, \\ \varphi_a \mathcal{O}^a & = & \begin{pmatrix} 0 & -\varphi_3 & \varphi_2 \\ \varphi_3 & 0 & -\varphi_1 \\ -\varphi_2 & \varphi_1 & 0 \end{pmatrix} \in \log \mathbf{SO}(3), \\ R(\vec{\varphi}) = \exp \varphi_a \mathcal{O}^a & \cong & \delta_{ab} \cos\varphi - \epsilon_{abc} \varphi_c \frac{\sin\varphi}{\varphi} + \varphi_a \varphi_b \frac{1 - \cos\varphi}{\varphi^2} \in \mathbf{SO}(3). \end{array}$$

The matrix $O(\frac{\vec{\varphi}}{\varphi})$ is determined up to 3-axial rotations $\mathbf{SO}(2)$. It is a representative of the class $O(\frac{\vec{\varphi}}{\varphi})\mathbf{SO}(2)$, which, by itself, is an element of the manifold $\mathbf{SO}(3)/\mathbf{SO}(2)$, therefore the double-element symbol $\Subset$.

All rotations (all angular momenta) can be parametrized with an abelian axial rotation subgroup $\mathbf{SO}(2)$ (from one abelian Lie subalgebra $\log \mathbf{SO}(2)$) and a rotation from the corresponding class $O \Subset \mathbf{SO}(3)/\mathbf{SO}(2) \cong \Omega^2$. This *Cartan factorization* (decomposition, diagonalization) of the rotation group into rotation angle and rotation axis will be denoted as follows:

$$\begin{pmatrix} \mathbf{SO}(3) \\ \log \mathbf{SO}(3) \end{pmatrix} \cong \begin{pmatrix} \mathbf{SO}(2) \\ \log \mathbf{SO}(2) \end{pmatrix} \circ \Omega^2.$$

In general, a Cartan factorization is not a manifold decomposition; here $\mathbf{SO}(3) \neq \Omega^1 \times \Omega^2$. A point on the 2-sphere Ω^2 gives the direction $\frac{\vec{\varphi}}{\varphi}$ of the rotation axis. The rotation $\mathbf{SO}(2)$ around this axis is determined by a point $\varphi \in [-\pi, \pi]$ on the sphere diameter for the possible rotation angles. Antipodal sphere points $\varphi = \pm\pi$ describe the same rotation.

1.2.2 Orientation Manifold of Scalar Products

Kepler's discovery of the elliptic planetary orbits put to an end the prejudice that circles were necessary for a "harmonious world." However, the operational characteriziation of circles and ellipses is not so different: Both are $\mathbf{SO}(2)$-orbits. The ellipse $\frac{x_1^2}{a^2} + \frac{x_2^2}{b^2} = 1$ with the metric (scalar product)

$$\begin{aligned} \sigma(a,b) &= \begin{pmatrix} \frac{1}{a^2} & 0 \\ 0 & \frac{1}{b^2} \end{pmatrix} = \frac{1}{ab} \begin{pmatrix} \frac{b}{a} & 0 \\ 0 & \frac{a}{b} \end{pmatrix} = h(a,b)^T \mathbf{1}_2 h(a,b), \\ \text{with } h(a,b) &= \begin{pmatrix} \frac{1}{a} & 0 \\ 0 & \frac{1}{b} \end{pmatrix} = \frac{1}{\sqrt{ab}} \begin{pmatrix} \sqrt{\frac{b}{a}} & 0 \\ 0 & \sqrt{\frac{a}{b}} \end{pmatrix}, \end{aligned}$$

is left invariant by the corresponding transformation group $\mathbf{SO}(2, \sigma)$

$$O(\varphi, \tfrac{a}{b}) = \begin{pmatrix} \cos\varphi & -\frac{b}{a}\sin\varphi \\ \frac{a}{b}\sin\varphi & \cos\varphi \end{pmatrix} = h(a,b) \begin{pmatrix} \cos\varphi & -\sin\varphi \\ \sin\varphi & \cos\varphi \end{pmatrix} h(a,b)^{-1}$$

with Lie algebra $\{\begin{pmatrix} 0 & -\frac{b}{a}\varphi \\ \frac{a}{b}\varphi & 0 \end{pmatrix} \bigm| 0 \le \varphi < 2\pi\}$. The ellipse area πab is unchanged for all transformations $h(a,b) \in \mathbf{SL}(\mathbb{R}^2)$ with unit determinant. The function $\tan\alpha = \frac{a}{b}$ characterizes the orthogonal triangle with the two axes of the ellipse.

Generalizing to a 3-dimensional position: There exist many scalar products σ for position translations and therefore many rotation groups $\mathbf{SO}(3, \sigma)$ in the position automorphisms $\mathbf{GL}(\mathbb{R}^3)$. With a fixed basis $\{\mathbf{p}^a\}_{a=1,2,3}$, the real 6-dimensional manifold of all scalar products is parametrizable by the symmetric positive definite 3×3 matrices (metric tensors)

$$\sigma(\mathbf{p}^a, \mathbf{p}^b)_{a,b=1,2,3} \cong (\sigma)_{\mathbf{p}} = \begin{pmatrix} \sigma^{11} & \sigma^{12} & \sigma^{13} \\ \sigma^{12} & \sigma^{22} & \sigma^{23} \\ \sigma^{13} & \sigma^{23} & \sigma^{33} \end{pmatrix}.$$

This scalar product manifold can be visualized by all 3-dimensional ellipsoids centered at the origin of Euclidean space $\mathbb{R}^3$ with a given basis. Starting from one scalar product σ, an automorphism $h \in \mathbf{GL}(\mathbb{R}^3)$, not from the invariance group $h \notin \mathbf{O}(3, \sigma)$, defines a different ellipsoid and scalar product with a different invariance rotation group

$$\mathbf{O}(3, h^T \circ \sigma \circ h) = h^{-1} \circ \mathbf{O}(3, \sigma) \circ h.$$

The manifold of all $\mathbb{S}$-bases (3-bein manifold) is the $\mathbf{GL}(\mathbb{R}^3)$-orbit of any basis

$$\mathbf{GL}(\mathbb{R}^3) \bullet \{\mathbf{p}^a\}_{a=1,2,3} \cong \mathbf{GL}(\mathbb{R}^3).$$

The manifold of all scalar products is the $\mathbf{GL}(\mathbb{R}^3)$-orbit of any scalar product σ. It is isomorphic to the noncompact 6-parametric quotient manifold $\mathbf{GL}(\mathbb{R}^3)/\mathbf{O}(3)$, the *3-bein manifold modulo to the rotation group.* This symmetric space is the product of the 1-parametric dilation group $\mathbf{D}(1)$ and the 5-parametric shear transformation manifold $\mathbf{SL}(\mathbb{R}^3)/\mathbf{SO}(3)$:

$$\mathbf{GL}(\mathbb{R}^3) \bullet \sigma \cong \mathbf{GL}(\mathbb{R}^3)/\mathbf{O}(3) \cong \mathbf{D}(\mathbf{1}_3) \times \mathbf{SL}(\mathbb{R}^3)/\mathbf{SO}(3).$$

In contrast to the "nature" of a scalar product, given by the dimensionality 3 and its positive definiteness, Weyl called, in this context, the manifold of the ellipsoids, i.e., of rotation groups $\mathbf{SO}(3)$ in $\mathbf{GL}(\mathbb{R}^3)$, the "orientations" of a scalar product. As an abstract Lie group, there is only one $\mathbf{SO}(3)$; it has many equivalent representations in the position automorphisms.

The six real operations in the 3-bein manifold: Since symmetric, any scalar product matrix $(\sigma)_\mathbf{p} = (\sigma)_\mathbf{p}^T$ can be transformed by an orthogonal matrix from the maximal compact subgroup $\mathbf{O}(3) \subset \mathbf{GL}(\mathbb{R}^3)$ (three parameters) to a matrix with the solutions of the characteristic polynomial of $(\sigma)_\mathbf{p}$ on the diagonal (transformation to principal axes of the ellipsoid):

$$\begin{aligned}(\sigma)_\mathbf{p} = & \; O^T \circ \begin{pmatrix} \frac{1}{\ell_1^2} & 0 & 0 \\ 0 & \frac{1}{\ell_2^2} & 0 \\ 0 & 0 & \frac{1}{\ell_3^2} \end{pmatrix} \circ O \text{ with } O^T = O^{-1}, \\ & \{\ell_{1,2,3}^{-2}\} = \{\xi \mid \det[(\sigma)_\mathbf{p} - \xi \mathbf{1}_3] = 0\}.\end{aligned}$$

The three diagonal elements with possibly different length units for the three dimensions are acted on by the maximal noncompact abelian subgroup, the three axis dilations $\mathbf{D}(1)^3 \subset \mathbf{GL}(\mathbb{R}^3)$. The positive discriminant $\frac{1}{\ell^6}$ with $\frac{4\pi}{3}\ell^3$ the volume of the ellipsoid defines the $\mathbf{SL}(\mathbb{R}^3)$-invariant $\mathbf{D}(\mathbf{1}_3)$-parameter of the scalar product as *length unit*, e.g., m (meters), with respect to the fixed basis $\{\mathbf{p}^a\}_{a=1,2,3}$:

$$\det(\sigma)_\mathbf{p} = \tfrac{1}{\ell^6}, \quad \ell^3 = \ell_1 \ell_2 \ell_3, \quad \text{length unit } \ell.$$

The double hyperbolic dilation group $\mathbf{D}(1)^2 \cong \mathbf{SO}_0(1,1)^2 \subset \mathbf{SL}(\mathbb{R}^3)$ allows the transition from a diagonal matrix to an equally normalized diagonal matrix (transformation from an ellipsoid to a sphere with equal volume)

$$(\sigma)_\mathbf{p} = O^T \circ d^T \circ \ell^{-2}\mathbf{1}_3 \circ d \circ O, \quad d = \begin{pmatrix} \frac{\ell}{\ell_1} & 0 & 0 \\ 0 & \frac{\ell}{\ell_2} & 0 \\ 0 & 0 & \frac{\ell}{\ell_3} \end{pmatrix} \in \mathbf{SO}_0(1,1)^2.$$

In an interpretation, not with different scalar products in one fixed basis but with one fixed scalar product considered in different bases, one can also say, For any scalar product σ there exist bases or adapted units such that its matrix is diagonal, even equally normalized (Euclidean bases).

In Euclidean bases the angular momenta have commutators with totally antisymmetric structure constants involving the totally antisymmetric Grassmann symbol ϵ^{abc} with $\epsilon^{123} = 1$:

$$\text{Euclidean bases:} \left\{ \begin{array}{l} \sigma = \frac{1}{\ell^2}\delta^{ab}\mathbf{x}_a \otimes \mathbf{x}_b, \quad \sigma^{-1} = \ell^2 \delta_{ab}\mathbf{p}^a \otimes \mathbf{p}^b, \\ \{\mathcal{O}^a = -\epsilon^{abc}\mathbf{p}^b \otimes \mathbf{x}_c \mid a = 1,2,3\}, \\ [\mathcal{O}^a, \mathcal{O}^b] = -\epsilon^{abc}\mathcal{O}^c \text{ with } \epsilon^{abc}\epsilon^{ade} = \delta^{bd}\delta^{ce} - \delta^{be}\delta^{cd}, \\ \operatorname{tr} \mathcal{O}^a \circ \mathcal{O}^b = -2\delta^{ab}, \text{ invariant: } \mathcal{O}_a \otimes \mathcal{O}^a. \end{array} \right.$$

The sum of the dual metric tensors for the rotations is used as Hamiltonian H for the isotropic 3-dimensional harmonic oscillator (short notation $\vec{\mathbf{p}}^2 =$

$\delta_{ab}\mathbf{p}^a \otimes \mathbf{p}^b$, etc.):

$$\begin{aligned} \frac{\sigma^{-1} \oplus \sigma}{2} &= \frac{\ell^2}{2\hbar}\vec{\mathbf{p}}^2 \oplus \frac{\hbar}{2\ell^2}\vec{\mathbf{x}}^2 = \frac{H}{\mu} \\ \text{with } H &= \frac{1}{2M}\vec{\mathbf{p}}^2 \oplus \frac{k}{2}\vec{\mathbf{x}}^2, \quad \left\{ \begin{array}{ll} \text{frequency unit:} & \mu = \sqrt{\frac{k}{M}}, \\ \text{length unit:} & \ell^2 = \frac{\hbar}{\sqrt{kM}}. \end{array} \right. \end{aligned}$$

1.3 Spacetime Translations

Nontrivial scalar products of time and position translations can be transferred to the direct vector space sum $\mathbb{T} \oplus \mathbb{S} \cong \mathbb{R} \oplus \mathbb{R}^3$. A relative sign remains free for the induced inner product

$$\begin{aligned} g_\pm(x_0\mathbf{p}^0 + x_a\mathbf{p}^a, y_0\mathbf{p}^0 + y_b\mathbf{p}^b) &= \tau(x_0\mathbf{p}^0, y_0\mathbf{p}^0) \pm \sigma(x_a\mathbf{p}^a, y_b\mathbf{p}^b) \\ &= x_0 y_0 \mu^2 \pm x_a y_a \ell^{-2}. \end{aligned}$$

The *nontrivial order structure of time (causal structure)* can be embedded into spacetime *only with a relative negative sign* for the scalar products of time and position translations

$$g_-^{jk} = \begin{pmatrix} \mu^2 & 0 \\ 0 & -\ell^{-2}\mathbf{1}_3 \end{pmatrix}.$$

A definite scalar product g_+ is compatible only with the trivial order.

However, time and space translations do not come as a direct sum with a fixed time and position. Time and position are relativized (special relativity): *Minkowski space* $\mathbb{M}$ is a real 4-dimensional vector space with a symmetric bilinear form g of signature $(1,3)$, called the *Lorentz (inner) product or (pseudo)metric*:

$$g(\ ,\ ) : \mathbb{M} \times \mathbb{M} \longrightarrow \mathbb{R}, \quad \left\{ \begin{array}{l} g(x,y) = g(y,x) = x_i y_j g^{ij}, \\ \operatorname{sign} g = (1,3). \end{array} \right.$$

Spacetime vectors should be considered as operations. "Spacetime" in this chapter is always understood as a vector space containing the spacetime translations, synonymous with Minkowski space.[4] The spaces on which spacetime translations act will be discussed in the chapters "Lorentz Symmetry" and "Harmonic Analysis."

Almost everywhere the generalization of these and the following structures on a spacetime $\mathbb{M} \cong \mathbb{R}^{1+s}$ is obvious, the most interesting question asks what distinguishes a special position dimension, e.g., $s = 3$.

The Lorentz-form-induced dual isomorphism between a Minkowski spacetime and its dual *energy-momentum space* $\mathbb{M}^T$ with the spacetime translation eigenvalues is used in the raising and lowering of indices:

$$\begin{aligned} \mathbb{M} &= \{x = x_i\mathbf{p}^i \mid x_i \in \mathbb{R}\} \cong \mathbb{R}^4, \quad \text{basis: } \{\mathbf{p}^i\}_{i=0,1,2,3}, \\ \mathbb{M}^T &= \{p = p^i\mathbf{x}_i \mid p^i \in \mathbb{R}\} \cong \mathbb{R}^4, \quad \text{dual bases: } \langle \mathbf{x}_j, \mathbf{p}^i \rangle = \delta^i_j, \\ g :\ \mathbb{M} &\longrightarrow \mathbb{M}^T, \quad g(x) = g(x,\ ), \quad g(x_i\mathbf{p}^i) = x_i g^{ij}\mathbf{x}_j = x^j\mathbf{x}_j, \\ g &= g^{ij}\mathbf{x}_i \otimes \mathbf{x}_j, \quad g^{-1} = g_{ij}\mathbf{p}^i \otimes \mathbf{p}^j. \end{aligned}$$

[4] A Minkowski manifold allows at each point a tangent Minkowski space.

The invariance group of a Lorentz isomorphism (metric) g is the associated *Lorentz group* $\mathbf{O}(1,3;g)$, sometimes written $\mathbf{O}(1,3)$ for short, a real 6-dimensional noncompact Lie group:

$$\begin{array}{ccc} \mathbb{M} & \stackrel{\Lambda}{\longrightarrow} & \mathbb{M} \\ g\downarrow & & \downarrow g \\ \mathbb{M}^T & \underset{\Lambda^{-1T}}{\longrightarrow} & \mathbb{M}^T \end{array}, \quad \begin{array}{l} \Lambda^T \circ g \circ \Lambda = g, \quad g(\Lambda.x, \Lambda.y) = g(x,y), \\ \Lambda = \Lambda_j^k \, \mathbf{p}^j \otimes \mathbf{x}_k \Rightarrow \Lambda_i^j g^{il} \Lambda_l^k = g^{jk}, \end{array}$$

$$\begin{array}{l} \mathbf{O}(1,3) = \{\Lambda \in \mathbf{GL}(\mathbb{R}^4) \mid \Lambda^T \circ g \circ \Lambda = g\} \in \underline{\mathbf{lgrp}}_{\mathbb{R}}, \\ \dim_{\mathbb{R}} \mathbf{O}(1,3) = 6. \end{array}$$

Time and position vectors are elements of their Lorentz group orbits. In a space with the action of a group it does not make sense to distinguish individual points; one has to "think in group orbits." The orbits $\mathbb{M}/\mathbf{O}(1,3)$ are characterized by the sign of the Lorentz square. Spacetime translations are called

timelike:	$x \in \mathbb{M}^{\text{time}} \iff g(x,x) > 0$ or $x = 0$,
lightlike:	$x \in \mathbb{M}^{\text{light}} \iff g(x,x) = 0$ and $x \neq 0$,
causal:	$x \in \mathbb{M}^{\text{time}} \cup \mathbb{M}^{\text{light}} \iff g(x,x) \geq 0$,
spacelike (generalized present):	$x \in \mathbb{M}^{\text{position}} \iff g(x,x) < 0$ or $x = 0$,
strictly present:	$x = 0$.

The lightlike vectors $\mathbb{M}^{\text{light}}$ are singular in spacetime; they do not arise in time or position.

The Lorentz group classes with respect to its *special Lorentz transformations* as normal subgroup constitute a group, isomorphic to the discrete reflection group $\mathbb{I}(2)$. The group $\mathbf{O}(1,3)$ has many $\mathbb{I}(2)$-isomorphic subgroups; none is a direct factor for $\mathbf{SO}(1,3)$. One obtains semidirect group decompositions[5]

$$\mathbf{O}(1,3) \cong \mathbb{I}(2) \vec{\times} \mathbf{SO}(1,3), \quad \mathbf{SO}(1,3) = \{\Lambda \in \mathbf{O}(1,3) \mid \det \Lambda = 1\}.$$

For any reflection group $\mathbb{I}(2) \cong \{\mathtt{R}, \mathtt{R}^2 = \mathrm{id}_{\mathbb{M}}\} \subset \mathbf{O}(1,3)$ with $\det \mathtt{R} = -1$ the semidirect product is given by

$$\begin{array}{l} \mathbb{I}(2) \times \mathbf{SO}(1,3) \longrightarrow \mathbf{SO}(1,3), \quad (\mathtt{I}, \Lambda) \longmapsto \mathtt{I} \circ \Lambda \circ \mathtt{I}, \\ \mathbb{I}(2) \vec{\times} \mathbf{SO}(1,3): \quad (\mathtt{I}_1, \Lambda_1) \circ (\mathtt{I}_2, \Lambda_2) = (\mathtt{I}_1 \circ \mathtt{I}_2, \Lambda_1 \circ \mathtt{I}_1 \circ \Lambda_2 \circ \mathtt{I}_1). \end{array}$$

Under special Lorentz transformations the nontrivial spacetime translations move in their 3-dimensional $\mathbf{SO}(1,3)$-orbits, the spacelike ones on a one-shell hyperboloid, the timelike ones on a two-shell hyperboloid, the lightlike ones on the skin of a vertexless double cone. The trivial vector $x = 0$ remains

[5] The semidirect decomposition of the real $\binom{1+s}{2}$-dimensional Lie group $\mathbf{O}(1,s) \cong \mathbb{I}(2) \vec{\times} \mathbf{SO}(1,s)$ is a direct decomposition for even position dimension $s \in 2\mathbb{N}$.

fixed. Therefore, the product group $\mathbf{D}(1) \times \mathbf{SO}(1,3)$ with the dilation group $\mathbf{D}(1) : x \longmapsto e^\lambda x$ decomposes Minkowski spacetime into four disjoint orbits:

$$\begin{aligned} \mathbb{M} &= \mathbb{M}^{\text{light}} \uplus (\mathbb{M}^{\text{time}} \cup \mathbb{M}^{\text{position}}) \\ &= \mathbb{M}^{\text{light}} \uplus \{0\} \uplus (\mathbb{M}^{\text{time}} \setminus \{0\}) \uplus (\mathbb{M}^{\text{position}} \setminus \{0\}). \end{aligned}$$

Lorentz transformations keep spacelike and timelike translations strictly apart; the main difference to a nonrelativistic framework is not a mixture of time and position, but the fact that the Lorentz-transformation-stable translations $\mathbb{M}^{\text{time}}$ and $\mathbb{M}^{\text{position}}$ are no longer linear spaces, in contrast to the linear spaces in a Lorentz-transformation-incompatible decomposition $\mathbb{M} \cong \mathbb{R} \oplus \mathbb{R}^3$ (more below).

A special Lorentz group $\mathbf{SO}(1,3;g)$ is the direct product of the metric-independent *spacetime reflection group* $\mathbb{I}(2)$, its center, and its *orthochronous Lorentz group* $\mathbf{SO}_0(1,3;g)$:

$$\mathbf{SO}(1,3) \cong \mathbb{I}(2) \times \mathbf{SO}_0(1,3), \quad \mathbb{I}(2) \cong \{\mathrm{id}_{\mathbb{M}}, -\mathrm{id}_{\mathbb{M}}\}.$$

$\mathbf{SO}_0(1,3)$ is the connection component of the unit in a Lorentz group:

$$\mathbf{O}(1,3) \cong \mathbb{I}(2) \,\vec{\times}\, [\mathbb{I}(2) \times \mathbf{SO}_0(1,3)].$$

A full discrete *spacetime reflection group* (a four-element Klein group) contains two Lorentz-invariant elements, the identity and its negative, the Minkowski space reflection, and two non-Lorentz-invariant reflections with negative determinant, one of which can be chosen arbitrarily, e.g., a time reflection $\mathtt{T}$, defining as its negative $\mathtt{P}$ as position reflection. The four reflections have the multiplication table

$\mathbf{1}_4$	$-\mathbf{1}_4$	$\mathtt{T}$	$\mathtt{P}$
$-\mathbf{1}_4$	$\mathbf{1}_4$	$\mathtt{P}$	$\mathtt{T}$
$\mathtt{T}$	$\mathtt{P}$	$\mathbf{1}_4$	$-\mathbf{1}_4$
$\mathtt{P}$	$\mathtt{T}$	$-\mathbf{1}_4$	$\mathbf{1}_4$

$\mathtt{T} \in \mathbf{O}(1,3)/\mathbf{SO}(1,3)$,
$\det \mathtt{T} = -1, \ \mathtt{P} = -\mathtt{T}$,

$$\mathbf{O}(1,3)/\mathbf{SO}_0(1,3) \cong \mathbb{I}(2) \times \mathbb{I}(2) \cong \{\pm\mathbf{1}_4, \mathtt{T}, \mathtt{P}\}.$$

With the experimentally observed violation of position reflection invariance, only the orthochronous group can be a basic symmetry.

The real 6-dimensional *Lie algebra* $\log \mathbf{SO}_0(1,3)$ *of a Lorentz group*

$$\begin{array}{ccc} \mathbb{M} & \xrightarrow{\mathcal{L}} & \mathbb{M} \\ g \downarrow & & \downarrow g \\ \mathbb{M}^T & \xrightarrow[-\mathcal{L}^T]{} & \mathbb{M}^T \end{array}, \qquad \begin{array}{l} g \circ \mathcal{L} = -\mathcal{L}^T \circ g, \quad g(x, \mathcal{L}.y) = -g(\mathcal{L}.x, y), \\ \mathcal{L} = \mathcal{L}^j_k \, \mathbf{p}^k \otimes \mathbf{x}_j \Rightarrow g_{ij}\mathcal{L}^j_k = -\mathcal{L}^j_i g_{jk}, \end{array}$$

$$\begin{aligned} \log \mathbf{SO}_0(1,3) &= \{\mathcal{L} \in \mathbf{AL}(\mathbb{R}^4) \bigm| g \circ \mathcal{L} = -\mathcal{L}^T \circ g\} \\ &= \{x_\alpha \otimes g(y_\alpha) - y_\alpha \otimes g(x_\alpha) \bigm| x_\alpha, y_\alpha \in \mathbb{M}\} \in \underline{\mathbf{lag}}_{\mathbb{R}}, \end{aligned}$$

has, with a dual basis for spacetime translations and energy-momenta, as basis for the g-antisymmetric generators

$$\text{basis of } \log \mathbf{SO}_0(1,3): \begin{cases} \{\mathcal{L}_i^j = -\mathbf{p}^j \otimes \mathbf{x}_i + g_{ik}\mathbf{p}^k \otimes \mathbf{x}_m g^{mj} \mid i,j = 0,1,2,3\}, \\ x \otimes p - g^{-1}(p) \otimes g(x) = -x_j p^i \mathcal{L}_i^j, \end{cases}$$
$$\mathcal{L}^{jk} = \mathcal{L}_i^j g^{ik} = -\mathbf{p}^j \otimes \mathbf{x}_i g^{ik} + \mathbf{p}^k \otimes \mathbf{x}_i g^{ij} = -\mathcal{L}^{kj},$$
$$[\mathcal{L}^{jk}, \mathcal{L}^{nm}] = g^{jn}\mathcal{L}^{km} - g^{kn}\mathcal{L}^{jm} - g^{jm}\mathcal{L}^{kn} + g^{km}\mathcal{L}^{jn}.$$

The dual isomorphism g allows an antisymmetric representation in $\mathbb{M} \otimes \mathbb{M}$.

The double trace defines an $\mathbf{O}(1,3)$-invariant inner product with signature $(3,3)$, the Killing form of the Lorentz Lie algebra:

$$\begin{aligned} &\log \mathbf{SO}_0(1,3) \times \log \mathbf{SO}_0(1,3) \longrightarrow \mathbb{R}, \quad \langle \mathcal{L} | \mathcal{L}' \rangle = \operatorname{tr} \mathcal{L} \circ \mathcal{L}', \\ &\operatorname{tr} [\mathcal{L}'', \mathcal{L}] \circ \mathcal{L}' + \operatorname{tr} \mathcal{L} \circ [\mathcal{L}'', \mathcal{L}'] = 0, \\ &\operatorname{tr} \mathcal{L}^{jk} \circ \mathcal{L}^{mn} = -2 \det \begin{pmatrix} g^{jm} & g^{jn} \\ g^{km} & g^{kn} \end{pmatrix}, \quad \text{invariant: } \mathcal{L}_{jk} \otimes \mathcal{L}^{jk}. \end{aligned}$$

Using the dual isomorphism g, the volume elements of the spacetime translations define a second type of invariant bilinear forms; this is possible only for dimension $1 + s = 4$

$$\begin{aligned} \epsilon(\ ,\ ): \ &\log \mathbf{SO}_0(1,3) \times \log \mathbf{SO}_0(1,3) \longrightarrow \mathbb{R}, \\ &\epsilon([\mathcal{L}'', \mathcal{L}], \mathcal{L}') + \epsilon(\mathcal{L}, [\mathcal{L}'', \mathcal{L}']) = 0, \\ &\epsilon(\mathcal{L}^{jk}, \mathcal{L}^{mn}) = \epsilon^{jkmn}, \quad \text{invariant: } \epsilon_{jkmn}\mathcal{L}^{jk} \otimes \mathcal{L}^{mn}. \end{aligned}$$

The signature for these bilinear forms is also $(3,3)$. The volume elements $\epsilon : \mathbb{M}^4 \longrightarrow \mathbb{R}$ of Minkowski spacetime cannot be expressed by the metric $g : \mathbb{M}^2 \longrightarrow \mathbb{R}$. The Lorentz Lie algebra has two independent types of invariant bilinear forms, which can be related to the rank 2 of the Lorentz Lie algebra (chapter "Lorentz Symmetry").

A *Poincaré group* $\mathbf{SO}_0(1,3) \vec{\times} \mathbb{R}^4$ with 10 real parameters is the semidirect product of the spacetime translations as abelian normal subgroup and homogeneously acting Lorentz transformations

$$\mathbf{O}(1,3) \vec{\times} \mathbb{R}^4 : \quad (\Lambda_1, x_1) \circ (\Lambda_2, x_2) = (\Lambda_1 \circ \Lambda_2, x_1 + \Lambda_1.x_2).$$

The Poincaré group representations as used for relativistic free particles are discussed in the chapters "Massive Particle Quantum Fields," "Massless Quantum Fields" and "Harmonic Analysis." The semidirect *Lie algebra of a Poincaré group* is the direct vector space sum of a Lorentz Lie algebra and the maximal abelian ideal with the translations

$$\begin{aligned} \log \mathbf{SO}_0(1,3) \ \vec{\oplus}\ \mathbb{R}^4, &\quad \log \mathbf{SO}_0(1,3) \cong \mathbb{R}^6, \\ [\log \mathbf{SO}_0(1,3), \mathbb{R}^4] = \mathbb{R}^4, &\quad [\mathcal{L}^{jk}, \mathbf{p}^n] = \mathcal{L}^{jk}(\mathbf{p}^n) = g^{jn}\mathbf{p}^k - g^{kn}\mathbf{p}^j, \\ [\mathbb{R}^4, \mathbb{R}^4] = \{0\}, &\quad [\mathbf{p}^j, \mathbf{p}^n] = 0. \end{aligned}$$

1.3.1 Order of Minkowski Space

In contrast to Euclidean spaces with compact action group $\mathbf{O}(s)$, Minkowski spacetimes with noncompact $\mathbf{O}(1,s)$-action have a nontrivial order (causal) structure.

Under orthochronous transformations $\mathbf{SO}_0(1,3)$ the nontrivial time- and lightlike vectors no longer move on two-shell hyperboloids. Time and light can be decomposed into three and two orbits, respectively, of $\mathbf{D}(1) \times \mathbf{SO}_0(1,3)$:

$$\begin{aligned} \mathbb{M}^{\text{time}} &= \{0\} \uplus \mathbb{M}_+^{\text{time}} \uplus \mathbb{M}_-^{\text{time}}, \\ \mathbb{M}^{\text{light}} &= \mathbb{M}_+^{\text{light}} \uplus \mathbb{M}_-^{\text{light}}. \end{aligned}$$

A distinction between future and past becomes possible using the bilinear form with which the Lorentz group $\mathbf{O}(1,3,g)$ is defined, and, in addition, any nontrivial causal translation $c \neq 0$, $g(c,c) \geq 0$ with respect to g and c:

$$\begin{aligned} \mathbb{M}_\pm^{\text{time}} &= \{x \,\big|\, g(x,x) > 0, \ \ \pm g(c,x) \geq 0\}, \\ \mathbb{M}_\pm^{\text{light}} &= \{x \neq 0 \,\big|\, g(x,x) = 0, \ \ \pm g(c,x) \geq 0\}, \\ (g,c)\text{-}order\text{: } x \succeq_c 0 &\iff x \in \{0\} \cup \mathbb{M}_+^{\text{time}} \cup \mathbb{M}_+^{\text{light}}. \end{aligned}$$

The c-dependence of the spacetime order involves only the unavoidable reflection: Two (g,c)-orders are either identical or reflected to each other, depending on the sign of $g(c_1,c_2)$:

$$\epsilon g(c_1,c_2) \geq 0 \text{ with } \epsilon = \pm 1 \iff (x \succeq_{c_1} 0 \iff \epsilon x \succeq_{c_2} 0).$$

A nontrivial Minkowski-order pair $(\succeq,\preceq)_g$ is natural, i.e., induced by the order of the real numbers as follows: Any $\mathbb{R}$-induced order pair $(\succeq,\preceq)$ of a real vector space will be characterized via $\mathbb{R}$-positivity of even-multilinear $\mathbb{R}$-forms $\{\gamma^2,\gamma^4,\dots\}$ (bilinear, quadrilinear, etc.), i.e., by conditions for the spacetime translations

$$x \succeq 0 \iff \gamma^{2n}(x,\dots,x) \geq 0.$$

The even-linear nontrivial forms start with symmetric bilinear forms $\gamma^2(x,y) = \gamma^2(y,x)$, which, if nondegenerate, are characterized on $\mathbb{R}^4$ by the signatures $(4,0)$, $(1,3)$, and $(2,2)$ and corresponding invariance groups $\mathbf{O}(p,q)$. Both the $\mathbf{O}(2,2)$- and $\mathbf{O}(4)$-invariant inner products are compatible only with the trivial order structure: each vector $x \in \mathbb{M}$ has itself $\{x\}$ as causal set. A bilinear $\mathbf{O}(1,3)$-form g defines a unique nontrivial order pair $(\succeq,\preceq)_g$.

1.3.2 Spacetime Topology

Like any $\mathbb{R}^n$-isomorphic topological vector space, a Minkowski space $\mathbb{M}$ also has a unique Hausdorff topology, the *natural topology*, induced by the scalars $\mathbb{R}$. It can be generated by a basis of open neighborhoods for 0 ("here-now"). The open 4-spheres or open 4-cubes are not Lorentz compatible structures: An indefinite inner product does not define a scalar product (length) topology.

An order-induced basis for the natural topology consists of open *"diamonds" (causal double cones)*: Each nontrivial causal vector $c \in \mathbb{M}$ defines an open 0-neighborhood $O_c(0)$ as intersection of the strict future for $-c$ with the strict past for c (or vice versa):

$$c \succ 0, \quad O_c(0) = \{x \in \mathbb{M} \bigm| -c \prec x \prec c\} = O_{-c}(0).$$

In the chapter "Spacetime as Unitary Operation Classes" the causality-related order topology of spacetime will be given by a norm (C*-algebra norm) via spectral values of translations.

All the diamonds define a basis $\mathcal{O}(0)$ of open 0-neighborhoods, which is not empty and which does not contain the empty set. Minkowski spacetime is *directed*, i.e., any two vectors have a common nonempty past and future by the intersection of the corresponding cones.[6] Therefore, the intersection of two diamonds always contains another one

$$\begin{array}{l} \mathcal{O}(0) = \{O_c(0) \bigm| c \in \mathbb{M},\ c \succ 0\}, \\ c_1 \succ 0,\ c_2 \succ 0 \Rightarrow \text{There exists } c \succ 0 \text{ with } O_c(0) \subseteq O_{c_1}(0) \cap O_{c_2}(0). \end{array}$$

1.3.3 Orientation Manifold of Lorentz Metrics

To repeat the $\mathbf{SO}(2)$-class property of circles and ellipses, now for hyperbolas: The two branches of the hyperbola $\frac{x_0^2}{a^2} - \frac{x_3^2}{b^2} = 1$ with "metric" g:

$$\begin{aligned} g(a,b) &= \begin{pmatrix} \frac{1}{a^2} & 0 \\ 0 & -\frac{1}{b^2} \end{pmatrix} = \frac{1}{ab}\begin{pmatrix} \frac{b}{a} & 0 \\ 0 & -\frac{a}{b} \end{pmatrix} = h^T(a,b)\eta_2 h(a,b), \\ \text{with } h(a,b) &= \begin{pmatrix} \frac{1}{a} & 0 \\ 0 & \frac{1}{b} \end{pmatrix} = \frac{1}{\sqrt{ab}}\begin{pmatrix} \sqrt{\frac{b}{a}} & 0 \\ 0 & \sqrt{\frac{a}{b}} \end{pmatrix}, \end{aligned}$$

are invariant under the transformation group $\mathbf{SO}_0(1,1,g)$

$$\Lambda(\psi, \tfrac{a}{b}) = \begin{pmatrix} \cosh\psi & \frac{b}{a}\sinh\psi \\ \frac{a}{b}\sinh\psi & \cosh\psi \end{pmatrix} = h(a,b)\begin{pmatrix} \cosh\psi & \sinh\psi \\ \sinh\psi & \cosh\psi \end{pmatrix} h(a,b)^{-1}$$

with Lie algebra $\{\begin{pmatrix} 0 & \frac{b}{a}\psi \\ \frac{a}{b}\psi & 0 \end{pmatrix} \bigm| \psi \in \mathbb{R}\}$. The area ab of any parallelogram with one corner on the hyperbola, one the intersection point of the asymptotes and sides parallel to the asymptotes is unchanged for all transformations $h(a,b) \in \mathbf{SL}(\mathbb{R}^2)$. Here $\frac{a}{b}$ characterizes the angle between the two asymptotes. Hyperbolas with equal sides $a = b$ are the analogue to the circle.

A real 4-dimensional vector space has many Lorentz metrics g. All bases and all Lorentz metrics arise from a fixed basis $\{\mathbf{p}^j\}_{j=0}^3$ and a fixed metric g:

$$\begin{aligned} \mathbf{GL}(\mathbb{R}^4) \bullet \{\mathbf{p}^j\}_{j=0}^3 &\cong \mathbf{GL}(\mathbb{R}^4), \\ \mathbf{GL}(\mathbb{R}^4) \bullet g &\cong \mathbf{GL}(\mathbb{R}^4)/\mathbf{O}(1,3). \end{aligned}$$

The real 10-parameter *tetrad manifold modulo the Lorentz group*, i.e., the orientation manifold of Lorentz groups $\mathbf{O}(1,3)$ in the spacetime automorphisms $\mathbf{GL}(\mathbb{R}^4)$, can be decomposed with a dilation group

$$\mathbf{GL}(\mathbb{R}^4)/\mathbf{O}(1,3) \cong \mathbf{D}(\mathbf{1}_4) \times \mathbf{SL}_0(\mathbb{R}^4)/\mathbf{SO}_0(1,3)$$

[6] $\mathbb{M} \cong \mathbb{R}^2$ is even a lattice, i.e., two vectors have a unique earliest future element and latest past element.

involving the unit connection component $\mathbf{SL}_0(\mathbb{R}^4) \cong \mathbf{SL}(\mathbb{R}^4)/\mathbb{I}(2)$. Different Lorentz metrics have different invariance groups:

$$h \in \mathbf{GL}(\mathbb{R}^4): \quad \mathbf{O}(1,3; h^T \circ g \circ h) = h^{-1} \circ \mathbf{O}(1,3;g) \circ h.$$

With a fixed basis the orientation manifold of Lorentz metrics is given by the symmetric 4×4 matrices of signature $(1,3)$ and can be visualized - easier for $\mathbf{O}(1,2)$ and $\mathbb{R}^{1+2}$ - by all timelike two-shell metric hyperboloids centered at the origin of a space $\mathbb{R}^4$ with fixed basis. The 10 operations of the tetrad manifold: The metric matrices can be diagonalized (transformation to the principal axes, one timelike and three spacelike ones) with an orthogonal matrix from the maximal compact group $\mathbf{O}(4) \subset \mathbf{GL}(\mathbb{R}^4)$ (six parameters). The four roots of the characteristic polynomial characterize the units in the basic directions for the maximal abelian noncompact group, the dilation group $\mathbf{D}(1)^4 \subset \mathbf{GL}(\mathbb{R}^4)$ with the subgroup $\mathbf{D}(1) \times \mathbf{D}(\mathbf{1}_3) \cong \mathbf{D}(\mathbf{1}_4) \times \mathbf{SO}_0(1,\mathbf{1}_3)$:

$$\begin{aligned}(g)_{\mathbf{P}} &= O^T \circ d^T \circ \begin{pmatrix} \mu^2 & 0 \\ 0 & -\ell^{-2}\mathbf{1}_3 \end{pmatrix} \circ d \circ O & &\text{with } \left\{ \begin{array}{ll} O & \in \mathbf{O}(4), \\ d & \in \mathbf{SO}_0(1,1)^2 \end{array} \right. \\ &= O^T \circ d^T \circ D^T \circ \sqrt{\tfrac{\mu}{\ell^3}} \begin{pmatrix} 1 & 0 \\ 0 & -\mathbf{1}_3 \end{pmatrix} \circ D \circ d \circ O & &\text{with } \quad D \in \mathbf{SO}_0(1,\mathbf{1}_3).\end{aligned}$$

The product of the time and position translation units for $\mathbf{D}(1) \times \mathbf{D}(\mathbf{1}_3)$ gives the $\mathbf{D}(\mathbf{1}_4)$-parameter, the volume of the 4-dimensional "rotated parallelogram." The quotient as *velocity unit* depends on a decomposition into time and position for

$$\begin{aligned}\mathbf{D}(\mathbf{1}_4): \quad &-\det(g)_{\mathbf{P}} = \tfrac{\mu^2}{\ell^6}, \quad \text{for } \mathbf{SO}_0(1,\mathbf{1}_3): \ c = \mu\ell, \\ &\begin{pmatrix} \mu^2 & 0 \\ 0 & -\ell^{-2}\mathbf{1}_3 \end{pmatrix} = \sqrt{\tfrac{\mu}{\ell^3}} \begin{pmatrix} \sqrt{c^3} & 0 \\ 0 & -\frac{1}{\sqrt{c}}\mathbf{1}_3 \end{pmatrix}.\end{aligned}$$

Einstein's general relativity "dynamizes" the ten operations from the tetrad manifold: Lorentz groups are defined by a signature $(1,3)$-metric hyperboloid $g_{\mu\nu}(P)$ assumed for the tangent translations at each point $P \in M$ of the real 4-dimensional spacetime manifold. The tetrad field $h(P) \in \mathbf{GL}(\mathbb{R}^4)/\mathbf{O}(1,3)$ diagonalizes at each spacetime point the translation metric $g_{\mu\nu}(P) = h^j_\mu(P) \eta_{jk} h^k_\nu(P)$. General relativity relates the tetrad h via the derivatives containing spacetime curvature to the energy-momentum tensor for the distribution of matter.

1.4 Decompositions of Spacetime

Time- or spacelike translations do not constitute vector spaces. Vector space decompositions of spacetime translations require additional structures.

There exist direct decompositions $\oplus$ of spacetime translations into maximally four vector subspaces:

$$\mathbb{M} \cong \bigoplus_{a=1}^{A} \mathbb{R}^{n_a}, \quad 1 \leq n_a \leq 3, \quad \sum_{a=1}^{A} n_a = 4.$$

For example, any basis decomposes spacetime into four 1-dimensional subspaces $\mathbb{M} \cong \bigoplus_{j=0}^{3} \mathbb{R}\mathbf{p}^j$. The direct summands may or may not be orthogonal $\perp$ with respect to the Lorentz metric g. A decomposition is invariant under a Lorentz subgroup. In the next subsections the following three decompositions are considered:

$$\mathbb{M} \cong \begin{cases} \mathbb{T} \perp \mathbb{S}^3 & \text{(time and position)}, \\ [\mathbb{L}_+ \oplus \mathbb{L}_-] \perp \mathbb{S}^2 & \text{(light and position)}, \\ \mathbb{L}_1 \oplus \mathbb{L}_2 \oplus \mathbb{L}_3 \oplus \mathbb{L}_4 & \text{(light)}. \end{cases}$$

The first two decompositions are relevant for the definition and description of massive and massless particles (chapters "Massive Particle Quantum Fields" and "Massless Quantum Fields").

1.4.1 Decompositions into Time and Position

Lorentz metrics $g \cong \eta$ in *Sylvester bases* come with the distinction of one nontrivial timelike translation in spacetime or the distinction of one energylike vector (rest system) in energy-momentum space $\mathbb{M}^T$. One thereby obtains a nonrelativistic orthogonal $\mathbb{M}$-decomposition into a 1-dimensional time $\mathbb{T}$ and its 3-dimensional position $\mathbb{S}$:

$$\text{Sylvester bases:} \begin{cases} \mathbb{M} \cong \mathbb{T} \perp \mathbb{S}, \quad \{\mathbf{p}^j\}_{j=0,1,2,3}, \\ g^{jk} = g(\mathbf{p}^j, \mathbf{p}^k) = \left(\begin{array}{c|c} 1 & 0 \\ \hline 0 & -\mathbf{1}_3 \end{array}\right) = \eta^{jk}, \\ g(x,y) = xy = x_0 y_0 - x_a y_a. \end{cases}$$

The invariance group in the orthochronous group $\mathbf{SO}_0(1,3)$ of this decomposition into two summands is a maximal compact rotation group $\mathbf{SO}(3)$.

The dual isomorphism between $\mathbb{M}$ and $\mathbb{M}^T$ uses for position $g^{ab} = -\delta^{ab}$

$$\text{Sylvester bases: } g(\mathbf{p}^0) = \mathbf{x}_0, \quad g(\mathbf{p}^a) = -\mathbf{x}_a.$$

The time reflection $\mathsf{T} = \begin{pmatrix} -1 & 0 \\ 0 & \mathbf{1}_3 \end{pmatrix}$ generates a reflection group $\mathbb{I}(2)$ for a semidirect group decomposition $\mathbf{O}(1,3) \cong \mathbb{I}(2) \vec{\times} \mathbf{SO}(1,3)$ and defines the position reflection $\mathsf{P} = \begin{pmatrix} 1 & 0 \\ 0 & -\mathbf{1}_3 \end{pmatrix}$. A causal spacetime vector is positive with a positive time component:

$$x \succeq 0 \iff x^2 \geq 0 \text{ and } x_0 \geq 0.$$

Noncompact properties of the Lorentz operations can be studied without the position rotations using a maximal abelian noncompact group, i.e., a real 1-dimensional subgroup $\mathbf{SO}_0(1,1)$ (Lorentz dilations) for $\mathbb{M} \cong \mathbb{R}^2$ with a 1-dimensional position space (trivial rotation group $\mathbf{SO}(1) = \{1\}$) or with two fixed position space directions. A Lorentz transformation can be parametrized in Sylvester bases with energy $p^0 > 0$ and momenta p^3, i.e., with two (pseudo)orthonormal basis vectors in the matrix columns (lines)

$$\eta = \begin{pmatrix} 1 & 0 \\ 0 & -1 \end{pmatrix}, \quad \Lambda(\tfrac{p}{m})\begin{pmatrix} m \\ 0 \end{pmatrix} = \begin{pmatrix} p^0 \\ p^3 \end{pmatrix} \Rightarrow \quad \Lambda(\tfrac{p}{m}) = \tfrac{1}{m}\begin{pmatrix} p^0 & p^3 \\ p^3 & p^0 \end{pmatrix}, \\ \text{with } (p^0)^2 - (p^3)^2 = m^2.$$

The hyperbolic normalization of the basic vectors motivates the parametrization with *rapidity* ψ:

$$\tfrac{p^3}{p^0} = \tanh\psi \Rightarrow \Lambda(\tfrac{p}{m}) = \Lambda(\psi) = \begin{pmatrix} \cosh\psi & \sinh\psi \\ \sinh\psi & \cosh\psi \end{pmatrix} = \cosh\psi \begin{pmatrix} 1 & \tanh\psi \\ \tanh\psi & 1 \end{pmatrix}.$$

The transformations $\Lambda(\psi)$ with imaginary "rotation angle" ("hyperbolic angle") $\psi = i\varphi$ may be called "axial boosts." Since the hyperbolic tangent is bounded, the Lie group is parametrizable also with a *bounded velocity* $\frac{v}{c}$ in the open interval $]-1,1[$

$$\left.\begin{array}{l} -1 < \tanh\psi = \frac{v(\psi)}{c} < 1, \\ (p^0, p^3) = \frac{m}{\sqrt{1-\frac{v(\psi)^2}{c^2}}}(1, \frac{v(\psi)}{c}) \end{array}\right\} \Rightarrow \Lambda(\tfrac{p}{m}) = \Lambda(\tfrac{v}{c}) = \frac{1}{\sqrt{1-\frac{v^2}{c^2}}}\begin{pmatrix} 1 & \frac{v}{c} \\ \frac{v}{c} & 1 \end{pmatrix}.$$

The *maximal* parameter c as upper limit for the Lie parameter modulus $|\vec{v}|$ is a consequence of the nontrivial signature $(1,3)$ of the Lorentz form g. Velocities $\vec{v}$ with $|\vec{v}| \geq c$ do not make operational sense:

$$\mathbf{SO}_0(1,1)\text{-Lie parameter } \psi \in \mathbb{R} \iff \tfrac{v}{c} \in]-1,1[,$$

c is the *natural velocity unit* for theories with Lorentz symmetry. With an arbitrary time unit $\frac{1}{\mu}$, e.g., second, the maximal velocity c defines the *naturally associated* length unit $\ell = c\mu$, e.g., meter, by

$$c = 299\ 792\ 458 \tfrac{\text{m}}{\text{s}} \text{ (meter definition since 1983)}.$$

The Lie algebra of the nonabelian group $\mathbf{O}(1,1) \cong \mathbb{I}(2) \vec{\times} [\mathbb{I}(2) \times \mathbf{SO}_0(1,1)]$ is real 1-dimensional. The orthochronous transformations $\Lambda(\psi) \in \mathbf{SO}_0(1,1)$ (abelian) can be obtained by exponentiation:

$$\begin{aligned} \log \mathbf{SO}_0(1,1) = & \ \{\psi\mathcal{B} = \psi\begin{pmatrix} 0 & 1 \\ 1 & 0 \end{pmatrix} \mid \psi = \operatorname{artanh}\tfrac{p^3}{p^0}\}, \\ \Lambda(\psi) = & \ \exp\psi\mathcal{B} = \sum_{k\geq 0} \tfrac{(\psi\mathcal{B})^k}{k!} = \tfrac{1}{m}\begin{pmatrix} p^0 & p^3 \\ p^3 & p^0 \end{pmatrix} \in \mathbf{SO}_0(1,1). \end{aligned}$$

Only with a decomposition into position and time does "velocity" as a nonrelativistic concept make sense. The linear addition of the hyperbolic "angle" (rapidity) in $\Lambda(\psi_1) \circ \Lambda(\psi_2) = \Lambda(\psi_1 + \psi_2)$ becomes the nonlinear velocity "addition":

$$\text{for } \mathbf{SO}_0(1,1): \quad \tfrac{v(\psi_1+\psi_2)}{c} = \tanh(\psi_1+\psi_2) = \tfrac{\tanh\psi_1 + \tanh\psi_2}{1+\tanh\psi_1\tanh\psi_2} = \frac{\frac{v(\psi_1)}{c} + \frac{v(\psi_2)}{c}}{1+\frac{v(\psi_1)v(\psi_2)}{c^2}},$$

$$\text{for } \mathbf{SO}_0(1,3)/\mathbf{SO}(3) \text{ (no group)}: \quad \tfrac{\vec{v}_{1+2}}{c} = \frac{\frac{\vec{v}_1}{c} + \frac{\vec{v}_2}{c}}{1+\frac{\vec{v}_1\vec{v}_2}{c^2}}.$$

The orthochronous Lorentz transformations $\mathbf{SO}_0(1,3)$ contain, in addition to the compact position rotation group, the classes of the noncompact 3-parameter *proper Lorentz transformations* (Sylvester manifold, no group)

$$\mathbf{SO}_0(1,3) \cong \mathbf{SO}(3) \times \mathcal{Y}^3, \quad \mathcal{Y}^3 \cong \mathbf{SO}_0(1,3)/\mathbf{SO}(3).$$

The orientation manifold $\mathcal{Y}^3$ of rotation groups in a Lorentz group is a 3-dimensional one-shell hyperboloid.

Any Lorentz transformation has a unique *polar decomposition* into a rotation and a proper Lorentz transformation

$$\mathbf{SO}_0(1,3) \ni \Lambda = O_\Lambda \circ |\Lambda| \text{ with } O_\Lambda \in \mathbf{SO}(3), \quad |\Lambda| = |\Lambda|^T \in \mathbf{SO}_0(1,3).$$

$|\Lambda|$ is computed by diagonalizing - with an orthogonal matrix R, in general different from O_Λ - the symmetric product $\Lambda^T \circ \Lambda$ and taking the positive roots of the positive diagonal elements:

$$|\Lambda|^2 = \Lambda^T \circ \Lambda = R_{|\Lambda|} \circ \operatorname{diag}(\Lambda^T \circ \Lambda) \circ R^T_{|\Lambda|} \Rightarrow |\Lambda| = R_{|\Lambda|} \circ \sqrt{\operatorname{diag}(\Lambda^T \circ \Lambda)} \circ R^T_{|\Lambda|},$$
$$O_\Lambda = \Lambda \circ |\Lambda|^{-1}.$$

In the corresponding Lie algebra decomposition[7]

$$\begin{aligned} \log \mathbf{SO}(3) &\cong \mathbb{R}^3 \cong \log \mathbf{SO}_0(1,3)/\log \mathbf{SO}(3), \\ \log \mathbf{SO}_0(1,3) &\ni \mathcal{L}(\vec{\varphi}, \vec{\psi}) = \tfrac{\omega_{jk}}{2} \mathcal{L}^{jk} = \vec{\varphi}\vec{\mathcal{O}} + \vec{\psi}\vec{\mathcal{B}} = \left(\begin{array}{c|ccc} 0 & \psi_1 & \psi_2 & \psi_3 \\ \hline \psi_1 & 0 & \varphi_3 & -\varphi_2 \\ \psi_2 & -\varphi_3 & 0 & \varphi_1 \\ \psi_3 & \varphi_2 & -\varphi_1 & 0 \end{array}\right), \end{aligned}$$

the angular momenta $\{\mathcal{L}^b_a\}_{a,b=1,2,3}$ with $\mathcal{L}^b_a = -\mathcal{L}^a_b$ for the rotations

$$\mathcal{L}^{bc} = \mathbf{p}^b \otimes \mathbf{x}_c - \mathbf{p}^c \otimes \mathbf{x}_b = \epsilon^{bca}\mathcal{O}^a, \quad \mathcal{O}^a = \tfrac{\epsilon^{abc}}{2}\mathcal{L}^{bc}$$

are paired with three boosts $\{\mathcal{L}^0_a = \mathcal{B}_a\}_{a=1,2,3}$ for the proper Lorentz transformations. The decomposition is compatible with the action of the rotations

$$[\mathcal{O}^a, \mathcal{O}^b] = -\epsilon^{abc}\mathcal{O}^c, \quad [\mathcal{O}^a, \mathcal{B}^b] = -\epsilon^{abc}\mathcal{B}^c, \quad [\mathcal{B}^a, \mathcal{B}^b] = \epsilon^{abc}\mathcal{O}^c,$$

with the two invariant inner products given by

$$\operatorname{tr}\begin{pmatrix} \mathcal{O}^a \circ \mathcal{O}^b & \mathcal{O}^a \circ \mathcal{B}^b \\ \mathcal{B}^a \circ \mathcal{O}^b & \mathcal{B}^a \circ \mathcal{B}^b \end{pmatrix} = 2\begin{pmatrix} -\mathbf{1}_3 & 0 \\ 0 & \mathbf{1}_3 \end{pmatrix}, \quad \begin{pmatrix} \epsilon(\mathcal{O}^a, \mathcal{O}^b) & \epsilon(\mathcal{O}^a, \mathcal{B}^b) \\ \epsilon(\mathcal{B}^a, \mathcal{O}^b) & \epsilon(\mathcal{B}^a, \mathcal{B}^b) \end{pmatrix} = \begin{pmatrix} 0 & \mathbf{1}_3 \\ \mathbf{1}_3 & 0 \end{pmatrix}.$$

Again with the dual isomorphism $g : \mathbb{M} \longrightarrow \mathbb{M}^T$, spacetime translations and energy-momenta can be visualized in one vector space $\mathbb{R}^4$. However, in contrast to angular momenta, boosts, positions, and momenta, which can all be brought into one space $\mathbb{R}^3$, the real 6-dimensional Lorentz transformations $\mathcal{L}^{ij} \in \mathbb{R}^6$ have no isomorphic image in a Minkowski space. The axial vectors $\vec{\mathcal{O}} \in \mathbb{R}^3$ for rotations are not embedded into 4-dimensional vectors. The axial

[7] A vector space isomorphism between $\log \mathbf{SO}(s)$ and $\log \mathbf{SO}_0(1,s)/\log \mathbf{SO}(s)$ requires equal dimensions $\binom{1+s}{2} = 2\binom{s}{2}$, i.e., it holds only for $s = 3$.

vector partners for Minkowski vectors $\mathbf{p}^j \in \mathbb{R}^4$ are the 3-dimensional volume elements $\{\mathbf{p}^5_j = \frac{\epsilon_{jklm}}{3!}\mathbf{p}^k \otimes \mathbf{p}^l \otimes \mathbf{p}^m\}_{j=0}^3 \in \mathbb{R}^4$.

Any rotation and Lorentz boost can be both rotated to a third axis rotation in $\mathbf{SO}(2)$ and a third directional boost in $\mathbf{SO}_0(1,1)$ with two, in general different, position rotations $O(\mathbf{e}) \in \mathbf{SO}(3)/\mathbf{SO}(2)$ as used above:

$$\left.\begin{array}{l} \vec{\varphi}\vec{\mathcal{O}} = O_4(\frac{\vec{\varphi}}{\varphi}) \circ \varphi\mathcal{O}^3 \circ O_4(\frac{\vec{\varphi}}{\varphi})^T, \mathcal{O}^3 = \left(\begin{array}{c|cc|c} 0 & 0 & 0 & 0 \\ \hline 0 & 0 & -1 & 0 \\ 0 & 1 & 0 & 0 \\ \hline 0 & 0 & 0 & 0 \end{array}\right) \\ \vec{\psi}\vec{\mathcal{B}} = O_4(\frac{\vec{\psi}}{\psi}) \circ \psi\mathcal{B}^3 \circ O_4(\frac{\vec{\psi}}{\psi})^T, \mathcal{B}^3 = \left(\begin{array}{c|cc|c} 0 & 0 & 0 & 1 \\ \hline 0 & 0 & 0 & 0 \\ 0 & 0 & 0 & 0 \\ \hline 1 & 0 & 0 & 0 \end{array}\right) \end{array}\right\}, O_4(\mathbf{e}) = \left(\begin{array}{c|c} 1 & 0 \\ \hline 0 & O(\mathbf{e}) \end{array}\right).$$

This gives the Cartan factorization (diagonalization) of the Lorentz operations into two abelian operations and two directions (2-spheres):

$$\begin{aligned} \mathbf{SO}_0(1,3) &\cong [\mathbf{SO}(2) \circ \Omega^2] \times [\mathbf{SO}_0(1,1) \circ \Omega^2], \\ \Lambda &= R_O \circ \operatorname{diag} O_\Lambda \circ R_O^{-1} \circ R_{|\Lambda|} \circ \operatorname{diag} |\Lambda| \circ R_{|\Lambda|}^{-1}, \\ & \qquad R_O, R_{|\Lambda|} \in \mathbf{SO}(3)/\mathbf{SO}(2) \cong \Omega^2, \\ \mathbf{SO}_0(1,3) &= \mathbf{SO}(3) \circ \mathbf{SO}_0(1,1) \circ \mathbf{SO}(3), \\ \Lambda &= R_1 \circ \operatorname{diag} |\Lambda| \circ R_2, \\ & \qquad R_{1,2} \in \mathbf{SO}(3). \end{aligned}$$

A boost basis has as Lie parameters the position-related momenta $\frac{\vec{p}}{m}$ of an energylike Lorentz orbit

$$\begin{aligned} \{\vec{\psi}\vec{\mathcal{B}} &= \psi\left(\begin{array}{c|c} 0 & \frac{\vec{p}^T}{|\vec{p}|} \\ \hline \frac{\vec{p}}{|\vec{p}|} & 0 \end{array}\right) \Big| \, \vec{p} \in \mathbb{R}^3, \ (p^0)^2 - \vec{p}^2 = m^2\}, \\ \psi &= \operatorname{artanh}\frac{|\vec{p}|}{p^0}, \quad \frac{\vec{p}}{p^0} = \frac{\vec{v}}{c}, \\ \Lambda(\tfrac{p}{m}) &= \exp \vec{\psi}\vec{\mathcal{B}} = \frac{1}{m}\left(\begin{array}{c|c} p^0 & p^a \\ \hline p^b & m\delta^{ab} + \frac{p^a p^b}{p^0+m} \end{array}\right) \in \mathbf{SO}_0(1,3)/\mathbf{SO}(3) \cong \mathcal{Y}^3(m), \\ \Lambda(\tfrac{p}{m})\binom{m}{0} &= \binom{p^0}{\vec{p}}. \end{aligned}$$

Both the four columns and rows of the matrix $\Lambda(\frac{p}{m})$ are a Sylvester basis for energy-momentum space for each p with $p^2 = m^2 > 0$. *A rest system for* m^2 is determined up to rotations $\mathbf{SO}(3)$ by $p^0 = m > 0$.

The Euclidean group $\mathbf{SO}(s) \vec{\times} \mathbb{R}^s$ has the same dimension $\binom{1+s}{2}$ as the orthogonal groups $\mathbf{SO}(p,q)$ with $p + q = 1 + s$. It can be obtained from the noncompact group $\mathbf{SO}_0(1,s)$ as a *contraction* limit, where s noncompact orthochronous transformations (one-branch hyperbolas) are flattened into s translations $\mathbf{SO}_0(1,1) \cong \mathbb{R}$. With a decomposition into position and time, the velocity-parametrized proper Lorentz transformations read

$$\begin{aligned} \binom{x_0}{\vec{x}} \longmapsto \Lambda(\tfrac{v}{c})\binom{x_0}{\vec{x}} \text{ with } \Lambda(\tfrac{v}{c}) &= \left(\begin{array}{c|c} C_\psi & C_\psi \frac{\vec{v}^T}{c} \\ \hline C_\psi \frac{\vec{v}}{c} & \mathbf{1}_3 + \frac{C_\psi^2}{1+C_\psi}\frac{\vec{v}\otimes\vec{v}^T}{c^2} \end{array}\right), \\ C_\psi &= \cosh\psi = \frac{1}{\sqrt{1-\frac{\vec{v}^2}{c^2}}}. \end{aligned}$$

After renormalizing the time component x_0 with the maximal velocity to $t = \frac{x_0}{c}$,

$$\begin{pmatrix} \frac{1}{c^2} & 0 \\ 0 & -\mathbf{1}_s \end{pmatrix} = h(\tfrac{1}{c})^T \eta_{1+s} h(\tfrac{1}{c}), \quad h(\tfrac{1}{c}) = \begin{pmatrix} \frac{1}{c} & 0 \\ 0 & \mathbf{1}_s \end{pmatrix},$$
$$\begin{pmatrix} t \\ \vec{x} \end{pmatrix} \longmapsto \Lambda(v,c)\begin{pmatrix} t \\ \vec{x} \end{pmatrix} \text{ with } \Lambda(v,c) = h(\tfrac{1}{c})\Lambda(\tfrac{v}{c})h(\tfrac{1}{c})^{-1} = \left(\begin{array}{c|c} C_\psi & C_\psi \frac{\vec{v}^T}{c^2} \\ \hline C_\psi \vec{v} & \mathbf{1}_3 + \frac{C_\psi^2}{1+C_\psi} \frac{\vec{v}\otimes\vec{v}^T}{c^2} \end{array}\right),$$

the contraction limit

$$c \to \infty: \quad \begin{cases} \mathbf{SO}_0(1,s)/\mathbf{SO}(s) & \longrightarrow \mathbb{R}^s, \\ \Lambda(v,c) & \longmapsto \left(\begin{array}{c|c} 1 & 0 \\ \hline \vec{v} & \mathbf{1}_3 \end{array}\right), \\ \mathbf{SO}_0(1,s) & \longrightarrow \mathbf{SO}(s) \vec{\times} \mathbb{R}^s, \end{cases}$$

describes for $s = 3$ the Inönü-Wigner contraction of the Lorentz group to the nonrelativistic *Galileo group*

$$\begin{aligned} \mathbf{SO}(3) \vec{\times} \mathbb{R}^3 \ni (O, \vec{v}) &\cong \left(\begin{array}{c|c} 1 & 0 \\ \hline \vec{v} & O \end{array}\right), \\ (O_1, \vec{v}_1) \circ (O_2, \vec{v}_2) &= (O_1 \circ O_2, \vec{v}_1 + O_1.\vec{v}_2). \end{aligned}$$

To visualize the contraction, think of $\mathbf{SO}_0(1,2)$ in $\mathbb{R}^3$ operating on the future like hyperboloids $\mathcal{Y}^2$ by axial rotations $\mathbf{SO}(2)$ and by hyperbolic stretchings $\mathbf{SO}_0(1,1)$. For $c \to \infty$ a hyperboloid is flattened to a plane $\mathcal{Y}^2 \to \mathbb{R}^2$, the rotations $\mathbf{SO}(2)$ remain rotations, and the hyperbolic stretchings become translations in the limiting plane.

Taking also into account time and position translations, one obtains the Inönü-Wigner contraction of the Poincaré group

$$\text{for } c \to \infty: \quad \mathbf{SO}_0(1,3) \vec{\times} \mathbb{R}^4 \longrightarrow [\mathbf{SO}(3) \vec{\times} \mathbb{R}^3] \vec{\times} [\mathbb{R} \oplus \mathbb{R}^3]$$

with the complicated composition law of a double semidirect product

$$[\mathbf{SO}(3) \vec{\times} \mathbb{R}^3] \vec{\times} [\mathbb{R} \oplus \mathbb{R}^3] \ni (O, \vec{v}; t, \vec{x}) \cong \left(\begin{array}{c|c|c} 1 & 0 & t \\ \hline \vec{v} & O & \vec{x} \\ \hline 0 & 0 & 1 \end{array}\right),$$
$$(O_1, \vec{v}_1; t_1, \vec{x}_1) \circ (O_2, \vec{v}_2; t_2, \vec{x}_2) = (O_1 \circ O_2, \vec{v}_1 + O_1.\vec{v}_2; t_1 + t_2, \vec{x}_1 + \vec{v}_1 t_2 + O_1.\vec{x}_2).$$

1.4.2 Decompositions into Position and Light

A *Cartan basis* of a 2-dimensional spacetime $\mathbb{M} \cong \mathbb{R}^2$ consists of eigenvectors of the abelian Lorentz group $\mathbf{SO}_0(1,1)$. In contrast to the real nondiagonalizable compact axial rotations $\mathbf{SO}(2)$, the noncompact stretching group $\mathbf{SO}_0(1,1)$ has real eigenvectors, seen explicitly in the isomorphism to the dilation group $\mathbf{D}(1)$, arising in two 1-dimensional representations. With the dilations $e^{\pm\psi}$ one eigenvector becomes "shorter", the other one "longer" ("Procrustean" transformations, self-dual dilations) with constant nondiagonal Lorentz metric

$$\text{for } \mathbf{SO}_0(1,1): \begin{cases} \mathbf{p}^- = \frac{\mathbf{p}^0 - \mathbf{p}^3}{\sqrt{2}}, \ \mathbf{p}^+ = \frac{\mathbf{p}^0 + \mathbf{p}^3}{\sqrt{2}}, \ g^{\mu\nu} = \begin{pmatrix} 0 & 1 \\ 1 & 0 \end{pmatrix}, \\ \Lambda(\psi) = \begin{pmatrix} e^{-\psi} & 0 \\ 0 & e^{\psi} \end{pmatrix}, \ e^{\pm\psi} = \frac{p^0 \pm p^3}{m} = \sqrt{\frac{c \pm v(\psi)}{c \mp v(\psi)}}, \end{cases}$$

The two 1-dimensional subspaces $\mathbb{L}_\pm$, spanned by $\mathbf{p}_\pm$ (light cone coordinates) contain only singular vectors. Sylvester and Cartan-Witt bases are denoted in this section by Latin and Greek indices respectively.

The Lorentz metric $g \cong \iota$ in *Witt bases* for 4-dimensional Minkowski spacetime is associated with a decomposition into three subspaces, two 1-dimensional lightlike nonorthogonal nilspaces $\mathbb{L}_\pm$, and a 2-dimensional position space $\mathbb{S}^2$, orthogonal to the nilspaces

$$\text{Witt bases:} \quad \begin{cases} \mathbb{M} & \cong [\mathbb{L}_+ \oplus \mathbb{L}_-] \perp \mathbb{S}^2, \quad \{\mathbf{p}^\mu\}_{\mu=+,1,2,-}, \\ g^{\mu\nu} & = g(\mathbf{p}^\mu, \mathbf{p}^\nu) = \left(\begin{array}{c|c|c} 0 & 0 & 1 \\ \hline 0 & -\mathbf{1}_2 & 0 \\ \hline 1 & 0 & 0 \end{array}\right) = \iota^{\mu\nu}, \\ g(x,y) & = xy = x_\mu \iota^{\mu\nu} y_\nu \\ & = x_+ y_- - x_1 y_1 - x_2 y_2 + x_- y_+. \end{cases}$$

Such a decomposition is possible with the distinction of a time translation and one position translation (rest system with rotation axis) or with the distinction of two linearly independent lightlike translations. The invariance group in $\mathbf{SO}_0(1,3)$ for such a decomposition is a maximal abelian subgroup $\mathbf{SO}_0(1,1) \times \mathbf{SO}(2)$.

With a tetrad as a representative of a class $\mathbf{SL}_0(\mathbb{R}^2)/\mathbf{SO}_0(1,1) \subset \mathbf{SL}_0(\mathbb{R}^4)/\mathbf{SO}_0(1,3)$ two lightlike vectors can be transformed from components in a Witt basis to components in a Sylvester basis with metric $g^{jk} = \eta^{jk}$: In a first step w leads from a Witt to a Sylvester basis:

$$\mathbb{L}_+ \oplus \mathbb{L}_- \cong \mathbb{T} \perp \mathbb{S}^1,$$
$$w\begin{pmatrix} 1 \\ 0 \\ 0 \\ 0 \end{pmatrix} = \frac{1}{\sqrt{2}}\begin{pmatrix} 1 \\ 0 \\ 0 \\ 1 \end{pmatrix}, \quad w\begin{pmatrix} 0 \\ 0 \\ 0 \\ -1 \end{pmatrix} = \frac{1}{\sqrt{2}}\begin{pmatrix} 1 \\ 0 \\ 0 \\ -1 \end{pmatrix}, \quad w = \left(\begin{array}{c|c|c} \frac{1}{\sqrt{2}} & 0 & -\frac{1}{\sqrt{2}} \\ \hline 0 & \mathbf{1}_2 & 0 \\ \hline \frac{1}{\sqrt{2}} & 0 & \frac{1}{\sqrt{2}} \end{array}\right).$$

Then, an $\mathbf{SO}(3)$-rotation $O(\frac{\vec{p}}{|\vec{p}|})$, given above, transforms to general Sylvester components

$$O_4(\tfrac{\vec{p}}{|\vec{p}|})\begin{pmatrix} 1 \\ 0 \\ 0 \\ \pm 1 \end{pmatrix} = \frac{1}{|\vec{p}|}\begin{pmatrix} |\vec{p}| \\ \pm\vec{p} \end{pmatrix}, \quad p^2 = 0, \quad p \neq 0, \quad p = (|\vec{p}|, \vec{p}).$$

In the four columns of the tetrad representing a class in $\mathbf{SL}_0(\mathbb{R}^4)/\mathbf{SO}_0(1,3)$,

$$O_4(\tfrac{\vec{p}}{|\vec{p}|}) \circ w = \frac{1}{|\vec{p}|}\left(\begin{array}{c|cc|c} \frac{|\vec{p}|}{\sqrt{2}} & 0 & 0 & -\frac{|\vec{p}|}{\sqrt{2}} \\ \hline \frac{p^1}{\sqrt{2}} & |\vec{p}| - \frac{(p^1)^2}{|\vec{p}|+p^3} & -\frac{p^1 p^2}{|\vec{p}|+p^3} & \frac{p^1}{\sqrt{2}} \\ \frac{p^2}{\sqrt{2}} & -\frac{p^1 p^2}{|\vec{p}|+p^3} & |\vec{p}| - \frac{(p^2)^2}{|\vec{p}|+p^3} & \frac{p^2}{\sqrt{2}} \\ \hline \frac{p^3}{\sqrt{2}} & -p^1 & -p^2 & \frac{p^3}{\sqrt{2}} \end{array}\right),$$

one has for each $p \neq 0$ with $p^2 = 0$ the Sylvester components of a Witt basis, for $p^3 = |\vec{p}|$ *the light system associated to* $|\vec{p}|$, determined up to dilations $\mathbf{SO}_0(1,1)$ and axial rotations $\mathbf{SO}(2)$.

1.4.3 Lightlike Bases

A Lorentz metric $g \cong \chi$ in a *Finkelstein basis* is associated with a decomposition of the spacetime translations into four 1-dimensional lightlike, not orthogonal nilspaces:

$$\text{Finkelstein bases:} \left\{ \begin{array}{l} \mathbb{M} \cong \mathbb{L}_1 \oplus \mathbb{L}_2 \oplus \mathbb{L}_3 \oplus \mathbb{L}_4, \quad \{\mathbf{p}^\mu\}_{\mu=0}^3, \\ g(\mathbf{p}^\mu, \mathbf{p}^\nu) = \begin{pmatrix} 0 & 1 & 1 & 1 \\ 1 & 0 & 1 & 1 \\ 1 & 1 & 0 & 1 \\ 1 & 1 & 1 & 0 \end{pmatrix} = \chi^{\mu\nu}, \\ g(\mathbf{p}^\mu, \mathbf{p}^\nu)^{-1} = \frac{1}{3} \begin{pmatrix} -2 & 1 & 1 & 1 \\ 1 & -2 & 1 & 1 \\ 1 & 1 & -2 & 1 \\ 1 & 1 & 1 & -2 \end{pmatrix} = \chi_{\mu\nu}. \end{array} \right.$$

The matrices for the dual metrics are different. As invariance group in $\mathbf{SO}_0(1,3)$ of this fourfold decomposition there remains only the trivial subgroup $\{\mathbf{1}_4\}$.

A representative from the tetrad manifold $\mathbf{GL}(\mathbb{R}^4)/\mathbf{SO}_0(1,3)$ for the transformation to a Sylvester basis $\{\mathbf{p}^j\}_{j=0}^3$ reads

$$\chi^{\mu\nu} = h^\mu_j \eta^{jk} h^\nu_k, \; \mathbf{p}^\mu = h^\mu_j \mathbf{p}^j \text{ with } h^\mu_j = \frac{1}{\sqrt{2}} \begin{pmatrix} \sqrt{\frac{3}{2}} & 1 & \frac{1}{\sqrt{3}} & \frac{1}{\sqrt{6}} \\ \sqrt{\frac{3}{2}} & -1 & \frac{1}{\sqrt{3}} & \frac{1}{\sqrt{6}} \\ \sqrt{\frac{3}{2}} & 0 & -\frac{2}{\sqrt{3}} & \frac{1}{\sqrt{6}} \\ \sqrt{\frac{3}{2}} & 0 & 0 & -\frac{3}{\sqrt{6}} \end{pmatrix} = \frac{1}{\sqrt{2}} \begin{pmatrix} \sqrt{\frac{3}{2}}, & \vec{e}^{\,0} \\ \sqrt{\frac{3}{2}}, & \vec{e}^{\,1} \\ \sqrt{\frac{3}{2}}, & \vec{e}^{\,2} \\ \sqrt{\frac{3}{2}}, & \vec{e}^{\,3} \end{pmatrix}.$$

The four position vectors $\{\vec{e}^{\,j}\}_{j=0}^3$ in the basis direct to the corners of a regular tetrahedron.

Finkelstein bases exist in general for orthochronous groups $\mathbf{SO}_0(1,s)$ with $s \geq 1$. Up to a dilation factor, the matrix for the Lorentz metric has trivial diagonal and 1's everywhere else. It reads with the unit matrix $\mathbf{1}_{1+s}$ and the matrix $\mathbf{E}_{1+s} = \begin{pmatrix} 1 & 1 \\ 1 & \mathbf{E}_s \end{pmatrix}$, full with $1 = \mathbf{E}_1$,

$$\begin{array}{l} \chi_{1+s} = \mathbf{E}_{1+s} - \mathbf{1}_{1+s} \Rightarrow \chi_{1+s}^{-1} = \frac{1}{s}\mathbf{E}_{1+s} - \mathbf{1}_{1+s}, \\ \det \chi_{1+s} = (-1)^s s, \quad \mathbf{E}_{1+s}^2 = (1+s)\mathbf{E}_{1+s}. \end{array}$$

In the transformation to a Sylvester basis with diagonal inner product η_{1+s},

$$\chi_{1+s} = h_{1+s}\eta_{1+s}h_{1+s}^T \text{ with } \eta_{1+s} = \begin{pmatrix} 1 & 0 \\ 0 & -\mathbf{1}_s \end{pmatrix}, \quad h_{1+s} = \frac{1}{\sqrt{2}} \begin{pmatrix} N(s) & \vec{e}^{\,0} \\ N(s) & \vec{e}^{\,1} \\ \dots & \dots \\ N(s) & \vec{e}^{\,s} \end{pmatrix},$$

the vectors $\{\vec{e}^{\,j}\}_{j=0}^s$ span a regular $(1+s)$-simplex in position $\mathbb{S}^s \cong \mathbb{R}^s$, centered at the origin:

$$\begin{pmatrix} \vec{e}^{\,0} \\ \vec{e}^{\,1} \\ \dots \\ \vec{e}^{\,s} \end{pmatrix} = \begin{pmatrix} 1 & \frac{1}{\sqrt{3}} & \frac{1}{\sqrt{6}} & \dots & \frac{1}{\sqrt{\binom{1+s}{2}}} \\ -1 & \frac{1}{\sqrt{3}} & \frac{1}{\sqrt{6}} & \dots & \frac{1}{\sqrt{\binom{1+s}{2}}} \\ 0 & -\frac{2}{\sqrt{3}} & \frac{1}{\sqrt{6}} & \dots & \frac{1}{\sqrt{\binom{1+s}{2}}} \\ 0 & 0 & -\frac{3}{\sqrt{6}} & \dots & \frac{1}{\sqrt{\binom{1+s}{2}}} \\ & \dots & & \dots & \dots \\ & \dots & & \dots & \dots \\ 0 & 0 & 0 & \dots & -\frac{s}{\sqrt{\binom{1+s}{2}}} \end{pmatrix} \text{ with } \left\{ \begin{array}{l} \sum_{j=0}^{s} \vec{e}^{\,s} = 0 \\ (\vec{e}^{\,j})^2 = N(s)^2 = \frac{2s}{1+s} \\ \cos(\vec{e}^{\,j}, \vec{e}^{\,k}) = -\frac{1}{s}, \quad j \neq k \end{array} \right. .$$

1.5 Summary

Time translations $\mathbb{T}$ and position translations $\mathbb{S}$ are vector spaces, isomorphic to $\mathbb{R}$ and $\mathbb{R}^3$ respectively. Time carries the natural order. Time and position come with a scalar product, which, for position, defines a preorder (polar coordinates). The dual energy and momentum spaces carry the induced properties. A Minkowski space $\mathbb{M} \cong \mathbb{R}^4$ with spacetime translations and its dual energy-momentum space comes with a symmetric inner product g with signature $(1,3)$. A Lorentz bilinear form g defines, up to the unavoidable reflection, a unique natural order structure with the orthochronous Lorentz group and a compatible diamond basis of 0-neighborhoods for the natural topology. Spacetime translations are a normal abelian subgroup in the Poincaré group $\mathbf{O}(1,3) \vec{\times} \mathbb{R}^4$ with subgroups $\mathbf{O}(1) \vec{\times} \mathbb{R}$ for the time translations and Euclidean subgroups $\mathbf{O}(3) \vec{\times} \mathbb{R}^3$ for position translations. There is a real 10-dimensional tetrad manifold $\mathbf{GL}(\mathbb{R}^4)/\mathbf{O}(1,3)$ of Lorentz groups or Lorentz "metrics." A decomposition $\mathbb{M} \cong \mathbb{T} \perp \mathbb{S}$ into a nonrelativistic time and position is compatible with a maximal compact group $\mathbf{SO}(3)$ (position rotations) and distinguishes a natural ratio (maximal velocity c) between time and length unit. A decomposition $\mathbb{M} \cong [\mathbb{L}_+ \oplus \mathbb{L}_-] \perp \mathbb{S}^2$ into two lightlike nilspaces and a definite 2-dimensional position space is compatible with the action of a maximal abelian subgroup $\mathbf{SO}_0(1,1) \times \mathbf{SO}(2)$ (Lorentz dilations and axial rotations).

translations vector space $V \cong \mathbb{R}^n$	**time** $\mathbb{T} \cong \mathbb{R}$	**position** $\mathbb{S} \cong \mathbb{R}^3$	**spacetime** $\mathbb{M} \cong \mathbb{R}^4$
translation eigenvalues linear forms V^T	frequencies (energies)	momenta	energy-momenta
manifold of bases $\mathbf{GL}(\mathbb{R}^n)$	$\dim_{\mathbb{R}} \mathbf{GL}(\mathbb{R}) = 1$	$\dim_{\mathbb{R}} \mathbf{GL}(\mathbb{R}^3) = 9$	$\dim_{\mathbb{R}} \mathbf{GL}(\mathbb{R}^4) = 16$
inner products invariance group $\mathbf{O}(n-s,s)$ invariance Lie algebra	natural products $\tau(x,y)$ $\mathbf{O}(1) \cong \mathbb{I}(2)$ $\{0\}$	scalar products $\sigma(x,y)$ $\mathbf{O}(3) \cong \mathbb{I}(2) \times \mathbf{SO}(3)$ angular momenta	Lorentz "metrics" $g(x,y)$ $\mathbf{O}(1,3) \cong \mathbb{I}(2) \vec{\times} [\mathbb{I}(2) \times \mathbf{SO}_0(1,3)]$ angular momenta, boosts
order pair $(\succeq, \preceq)$	natural, unique $(\geq, \leq)$	natural preorder $(\succeq, \preceq)_\sigma$	natural $(\succeq, \preceq)_g$
orientation manifold of inner products ("metrics")	distances $\dim_{\mathbb{R}} \mathbf{D}(1)$ $= 1$	ellipsoids $\dim_{\mathbb{R}} \mathbf{D}(1) \times \mathbf{SL}(\mathbb{R}^3)/\mathbf{SO}(3)$ $= 6$	2-shell hyperboloids $\dim_{\mathbb{R}} \mathbf{D}(1) \times \mathbf{SL}_0(\mathbb{R}^4)/\mathbf{SO}_0(1,3)$ $= 10$
open 0-basis for natural topology	intervals $\{\{-\alpha < x < \alpha\} \mid \alpha > 0\}$	spheres $\{\{x \mid \lVert x \rVert_\sigma < \alpha\} \mid \alpha > 0\}$	causal double cones $\{\{-c \prec x \prec c\} \mid c \succ 0\}$

space and time translations

MATHEMATICAL TOOLS

1.6 Relations and Mappings

A *relation between two sets S and T* is a subset ρ of the product set $S \times T$:

$$x \rho y \iff (x,y) \in \rho \subseteq S \times T.$$

With the *power set* of a set S denoted by 2^S, a relation is an element of $2^{S \times T}$.

Relations have a *reflection*

$$\rho \leftrightarrow \rho^* \text{ with } \rho^* = \{(x,y) \mid (y,x) \in \rho\} \subseteq T \times S, \quad \rho = \rho^{**}.$$

Relations can be composed ("multiplied") (*concatenation, composition*), sometimes with the empty set as result;

$$\begin{aligned}
&\rho_i \subseteq S_i \times T_i, \quad i = 1,2 \Rightarrow \rho_2 \circ \rho_1 \subseteq S_1 \times T_2,\\
&\rho_2 \circ \rho_1 = \{(x,y) \mid \text{There exists } z \in T_1 \cap S_2 \text{ with } (x,z) \in \rho_1 \text{ and } (z,y) \in \rho_2\},\\
&(\rho_2 \circ \rho_1)^* = \rho_1^* \circ \rho_2^*.
\end{aligned}$$

The *identities* are neutral factors:

$$\begin{aligned}
&\mathrm{id}_S = \{(x,x) \mid x \in S\} \subseteq S \times S,\\
&\mathrm{id}_{T_1} \circ \rho_1 = \rho_1, \quad \rho_2 \circ \mathrm{id}_{S_2} = \rho_2.
\end{aligned}$$

A relation $\rho \subseteq S \times T$ that is *surjective in the first place (domain)* and *injective in the second place (range)* is called a mapping

$$\begin{aligned}
x \in S &\Rightarrow \text{There exists a } y \in T \text{ with } x\rho y,\\
x\rho y \text{ and } x\rho z &\Rightarrow y = z,\\
\rho &= \bigcup_{x \in S} (x, \rho(x)) \subseteq S \times T.
\end{aligned}$$

It is denoted as follows:

$$\rho : S \longrightarrow T \text{ with } x\rho y \iff x \stackrel{\rho}{\longmapsto} y \iff x \longmapsto \rho(x) = \rho.x = y.$$

All mappings (set morphisms) are denoted by $\mathbf{set}(S,T) = T^S$. Surjective mappings ("onto") and injective mappings ("one-to-one") have those properties in the second and first places respectively.

Any relation $\rho \subseteq S \times T$ defines a mapping $\underline{\rho} : \underline{S} \longrightarrow 2^T$ from the nontrivially used subset $\underline{S} \subseteq S$ into the power set $\underline{\rho}(x) = \{y \in T \mid x\rho y\}$.

1.7 Equivalence and Order

A *binary (self)relation* $\rho \subseteq M \times M$ can be reflected and multiplied by the equality (identity) id_M as neutral element. A binary relation with $\rho \cup \rho^* = M \times M$ is *total.*

A binary relation is called a *preorder* with the two properties

$$\begin{aligned}
&\textit{reflexive:} && x\rho x \text{ for all } x, && \text{i.e., } \mathrm{id}_M \subseteq \rho,\\
&\textit{transitive:} && x\rho y \text{ and } y\rho z \Rightarrow x\rho z, && \text{i.e., } \rho \circ \rho \subseteq \rho.
\end{aligned}$$

If the relation is also, with respect to reflection,

$$\begin{array}{lll} \textit{antisymmetric:} & x\rho y \text{ and } y\rho x \Rightarrow x = y, & \text{i.e., } \rho \cap \rho^* = \mathrm{id}_M, \\ \text{or } \textit{symmetric:} & x\rho y \Rightarrow y\rho x, & \text{i.e., } \rho = \rho^*, \end{array}$$

it is called an *order relation* or respectively an *equivalence relation.*

Reflexivity, transitivity, and antisymmetry also characterize the logical conclusion structure of propositions $x, y, \ldots$ with the implication $x\rho y : x \Rightarrow y$.

The set inclusion of $M \times M$ gives an order to binary relations. id_M is the finest equivalence and the finest order relation.

An equivalence relation $\sim$ defines *equivalence classes* $[x]_\sim$. Two classes are either disjoint or identical. The fibered (classified) set $M/\sim$ has as elements these classes, which are subsets of M. The associated projection $\pi_\sim$ maps onto the classes in the quotient:

$$\begin{aligned} [x]_\sim &= \{y \in M \mid y \sim x\}, \quad M = \biguplus_{\text{representatives } x_r} [x_r]_\sim, \\ M/\sim &= \{[x_r]_\sim \mid \text{representatives } x_r\} \subseteq 2^M, \\ \pi_\sim : M &\longrightarrow M/\sim, \quad \pi_\sim(x) = [x]_\sim. \end{aligned}$$

Given a binary relation $\rho \subseteq N \times N$, a mapping $f : M \longrightarrow N$ induces a binary relation on the domain M,

$$x\rho_f y \iff f(x)\rho f(y), \text{ i.e., } \rho_f = (f \times f)^{-1}[\rho].$$

The mapping $[f]$ induced by the class projections is injective:

$$\begin{array}{ccc} M & \xrightarrow{f} & N \\ \pi_M \downarrow & & \downarrow \pi_N, \\ M/\rho_f & \xrightarrow[{[f]}]{} & N/\rho \end{array} \qquad \begin{array}{l} [f] : M/\rho_f \longrightarrow N/\rho \\ \quad [x] \longmapsto [f(x)] \end{array}.$$

For example, the equality of mapping values on the range (with $N \cong N/=$) induces an equivalence relation on the domain $x \overset{f}{\sim} y \iff f(x) = f(y)$ with the mapping image isomorphic to the equivalence classes $f[M] \cong M/\overset{f}{\sim}$.

A preorder ρ leads to an equivalence relation $\overset{\rho}{\sim}$:

$$x \overset{\rho}{\sim} y \iff x\rho y \text{ and } y\rho x, \text{ i.e., } \overset{\rho}{\sim} = \rho \cap \rho^*,$$

and an order for the classes

$$[x] \succeq_\rho [y] \iff x\rho y.$$

For an order relation ("larger," "later," etc.) one writes $x \succ y$ if $x \succeq y$ and $x \neq y$.

Because of the antisymmetry, order relations come in pairs, (ρ, ρ^*), identical only for the trivial order, the identity: For $\preceq$ the *reflected order relation* $\succeq$ ("smaller", "earlier") is defined by

$$y \succeq x \iff x \preceq y, \text{ i.e., } \succeq^* = \preceq .$$

If $(x, y) \in M \times M$ induces $x \preceq y$ or $x \succeq y$, then M is *totally (linearly) ordered*, i.e., $\succeq \cup \succeq^* = M \times M$.

If any two elements have a larger (smaller) one, M is called *directed from above (from below)*. If there is exactly one smallest larger element (*supremum* $\sup(x, y)$) and largest smaller one (*infimum* $\inf(x, y)$), M is a *lattice*.

An element of an ordered set $x \in M$ leads, with physical terminology, to the following subsets of M

$$\begin{array}{rll}
\textit{future of } x: & [x]^{\succeq} = \{y \mid x \preceq y\}, & \\
\textit{past of } x: & [x]^{\preceq} = \{y \mid x \succeq y\}, & \\
\textit{strict presence of } x: & [x]^{\preceq} \cap [x]^{\succeq} = \{x\}, & \\
\textit{causal set of } x: & [x]^{\text{caus}} = [x]^{\preceq} \cup [x]^{\succeq} & (= M, \text{ if total order}), \\
\textit{non-causal set of } x: & M \setminus [x]^{\text{caus}} & (= \emptyset, \text{ if total order}), \\
\textit{generalized presence of } x: & M \setminus [x]^{\text{caus}} \cup \{x\} & (= \{x\}, \text{ if total order}).
\end{array}$$

An order-compatible mapping $f : M \longrightarrow N$ for ordered sets M, N is called *monotonic* (or *contramonotonic*)

$$x \preceq_M y \Rightarrow f(x) \preceq_N f(y) \quad (\text{or} \quad f(x) \succeq_N f(y)).$$

A *vector space* $V \in \underline{\mathbf{vec}}_{\mathbb{K}}$, an additive group compatible with the action of a number field $\mathbb{K} = \mathbb{R}$ or $\mathbb{K} = \mathbb{C}$, is *ordered* if the order is compatible with addition and scalar multiplication. It defines by the *positive and negative* vectors the *future and past cone (of* 0*)* and the causal double cone

$$\begin{aligned}
V_+ &= \{v \succeq 0\} : \quad \begin{cases} v, w \in V_+, \ \alpha \geq 0 \Rightarrow \alpha v, \ v + w \in V_+, \\ v \in V_+ \text{ and} - v \in V_+ \iff v = 0, \end{cases} \\
V_- &= \{v \preceq 0\} = -V_+, \quad V^{\text{caus}} = V_+ \cup V_-.
\end{aligned}$$

One has as *characteristic order functions* (notation $\vartheta(x) = \vartheta_+(x)$)

$$\vartheta_\pm : V \longrightarrow \mathbb{R}, \quad \vartheta_+(x) = \vartheta_-(-x) = \begin{cases} 1, & x \succeq 0, \\ 0, & \text{elsewhere}. \end{cases}$$

In general, a *cone* is defined as an ordered additive monoid (below) with an order-compatible action of an ordered abelian ring (below).

1.8 Numbers

With von Neumann, the *natural numbers* $\mathbb{N}_0$ are recursively definable as cardinality equivalence classes by the empty set and the set containing the empty set, etc.

$$0 = \emptyset, \quad 1 = \{0\}, \quad 2 = \{0, 1\}, \quad \ldots, \quad n + 1 = \{0, \ldots, n\}.$$

The natural numbers are both an additive and a multiplicative monoid (below) with total order

$$n \geq m \iff \text{There exists } k \in \mathbb{N}_0 \text{ with } n = m + k.$$

Morphisms of $\mathbb{N}$ define the ring (below) of *integers* as additive equivalence classes:

$$\mathbb{Z} = \mathbb{N} \times \mathbb{N} / \overset{+}{\sim} \text{ with } (n,m) \overset{+}{\sim} (n',m') \iff n + m' = m + n'.$$

Transitivity follows with the cancellation rule $n + k = m + k \Rightarrow n = m$. The $\mathbb{N}_0$-order defines the $\mathbb{Z}$-norm

$$\mathbb{Z} \longrightarrow \mathbb{N}_0, \quad z \longmapsto |z| = \begin{cases} z, & z \in \mathbb{N}_0, \\ -z, & -z \in \mathbb{N}_0. \end{cases}$$

Similarly, morphisms of $\mathbb{Z}$ define the *rational field* as multiplicative equivalence classes

$$\mathbb{Q} = \mathbb{Z} \times [\mathbb{Z} \setminus \{0\}] / \overset{\cdot}{\sim} \text{ with } (z;u) \overset{\cdot}{\sim} (z';u') \iff zu' = uz'.$$

$\mathbb{Q}$ has the induced order and norm.

The algebraic closure of $\mathbb{Q}$ is given by the *algebraic field* $\mathbb{A}$ as the set of all solutions of all polynomials with integer coefficients.

The *real field* $\mathbb{R}$ consists of mappings, defined with series $\mathbb{Q}^{\mathbb{N}_0}$ of rational numbers, wherein Cauchy series, using the $\mathbb{Q}$-norm, are distinguished and collected into equivalence classes. The reals inherit order and norm. The *complex field* $\mathbb{C}$ arises by adjoining the solutions $\{\pm i = \pm\sqrt{-1}\}$ of the irreducible (over $\mathbb{R}$) polynomial $X^2 + 1 = (X - i)(X + i)$.

Altogether one has the extensions from the naturals

$$\begin{array}{ccccccc} \mathbb{N} & \longrightarrow & \mathbb{Z} & \longrightarrow & \mathbb{Q} & \longrightarrow & \mathbb{R} & & \\ & & & & & & & \searrow & \\ & & \downarrow & & \downarrow & & & & \mathbb{C} = \mathbb{R} + i\mathbb{R}. \\ & & & & & & & \nearrow & \\ & & \mathbb{Z} + i\mathbb{Z} & \longrightarrow & \mathbb{Q} + i\mathbb{Q} & \longrightarrow & \mathbb{A} & & \end{array}$$

$\mathbb{A}$ is algebraically closed, but not Cauchy complete, whereas $\mathbb{R}$ is Cauchy complete, but not algebraically closed. The minimal $\mathbb{C}$, both algebraically closed and Cauchy complete, is also *exponentially closed* with the transcendental Euler number $e = \exp 1 = \sum_{k=0}^{\infty} \frac{1}{k!} \notin \mathbb{A}$,

$$\mathbb{C}^\diamond = \mathbb{C} \setminus \{0\} = \exp \mathbb{C},$$

which is of paramount importance for the Lie group-Lie algebra relation.

The *natural order* of the numbers can be defined by a scalar product (below)

$$\begin{array}{l} \alpha, \beta \in \mathbb{C}, \;\; \alpha \geq \beta \iff \alpha - \beta \geq 0, \\ \gamma \in \mathbb{C}, \;\; \gamma \geq 0 \iff \gamma = \overline{\delta}\delta = |\delta|^2, \;\; \delta \in \mathbb{C}, \\ \mathbb{C} \times \mathbb{C} \longrightarrow \mathbb{C}, \;\; (\alpha, \beta) \longmapsto \langle \alpha | \beta \rangle = \overline{\alpha}\beta; \end{array}$$

$\delta \longmapsto \overline{\delta}$ is the canonical conjugation, the unique nontrivial field automorphism of the complex numbers $\mathbb{C}$ leaving the reals invariant.

1.9 Monoids and Groups

A set M with an internal binary composition (mapping) $(x, y) \longmapsto xy \in M$ is called a *magma*. Two magma structures $\{\diamond, \bullet\}$ for a set S are called *equivalent* if they are compatible with a set bijection

$$\begin{array}{ccc} S \times S & \xrightarrow{\diamond} & S \\ {\scriptstyle g\times g}\downarrow & & \downarrow{\scriptstyle g} \\ S \times S & \xrightarrow[\bullet]{} & S \end{array}, \quad g(x \diamond y) = g(x) \bullet g(y).$$

$g : S \longrightarrow S$ is a magma isomorphism for the possibly different magmas $(S, \diamond)$ and $(S, \bullet)$ with equal underlying set.

A magma composition is *abelian* with $xy = yx$ for all $x, y \in M$. An associative magma, i.e., $(xy)z = x(yz)$, is called a *semigroup*. A magma with neutral element 1_M (or e or 0), i.e., $1_M x = x 1_M = x$, is called *unital*. A *monoid* $M \in \underline{\mathbf{mon}}$ is a unital associative magma. If not stated otherwise, "unital" includes "associative." The *regular group* of a monoid M consists of the invertible elements

$$M^\diamond = \{g \in M \mid \text{There exists } h \in M \text{ with } gh = hg = 1_M\} \in \underline{\mathbf{grp}}.$$

A finite group of order $1 + n$ is characterizable by its $(1 + n) \times (1 + n)$ *multiplication table*

1	g_1	$\cdots$	g_n
g_1	$g_1 g_1$	$\cdots$	$g_1 g_n$
$\cdot$		$\cdots$	
$\cdot$		$\cdots$	
g_n	$g_n g_1$	$\cdots$	$g_n g_n$

where with the bijective left and right multiplication, $h \longmapsto gh$ and $h \longmapsto hg$, each row and each column contain exactly all group elements.

The binary relations $(2^{S\times S}, \circ, \mathrm{id}_S)$ of a set S are a monoid, the endomorphisms a submonoid $\mathbf{A}(S)$ (mappings, arrows) with the bijections $\mathbf{G}(S)$ as regular group, e.g., the permutation group $\mathbf{G}(n)$ of n elements:

$$S \in \underline{\mathbf{set}} : \begin{cases} 2^{S\times S} \in \underline{\mathbf{mon}}, \\ \mathbf{A}(S) = S^S = \underline{\mathbf{set}}(S, S) \in \underline{\mathbf{mon}}, \\ \mathbf{G}(S) = \mathbf{A}(S)^\diamond = S! = \overset{\circ}{\underline{\mathbf{set}}}\,(S, S) \in \underline{\mathbf{grp}}, \\ S! \subseteq S^S \subseteq 2^{S\times S}. \end{cases}$$

Normal (invariant) subgroups $N \subseteq G$ are *kernels* of group morphisms $f : G \longrightarrow G'$:

$$\begin{aligned} &N = \mathrm{kern}\, f = \{g \in G \mid f(g) = e\} \\ &\big(\text{normal } N \subseteq G\big) \iff \big(g \in G \Rightarrow gNg^{-1} = N\big). \end{aligned}$$

With a group morphism $f \in \mathbf{grp}(G, G')$ and a normal subgroup $N' \subseteq G'$ one gets an injective group morphism for the quotient groups with the classes for $g \sim g' \iff g'g^{-1} \in N'$ and an isomorphism $[f]$ between the domain classes and the image classes:

$$\begin{array}{ccccc} & G & \stackrel{f}{\longrightarrow} & G' & \\ \pi & \downarrow & & \downarrow & \pi' . \\ & G/f^{-1}[N'] & \underset{[f]}{\longrightarrow} & G'/N' & \end{array}$$

As a special case, f induces an isomorphism between the image and the classes of the domain with respect to the kernel:

$$f : G \longrightarrow G' \Rightarrow f[G] \cong G/\operatorname{kern} f.$$

The *alternating (special) permutation subgroup* $\mathbf{G}(n)_+$ contains the even permutations. Being the kernel of the signature morphism into $\mathbf{G}(2)$ (identity and transposition), surjective for $n \geq 2$, it is a normal subgroup:

$$\begin{array}{l} \operatorname{sign} : \mathbf{G}(n) \longrightarrow \{\pm 1\} = \mathbb{I}(2) \cong \mathbf{G}(2), \\ \mathbf{G}(n)_+ = \{\pi \in \mathbf{G}(n) \mid \operatorname{sign} \pi = 1\}, \\ n \geq 2 : \quad \operatorname{card} \mathbf{G}(n)_+ = \frac{n!}{2}, \quad \mathbf{G}(2) \cong \mathbf{G}(n)/\mathbf{G}(n)_+. \end{array}$$

1.9.1 Products of Groups

The powers of a group element define a *cyclic subgroup*

$$g \in G \in \underline{\mathbf{grp}} \Rightarrow g^{\mathbb{N}_0} = \{g^n \mid n = 0, 1, \ldots \text{ with } g^0 = 1\} \in \underline{\mathbf{grp}}.$$

A finite cyclic group is isomorphic to a *unit root group (cyclotomic group)*, which can be written as multiplicative group $\mathbb{I}(n)$ or additive group $\mathbb{Z}_n$ (addition modulo n):

$$\begin{aligned} n \in \mathbb{N} : \quad \mathbb{I}(n) &= \{z \in \mathbb{C} \mid z^n = 1\} = \{e^{\frac{2\pi i k}{n}} \mid k = 0, \ldots, n-1\} \\ &\cong \mathbb{Z}_n = \mathbb{Z}/n\mathbb{Z} = \{[0], [1], \ldots, [n-1]\}. \end{aligned}$$

Subgroups of the reals are $\mathbb{I}(1) = \{1\}$ and $\mathbb{I}(2) = \{\pm 1\}$. All groups of prime order p are isomorphic to $\mathbb{I}(p)$ and of order p^2 either to the cyclic group $\mathbb{I}(p^2)$ or to the direct product (below) bicyclic group $\mathbb{I}(p) \times \mathbb{I}(p)$. After the smallest groups $\mathbb{I}(1), \mathbb{I}(2), \mathbb{I}(3), \mathbb{I}(4), \mathbb{I}(2) \times \mathbb{I}(2)$, and $\mathbb{I}(5)$ the nonabelian groups have at least six elements, e.g., $\mathbf{G}(3)$.

A normal subgroup $N \subseteq G$ and a subgroup $H \subseteq G$ define the *product group* $HN = NH \subseteq G$. If one restricts the canonical projection on G/N to the subgroup H, there arises the following isomorphism:

$$\left.\begin{array}{l} \pi : G \longrightarrow G/N \\ \pi_H : H \longrightarrow HN/N \end{array}\right\} \Rightarrow HN/N \cong H/H \cap N,$$

since HN/N is the image of the restriction and $H \cap N$ its kernel.

Two groups $N_{1,2}$ define the *direct product* $N_1 \times N_2$ with independent composition. The two factors $N_1 \cong (N_1, e_2)$ and $N_2 \cong (e_1, N_2)$ with neutral elements $e_{1,2}$ are embedded as normal subgroups.

If the elements of a group G are uniquely factorizable with two normal subgroups, i.e., $G = N_1 N_2$ and $N_1 \cap N_2 = \{e\}$, then G is isomorphic to the direct product $G \cong N_1 \times N_2$.

The realization of a group H with neutral element e_H (product notation hh') in the automorphisms of a group N with neutral element e_N (product notation $v \diamond w$),

$$\begin{array}{l} H \longrightarrow \overset{\circ}{\mathbf{grp}}\,(N,N), \quad h \longmapsto h\bullet, \quad \left\{\begin{array}{l} e_H\bullet = \mathrm{id}_N, \\ (hh')\bullet = (h\bullet) \circ (h'\bullet), \end{array}\right. \\ H \times N \longrightarrow N, \quad (h,v) \longrightarrow h \bullet v, \\ h\bullet : N \longrightarrow N, \quad \left\{\begin{array}{l} h \bullet e_N = e_N, \\ h \bullet (v \diamond w) = (h \bullet v) \diamond (h \bullet w), \end{array}\right. \end{array}$$

induces the action of the product set $H \times N$ on N,

$$(h,v) : N \longrightarrow N, \quad w \longmapsto v \diamond (h \bullet w),$$

and defines the *semidirect product group* $H \vec{\times} N$, a subgroup of the bijections $\mathbf{G}(N)$ (in general not of the group automorphisms $\overset{\circ}{\mathbf{grp}}\,(N,N)$):

$$\begin{array}{l} H \vec{\times} N = \{(h,v) \bigm| h \in H, \;\; v \in N\} \in \underline{\mathbf{grp}}, \quad H \vec{\times} N \subseteq \mathbf{G}(N), \\ \text{composition: } (h,v) \circ (h',v') = (hh', v \diamond (h \bullet \; v')), \\ \text{neutral element: } (e_H, e_N) \circ (h,v) = (h,v) \circ (e_H, e_N) = (h,v), \\ \text{inverse: } (h,v)^{-1} = (h^{-1}, h^{-1} \bullet v^{-1}). \end{array}$$

In addition to the notation $H \vec{\times} N$ and $N \overleftarrow{\times} H$, indicating the action of H on N, there is the notation $N \circ H$, indicating the possible factorization of the elements

$$H \vec{\times} N = N \overleftarrow{\times} H = N \circ H = \{(h,v) = (e_H, v) \circ (h, e_N) \bigm| h \in H, \;\; v \in N\}.$$

The group N is embedded as a normal subgroup (e_H, N),

$$(h,v) \circ (e_H, w) \circ (h,v)^{-1} = (e_H, v \diamond (h \bullet w) \diamond v^{-1}),$$

the group $H \cong (H, e_N)$ in general only as a subgroup. Therefore the equivalence classes with respect to N are group isomorphic to H, those with respect to H in general only bijective to N:

$$\begin{array}{l} H \vec{\times} N/N = \{(h,N) \bigm| h \in H\} \cong H \text{ as a group}, \\ H \vec{\times} N/H = \{(H,v) \bigm| v \in N\} \cong N \text{ as a set}. \end{array}$$

The direct product $H \times N$ is the special case with the trivial H-realization, $h\bullet = \mathrm{id}_N$ for all $h \in H$.

It may be useful to indicate the H-realization, e.g., $H \vec{\times}_\Lambda N$, then $h \bullet v = \Lambda(h)(v)$.

If the elements of a group G are uniquely factorizable with a normal subgroup $N \subseteq G$ and a subgroup $H \subseteq G$, i.e., $G = N \circ H$ and $N \cap H = \{e_G\}$, the group G is isomorphic to a semidirect product

$$\begin{array}{l} H \times N \longrightarrow N, \quad (h,n) \longmapsto h \bullet n = \mathrm{Int}\, h(n) = h \circ n \circ h^{-1}, \\ H \vec{\times} N \cong G \text{ with } (h,n) \cong n \circ h. \end{array}$$

For example, $\mathbf{G}(n) \cong \mathbf{G}(2) \vec{\times} \mathbf{G}(n)_+$.

Extensions generalize (semi)direct products: In an *exact group sequence* (only groups and group morphisms, injective ι and surjective π),

$$\text{in } \underline{\mathbf{grp}}: \quad N \xrightarrow{\iota} G \xrightarrow{\pi} H = \pi[G] \text{ with } N \cong \iota[N] = \mathrm{kern}\, \pi,$$

the group G is an *extension of the group* $H \cong G/N$ *by the group* N. It is *central* for $\iota[N] \subseteq \mathrm{centr}\, G$ and

$$G \cong \left\{ \begin{array}{ll} H \vec{\times} N & \textit{inessential extension} \text{ (semidirect product)}, \\ H \times N & \textit{trivial extension} \text{ (direct product)}. \end{array} \right.$$

1.10 Vector Space Duality

A $\mathbb{K}$-vector space V has bases $\{e^\iota \mid \iota \in I\}$ with $V = \bigoplus_{\iota \in I} \mathbb{K} e^\iota$ and the cardinality of I its dimension. One has to be aware, both mathematically and physically, that the distinction of a basis introduces new structures in a vector space.

One has for vector subspaces $W, W' \subseteq V \in \underline{\mathbf{vec}}_\mathbb{K}$ and linear mappings

$$\begin{array}{l} \dim_\mathbb{K} W + \mathrm{codim}_\mathbb{K} W = \dim_\mathbb{K} V, \quad \mathrm{codim}_\mathbb{K} W = \dim_\mathbb{K} V/W, \\ \dim_\mathbb{K}(W + W') + \dim_\mathbb{K}(W \cap W') = \dim_\mathbb{K} W + \dim_\mathbb{K} W', \\ f : V \longrightarrow V', \quad \mathrm{rank}_\mathbb{K} f = \mathrm{codim}_\mathbb{K}\, \mathrm{kern}\, f = \dim_\mathbb{K} V/\mathrm{kern}\, f. \end{array}$$

The *dual space* $V^T = \mathbf{vec}_\mathbb{K}(V, \mathbb{K})$ for a vector space $V \in \underline{\mathbf{vec}}_\mathbb{K}$ is constituted by the morphisms into the distinguished vector space $\mathbb{K}$: V^T consists of the linear V-forms with the bilinear *dual product*

$$\langle\ ,\ \rangle : V^T \times V \longrightarrow \mathbb{K}, \quad \langle \theta, v \rangle = \theta(v).$$

The definition of the "exchanged" dual product allows a *transposition sign* with $\epsilon_V^2 = 1 \in \mathbb{K}$, i.e., $\epsilon_V = \pm 1$, called the *Fermi and Bose signs* respectively:

$$\langle\ ,\ \rangle : V \times V^T \longrightarrow \mathbb{K}, \quad \langle v, \theta \rangle = \epsilon_V \langle \theta, v \rangle.$$

For all vector spaces one has with respect to duality two possibilities: $(V, V^T, +)$ (Fermi duality) and $(V, V^T, -)$ (Bose duality). The introduction of the transposition sign complicates the following formulas. For simplicity, one can take everywhere $\epsilon_V = +1$. However, if quantum structures are introduced via duality (chapter "Quantum Algebra"), the introduction of $\epsilon = \pm 1$ with this origin will prove useful.

Many operations on and properties of the vector space V can be "dually rolled over" to the dual space V^T via the dual product. E.g., for vector spaces with equal transposition signs, the *transposed mappings* are defined as follows:

$$\epsilon_V = \epsilon_W : \begin{cases} f : V \longrightarrow W \\ f^T : W^T \longrightarrow V^T \end{cases} \text{with } \langle \theta, f(v) \rangle = \langle f^T(\theta), v \rangle.$$

For the contravariant *transposition functor* all Fermi and all Bose vector spaces over $\mathbb{K}$ are collected in the two categories $\underline{\mathbf{vec}}_{\mathbb{K}}^{\pm}$:

$$^T : \underline{\mathbf{vec}}_{\mathbb{K}}^{\epsilon} \longrightarrow \underline{\mathbf{vec}}_{\mathbb{K}}^{\epsilon}, \quad f \begin{array}{c} V \\ \downarrow \\ W \end{array} \longmapsto \begin{array}{c} V^T \\ \uparrow \\ W^T \end{array} f^T,$$

$$(\mathrm{id}_V)^T = \mathrm{id}_{V^T}, \quad (f \circ g)^T = g^T \circ f^T.$$

V is naturally embedded in the bidual V^{TT}

$$V \longrightarrow V^{TT}, \quad v \longmapsto v^{TT} \text{ with } \langle v^{TT}, \omega \rangle = \langle \omega, v \rangle.$$

For *finite-dimensional vector spaces* $V \cong \mathbb{K}^n$ there is the isomorphism $V \cong V^T$ without natural isomorphism and the natural isomorphism $V^{TT} \cong V$, $f^{TT} \cong f$. In this case, there are *dual bases* $\{e^k\}_{k=1}^n$ of V and $\{\check{e}_k\}_{k=1}^n$ of V^T:

$$\langle \check{e}_j, e^k \rangle = \delta_j^k = \epsilon_V \langle e^k, \check{e}_j \rangle.$$

A basis determines a unique dual basis and thus a dual isomorphism $e^j \longmapsto \check{e}_j$.

With dual bases one has corresponding components of a vector v and of a form ω:

$$\begin{array}{llll} v &= v_i e^i \in V, & v_i &= \langle \check{e}_i, v \rangle \in \mathbb{K}, \quad v \cong \begin{pmatrix} v_1 \\ v_2 \\ \dots \\ v_n \end{pmatrix}, \quad e^1 \cong \begin{pmatrix} 1 \\ 0 \\ \dots \\ 0 \end{pmatrix}, \dots, \\ \omega &= \omega^i \check{e}_i \in V^T, & \omega^i &= \langle \omega, e^i \rangle \in \mathbb{K}, \quad \begin{cases} \omega & \cong (\omega^1, \omega^2, \dots, \omega^n), \\ \check{e}_1 & \cong (1, 0, \dots, 0), \dots . \end{cases} \end{array}$$

1.11 Bilinearity and Tensor Product

Any bilinear mapping f involving three vector spaces $V, W, U \in \underline{\mathbf{vec}}_{\mathbb{K}}$ can be factorized into a *universal bilinear injection* ι and a unique individual linear mappping $\tilde{f}$ defining a *tensor product* $W \otimes V$, uniquc up to vector space isomorphism:[8]

$$\begin{array}{ccc} V\times W & \stackrel{\iota}{\longrightarrow} & W\otimes V \\ f\Big\downarrow & & \Big\downarrow\tilde f \\ U & \underset{\mathrm{id}_U}{\longrightarrow} & U \end{array}, \qquad \begin{array}{l} \iota(v,w)=w\otimes v, \\ \tilde f(w\otimes v)=f(v,w). \end{array}$$

The definition of $\tilde f$ for *decomposable tensors* $w \otimes v$ has to be extended by linearity.

The tensor product can be defined also by imposing distributivity in the vector space $V \times W$ via classes with respect to the vector subspace generated by the appropriate vector combinations in $V \times W$:

$$\begin{aligned} &\mathbb{K}\text{-span of } \{(v_1+v_2,w)-(v_1,w)-(v_2,w),\ \ (v,w_1+w_2)-(v,w_1)-(v,w_2), \\ &\qquad\qquad (\alpha v,w)-\alpha(v,w),\ \ (v,\alpha w)-\alpha(v,w) \,\big|\, v\in V,\ w\in W, \alpha\in\mathbb{K}\}, \\ &V\otimes W \cong V\times W/\mathbb{K}\text{-span of } \{\ldots\}. \end{aligned}$$

The related *bifunctor of bilinearity* is biadditive and bicovariant:

$$\otimes : \underline{\mathbf{vec}}_{\mathbb{K}} \times \underline{\mathbf{vec}}_{\mathbb{K}} \longrightarrow \underline{\mathbf{vec}}_{\mathbb{K}}, \quad f\times g \begin{array}{c} V_1\times W_1 \\ \Big\downarrow \\ V_2\times W_2 \end{array} \longmapsto \begin{array}{c} W_1\otimes V_1 \\ \Big\downarrow \\ W_2\otimes V_2 \end{array} g\otimes f,$$

$$\mathrm{id}_W\otimes\mathrm{id}_V=\mathrm{id}_{W\otimes V}, \quad (g\otimes f)\circ(g'\otimes f')=(g\circ g')\otimes(f\circ f').$$

Especially, any linear mapping $f \in \mathbf{vec}_{\mathbb{K}}(V, W^T)$ is equivalent to an *associate bilinear form* $f(\ ,\ )$ *of* $V \times W$ *and to a linear form* $\tilde f$ *of the tensor product* $W \otimes V$:

$$\begin{array}{rl} \mathbf{vec}_{\mathbb{K}}(V,W^T) & \cong \mathbf{bilin}_{\mathbb{K}}(V\times W,\mathbb{K}) \cong (W\otimes V)^T, \\ f: V\longrightarrow W^T & \iff f(\ ,\ ): V\times W\longrightarrow\mathbb{K} \\ & \iff \tilde f: W\otimes V\longrightarrow\mathbb{K}, \\ \langle f(v),w\rangle & = f(v,w)=\tilde f(w\otimes v). \end{array}$$

For finite-dimensional vector spaces $V \cong \mathbb{K}^n$ there exists a natural isomorphism between the linear mappings and the tensor product (duality and tensor product commute):

$$\begin{array}{l} V\cong\mathbb{K}^n \Rightarrow \mathbf{vec}_{\mathbb{K}}(V,W^T)\cong(W\otimes V)^T\cong W^T\otimes V^T, \\ \omega\otimes\theta: V\longrightarrow W^T, \quad (\omega\otimes\theta)(v)=\omega\langle\theta,v\rangle. \end{array}$$

In the following, corresponding formulas with tensor expressions are valid only for finite dimensions.

[8]The exchange in ι is formally useful in the following, but basically irrelevant because of the natural isomorphism $V\otimes W\cong W\otimes V$.

If both vector spaces are finite-dimensional and have equal transposition signs $\epsilon_V = \epsilon_W = \epsilon$, then one has tensor expressions for transposed endomorphisms and a natural isomorphism

$$V \cong \mathbb{K}^n, \quad W \cong \mathbb{K}^m, \quad \begin{cases} V \otimes W \cong \mathbf{vec}_{\mathbb{K}}(W^T, V) \cong \mathbb{K}^{nm}, \\ W \otimes V \cong \mathbf{vec}_{\mathbb{K}}(V^T, W) \cong \mathbb{K}^{nm}, \end{cases}$$
$$^T : V \otimes W \longrightarrow W \otimes V, \quad \begin{cases} f \longmapsto f^T, \\ (w \otimes v)^T = \epsilon v \otimes w, \end{cases}$$

since with $v \in V, w \in W, \omega \in V^T, \theta \in W^T$,

$$\langle \omega, (w \otimes v)^T(\theta) \rangle = \langle w \otimes v(\omega), \theta \rangle = \langle v, \omega \rangle \langle w, \theta \rangle = \epsilon \langle \omega, (v \otimes w)(\theta) \rangle.$$

1.12 Algebras

A $\mathbb{K}$-*algebra* $A \in \underline{\mathbf{ag}}_{\mathbb{K}}$ is a $\mathbb{K}$-vector space with a bilinear product $\diamond$ for the vectors, characterizable by *structure constants with respect to a basis* $\{e^i\}_{i \in I}$:

$$A \times A \longrightarrow A, \quad \begin{cases} (a, b) \longmapsto a \diamond b, \\ e^i \diamond e^j = \alpha_k^{ij} e^k, \quad \alpha_k^{ij} \in \mathbb{K}, \quad i, j, k \in I. \end{cases}$$

Two algebra structures for a vector space V are called *linearily equivalent* if they are compatible with a vector space automorphism of V.

Depending on the three-element product behavior one defines different algebra types: For an *associative algebra* $A \in \underline{\mathbf{aag}}_{\mathbb{K}}$ one has $a(bc) = (ab)c$; for a *Lie algebra* $L \in \underline{\mathbf{lag}}_{\mathbb{K}}$ one has the *individual quadratic nilpotency* (antisymmetry, negative commutativity) of the product (Lie bracket)

$$[l, l] = 0, \quad l \in L \iff [m, n] = -[n, m], \quad m, n \in L,$$

and the compatible *Leibniz product rule (Jacobi identity)*

$$\begin{aligned} & [l, [m, n]] = [[l, m], n] + [m, [l, n]] \\ \iff \quad & [l, [m, n]] + [m, [n, l]] + [n, [l, m]] = 0, \quad l, m, n \in L, \end{aligned}$$

e.g., the real 3-dimensional *Heisenberg Lie algebra* $\log \mathbf{H}(1)$ for one position-momentum pair $(\mathbf{x}, \mathbf{p})$ and its bracket $\mathbf{I}$,

$$\begin{aligned} \log \mathbf{H}(1) \quad & = \{q\mathbf{x} + y\mathbf{p} + t\mathbf{I} \mid q, y, t \in \mathbb{R}\} \cong \mathbb{R}^3, \\ \text{with} \quad & [\mathbf{x}, \mathbf{p}] = \mathbf{I}, \ [\mathbf{I}, \mathbf{x}] = 0 = [\mathbf{I}, \mathbf{p}]. \end{aligned}$$

Each associative algebra carries a *natural Lie algebra structure* via the *commutator* $[a, b] = ab - ba$. This defines the covariant functor $\underline{\mathbf{aag}}_{\mathbb{K}} \longrightarrow \underline{\mathbf{lag}}_{\mathbb{K}}$.

Associative and Lie algebras are subcategories of *"general" algebras.* There exist other types of algebras. Without scalar multiplication one has *"general" rings, associative rings* $\underline{\mathbf{rng}}$, *and Lie rings.* A *field* is a unital ring with invertible nontrivial elements.

For a possibly nonunital $A \in \underline{\mathbf{aag}}_{\mathbb{K}}$ the vector space $\mathbb{K} \times A$ has a unital algebra structure with the product $(\alpha, a)(\beta, b) = (\alpha\beta, ab + \alpha b + \beta a)$.

A subset $E \subseteq A \in \underline{\mathbf{aag}}_{\mathbb{K}}$ is called a *generating system* if the finite $\mathbb{K}$-linear combinations of the finite products in E, the associative $\mathbb{K}$-subalgebra

$$\{\alpha_i a_i \mid a_i = b_{i_1} \cdots b_{i_m};\ b_{i_j} \in E,\ \alpha_i \in \mathbb{K};\ n, m \in \mathbb{N}\},$$

constitute the full algebra A.

A subalgebra $B \subseteq A \in \underline{\mathbf{ag}}_{\mathbb{K}}$ is called a *left, right, or (two-sided) ideal* if $A \diamond B \subseteq B$, $B \diamond A \subseteq B$, or $A \diamond B \cup B \diamond A \subseteq B$ respectively. In an abelian algebra all ideals are two-sided. The classes with respect to an ideal constitute the *quotient algebra* A/I with a natural $\mathbb{K}$-algebra structure (with unit 1 if $1 \in A \in \underline{\mathbf{aag}}_{\mathbb{K}}$). The elements of I "vanish" in A/I, i.e., they belong to the class of 0. The kernel of an algebra morphism is an ideal.

An element $s \in A \in \underline{\mathbf{aag}}_{\mathbb{K}}$ defines the *left and right principal ideals As* and sA respectively. Each subset $S \subseteq A$ defines a unique minimal ideal $I(S) \supseteq S$, e.g., for a unital algebra A,

$$I(S) = \{\sum_{i=1}^{n} a_i s_i b_i \mid s_i \in S,\ a_i, b_i \in A,\ n \in \mathbb{N}\}.$$

In the "negative commutative" Lie algebras there are only two-sided ideals, given by the Lie subalgebras $I \subseteq L$ with $[L, I] \subseteq I$; L/I carries a natural Lie algebra structure.

1.12.1 Vector Space Endomorphisms

The vector space endomorphisms with the composition product constitute the associative *endomorphism algebra* $\mathbf{AL}(V)$ ("linear arrows") with $\mathrm{id}_V = \mathbf{1}_V$ as unit. The *automorphism group (linear group)* $\mathbf{GL}(V)$ is its regular group

$$V \in \underline{\mathbf{vec}}_{\mathbb{K}} : \begin{cases} \mathbf{AL}(V) = \mathbf{vec}_{\mathbb{K}}(V, V) \in \underline{\mathbf{aag}}_{\mathbb{K}}, \\ \mathbf{GL}(V) = \overset{\circ}{\mathbf{vec}}_{\mathbb{K}}(V, V) \in \underline{\mathbf{grp}}, \\ \mathbf{GL}(V) = \mathbf{AL}(V)^{\circ} \subset \mathbf{AL}(V). \end{cases}$$

For a finite-dimensional vector space $V \cong \mathbb{K}^n$ the endomorphism algebra ($n \times n$ matrices with matrix product) is naturally isomorphic to the tensor product:

$$\begin{aligned} \mathbf{vec}_{\mathbb{K}}(V, V) &= \mathbf{AL}(V) \cong V \otimes V^T \cong \mathbb{K}^{n^2} \in \underline{\mathbf{aag}}_{\mathbb{K}} \\ &\text{with } (w \otimes \theta) \circ (v \otimes \omega) = \langle \theta, v \rangle w \otimes \omega, \\ \mathbf{vec}_{\mathbb{K}}(V^T, V^T) &= \mathbf{AL}(V^T) \cong V^T \otimes V \cong \mathbb{K}^{n^2} \in \underline{\mathbf{aag}}_{\mathbb{K}} \\ &\text{with } (\theta \otimes w) \circ (\omega \otimes v) = \langle w, \omega \rangle \theta \otimes v, \\ &g, f, v \otimes \omega \in V \otimes V^T : \quad f \circ (v \otimes \omega) \circ g = f(v) \otimes g^T(\omega). \end{aligned}$$

In the action on the vector spaces the form "eats" the vector in the dual product and vice versa:

$$\begin{aligned} (V \otimes V^T) \times V &\longrightarrow V, & (v \otimes \omega)(u) &= \langle \omega, u \rangle v, \\ (V^T \otimes V) \times V^T &\longrightarrow V^T, & (\omega \otimes v)(\theta) &= \langle v, \theta \rangle \omega = \epsilon \langle \theta, v \rangle \omega. \end{aligned}$$

The action of the automorphisms on the vector space defines the *full affine group* as the semidirect product with the vectors (translations):

$$\begin{array}{l}\mathbf{GL}(V)\,\vec{\times}\,V = V\circ\mathbf{GL}(V) = \{(g,v)\bigm| g\in\mathbf{GL}(V),\quad v\in V\}\in\underline{\mathbf{grp}},\\ (g_1,v_1)\circ(g_2,v_2) = (g_1\circ g_2, v_1+g_1.v_2),\quad (g,v) = (\mathrm{id}_V,v)\circ(g,0).\end{array}$$

It can be represented projectively in its action on $w\in V$:

$$\begin{array}{rl} w\longmapsto (g,v)\bullet w = v+g.w \quad \text{by} & \begin{pmatrix}1\\w\end{pmatrix} \longmapsto \left(\begin{array}{c|c}1&0\\\hline v&g\end{array}\right)\begin{pmatrix}1\\w\end{pmatrix} = \begin{pmatrix}1\\v+g.w\end{pmatrix},\\ (g,v) & \cong \left(\begin{array}{c|c}1&0\\\hline v&g\end{array}\right) \cong \left(\begin{array}{c|c}g&v\\\hline 0&1\end{array}\right).\end{array}$$

The transposed endomorphisms are as Lie algebras, not as associative algebras for $n\geq 2$, naturally isomorphic to each other via the *negative transposition*

$$\begin{array}{l}\check{\ }:\mathbf{AL}(V)\longrightarrow\mathbf{AL}(V^T),\quad \begin{cases} l\longmapsto \check{l} = -l^T,\\ (v\otimes\omega)\check{\ } = -(v\otimes\omega)^T = -\epsilon\omega\otimes v,\end{cases}\\ [l,f]\check{\ } = -[l,f]^T = [-l^T,-f^T] = [\check{l},\check{f}],\\ [v\otimes\omega, u\otimes\theta]\check{\ } = [-\epsilon\omega\otimes v, -\epsilon\theta\otimes u].\end{array}$$

Via *inverse transposition*, transition to the *contragredient* automorphism, the automorphism groups of dual vector spaces are naturally isomorphic to each other:

$$\check{\ }:\mathbf{GL}(V)\longrightarrow\mathbf{GL}(V^T),\quad g\longmapsto\check{g} = g^{-1T},\quad (g\circ h)\check{\ } = \check{g}\circ\check{h}.$$

The same hat symbol for Lie algebra $\check{l}$ and group $\check{g}$ should not lead to any confusion.

The *trace for endomorphisms* of $V\cong\mathbb{K}^n$ is defined by the dual product. It is a transposition-invariant linear form of the endomorphisms:

$$\mathrm{tr}\,:\mathbf{AL}(\mathbb{K}^n)\longrightarrow\mathbb{K},\quad \left\{\begin{array}{l}\mathrm{tr}\,v\otimes\omega = \langle\omega,v\rangle = \epsilon\langle v,\omega\rangle = \epsilon\,\mathrm{tr}\,\omega\otimes v,\\ \mathrm{tr}\,(f_1+f_2) = \mathrm{tr}\,f_1+\mathrm{tr}\,f_2,\quad \mathrm{tr}\,\alpha f = \alpha\,\mathrm{tr}\,f,\\ \mathrm{tr}\,\mathbf{1}_n = n,\quad \mathrm{tr}\,f = \mathrm{tr}\,f^T.\end{array}\right.$$

The dual product is the bilinear form of $V\times V^T$ associated to the identity id_V; the trace on $V^T\otimes V$ is the associated linear form

$$\begin{array}{rl}\mathrm{id}_V: V\longrightarrow V & \iff \langle\ ,\ \rangle: V\times V^T\longrightarrow\mathbb{K}\\ & \iff \mathrm{tr}\,:V^T\otimes V\longrightarrow\mathbb{K},\\ \langle\,\mathrm{id}_V(v),\omega\rangle & = \langle v,\omega\rangle = \mathrm{tr}\,\omega\otimes v\\ & = \epsilon\,\mathrm{tr}\,v\otimes\omega = \langle\epsilon\,\mathrm{id}_V^T(\omega),v\rangle = \epsilon\langle\omega,v\rangle.\end{array}$$

The (bi)linear form associated to the transposition $f\longmapsto f^T$ of the endomorphisms is the symmetric "double" trace

$$\begin{array}{rl}\mathbf{AL}(V)\longrightarrow\mathbf{AL}(V^T) & \iff T(\ ,\ ):\mathbf{AL}(V)\times\mathbf{AL}(V)\longrightarrow\mathbb{K}\\ & \iff \tilde{T}:\mathbf{AL}(V)\otimes\mathbf{AL}(V)\longrightarrow\mathbb{K},\\ \langle f^T,g\rangle & = T(f,g) = \tilde{T}(g\otimes f) = \mathrm{tr}\,f\circ g = \mathrm{tr}\,g\circ f,\\ \langle\epsilon\omega\otimes v, u\otimes\theta\rangle & = T(v\otimes\omega,u\otimes\theta) = \tilde{T}(u\otimes\theta\otimes v\otimes\omega) = \langle\omega,u\rangle\langle\theta,v\rangle.\end{array}$$

The trace of a tensor product is the product of the traces:

$$f \otimes g : V \otimes W \longrightarrow V \otimes W, \quad \mathrm{tr}\, f \otimes g = \mathrm{tr}\, f\, \mathrm{tr}\, g.$$

With dual bases one has the following tensor expressions (matrices) for endomorphisms $f \in V \otimes V^T$:

$$\begin{array}{ll} f = f^j_k\, e^k \otimes \check{e}_j,\ f(e^j) = f^j_k e^k, & f^T = f^j_k\, \epsilon\check{e}_j \otimes e^k,\ f^T(\check{e}_k) = f^j_k \check{e}_j, \\ \mathrm{id}_V = e^i \otimes \check{e}_i, & \mathrm{id}_{V^T} = \epsilon\check{e}_i \otimes e^i, \\ f^j_k = \langle e^j \otimes \check{e}_k, f\rangle = \mathrm{tr}\,(e^j \otimes \check{e}_k) \circ f = & \mathrm{tr}\, f^T \circ (\epsilon\check{e}_k \otimes e^j) = \langle f^T, \epsilon\check{e}_k \otimes e^j\rangle, \\ f \cong \begin{pmatrix} f^1_1 & \dots & f^n_1 \\ \vdots & \dots & \vdots \\ f^1_n & \dots & f^n_n \end{pmatrix}, & f^T \cong \begin{pmatrix} f^1_1 & \dots & f^1_n \\ \vdots & \dots & \vdots \\ f^n_1 & \dots & f^n_n \end{pmatrix}. \end{array}$$

For finite dimension $V \cong \mathbb{K}^n$, the determinant is a monoid morphism:

$$\det : \mathbf{AL}(\mathbb{K}^n) \longrightarrow \mathbb{K}, \quad \begin{cases} \det f_1 \circ f_2 = \det f_1 \det f_2, & \det \mathbf{1}_n = 1, \\ \det f^T = \det f. \end{cases}$$

The determinant of a tensor product with $V \otimes W \cong \mathbb{K}^n \otimes \mathbb{K}^m$ is the dimension-powered product of the determinants:

$$f \otimes g : V \otimes W \longrightarrow V \otimes W, \quad \begin{array}{l} \det f \otimes g = (\det f)^m (\det g)^n, \\ \det \alpha \mathbf{1}_n = \alpha^n. \end{array}$$

The automorphism group $\mathbf{GL}(\mathbb{K}^n)$ contains as normal subgroups the *unilinear group* $\mathbf{UL}(\mathbb{K}^n)$ with determinant modulus 1 and the *special automorphisms* $\mathbf{SL}(\mathbb{K}^n)$ with determinant 1:

$$\begin{array}{rl} \mathbf{UL}(\mathbb{K}^n) & = \{g \in \mathbf{GL}(\mathbb{K}^n) \,\big|\, |\det g| = 1\} \in \underline{\mathbf{lgrp}}_{\mathbb{K}}, \\ \mathbf{SL}(\mathbb{K}^n) & = \{g \in \mathbf{GL}(\mathbb{K}^n) \,\big|\, \det g = 1\} \in \underline{\mathbf{lgrp}}_{\mathbb{K}}, \\ \det : & \mathbf{GL}(\mathbb{K}^n) \longrightarrow \mathbb{K}^\diamond, \quad \begin{cases} \mathbb{R}^\diamond & = \mathbf{D}(1) \times \mathbb{I}(2), \\ \mathbb{C}^\diamond & = \mathbf{D}(1) \times \mathbf{U}(1), \end{cases} \\ \mathbf{GL}(\mathbb{K}^n) & = \mathbf{D}(\mathbf{1}_n) \times \mathbf{UL}(\mathbb{K}^n) = \mathbb{K}^\diamond \mathbf{1}_n \circ \mathbf{SL}(\mathbb{K}^n). \end{array}$$

In general, the product is not direct: For the complex field, a decomposition

$$\mathbf{GL}(\mathbb{K}^n) \ni g = (\det g)^{\frac{1}{n}} \mathbf{1}_n \circ \frac{g}{(\det g)^{\frac{1}{n}}}$$

has the freedom of the cyclotomic group $\mathbb{I}(n)$. For real odd-dimensional vector spaces, the product of the special automorphisms with the nontrivial numbers is direct:

$$\mathbb{K}^\diamond \mathbf{1}_n \cap \mathbf{SL}(\mathbb{K}^n) = \begin{cases} \{\mathbf{1}_n\}, & \mathbb{K} = \mathbb{R}, \quad n = 1, 3, \dots, \\ \mathbb{I}(2)\mathbf{1}_n, & \mathbb{K} = \mathbb{R}, \quad n = 2, 4, \dots, \\ \mathbb{I}(n)\mathbf{1}_n, & \mathbb{K} = \mathbb{C}. \end{cases}$$

The normal subgroup $\mathbf{SL}_0(\mathbb{R}^n)$ is the unit connection component in $\mathbf{SL}(\mathbb{R}^n)$.

The endomorphism Lie algebra $\mathbf{AL}(V)$ of a finite-dimensional vector space $V \cong \mathbb{K}^n$ is the direct sum of two Lie algebra ideals, the traceless endomorphisms $\mathbf{AL}(\mathbb{K}^n)_0$, and the trace values $\mathbb{K}\mathbf{1}_n$ isomorphic to the abelian scalars $\mathbb{K}$:

$$\begin{aligned} \mathbf{AL}(\mathbb{K}^n) &= \log \mathbf{GL}(\mathbb{K}^n) = \mathbb{K}\mathbf{1}_n \oplus \mathbf{AL}(\mathbb{K}^n)_0 \in \underline{\mathbf{aag}}_{\mathbb{K}}, \\ f &= \tfrac{\operatorname{tr} f}{n}\mathbf{1}_n + (f - \tfrac{\operatorname{tr} f}{n})\mathbf{1}_n, \\ \mathbf{AL}(\mathbb{K}^n)_0 &= \log \mathbf{SL}(\mathbb{K}^n) = \{f \in \mathbf{AL}(\mathbb{K}^n) \mid \operatorname{tr} f = 0\} \in \underline{\mathbf{lag}}_{\mathbb{K}}, \\ \mathbb{K} &= \log \mathbf{GL}(\mathbb{K}) \in \underline{\mathbf{aag}}_{\mathbb{K}}. \end{aligned}$$

Lie algebras of discrete groups D are trivial: $\log D = \{0\}$.

1.12.2 Products of Lie Algebras

The representation of a Lie algebra M in the endomorphisms of a Lie algebra I,

$$\begin{aligned} &\mathcal{D} : M \longrightarrow \mathbf{lag}_{\mathbb{K}}(I, I), \quad m \longmapsto \mathcal{D}_m, \quad \mathcal{D}_{[m,m']} = [\mathcal{D}_m, \mathcal{D}_{m'}], \\ &\mathcal{D}_m : I \longrightarrow I, \quad \mathcal{D}_m([k,l]) = [\mathcal{D}_m(k), \mathcal{D}_m(l)], \end{aligned}$$

defines the action of the direct vector space sum $M \oplus I$ on I,

$$m \oplus k : I \longrightarrow I, \quad l \longmapsto \mathcal{D}_m(l) + k,$$

and hence the *semidirect Lie algebra* $M \vec{\oplus} I$:

$$\text{Lie bracket of } m \vec{\oplus} I : \begin{cases} [m, m'] \text{ as in } M, \\ [k, k'] \text{ as in } I, \\ [m, k] = \mathcal{D}_m(k). \end{cases}$$

$M \vec{\oplus} I$ is a Lie subalgebra of the linear mappings $\mathbf{vec}_{\mathbb{K}}(I, I)$, in general not of the Lie-algebra morphisms $\mathbf{lag}_{\mathbb{K}}(I, I)$. I is embedded as an ideal in $M \vec{\oplus} I$.

The action of the endomorphism algebra on the vector space defines the *full affine Lie algebra* as direct sum and semidirect Lie bracket with the vectors

$$\begin{aligned} \mathbf{AL}(V) \vec{\oplus} V &= \{l + v \mid l \in \mathbf{AL}(V), \ v \in V\} \in \underline{\mathbf{lag}}_{\mathbb{K}}, \\ [l_1 + v_1, l_2 + v_2] &= l_1 \circ l_2 - l_2 \circ l_1 + l_1(v_2) - l_2(v_1), \\ l + v &\cong \left(\begin{array}{c|c} 0 & 0 \\ \hline v & l \end{array}\right) \cong \left(\begin{array}{c|c} l & v \\ \hline 0 & 0 \end{array}\right). \end{aligned}$$

If the elements of a Lie algebra L are uniquely decomposable $L = M \oplus I$ with an ideal $I \subseteq L$ and a Lie subalgebra $M \subseteq L$, then L is isomorphic to a *semidirect product of Lie algebras*:

$$\begin{aligned} &\mathcal{D} : M \longrightarrow \mathbf{lag}_{\mathbb{K}}(I, I), \quad \mathcal{D}_m = \operatorname{ad} m \text{ with } \operatorname{ad} m(k) = [m, k], \\ &L \cong M \vec{\oplus} I. \end{aligned}$$

The *direct product of Lie algebras* M and I uses the trivial representation $\mathcal{D}_m = 0$ for all $m \in M$. M and I are embedded as ideals.

Embedding the concept of (semi)direct products: In an *exact Lie algebra sequence*

$$\text{in } \underline{\mathbf{lag}}_{\mathbb{K}} : \ I \stackrel{\iota}{\longrightarrow} L \stackrel{\pi}{\longrightarrow} M = \pi[L] \text{ with } I \cong \iota[I] = \operatorname{kern} \pi.$$

L is an *extension of* $M \cong L/I$ *by* I, called *central* for $\iota[I] \subseteq \operatorname{centr} L$ and

$$L \cong \begin{cases} M \vec{\oplus} I & \textit{inessential} \text{ (semidirect product)}, \\ M \oplus I & \textit{trivial} \text{ (direct product)}. \end{cases}$$

1.13 Reflections (Conjugations)

An *involutive contra-automorphism* of a magma M is called a *reflection*, a realization of the reflection group $\mathbb{I}(2) = \{\pm 1\}$:

$$* : M \longrightarrow M, \quad \begin{cases} g^{**} = g & \text{(involutive, for set)}, \\ (gh)^* = h^* g^* & \text{(contra-, for magma)}. \end{cases}$$

It defines the *symmetric* elements

$$M^*_= = \{g \in M \mid g = g^*\}.$$

Any element has its *symmetric domain*

$$\Delta : M \longrightarrow M^*_=, \quad \Delta(g) = g^* g = \Delta(g)^*.$$

Elements with equal domains for reflected elements are called $*$-*normal*:

$$\Delta(g) = g^* g = g g^* = \Delta(g^*).$$

For an abelian composition the concept is trivial.

In a group with reflection $G \in *\underline{\mathbf{grp}}$, the concatenation of the reflection with the inversion is an involutive automorphism:

$$\hat{} : G \longrightarrow G, \quad \hat{g} = (g^{-1})^* = (g^*)^{-1}.$$

The elements where the reflection coincides with the *inversion as canonical group reflection* are called *unitary* for a multiplicative notation and *antisymmetric* for an additive one:

$$\begin{aligned} U(G,*) &= \{u \in G \mid u^{-1} = u^*\} \in \underline{\mathbf{grp}}, \\ A(G,*) &= \{v \in G \mid -v = v^*\} \in \underline{\mathbf{grp}}. \end{aligned}$$

A (conjugate) linear[9] involutive automorphism is called a *(conjugate) linear reflection of the vector space* $V \in *\underline{\mathbf{vec}}_{\mathbb{K}}$:

$$\begin{aligned} * : V \longrightarrow V, \quad & v^{**} = v, \quad (v+w)^* = v^* + w^*, \\ (\alpha v)^* = & \begin{cases} \alpha v^*, & \mathbb{K} = \mathbb{R}, \mathbb{C} \quad \text{linear}, \\ \overline{\alpha} v^*, & \mathbb{K} = \mathbb{C} \quad \text{conjugate linear}. \end{cases} \end{aligned}$$

[9] In general, the property "linear" is not mentioned explicitly, in contrast to "conjugate linear" or "antilinear."

A conjugate linear reflection is also called a *conjugation.*

The inversion $v \longmapsto -v$ is the *canonical linear reflection of* V. A (conjugate) linear reflection decomposes the vector space V (as a real vector space) into two (real) subspaces:

$$\begin{array}{l} V = V_+ \oplus V_-, \quad v = \frac{v+v^*}{2} + \frac{v-v^*}{2}, \quad V_+ = V^*_=, \quad V_- = A(V,*), \\ V_\pm = \{v = \pm v^* \mid v \in V\} \in \left\{ \begin{array}{ll} \underline{\mathbf{vec}}_{\mathbb{K}} & \text{linear,} \\ \underline{\mathbf{vec}}_{\mathbb{R}} & \text{conjugate linear.} \end{array} \right. \end{array}$$

For a conjugate linear reflection, the real subspaces are isomorphic via i-multiplication:

$$V \in \underline{\mathbf{vec}}_{\mathbb{C}}, \quad V_+ \longrightarrow V_-, \quad v_+ \longmapsto iv_+, \quad V_+ \cong V_- \in \underline{\mathbf{vec}}_{\mathbb{R}}.$$

They are called *real forms of* V.

With a vector space also the dual space carries a (conjugate) linear reflection. With ω also the reflected ω^* is linear:

$$\begin{array}{ccc} V & \stackrel{*}{\longrightarrow} & V \\ {\scriptstyle \omega}\Big\downarrow & & \Big\downarrow{\scriptstyle \omega^*}, \\ \mathbb{K} & \underset{-}{\longrightarrow} & \mathbb{K} \end{array} \qquad * : V^T \longrightarrow V^T,$$

$$\omega^*(v) = \langle \omega^*, v \rangle = \left\{ \begin{array}{ll} \omega(v^*) = \langle \omega, v^* \rangle & \text{linear,} \\ \overline{\omega(v^*)} = \overline{\langle \omega, v^* \rangle} & \text{conjugate linear.} \end{array} \right.$$

A (conjugate) linear involutive contra-automorphism is called a *(conjugate) linear reflection of the algebra* $A \in *\underline{\mathbf{ag}}_{\mathbb{K}}$, i.e., a vector space reflection and

$$* : A \longrightarrow A, \quad (a \diamond b)^* = b^* \diamond a^*.$$

A reflection-stable ideal $I = I^*$ yields a quotient algebra A/I with reflection.

For a Lie algebra the inversion $l \longmapsto -l$ is the *canonical linear reflection of* L. In the nonabelian case only the *antisymmetric* subspace L_- is even a real Lie subalgebra of $L \in *\underline{\mathbf{lag}}_{\mathbb{K}}$

$$[L_-, L_-] \subseteq L_- \in \underline{\mathbf{lag}}_{\mathbb{R}}, \quad [L_+, L_+] \subseteq L_-, \quad [L_-, L_+] \subseteq L_+.$$

A normal element in an algebra is characterized by being the sum of a symmetric and an antisymmetric element that commute with each other:

$$\begin{array}{rcl} \text{normal } a \in A \in *\underline{\mathbf{ag}}_{\mathbb{K}} & \iff & a \diamond a^* = a^* \diamond a \\ & \iff & a = a_+ + a_-, \quad a_\pm = \pm a^*_\pm \\ & & \text{with } a_+ \diamond a_- = a_- \diamond a_+, \\ \text{for } A \in *\underline{\mathbf{aag}}_{\mathbb{K}} : & & [a_+, a_-] = 0, \\ \text{for } A \in *\underline{\mathbf{lag}}_{\mathbb{K}} : & & [a, a^*] = 0, \ [a_+, a_-] = 0, \\ & \Rightarrow & a \diamond a^* = a^* \diamond a = a_+^2 - a_-^2, \quad a_\pm = \frac{a \pm a^*}{2}. \end{array}$$

1.13.1 Inner Products

Since there does not exist a natural (iso)morphism between a vector space $V(\cong \mathbb{K}^n)$ and its dual space V^T, in contrast to the natural (iso)morphisms between V and its bidual V^{TT}, a dual (iso)morphism is an important V-characterizing structure.

A *(conjugate) linear dual morphism* ζ,

$$\zeta : V \longrightarrow V^T, \quad w \longmapsto \zeta(w), \quad w_j e^j \longmapsto \overline{w_j} \zeta^{ji} \check{e}_i \quad \text{if } V \cong \mathbb{K}^n$$

(with $\{e^i, \check{e}_i\}_{i=1}^n$ dual bases of V and V^T respectively if $V \cong \mathbb{K}^n$) defines an associate *inner product on* V, i.e., a *bilinear form* (for ζ linear $\mathbb{K} = \mathbb{R}, \mathbb{C}$) and, for ζ conjugate linear $\mathbb{K} = \mathbb{C}$, a *sesquilinear form* (in the first variable conjugate linear, in the second one linear)

$$\zeta(\, , \,) : V \times V \longrightarrow \mathbb{C}, \quad \begin{cases} \zeta(w, v) = \langle \zeta(w), v \rangle = \operatorname{tr} v \otimes \zeta(w) = \overline{w_j} \zeta^{ji} v_i, \\ \zeta(e^j, e^i) = \zeta^{ji}. \end{cases}$$

For the linear case, one has to omit the complex conjugation for the scalars $\mathbb{K} = \mathbb{C}$ in the corresponding formulas here and in the following sections. With a given inner product $V \times V \longrightarrow \mathbb{K}$, a (conjugate) linear dual morphism is defined by $v \longmapsto \overline{\zeta(v, \,)}$.

Bilinear forms for the finite-dimensional case induce tensors of power 2, $\zeta = \zeta^{jk} \check{e}_j \otimes \check{e}_k \in V^T \otimes V^T$; sesquilinear forms have no such tensor expressions (chapter "Antistructures: The Real in the Complex").

A dual morphism ζ with trivial kernel is injective and the inner product $\zeta(\, , \,)$ is *nondegenerate*:

$$\text{kern } \zeta = 0 \quad \text{or } \zeta(V, v) = \{0\} \iff v = 0.$$

A nondegenerate inner product of a *finite dimensional* vector space $V \cong \mathbb{K}^n$ defines a *(conjugate) linear dual isomorphism.* With a dual isomorphism *Dirac's bra and ket notation* can be used (conjugate linear for a sesquilinear ζ):

$$\zeta : V \longrightarrow V^T, \quad |w\rangle \longmapsto \langle w|, \quad \zeta(v, w) = \langle v|w\rangle.$$

Two dual isomorphisms, both (conjugate) linear, define an automorphism $\zeta^{-1} \circ \eta : V \longrightarrow V$, always linear.

The inner product ζ of $V \cong \mathbb{K}^n$ is extendable as inner product $\bigwedge^n \zeta$ of the 1-dimensional vector space $\bigwedge^n V \cong \mathbb{K}$ (Grassmann power, chapter "Spin, Rotations and Position") with the parallel epipeds $v_1 \wedge \cdots \wedge v_n$ as elements that have as bases the volume elements $e^1 \wedge \cdots \wedge e^n$. An inner product ζ associates a *volume unit* ζ_e to any basis:

$$\bigwedge^n \zeta : \bigwedge^n V \times \bigwedge^n V \longrightarrow \mathbb{K}, \quad \zeta^{2n}(e^1 \wedge \cdots \wedge e^n) = (\zeta_e)^{2n} \in \mathbb{K}.$$

The *discriminant* of an inner product of $V \cong \mathbb{K}^n$ associates to n vectors a determinant,

$$\{v_1, \dots, v_n\} \longmapsto \zeta^{2n}(v_1 \wedge \cdots \wedge v_n) = \det \begin{pmatrix} \zeta(v_1, v_1) & \dots & \zeta(v_1, v_n) \\ \dots & \dots & \dots \\ \zeta(v_n, v_1) & \dots & \zeta(v_n, v_n) \end{pmatrix};$$

ζ is nondegenerate iff the discriminant is nontrivial for a V-basis $\{e^i\}_{i=1}^n$ and then, because of the transformation behavior with $f \in \mathbf{AL}(V)$,

$$\zeta^{2n}(f.e^1 \wedge \cdots \wedge f.e^n) = \begin{cases} (\det f)^2 \zeta^{2n}(e^1 \wedge \cdots \wedge e^n) \text{ (bilinear form)}, \\ |\det f|^2 \zeta^{2n}(e^1 \wedge \cdots \wedge e^n) \text{ (sesquilinear form)}, \end{cases}$$

nontrivial for all bases $\mathbf{GL}(V) \bullet \{e^i\}_{i=1}^n$.

The inverse of a (conjugate) linear dual isomorphism yields a nondegenerate *inner product of the dual space* V^T (with transposition sign ϵ):

$$\begin{aligned} &\zeta^{-1} : V^T \longrightarrow V, \quad \omega^i \check{e}_i \longmapsto \overline{\omega^i} \zeta_{ik} e^k, \quad \overline{\zeta^{ji}} \zeta_{ik} = \delta^j_k, \\ &\zeta^{-1}(\, , \,) : V^T \times V^T \longrightarrow \mathbb{C}, \begin{cases} \zeta^{-1}(\omega, \theta) = \langle \zeta^{-1}.\omega, \theta \rangle \\ \quad = \operatorname{tr} \theta \otimes \zeta^{-1}(\omega) = \epsilon \overline{\omega^i} \zeta_{ik} \theta^k, \\ \zeta^{-1}(\check{e}_k, \check{e}_i) = \epsilon \zeta_{ik}. \end{cases} \end{aligned}$$

One has the following connection between a dual isomorphism and its nondegenerate inner products:

$$\zeta(w, v) = \langle \zeta.w, v \rangle = \epsilon \zeta^{-1}(\zeta.v, \zeta.w).$$

With a dual isomorphism the *direct sum of dual vector spaces* $V \oplus V^T$ becomes a vector space with (conjugate) linear reflection (conjugation):

$$\zeta \oplus \zeta^{-1} : V \oplus V^T \longrightarrow V \oplus V^T, \quad \zeta(v) + \zeta^{-1}(\omega) = v^* + \omega^*.$$

A (conjugate) linear dual morphism ζ is called *symmetric or antisymmetric* if the associate inner product has this property:

$$\begin{aligned} &\zeta : V \longrightarrow V^T, \quad \zeta(v) = v^*, \\ &\text{linear, } \zeta = \pm\epsilon\zeta^T : \begin{cases} \langle \zeta.v, u \rangle = \langle v^*, u \rangle = \zeta(v, u) \\ = \pm\zeta(u, v) = \pm\langle u^*, v \rangle = \pm\langle \zeta.u, v \rangle = \pm\epsilon\langle \zeta^T.v, u \rangle, \\ \zeta = \zeta^{jk} \, \check{e}_j \otimes \check{e}_k, \quad \zeta^T = \epsilon\zeta^{kj} \, \check{e}_j \otimes \check{e}_k, \end{cases} \\ &\text{conjugate linear, } \zeta = \pm\zeta^* : \begin{cases} \langle \zeta.v, u \rangle = \langle v^*, u \rangle = \zeta(v, u) \\ = \pm\overline{\zeta(u, v)} = \pm\overline{\langle u^*, v \rangle} \\ = \pm\overline{\langle \zeta.u, v \rangle} = \pm\langle \zeta^*.v, u \rangle. \end{cases} \end{aligned}$$

This is the definition of ζ^* for the conjugate linear case. Here one has to consider, because of $\zeta = -\zeta^* \iff i\zeta = (i\zeta)^*$, only symmetric dual morphisms $\zeta = \zeta^*$.

Two subsets $A, B \subseteq V$ are called ζ-*orthogonal* if $\zeta(A, B) = \{0\}$orthogonal. The ζ- *orthogonal space* for a vector subspace U is the vector space

$$U_\zeta^\perp = \{v \in V \mid \zeta(U, v) = \{0\}\}.$$

The ζ-orthogonal space of V shows whether ζ is degenerate and defines, in the case of finite dimension, the *rank of* ζ:

$$\begin{array}{l}\zeta \text{ nondegenerate } \iff V_\zeta^\perp = \{0\},\\ \operatorname{rank}_{\mathbb{K}}\zeta = \dim_{\mathbb{K}} V - \dim_{\mathbb{K}} V_\zeta^\perp = \dim_{\mathbb{K}} V/V_\zeta.\end{array}$$

The (anti-) symmetric inner product induced on the quotient vector space $V/V_\zeta^\perp$ is nondegenerate.

If ζ is nondegenerate and (anti-) symmetric, then ζ^{-1} also has those properties.

1.13.2 Endomorphism Reflections

A (conjugate) linear (anti-) symmetric dual isomorphism $\zeta = \pm\epsilon\zeta^T$ or $\zeta = \pm\zeta^*$ leads together with the transposition to a *(conjugate) linear reflection of the endomorphism algebra*:

$$* : \mathbf{AL}(V) \longrightarrow \mathbf{AL}(V), \quad \begin{cases} f^* = \zeta^{-1} \circ f^T \circ \zeta,\\ f^{**} = f, \quad (f \circ g)^* = g^* \circ f^*,\end{cases}$$

$$* : V \otimes V^T \longrightarrow V \otimes V^T, \quad \begin{cases} (v \otimes \omega)^* = \pm\zeta^{-1}(\omega) \otimes \zeta(v) = \pm\omega^* \otimes v^*,\\ (v \otimes \omega)^{**} = v \otimes \omega,\\ \operatorname{tr} f^* = \begin{cases} \operatorname{tr} f \text{ (linear)},\\ \overline{\operatorname{tr} f} \text{ (conjugate linear)},\end{cases}\end{cases}$$

$$\text{with } (v \otimes \omega)^*(u) = \zeta^{-1}(\epsilon\, \omega\langle v, \zeta.u\rangle) = \pm\zeta^{-1}(\omega\overline{\langle \zeta.v, u\rangle}) = \pm\zeta^{-1}(\omega)\langle \zeta.v, u\rangle.$$

A more exact notation is $f^* = f^\zeta$. Two reflections, both linear or conjugate linear, are related to each other by an isomorphism, always linear:

$$f^\eta = (\zeta^{-1} \circ \eta)^{-1} \circ f^\zeta \circ (\zeta^{-1} \circ \eta).$$

With a reflection, the regular group $\mathbf{GL}(V)$ and the endomorphism Lie algebra are reflected:

$$\begin{array}{ll}\hat{\ } : \mathbf{GL}(V) \longrightarrow \mathbf{GL}(V), & \hat{g} = g^{-1*} = \zeta^{-1} \circ g^{-1T} \circ \zeta, \quad \hat{\hat{g}} = g,\\ \hat{\ } : \mathbf{AL}(V) \longrightarrow \mathbf{AL}(V), & \hat{l} = -l^* = -\zeta^{-1} \circ l^T \circ \zeta, \quad \hat{\hat{l}} = l.\end{array}$$

In general, one defines for an injective (conjugate) linear (anti-) symmetric dual morphism, i.e., for a nondegenerate inner product, the linear mapping f^*, ζ-*adjoint* to the endomorphism $f \in \mathbf{AL}(V)$ with the domain V_{f^*}, where $f^T \circ \zeta$ remains in the image of ζ:

$$\begin{array}{l} V_{f^*} = \{v \in V \mid \text{There exists } w \in V \text{ with } \zeta(w, u) = \zeta(v, f.u) \text{ for all } u \in V\},\\ f^* : V_{f^*} \longrightarrow V, \quad \zeta(f^*.v, u) = \zeta(v, f.u).\end{array}$$

If a dual isomorphism ζ is (conjugate) linear (anti-) symmetric, then the (conjugate) linear dual isomorphism $Z = \zeta \otimes \zeta^{-1}$ on $\mathbf{AL}(V)$ is always symmetric and vice versa:

$$\begin{array}{l} Z(\ ,\ ) : \mathbf{AL}(V) \times \mathbf{AL}(V) \longrightarrow \mathbb{K},\\ Z(f, g) = \operatorname{tr} f^* \circ g = \overline{Z(g, f)}, \quad \begin{cases} \zeta = \pm\epsilon\zeta^T \iff Z = Z^T,\\ \zeta = \pm\zeta^* \iff Z = Z^*,\end{cases}\\ Z(v \otimes \omega, u \otimes \theta) = \langle \zeta.v, u\rangle\langle \theta, \zeta^{-1}.\omega\rangle = \zeta(v, u)\zeta^{-1}(\omega, \theta).\end{array}$$

1.13.3 Quadratic and Positive Forms ("Squares")

A *quadratic form ("square")* d maps a real or complex vector space to the real numbers with the properties

$$\begin{array}{rl} d: & V \longrightarrow \mathbb{R},\quad d(\alpha v) = |\alpha|^2 d(v), \\ V \in \underline{\mathbf{vec}}_{\mathbb{R}}: & \left\{ \begin{array}{l} d(v+w) - d(v-w) \\ \text{is linear in } v \text{ and } w, \end{array} \right. \\ V \in \underline{\mathbf{vec}}_{\mathbb{C}}: & \left\{ \begin{array}{l} d(v+w) - d(v-w) + id(v-iw) - id(v+iw) \\ \text{is linear in } w \text{ and conjugate linear in } v. \end{array} \right. \end{array}$$

The conditions are chosen exactly in such a way that quadratic forms lead to *symmetric* inner products by the definitions

$$\begin{array}{l} \zeta(\ ,\ ) : V \times V \longrightarrow \mathbb{K}, \\ V \in \underline{\mathbf{vec}}_{\mathbb{R}}: \quad 4\zeta(v,w) = d(v+w) - d(v-w) = 4\zeta(w,v), \\ V \in \underline{\mathbf{vec}}_{\mathbb{C}}: \quad 4\zeta(v,w) = \left\{ \begin{array}{l} d(v+w) - d(v-w) \\ +id(v-iw) - id(v+iw) \end{array} \right\} = 4\overline{\zeta(w,v)}. \end{array}$$

In the other direction, each symmetric inner product defines its quadratic form by

$$d(v) = \zeta(v,v);$$

d is called degenerate if $\zeta(\ ,\ )$ is degenerate, etc.

Vectors with trivial square $d(v) = \zeta(v,v) = 0$ are called *singular*.

Quadratic forms satisfy the *parallelogram equation*

$$d(v+w) + d(v-w) = 2d(v) + 2d(w).$$

A quadratic form d is called *positive (prescalar product)* , $d \succeq 0$, if

$$d : V \longrightarrow \mathbb{R}_+,\quad d(v) = \zeta(v,v) \geq 0.$$

If d is negative, then $-d$ is positive. The positivity of a sesquilinear form involves its symmetry $\zeta(v,v) \geq 0 \Rightarrow \zeta(v,w) = \overline{\zeta(w,v)}$.

A form ζ is positive if and only if it has positive discriminants for $\zeta(v^i, v^j)_{i,j=1}^N$ for all vectors and all $N = 1, \ldots, n$. Therefore one has, for $N = 2$, the multiplicative *Cauchy-Schwarz inequality* and the additive *Minkowski inequality*:

$$d \succeq 0 \iff \left\{ \begin{array}{l} \zeta(v,v) \geq 0, \\ \det \begin{pmatrix} \zeta(v,v) & \zeta(v,w) \\ \zeta(w,v) & \zeta(w,w) \end{pmatrix} \geq 0 \Rightarrow \left\{ \begin{array}{l} |\zeta(v,w)|^2 \leq \zeta(v,v)\zeta(w,w), \\ \sqrt{d(v+w)} \leq \sqrt{d(v)} + \sqrt{d(w)}. \end{array} \right. \end{array} \right.$$

Positivity of ζ is also characterizable, as for the complex numbers, by a product

$$d \succeq 0 \iff d = \xi^\star \circ \xi$$

of a complex $(n \times n)$-matrix ξ and its Euclidean conjugate $\xi^\star$ (transposed matrix with complex conjugate elements).

A prescalar product defines a *prenorm* $\| \ \|_\zeta$ and a *premetric* $d_\zeta(\ , \ )$:

$$\begin{aligned} \| \ \|_\zeta &: V \longrightarrow \mathbb{R}_+, & \| v \|_\zeta^2 &= \zeta(v,v) = d(v), \\ d_\zeta(\ , \ ) &: V \times V \longrightarrow \mathbb{R}, & d_\zeta(v,w) &= \| v - w \|_\zeta \, . \end{aligned}$$

A *scalar product* is strictly positive:

$$\zeta(v,v) > 0 \iff v \neq 0.$$

Such a nondegenerate form leads to a norm and a metric and defines a pre-Hilbert space (chapter "Quantum Probability").

1.14 Equivalent Vector Space Bases

A linear group $G \subseteq \mathbf{GL}(V)$ acting on a vector space $V \cong \mathbb{K}^n$ defines G-*equivalent bases* by

$$\begin{aligned} \{e^i\}_{i=1}^n \overset{G}{\sim} \{f^i\}_{i=1}^n &\iff \text{There exists a } g \in G \text{ with } f^i = g(e^i) \text{ for all } i \\ &\iff f^i \otimes \check{e}_i \in G. \end{aligned}$$

All bases (n-*beins*) are $\mathbf{GL}(\mathbb{K}^n)$-equivalent. The equivalence classes $\mathbf{GL}(V)/G$ constitute the *n-bein manifold modulo G*. The columns in the matrix of g with respect to a basis $\{e^i\}_{i=1}^n$ contain the components of the equivalent basis $\{f^i = g(e^i)\}_{i=1}^n$:

$$g = f^i \otimes \check{e}_i = g(e^i) \otimes \check{e}_i = g^i_j e^j \otimes \check{e}_i \cong \begin{pmatrix} g^1_1 & \cdots & g^n_1 \\ & \cdots & \\ g^1_n & \cdots & g^n_n \end{pmatrix}.$$

If a group is the invariance group of an inner product ζ,

$$G = \{g \in \mathbf{GL}(V) \bigm| \zeta(g.v, g.u) = \zeta(v,u) \text{ for all } v, u \in V\},$$

then G-equivalent bases are characterized by equal ζ-matrices:

$$\begin{aligned} \{e^i\}_{i=1}^n \overset{G}{\sim} \{f^i\}_{i=1}^n &\iff f^i \otimes \check{e}_i = g \in G \\ &\iff \zeta(e^i, e^j) = \zeta(f^i, f^j) \text{ for all } i, j. \end{aligned}$$

Obviously, G-equivalent bases have equal ζ-matrices. For equal ζ-matrices one has for $\gamma = f^i \otimes \check{e}_i$ the G-condition $\zeta(\gamma.e^i, \gamma.e^j) = \zeta(e^i, e^j)$.

1.14.1 Equivalent Inner Products

Two inner products and their dual morphisms are called *linearily equivalent*, $\zeta_1 \overset{\mathbf{GL}(V)}{\sim} \zeta_2$, if they are connected by an automorphism $g \in \mathbf{GL}(V)$:

$$\begin{array}{ccc} V \times V & \overset{g \times g}{\longrightarrow} & V \times V \\ \zeta_2(\ , \ ) \downarrow & & \downarrow \zeta_1(\ , \ ), \\ \mathbb{K} & \underset{\mathrm{id}_\mathbb{K}}{\longrightarrow} & \mathbb{K} \end{array} \qquad \begin{array}{ccc} V & \overset{g}{\longrightarrow} & V \\ \zeta_2 \downarrow & & \downarrow \zeta_1 , \\ V^T & \underset{g^{-1T}}{\longrightarrow} & V^T \end{array}$$

$$\zeta_1(v, g.u) = \zeta_2(g^{-1}.v, u), \qquad \zeta_1 = g^T \circ \zeta_2 \circ g.$$

Symmetry is a property of the equivalence classes,

$$\zeta_1 \overset{\mathbf{GL}(V)}{\sim} \zeta_2 : \quad \begin{cases} \zeta_1 = \pm\epsilon\zeta_1^T & \iff \zeta_2 = \pm\epsilon\zeta_2^T \quad \text{(linear)}, \\ \zeta_1 = \pm\zeta_1^* & \iff \zeta_2 = \pm\zeta_2^* \quad \text{(conjugate linear)}, \end{cases}$$

as well as rank and signature.

Any *symmetric inner product* ζ *of* $V \cong \mathbb{K}^n$ leads to direct *Sylvester decompositions* of V into three vector subspaces - two orthogonal subspaces with strictly definite squares and the orthogonal

$$\begin{aligned} &V \cong V_\zeta^+ \oplus V_\zeta^- \oplus V_\zeta^\perp \cong \mathbb{K}^p \oplus \mathbb{K}^q \oplus \mathbb{K}^{n_\perp}, \\ &0 \neq v_\pm \in V_\zeta^\pm : \quad \zeta(v_+, v_+) > 0, \quad \zeta(v_-, v_-) < 0, \end{aligned}$$

where $r = p + q$ is the *rank of* ζ. $V_\zeta^\perp$ with dimension $n_\perp = n - r$ is ζ-orthogonal to all vectors. The dimensions, equal for all linearily equivalent forms, define the *signature* (p, q) and the *definite character* $J = |p - q|$ of ζ. (q, p) is the signature of $-\zeta$. For complex bilinear inner products the signature is meaningless.

For any such form ζ, there exist *Sylvester bases* $\{e^j\}_{j=1}^n$, defined by the ζ-diagonality and equal normalization for positive and negative elements:

$$\zeta \simeq e^{2\lambda} \begin{pmatrix} e^{2\rho}\mathbf{1}_p & 0 & 0 \\ 0 & -e^{-2\rho}\mathbf{1}_q & 0 \\ 0 & 0 & 0_{n_\perp} \end{pmatrix}, \quad \lambda, \rho \in \mathbb{R}.$$

Sylvester bases for a nondegenerate ζ are also called *orthogonal bases.* e^λ is the *normalization* of the Sylvester basis, $e^{\pm\rho}$ the *signature normalization.*

In analogy, there exist direct *Witt decompositions* for ζ into four vector subspaces

$$V \cong V_\zeta^{\text{nil}} \oplus V_\zeta^{\text{nil}'} \oplus V_\zeta^{\text{def}} \oplus V_\zeta^\perp \cong \mathbb{K}^I \oplus \mathbb{K}^I \oplus \mathbb{K}^J \oplus \mathbb{K}^{n_\perp}$$

two *nilspaces* $V_\zeta^{\text{nil}}, V_\zeta^{\text{nil}'} \cong \mathbb{K}^I$, where $I = \min\{p, q\}$ is the *index* of ζ, the *definite space* $V_\zeta^{\text{def}} \cong \mathbb{K}^J$, $J = |p - q|$, and the orthogonal. There exist *Witt bases* $\{e^\nu\}_{\nu=1}^n$ that contain I Witt pairs $\{e^{2I+1-\nu}, e^\nu\}_{\nu=1}^I$ for the nilspaces

$$\begin{aligned} \zeta &\simeq \epsilon e^{2\lambda} \begin{pmatrix} 0 & \overline{\mu}\mathbf{z}_I & 0 & 0 \\ \mu\mathbf{z}_I & 0 & 0 & 0 \\ 0 & 0 & \mathbf{1}_J & 0 \\ 0 & 0 & 0 & 0_{n_\perp} \end{pmatrix}, \quad \lambda \in \mathbb{R}, \quad \mu \in \mathbb{K}^\diamond, \quad \epsilon = \operatorname{sign}(p - q), \\ \mathbf{z}_I &= \begin{pmatrix} 0 & 0 & \dots & 0 & 1 \\ 0 & 0 & \dots & 1 & 0 \\ & & \dots & & \\ 0 & 1 & \dots & 0 & 0 \\ 1 & 0 & \dots & 0 & 0 \end{pmatrix} \in \mathbb{K}^{I^2}, \quad \begin{pmatrix} 0 & \mathbf{z}_I \\ \mathbf{z}_I & 0 \end{pmatrix} \cong \begin{pmatrix} \mathbf{z}_2 & 0 & \dots & 0 \\ 0 & \mathbf{z}_2 & \dots & 0 \\ & & \dots & \\ 0 & 0 & \dots & \mathbf{z}_2 \end{pmatrix}; \end{aligned}$$

e^λ and μ are called the *Witt normalization* and *Witt parameter.*

The transformation from a Witt to a Sylvester basis is characterizable in the 2-dimensional case by a 2×2 matrix w

$$\begin{aligned} &\zeta(e^j, e^k) \simeq \begin{pmatrix} e^{2\rho} & 0 \\ 0 & -e^{-2\rho} \end{pmatrix}, \quad \zeta(e^\mu, e^\nu) \simeq \begin{pmatrix} 0 & 1 \\ 1 & 0 \end{pmatrix}, \\ &w = \frac{1}{\sqrt{2}} \begin{pmatrix} e^{-\rho} & -e^\rho \\ e^{-\rho} & e^\rho \end{pmatrix}, \quad w_j^\mu \zeta^{jk} w_k^\nu = \zeta^{\mu\nu}. \end{aligned}$$

Strictly positive linear and conjugate linear dual isomorphisms δ_n of finite-dimensional spaces $V \cong \mathbb{R}^n$ and $V \cong \mathbb{C}^n$ respectively (signature $(n,0)$) are called *Euclidean*. There exist *Euclidean bases* defined by diagonality with equal normalization from which the isomorphisms arise by linear and conjugate linear extension

$$\delta_n : V \longrightarrow V^T, \quad \delta_n(e^j) = e^{2\lambda}\delta^{jk}\check{e}_k.$$

The associate (conjugate) linear *Euclidean reflection* is the transposition of the $n \times n$ matrix for linear δ_n and the familiar hermitian conjugation for conjugate linear δ_n. A Euclidean conjugation of $V \oplus V^T \cong \mathbb{C}^{2n}$ and the induced endomorphism conjugation of $\mathbf{AL}(V) \cong \mathbb{C}^{n^2}$ is always denoted by the *Euclidean five pointed star* $\star$

$$\begin{array}{l} \star : V \longrightarrow V^T, \quad (e^j)^\star = e^{2\lambda}\delta^{jk}\check{e}_k, \quad (\check{e}_k)^\star = e^{-2\lambda}\delta_{kj}e^j, \quad (\alpha v)^\star = \overline{\alpha} v^\star, \\ \star : \mathbf{AL}(V) \longrightarrow \mathbf{AL}(V), \quad \left\{ \begin{array}{l} (v \otimes \omega)^\star = \omega^\star \otimes v^\star, \\ (e^j \otimes \check{e}_k)^\star = \delta_{kl} e^l \otimes \check{e}_i \delta^{ij}. \end{array} \right. \end{array}$$

For *symmetric bilinear forms* ζ *of* $V \cong \mathbb{C}^n$ there exist orthogonal bases with only $+1$ or 0 on the diagonal.

For *linear antisymmetric* dual morphisms ζ of $V \cong \mathbb{K}^n$ there exist decompositions $V \cong V_\zeta^\epsilon \oplus V_\zeta^\perp \cong \mathbb{K}^{2d} \oplus \mathbb{K}^{n_\perp}$ and *symplectic pair bases* with real antisymmetric 2×2 matrix block diagonals or skew-diagonal ϵ-form for ζ, nontrivial in the $2d \times 2d$ uppermost left corner:

$$\zeta \simeq \begin{pmatrix} \epsilon_{2d} & 0 \\ 0 & 0_{n_\perp} \end{pmatrix} \cong \begin{pmatrix} \epsilon_2 & 0 & \dots & 0 & 0 \\ & & \dots & & \\ 0 & 0 & \dots & \epsilon_2 & 0 \\ 0 & 0 & \dots & 0 & 0_{n_\perp} \end{pmatrix}, \quad \epsilon_{2d} = \begin{pmatrix} 0 & \mathbf{z}_d \\ -\mathbf{z}_d & 0 \end{pmatrix} \in \mathbb{K}^{(2d)^2}.$$

The rank of ζ is $2d$.

1.14.2 Invariance Groups and Lie Algebras of Inner Products

The *invariance group* $\mathbf{UL}(V,\zeta)$ *of a (conjugate) linear and (anti-) symmetric dual isomorphism* ζ, i.e., of an (anti-) symmetric nondegenerate inner product $\zeta(\ ,\ )$, is

$$\mathbf{UL}(V,\zeta) = \{g \in \mathbf{GL}(V) \,\big|\, g^* = \zeta^{-1} \circ g^T \circ \zeta = g^{-1}\} \in \underline{\mathbf{grp}}.$$

$$\begin{array}{ccc} V & \xrightarrow{g} & V \\ \zeta \downarrow & & \downarrow \zeta \\ V^T & \xrightarrow[g^{-1T}]{} & V^T \end{array}, \qquad \begin{array}{ccc} V \times V & \xrightarrow{g \times g} & V \times V \\ \zeta(\ ,\ ) \downarrow & & \downarrow \zeta(\ ,\ ) \\ \mathbb{K} & \xrightarrow[\mathrm{id}_\mathbb{K}]{} & \mathbb{K} \end{array},$$

$$\zeta = g^T \circ \zeta \circ g, \qquad \zeta(v, g.u) = \zeta(g^{-1}.v, u).$$

For the ζ-induced endomorphism reflection, the elements of the invariance group go to their inverses, the invariance group is $*$-*unitary*; one can write $\mathbf{UL}(V,\zeta) = \mathbf{U}(\mathbf{GL}(V), *)$.

The invariance groups and also their elements are called

$$\mathbf{UL}(V,\zeta) = \left\{ \begin{array}{rl} \mathbf{O}(V,\zeta),\ \textit{orthogonal}, & V \in \underline{\mathbf{vec}}_{\mathbb{K}},\ \ \zeta = \epsilon\zeta^T \\ & \text{(linear symmetric)}, \\ \mathbf{Sp}(V,\zeta),\ \textit{symplectic}, & V \in \underline{\mathbf{vec}}_{\mathbb{K}},\ \ \zeta = -\epsilon\zeta^T \\ & \text{(linear antisymmetric)}, \\ \mathbf{U}(V,\zeta),\ \textit{unitary}, & V \in \underline{\mathbf{vec}}_{\mathbb{C}},\ \ \zeta = \zeta^* \\ & \text{(conjugate linear symmetric)}. \end{array} \right.$$

For finite-dimensional spaces $V \cong \mathbb{K}^n$ all determinants have modulus 1; one has a *unilinear invariance group* $\mathbf{UL}(\mathbb{K}^n,\zeta)$. The normal subgroup with determinant $+1$ elements is called the *special invariance group* $\mathbf{SL}(\mathbb{K}^n,\zeta)$:

$$\begin{array}{l} g \in \mathbf{UL}(\mathbb{K}^n,\zeta) \Rightarrow |\det g| = 1, \\ \mathbf{UL}(\mathbb{K}^n,\zeta) \cong \mathbf{U}(\mathbb{K},\zeta) \circ \mathbf{SL}(\mathbb{K}^n,\zeta) \text{ in general no direct product}, \\ \mathbf{SL}(\mathbb{K}^n,\zeta) = \{g \in \mathbf{GL}(\mathbb{K}^n) \mid g^* = \zeta^{-1} \circ g^T \circ \zeta = g^{-1},\quad \det g = 1\}. \end{array}$$

The *phase group* $\mathbf{U}(\mathbb{K},\zeta) \subseteq \mathbf{GL}(\mathbb{K})$ has, in general, no isomorphic normal subgroup in $\mathbf{UL}(\mathbb{K}^n,\zeta)$:

$$\begin{array}{l} \mathbf{U}(\mathbb{K},\zeta) = \{\det g \mid g \in \mathbf{UL}(\mathbb{K}^n,\zeta)\}, \\ \mathbf{U}(\mathbb{K},\zeta) \cong \left\{ \begin{array}{l} \mathbb{I}(2) = \{1,-1\},\quad \text{for } g \in \mathbf{O}(\mathbb{K}^n,\zeta) \Rightarrow (\det g)^2 = 1, \\ \{1\},\quad \text{for } g \in \mathbf{Sp}(\mathbb{K}^n,\zeta) \Rightarrow \det g = 1, \\ \mathbf{U}(1) = \{e^{i\alpha} \mid \alpha \in \mathbb{R}\},\quad \text{for } g \in \mathbf{U}(\mathbb{C}^n,\zeta) \Rightarrow |\det g|^2 = 1. \end{array} \right. \end{array}$$

The corresponding *invariance Lie algebra* $\log \mathbf{UL}(V,\zeta)$ *of a (conjugate) linear and (anti-) symmetric dual isomorphism* ζ is

$$\log \mathbf{UL}(V,\zeta) = \{l \in \mathbf{AL}(V) \mid l^* = \zeta^{-1} \circ l^T \circ \zeta = -l\} \in \underline{\mathbf{lag}}_{\mathbb{K}},$$

$$\begin{array}{ccc} V & \stackrel{l}{\longrightarrow} & V \\ \zeta \downarrow & & \downarrow \zeta \\ V^T & \underset{-l^T}{\longrightarrow} & V^T \end{array}, \qquad \begin{array}{ccc} V \times V & \stackrel{l \times \mathrm{id}_V + \mathrm{id}_V \times l}{\longrightarrow} & V \times V \\ \zeta(\ ,\ ) \downarrow & & \downarrow \zeta(\ ,\ ) \\ \mathbb{K} & \underset{0}{\longrightarrow} & \mathbb{K} \end{array},$$

$$\zeta \circ l = -l^T \circ \zeta, \qquad \zeta(v, l.u) = -\zeta(l.v, u).$$

The ζ-induced endomorphism reflection maps the elements of the invariance Lie algebra into their negatives: the invariance Lie algebra is $*$-*antisymmetric,* $\log \mathbf{UL}(V,\zeta) = A(\mathbf{AL}(V), *)$. V is a vector space with $\log \mathbf{UL}(V,\zeta)$ as operator Lie algebra. The Lie algebras are called orthogonal, symplectic, or unitary, as above.

For finite-dimensional spaces $V \cong \mathbb{K}^n$ the invariance Lie algebra is directly decomposable into the *traceless* ζ-antisymmetric endomorphisms and the *trace part* with the imaginary trace $\operatorname{tr} l = -\overline{\operatorname{tr} l}$:

$$\begin{array}{l} \log \mathbf{UL}(\mathbb{K}^n,\zeta) = \log \mathbf{U}(\mathbb{K},\zeta) \oplus \log \mathbf{SL}(\mathbb{K}^n,\zeta), \\ \log \mathbf{SL}(\mathbb{K}^n,\zeta) = \{l \in \mathbf{AL}(\mathbb{K}^n) \mid l^* = \zeta^{-1} \circ l^T \circ \zeta = -l,\quad \operatorname{tr} l = 0\}, \\ \log \mathbf{U}(\mathbb{K},\zeta) \cong \left\{ \begin{array}{r} \{0\},\ \ \zeta \text{ linear},\quad \operatorname{tr} l = 0, \\ i\mathbb{R},\ \ \mathbb{K} = \mathbb{C},\ \ \zeta \text{ conjugate linear}. \end{array} \right. \end{array}$$

Using the dual isomorphism $\zeta : V \longrightarrow V^T$ the orthogonal and the symplectic Lie algebra can be defined as follows (one has "opposite" symmetry properties):

$$\begin{aligned}\log \mathbf{SO}(V,\zeta) &= \{v_m \otimes \zeta(u_m) - u_m \otimes \zeta(v_m) \,\big|\, v_m, u_m \in V\}, \quad \zeta = +\epsilon\zeta^T,\\ \log \mathbf{Sp}(V,\zeta) &= \{v_m \otimes \zeta(u_m) + u_m \otimes \zeta(v_m) \,\big|\, v_m, u_m \in V\}, \quad \zeta = -\epsilon\zeta^T.\end{aligned}$$

If an invariance is characterized by (anti-) symmetry and signature, one uses the following notation for the special invariance Lie groups and the associate Lie algebras:

$$\begin{array}{ll}\text{linear and } \mathbb{K} = \mathbb{R} \\ \text{(real Lie symmetries)}\end{array} \left\{\begin{array}{l}\mathbf{SO}(\mathbb{R}^n, p, q) = \mathbf{SO}(p,q), \quad \log \mathbf{SO}(p,q) \cong \mathbb{R}^{\binom{n}{2}},\\ \mathbf{Sp}(\mathbb{R}^{2n}) = \mathbf{Sp}(2n), \quad \log \mathbf{Sp}(2n) \cong \mathbb{R}^{\binom{1+2n}{2}};\end{array}\right.$$

$$\begin{array}{ll}\text{linear and } \mathbb{K} = \mathbb{C} \\ \text{(complex Lie symmetries)}\end{array} \left\{\begin{array}{l}\mathbf{SO}(\mathbb{C}^n), \quad \log \mathbf{SO}(\mathbb{C}^n) \cong \mathbb{C}^{\binom{n}{2}},\\ \mathbf{Sp}(\mathbb{C}^{2n}), \quad \log \mathbf{Sp}(\mathbb{C}^{2n}) \cong \mathbb{C}^{\binom{1+2n}{2}};\end{array}\right.$$

$$\begin{array}{ll}\text{conjugate linear, } \mathbb{K} = \mathbb{C} \\ \text{(real Lie symmetries)}\end{array} \left\{\begin{array}{l}\mathbf{SU}(\mathbb{C}^n, p, q) = \mathbf{SU}(p,q),\\ \log \mathbf{SU}(p,q) \cong \mathbb{R}^{n^2-1}.\end{array}\right.$$

$\mathbf{SO}(p,q)$, $\mathbf{Sp}(2n)$, and also the unitary $\mathbf{SU}(p,q)$ in complex automorphisms are real Lie groups with real Lie algebras.

1.15 Matrix Diagonalization and Orientation Manifolds

A complex $(n \times n)$ matrix f is *unitarily diagonalizable* and a real $(n \times n)$ matrix f is *orthogonally box-diagonalizable* if and only if it is $*$-normal:

$$\begin{aligned} f \circ f^* = f^* \circ f \iff & f = f_+ + f_- \\ & \text{with } f_\pm^* = \pm f_\pm, \ [f_+, f_-] = 0, \\ \iff & \left\{\begin{array}{ll} f \ \mathbf{U}\text{-diagonalizable}, & \mathbb{K} = \mathbb{C}, \\ f \ \mathbf{O} \text{ box-diagonalizable}, & \mathbb{K} = \mathbb{R}, \end{array}\right. \\ U^* \circ f \circ U = \operatorname{diag} f \ & \text{with } U^* = U^{-1}. \end{aligned}$$

The definition of normality with diagonalization requires a reflection $*$, i.e., a bilinear form for the real and a sesquilinear form for the complex case (more below). In the complex case the resulting matrix $\operatorname{diag} f$ is diagonal, in the real case only box-diagonal: Its diagonal has real elements (1×1 matrices) or nondiagonalizable antisymmetric 2×2 matrices in $\mathbf{D}(1) \times \mathbf{O}(2)$-form:

$$\pm e^\lambda O(\varphi), \quad O(\varphi) = \begin{pmatrix} \cos\varphi & -\sin\varphi \\ \sin\varphi & \cos\varphi \end{pmatrix}, \quad \lambda \in \mathbb{R}, \quad 0 < \varphi < 2\pi.$$

A symmetric matrix $\zeta = \zeta^*$ is diagonalizable, in the real case the symmetry is even necessary:

$$\mathbb{K} = \mathbb{R}: \quad \zeta = \zeta^* \iff \zeta \text{ is } \mathbf{O}\text{-diagonalizable}.$$

In general, there exist different reflections $*$ and thus different types of "normality" and "unitarity" ("orthogonality"). The positive Euclidean matrix reflection $\star$ ($\mathbf{O}(n)$ and $\mathbf{U}(n)$-reflection) uses the matrix transposition in the real case and the conjugate transposition in the complex case:

$$f^\star = \begin{cases} f^T & \text{for } \mathbb{R}, \\ \overline{f}^T & \text{for } \mathbb{C}. \end{cases}$$

The orthogonal group $\mathbf{SO}(n)$ has $\binom{n}{2}$ independent axial rotation axes with associated 2-planes, therefore $\dim_{\mathbb{R}} \mathbf{SO}(n) = \binom{n}{2}$. A rotation is box-diagonalizable with r axial rotations for $n = 2r$ and one additional "isolated" diagonal element 1 for $n = 1 + 2r$. This defines the *rank* for the orthogonal groups. Analogously one obtains the rank $n-1$ for special unitary groups $\mathbf{SU}(n)$ with $n-1$ independent diagonal $\mathbf{U}(1)$-phases:

$$\begin{aligned} \mathbf{SO}(2r) \ni O \cong\ & R \circ \operatorname{diag}(O(\varphi_1), ..., O(\varphi_r)) \circ R^T, \\ & R \in \mathbf{SO}(2r)/\mathbf{SO}(2)^r, \\ \mathbf{SO}(1+2r) \ni O \cong\ & R \circ \operatorname{diag}(O(\varphi_1), ..., O(\varphi_r), 1) \circ R^T, \\ & R \in \mathbf{SO}(1+2r)/\mathbf{SO}(2)^r, \\ \mathbf{SU}(n) \ni U \cong\ & V \circ \operatorname{diag}(e^{i\varphi_1}, ..., e^{i\varphi_n}) \circ V^\star, \\ & \varphi_1 + \ldots \varphi_n = 0, \quad V \in \mathbf{SU}(n)/\mathbf{U}(1)^{n-1}. \end{aligned}$$

Similar expressions hold for all orthogonal groups $\mathbf{SO}_0(p,q)$ with diagonal $\mathbf{SO}(2)$'s, $\mathbf{SO}_0(1,1)$'s, and 1's and for all unitary groups $\mathbf{SU}(p,q)$ with diagonal $\mathbf{U}(1)$'s and $\mathbf{D}(1)$'s.

The $\mathbf{O}(p,q)$ and $\mathbf{U}(p,q)$-reflections $*$ for matrices, indefinite in the case $pq \neq 0$, are obtained by a left-right multiplication with the signature matrix in addition to the positive reflection $\star$:

$$f^* = \begin{pmatrix} \mathbf{1}_p & 0 \\ 0 & -\mathbf{1}_q \end{pmatrix} \circ f^\star \circ \begin{pmatrix} \mathbf{1}_p & 0 \\ 0 & -\mathbf{1}_q \end{pmatrix}.$$

All nondegenerate matrices, symmetric with the positive conjugation $\zeta = \zeta^\star$ and therefore $\mathbf{O}(n)$ or $\mathbf{U}(n)$-diagonalizable $U^\star \circ \zeta \circ U = \operatorname{diag}\zeta$ to a real diagonal with signature (p,q), parametrize the *orientation manifolds* $\mathbf{GL}(\mathbb{R}^n)/\mathbf{O}(p,q)$ and $\mathbf{GL}(\mathbb{C}^n)/\mathbf{U}(p,q)$ for the inner product ζ. Here, two reflections are involved: the ζ-defined reflection with signature (p,q) and the definite Euclidean one. The p strictly positive and q strictly negative numbers in $\operatorname{diag}\zeta$ define the dilation coordinates (normalizations of the axes) in $\mathbf{D}(1)^n$:

$$\begin{aligned} \mathbf{GL}(\mathbb{R}^n)/\mathbf{O}(p,q) &\cong \mathbf{D}(1)^n \circ \mathbf{SO}(n), \\ \mathbf{GL}(\mathbb{C}^n)/\mathbf{U}(p,q) &\cong \mathbf{D}(1)^n \circ \mathbf{U}(n)/\mathbf{U}(1)^n, \end{aligned}$$

with the determinant $\det\zeta$ the $\mathbf{SL}(\mathbb{K}^n)$-invariant $\mathbf{D}(\mathbf{1}_n)$-coordinate (overall normalization).

1.16 Reflections in Orthogonal Groups

A real linear reflection $\mathsf{R} \cong \begin{pmatrix} \mathbf{1}_m & 0 \\ 0 & -\mathbf{1}_{n-m} \end{pmatrix}$ of a vector space $V \cong \mathbb{R}^n$ is an element of all orthogonal groups $\mathbf{O}(p,q)$ with $p+q=n$. A positively oriented reflection,

$\det \mathtt{R} = 1$, is an element even of the special orthogonal group, $\mathtt{R} \in \mathbf{SO}(p,q)$, $p+q \geq 1$.

Orthogonal groups have discrete reflection subgroups $\mathbb{I}(2)$ (parity) as (semi) direct factors, as seen in the simplest compact and noncompact examples:

$$\begin{aligned} \mathbf{O}(2) &\ni \begin{pmatrix} \epsilon\cos\varphi & \epsilon\sin\varphi \\ -\sin\varphi & \cos\varphi \end{pmatrix}, \quad \epsilon \in \mathbb{I}(2) = \{\pm 1\}, \quad \varphi \in [0, 2\pi[, \\ \mathbf{O}(1,1) &\ni \epsilon' \begin{pmatrix} \epsilon\cosh\psi & \epsilon\sinh\psi \\ \sinh\psi & \cosh\psi \end{pmatrix}, \quad \epsilon, \epsilon' \in \mathbb{I}(2), \quad \psi \in \mathbb{R}. \end{aligned}$$

The classes of a real orthogonal group with respect to its special normal subgroup constitute a reflection group

$$\mathbf{O}(p,q)/\mathbf{SO}(p,q) \cong \mathbb{I}(2).$$

For real odd-dimensional spaces V, e.g., for position $\mathbb{R}^3$, one has direct products of the special groups with the central reflection group, whereas for even-dimensional spaces, e.g., a Minkowski space $\mathbb{R}^4$, there arise semidirect products of the special group with a reflection group that can be generated by any negatively oriented reflection

$$\mathbf{O}(p,q) \cong \begin{cases} \mathbb{I}(2) \times \mathbf{SO}(p,q), & p+q = 1,3,\ldots, \\ & \mathbb{I}(2) \cong \{\pm \mathrm{id}_V\}, \\ \mathbb{I}(2) \vec{\times} \mathbf{SO}(p,q), & p+q = 2,4,\ldots, \\ & \mathbb{I}(2) \cong \{\mathtt{R}, \mathrm{id}_V\} \text{ with } \det \mathtt{R} = -1. \end{cases}$$

In the semidirect case the product is given as follows:

$$(\mathtt{I}, \Lambda) \in \mathbb{I}(2) \vec{\times} \mathbf{SO}(p,q) \Rightarrow (\mathtt{I}_1, \Lambda_1)(\mathtt{I}_2, \Lambda_2) = (\mathtt{I}_1 \circ \mathtt{I}_2, \Lambda_1 \circ \mathtt{I}_1 \circ \Lambda_2 \circ \mathtt{I}_1).$$

Obviously, in the semidirect case the reflection group $\mathbb{I}(2)$ is not compatible with the action of the (special) orthogonal group.

$$p+q = 2,4,\ldots, \quad \det \mathtt{R} = -1 \Rightarrow [\mathtt{R}, \mathbf{SO}(p,q)] \neq \{0\}.$$

For example, the group $\mathbf{O}(2)$ is nonabelian, and a space reflection and a time reflection of Minkowski space is not Lorentz group $\mathbf{SO}(1,3)$-compatible.

For noncompact orthogonal groups there is another discrete reflection group: The subgroup G_0 (unit connection component and Lie algebra exponent) of a Lie group G is normal with a discrete quotient group G/G_0. The connected components of the full orthogonal groups are those of the special groups $\mathbf{O}_0(p,q) = \mathbf{SO}_0(p,q)$. For the compact case these are the special groups, for the noncompact ones one has two components:

$$\begin{aligned} &\mathbf{SO}_0(n) = \mathbf{SO}(n), \\ pq \geq 1 \Rightarrow \ &\mathbf{SO}(p,q)/\mathbf{SO}_0(p,q) \cong \mathbb{I}(2). \end{aligned}$$

For $\mathbf{SO}(2n)$ with nontrivial center $\mathbb{I}(2)$ there is no subgroup isomorphic to $\mathbf{SO}(2n)/\mathbb{I}(2)$.

Summarizing: A compact orthogonal group gives rise to a reflection group $\mathbb{I}(2)$,

$$\begin{aligned} \mathbf{O}(n) &\cong \begin{cases} \{\pm \mathbf{1}_n\} \times \mathbf{SO}(n), & n = 1,3,\ldots, \\ \mathbb{I}(2) \vec{\times} \mathbf{SO}(n), & n = 2,4,\ldots, \end{cases} \\ \text{with } \mathbb{I}(2) &\cong \{\mathtt{R}, \mathbf{1}_n\}, \quad \det \mathtt{R} = -1, \end{aligned}$$

a noncompact one to a reflection Klein group $\mathbb{I}(2) \times \mathbb{I}(2)$,

$$\begin{array}{rl} pq \geq 1: \quad \mathbf{O}(p,q) & \cong \left\{ \begin{array}{ll} \{\pm \mathbf{1}_{p+q}\} \times [\mathbb{I}(2) \vec{\times} \mathbf{SO}_0(p,q)], & p+q = 3,5,\dots, \\ \mathbb{I}(2) \vec{\times} [\{\pm \mathbf{1}_{p+q}\} \times \mathbf{SO}_0(p,q)], & p+q = 2,4,\dots, \end{array} \right. \\ \text{with } \mathbb{I}(2) & \cong \{\mathtt{R}, \mathbf{1}_n\}, \quad \det \mathtt{R} = -1. \end{array}$$

Also, the connected subgroup $\mathbf{SO}_0(p,q)$ may contain positively oriented reflections, which are called continuous since they can be written as exponentials $\mathtt{R} = e^l$ with an element of the orthogonal Lie algebra, $l \in \log \mathbf{SO}_0(p,q)$. For example,the central reflections $-\mathbf{1}_{2n} \in \mathbf{SO}(2n)$ in even-dimensional Euclidean spaces, e.g., in the Euclidean 2-plane. A negatively oriented reflection $\mathtt{R}$ of a space V can be embedded as a reflection $\mathtt{R} \oplus \mathtt{S}$ with any orientation of a strictly higher-dimensional space $V \oplus W$:

$$\begin{array}{rl} V \xleftrightarrow{\mathtt{R}} V, & \det \mathtt{R} = -1, \\ V \oplus W \xleftrightarrow{\mathtt{R} \oplus \mathtt{S}} V \oplus W, & \det(\mathtt{R} \oplus \mathtt{S}) = -\det \mathtt{S}, \end{array}$$

where, for compact orthogonal groups on V and $V \oplus W$, a reflection $\mathtt{R} \oplus \mathtt{S}$ with $\det \mathtt{S} = -1$ is a continuous reflection, i.e., a rotation. There are the familiar examples for $\mathbf{O}(n) \hookrightarrow \mathbf{SO}(1+n)$: Two noodles in letter L-form, lying with opposite helicity on the kitchen table, can be 3-space rotated into each other, and a left- and a right-hand glove are identical up to Euclidean 4-space rotations. The embedding of the central position space reflection into Minkowski spacetime can go into a positively or negatively oriented reflection, both of which not continuous, i.e., they are in a discrete Klein reflection group:

$$-\mathbf{1}_3 \hookrightarrow \begin{pmatrix} \pm 1 & 0 \\ 0 & -\mathbf{1}_3 \end{pmatrix}, \quad \{\mathtt{S}, -\mathbf{1}_4\} \subset \mathbf{O}(1,3)/\mathbf{SO}_0(1,3).$$

Bibliography

[1] N. Bourbaki, *Théorie des Ensembles, Chapitre 4* (Structures) (1957), Hermann, Paris.

[2] N. Bourbaki, *Algebra I, Chapters 1-3* (1989), Springer, Berlin, Heidelberg, New York, London, Paris, Tokyo.

[3] N. Bourbaki, *Algèbre, Chapitre 9* (Formes sesquilineaires et formes quadratiques) (1959), Hermann, Paris.

[4] R. Gilmore, *Lie Groups, Lie Algebras and Some of Their Applications* (1974), John Wiley & Sons, New York, London, Sidney, Toronto.

[5] H. Reichenbach, *The Philosophy of Space and Time* (1927), Dover, New York.

[6] H. Weyl, *Raum, Zeit, Materie* (1923), Wissenschaftliche Buchgesellschaft, Darmstadt.

2

TIME REPRESENTATIONS

A dynamics can be characterized and will be understood as an action of time and position which, together with time and position translations as their tangent structures, are modeled by operations from real Lie groups and Lie algebras respectively. Time and position operations come in realizations and representations acting on sets and vector spaces. The solution of a dynamics is the decomposition of the time and position representations involved into nondecomposable, perhaps even irreducible, representations. Representations are characterized by invariants (masses, spins, etc.) and eigenvalues (energies, momenta, helicity, etc.) for the operations that define the properties of physical objects. With this program, physical theories become to a great extent applied representation and realization theory.

In classical dynamics the time realizations are visible in the mass point orbits $t \longmapsto \mathbf{x}(t)$ in position as solutions of the equations of motion with a fixed energy, imposed by initial or boundary conditions. In quantum mechanics time and position orbits are information-valued ("probability amplitudes"). They are given, e.g., by Schrödinger wave functions $(t, \vec{x}) \longmapsto \Psi(t, \vec{x}) = e^{iEt}\psi_E(\vec{x})$, which are time and position orbits in a complex Hilbert space with probability interpretation (chapter "The Kepler Factor").

In this chapter, the complex finite-dimensional representations of the "simplest" nontrivial real 1-dimensional Lie group $\mathbf{D}(1) = \exp \mathbb{R}$ and its Lie algebra $\log \mathbf{D}(1) = \mathbb{R}$ are considered. Its structures can be formulated in a time-related language: The simplest, and also characteristic, examples are given by the free Newtonian mass point with mass M and the harmonic oscillator with string constant k with Hamiltonians $H_{1,0}$ and the equations of motion for position-momentum $(\mathbf{x}, \mathbf{p})$:

$$\begin{aligned} H_1 &= \tfrac{\mathbf{p}^2}{2M} & &\Rightarrow \tfrac{d\mathbf{x}}{dt} = \tfrac{\mathbf{p}}{M}, \quad \tfrac{d\mathbf{p}}{dt} = 0, \\ H_0 &= \tfrac{\mathbf{p}^2}{2M} + \tfrac{k}{2}\mathbf{x}^2 & &\Rightarrow \tfrac{d\mathbf{x}}{dt} = \tfrac{\mathbf{p}}{M}, \quad \tfrac{d\mathbf{p}}{dt} = -k\mathbf{x}. \end{aligned}$$

$\mathbf{D}(1)$ will be called the time group or, in special relativity, eigentime group. In larger operation groups, e.g., in spacetime, the general name "causal group" is appropriate.

To represent real Lie operations on complex vector spaces, such a space has to come with a conjugation, definite or indefinite: Real groups have to be represented in the complex by unitary automorphisms.[1] The conjugation for a time representation implements the time reflection and determines an inner product of the complex representation space, which, for a positive conjugation, is the origin for the scalar product, leading in quantum theories to "probability amplitudes" for the interpretation of experiments (chapter "Quantum Probabilities").

Obviously from a general group theoretical point of view, the representations of the two real 1-dimensional abelian Lie groups $\mathbf{D}(1) = \exp \mathbb{R}$ and $\mathbf{U}(1) = \exp i\mathbb{R}$ as noncompact and compact subgroups of the complex group $\mathbf{GL}(\mathbb{C}) = \mathbb{C}^\diamond = \exp \mathbb{C}$ are the basic ingredients for the representations of all real and complex Lie groups that contain "many" $\mathbf{D}(1)$ and $\mathbf{U}(1)$-isomorphic subgroups.

2.1 The Time Group

The totally ordered noncompact real 1-dimensional Lie group $\mathbf{D}(1)$ is used as a time model. This abelian group can be written multiplicatively as $\mathbf{D}(1)$ or additively as $\mathbb{R}$:

$$\mathbf{D}(1) = \exp \mathbb{R} = \{e^t \mid t \in \mathbb{R}\} \cong \mathbb{R} = \log \mathbf{D}(1).$$

The Lie group isomorphism is given by the exponential and the logarithm.

The classes with respect to the discrete subgroup of the integers $\exp \mathbb{Z} \cong \mathbb{Z}$ constitute the quotient group $\exp \mathbb{R}/\exp \mathbb{Z}$, isomorphic to the *compact* additive group $\mathbb{R}/\mathbb{Z}$ of the reals modulo the integers, i.e., to the 1-dimensional torus. As a real Lie group it is isomorphic to the unit circle $\mathbf{U}(1)$, the phase group of the complex numbers:

$$\mathbf{U}(1) = \exp i\mathbb{R} = \{e^{i\alpha} \mid 0 \leq \alpha < 2\pi\} \cong \mathbb{R}/\mathbb{Z}, \quad \log \mathbf{U}(1) = i\mathbb{R}.$$

All connected Lie groups with $\mathbb{R}$-isomorphic Lie algebra arise from the simply connected Lie group $\mathbf{D}(1)$ by the classes with respect to the discrete normal subgroups. Since $\exp \mathbb{Z} \cong \mathbb{Z}$ is, up to isomorphism, the only nontrivial closed subgroup, only $\mathbf{D}(1)$ and $\mathbf{U}(1)$ occur as images of nontrivial time $\mathbf{D}(1)$-representations.

Thus three types of time orbits are possible: the trivial representation with an 1-elementic orbit $\mathbb{R}/\mathbb{R} \cong \{1\}$ and two 1-dimensional orbits, as manifolds isomorphic either to the circle $\mathbb{R}/\mathbb{Z} \cong \mathbf{U}(1)$ or to the real line $\mathbb{R}$. This is illustrated (chapter "The Kepler Factor") in a classical description by the solar system, with hyperbolic orbits for never-returning comets, elliptic ones for planets, and the trivial orbit for the sun (more exactly, for the center of mass) with trivially represented $\mathbb{R}$-subgroups $\{0\}$, $\mathbb{Z}$, and $\mathbb{R}$. Quantum-mechanical energy eigenstates are $\mathbf{U}(1)$-orbits in the Hilbert space under question.

[1] Orthogonality and unitarity can be definite or indefinite, e.g., $\mathbf{O}(1,3)$ or $\mathbf{O}(4)$ and $\mathbf{U}(1,3)$ or $\mathbf{U}(4)$.

The compact group $\mathbf{U}(1)$ plays a decisive role in characterizing particles by rational number properties, e.g., by integer electromagnetic charge numbers (internal operation group $\mathbf{U}(1)$) or integer and half-integer spin numbers (external operation group $\mathbf{U}(1) \cong \mathbf{SO}(2) \subset \mathbf{SU}(2)$).

2.2 Representations of the Complex Numbers

The nondecomposable, finite-dimensional representations of the complex 1-dimensional Lie group $\mathbb{C}^\diamond$ are characterized by the dimension of the representation space $V \cong \mathbb{C}^{1+N}$ and one *complex eigenvalue (invariant)* $\zeta \in \mathbb{C}$. They contain an endomorphism n_N nilpotent to the power $1+N$:

$$D_N^\zeta : \mathbb{C}^\diamond \longrightarrow \mathbf{GL}(V), \quad e^z \longmapsto D_N^\zeta(e^z) = e^{(\zeta\, \mathrm{id}_V + n_N)z} = e^{\zeta z} \sum_{k=0}^{N} \frac{(n_N z)^k}{k!},$$
$$n_0 = 0, \quad \text{for } N \geq 1: \ (n_N)^N \neq 0, \ (n_N)^{1+N} = 0;$$

N is called the *nildimension* of the representation. The two factors involved are representations themselves: the ζ-power of the group and a special factor

$$D_N^\zeta = D_0^\zeta \otimes D_N^0, \quad \begin{cases} D_0^\zeta : & e^z \longmapsto (e^z)^\zeta \in \mathbf{GL}(\mathbb{C}), \\ D_N^0 : & e^z \longmapsto e^{n_N z} \in \mathbf{SL}(\mathbb{C}^{1+N}). \end{cases}$$

The invariant ζ and the dimension $(1+N)$ determine the minimal polynomial $(X - e^{\zeta z})^{1+N}$ of the representation D_N^ζ.

The irreducible representations $e^z \longmapsto D(z)$ of the abelian Lie group $\mathbb{C}^\diamond$ have to be complex 1-dimensional and fulfill, by derivation of $D(z_1 + z_2) = D(z_1)D(z_2)$, the differential equation $D'(z) = D'(0)D(z)$, i.e., $D(z) = e^{\zeta z}$ with invariant $D'(0) = \zeta \in \mathbb{C}$. The nilpotent endomorphisms can be written as a nilcyclic Jordan matrix with *Jordan bases*:

$$\text{dual bases: } \langle \check{e}_B, e^A \rangle = \delta_B^A, \quad A, B = 0, \dots, N, \quad \mathrm{id}_V = e^A \otimes \check{e}_A,$$
$$n_N = \nu(e^0 \otimes \check{e}_1 + \dots + e^{N-1} \otimes \check{e}_N) \simeq \nu \begin{pmatrix} 0 & 1 & 0 & \dots & 0 & 0 \\ 0 & 0 & 1 & \dots & 0 & 0 \\ & & & \dots & & \\ 0 & 0 & 0 & \dots & 0 & 1 \\ 0 & 0 & 0 & \dots & 0 & 0 \end{pmatrix},$$
$$e^{(\zeta\, \mathrm{id}_V + n_N)z} \simeq e^{\zeta z} \begin{pmatrix} 1 & \nu z & \frac{(\nu z)^2}{2} & \dots & \frac{(\nu z)^N}{N!} \\ 0 & 1 & \nu z & \dots & \frac{(\nu z)^{N-1}}{(N-1)!} \\ & \dots & & \dots & \\ 0 & \dots & & 1 & \nu z \\ 0 & \dots & & 0 & 1 \end{pmatrix}.$$

In contrast to ζ, the *nilconstant* ν is not an invariant; it is basis-dependent, as illustrated for $N = 1$ by

$$\begin{pmatrix} e^\gamma & 0 \\ 0 & e^{-\gamma} \end{pmatrix} \begin{pmatrix} 0 & \nu \\ 0 & 0 \end{pmatrix} \begin{pmatrix} e^{-\gamma} & 0 \\ 0 & e^\gamma \end{pmatrix} = \begin{pmatrix} 0 & \nu e^{2\gamma} \\ 0 & 0 \end{pmatrix}.$$

The nilconstant has to be nontrivial for $N \geq 1$. It plays a role as "gauge fixing constant" in the theory of gauge fields (chapter "Massless Quantum Fields").

The representation, *dual* to D_N^ζ, acts on the linear forms $V^T \cong \mathbb{C}^{1+N}$:

$$\check{D}_N^\zeta : \mathbb{C}^\diamond \longrightarrow \mathbf{GL}(V^T), \quad e^z \longmapsto \check{D}_N^\zeta(e^z) = e^{-(\zeta\, \mathrm{id}_V + n_N)^T z}.$$

Only for trivial nildimension $N = 0$ are all vectors with $\mathbb{C}^\diamond$-action eigenvectors. Only these 1-dimensional representations are *irreducible (simple)*. The nondecomposable representations for $N \geq 1$ are *reducible (multiple)* with a 1-dimensional invariant eigenvector subspace $V_\zeta, V_\zeta^T \cong \mathbb{C}$, in the basis above given by $\mathbb{C}e^0$ and $\mathbb{C}\check{e}_N$ respectively.

2.3 Time Representations and Unitarity

The action of time (dynamics) is realizable by *complex* representations of the *real* time group $\mathbf{D}(1)$. It can be obtained in a product with the phase group from the complex numbers by imposing the canonical conjugation

$$\mathbb{C}^\diamond = \mathbf{U}(1) \times \mathbf{D}(1), \quad e^z = e^{i\alpha}e^t.$$

In the following an additive notation will be used for the time group representations

$$D : \mathbf{D}(1) \longrightarrow \mathbf{GL}(V), \quad e^t \longmapsto D(t), \quad D(t+s) = D(t) \circ D(s),$$

possible with the multiplicative-additive Lie group isomorphism $(\exp \mathbb{R}, \cdot) \cong (\mathbb{R}, +)$.

To obtain from $\mathbb{C}^\diamond$-representations the complex time $\mathbf{D}(1)$-representations, the vector spaces need a conjugation: *Complex time representations have to be unitary*, i.e., conjugate linear self-dual with a $\mathbf{D}(1)$-invariant nondegenerate inner product of the representation space. A *conjugate linear* symmetric isomorphism $\mathbf{z}$ between a finite-dimensional complex $\mathbb{C}^\diamond$-representation space V and its linear forms V^T,

$$\begin{array}{ll} \mathbf{z} : V \longrightarrow V^T, \ \mathbf{z}(v) & = v^*, \ \mathbf{z}^{-1}(\theta) = \theta^*, \\ \text{Dirac notation: } |v\rangle & \overset{*}{\leftrightarrow} \langle v|, \end{array}$$

induces a conjugation also for endomorphisms, etc., denoted in the following by $*$. With the conjugation of the representation space endomorphisms

$$\begin{array}{l} * : \mathbf{AL}(V) \longrightarrow \mathbf{AL}(V), \ f \longmapsto f^* = \mathbf{z}^{-1} \circ f^T \circ \mathbf{z}, \\ \qquad\qquad\qquad\qquad\qquad |v\rangle\langle w| \overset{*}{\leftrightarrow} |w\rangle\langle v| \end{array}$$

the time $\mathbf{D}(1)$-representations in the $\mathbb{C}^\diamond$-representations are $*$-*unitary*

$$D(t)^* = \mathbf{z}^{-1} \circ D(t)^T \circ \mathbf{z} = D(t)^{-1} = D(-t).$$

Therefore the represented Lie algebra $\mathbb{R}$ is $*$-antisymmetric with a basis $iH \in \mathbf{AL}(V)$ involving a symmetric Hamiltonian H (more below):

$$D(t) = e^{iHt} \Rightarrow H^* = H.$$

The representation space conjugation implements the *time reflection*

$$\begin{array}{ccc} V & \xrightarrow{D(t)} & V \\ {\scriptstyle *}\downarrow & & \downarrow{\scriptstyle *} \\ V^T & \underset{D(-t)^T}{\longrightarrow} & V^T \end{array}, \qquad \begin{array}{rcl} t & \overset{\mathrm{T}}{\leftrightarrow} & -t, \\ |v\rangle & \overset{\mathrm{T}}{\leftrightarrow} & \langle v|, \\ D(t)^* & = & D(-t). \end{array} \quad .$$

The nondecomposable time representations are

$$\begin{aligned} D_N^{im} : \mathbf{D}(1) \longrightarrow \mathbf{GL}(V), \quad e^t \longmapsto D_N^{im}(t) &= e^{i(m\,\mathrm{id}_V + \mathcal{N}_N)t}, \\ \text{dual: } \check{D}_N^{im}(t) &= e^{-i(m\,\mathrm{id}_V + \mathcal{N}_N)^T t}. \end{aligned}$$

The conjugation properties, i.e., the unitarity

$$\overline{e^{\zeta t}} = e^{-\zeta t}, \quad \mathbf{z}_N^{-1} \circ n_N^T \circ \mathbf{z}_N = -n_N,$$

require a $*$-antisymmetric nilpotent endomorphism and an *imaginary invariant*, for the time action i times a real *frequency (energy)*:

$$\text{for } \mathbf{D}(1): \quad \zeta = im, \quad m \in \mathbb{R}, \quad n_N = i\mathcal{N}_N, \quad \mathcal{N}_N = \mathcal{N}_N^*.$$

The energies $m \in \mathbb{R}$ as invariants constitute the dual vector space for the time translations (chapter "Spacetime Translations"). The nilpotent endomorphism can be written with the nilcyclic matrix above and a real nilconstant $\nu \in \mathbb{R}^\diamond$.

The 1-dimensional time representations in $\mathbf{U}(1)$ are given by the imaginary continuous powers $(e^t)^{im}$ of the group elements. They are irreducible, but not faithful. The faithful representations with $N \geq 1$ are reducible, but nondecomposable. Faithful definite unitary representations of the time group $\mathbf{D}(1)$ have to be infinite-dimensional (chapter "Harmonic Analysis").

Eigenvectors of causal group representations in $\mathbf{U}(1)$ will be related to objects (stable states and particles). They have a nontrivial periodic time property via the complex phase $e^{imt} \in \mathbf{U}(1)$ for $m \neq 0$. Their norm via the $\mathbf{U}(1)$-scalar product is time independent. Obviously, additional properties have to be considered to define physical objects in spacetime, e.g., eigenvalues for position translations and rotations and for charge operations.

There exist irreducible 1-dimensional representations of $\mathbf{D}(1)$ with complex eigenvalue

$$\mathbf{D}(1) \ni e^t \longmapsto e^{(im-\Gamma)t} \in \mathbf{U}(1) \times \mathbf{D}(1), \quad m, \Gamma \in \mathbb{R}.$$

They are time representation coefficients in infinite-dimensional Hilbert spaces for $\Gamma > 0$. To obtain a finite-dimensional unitary representation the eigenvalue has to come in reflected pairs $m \pm i\Gamma$, e.g.,

$$\mathbf{D}(1) \ni e^t \longmapsto e^{imt}\begin{pmatrix} e^{-\Gamma t} & 0 \\ 0 & e^{\Gamma t} \end{pmatrix} \in \mathbf{U}(1,1).$$

Those representations will not be discussed in this chapter. They play a role for unstable particles with an energy width Γ and, for $m = 0$, as representation of

Lorentz boosts $\mathbf{SO}_0(1,1)$. Their Hilbert space interpretation requires infinite-dimensional spaces (chapter "Harmonic Analysis").

Jordan bases for reducible, but nondecomposable time representation spaces are *Witt bases* for the inner product

$$\mathbf{z}_N(e^A) = \check{e}_{N-A},\ \mathbf{z}_N = \langle\ |\ \rangle_N \cong \begin{pmatrix} 0 & \cdots & 0 & 1 \\ 0 & \cdots & 1 & 0 \\ & & \cdots & \\ 1 & 0 & \cdots & 0 \end{pmatrix}.$$

The skew-diagonal matrix for the conjugate linear $\mathbf{z}_N$ is the matrix for the linear isomorphism $\star \circ \mathbf{z}_N$ arising as product with a Euclidean conjugation $\star$. The unitarity is explicitly shown for the simplest faithful representation with dimension $1 + N = 2$

$$\begin{aligned} \mathbf{z}_2^{-1} \circ D_1^{im}(t)^T \circ \mathbf{z}_2 &\cong \begin{pmatrix} 0 & 1 \\ 1 & 0 \end{pmatrix} e^{-i\overline{m}t} \begin{pmatrix} 1 & 0 \\ -i\overline{\nu}t & 1 \end{pmatrix} \begin{pmatrix} 0 & 1 \\ 1 & 0 \end{pmatrix} = e^{-i\overline{m}t} \begin{pmatrix} 1 & -i\overline{\nu}t \\ 0 & 1 \end{pmatrix} \\ &= D_1^{im}(-t) \text{ for } m, \nu \in \mathbb{R}. \end{aligned}$$

The conjugation $\mathbf{z}_N$ is invariant under the real $(1+N)^2$-dimensional unitary groups

$$\mathbf{U}_N(\mathbb{R}) = \begin{cases} \mathbf{U}(1 + \frac{N}{2}, \frac{N}{2}), & N = 0, 2, \ldots, \\ \mathbf{U}(\frac{1+N}{2}, \frac{1+N}{2}), & N = 1, 3, \ldots. \end{cases}$$

It contains the represented time as subgroup

$$\mathbf{U}_N(\mathbb{R}) \supseteq D_N^{im}[\mathbb{R}] \cong \begin{cases} \{1\}, & N = 0, \quad m = 0, \\ \mathbf{U}(1), & N = 0, \quad m \neq 0, \\ \mathbf{D}(1), & N \geq 1. \end{cases}$$

The conjugation associate inner product

$$\begin{aligned} \mathbf{z}_N(\ ,\ ) : V \times V \longrightarrow \mathbb{C}, \quad & \mathbf{z}_N(v,u) = \langle v^*, u\rangle = \langle v|u\rangle = \overline{\langle u|v\rangle}, \\ & \langle v|u\rangle \stackrel{*,\mathrm{T}}{\longleftrightarrow} \langle u|v\rangle, \end{aligned}$$

is a scalar product only for irreducible time representations. In the case $N \geq 1$ ("indefinite metric"), there arise $\frac{1+N}{2}$ positive-negative basis vector pairs (Witt pairs) for even dimension $1 + N$ (index $\frac{1+N}{2}$ and character 0). For odd dimension, there is one unpaired vector, positive in the skew-diagonal matrix above (index $\frac{N}{2}$ and character 1).

2.4 Causal Time Representations

Also the natural order of time can be represented. The representations of the *abelian monoids future and past* $\mathbf{D}(1)_\pm \cong \mathbb{R}_\pm$ are obtained from $\mathbf{D}(1)$-representations by multiplication with the characteristic order function

$$\mathbf{D}(1)_\pm \longrightarrow \mathbf{GL}(V), \quad D(t)_\pm = \vartheta(\pm t) D(\pm t) = D(-t)_\mp.$$

The trivial representations are the order functions $D_0^0(t)_\pm = \vartheta(\pm t)$.

The sum and the difference are *causal representations of time*, not as a group, but as the bimonoid $\mathbf{D}(1)_+ \cup \mathbf{D}(1)_- \cong \mathbb{R}_+ \cup \mathbb{R}_- \cong \mathbb{R}$:

$$\mathbf{D}(1) \longrightarrow \mathbf{GL}(V), \quad \begin{cases} D(t)_{\text{caus}} &= D(t)_+ + D(t)_- &= D(|t|), \\ \epsilon(t)D(t)_{\text{caus}} &= D(t)_+ - D(t)_- &= \epsilon(t)D(|t|), \end{cases}$$

with the examples for a compact and noncompact causal representation

$$D_0^{im}(t)_\pm \cong \vartheta(\pm t)e^{\pm imt}, \; D_0^{im}(t)_{\text{caus}} \cong e^{im|t|}, \; D_1^{im}(t)_{\text{caus}} \cong e^{im|t|}\begin{pmatrix} 1 & i\nu|t| \\ 0 & 1 \end{pmatrix}.$$

Causal time representations are used in Feynman propagators (chapter "Propagators").

2.5 Nondecomposable Hamiltonians

As familiar from differential equations in classical physics, a dynamics can be formulated infinitesimally by expanding the time group actions with its Lie algebra (time translations). If $\mathbb{R}$ is used to parametrize both the time group $\mathbf{D}(1) = \exp \mathbb{R} \cong \mathbb{R} \in \underline{\mathbf{lgrp}}_{\mathbb{R}}$ and its Lie algebra (time translations) $\mathbb{R} \in \underline{\mathbf{lag}}_{\mathbb{R}}$, a mixing of both structures on $\mathbb{R}$ has to be avoided, e.g., only the Lie algebra structure allows a scalar multiplication $t \longmapsto \alpha t$.

The representations D_N^ζ of the complex group $\mathbb{C}^\diamond$ and the unitary representations D_N^{im} of the real time group $\mathbf{D}(1)$ have the associate representations $\mathcal{D}_N^\zeta$ and $\mathcal{D}_N^{im}$ for the Lie algebra $\log \mathbb{C}^\diamond = \mathbb{C}$ and for $\log \mathbf{D}(1) = \mathbb{R}$ for the time translations. With a factor i the *Hamiltonian* H_N (in a basis called a Hamiltonian matrix) represents a time translation basis $\mathbf{p}^0$:

$$\begin{aligned} \mathcal{D}_N^{im} : \quad & \log \mathbf{D}(1) \longrightarrow \mathbf{AL}(V), \;\; \mathcal{D}_N^{im}(\mathbf{p}^0) = iH_N = i(m\,\text{id}_V + \mathcal{N}_N), \\ & [H_N, \mathcal{N}_N] = 0, \;\; D_N^{im}(t) = e^{iH_N t}, \\ & \text{minimal polynomial: } p_{iH_N}(X) = (X - im)^{1+N}. \end{aligned}$$

The representation invariant is given by the trace

$$\frac{1}{1+N} \operatorname{tr} iH_N = im.$$

The dual Lie algebra representation arises by negative transposition:

$$\begin{aligned} \check{\mathcal{D}}_N^{im} : \log \mathbf{D}(1) \longrightarrow \mathbf{AL}(V^T), \; \check{\mathcal{D}}_N^{im}(\mathbf{p}^0) &= -i(m\,\text{id}_V + \mathcal{N}_N)^T \\ = i\check{H}_N &= -iH_N^T. \end{aligned}$$

The conjugation $\mathbf{z}_N$ above reflects the basis $\mathbf{p}^0 \overset{\text{T}}{\leftrightarrow} -\mathbf{p}^0$, $iH_N \overset{\text{T}}{\leftrightarrow} -iH_N$. The Hamiltonian is $*$-symmetric:

$$\begin{aligned} H_N^* &= \mathbf{z}_N^{-1} \circ H_N^T \circ \mathbf{z}_N = H_N, \\ \text{e.g., } H_1 &\simeq \begin{pmatrix} m & \nu \\ 0 & m \end{pmatrix} = \begin{pmatrix} 0 & 1 \\ 1 & 0 \end{pmatrix}\begin{pmatrix} m & 0 \\ \nu & m \end{pmatrix}\begin{pmatrix} 0 & 1 \\ 1 & 0 \end{pmatrix}, \quad m, \nu \in \mathbb{R}. \end{aligned}$$

The reducible, nondecomposable representations for $N \geq 1$ have a nontrivial *nil-Hamiltonian* $\mathcal{N}_N$. There exist *nilvectors* of the Hamiltonian. The 1-dimensional *time eigenspace* $V_{im} \subseteq V$ is characterized by its invariance under action with the nil-Hamiltonian $\mathcal{N}_N$:

$$\mathcal{N}_N(v) = 0 \iff H_N(v) = mv, \quad \mathcal{N}_N^T(\theta) = 0 \iff H_N^T(\theta) = m\theta.$$

The action of the nil-Hamiltonian will play a role for gauge transfomations in gauge theories (chapter "Massless Quantum Fields").

Only for the irreducible case $N = 0$ do the time eigenvectors $V_{im} = (\mathcal{N}_n)^N[V]$ and eigenforms $V_{im}^T = (\mathcal{N}_n^T)^N[V^T]$ constitute dual spaces. For $N \geq 1$, they are orthogonal to each other

$$\langle \theta_{im}, v_{im} \rangle = (-1)^N \langle \theta, (\mathcal{N}_N)^{2N}.v \rangle = \begin{cases} \langle \theta, v \rangle, & \text{for } N = 0,\ V_{im}^T = (V_{im})^T, \\ 0, & \text{for } N \geq 1,\ V_{im}^T \neq (V_{im})^T. \end{cases}$$

The vectors of a Jordan-Witt basis for $N \geq 1$ have a well defined nildimension:

$$(\mathcal{N}_N)^A.e^N = \nu^A e^{N-A}, \quad A = 0, \ldots, N.$$

The nil-Hamiltonian generates a basis of the representation space from a nilvector with maximal nildimension N, e.g., from e^N or the dual basis from $\check{e}_0$

$$v \in V \text{ is cyclic principal} \iff \{(\mathcal{N}_N)^A.v \mid A = 0, \ldots, N\} \text{ is a } V\text{-basis}.$$

2.6 Time Orbits and Equations of Motion

Equations of motion express the time Lie algebra action by derivation with respect to the Lie parameter $t \in \mathbb{R}$.

Dual *time orbits* $v(t) \in V$ and $\omega(t) \in V^T$ for a $\mathbf{D}(1)$-representation are "fields on time" with values in the dual representation spaces

$$\begin{array}{lll} v(\ ) : \mathbf{D}(1) \longrightarrow V, & v(t) = D_N^{im}(t)(v), & v(0) = v, \\ \omega(\ ) : \mathbf{D}(1) \longrightarrow V^T, & \omega(t) = \check{D}_N^{im}(t)(\omega), & \omega(0) = \omega, \\ \text{with} & \langle \omega(t), v(t) \rangle = \langle \omega, v \rangle. & \end{array}$$

In dual Jordan-Witt bases the time orbits $e^A(t)$ and $\check{e}_A(t)$ are the rows of the $D_N^{im}(t)$-matrix above:

$$\begin{aligned} &e^0(t) = D_N^{im}(t)(e^0) = e^{imt}e^0, \\ &e^1(t) = D_N^{im}(t)(e^1) = e^{imt}(e^1 + i\nu t e^0), \\ &\ldots \\ &e^N(t) = D_N^{im}(t)(e^N) = e^{imt}(e^N + i\nu t e^{N-1} + \cdots + \tfrac{(i\nu t)^N}{N!} e^0). \end{aligned}$$

Time representations can be written as *two-point products of time orbits* for dual bases with transposition sign $\epsilon = \pm 1$:

$$\begin{array}{lll} D_N^{im}(t) &= D_N^{im}(t)_B^A e^B \otimes \check{e}_A &= e^A(t_1) \otimes \check{e}_A(t_2) \text{ with } t_1 - t_2 = t, \\ \check{D}_N^{im}(t) &= D_N^{im}(-t)_B^A \epsilon \check{e}_A \otimes e^B &= \epsilon \check{e}_A(t_1) \otimes e^A(t_2), \end{array}$$

with the *matrix elements of time representations*

$$\langle \check{e}_B(t_2), e^A(t_1)\rangle = D_N^{im}(t)_B^A = \frac{(i\nu t)^{A-B}}{(A-B)!} e^{imt} = \frac{1}{(A-B)!}\left(\frac{\partial}{\partial m}\right)^{A-B} e^{imt}.$$

The group parameter occurs as the time difference. Time representation matrix elements are embedded into propagators with quantum particle fields as orbits of spacetime translations (chapter "Propagators").

The *equations of motion* at the neutral time element describe the action of the time translations ($d_t = \frac{d}{dt}$):

$$\begin{array}{llll} v \in V: & \mathbf{p}^0 \bullet v &= \mathcal{D}_N^{im}(\mathbf{p}^0)(v) &= iH_N(v) &= d_t|_{t=0} v(t), \\ \omega \in V^T: & \mathbf{p}^0 \bullet \omega &= \check{\mathcal{D}}_N^{im}(\mathbf{p}^0)(\omega) &= -iH_N^T(\omega) &= d_t|_{t=0} \omega(t). \end{array}$$

Because of the abelian structure for time, the equations of motion hold for all times, in a Jordan-Witt basis:

$$\left.\begin{array}{ll} iH_N(e^0) &= d_t e^0(t) = ime^0(t), \\ iH_N(e^1) &= d_t e^1(t) = ime^1(t) + i\nu e^0(t), \\ & \dots \\ iH_N(e^N) &= d_t e^N(t) = ime^N(t) + i\nu e^{N-1}(t) \end{array}\right\} \Rightarrow (d_t - im)^{1+A} e^A(t) = 0.$$

2.7 Self-Dual Time Representations

The conjugate self-duality of the complex representations of real time operations allows a formulation with real representation spaces and orthogonal inner products. In general, however, in real vector spaces there do not exist time action eigenvectors, which is related to the real nondiagonalizability of the (2×2) matrix $\begin{pmatrix} 0 & 1 \\ -1 & 0 \end{pmatrix}$ with the eigenvalues $\{\pm i\}$. To define objects (states, particles), complex vector spaces are needed.

For the complicated-looking general formalism in this section, the simplest and most relevant examples with nildimensions $N = 0, 1$ are treated in the following sections.

The direct sum $\mathbf{V} = V \oplus V^T \cong \mathbb{C}^{2(1+N)}$ of dual vector spaces (vectors and their linear forms) carries the dual-product-induced canonical bilinear form, either symmetric or antisymmetric, depending on the transposition sign $\epsilon = 1$ (called Fermi)[2] and $\epsilon = -1$ (called Bose):

$$\begin{array}{l} v, w \in V, \;\; \omega, \theta \in V^T: \;\; \langle \omega, v\rangle = \epsilon\langle v, \omega\rangle, \;\; \langle \omega, \theta\rangle = 0 = \langle v, w\rangle, \\ \text{dual bases } \{e^A, \check{e}_A\}_{A=0}^N: \;\; \left\{\begin{array}{lll} \langle \check{e}_B, e^A\rangle &= \delta_B^A = & \epsilon\langle e^A, \check{e}_B\rangle, \\ \langle \check{e}_A, \check{e}_B\rangle &= 0 = & \langle e^A, e^B\rangle. \end{array}\right. \end{array}$$

[2]The (anti)symmetric canonical form is the starting point for the quantization in the chapter "Quantum Algebras."

The time-representation-related conjugation $\mathbf{z}_N \oplus \mathbf{z}_N^{-1} = *$ of the doubled space $\mathbf{V}$ allows the construction of *real time representations* on it. $\mathbf{V}$ can be decomposed into two isomorphic real subspaces. Dual Jordan-Witt bases give a $*$-symmetric $\mathbf{V}$-basis:

$$\begin{array}{l} \mathbf{V} = \mathbf{V}^* = \mathbf{V}_+ \oplus \mathbf{V}_-, \quad \mathbf{V}_+ \cong \mathbf{V}_- \cong \mathbb{R}^{2(1+N)}, \\ \begin{array}{l} \mathbf{e}_+^A = (\mathbf{e}_+^A)^* = \frac{e^A + \check{e}_{N-A}}{\sqrt{2}} \\ \mathbf{e}_-^A = (\mathbf{e}_-^A)^* = \frac{e^A - \check{e}_{N-A}}{i\sqrt{2}} \end{array} \Rightarrow \left\{ \begin{array}{lll} \langle \mathbf{e}_+^A, \mathbf{e}_+^B \rangle & = \langle \mathbf{e}_-^A, \mathbf{e}_-^B \rangle & = \frac{1+\epsilon}{2} \delta_N^{A+B}, \\ \langle \mathbf{e}_+^B, i\mathbf{e}_-^A \rangle & = -\langle i\mathbf{e}_-^A, \mathbf{e}_+^B \rangle & = \frac{1-\epsilon}{2} \delta_N^{A+B}. \end{array} \right. \end{array}$$

The Bose case differs from the Fermi case with respect to the dual structure: For Fermi, duality respects the decomposition into real-imaginary, i.e., the real vectors $\mathbf{e}_+$ and $\mathbf{e}_-$ are self-dual

$$\epsilon = +1: \quad (\mathbf{V}_\pm)^T = \mathbf{V}_\pm, \quad \langle \mathbf{e}_+^A, \mathbf{e}_+^B \rangle = \langle \mathbf{e}_-^A, \mathbf{e}_-^B \rangle = \delta_N^{A+B}.$$

For Bose, duality joins real and imaginary, i.e., the real vector $\mathbf{e}_+$ is the dual of the imaginary vector $i\mathbf{e}_-$:

$$\epsilon = -1: \quad (\mathbf{V}_\pm)^T = \mathbf{V}_\mp, \quad \langle \mathbf{e}_+^B, i\mathbf{e}_-^A \rangle = \langle -i\mathbf{e}_-^A, \mathbf{e}_+^B \rangle = \delta_N^{A+B}.$$

The Bose duality structure arises in quantum mechanics for position-momentum pairs and is expressed by $[i\mathbf{p}, \mathbf{x}] = 1$, combining real with imaginary.

The complex decomposable self-dual representations on the direct sum $\mathbf{V}$ can be written with the symmetric bases, for the time group with $\mathbf{D} = D \oplus \check{D}$,

$$\begin{array}{rl} \mathbf{D}(1) & \longrightarrow \mathbf{GL}(\mathbf{V}), \quad e^t \longmapsto \mathbf{D}_N^{im}(t), \\ \mathbf{D}_N^{im}(t) & = e^A(t_1) \otimes \check{e}_A(t_2) + \epsilon \check{e}_A(t_1) \otimes e^A(t_2), \quad t = t_1 - t_2 \\ & = \left\{ \begin{array}{ll} \mathbf{e}_-^A(t_1) \otimes \mathbf{e}_-^{N-A}(t_2) + \mathbf{e}_+^A(t_1) \otimes \mathbf{e}_+^{N-A}(t_2), & \epsilon = +1, \\ i\Big[\mathbf{e}_-^A(t_1) \otimes \mathbf{e}_+^{N-A}(t_2) + \mathbf{e}_+^A(t_1) \otimes \mathbf{e}_-^{N-A}(t_2)\Big], & \epsilon = -1. \end{array} \right. \end{array}$$

and for the time Lie algebra with the Hamiltonian $\mathbf{H}_N = H_N - H_N^T$

$$\begin{array}{l} \log \mathbf{D}(1) \longrightarrow \mathbf{AL}(\mathbf{V}), \quad \mathbf{p}^0 \longmapsto i\mathbf{H}_N \\ \mathbf{H}_N = m[e^A, \check{e}_A]_{-\epsilon} + \nu[e^A, \check{e}_{A+1}]_{-\epsilon} \\ = \left\{ \begin{array}{ll} im(\mathbf{e}_-^A \otimes \mathbf{e}_+^{N-A} - \mathbf{e}_+^A \otimes \mathbf{e}_-^{N-A}) + i\nu(\mathbf{e}_-^A \otimes \mathbf{e}_+^{N-1-A} - \mathbf{e}_+^A \otimes \mathbf{e}_-^{N-1-A}), & \epsilon = +1, \\ m(\mathbf{e}_-^A \otimes \mathbf{e}_-^{N-A} + \mathbf{e}_+^A \otimes \mathbf{e}_+^{N-A}) + \nu(\mathbf{e}_-^A \otimes \mathbf{e}_-^{N-1-A} + \mathbf{e}_+^A \otimes \mathbf{e}_+^{N-1-A}), & \epsilon = -1. \end{array} \right. \end{array}$$

Here tensor (anti-) commutators are used, defined in the endomorphism algebra $\mathbf{AL}(\mathbf{V}) \cong \mathbf{V} \otimes \mathbf{V}$ of the self-dual space by $[a, b]_\epsilon = a \otimes b + \epsilon b \otimes a$. The minimal polynomial for the Hamiltonian contains the factor $X^2 + m^2$, irreducible in the reals

$$p_{i\mathbf{H}_N}(X) = (X^2 + m^2)^{1+N}.$$

The equations of motion for the conjugation symmetric time orbits read

$$\left. \begin{array}{ll} d_t \mathbf{e}_\pm^0 = \mp m \mathbf{e}_\mp^0, & (d_t^2 + m^2)\mathbf{e}_\pm^0 = 0, \\ d_t \mathbf{e}_\pm^1 = \mp (m \mathbf{e}_\mp^1 + \nu \mathbf{e}_\mp^0), & (d_t^2 + m^2)\mathbf{e}_\pm^1 = \mp 2m\nu \mathbf{e}_\pm^0, \\ \dots & \dots \\ d_t \mathbf{e}_\pm^N = \mp (m \mathbf{e}_\mp^N + \nu \mathbf{e}_\mp^{N-1}), & (d_t^2 + m^2)\mathbf{e}_\pm^N = \mp 2m\nu \mathbf{e}_\pm^{N-1} \end{array} \right\} \Rightarrow (d_t^2 + m^2)^{1+A} \mathbf{e}_\pm^A = 0.$$

The representations leave the real subspaces $\mathbf{V}_\pm \cong \mathbb{R}^{2(1+N)}$ invariant. Therefore the group $\mathbf{D}(1)$ is represented as a subgroup of a real orthogonal group in $\mathbf{GL}(\mathbb{C}^{2(1+N)})$:

$$D_N^{im}[\mathbb{R}] \subseteq \begin{cases} \mathbf{SO}_0(2+N,N), & N=0,2,\dots, \\ \mathbf{SO}_0(1+N,1+N), & N=1,3,\dots, \end{cases}$$

and leaves invariant the corresponding orthogonal inner product.

2.8 Compact Time Representations

The complex 1-dimensional irreducible $\mathbb{R}$-representations in $\mathbf{U}(1)$, called Hilbert representations, are the building blocks for Fourier analysis, Fourier integrals, etc. (chapter "Harmonic Analysis"). In the case of time translations $\mathbb{R}$, they are relevant for bound state vectors and particles.

The irreducible representations D_0^{im} have a real frequency (energy) m:

$$\mathbf{D}(1) \ni e^t \longmapsto \begin{cases} e^{imt} \in \mathbf{U}(1) & \subset \mathbf{GL}(V), & V &= \mathbb{C}\mathrm{u}, \\ e^{-imt} \in \mathbf{U}(1) & \subset \mathbf{GL}(V^T), & V^T &= \mathbb{C}\check{\mathrm{u}}. \end{cases}$$

They come with a Euclidean conjugation $\mathbf{z}_0 = \star$:

$$V \overset{\star}{\leftrightarrow} V^T, \quad |\mathrm{u}\rangle = \mathrm{u} \overset{\star}{\leftrightarrow} \check{\mathrm{u}} = \mathrm{u}^\star = \langle \mathrm{u}|.$$

The definite group $\mathbf{U}(1)$ with the time action gives a *scalar product* of the representation space

$$\mathbf{z}_0(\mathrm{u},\mathrm{u}) = \langle \mathrm{u}^\star, \mathrm{u}\rangle = \langle \mathrm{u}|\mathrm{u}\rangle = 1.$$

In quantum theory, the conjugation-related dual basis $(\mathrm{u},\mathrm{u}^\star)$ is called the creation and annihilation operator. The scalar product leads to probabilities (chapters "Quantum Algebra" and "Quantum Probability").

The (anti)symmetric combinations give a basis of the self-dual space $\mathbf{V} = V \oplus V^T \cong \mathbb{C}^2$

$$\mathbf{u}_+ = \mathbf{u}_+^\star = \tfrac{\mathrm{u}+\mathrm{u}^\star}{\sqrt{2}}, \quad \mathbf{u}_- = \mathbf{u}_-^\star = \tfrac{\mathrm{u}-\mathrm{u}^\star}{i\sqrt{2}}.$$

Special notation is used, a real position-momentum pair $(\mathbf{x},\mathbf{p}) = (\mathbf{x}^\star,\mathbf{p}^\star)$ for Bose and a real-imaginary pair $(\mathbf{r},\mathbf{l}) = (\mathbf{r}^\star, -\mathbf{l}^\star)$ for Fermi:

$$(\mathbf{u}_+,\mathbf{u}_-) = \begin{cases} (\mathbf{r}, -i\mathbf{l}), & \epsilon = +1 \text{ with } \begin{pmatrix} \langle \mathbf{r},\mathbf{r}\rangle & \langle \mathbf{l},\mathbf{r}\rangle \\ \langle \mathbf{r},\mathbf{l}\rangle & -\langle \mathbf{l},\mathbf{l}\rangle \end{pmatrix} = \begin{pmatrix} 1 & 0 \\ 0 & 1 \end{pmatrix}, \\ \\ (\frac{1}{\ell}\mathbf{x}, -\ell\mathbf{p}), & \epsilon = -1 \text{ with } \begin{pmatrix} \langle i\mathbf{p},\mathbf{x}\rangle & \langle \mathbf{x},\mathbf{x}\rangle \\ \langle \mathbf{p},\mathbf{p}\rangle & -\langle \mathbf{x}, i\mathbf{p}\rangle \end{pmatrix} = \begin{pmatrix} 1 & 0 \\ 0 & 1 \end{pmatrix}. \end{cases}$$

In contrast to the Fermi case the real combinations in the Bose case allow nontrivial inverse normalization factors $\ell, \frac{1}{\ell}$ for dual bases.

The bounded time orbits obey equations of motion of first and second order for the complex and real bases respectively (harmonic oscillator with cyclic orbits):

$$\begin{aligned} d_t\mathrm{u} = im\mathrm{u}, \quad \mathrm{u}(t) = e^{imt}\mathrm{u}, \quad d_t\mathrm{u}^\star &= -im\mathrm{u}^\star, \quad \mathrm{u}^\star(t) = e^{-imt}\mathrm{u}^\star, \\ d_t(\mathbf{u}_+,\mathbf{u}_-) &= (\mathbf{u}_+,\mathbf{u}_-)h_0(m), \quad (d_t^2+m^2)\mathbf{u}_\pm = 0, \\ (\mathbf{u}_+,\mathbf{u}_-)(t) &= (\mathbf{u}_+,\mathbf{u}_-)e^{h_0(m)t}, \\ h_0(m) &= m\begin{pmatrix} 0 & 1 \\ -1 & 0 \end{pmatrix}, \quad e^{h_0(m)t} = \begin{pmatrix} \cos mt & \sin mt \\ -\sin mt & \cos mt \end{pmatrix} \in \mathbf{SO}(2); \end{aligned}$$

$h_0(m)$ is the antisymmetric *Hamiltonian matrix* . The first and second order equations of motion reflect the irreducible complex and real minimal polynomials of degree 1 and 2, given by $(X \mp im)$ and $(X^2 + m^2)$ respectively.

For $m \neq 0$, the real 2-dimensional time representation is irreducible since $\begin{pmatrix} 0 & 1 \\ -1 & 0 \end{pmatrix}$ is not diagonalizable over the reals. The symmetric combinations $(\mathbf{u}_+,\mathbf{u}_-)$, e.g., position and momentum $(\mathbf{x},\mathbf{p})$, are not time eigenvectors, in contrast to the eigenvectors $(\mathrm{u},\mathrm{u}^\star)$.

The self-dual group representation acts on the direct sum $\mathbf{V}$:

$$\begin{aligned} \mathbf{D}_0^{im}(t) &= e^{imt}\,\mathrm{id}_V + e^{-imt}\,\mathrm{id}_{V^T} = e^{imt}\mathrm{u}\otimes\mathrm{u}^\star + e^{-imt}\epsilon\mathrm{u}^\star\otimes\mathrm{u} \cong \begin{pmatrix} e^{imt} & 0 \\ 0 & e^{-imt} \end{pmatrix} \\ &= \begin{cases} \cos mt(\mathbf{r}\otimes\mathbf{r} - \mathbf{l}\otimes\mathbf{l}) + i\sin mt(\mathbf{l}\otimes\mathbf{r} - \mathbf{r}\otimes\mathbf{l}) \\ \quad \cong \begin{pmatrix} \cos mt & i\sin mt \\ i\sin mt & \cos mt \end{pmatrix} \in \mathbf{SO}(2), \quad \epsilon = +1, \\ \\ \cos mt(\mathbf{x}\otimes i\mathbf{p} - i\mathbf{p}\otimes\mathbf{x}) + i\sin mt(\ell^2\mathbf{p}\otimes\mathbf{p} + \frac{\mathbf{x}\otimes\mathbf{x}}{\ell^2}) \\ \quad \cong \begin{pmatrix} \cos mt & i\ell^2\sin mt \\ \frac{i}{\ell^2}\sin mt & \cos mt \end{pmatrix} \in \mathbf{SO}(2), \quad \epsilon = -1. \end{cases} \end{aligned}$$

The time Lie algebra is spanned by the self-dual Hamiltonian

$$\begin{aligned} \mathbf{H}_0 &= m(\,\mathrm{id}_V - \,\mathrm{id}_{V^T}) = m[\mathrm{u},\mathrm{u}^\star]_{-\epsilon} \cong m\begin{pmatrix} 1 & 0 \\ 0 & -1 \end{pmatrix} \\ &= \begin{cases} m(\mathbf{l}\otimes\mathbf{r} - \mathbf{r}\otimes\mathbf{l}) & \cong m\begin{pmatrix} 0 & 1 \\ 1 & 0 \end{pmatrix}, \quad \epsilon = +1, \\ m(\ell^2\mathbf{p}\otimes\mathbf{p} + \frac{\mathbf{x}\otimes\mathbf{x}}{\ell^2}) & \cong m\begin{pmatrix} 0 & \ell^2 \\ \frac{1}{\ell^2} & 0 \end{pmatrix}, \quad \epsilon = -1. \end{cases} \end{aligned}$$

Comparing with the harmonic oscillator $H = \frac{\mathbf{p}^2}{2M} + k\frac{\mathbf{x}^2}{2}$ with mass M and spring constant k, one has the frequency $m^2 = \frac{k}{M}$ and the intrinsic length $\ell^4 = \frac{1}{kM}$ as normalization factor for the metric.

2.9 Noncompact Time Representations

In contrast to the irreducible time representations in $\mathbf{U}(1)$, parametrizable by eigenvectors, the noncompact ones are not so familiar. Their orbits are seen in the motion of a free Newtonian mass point. Also relativistic massless fields involve noncompact time representations without particle interpretation, e.g., the Coulomb degree of freedom in the electromagnetic gauge field (chapter "Massless Quantum Fields").

The complex 2-dimensional nondecomposable, but reducible time representations D^1_{im},

$$\mathbf{D}(1) \ni e^t \longmapsto \left\{ \begin{array}{ll} e^{imt}\begin{pmatrix} 1 & i\nu t \\ 0 & 1 \end{pmatrix} \in \mathbf{U}(1,1) & \subset \mathbf{GL}(V), \quad V \cong \mathbb{C}^2, \\ e^{-imt}\begin{pmatrix} 1 & 0 \\ -i\nu t & 1 \end{pmatrix} \in \mathbf{U}(1,1) & \subset \mathbf{GL}(V^T), \quad V^T \cong \mathbb{C}^2, \end{array} \right.$$

are in the indefinite group $\mathbf{U}(1,1)$ with conjugation $\mathbf{z}_1 = \times$:

$$\begin{array}{l} \mathbf{z}_1 \simeq \begin{pmatrix} 0 & 1 \\ 1 & 0 \end{pmatrix}, \quad V \overset{\times}{\leftrightarrow} V^T, \quad \left\{ \begin{array}{l} e^0 = \mathrm{g} \overset{\times}{\leftrightarrow} \check{e}_1 = \mathrm{g}^\times, \\ e^1 = \mathrm{b} \overset{\times}{\leftrightarrow} \check{e}_0 = \mathrm{b}^\times, \end{array} \right. \\ \langle \mathrm{b}^\times, \mathrm{g}\rangle = 1 = \epsilon\langle \mathrm{g}, \mathrm{b}^\times\rangle, \quad \langle \mathrm{g}^\times, \mathrm{b}\rangle = 1 = \epsilon\langle \mathrm{b}, \mathrm{g}^\times\rangle. \end{array}$$

They are faithful. Time eigenvectors are symbolized with g ("good"), time nilvectors with b ("bad"). The direct sum of the dual representation spaces $\mathbf{V} = V \oplus V^T \cong \mathbb{C}^4$ has as a conjugation symmetric basis

$$\begin{array}{ll} \mathbf{b}_+ = \frac{\mathrm{b}+\mathrm{b}^\times}{\sqrt{2}}, & \mathbf{g}_+ = \frac{\mathrm{g}+\mathrm{g}^\times}{\sqrt{2}} \\ \mathbf{b}_- = \frac{\mathrm{b}-\mathrm{b}^\times}{i\sqrt{2}}, & \mathbf{g}_- = \frac{\mathrm{g}-\mathrm{g}^\times}{i\sqrt{2}} \end{array}, \quad \left\{ \begin{array}{lll} \langle \mathbf{b}_\pm, \mathbf{g}_\pm\rangle & = \langle \mathbf{g}_\pm, \mathbf{b}_\pm\rangle & = \frac{1+\epsilon}{2}, \\ \langle \mathbf{b}_+, i\mathbf{g}_-\rangle & = -\langle i\mathbf{g}_-, \mathbf{b}_+\rangle & \\ = \langle \mathbf{g}_+, i\mathbf{b}_-\rangle & = -\langle i\mathbf{b}_-, \mathbf{g}_+\rangle & = \frac{1-\epsilon}{2}. \end{array} \right.$$

The dual Hamiltonians with sum in $\mathbf{AL}(\mathbf{V})$,

$$\begin{array}{ll} H_1 & = m(\mathrm{g} \otimes \mathrm{b}^\times + \mathrm{b} \otimes \mathrm{g}^\times) + \nu \mathrm{g} \otimes \mathrm{g}^\times \in \mathbf{AL}(V), \\ -H_1^T & = -\epsilon m(\mathrm{b}^\times \otimes \mathrm{g} + \mathrm{g}^\times \otimes \mathrm{b}) - \epsilon\nu \mathrm{g}^\times \otimes \mathrm{g} \in \mathbf{AL}(V^T), \\ i\mathbf{H}_1 & = \left\{ \begin{array}{ll} m\Big([\mathbf{b}_+, \mathbf{g}_-] + [\mathbf{g}_+, \mathbf{b}_-]\Big) + \nu[\mathbf{g}_+, \mathbf{g}_-], & \epsilon = +1, \\ im\Big(\{\mathbf{b}_-, \mathbf{g}_-\} + \{\mathbf{g}_+, \mathbf{b}_+\}\Big) + i\nu(\mathbf{g}_- \otimes \mathbf{g}_- + \mathbf{g}_+ \otimes \mathbf{g}_+), & \epsilon = -1, \end{array} \right. \end{array}$$

give the equations of motion for the time orbits in the complex formulation

$$\begin{array}{llll} d_t\begin{pmatrix} \mathrm{b} \\ \mathrm{g} \end{pmatrix} & = i\begin{pmatrix} m & \nu \\ 0 & m \end{pmatrix}\begin{pmatrix} \mathrm{b} \\ \mathrm{g} \end{pmatrix}, & d_t(\mathrm{g}^\times, \mathrm{b}^\times) & = -i(\mathrm{g}^\times, \mathrm{b}^\times)\begin{pmatrix} m & \nu \\ 0 & m \end{pmatrix}, \\ \begin{pmatrix} \mathrm{b}(t) \\ \mathrm{g}(t) \end{pmatrix} & = e^{imt}\begin{pmatrix} 1 & i\nu t \\ 0 & 1 \end{pmatrix}\begin{pmatrix} \mathrm{b} \\ \mathrm{g} \end{pmatrix}, & \begin{pmatrix} 1 & i\nu t \\ 0 & 1 \end{pmatrix} e^{imt} & = \begin{pmatrix} 1 & \nu\frac{d}{dm} \\ 0 & 1 \end{pmatrix} e^{imt} \in \mathbf{U}(1,1); \end{array}$$

b and $\mathrm{b}^\times$ are cyclic principal vectors. The characteristic matrix element $i\nu t e^{imt}$ arises by derivation of the matrix element e^{imt} of an irreducible representation with respect to the characterizing energy (frequency) m.

The equations of motion for the symmetric combinations,

$$\begin{array}{l} d_t(\mathbf{g}_+, \mathbf{g}_-, \mathbf{b}_+, \mathbf{b}_-) = (\mathbf{g}_+, \mathbf{g}_-, \mathbf{b}_+, \mathbf{b}_-)h_1(m, \nu), \\ \left.\begin{array}{ll} (d_t^2 + m^2)\mathbf{b}_\pm & = \mp m\nu \mathbf{g}_\pm \\ (d_t^2 + m^2)\mathbf{g}_\pm & = 0 \end{array}\right\} \Rightarrow (d_t^2 + m^2)^2 \mathbf{b}_\pm = 0, \\ h_1(m, \nu) = \left(\begin{array}{cc|cc} 0 & m & 0 & \nu \\ -m & 0 & -\nu & 0 \\ \hline 0 & 0 & 0 & m \\ 0 & 0 & -m & 0 \end{array}\right) = \begin{pmatrix} 1 & \nu\frac{d}{dm} \\ 0 & 1 \end{pmatrix} h_0(m), \end{array}$$

are solved by the noncompact spiraling orbits for a $\mathbf{D}(1)$-isomorphic subgroup of $\mathbf{SO}_0(2,2)$:

$$(\mathbf{g}_+, \mathbf{g}_-, \mathbf{b}_+, \mathbf{b}_-)(t) = (\mathbf{g}_+, \mathbf{g}_-, \mathbf{b}_+, \mathbf{b}_-)e^{h_1(m,\nu)t},$$

$$e^{h_1(m,\nu)t} = \begin{pmatrix} 1 & \nu\frac{d}{dm} \\ 0 & 1 \end{pmatrix} e^{h_0(m)t} = \left(\begin{array}{cc|cc} \cos mt & \sin mt & -\nu t \sin mt & \nu t \cos mt \\ -\sin mt & \cos mt & -\nu t \cos mt & -\nu t \sin mt \\ \hline 0 & 0 & \cos mt & \sin mt \\ 0 & 0 & -\sin mt & \cos mt \end{array}\right) \in \mathbf{SO}_0(2,2);$$

$h_1(m,\nu)$ is the *Hamiltonian matrix*. For $m \neq 0$, these real 4-dimensional time representations are nondecomposable.

Also, a trivial frequency $m = 0$ gives a nontrivial time representation, decomposable into two real 2-dimensional representations:

$$m = 0: \quad \begin{cases} (\mathbf{g}_+, \mathbf{b}_-)(t) = (\mathbf{g}_+, \mathbf{b}_- + \nu t \mathbf{g}_+), \\ (\mathbf{g}_-, \mathbf{b}_+)(t) = (\mathbf{g}_-, \mathbf{b}_+ - \nu t \mathbf{g}_-). \end{cases}$$

The *Bose case* gives the motion of two independent free Newtonian mass points with position-momentum pairs $(\mathbf{x}, \mathbf{p})$, $(\mathbf{x}', \mathbf{p}')$:

$$m = 0,\ \epsilon = -1: \quad \begin{cases} \mathbf{b}_+ & = \mathbf{x},\ \mathbf{g}_+ = \mathbf{p}',\ \mathbf{b}_- = \mathbf{x}',\ \mathbf{g}_- = -\mathbf{p}, \\ \langle i\mathbf{p}, \mathbf{x}\rangle & = -\langle \mathbf{x}, i\mathbf{p}\rangle = 1 = \langle i\mathbf{p}', \mathbf{x}'\rangle = -\langle \mathbf{x}', i\mathbf{p}'\rangle, \\ \mathbf{H}_1 & = \nu(\mathbf{p}\otimes\mathbf{p} + \mathbf{p}'\otimes\mathbf{p}') = H \oplus H', \\ H, H' & \cong \begin{pmatrix} 0 & \nu \\ 0 & 0 \end{pmatrix},\ e^{iHt} \cong \begin{pmatrix} 1 & i\nu t \\ 0 & 1 \end{pmatrix}. \end{cases}$$

The momenta $\{\mathbf{p}, \mathbf{p}'\}$ are eigenvectors with trivial eigenvalue. The positions $\{\mathbf{x}, \mathbf{x}'\}$ are cyclic principal vectors, not eigenvectors

$$d_t\begin{pmatrix} \mathbf{x} \\ \mathbf{p} \end{pmatrix} = \begin{pmatrix} 0 & \nu \\ 0 & 0 \end{pmatrix}\begin{pmatrix} \mathbf{x} \\ \mathbf{p} \end{pmatrix}, \quad d_t\begin{pmatrix} \mathbf{x}' \\ \mathbf{p}' \end{pmatrix} = \begin{pmatrix} 0 & \nu \\ 0 & 0 \end{pmatrix}\begin{pmatrix} \mathbf{x}' \\ \mathbf{p}' \end{pmatrix}.$$

The eigenvectors have trivial norm since $\langle \mathrm{g}|\mathrm{g}\rangle = 0$. Comparing with the free mass point $H = \frac{\mathbf{p}^2}{2M}$, the nilconstant is the inverse mass $\nu = \frac{1}{M}$.

For the Fermi case, the time development cannot be decomposed for $m = 0$. It always couples partners from nondual pairs, e.g., $\langle \mathbf{b}_+, \mathbf{g}_+\rangle = 1$.

The Bose structure arises for gauge fields, the Fermi structure for Fadeev-Popov fields (chapter "Massless Quantum Fields").

2.10 Invariants and Weights

The equivalence classes $(N|\zeta)$ of the nondecomposable representations D_N^ζ of the complex Lie group $\exp\mathbb{C} = \mathbb{C}^\diamond$ and $\mathcal{D}_N^\zeta$ for its Lie algebra $\mathbb{C}$ with nildimension N and eigenvalue ζ constitute the abelian *representation monoid*

$$\mathbf{ndecrep}\,\mathbb{C}^\diamond = \{(N|\zeta)\} \cong \mathbb{N}_0 \times \mathbb{C} \in \underline{\mathbf{mon}}.$$

The product of two representations can be decomposed as follows:

$$(N_1|\zeta_1) \otimes (N_2|\zeta_2) = \bigoplus_{N=|N_1-N_2|}^{N_1+N_2} (N|\zeta_1 + \zeta_2).$$

This defines the monoid composition $\vee$, where the highest dimensional representation is taken: Nildimensions and eigenvalues are added. The trivial representation class is the neutral element for the monoid

$$(N_1|\zeta_1) \vee (N_2|\zeta_2) = (N_1 + N_2|\zeta_1 + \zeta_2), \quad \text{neutral element: } (0|0).$$

The complex 2-dimensional representations $\{(1|\zeta) \mid \zeta \in \mathbb{C}\}$ are *cyclic fundamental*: Their totally symmetrized products $\bigvee$ give all finite-dimensional faithful nondecomposable $\mathbb{C}^\diamond$-representations

$$\bigvee^N (1|\zeta) = (N|N\zeta), \quad N = 0, 1, 2 \ldots.$$

The antisymmetric square gives the irreducible representations

$$(1|\zeta) \wedge (1|\zeta) = (0|2\zeta).$$

The one invariant of a $\mathbb{C}^\diamond$-representation is also its eigenvalue, called the *weight*. For the abelian group $\mathbb{C}^\diamond$, the weight set is isomorphic to the equivalence classes of the irreducible representations, called the *dual group* of $\mathbb{C}^\diamond$:

$$\textbf{ndecrep}\,\mathbb{C}^\diamond \supset \textbf{irrep}\,\mathbb{C}^\diamond = \{(0|\zeta)\} \cong \textbf{weights}\,\mathbb{C}^\diamond = \{\zeta\} = \mathbb{C} \in \underline{\textbf{grp}}.$$

The nondecomposable complex representations of the real 1-dimensional Lie groups

$$\begin{array}{ll} \mathbf{D}(1) \ni e^t & \longmapsto e^{i(m\,\mathrm{id}_V + \mathcal{N}_N)t}, \\ \mathbf{U}(1) \ni e^{i\alpha} & \longmapsto e^{Zi\alpha}, \end{array}$$

have a conjugation and an inner product for the representation space.

The equivalence classes of the nondecomposable unitary $\mathbf{D}(1)$-representations with an imaginary invariant (weight) im constitute the *representation monoid*

$$\textbf{ndecrep}\,\mathbf{D}(1) = \{(N|im)\} \cong \mathbb{N}_0 \times i\mathbb{R} \in \underline{\textbf{mon}}.$$

Again, the imaginary weight group is isomorphic to the *dual group of* $\mathbf{D}(1)$ with the irreducible representation classes

$$\textbf{ndecrep}\,\mathbf{D}(1) \supset \textbf{irrep}\,\mathbf{D}(1) = \{(0|im)\} \cong \textbf{weights}\,\mathbf{D}(1) = \{im\} = i\mathbb{R} \in \underline{\textbf{grp}}.$$

The representations of the compact group $\mathbf{U}(1)$ have to take into account the periodicity $e^{i\alpha} = e^{i(\alpha+2\pi)}$. They are decomposable into irreducible complex 1-dimensional representations with integer invariant *winding number* (weight):

$$\begin{array}{l} \mathbf{U}(1) \longrightarrow \mathbf{U}(1), \; e^{i\alpha} \longmapsto e^{i\alpha Z}, \quad Z \in \mathbb{Z}, \\ \mathbf{U}(1) \cong \mathbf{U}(1)/\mathbb{I}(n), \quad n = 1, 2, \ldots. \end{array}$$

They are characterized by their kernel, the cyclotomic group $\mathbb{I}(Z) = \mathbb{I}(-Z)$. The nontrivial ones have positive or negative orientation $\epsilon(Z) = \pm 1$. Only $|Z| - 1$ gives faithful representations. The *dual group of* $\mathbf{U}(1)$ with the equivalence classes of the irreducible $\mathbf{U}(1)$-representations is isomorphic to the $\mathbf{U}(1)$-weight group, containing the oriented winding numbers Z

$$\textbf{irrep}\,\mathbf{U}(1) \cong \textbf{weights}\,\mathbf{U}(1) = \{Z\} = \mathbb{Z} \in \underline{\textbf{grp}}.$$

The $\mathbf{U}(1)$-weights are used, e.g., for integer charge numbers. They are isomorphic to the discrete subgroup $\mathbb{Z} \subseteq \mathbb{R}$ of the $\mathbf{D}(1)$-weights. The $\mathbf{D}(1)$-weights constitute, in the language of the chapter "Spacetime Translations", up to i the dual space (frequencies, energies) of the time translations.

2.11 Summary

The totally ordered simply connected noncompact real group $\mathbf{D}(1) = \exp \mathbb{R} \cong \mathbb{R}$ (causal group) is used as Lie group model for time with the time translation Lie algebra $\log \mathbf{D}(1) = \mathbb{R}$. The nondecomposable complex finite-dimensional $\mathbf{D}(1)$-representations $D_N^{im}(t) = e^{i(m\,\mathrm{id}_V + \mathcal{N}_N)t}$ are characterized by their dimensionality $1+N$ and one invariant (eigenvalue) im with a real energy (frequency) m. For nildimension $N = 0$ they are irreducible (simple), but not faithful; for $N \geq 1$ they are faithful, but reducible (multiple) with a nontrivial nilpotent contribution $\mathcal{N}_N$; $iH_N = i(m\,\mathrm{id}_V + \mathcal{N}_N)$ with the Hamiltonian H_N represents a time translation basis. Self-duality of the complex nondecomposable time representations, i.e., real images for the real time operations, requires a conjugation $\mathbf{z}_N$ of the representation space that determines an inner product. With respect to this conjugation the nondecomposable representations of the time group are unitary; those of the time Lie algebra are antisymmetric. In contrast to the irreducible time representations in $\mathbf{U}(1)$ (with probability interpretation in quantum theory), the faithful representations in the groups $\mathbf{U}(1+\frac{N}{2}, \frac{N}{2})$ and $\mathbf{U}(\frac{1+N}{2}, \frac{1+N}{2})$ for $N \geq 1$ have no definite scalar product. The conjugation implements the time reflection. Vector spaces with compact time representations can be spanned by time eigenvectors, noncompact time representations also contain time nilvectors; they lead to bounded and unbounded time orbits.

All representations of the compact group $\mathbf{U}(1) \cong \mathbb{R}/\mathbb{Z}$ are decomposable into the complex 1-dimensional irreducible ones $D^Z(i\alpha) = e^{i\alpha Z}$, characterized by an integer winding number Z as eigenvalue (invariant).

$\mathbf{ndecrep}\,\mathbb{C}^\diamond \cong \mathbb{N}_0 \times \mathbb{C}$	$\mathbf{irrep}\,\mathbb{C}^\diamond \cong \mathbb{C}$
$e^z \longmapsto e^{(\zeta+n_N)z} \in \mathbf{GL}(\mathbb{C}^{1+N})$	$e^{\zeta z} \in \mathbb{C}^\diamond$
	$\mathbf{weights}\,\mathbb{C}^\diamond = \mathbb{C}$

representations and weights of $\mathbb{C}^\diamond = \exp \mathbb{C}$

$\mathbf{ndecrep}\,\mathbf{D}(1) \cong \mathbb{N}_0 \times i\mathbb{R}$	$\mathbf{irrep}\,\mathbf{D}(1) \cong i\mathbb{R}$	$\mathbf{irrep}\,\mathbf{U}(1) \cong \mathbb{Z}$
$e^{iH_N t} \in \mathbf{U}(1+\frac{N}{2}, \frac{N}{2}), \mathbf{U}(\frac{1+N}{2}, \frac{1+N}{2})$	$e^{imt} \in \mathbf{U}(1)$	$e^{i\alpha Z} \in \mathbf{U}(1)$
	$\mathbf{weights}\,\mathbf{D}(1) = i\mathbb{R}$	$\mathbf{weights}\,\mathbf{U}(1) = \mathbb{Z}$

representations and weights of $\mathbf{D}(1) = \exp \mathbb{R}$ **and** $\mathbf{U}(1) = \exp i\mathbb{R}$

MATHEMATICAL TOOLS

2.12 Group Realizations and Klein Spaces

A set S defines a monoid by its endomorphisms $\mathbf{A}(S) = \{f : S \longrightarrow S\}$ (arrow monoid, self-mappings) with the associative concatenation $f \circ g$ and the identity id_S. A monoid realized in a mapping monoid $M \longrightarrow \mathbf{A}(S)$ can be interpreted by operations acting on the set S. If the realized monoid M has additional structures (linearity, topology, analyticity, etc.), the set S acted on by the monoid $\mathbf{A}(S)$ should be also enriched correspondingly with a vector space, topological space, analytic manifold, etc., structure. A group with neutral element $e \in G \in \underline{\mathbf{grp}}$ is *realized on a set* S by a group morphism in the automorphism group $R^S : G \longrightarrow \mathbf{G}(S)$ (self-bijections or permutation group). A set with a G-action $x \longmapsto g \bullet x = R^S(g).x$ is called a *G-set* with the category $\underline{\mathbf{set}}_G \ni S$. A set acted on by different groups belongs to different categories, e.g., the plane translations $\mathbb{R}^2 \in \underline{\mathbf{vec}}_{\mathbb{R}}$ or the plane translations with the action of rotations $\mathbb{R}^2 \in \underline{\mathbf{set}}_{\mathbf{SO}(2)}$ or the plane translations with Lorentz transformations $\mathbb{R}^2 \in \underline{\mathbf{set}}_{\mathbf{SO}_0(1,1)}$.

A manifold with a Lie group action $G \bullet M$ is called a *Klein space*. If a (Lie) group or Lie algebra acts on a manifold, only *compatible, especially invariant and covariant*, structures make sense (*Erlanger Programm of Felix Klein* (1872)), e.g., inner products or order and irreducible realizations or representations.

Most of the following concepts given for group realizations are valid also for monoid realizations $M \longrightarrow \mathbf{A}(S)$ and can be formulated correspondingly.

The image of a realized group is isomorphic to the quotient group with kernel

$$R^S[G] \cong G/\operatorname{kern} R^S \in \underline{\mathbf{grp}}.$$

Injective realizations are called *faithful*, in this case G operates *effectively*. A *simple group* G, defined by $G \neq \{e\}$ and without proper normal subgroup, has only trivial or faithful realizations. An element $x \in S \in \underline{\mathbf{set}}_G$ with $G \bullet x = \{x\}$ is called *G-invariant*. It belongs to the *invariance set for the group* G:

$$\mathrm{INV}_G S = \{x \in S \mid G \bullet x = \{x\}\} \in \underline{\mathbf{set}}.$$

The power set inherits a G-action, i.e., $S \in \underline{\mathbf{set}}_G \Rightarrow 2^S \in \underline{\mathbf{set}}_G$:

$$G \times 2^S \longrightarrow 2^S, \quad g \bullet T = \{g \bullet x \mid x \in T\}.$$

With two G-sets also their product is again a G-set

$$\begin{aligned} &G \times (S \times T) \longrightarrow S \times T, \quad g \bullet (x,y) = (g \bullet x, g \bullet y),\\ &R^{S \times T} = R^S \times R^T. \end{aligned}$$

Therefore, the commutativity of the following diagram defines the G-action on the mappings $T^S = \mathbf{set}(S,T)$ between two G-sets. i.e., $T, S \in \underline{\mathbf{set}}_G \Rightarrow T^S \in \underline{\mathbf{set}}_G$:

$$\begin{array}{ccc} S & \xrightarrow{g\bullet} & S \\ {\scriptstyle f}\downarrow & & \downarrow{\scriptstyle {}_gf} \\ T & \xrightarrow[g\bullet]{} & T \end{array} \quad {}_gf(x) = g \bullet f(g^{-1} \bullet x),$$

$$G \times T^S \longrightarrow T^S, \quad (g,f) \longmapsto {}_gf = R^T(g) \circ f \circ R^S(g^{-1}).$$

Morphisms for G-sets, called *intertwiners*, are defined by the G-invariant mappings in all mappings T^S, i.e., by the commutativity of the diagram above with ${}_gf = f$, $f(g \bullet v) = g \bullet f(v)$:

$$\mathbf{set}_G(S,T) = \mathrm{INV}_G T^S = \{f : S \longrightarrow T \mid {}_gf = f \text{ for all } g \in G\}.$$

For an isomorphism f, R^S and R^T are called *equivalent (isomorphic) realizations.* The intertwining endomorphisms for a G-set constitute a submonoid in all endomorphisms, called the *intertwining monoid for* $S \in \underline{\mathbf{set}}_G$:

$$\mathbf{set}_G(S,S) = \mathbf{A}_G(S) = \mathrm{INV}_G S^S = \{f \in \mathbf{A}(S) \mid R(g) \circ f = f \circ R(g)\} \in \underline{\mathbf{mon}}.$$

The equivalence classes of the realizations of a group G define the abelian *realization monoid* with the multiplication via the set product and the neutral element by the trivial G-representation on 1-elementic sets, e.g., on $\{\emptyset\}$,

$$\mathbf{real}\, G = \{[R^S] \mid S \in \underline{\mathbf{set}}_G\}, \quad \begin{cases} [R^S] \times [R^T] & = [R^S \times R^T] = [R^{S\times T}], \\ G \ni g & \longmapsto \mathrm{id}_\emptyset \in \mathbf{G}(\emptyset), \\ & \text{trivial } G\text{-realization.} \end{cases}$$

This defines the contravariant *functor of the group realizations* to the abelian realization monoids

$$\mathbf{real} : \underline{\mathbf{grp}} \longrightarrow \underline{\mathbf{mon}}, \quad f \begin{array}{c} G_1 \\ \big\downarrow \\ G_2 \end{array} \circ f \longmapsto \begin{array}{c} \mathbf{real}\, G_1 \\ \big\uparrow \\ \mathbf{real}\, G_2 \end{array},$$
$$R_2 : G_2 \longrightarrow \mathbf{G}(S) \Rightarrow R_2 \circ f : G_1 \longrightarrow \mathbf{G}(S).$$

For example, the subgroup embedding $H \hookrightarrow G$ restricts the G-realizations R to H-realizations $R|_H$.

2.12.1 Self-Realizations of a Group

The group product can be interpreted in terms of realizations of the group: *Left and right translations (multiplications)* are faithful self-realizations in the bijection (permutation) group $\mathbf{G}(G) = \overset{\circ}{\mathbf{set}}\,(G,G)$. The inversion intertwines both realizations

$$\begin{array}{l} L, R : G \longrightarrow \mathbf{G}(G), \\ L_k(g) = kg, \ R_k(g) = gk^{-1}, \end{array} \quad \begin{array}{ccc} G & \xrightarrow{L_k} & G \\ \text{inversion} \big\downarrow & & \big\downarrow \text{inversion}, \\ G & \xrightarrow[R_k]{} & G \end{array} \quad \begin{array}{ccc} g & \longmapsto & kg \\ \big\downarrow & & \big\downarrow \\ g^{-1} & \longmapsto & g^{-1}k^{-1} \end{array}.$$

The left-right realization of the squared group,

$$L \times R : G \times G \longrightarrow \mathbf{G}(G), \quad L_{k_1} \circ R_{k_2}(g) = R_{k_2} \circ L_{k_1}(g) = k_1 g k_2^{-1},$$

contains the *inner automorphisms* for the group $G \cong (G \times G)_\Delta$ as the diagonal elements:

$$\begin{array}{rl} \mathrm{Int} : G \longrightarrow \overset{\circ}{\mathbf{grp}}\,(G,G), & k \longmapsto \mathrm{Int}\, k = L_k \circ R_k = R_k \circ L_k, \\ \text{with} & \mathrm{Int}\, k(g) = kgk^{-1}. \end{array}$$

They constitute a normal subgroup in all group bijections and are trivial for an abelian group. Normal subgroups are invariant under inner automorphisms.

Any finite group $\operatorname{card} G = N$ is isomorphic to a subgroup of the permutation group $\mathbf{G}(n)$ as seen with the left multiplication realization $G \ni k \longmapsto L_k \in \mathbf{G}(N)$.

The *adjoint group* $\operatorname{Int} G$ *of a group* G is the image with respect to inner automorphisms, i.e., the quotient of the group with its center

$$\operatorname{Int} G = G/\operatorname{centr} G.$$

The *adjoint group square* is defined as the semidirect product of the adjoint group with the group:

$$\begin{aligned} &\operatorname{Int} G \ \vec{\times}\ G = \{(k,g) \mid k, g \in G\}, \\ &(k_1, g_1) \circ (k_2, g_2) = (k_1 k_2, g_1 \operatorname{Int} k_1(g_2)). \end{aligned}$$

For a unital algebra A the *inner automorphisms* are defined for the regular elements $g \in A^\diamond$

$$\begin{aligned} \operatorname{Ad} : A^\diamond \longrightarrow \overset{\circ}{\mathbf{aag}}_{\mathbb{K}}(A, A), \quad &\operatorname{Ad} k : A \longrightarrow A, \quad \operatorname{Ad} k(a) = kak^{-1}, \\ &\operatorname{Ad} k\Big|_{A^\diamond} = \operatorname{Int} k\Big|_{A^\diamond}. \end{aligned}$$

2.12.2 Fix- and Stabilgroups

For a group action $G \times S \longrightarrow S$ each subset has its *fixgroup*

$$T \subseteq S: \quad G_T = \{g \in G \mid g \bullet x = x \text{ for all } x \in T\} \in \underline{\mathbf{grp}}.$$

The fixgroup of a set is the intersection of the fixgroups $G_{\{x\}} = G_x$ of its elements. The fixgroup of the whole set is a normal subgroup, since it is the kernel of the G-realization $R : G \longrightarrow \mathbf{G}(S)$

$$\begin{aligned} &\{g \in G \mid g \bullet x = x \text{ for all } x \in S\} = \bigcap_{x \in S} G_x \\ &= \operatorname{kern} R = \{h \in G \mid R(h) = \operatorname{id}_S\}. \end{aligned}$$

The fixgroup of a group subset with respect to the inner automorphisms is its *centralizer*

$$X \subseteq G: \quad G_X^{\operatorname{Int}} = \{k \in G \mid kgk^{-1} = g \text{ for all } g \in X\} \in \underline{\mathbf{grp}}.$$

The centralizer of the whole group is the group center.

The *stabilgroup* of a subset $T \subseteq S$ consists of those group elements, which keep T stable, i.e., the T-bijections in G:

$$T \subseteq S: \quad G_{\{T\}} = \{g \in G \mid g \bullet T = T\} \in \underline{\mathbf{grp}}.$$

For one element $x \in S$ the stabilgroup coincides with the fixgroup. The stabilgroup with respect to the G-action in S is a fixgroup with respect to the G-action in the power set of S.

The *normalizer of a subset X of a group G* is its stabilgroup with respect to inner automorphisms:

$$X \subseteq G: \quad G^{\mathrm{Int}}_{\{X\}} = \{g \in G \mid gXg^{-1} = X\} \in \underline{\mathbf{grp}}.$$

For a normal subgroup X the full group G is its normalizer.

2.12.3 Group Orbits as Irreducible Realizations

A set S with G-action is called *G-irreducible* (also the realization) if it has no proper G-invariant subsets:

$$T \subseteq S: \quad G \bullet T = T \Rightarrow T = S \text{ or } \emptyset.$$

Equivalent: G acts *transitively* on S, i.e., each element $x \in S$ is *cyclic*:

$$x \in S \Rightarrow G \bullet x = S.$$

A G-realization defines *G-orbits* in the set S as images of the *orbit mappings*

$$\begin{array}{l} x(\ ): G \longrightarrow S, \quad x(g) = R(g)(x), \quad x(e) = x, \\ G \bullet x = x[G] \subseteq S. \end{array}$$

Orbit mappings are G-intertwiners between group and action set:

$$\begin{array}{ccc} G & \xrightarrow{L_k} & G \\ {\scriptstyle x(\ )}\downarrow & & \downarrow{\scriptstyle x(\ )}, \\ S & \xrightarrow[R(k)]{} & S \end{array} \qquad \begin{array}{rl} x(\ ) & \in \mathbf{set}_G(G, S), \\ x(kg) & = R(k).x(g) = k \bullet x(g). \end{array}$$

A group action decomposes the set S into disjoint orbits $G \bullet x$ (*orbit decomposition*). These orbits are G-irreducible sets (transitivity subsets of S)

$$x \sim_G y \iff G \bullet x = G \bullet y \Rightarrow \left\{ \begin{array}{rl} S & = \biguplus_{\text{repr } x_r} G \bullet x_r, \\ S/G & = \{G \bullet x \mid x \in S\} \\ & \cong \{x_r \mid \text{representatives}\}. \end{array} \right.$$

Therefore, each group realization is decomposable into irreducible realizations $G \times (G \bullet x) \longrightarrow G \bullet x$.

The fixgroups of the elements of one orbit are isomorphic by an inner G-automorphism

$$x, y \in S,\ G \bullet x = G \bullet y : \begin{cases} h \in G_x \iff h \bullet x = x \\ \Rightarrow \text{There exists } g \in G \text{ with } ghg^{-1} \in G_y \\ \Rightarrow G_x = gG_y g^{-1} \cong G_y. \end{cases}$$

The orbits of the inner automorphisms decompose the group into *conjugacy classes* that have characteristic centralizers (fixgroups)

$$\operatorname{Int} G(g) = \{kgk^{-1} \mid k \in G\} \cong G/G_g^{\operatorname{Int}},\quad G = \biguplus_r \operatorname{Int} G(g_r),$$

with $G_e^{\operatorname{Int}} = G$. For an abelian group $\{g\} = \operatorname{Int} G(g) \cong G/G$, e.g., for the cyclic groups $\mathbb{I}(n) = \biguplus_{k=1}^{n} \{e^{\frac{2i\pi k}{n}}\}$. The smallest nonabelian permutation group $\mathbf{G}(3)$ gives as centralizers the full group, the permutation subgroup $\mathbf{G}(2)$, and the alternating permutation normal subgroup $\mathbf{G}(3)_+$:

1	(32)	(31)	(21)	(312)	(231)
(32)	**1**	(312)	(231)	(31)	(21)
(31)	(231)	**1**	(312)	(21)	(32)
(21)	(312)	(231)	**1**	(32)	(31)
(312)	(21)	(32)	(31)	(231)	**1**
(231)	(31)	(21)	(32)	**1**	(312)

$$\mathbf{G}(3) \cong \mathbf{G}(2) \vec{\times} \mathbf{G}(3)_+ \cong \mathbb{I}(2) \vec{\times} \mathbb{I}(3)$$

$$\Rightarrow \begin{cases} \operatorname{Int} \mathbf{G}(3)(\mathbf{1}) & = \{\mathbf{1}\} & \cong \mathbf{G}(3)/\mathbf{G}(3) & \cong \mathbf{G}(1), \\ \operatorname{Int} \mathbf{G}(3)((32)) & = \{(32),(31),(21)\} & \cong \mathbf{G}(3)/\mathbf{G}(2), & \\ \operatorname{Int} \mathbf{G}(3)((312)) & = \{(312),(231)\} & \cong \mathbf{G}(3)/\mathbf{G}(3)_+ & \cong \mathbf{G}(2). \end{cases}$$

For a *finite group, the conjugacy classes characterize the equivalence classes of the irreducible group representations* (more below).

Larger equivalence classes in the set S, called *strata*, are obtained by elements with isomorphic fixgroups; this defines the *strata decomposition* (isotypical orbit decomposition) of a G-set. All elements $y \in S$ arising from $x \in S$ by transformation with a mapping from the centralizer of the fixgroup G_x in all bijections $\mathbf{G}(S)$, have equal fixgroup

$$\begin{aligned} & \{f \in \mathbf{G}(S) \mid f \circ R(g) \circ f^{-1} = R(g) \text{ for all } g \in G_x\}, \\ y = f(x) \quad & \iff f^{-1}(y) = x = g \bullet x = g \bullet (f^{-1}(y)) = f^{-1}(g \bullet y) \\ & \iff y = g \bullet y. \end{aligned}$$

2.12.4 Left and Right Cosets

Each subgroup $H \subseteq G$ defines two equivalence relations for the full group by equal left or right orbits of the subgroup. The equivalence classes G/H and $H \setminus G$ are subset families of the full group. They are called the *H left and H right cosets (subgroup classes) of* G, their cardinality $\operatorname{card} G/H$ is the *index of the subgroup* H *in* G

$$g \sim_L k \iff g^{-1}k \in H \iff gH = kH \Rightarrow \begin{cases} G = \biguplus_{\text{repr } g_r} g_r H, \\ G/H = \{gH \mid g \in G\}, \end{cases}$$
$$g \sim_R k \iff kg^{-1} \in H \iff Hg = Hk \Rightarrow \begin{cases} G = \biguplus_{\text{repr } g_r} Hg_r, \\ H\backslash G = \{Hg \mid g \in G\}. \end{cases}$$

Left and right subgroup classes are isomorphic with the inversion $g \leftrightarrow g^{-1}$; therefore in the following mostly left classes are used. Precisely for a normal subgroup N left and right classes are identical $G/N = N\backslash G$, since $gNg^{-1} = N$ for all $g \in G$.

The subgroup classes carry a canonical G-action and are G-irreducible:

$$\begin{array}{ll} G \times G/H \longrightarrow G/H, & k \bullet gH = kgH, \\ G \times H\backslash G \longrightarrow H\backslash G, & k \bullet Hg = Hgk^{-1}. \end{array}$$

The fixgroup of any element for the left G-action is isomorphic to the subgroup which determines the classes

$$gH \in G/H \Rightarrow G_{\{gH\}} = \{k \in G \mid kgH = gH\} = gHg^{-1} \cong H.$$

The following simple insight is basically very important for the understanding of group realizations and representations: The group orbit $G \bullet x$ of a point $x \in S$ is, as a set with G action, isomorphic to the fixgroup classes G/G_x of the group G:

$$G \bullet x \cong G/G_x \in \underline{\mathbf{set}}_G,$$

$$\begin{array}{ccc} G \bullet x & \stackrel{k\bullet}{\longrightarrow} & G \bullet x \\ \cong \downarrow & & \downarrow \cong \\ G/G_x & \underset{k}{\longrightarrow} & G/G_x \end{array}, \quad \begin{array}{ccc} g \bullet x & \longmapsto & kg \bullet x \\ \updownarrow & & \updownarrow \\ gG_x & \longmapsto & kgG_x \end{array}.$$

Therefore, one has, up to isomorphism, *all irreducible realizations of a group* G, i.e., all equivalence classes of irreducible G-sets $\underline{\mathbf{set}}_G$, if one knows *all isomorphism classes of* G*-subgroups*

$$\mathbf{irreal}\ G \cong \{G/H \mid \text{subgroup } H \subseteq G\} \cong \{H\backslash G\}.$$

2.13 Group and Lie Algebra Representations

A group G acts on a *vector space* $V \in \underline{\mathbf{vec}}_{\mathbb{K}}$ via a group morphism into the nonsingular *linear* mappings $\mathbf{GL}(V)$, called a *group representation*:

$$D^V : G \longrightarrow \mathbf{GL}(V), \quad D^V(gh) = D^V(g) \circ D^V(h), \quad D^V(e) = \mathrm{id}_V.$$

$D^V[G] \vec{\times} V \subseteq \mathbf{GL}(V) \vec{\times} V$ is an affine subgroup.

A vector space V defines by its endomorphisms $\mathbf{AL}(V)$ (linear arrows) a unital algebra with a natural Lie algebra structure (commutators). On a vector space V a Lie algebra L acts via a Lie algebra morphism into the endomorphism algebra $\mathbf{AL}(V)$ with its natural Lie algebra structure (*Lie algebra representation*):

$$\begin{array}{l} \mathcal{D}^V : L \longrightarrow \mathbf{AL}(V), \quad \mathcal{D}^V([l,m]) = [\mathcal{D}^V(l), \mathcal{D}^V(m)] \\ \text{if } V \text{ and } L \text{ are finite-dimensional: } \begin{cases} \mathcal{D}^V(l) \in V \otimes V^T, \\ \mathcal{D}^V \in V \otimes V^T \otimes L^T; \end{cases} \end{array}$$

$\mathcal{D}^V[L] \vec{\oplus} V \subseteq \mathbf{AL}(V) \vec{\oplus} V$ is an affine Lie subalgebra.

A vector space with a G or L-action is called a *G- or L-module*, $V \in \underline{\mathbf{mod}}_G, V \in \underline{\mathbf{mod}}_L$. More about modules below.

The image of a Lie algebra is isomorphic to the quotient Lie algebra with kernel

$$\mathcal{D}^V[L] \cong L/\,\mathrm{kern}\ \mathcal{D}^V \in \underline{\mathbf{lag}}_{\mathbb{K}}.$$

A *simple Lie algebra* L, defined as nonabelian and without proper ideals, has only trivial or faithful representations.

Any finite-dimensional $\mathbb{K}$-Lie algebra is isomorphic to an endomorphism subalgebra of a finite-dimensional $\mathbb{K}$-vector space (*theorem of Ado*). For example, the real Heisenberg Lie algebra $\log \mathbf{H}(1) \cong \mathbb{R}^3$ for one position-momentum pair has the 3-dimensional faithful representation by nilpotent matrices:

$$\begin{array}{l} [\mathbf{x}, \mathbf{p}] = \mathbf{I}, \ [\mathbf{x}, \mathbf{I}] = 0 = [\mathbf{p}, \mathbf{I}], \\ \log \mathbf{H}(1) \ni q\mathbf{x} + y\mathbf{p} + t\mathbf{I} \longmapsto \begin{pmatrix} 0 & q & t \\ 0 & 0 & y \\ 0 & 0 & 0 \end{pmatrix} \in \mathbf{AL}(\mathbb{R}^3). \end{array}$$

An element $v \in V \in \underline{\mathbf{mod}}_L$ is called *L-invariant* with $L \bullet v = \{0\}$. The vector subspace with the *L-invariants of V* is

$$\mathrm{INV}_L V = \{v \in V \mid L \bullet v = \{0\}\} \in \underline{\mathbf{vec}}_{\mathbb{K}}.$$

The morphisms for the categories $\underline{\mathbf{mod}}_G$ and $\underline{\mathbf{mod}}_L$, i.e., the linear intertwiners for vector spaces with group and Lie algebra action, are characterized by their compatibility, i.e., by the invariant sets in the mappings

$$\begin{array}{rl} \mathbf{mod}_G(V, W) & = \mathrm{INV}_G \mathbf{vec}_{\mathbb{K}}(V, W) \\ & = \{f \mid D^V(g) \circ f = f \circ D^W(g) \text{ for all } g \in G\}, \\ \mathbf{mod}_L(V, W) & = \mathrm{INV}_L \mathbf{vec}_{\mathbb{K}}(V, W) \\ & = \{f \mid \mathcal{D}^V(l) \circ f = f \circ \mathcal{D}^W(l) \text{ for all } l \in L\}. \end{array}$$

The dimensions of those vector spaces define the *intertwining dimensions*. With an isomorphism f, one has *equivalent (isomorphic) representations.*

An intertwiner $f : V \longrightarrow W$ gives an intertwining isomorphism $\overline{f}$ between the kernel classes and the range (on both vector spaces the representation is well-defined):

$$\begin{array}{ccc} V/\operatorname{kern} f & \xrightarrow{\overline{D}^V(g),\overline{\mathcal{D}}^V(l)} & V/\operatorname{kern} f \\ \overline{f}\ \Big\downarrow & & \Big\downarrow\ \overline{f} \\ f[V] & \xrightarrow[\underline{D}^W(g),\underline{\mathcal{D}}^W(l)]{} & f[V] \end{array}.$$

The intertwining endomorphisms for a group or a Lie algebra vector space constitute subalgebras of all endomorphisms, the *intertwining algebra* with all G and L-invariant endomorphisms

$$\begin{aligned} \mathbf{mod}_G(V,V) &= \mathbf{AL}_G(V) = \mathrm{INV}_G\mathbf{vec}_{\mathbb{K}}(V,V) \\ &= \{f \in \mathbf{AL}(V) \mid [D(g), f] = 0\} \quad \in \underline{\mathbf{aag}}_{\mathbb{K}}, \\ \mathbf{mod}_L(V,V) &= \mathbf{AL}_L(V) = \mathrm{INV}_L\mathbf{vec}_{\mathbb{K}}(V,V) \\ &= \{f \in \mathbf{AL}(V) \mid [\mathcal{D}(l), f] = 0\} \quad \in \underline{\mathbf{aag}}_{\mathbb{K}}. \end{aligned}$$

It is also called the *commutant or centralizer* of the (G, L)-representation $(D, \mathcal{D})$.

2.13.1 Sum and Product Representations

If a group or a Lie algebra acts on two vector spaces V, W, the action is explained on the *direct sum* $V \oplus W$ by the direct sum representations $D^V \oplus D^W$ and $\mathcal{D}^V \oplus \mathcal{D}^W$ and on the *linear mappings* by

$$D(g), \mathcal{D}(l) : \mathbf{vec}_{\mathbb{K}}(V,W) \longrightarrow \mathbf{vec}_{\mathbb{K}}(V,W), \begin{cases} D(g)(f) &= D^W(g) \circ f \circ D^V(g^{-1}), \\ \mathcal{D}(l)(f) &= \mathcal{D}^W(l) \circ f - f \circ \mathcal{D}^V(l). \end{cases}$$

Also, the *tensor product* inherits the action, first for a group

$$\begin{aligned} D^{V\otimes W}(g) : V \otimes W &\longrightarrow V \otimes W, \quad D^{V\otimes W}(g) = D^V(g) \otimes D^W(g), \\ v \otimes w &\longmapsto (g \bullet v) \otimes (g \bullet w). \end{aligned}$$

For a Lie algebra one induces the derivation property with the Leibniz rule

$$\begin{aligned} \mathcal{D}^{V\otimes W}(l) : V \otimes W &\longrightarrow V \otimes W, \quad \mathcal{D}^{V\otimes W}(l) = \mathcal{D}^V(l) \otimes \mathrm{id}_W + \mathrm{id}_V \otimes \mathcal{D}^W(l), \\ v \otimes w &\longmapsto (l \bullet v) \otimes w + v \otimes (l \bullet w). \end{aligned}$$

2.13.2 Scalar and Dual Representations

A *scalar representation* of a group G or Lie algebra L,

$$G \longrightarrow \mathbf{GL}(\mathbb{K}) \cong \mathbb{K}^\diamond, \quad L \longrightarrow \mathbf{AL}(\mathbb{K}) \cong \mathbb{K},$$

is trivial for $G \longrightarrow \{1\}$ and $L \longrightarrow \{0\}$. The determinant and the trace of finite-dimensional representations are the *associated scalar representations*

$$V \cong \mathbb{K}^n: \begin{array}{rll} \det D: & G \longrightarrow \mathbf{GL}(\mathbb{K}), & \det D(g_1 g_2) = \det D(g_1) \det D(g_2), \\ \operatorname{tr} \mathcal{D}: & L \longrightarrow \mathbf{AL}(\mathbb{K}), & \operatorname{tr} \mathcal{D}([l_1, l_2]) = 0 = \big[\operatorname{tr} \mathcal{D}(l_1), \operatorname{tr} \mathcal{D}(l_2)\big]. \end{array}$$

With the trivial representation on the scalars $\mathbb{K}$ one obtains *dual representations* on dual spaces (V, V^T): They use the natural morphisms of the automorphism groups and endomorphism Lie algebras for dual vector spaces. They are characterizable by the multiplicative and additive invariance of the *dual product* and are given by the inverse transposed (contragredient) and negative transposed mappings

$$\begin{array}{ccc} G & \xrightarrow{D^V} & \mathbf{GL}(V) \\ \text{inversion} \downarrow & & \downarrow \text{transposition}, \\ G & \xrightarrow[D^{V^T}]{} & \mathbf{GL}(V^T) \end{array} \qquad \begin{array}{ccc} L & \xrightarrow{\mathcal{D}^V} & \mathbf{AL}(V) \\ \text{inversion} \downarrow & & \downarrow \text{transposition}, \\ L & \xrightarrow[\mathcal{D}^{V^T}]{} & \mathbf{AL}(V^T) \end{array}$$

$$\begin{array}{ll} D^{V^T}(g) = D^V(g^{-1})^T, & \mathcal{D}^{V^T}(l) = \mathcal{D}^V(-l)^T, \\ \langle g \bullet \omega, g \bullet v\rangle = \langle \omega, v\rangle, & \langle l \bullet \omega, v\rangle + \langle \omega, l \bullet v\rangle = 0. \end{array}$$

Dual representations are denoted in the following shortly by $(D, \check{D} = D^{-1T})$ and $(\mathcal{D}, \check{\mathcal{D}} = -\mathcal{D}^T)$. For finite-dimensional representations one has natural isomorphisms (an involution for the submonoid with the finite-dimensional representations)

$$V \cong \mathbb{K}^n: \quad \left\{ \begin{array}{ll} D[G] \cong \check{D}[G], & \mathcal{D}[L] \cong \check{\mathcal{D}}[L], \\ \check{\check{D}} \cong D, & \check{\check{\mathcal{D}}} \cong \mathcal{D}. \end{array} \right.$$

The isomorphisms $D(g) \leftrightarrow \check{D}(g)$ and $\mathcal{D}(l) \leftrightarrow \check{\mathcal{D}}(l)$ do not presuppose an isomorphism $V \cong V^T$ for the dual representation spaces.

For a G-representation on vectors V and forms V^T, the dual product is constant on G:

$$\left. \begin{array}{ll} v(\ ): G \longrightarrow V, & v(g) = D(g)(v), \\ \omega(\ ): G \longrightarrow V^T, & \omega(g) = \check{D}(g)(\omega), \end{array} \right\} \quad \langle \omega(g), v(g)\rangle = \langle \omega, v\rangle.$$

Group representations can be written as *two-point products* of the orbits for dual bases $\{e^A, \check{e}_A\}_{A=1}^n$. They depend only on the quotient (or the difference for additive groups) of group elements:

$$\begin{array}{lll} D(g) & = D_A^B(g) e^A \otimes \check{e}_B & = e^A(g_1) \otimes \check{e}_A(g_2), \\ \check{D}(g) & = D_A^B(g^{-1}) \epsilon \check{e}_B \otimes e^A & = \epsilon \check{e}_A(g_1) \otimes e^A(g_2), \quad g = g_1 g_2^{-1}. \end{array}$$

The orbits of a vector $v \in V$ and a form $\omega \in V^T$ give *matrix elements or coefficients of the group representation* via the trace (if defined), it is a function from the group into the scalars:

$$\begin{array}{l} D_\omega^v : G \longrightarrow \mathbb{K}, \quad g \longmapsto D_\omega^v(g), \\ D_\omega^v(g) = \langle \omega(g_2), v(g_1)\rangle = \operatorname{tr} v(g_1) \otimes \omega(g_2) = \langle \omega, g \bullet v\rangle = \langle \omega, v(g)\rangle, \\ D_B^A(g) = \langle \check{e}_B(g_2), e^A(g_1)\rangle = \langle \check{e}_B, e^A(g)\rangle, \quad g = g_2^{-1} g_1. \end{array}$$

This explains the *duality* of V, V^T as G-vector spaces:

$$\langle g_2 \bullet \omega, g_1 \bullet v\rangle = \operatorname{tr} D(g_2^{-1}g_1) \circ (v \otimes \omega) = \langle D(g_2^{-1}g_1), v \otimes \omega\rangle.$$

Given a representation D of a group or $\mathcal{D}$ of a Lie algebra, one has the scalar representation on the scalars, the dual ones on the dual vector spaces, the representations on direct sums and tensor products. Continuing, the group or Lie algebra representation is determined on all co- and contravariant tensors, e.g., on the endomorphism algebra $V \otimes V^T$:

$$\begin{array}{ll} D^{V\otimes V^T}(g) : V \otimes V^T \longrightarrow V \otimes V^T, & \left\{\begin{array}{rl} v \otimes \omega & \longmapsto D(g).v \otimes D(g^{-1})^T.\omega, \\ f & \longmapsto D(g) \circ f \circ D(g^{-1}), \end{array}\right. \\ \mathcal{D}^{V\otimes V^T}(l) : V \otimes V^T \longrightarrow V \otimes V^T, & \left\{\begin{array}{rl} v \otimes \omega & \longmapsto \mathcal{D}(l).v \otimes \omega - v \otimes \mathcal{D}(l)^T.\omega, \\ f & \longmapsto [\mathcal{D}(l), f]. \end{array}\right. \end{array}$$

The squared group $G \times G$ is represented by

$$\mathbf{AL}(V) \longrightarrow \mathbf{AL}(V), \quad f \longmapsto D(k_1) \circ f \circ D(k_2^{-1}).$$

2.13.3 Representation Monoids

The equivalence classes of group and Lie algebra representations have a multiplication given by the tensor product and an addition by the direct sum

$$\mathbf{rep}\ G = \{[D^V] \mid V \in \underline{\mathbf{mod}}_G\}, \quad \left\{\begin{array}{rl} [D^V] \otimes [D^W] & = [D^{V\otimes W}], \\ \text{trivial: } G \ni g & \longmapsto 1 \in \mathbb{K}^\diamond, \\ [D^V] \oplus [D^W] & = [D^{V \oplus W}], \end{array}\right.$$

$$\mathbf{rep}\ L = \{[\mathcal{D}^V] \mid V \in \underline{\mathbf{mod}}_L\}, \quad \left\{\begin{array}{rl} [\mathcal{D}^V] \oplus [\mathcal{D}^W] & = [\mathcal{D}^{V \oplus W}], \\ \text{trivial: } L \ni l & \longmapsto 0 \in \{0\}, \\ [\mathcal{D}^V] \otimes [\mathcal{D}^W] & = [\mathcal{D}^{V\otimes W}]. \end{array}\right.$$

$(\mathbf{rep}\, G, \otimes, 1)$ is an abelian monoid with the trivial representation class as neutral element and a semigroup $(\mathbf{rep}\, G, \oplus)$. $(\mathbf{rep}\, L, \oplus, 0, \otimes)$ is an abelian ring with the trivial representation class as neutral element.

Therefore one obtains the contravariant *representation functors for groups and Lie algebras*

$$\mathbf{rep} : \underline{\mathbf{grp}} \longrightarrow \underline{\mathbf{mon}}, \quad f \Big\downarrow \begin{matrix} G_1 \\ \\ G_2 \end{matrix} \quad \circ f \longmapsto \Big\uparrow \begin{matrix} \mathbf{rep}\ G_1 \\ \\ \mathbf{rep}\ G_2 \end{matrix},$$

$$\mathbf{rep} : \underline{\mathbf{lag}_{\mathbb{K}}} \longrightarrow \underline{\mathbf{rng}}, \quad f \Big\downarrow \begin{matrix} L_1 \\ \\ L_2 \end{matrix} \quad \circ f \longmapsto \Big\uparrow \begin{matrix} \mathbf{rep}\ L_1 \\ \\ \mathbf{rep}\ L_2 \end{matrix}.$$

The dual representations define a reflection.

2.14 Invariant Inner Products and Self-Dual Representations

An (anti-)symmetric inner product *invariant with respect to a group G or Lie algebra L*,

$$\zeta(\ ,\ ) : V \times V \longrightarrow \mathbb{K}, \quad \begin{cases} \zeta(v, g \bullet w) &= \zeta(g^{-1} \bullet v, w), \quad g \in G, \\ \zeta(v, l \bullet w) &= -\zeta(l \bullet v, w), \quad l \in L, \end{cases}$$

defines for an invariant subspace the invariant ζ-orthogonal subspace

$$\begin{array}{l} W \subseteq V: \ \ G \bullet W \subseteq W, \ L \bullet W \subseteq W, \\ W_\zeta^\perp = \{v \in V \mid \zeta(W, v) = \{0\}\} \Rightarrow G \bullet W_\zeta^\perp \subseteq W_\zeta^\perp, \ \ L \bullet W_\zeta^\perp \subseteq W_\zeta^\perp. \end{array}$$

Therefore, if an irreducible nontrivial representation (irreducibility will be explained below) has a nontrivial invariant form, the inner product must be nondegenerate:

$$V \text{ irreducible} \Rightarrow \zeta \text{ nondegenerate} \iff V_\zeta^\perp = \{0\}.$$

With Schur's lemma (below) an irreducible, complex finite-dimensional representation of a group or Lie algebra can have, up to scalar multiples, only one invariant linear or only one invariant conjugate linear dual isomorphism ζ, possibly both or none. With two $\zeta_{1,2}$, one has $\zeta_1 \circ \zeta_2^{-1} = \alpha \, \mathrm{id}_V$. For a nondecomposable representation the uniqueness holds up to nilpotent contributions.

A product representation on $V_1 \otimes V_2$ with invariant forms $\zeta_{1,2}$ has $\zeta_1 \otimes \zeta_2$ as an invariant form.

If there exists a (conjugate) linear dual isomorphism ζ compatible with dual representations, the representations are called *(conjugate) linear self-dual*:

$$\begin{array}{ccc} V & \xrightarrow{D(g),\ \mathcal{D}(l)} & V \\ \zeta \big\downarrow & & \big\downarrow \zeta \\ V^T & \xrightarrow[\check{D}(g),\ \check{\mathcal{D}}(l)]{} & V^T \end{array}, \quad \begin{array}{rll} D(g^{-1}) &= \zeta^{-1} \circ D(g)^T \circ \zeta &= D(g)^*, \\ -\mathcal{D}(l) &= \zeta^{-1} \circ \mathcal{D}(l)^T \circ \zeta &= \mathcal{D}(l)^*. \end{array}$$

The properties of nondegenerate inner products determine the groups and the Lie algebras (for the Lie algebra, e.g., $\mathcal{D}[L] \subseteq \log \mathbf{O}(V, \zeta)$):

$$\zeta(v, u) = \begin{cases} \zeta(u, v) \\ -\zeta(u, v) \\ \overline{\zeta(u, v)} \end{cases} \Rightarrow D[G] \subseteq \begin{cases} \mathbf{O}(V, \zeta), & \text{orthogonally self-dual,} \\ \mathbf{Sp}(V, \zeta), & \text{symplectically self-dual,} \\ \mathbf{U}(V, \zeta), & \text{unitarily self-dual.} \end{cases}$$

A *complex* representation of a *real* Lie group $G_\mathbb{R}$ or Lie algebra $L_\mathbb{R}$ has to come with a conjugation of the representation space, i.e., it has to be unitarily self-dual:

$$\begin{array}{ll} D[G_\mathbb{R}] \subseteq \mathbf{U}(p, q) & \text{with } D(g^{-1}) = D(g)^*, \\ \mathcal{D}[L_\mathbb{R}] \subseteq \log \mathbf{U}(p, q) & \text{with } -\mathcal{D}(l) = \mathcal{D}(l)^*. \end{array}$$

A representation in compact $\mathbf{O}(n)$ or $\mathbf{U}(n)$ is *positive self-dual.* Then, the double trace is positive:

$$\operatorname{tr} \mathcal{D}(l)^\star \circ \mathcal{D}(l) = - \operatorname{tr} \mathcal{D}(l) \circ \mathcal{D}(l) \geq 0.$$

A definite unitary representation $D[G] \subseteq \mathbf{U}(V)$ is called a *Hilbert representation.*

If there exists an invariant nondegenerate inner product ζ, the induced conjugation $*$ *implements the inversion* of the represented group $g \longmapsto g^{-1}$ and Lie algebra $l \longmapsto -l$, e.g., for a group

$$\begin{array}{ccc} G & \overset{D}{\longrightarrow} & \mathbf{GL}(V) \\ {\scriptstyle \text{inversion}}\downarrow & & \downarrow * \\ G & \underset{D}{\longrightarrow} & \mathbf{GL}(V) \end{array}, \quad \begin{array}{ccc} g & \longmapsto & D(g) \\ \downarrow & & \downarrow \\ g^{-1} & \longmapsto & D(g^{-1}) = D(g)^* \end{array}.$$

There exist the natural isomorphisms (dual representations) for transposed endomorphism Lie algebras, $\mathbf{AL}(V) \cong \mathbf{AL}(V^T)$, $l \leftrightarrow -l^T$, and automorphism groups, $\mathbf{GL}(V) \cong \mathbf{GL}(V^T)$, $g \leftrightarrow g^{-1T}$. This does not imply that these isomorphisms have to arise as products $\zeta \otimes \zeta^{-1}$ with a dual isomorphism $V \overset{\zeta}{\leftrightarrow} V^T$.

The direct sum of a finite-dimensional vector space and its dual space is naturally isomorphic to its dual via the natural bidual isomorphism $V \cong V^{TT}$

$$\begin{array}{ll} \mathbb{K}^n & \cong V \cong V^{TT} \text{ natural bidual isomorphism}, \\ \mathbf{V} & = V \oplus V^T \Rightarrow \mathbf{V}^T = V^T \oplus V^{TT} \cong V^T \oplus V \cong \mathbf{V}. \end{array}$$

With transposition sign $\epsilon = \pm 1$ the *self-dual vector space sum* $\mathbf{V} \cong \mathbb{K}^{2n}$ carries an ϵ-symmetric nondegenerate bilinear form as an extension of the dual product, the *canonical bilinear form*:

$$\mathbf{V} \times \mathbf{V} \longrightarrow \mathbb{K}, \quad \begin{cases} \langle \omega, u \rangle = \epsilon \langle u, \omega \rangle, \quad \epsilon^2 = 1, \\ \langle v, u \rangle = 0, \quad \langle \theta, \omega \rangle = 0, \\ \langle \mathbf{w}, \mathbf{v} \rangle = \langle v + \omega, u + \theta \rangle = \langle \omega, u \rangle + \langle v, \theta \rangle = \epsilon \langle \mathbf{v}, \mathbf{w} \rangle, \end{cases}$$

$$\text{dual basis of } \mathbf{V} : \{e^A, \check{e}_A\}_{A=1}^n, \quad \begin{cases} \langle \check{e}_B, e^A \rangle = \epsilon \langle e^A, \check{e}_B \rangle = \delta_B^A, \\ \langle e^B, e^A \rangle = 0 = \langle \check{e}_A, \check{e}_B \rangle, \end{cases}$$

$$\langle \mathbf{V}, \mathbf{V} \rangle \sim \begin{pmatrix} 0 & \mathbf{1}_n \\ \epsilon \mathbf{1}_n & 0 \end{pmatrix}.$$

The canonical bilinear form on $\mathbf{V}$ does not define a dual isomorphism $V \cong V^T$. If there exists a dual isomorphism $V \overset{\zeta}{\leftrightarrow} V^T$, then $\mathbf{V} \overset{\zeta \oplus \zeta^{-1}}{\leftrightarrow} \mathbf{V}$ defines an additional inner product.

The endomorphism algebra of the self-dual sum $\mathbf{V} \cong \begin{pmatrix} V \\ V^T \end{pmatrix}$ has the two endomorphism subalgebras $\mathbf{AL}(V)$, $\mathbf{AL}(V^T)$ in the diagonal and two transposition related vector spaces in the skew-diagonal:

$$\begin{array}{l}\underline{\mathbf{aag}}_{\mathbb{K}} \ni \mathbf{AL(V)} \cong \mathbf{V}\otimes\mathbf{V} \simeq \begin{pmatrix} V\otimes V^T & V\otimes V \\ V^T\otimes V^T & V^T\otimes V\end{pmatrix} \in \begin{pmatrix} \underline{\mathbf{aag}}_{\mathbb{K}} & \underline{\mathbf{vec}}_{\mathbb{K}} \\ \underline{\mathbf{vec}}_{\mathbb{K}} & \underline{\mathbf{aag}}_{\mathbb{K}}\end{pmatrix},\\ [\mathbf{v},\mathbf{w}]_\epsilon = \mathbf{v}\otimes\mathbf{w} + \epsilon\mathbf{w}\otimes\mathbf{v} = \epsilon[\mathbf{w},\mathbf{v}]_\epsilon.\end{array}$$

The direct sum of dual representations on $\mathbf{V}$ is decomposable. According to the definition of "dual representations", it leaves the canonical bilinear form invariant

$$\begin{array}{rl} g\in G\in \underline{\mathbf{grp}}: & \mathbf{D}(g) = (D\ \oplus\ \check{D})(g) \simeq \begin{pmatrix} D(g) & 0 \\ 0 & D(g^{-1})^T\end{pmatrix},\\ \text{with} & \langle g\bullet \mathbf{w}, g\bullet\mathbf{v}\rangle = \langle \mathbf{w},\mathbf{v}\rangle,\\ l\in L\in \underline{\mathbf{lag}}_{\mathbb{K}}: & \boldsymbol{\mathcal{D}}(l) = (\mathcal{D}\ \oplus\ \check{\mathcal{D}})(l) = \mathcal{D}^B_A(l)\ [e^A,\check{e}_B]_{-\epsilon} \simeq \begin{pmatrix} \mathcal{D}(l) & 0 \\ 0 & -\mathcal{D}(l)^T\end{pmatrix},\\ \text{with} & \langle l\bullet\mathbf{w},\mathbf{v}\rangle + \langle\mathbf{w}, l\bullet\mathbf{v}\rangle = 0.\end{array}$$

Both for Fermi and for Bose one obtains orthogonal subgroups $D[G]\subseteq \mathbf{O}(\mathbb{K}^{2n})$.

2.15 Characters of Groups

The morphisms of a group G into a fixed abelian group A with unit e_A inherit pointwise the abelian group structure:

$$A\in\underline{\mathbf{abgrp}} \Rightarrow \mathbf{grp}(G,A)\in\underline{\mathbf{abgrp}}: \left\{\begin{array}{l}\chi,\chi'\in\mathbf{grp}(G,A),\\ \text{product: } (\chi\chi')(g) = \chi(g)\chi'(g),\\ \text{inverse: } \chi_{-1}(g) = \chi(g^{-1}),\\ \text{unit: } e_A(g) = e_A\in A.\end{array}\right.$$

In the category of abelian groups, each group $A\in\underline{\mathbf{abgrp}}$ defines the contravariant *A-dual functor*

$$\mathbf{abgrp}(\ ,A): \underline{\mathbf{abgrp}} \longrightarrow \underline{\mathbf{abgrp}}$$

$$\begin{array}{ccc} G & & \mathbf{abgrp}(G,A) = \hat{G}_A \\ f\downarrow & \longmapsto & \uparrow \hat{f}_A \\ G' & & \mathbf{abgrp}(G',A) = \hat{G}'_A\end{array},\quad \hat{f}_A(\chi')(g) = \chi'(f(g)).$$

An element $\chi\in\mathbf{abgrp}(G,A)$ is called an *A-character of the group G*. The A-characters constitute the *A-dual group* $\mathbf{abgrp}(G,A)$ *for G*. If the group A is fixed, the notation may be simplified: $\hat{G} = \hat{G}_A$, $\hat{f} = \hat{f}_A$, etc.

$\mathbb{K}$-vector spaces are abelian groups $(V,+,0)$ with the distinguished vector space $\mathbb{K}$: The linear forms (dual space) $q\in\hat{V}_{\mathbb{K}} = V^T = \mathbf{vec}_{\mathbb{K}}(V,\mathbb{K})$ with $q(x) = \langle q,x\rangle$ are the $\mathbb{K}$-characters of V.

$\mathbf{U}(1)$-representations $G\longrightarrow\mathbf{U}(1) = \exp i\mathbb{R}$ e.g., on vector spaces $\mathbb{R}^n$ with $x\longmapsto e^{\langle iq,x\rangle}$ as used for Fourier transforms, are called *characters*.

With the A-dual group one has the *A-valued dual product*

$$\hat{G}\times G\longrightarrow A,\quad \langle\!\langle\chi,g\rangle\!\rangle = \chi(g),\quad \left\{\begin{array}{l}\langle\!\langle\chi,gg'\rangle\!\rangle = \langle\!\langle\chi,g\rangle\!\rangle\langle\!\langle\chi,g'\rangle\!\rangle,\\ \langle\!\langle\chi\chi',g\rangle\!\rangle = \langle\!\langle\chi,g\rangle\!\rangle\langle\!\langle\chi',g\rangle\!\rangle,\end{array}\right.$$

$$f: G\longrightarrow G',\quad \hat{f}:\hat{G}'\longrightarrow\hat{G},\quad \langle\!\langle\hat{f}(\chi'),g\rangle\!\rangle = \langle\!\langle\chi',f(g)\rangle\!\rangle.$$

The *dual groups* (characters and $\mathbf{D}(1)$-characters) of the Lie groups $\mathbf{D}(1) \cong \mathbb{R}$ and $\mathbf{U}(1)$ are

$$\begin{array}{lll} \mathbf{ablgrp}_{\mathbb{R}}(\mathbf{D}(1), \mathbf{U}(1)) & \cong \mathbf{ablgrp}_{\mathbb{R}}(\mathbf{D}(1), \mathbf{D}(1)) & \cong \mathbb{R}, \\ \mathbf{ablgrp}_{\mathbb{R}}(\mathbf{U}(1), \mathbf{U}(1)) & \cong \mathbf{ablgrp}_{\mathbb{R}}(\mathbf{U}(1), \mathbf{D}(1)) & \cong \mathbb{Z}, \end{array}$$

and for the discrete group $\mathbb{Z}$ with $Z \longmapsto (e^{i\alpha})^Z$,

$$\begin{array}{lll} \mathbf{abgrp}(\mathbb{Z}, \mathbf{U}(1)) \cong & \mathbf{abgrp}(\mathbb{Z}, \mathbf{D}(1)) & \cong \mathbf{U}(1), \\ & \mathbf{abgrp}(\mathbb{Z}, \mathbb{Z}) & \cong \mathbb{Z}. \end{array}$$

The *character of a group representation* is defined by the trace:

$$D : G \longrightarrow \mathbf{GL}(V), \quad \chi_D : G \longrightarrow \mathbb{K}, \quad \chi_D(g) = \operatorname{tr} D(g).$$

For infinite-dimensional representations additional structures have to be used for the definition of a trace. For a nonabelian group, a representation character does not have to be a group representation; it is a function with equal values on the conjugacy classes:

$$k \in G \Rightarrow \chi_D(kgk^{-1}) = \chi_D(g).$$

The conjugacy classes for a finite group and therefore their character values characterize its irreducible representations. This property of representation characters is relevant not only for finite groups.

2.16 Representations of Ordered Monoids

The order of an abelian group $(M, +)$ can be characterized with properties for the neutral element 0:

$$\begin{array}{lll} \text{reflexive:} & 0 \succeq 0, & \\ \text{transitive:} & x \succeq 0 \text{ and } y \succeq 0 & \Rightarrow x + y \succeq 0, \\ \text{antisymmetric:} & x \succeq 0 \text{ and } -x \succeq 0 & \iff x = 0. \end{array}$$

M contains two monoids $(M_\pm, +)$ (future and past). A *representation of the monoids*

$$D_\pm : M_\pm \longrightarrow \mathbf{GL}(V), \quad D_\pm(0) = \mathrm{id}_V, \quad D_\pm(x + y) = D_\pm(x) \circ D_\pm(y)$$

is reflection-(anti)symmetric for $D_+(x) = \pm D_-(-x)$. The definition of dual representations for direct sums and tensor products is obvious.

The characteristic order functions give the *trivial representation*

$$\vartheta_\pm : M_\pm \longrightarrow \{1\}, \quad \left\{ \begin{array}{ll} \vartheta_+(x) & = \left\{ \begin{array}{ll} 1, & x \succeq 0, \\ 0, & x \notin M_+, \end{array} \right. \\ \vartheta_-(x) & = \left\{ \begin{array}{ll} 1, & x \preceq 0, \\ 0, & x \notin M_-, \end{array} \right. \end{array} \right. \quad \vartheta_-(x) = \vartheta_+(-x).$$

The order functions lead from a representation of the ordered abelian group M to monoid representations $D_\pm(x) = \vartheta_\pm(x)D(\pm x)$.

In general, the *bimonoid* $M_{\text{caus}} = M_+ \cup M_- \subseteq M$ is not a subgroup. The trivial (anti)symmetric bimonoid representation

$$M_{\text{caus}} \longrightarrow \{\pm 1\}, \quad \begin{cases} \vartheta(x) & = \vartheta_+(x) + \vartheta_-(x) = \vartheta(-x), \\ \epsilon(x) & = \vartheta_+(x) - \vartheta_-(x) = -\epsilon(-x), \end{cases}$$

e.g., for Minkowski spacetime

$$x \in \mathbb{M} \cong \mathbb{R}^4 : \quad \vartheta_\pm(x) = \vartheta(\pm x_0)\vartheta(x^2), \quad \begin{cases} \vartheta(x) & = \vartheta(x^2), \\ \epsilon(x) & = \epsilon(x_0)\vartheta(x^2), \end{cases}$$

is generalizable with a group representation to reflection (anti)symmetric bimonoid representations

$$M_{\text{caus}} \longrightarrow \mathbf{GL}(V), \ x \longmapsto \vartheta_+(x)D(x) \pm \vartheta_-(x)D(-x).$$

2.17 Minimal Polynomials

The following structures are relevant and will be used for vector space endomorphisms, their eigenvalues, their diagonalization and triagonalization.

The *polynomials in one indeterminate* X with pointwise multiplication constitute a commutative unital $\mathbb{K}$-algebra of countable dimension $\aleph_0$, isomorphic to the finite series $\mathbb{K}^{(\mathbb{N}_0)}$:

$$\mathbb{K}[X] = \{p(X) = \sum_{r=0}^{n} \alpha_r X^r \bigm| \alpha_r \in \mathbb{K}\} \in \underline{\mathbf{aag}}_{\mathbb{K}}.$$

A basis is given by the monomials $\{X^r\}_{r\geq 0}$. The invertible elements are the nontrivial constants $\alpha \in \mathbb{K}^\diamond = \mathbb{K}[X]^\diamond$. The highest-power nontrivial term $\alpha_n X^n$, $\alpha_n \neq 0$, of a polynomial defines its *degree*, $\deg p(X) = n$; a polynomial is called *unitary* for $\alpha_n = 1$. Any ideal of $\mathbb{K}[X]$ is generated by its unitary polynomial $p_I(X)$ of smallest degree, $I = \mathbb{K}[X]p_I(X)$, $\mathbb{K}[X]$ is a principal ideal ring.

A *zero (root)* $\alpha \in \mathbb{K}$ of a polynomial $p(X)$ satidsfies $p(\alpha) = 0$.

A nonconstant polynomial without proper polynomial divisors is called *irreducible* (simple, nonfactorizable, prime). In $\mathbb{C}$ (algebraically closed) the irreducible polynomials have degree 1, in $\mathbb{R}$ one has in addition degree-2 polynomials:

$$\text{irreducible in } \mathbb{K}[X] : \begin{cases} \mathbb{K} = \mathbb{R}, \mathbb{C} : & \{X - \alpha \bigm| \alpha \in \mathbb{K}\}, \\ \mathbb{K} = \mathbb{R} : & \{(X-\beta)^2 + \gamma^2 \bigm| \beta, \gamma \in \mathbb{R}, \ \gamma \neq 0\}. \end{cases}$$

The irreducible degree-1 polynomials $X - \alpha$ with $0 \neq \alpha \in \mathbb{K}^\diamond$ are related to the operations $\beta \longmapsto \alpha\beta$ with $\mathbb{R}^\diamond = \mathbb{I}(2) \times \mathbf{D}(1)$ and $\mathbb{C}^\diamond = \mathbf{U}(1) \times \mathbf{D}(1)$. The real irreducible degree-2 polynomials are the characteristic polynomials for the $\mathbf{O}(2) \times \mathbf{D}(1)$-mappings

$$\begin{pmatrix} \beta & \gamma \\ -\gamma & \beta \end{pmatrix} = \epsilon(\beta) e^{\lambda} \begin{pmatrix} \cos\varphi & -\sin\varphi \\ \sin\varphi & \cos\varphi \end{pmatrix} \in \mathbf{O}(2) \times \mathbf{D}(1), \quad \begin{cases} e^{\lambda} &= \sqrt{\beta^2 + \gamma^2}, \\ \tan\varphi &= \frac{\gamma}{\beta}, \end{cases}$$

where the imaginary unit is realized in the real as a 2×2 matrix

$$\begin{pmatrix} \cos\varphi & \sin\varphi \\ -\sin\varphi & \cos\varphi \end{pmatrix} = \exp\varphi \begin{pmatrix} 0 & 1 \\ -1 & 0 \end{pmatrix} \text{ with } \begin{pmatrix} 0 & 1 \\ -1 & 0 \end{pmatrix}^2 = -\begin{pmatrix} 1 & 0 \\ 0 & 1 \end{pmatrix}.$$

Some structures of the polynomial ring $\mathbb{K}[X]$ arise also for the principal ideal ring of the integers $\mathbb{Z}$ with the primes $\{2, 3, 5, \dots\}$ as irreducible positive elements and can be illustrated there.

Irreducible polynomials define maximal principal ideals. Any unitary nonconstant polynomial $p(X) \neq \alpha \in \mathbb{K}$ has a unique *extremal decomposition* into powers of unitary, distinct, irreducible polynomials. An analogue is the prime number decomposition of a natural number.

Elements of unital $\mathbb{K}$-algebras $A \in \underline{\mathbf{aag}}_{\mathbb{K}}$ are investigated with the polynomials $\mathbb{K}[X]$: The positive powers of an element $a \in A$ generate its *polynomial ring* $\mathbb{K}[a]_A$, a commutative unital subalgebra

$$A \supseteq \mathbb{K}[a]_A = \{p(a) = \sum_{r=0}^{n} \alpha_r a^r \mid \alpha_r \in \mathbb{K}, \quad a^0 = 1_A\} \in \underline{\mathbf{aag}}_{\mathbb{K}}.$$

The surjective *insertion ring morphism* from the principal ideal ring $\mathbb{K}[X]$ with the polynomials has as kernel an ideal

$$a_X : \mathbb{K}[X] \longrightarrow \mathbb{K}[a]_A, \; p(X) = \sum_r \alpha_r X^r \longmapsto p(a) = \sum_r \alpha_r a^r.$$

This ideal defines for each algebra element a unique generating unitary polynomial

$$\begin{array}{l} \operatorname{kern} a_X = p_a(X)\mathbb{K}[X], \;\; p_a(a) = 0, \;\; \mathbb{K}[a]_A \cong \mathbb{K}[X]/p_a(X)\mathbb{K}[X], \\ A \longrightarrow \mathbb{K}[X], \;\; a \longmapsto p_a(X). \end{array}$$

If the kernel is trivial, $p_a(X) = 0$, $\mathbb{K}[X] \cong \mathbb{K}[a]_A$, the algebra element $a \in A$ is called *transcendental*, a nontrivial $p_a(X)$ is called the *minimal polynomial of* $a \in A$, and the element a *algebraic*. The minimal polynomial divides each polynomial $p(X)$ with $p(a) = 0$. It is the smallest degree polynomial with $p_a(a) = 0$.

One has the following connection between dimension of the polynomial ring $\mathbb{K}[a]_A$ and the algebra:

$$\begin{array}{l} a \text{ transcendental} \iff p_a(X) = 0 \iff \dim_{\mathbb{K}} \mathbb{K}[a]_A = \aleph_0 = \dim_{\mathbb{K}} \mathbb{K}[X], \\ a \text{ algebraic} \iff p_a(a) = 0 \iff \dim_{\mathbb{K}} \mathbb{K}[a]_A = \deg p_a(X) < \aleph_0, \\ \dim_{\mathbb{K}} A < \aleph_0 \Rightarrow \text{ all } a \in A \text{ algebraic}, \;\; \deg p_a(X) \leq \dim_{\mathbb{K}} A. \end{array}$$

The zeros of the minimal polynomial $p_a(X)$ are called *eigenvalues of the algebra element* $a \in A$.

2.17.1 Algebraic Elements

An algebraic element $a \in A \in \underline{\mathbf{aag}}_{\mathbb{K}}$, e.g., an endomorphism (matrix) $a \in \mathbf{AL}(\mathbb{K}^n)$, can be characterized in more detail according to the extremal decomposition of its minimal polynomial. A *simple (irreducible)* algebra element has an irreducible minimal polynomial:

$$A \ni a \text{ simple} \iff p_a(X) = p_0(X) = \begin{cases} X - \alpha_0, & \mathbb{K} = \mathbb{C}, \mathbb{R}, \\ (X - \beta_0)^2 + \gamma_0^2, \quad \gamma_0 \neq 0, & \mathbb{K} = \mathbb{R}. \end{cases}$$

Nonsimple algebraic elements $a \in A$ are called *reducible* if their minimal polynomials are reducible.

A *semisimple* element has a semisimple minimal polynomial, i.e., a product of different irreducible polynomials $p_{0i}(X)$:

$$A \ni a \text{ semisimple} \iff p_a(X) = \prod_{i=1}^{m} p_{0i}(X).$$

An element is called *nondecomposable (monogeneous)* if it is a power of only one irreducible polynomial, a. For a power $N_0 > 1$, it is called *multiple*:

$$A \ni a \text{ nondecomposable} \iff p_a(X) = p_0(X)^{N_0}, \quad N_0 \geq 1.$$

An element with a minimal polynomial having only irreducible factors of first degree, possibly also with higher powers, is called *split*:

$$A \ni a \text{ split} \iff p_a(X) = \prod_{i=1}^{m} (X - \alpha_i)^{N_i}.$$

The power N_i is the *order of the zero* α_i. In the *complex* case, $\mathbb{K} = \mathbb{C}$, all algebraic elements are split; there one has the branched structure

$$\text{simple } (X - \alpha) \Rightarrow \left\{ \begin{matrix} \text{semisimple } \prod_{i=1}^{m} (X - \alpha_i) \\ \text{nondecomposable } (X - \alpha)^N \end{matrix} \right\} \Rightarrow \text{ algebraic } \prod_{i=1}^{m} (X - \alpha_i)^{N_i}.$$

Analogous structures for the natural numbers $\mathbb{N}$ are seen in the follwowing examples: 7 (prime, simple), $66 = 2 \cdot 3 \cdot 11$ (semisimple), $12 = 2^2 \cdot 3$ (decomposable), $81 = 3^4$ (nondecomposable), and the extremal decomposition of any natural number into powers of primes, unique up to ordering.

2.17.2 Projectors and Nilpotents; Jordan Bases

An element $\mathcal{P}$ of a unital algebra $A \in \underline{\mathbf{aag}}_{\mathbb{K}}$ is called *idempotent* for $\mathcal{P}^2 = \mathcal{P}$, if nontrivial, $\mathcal{P}$ is a *projector*. A projector is *decomposable* if it is the sum $\mathcal{P} = \mathcal{P}_1 + \mathcal{P}_2$ of two projectors. With a projector $\mathcal{P} \neq 1$ the unit is not primitive $1 = \mathcal{P} + (1 - \mathcal{P})$. A set of idempotents $\{\mathcal{P}_j\}_j$ with $\mathcal{P}_j\mathcal{P}_l = \delta_{jl}\mathcal{P}_l$ is called *orthogonal* and *complete* for $1 = \sum_j \mathcal{P}_j$. A dual basis defines a complete orthogonal family of nondecomposable projectors $\mathcal{P}_j$:

$$V \cong \mathbb{K}^n : \quad \mathrm{id}_V = \mathbf{1}_V = \sum_j \mathcal{P}_j = \sum_j e^j \otimes \check{e}_j, \quad \mathrm{tr}\, \mathcal{P}_j = 1.$$

Any idempotent defines an *involutor* $\mathcal{I}^2 = 1_A$ by $\mathcal{I} = 1_A - 2\mathcal{P}$ and vice versa. An algebra element with $\mathcal{N}^K = 0$, but $\mathcal{N}^{K-1} \neq 0$, $K \geq 1$, is *nilpotent with power* K. 0 is nilpotent with power 1. The associate minimal polynomials are

$$\begin{array}{rl} \alpha \in \mathbb{K} & \Rightarrow p_{\alpha 1_A}(X) = X - \alpha, \\ \mathcal{P} \neq 1 \text{ projector} & \Rightarrow p_{\mathcal{P}}(X) = X(X-1), \\ \mathcal{I} \neq 1 \text{ involutor} & \Rightarrow p_{\mathcal{I}}(X) = (X+1)(X-1), \\ \mathcal{N} \text{ nilpotent with power } K & \Rightarrow p_{\alpha 1_A + \mathcal{N}}(X) = (X-\alpha)^K. \end{array}$$

An algebraic element $a \in A$ is decomposable into a sum characterized by projectors and nilpotents: A polynomial $p(X) \in p_a(X)\mathbb{K}[X]$, i.e., $p(a) = 0$, has the extremal decomposition with irreducible polynomials

$$p(X) = \prod_{i=1}^{m} p_i(X)^{N_i}.$$

Since the "cofactors" $q_j(X)$,

$$q_j(X) = \frac{p(X)}{p_j(X)^{N_j}} = \prod_{i \neq j} p_i(X)^{N_i} \in \mathbb{K}[X], \quad j = 1, \dots, m,$$

have as their greatest common divisor a number, there exists a *Bézout decomposition* of 1 with polynomials $h_j(X) \in \mathbb{K}[X]$,

$$1 = \sum_{i=1}^{m} h_i(X) q_i(X).$$

The insertion morphism $\mathbb{K}[X] \longrightarrow \mathbb{K}[a]_A$ defines a complete orthogonal family with idempotents $\{\mathcal{P}_i(a)\}_{i=1}^m$ in the algebra A:

$$1_A = \sum_{i=1}^{m} \mathcal{P}_i(a), \ \mathcal{P}_i(a) = h_i(a) q_i(a) \left\{ \begin{array}{l} i \neq j: \ p(X) \text{ divides } q_i(X) q_j(X) \\ \Rightarrow q_i(a) \ q_j(a) = 0 \Rightarrow \mathcal{P}_i(a)\mathcal{P}_j(a) = 0 \\ \Rightarrow i = j: \ \mathcal{P}_i(a)\mathcal{P}_j(a) = \mathcal{P}_i(a). \end{array} \right.$$

The analogue for the numbers: Primes have a basic structure for the integers: An example: The prime factors $63 = 3^2 \cdot 7$ give a Bézout decomposition of the unit $1 = -2 \cdot 7 + 5 \cdot 3$.

Starting from the minimal polynomial $p_a(X)$, no idempotent can be trivial: If $\mathcal{P}_j(a) = 0$, then $1_A = \sum_{i \neq j} \mathcal{P}_i(a)$, and, hence, $q_j(a) = \sum_{i \neq j} h_i(a) \circ q_i(a) \ q_j(a) = 0$, but $\deg q_j(X) < \deg p_a(X)$. Summarizing: Any algebraic element $a \in A$ defines for each polynomial with $p(a) = 0$ a *complete orthogonal family*, for the minimal polynomial $p_a(X)$ even a *complete projector family.*

Three examples:

$$\begin{array}{rl} a = \alpha \in \mathbb{K}, \ p_a = X - \alpha & \Rightarrow q(X) = 1, \quad h(X) = 1, \quad \mathcal{P}(a) = 1_A, \\ a \text{ nilpotent}, \ p_a = X^K & \Rightarrow q(X) = 1, \quad h(X) = 1, \quad \mathcal{P}(a) = 1_A, \\ a \text{ projector}, \ p_a = X(X-1) & \Rightarrow \left\{ \begin{array}{l} q_1(X) = X, \quad q_2(X) = X - 1, \\ h_1(X) = 1, \quad h_2(X) = -1, \\ \mathcal{P}_1(a) = a, \quad \mathcal{P}_2(a) = 1_A - a. \end{array} \right. \end{array}$$

For a *split element* $a \in A$ with minimal polynomial $p_a(X) = \prod_{i=1}^{m}(X - \alpha_i)^{N_i}$ one constructs first the complete projector family $\mathcal{P}_i(a) \in \mathbb{K}[a]_A$. For $N_i \geq 2$ that is not enough: The difference $a - \alpha_i 1_A$ with the a-eigenvalue α_i multiplied by the corresponding projector defines an *element* $\mathcal{N}_i(a) \in \mathbb{K}[a]_A$, *nilpotent with power* N_i:

$$\mathcal{N}_i(a) = (a - \alpha_i 1_A)\, \mathcal{P}_i(a) \Rightarrow \begin{cases} \mathcal{P}_i(a)\, \mathcal{N}_j(a) = \delta_{ij}\mathcal{N}_j(a), \\ \mathcal{N}_i(a)\, \mathcal{N}_j(a) = \delta_{ij}\mathcal{N}_j(a)^2, \\ \mathcal{N}_i(a)^K = (a - \alpha_i 1_A)^K\, \mathcal{P}_i(a), \quad K \in \mathbb{N}, \\ \mathcal{N}_i(a)^{N_i} = (a - \alpha_i 1_A)^{N_i}\, h_i(a) q_i(a) \\ \qquad\qquad = h_i(a) p_a(a) = 0. \end{cases}$$

Therefore, any split element a can be decomposed uniquely into a projector part (semisimple) and a nilpotent element, its *spectral decomposition*:

$$a = \sum_{i=1}^{m} a\, \mathcal{P}_i(a) = h(a) + n(a), \quad \begin{cases} h(a) = \sum_{i=1}^{m} \alpha_i \mathcal{P}_i(a) \text{ semisimple}, \\ p_{h(a)} = \prod_{i=1}^{m}(X - \alpha_i), \\ n(a) = \sum_{i=1}^{m} \mathcal{N}_i(a) \text{ nilpotent}, \\ p_{n(a)} = X^{\max_i\{N_i - 1\}}. \end{cases}$$

All projectors and nontrivial powers of the nilpotents constitute a *Jordan basis* of the polynomial algebra whose elements, obviously, are a-polynomials:

$$\text{basis of } \mathbb{K}[a]_A \cong \mathbb{K}^N : \{\mathcal{P}_i(a), \mathcal{N}_i(a)^{K_i} \bigm| i = 1, \ldots, m; K_i = 1, \ldots, N_i - 1\}.$$

Semisimple elements a have a Jordan basis consisting only of projectors:

$$p_a(X) = \prod_{i=1}^{m}(X - \alpha_i), \quad q_j(X) = \prod_{i \neq j}(X - \alpha_i), \quad q_j(\alpha_k) = \delta_{jk} q_j(\alpha_j),$$

$$h_j(X) = \tfrac{1}{q_j(\alpha_j)} \in \mathbb{K}, \quad \sum_{i=1}^{m} h_i q_i(X) = 1, \text{ since} \begin{cases} \deg \sum_{i=1}^{m} h_i q_i(X) \leq m - 1, \\ \sum_{i=1}^{m} h_i q_i(\alpha_j) = 1, \end{cases}$$

$$\Rightarrow \mathcal{P}_j(a) = \tfrac{q_j(a)}{q_j(\alpha_j)}.$$

2.17.3 Exponential and Logarithm

In the following $A \cong \mathbb{C}^n$ is a unital algebra. Each $a \in A$ has a minimal polynomial and an associate Jordan basis

$$p_a(X) = \prod_{j=1}^{m}(X - \alpha_j)^{N_j}, \quad a = \sum_{j=1}^{m}[\alpha_j \mathcal{P}_j(a) + \mathcal{N}_j(a)].$$

An element $a \in A$ is regular iff all zeros α_j are nontrivial:

$$
\begin{aligned}
&a \in A^\diamond \iff \text{all } \alpha_j \neq 0,\\
&a = \sum_{j=1}^{m} \alpha_j \mathcal{P}_j(a)(1_A + \tfrac{\mathcal{N}_j(a)}{\alpha_j}) \Rightarrow a^{-1} = \sum_{j=1}^{m} \tfrac{1}{\alpha_j}\Big[\mathcal{P}_j(a) + \sum_{K_j=1}^{N_j-1} \Big(\tfrac{-\mathcal{N}_j(a)}{\alpha_j}\Big)^{K_j}\Big],\\
&\text{since } \mathcal{N}^N = 0 \Rightarrow (1_A + \mathcal{N})^{-1} = 1_A + \sum_{K=1}^{N-1} (-\mathcal{N})^K.
\end{aligned}
$$

The *exponential mapping* goes into the regular group $A^\diamond$:

$$
\exp : A \longrightarrow A^\diamond, \quad \begin{cases} \exp a = e^a = \sum_{K\geq 0} \frac{a^K}{K!},\\ \mathcal{N}^N = 0 \Rightarrow e^{\mathcal{N}} = 1_A + \sum_{K=1}^{N-1} \frac{\mathcal{N}^K}{K!},\\ e^{\pm a} = \sum_{j=1}^{m} e^{\pm\alpha_j}[\mathcal{P}_j(a) + \sum_{K_j=1}^{N_j-1} \frac{\mathcal{N}_j(a)^{K_j}}{K_j!}], \end{cases}
$$
$$
e^a e^{-a} = 1_A.
$$

The *logarithm mapping* is defined for regular elements by the formal series of $\log a = \log(1_A + a - 1_A)$:

$$
\log : A^\diamond \longrightarrow A/\log 1_A, \quad \begin{cases} \log a = -\sum_{K\geq 1} \frac{(1_A - a)^K}{K},\\ \mathcal{N}^N = 0 \Rightarrow \log(1_A + \mathcal{N}) = -\sum_{K=1}^{N-1} \frac{(-\mathcal{N})^K}{K}. \end{cases}
$$

Here $A/\log 1_A$ is the additive group that arises from A by the classes modulo $\log 1_A = \{n \in A \mid \exp n = 1_A\}$:

$$
\begin{aligned}
\log a &= \sum_{j=1}^{m} \Big[(\log \alpha_j)\mathcal{P}_j(a) - \sum_{K_j=1}^{N_j-1} \frac{(-\mathcal{N}_j(a))^{K_j}}{K_j}\Big] + \log 1_A,\\
\alpha_j &= |\alpha_j| e^{i\varphi_j}, \quad \log \alpha_j = \log|\alpha_j| + i\varphi_j + \log 1_A.
\end{aligned}
$$

Taking into account these classes, the exponential and logarithm mappings are inverse to each other:

$$
A \in \underline{\mathbf{aag}}_{\mathbb{C}}, \quad A \cong \mathbb{C}^n \Rightarrow A^\diamond = e^A, \quad \begin{cases} e^{\log a} = a,\\ \log e^a = a + \log 1_A. \end{cases}
$$

As an example, the regular group of a complex finite-dimensional endomorphism algebra is the exponential:

$$
V \cong \mathbb{C}^m \Rightarrow \begin{cases} \mathbf{GL}(\mathbb{C}^m) = \exp \mathbf{AL}(\mathbb{C}^m),\\ \mathbf{AL}(\mathbb{C}^m) = \log \mathbf{GL}(\mathbb{C}^m). \end{cases}
$$

2.18 The Hausdorff Product

In an *exponentially closed* algebra $A \in \underline{\mathbf{aag}}_{\mathbb{K}}$, i.e., each element has an exponent in the regular group of the algebra, e.g., the matrices $\mathbf{AL}(\mathbb{K}^n)$,

$$\exp : A \longrightarrow A^\diamond, \quad \exp a = \sum_{k\geq 0} \tfrac{a^k}{k!},$$

the *Hausdorff group product* $+\!\!+$, in general nonabelian, is defined up to $\log 1_A$ via the product of two exponents $\exp a \exp b = \exp(a +\!\!+ b)$:

$$\begin{aligned} &A/\log 1_A \times A/\log 1_A \longrightarrow A/\log 1_A, \quad a +\!\!+ b = \log(\exp a \exp b),\\ &(a +\!\!+ b) +\!\!+ c = a +\!\!+ (b +\!\!+ c), \quad a +\!\!+ 0 = 0 +\!\!+ a = 0, \quad a +\!\!+ (-a) = 0,\\ &a +\!\!+ b = a + b \iff [a, b] = 0. \end{aligned}$$

The neutral and inverse element are as in the additive group $a + b$. With the natural Lie algebra structure of A, the Hausdorff product can be expanded into "bracket polynomials" of fixed degree:

$$a +\!\!+ b = a + b + \tfrac{[a,b]}{2} + \tfrac{[a,[a,b]]+[b,[b,a]]}{2\cdot 3!} + \tfrac{[a,[[a,b],b]]}{4!} + \cdots .$$

It can be defined for a Lie algebra where the series makes sense. For a nilpotent Lie algebra structure, i.e., $[A, A] = \{0\}$ or $[A, [A, A]] = \{0\}$, etc., the series ends correspondingly.

2.19 (Semi)Simple and Decomposable Endomorphisms

A nontrivial vector space V with an endomorphism set S, e.g., one endomorphism $S = \{f\}$ or a represented group $S = D[G]$ or Lie algebra $S = \mathcal{D}[L]$, without a *proper* S-invariant vector subspace, i.e.,

$$U \subseteq V, \quad S.U \subseteq U \Rightarrow U = \{0\} \text{ or } U = V,$$

is called *S-irreducible (S-simple)* and *S-reducible* otherwise. For a represented group G or Lie algebra L one says G-irreducible or L-irreducible, etc.

For irreducible inequivalent representations on V, W, the intertwining transformations are trivial $\mathbf{mod}_L(V, W) = \{0\} = \mathbf{mod}_G(V, W)$.

The trivial representations on a 1-dimensional $\mathbb{K}$-isomorphic vector space $G \longrightarrow \{1\} \subset \mathbf{GL}(\mathbb{K})$ and $L \longrightarrow \{0\} \subset \mathbf{AL}(\mathbb{K})$ are irreducible.

With an S-invariant subspace $S.U \subseteq U \subseteq V$, the restriction $D[G]|_U$ is called a *subrepresentation*. Given a G-representation on a vector space U, a G-representation on a larger space $V \supseteq U$ is called a *suprepresentation*.

With a subrepresentation the action by the endomorphism S is well defined on the quotient space

$$S \times V/U \longrightarrow V/U, \quad s(v + U) = s(v) + U;$$

$D[G]|_{V/U}$ is called a *quotient representation*.

In the case of a *finite-dimensional* S-reducible vector space V with invariant subspace U, there exist V-decompositions with *block triagonal form (block Jordan form)* and *nilpotent* linear mappings $n(s)$:

$$\begin{array}{ll} V \cong \mathbb{K}^n, & S.U \subseteq U,\ V = U \oplus U',\ U' \cong V/U, \ \text{ in general } S.U' \not\subseteq U', \\ s \in S \Rightarrow & s = s_0 + n(s) \simeq \begin{pmatrix} s|_{U'} & n(s) \\ \mathbf{0} & s|_U \end{pmatrix}. \end{array}$$

An S-reducible vector space that allows even a decomposition into proper, invariant vector subspaces

$$V = U_1 \oplus U_2, \quad S.U_{1,2} \subseteq U_{1,2}, \quad S = S|_{U_1} \oplus S|_{U_2},$$

is called *S-decomposable*, and *S-nondecomposable* otherwise. If V is S-decomposable there exists projectors

$$\begin{array}{l} \mathrm{id}_V = \bigoplus\limits_{j\in I} \mathcal{P}_j, \quad V = \bigoplus\limits_j U_j, \quad s \in S,\ s.U_j \subseteq U_j \Rightarrow s \circ \mathcal{P}_j = \mathcal{P}_j \circ s \circ \mathcal{P}_j \\ \Rightarrow s = \bigoplus\limits_j \mathcal{P}_j \circ s \circ \bigoplus\limits_k \mathcal{P}_k = \bigoplus\limits_j \mathcal{P}_j \circ s \circ \mathcal{P}_j = \bigoplus\limits_j s_j. \end{array}$$

The linear extension of irreducibility for orbits G/H of a group acting on a set looks as follows: Any vector v in a space V with G-action defines by its orbit $G \bullet v \cong G/G_v$ a generating system whose span (and, for topological vector spaces, the closure) is the *cyclic space generated by* v, a vector subspace invariant under G-action:

$$G \bullet v \subseteq V \in \underline{\mathbf{mod}}_G \Rightarrow V \supseteq \overline{\mathbb{K}^{(G\bullet v)}} \in \underline{\mathbf{mod}}_G.$$

A representation with a cyclic vector $\mathbb{K}^{(G\bullet v)} = V$ is called cyclic:

$$\text{irreducible} \Rightarrow \left\{ \begin{array}{l} \text{cyclic,} \\ \text{nondecomposable.} \end{array} \right.$$

Cyclic spaces can be decomposable.

V is called an *S-semisimple (completely decomposable)* vector space for $V = \{0\}$, and for $V \neq \{0\}$, if V is decomposable into S-irreducible (S-simple) vector subspaces $V = \bigoplus\limits_{j\in I} U_j$. Exactly for this case, each invariant subspace has an invariant complement. If all simple subspaces are isomorphic to each other, $U_j \cong U$, the vector space is called *U-isotypical with respect to S.*

A *set* $S \subseteq \mathbf{AL}(V)$ generates a minimal endomorphism *subalgebra* $A(S)$ with unit id_V. With respect to the set and the generated algebra of endomorphisms the vector space has analogous properties,

$$\left. \begin{array}{c} V \text{ is } S\text{-(ir)reducible} \\ \text{or } S\text{-(semi)simple} \end{array} \right\} \iff \left\{ \begin{array}{c} V \text{ is } A(S)\text{-(ir)reducible} \\ \text{or } A(S)\text{-(semi)simple,} \end{array} \right.$$

V is called an *$A(S)$-module*, $V \in \underline{\mathbf{mod}}_{A(S)}$.

The tensor product $V_1 \otimes_{\mathbb{K}} V_2 \in \mathbf{\underline{mod}}_A$ (below) of two nondecomposable A-modules $V_{1,2}$ (with the A-action defined via the product representations for groups or Lie algebras) can be decomposable. If possible, a decomposition into A-invariant subspaces U_j (A-submodules) is called a *Clebsch-Gordan decomposition*:

$$\begin{array}{l} V_1 \otimes V_2 = \bigoplus_j U_j, \quad A \bullet U_j \subseteq U_j, \quad D^{V_1 \otimes V_2}(a) = \bigoplus_j D^{U_j}(a), \\ \mathcal{P}_j^{12} : V_1 \otimes V_2 \longrightarrow U_j. \end{array}$$

The *Clebsch-Gordan projectors* $\mathcal{P}_j^{12}$ build the decomposition isomorphism $\bigoplus_j \mathcal{P}_j^{12}$ of the product $V_1 \otimes V_2$. The matrix elements of $\mathcal{P}_j^{12}$ with respect to a basis are *Clebsch-Gordan coefficients*:

$$\begin{array}{l} \text{bases of } V_1, V_2, U_j : \ \{e_{(1)}^{m_1}\}_{m_1}, \ \{e_{(2)}^{m_2}\}_{m_2}, \ \{e_{(j)}^{m}\}_{m}, \\ \mathcal{P}_j^{12} : V_1 \otimes V_2 \longrightarrow U_j : \ \mathcal{P}_j^{12} \simeq \langle 1m_1, 2m_2 || jm \rangle. \end{array}$$

With dual (V, V^T)-bases, the coefficients of the Clebsch-Gordan projection $V \otimes V^T \longrightarrow \mathbb{K}\,\mathrm{id}_V$ to the subspace with trivial representation are δ_A^B. The representation coefficients $\mathcal{D}_A^{jB}$ of a Lie algebra L are the coefficients of the Clebsch-Gordan projection $V \otimes V^T \longrightarrow \mathcal{D}[L]$ on the L-image.

With an invariant dual isomorphism $\gamma : V \longrightarrow V^T$ (self-dual representation) one can go from $V \otimes V^T$ to $V \otimes V$ or to $V^T \otimes V^T$. For example, γ_{AB} and γ^{AB} are the coefficients for the projection on the 1-dimensional subspace in $V \otimes V$ and $V^T \otimes V^T$ respectively with trivial representation.

2.20 Representations of Compact (Finite) Groups

For compact groups reducibility entails decomposability.

With the discrete topology a *finite group* G is compact. Any reducible, finite-dimensional representation D is decomposable (semisimple) (*theorem of Maschke*): D has with representations D^U and $D^{U'}$, $U' \cong V/U$, a block-triagonal form:

$$\begin{array}{rl} D \text{ reducible} & \Rightarrow D(g) \simeq \begin{pmatrix} D^U(g) & N(g) \\ 0 & D^{U'}(g) \end{pmatrix}, \\ D^{U,U'} \text{ representations:} & \left\{ \begin{array}{ll} D^U(gh) & = D^U(g) \circ D^U(h), \\ D^{U'}(gh) & = D^{U'}(g) \circ D^{U'}(h), \end{array} \right. \\ \text{and} & N(gh) = D^U(g) \circ N(h) + N(g) \circ D^{U'}(h). \end{array}$$

D is even equivalent to a block-diagonal form by an automorphism f, which trivializes the nondiagonal part $N(g)$. f can be constructed explicitly:

$$\begin{array}{l} \text{ansatz: } f \simeq \begin{pmatrix} \mathrm{id}_U & B \\ \mathbf{0} & \mathrm{id}_{U'} \end{pmatrix} \Rightarrow f \circ D(g) \circ f^{-1} \simeq \begin{pmatrix} D^U(g) & \tilde{N}(g) \\ 0 & D^{U'}(g) \end{pmatrix}, \\ \text{with } \tilde{N}(g) = N(g) + B \circ D^{U'}(g) - D^U(g) \circ B, \\ \text{therefore } \tilde{N}(g) = 0 \iff -B \circ D^U(g) = -D^U(g) \circ B + N(g). \end{array}$$

The property of $N(gh)$ above yields the mapping $B : U \longrightarrow U'$ looked for after summation over all group elements:

$$\begin{aligned}
&\sum_{h\in G} N(gh)D^{U'}(h^{-1}) = D^U(g)\sum_{h\in G} N(h)D^{U'}(h^{-1}) + \operatorname{card} G N(g)\\
&\Rightarrow \tilde{N}(g) = 0 \text{ with } B = -\tfrac{1}{\operatorname{card} G}\sum_{h\in G} N(h)D^{U'}(h^{-1}).
\end{aligned}$$

All complex representations of compact groups are semisimple and decomposable into irreducible Hilbert representations that are all finite-dimensional (*theorem of Weyl*).

More general also for noncompact groups (chapter "Harmonic Analysis"): Hilbert representations of locally compact groups, e.g., real finite-dimensional Lie groups, are decomposable into a direct sum of cyclic ones and into a direct integral of irreducible ones.

2.21 Algebra Representations and Modules

A unital $\mathbb{K}$-algebra, i.e., $\mathbb{K}1_A \subseteq A$, is represented as an *operator algebra* in the endomorphism algebra of a $\mathbb{K}$-vector space V by a unital algebra morphism. A acts on V:

$$A \times V \longrightarrow V, \quad (a, v) \longmapsto a \bullet v.$$

Therefore V is called an *A-module*, $V \in \underline{\mathbf{mod}}_A$. For $A = \mathbb{K}$ with $\mathbb{K}$ represented by $\mathbb{K}\,\mathrm{id}_V$, the A-module V is simply the $\mathbb{K}$-vector space V.

A-invariant, (ir)reducible, etc., are meant with respect to the acting algebra A. An A-invariant vector subspace is an A-submodule. V is called *A-monogeneous* if there exists a *cyclic vector* $v \in V$, i.e., $V = A \bullet v$. For an irreducible A-module, i.e., without a proper A-submodule, each nontrivial vector is cyclic.

The A-modules constitute the category $\underline{\mathbf{mod}}_A$: *$A$-module morphisms* $f \in \mathbf{mod}_A(V_1, V_2)$ are vector space morphisms compatible with the algebra action (A-intertwiners as A-invariants in the linear mappings):

$$\begin{array}{ccc}
V_1 & \xrightarrow{a\bullet} & V_1\\
{\scriptstyle f}\downarrow & & \downarrow{\scriptstyle f}\\
V_2 & \underset{a\bullet}{\longrightarrow} & V_2
\end{array}, \quad
\begin{array}{l}
f(a \bullet v) = a \bullet f(v) \text{ for all } a \in A,\ v \in V_1,\\
f \in \mathbf{mod}_A(V_1, V_2) = \mathrm{INV}_A \mathbf{vec}_{\mathbb{K}}(V_1, V_2).
\end{array}$$

Images and inverse images of A-submodules and images of direct sums of A-submodules remain A-submodules, e.g., the image of a semisimple A-module is semisimple.

Therefore, irreducibility restricts strongly the A-intertwiners:

$$f \in \mathbf{mod}_A(V_1, V_2) : \left\{ \begin{array}{l}
A\text{-irreducible } V_2 \Rightarrow f = 0 \text{ or } f \text{ surjective},\\
A\text{-irreducible } V_1 \Rightarrow f = 0 \text{ or } f \text{ injective},\\
A\text{-irreducible } V_1 \text{ and } V_2 \Rightarrow \left\{ \begin{array}{l} f = 0 \text{ or}\\ f \text{ isomorphism}. \end{array} \right.
\end{array} \right.$$

The intertwining algebra for an A-irreducible vector space contains up to the trivial endomorphism only automorphisms; it is a possibly noncommutative field:

$$f \in \mathbf{mod}_A(V,V) = \mathbf{AL}_A(V): \ A\text{-irreducible } V \Rightarrow \begin{cases} f = 0 \text{ or} \\ f \text{ automorphism.} \end{cases}$$

The *Schur's lemma* characterizes the intertwining algebra (commutant) of a *complex finite-dimensional nondecomposable A-module V*: The minimal polynomial $p_f(X)$ of an A-intertwiner $f = \sum_{j=1}^{m} [\alpha_j \mathcal{P}_j(f) + \mathcal{N}_j(f)]$ has exactly one zero α_1, otherwise, $\bigoplus_{j=1}^{m} \mathcal{P}_j(f)[V]$ with $m \geq 2$ would give a V-decomposition. If V is *A-irreducible*, the nilpotent part is trivial:

$$V \cong \mathbb{C}^D \text{ is } \begin{cases} A\text{-nondecomposable} \Rightarrow \begin{cases} p_f(X) = (X - \alpha_1)^N, \ N \leq D, \\ f = \alpha_1 \operatorname{id}_V + \mathcal{N}_1(f) \cong \alpha_1 \mathbf{1}_D + \begin{pmatrix} 0 & \times \\ 0 & 0 \end{pmatrix}, \\ \dim_{\mathbb{C}} \mathbf{AL}_A(\mathbb{C}^D) \leq D, \end{cases} \\ \\ A\text{-irreducible} \Rightarrow \begin{cases} p_f(X) = X - \alpha_1, \\ f = \alpha_1 \operatorname{id}_V, \\ \mathbf{AL}_A(\mathbb{C}^D) \cong \mathbb{C}. \end{cases} \end{cases}$$

For an irreducible real representation space V, $\mathbf{AL}_A(\mathbb{R}^D)$ is isomorphic to $\mathbb{R}$ or to $\mathbb{C} \cong \mathbb{R}^2$ or to the quaternions $\mathbb{H} \cong \mathbb{R}^4$ (chapter "Quantum Algebras").

If the algebra A is *abelian*, the representation consists of intertwiners :

$$\text{abelian } D[A] \subseteq \mathbf{AL}_A(V) \Rightarrow D(b) = \alpha(b) \operatorname{id}_V + n(b) \text{ for all } b \in A,$$

where $\alpha(b) \in \mathbb{C}$ and the nilpotent endomorphism $n(b)$ have the corresponding representation properties. Therefore, any irreducible complex finite-dimensional representation of an abelian unital algebra A, of an abelian group G, and of an abelian Lie algebra L has to be 1-dimensional, $\mathbf{AL}_{A,G,L}(\mathbb{C}^D) \cong \mathbb{C}$.

All nondecomposable complex representations of an abelian Lie algebra are obtained from all commuting representations of all bases, i.e., they can be built with the nondecomposable representations of the Lie algebra $L \cong \mathbb{K}$:

$$L \cong \mathbb{K}l \in \underline{\mathbf{lag}}_{\mathbb{K}} \Rightarrow \begin{cases} \mathcal{D}_N(l) = \alpha \operatorname{id}_V + \mathcal{N}, \\ \alpha \in \mathbb{C}, \ \mathcal{N}^{N-1} \neq 0, \ \mathcal{N}^N = 0. \end{cases}$$

The left ideal orbit of a vector of an A-module is an A-submodule:

$$v \in V \in \underline{\mathbf{mod}}_A, \quad L \subseteq A \Rightarrow L \bullet v \in \underline{\mathbf{mod}}_A.$$

For irreducible V the orbits $L \bullet v$ have to be trivial or the full space V. Therefore the irreducible A-modules (A-representations) are isomorphic to minimal left ideals

$$\mathbf{irrep}\, A \cong \{L^{\min} \subseteq A\}.$$

For a group G, the left and right ideals correspond to the left and right classes (cosets) G/H and $H\backslash G$ with subgroups.

The *left multiplications (translations)* in a unital algebra A define the *left regular (adjoint) A-representation.* The right regular contrarepresentation is given by the right multiplications:

$$\begin{array}{ll} L : A \longrightarrow \mathbf{AL}(A), & a \longmapsto L_a : A \longmapsto A, \quad L_a(b) = ab, \\ & L_a \circ L_c = L_{ac}, \\ \check{R} : A \longrightarrow \mathbf{AL}(A), & a \longmapsto \check{R}_a : A \longmapsto A, \quad \check{R}_a(b) = ba, \\ & \check{R}_a \circ \check{R}_c = \check{R}_{ca}. \end{array}$$

The (minimal) left ideals $AL = L \subseteq A$ are the A-invariant (irreducible) subspaces of A under left A-multiplication. The intersection $L_1 \cap L_2$ and the (direct) sum $L_1 + L_2$ ($L_1 \oplus L_2$) of two left ideals is a left ideal. Two different minimal left ideals have the intersection $\{0\}$.

In an $n \times n$ matrix algebra $M \cong \mathbb{K}^{n^2}$ the columns are minimal left ideals, the rows minimal right ideals:

$$\begin{array}{l} L_k = \{\sum_{j=1}^{n} \alpha_j e^j \otimes \check{e}_k \mid k = 1, \ldots, N\} \cong \mathbb{K}^n, \quad M \circ L_k = L_k, \\ \qquad \begin{pmatrix} \times & \times \\ \times & \times \end{pmatrix} \begin{pmatrix} \times & 0 \\ \times & 0 \end{pmatrix} \subseteq \begin{pmatrix} \times & 0 \\ \times & 0 \end{pmatrix}, \\ R^j = \{\sum_{k=1}^{n} \alpha_k e^j \otimes \check{e}_k \mid j = 1, \ldots, N\} \cong \mathbb{K}^n, \quad R^j \circ M = R^j, \\ \qquad \begin{pmatrix} \times & \times \\ 0 & 0 \end{pmatrix} \begin{pmatrix} \times & \times \\ \times & \times \end{pmatrix} \subseteq \begin{pmatrix} \times & \times \\ 0 & 0 \end{pmatrix}. \end{array}$$

With the orthogonal projectors of an algebraic element a in an algebra A each A-module is decomposable into a-invariant subspaces. The polynomial algebra (ring) $\mathbb{K}[a]_A$ is decomposable into ideals:

$$\begin{array}{l} 1_A = \bigoplus_{i=1}^{m} \mathcal{P}_i(a), \quad V = \bigoplus_{i=1}^{m} V_i, \quad V_i = \mathcal{P}_i(a) \bullet V, \\ \mathbb{K}[a]_A = \bigoplus_{i=1}^{m} \mathbb{K}_i[a], \quad \mathbb{K}_i[a] = \mathcal{P}_i(a)\mathbb{K}[a]. \end{array}$$

A split element in the left regular representation has a matrix with $N_i \times N_i$ block diagonal matrices for the powers of first degree polynomials:

$$\begin{array}{rl} p_a(X) = \prod_{i=1}^{m} (X - \alpha_i)^{N_i}, & \mathbb{K}_i[a] \cong \mathbb{K}^{N_i}, \\ \text{Jordan basis of } \mathbb{K}_i[a]: & \{\mathcal{P}_i(a), \mathcal{N}_i(a)^K \mid K = 1, \ldots, N_i - 1\}, \\ \text{left regular representation:} & L_a = \sum_{i=1}^{m} [\alpha_i \mathcal{P}_i(a) + \mathcal{N}_i(a)], \\ & \mathcal{P}_i(a) = \mathbf{1}_{N_i}, \quad \mathcal{N}_i(a) \simeq \begin{pmatrix} 0 & 1 & 0 & \ldots & 0 & 0 \\ 0 & 0 & 1 & \ldots & 0 & 0 \\ & \ldots & & \ldots & & \\ 0 & 0 & 0 & \ldots & 0 & 1 \\ 0 & 0 & 0 & \ldots & 0 & 0 \end{pmatrix}. \end{array}$$

2.21.1 Group Algebra

Any group defines its *group algebra* with the covariant free functor (linear extension functor)

$$\mathbb{K}^{(\ )} : \underline{\mathbf{grp}} \longrightarrow \underline{\mathbf{aag}_{\mathbb{K}}},\ G \longmapsto \mathbb{K}^{(G)} = \{\alpha : G \longrightarrow \mathbb{K} \,\big|\, \text{finite support}\}.$$

The group algebra $\mathbb{K}^{(G)}$ (for finite groups also denoted by $\mathbb{K}^{G}$) contains the $\mathbb{K}$-valued functions on the group with finite support. It has the group elements as *canonical basis*:

$$\mathbb{K}^{(G)} \cong \bigoplus_{g\in G} \mathbb{K}g \ni \sum_{g\in G,\text{finite}} \alpha(g)g, \quad \dim_{\mathbb{K}} \mathbb{K}^{(G)} = \operatorname{card} G.$$

The vector space structure of the group algebra is pointwise inherited from the scalars $\mathbb{K}$; the multiplication uses the group product in the $\mathbb{K}$-linear combination of groups elements, in this context called *convolution product*. The *monoid algebra* $\mathbb{K}^{(M)}$ is universal: Any representation D of the monoid M in a unital algebra A can be factorized by the canonical injection ι and a unital algebra representation $\tilde{D}$:

$$\mathbb{K}^{(\ )} : \underline{\mathbf{mon}} \longrightarrow \underline{\mathbf{aag}_{\mathbb{K}}} \text{ (unital)},\ M \longmapsto \mathbb{K}^{(M)},$$

$$M, \iota, D \in \underline{\mathbf{mon}}, \quad \begin{array}{ccc} M & \stackrel{\iota}{\longrightarrow} & \mathbb{K}^{(M)} \\ {\scriptstyle D}\downarrow & & \downarrow{\scriptstyle \tilde{D}} \\ A & \underset{\mathrm{id}_A}{\longrightarrow} & A \end{array}, \quad \mathbb{K}^{(M)}, A, \tilde{D} \in \underline{\mathbf{aag}_{\mathbb{K}}}.$$

The isomorphic *left and right regular representations of a group* G and the *two-sided regular representation of the doubled group* $G \times G$ act on the group algebra $\mathbb{K}^{(G)}$:

$$G \times G \times \mathbb{K}^{(G)} \longrightarrow \mathbb{K}^{(G)},\ (k_1, k_2, a) \longmapsto L_{k_1} \circ R_{k_2}(a) = k_1 a k_2^{-1}.$$

For a finite group with order n, the multiplication table can be used to find the nonsingular $(n \times n)$ matrices for L_k and R_k that exhibit the permutations of the group elements: In permutation matrices each column and each row contains in addition to 0's exactly one 1.

The fixgroup of the doubled group action is, as seen at the neutral element, the diagonal group $(G \times G)_{\Delta} \cong G$ that defines the *inner automorphism*

$$G \times \mathbb{K}^{(G)} \longrightarrow \mathbb{K}^{(G)},\ (k, a) \longmapsto \begin{cases} L_k(a) &= ka, \\ R_k(a) &= ak^{-1}, \\ \operatorname{Ad} k(a) &= kak^{-1}. \end{cases}$$

The group algebra is especially important for finite groups; for continuous groups there are several different generalizations (chapter "Harmonic Analysis").

2.21.2 (Semi)Simple Associative Algebras

A unital algebra is called *semisimple* if it is semisimple as an A-module, i.e., with respect to left multiplication. Then each left ideal can be complemented, i.e., A is the direct sum of minimal nontrivial left ideals, $A = \bigoplus_{j\in J} L_j^{\min}$. "Left" can be exchanged, equivalently, with "right." A is semisimple iff each A-module is A-semisimple, i.e., isomorphic to a direct sum of minimal nontrivial left ideals of the algebra.

A full matrix algebra $M = \mathbb{K}^{n^2}$ is semisimple as the direct sum of its columns (rows), $M = \bigoplus_{k=1}^{n} L_k^{\min}$. Columns and rows correspond to the minimal left and right ideals. Maschke's theorem implies that the group algebra $\mathbb{K}^G \cong \mathbb{K}^{\operatorname{card} G}$ of a finite group is semisimple.

Each nontrivial minimal left ideal $L_j^{\min}$ of a semisimple algebra is monogeneous with a projector $L_j^{\min} = A\mathcal{P}_j$. The orthogonal projectors $\{\mathcal{P}_j\}_{j\in J}$ can be taken from a decomposition of the unit $1_A \in A$, leading to left ideals $L_j^{\min} L_k^{\min} = \delta_{jk} L_k^{\min}$.

In a semisimple algebra A the direct sum of isomorphic minimal nontrivial left ideals is a minimal two-sided ideal $A_z^{\min}$, a maximal isotypical direct summand. A semisimple algebra is the direct sum of minimal ideals; there exists only a *finite* number of them:

$$A = \bigoplus_{j\in J} L_j^{\min} = \bigoplus_{z=1}^{n} A_z^{\min}, \quad A_z^{\min} = \bigoplus_{j\in J_z} L_{jz}^{\min}, \quad L_{kz}^{\min} \cong L_{jz}^{\min}.$$

If in a semisimple algebra A all minimal left ideals are isomorphic, i.e., A is isotypical, then A is called *simple*. Or equivalently, a simple algebra A has no proper ideal. The minimal ideals $A_z^{\min}$ above are simple algebras.

With the irreducible polynomials, an *abelian simple* algebra is 1- or 2-dimensional for the real case, for $\mathbb{K} = \mathbb{C}$ it is 1-dimensional. For a semisimple algebra element a (no nilpotents), the polynomial algebra $\mathbb{K}[a]_A$ is semisimple. Its ideals are minimal, $\mathbb{C}[a]^{\min} \cong \mathbb{C}$ and $\mathbb{R}[a]^{\min} \cong \mathbb{R}$ or $\mathbb{R}^2$.

The endomorphism algebras (matrix algebras) $M = \mathbb{K}^{n^2}$ are simple. Conversely: Each *complex* finite-dimensional simple algebra is isomorphic to an endomorphism algebra $A = L \otimes L^T$ (full matrix algebra) with the columns $L \cong \mathbb{C}^n$ (rows $L^T = R \cong \mathbb{C}^n$) as irreducible left (right) ideals (*theorem of Burnside and Wedderburn*).

Therefore: The group algebra of a finite group G is a direct sum of simple algebras, for the complex case the direct sum of full matrix algebras:

$$\begin{aligned}\mathbb{C}^G &\cong \bigoplus_{z=1}^{n} A_z^{\min}, \quad A_z^{\min} = L_z^{\min} \otimes (L_z^{\min})^T \cong \mathbb{C}^{n_z} \otimes \mathbb{C}^{n_z},\\ \operatorname{card} G &= \sum_{z=1}^{n} n_z^2.\end{aligned}$$

The n types of minimal left ideals $L_z^{\min} \cong \mathbb{C}^{n_z}$ characterize all irreducible complex representation vector spaces, the minimal two sided ideals characterize all irreducible group algebra representations $\mathbb{C}^G \longrightarrow A_z^{\min}$. Here n is the number of conjugacy classes of the group G, i.e., the disjoint orbits under inner group automorphisms:

$$G = \biguplus_{z=1}^{n} \operatorname{Int} G(g_z) \cong \biguplus_{z=1}^{n} G/G_{g_z}^{\mathrm{Int}}, \quad \operatorname{card} G = \sum_{z=1}^{n} \frac{\operatorname{card} G}{\operatorname{card} G_{g_z}^{\mathrm{Int}}}.$$

Different conjugacy classes may have isomorphic centralizers (fixgroups).

For finite abelian groups $n = \operatorname{card} G$. For example, the algebra for the cyclic group $\mathbb{C}^{\mathbb{I}(n)} \cong \mathbb{C}^n$ has n minimal ideals $\mathbb{C}$ with the irreducible representations

$$\{\mathbb{I}(n) \ni e^{\frac{2\pi i}{n}} \longmapsto e^{\frac{2\pi i k}{n}} \in \mathbf{GL}(\mathbb{C}) \mid k = 1, \ldots, n\}.$$

The representations for different k are inequivalent.

In the permutation group algebra $\mathbb{C}^{\mathbf{G}(N)}$ with dimension $N!$, the n minimal *Young ideals* $A_z^{\min}$ characterize the irreducible representations (conjugacy classes) of the permutations. A Young frame for $\mathbf{G}(N)$ left-aligns N boxes, vertically not increasing; n is the number of the possible related partitions of N. Each frame has n_z possibilities to fill it with the numbers $\{1, \ldots, N\}$ defining n_z basic vectors: The numbers have to increase both horizontally (symmetrization) and vertically (antisymmetrization), e.g.,

$$\mathbf{G}(1) = \{1\}: \quad 1! = 1,$$

$$\mathbf{G}(2) = \mathbb{I}(2): \quad \begin{cases} \mathbf{G}(2) = & \{1\} & \uplus & \{-1\}, \\ 2! = & 1 & + & 1, \\ & \begin{array}{|c|}\hline 1 \\ \hline 2 \\ \hline \end{array} & \oplus & \begin{array}{|c|c|}\hline 1 & 2 \\ \hline \end{array}, \\ 2! = & 1^2 & + & 1^2, \end{cases}$$

$$\begin{aligned}\mathbf{G}(3) &\cong \mathbf{G}(2) \vec{\times} \mathbf{G}(3)_+ \\ &\cong \mathbb{I}(2) \vec{\times} \mathbb{I}(3)\end{aligned} \quad \begin{cases} \mathbf{G}(3) \cong \frac{\mathbf{G}(3)}{\mathbf{G}(3)} & \uplus & \frac{\mathbf{G}(3)}{\mathbb{I}(2)} & \uplus & \frac{\mathbf{G}(3)}{\mathbb{I}(3)}, \\ 3! = \quad 1 & + & 3 & + & 2, \\ \begin{array}{|c|}\hline 1 \\ \hline 2 \\ \hline 3 \\ \hline \end{array} & \oplus & \begin{array}{c} \begin{array}{|c|c|}\hline 1 & 2 \\ \hline 3 & \\ \cline{1-1} \end{array} \\ \\ \begin{array}{|c|c|}\hline 1 & 3 \\ \hline 2 & \\ \cline{1-1} \end{array} \end{array} & \oplus & \begin{array}{|c|c|c|}\hline 1 & 2 & 3 \\ \hline \end{array}, \\ 3! = \quad 1^2 & + & 2^2 & + & 1^2, \end{cases}$$

and permuting four elements

$$\mathbf{G}(4) \cong \mathbf{G}(2) \vec{\times} \mathbf{G}(4)_+, \quad \mathbf{G}(4)_+ \cong \mathbf{G}(3)_+ \vec{\times} [\mathbf{G}(2) \times \mathbf{G}(2)],$$

$$\begin{array}{rcccccccccc} \mathbf{G}(4) = & \frac{\mathbf{G}(4)}{\mathbf{G}(4)} & \uplus & \frac{\mathbf{G}(4)}{[\mathbb{I}(2)]^2} & \uplus & \frac{\mathbf{G}(4)}{\mathbb{I}(3)} & \uplus & \frac{\mathbf{G}(4)}{\mathbb{I}(4)} & \uplus & \frac{\mathbf{G}(4)}{[\mathbb{I}(2)]^3}, \\ 4! = & 1 & + & 6 & + & 8 & + & 6 & + & 3, \end{array}$$

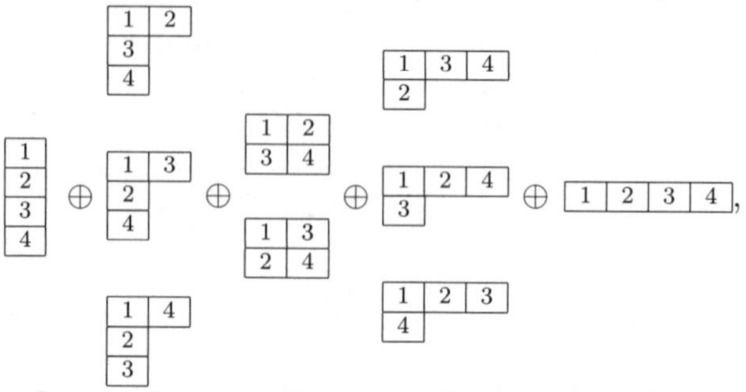

$$4! = 1^2 + 3^2 + 2^2 + 3^2 + 1^2.$$

The two 1-dimensional representations for $N \geq 2$ are the trivial one $\mathbf{G}(N) \longrightarrow \{1\}$ and the signature representation $\mathbf{G}(N) \longrightarrow \{\pm 1\}$.

2.22 Characteristic and Minimal Polynomial

An endomorphism f of a finite-dimensional vector space $V \cong \mathbb{K}^D$ defines by its polynomials $\mathbb{K}[f] \subseteq \mathbf{AL}(V)$ a unital subalgebra. V is a module with respect to this algebra. f is algebraic with minimal polynomial $p_f(X)$, which divides the *characteristic polynomial*

$$\begin{aligned} c_f(X) = \det(X \, \mathrm{id}_V - f) &= \sum_{k=0}^{D} (-1)^{D-k} X^k t_{D-k}(f) \\ &= X^D - X^{D-1} \operatorname{tr} f + \cdots + (-1)^D \det f, \end{aligned}$$

e.g.,

$$f = \begin{pmatrix} 2 & 1 & 0 \\ 4 & 2 & 0 \\ 0 & 0 & 4 \end{pmatrix} = \frac{f \circ f}{4} \Rightarrow \begin{cases} p_f(X) &= X(X-4), \\ c_f(X) &= X(X-4)^2. \end{cases}$$

If a vector space $V \cong \mathbb{K}^D$ is monogeneous with an endomorphism f, i.e., $V = \mathbb{K}[f](v)$ with a cyclic vector $v \in V$, then there exists a basis $\{f^k(v) \bigm| k = 0, \ldots, D-1\}$ with a matrix for f displaying the coefficients of the characteristic polynomial (since $f^D(v)$ is a linear combination of the basis)

$$f \cong \begin{pmatrix} 0 & 1 & 0 & \ldots & 0 & 0 \\ 0 & 0 & 1 & \ldots & 0 & 0 \\ & & & \ldots & & \\ 0 & 0 & 0 & \ldots & 0 & 1 \\ -\beta_0 & -\beta_1 & -\beta_2 & \ldots & & -\beta_{D-1} \end{pmatrix}, \quad \begin{aligned} &\det(f - X \, \mathrm{id}_V) = \sum_{k=0}^{D} \beta_k X^k, \\ &\beta_k \in \mathbb{K}, \quad \beta_D = 1. \end{aligned}$$

As an example, the characteristic and minimal polynomials of a split endomorphism for different values $\{\alpha_j\}$:

$$f \text{ split} \iff \begin{cases} c_f(X) = \det(X\,\mathrm{id}_V - f) = \prod_{j=1}^{m}(X - \alpha_j)^{D_j}, \\ p_f(X) = \prod_{j=1}^{m}(X - \alpha_j)^{N_j},\ N_j \leq D_j. \end{cases}$$

2.22.1 Triagonalization and Diagonalization

An endomorphism is

$$\begin{matrix} \textit{triagonalizable}, & \textit{niltriagonalizable}, & \textit{diagonalizable} \\ \begin{pmatrix} \times & \times \\ 0 & \times \end{pmatrix}, & \begin{pmatrix} 0 & \times \\ 0 & 0 \end{pmatrix}, & \begin{pmatrix} \times & 0 \\ 0 & \times \end{pmatrix} \end{matrix}$$

if there exist bases whose matrices have upper (lower) triagonal form, upper (lower) triagonal form with trivial diagonal, and diagonal form respectively, e.g., $f = g \circ \mathrm{diag}\, f \circ g^{-1}$ with $g \in \mathbf{GL}(\mathbb{K}^D)$. For a *block-triagonalizable and diagonalizable* real endomorphism there may arise in the block diagonal, in addition to numbers, (2×2)-matrices (irreducible polynomials)

$$\pm e^{\lambda} \begin{pmatrix} \cos\varphi & -\sin\varphi \\ \sin\varphi & \cos\varphi \end{pmatrix}, \quad \sin\varphi \neq 0.$$

The minimal polynomial of a split endomorphism $f \in \mathbf{AL}(\mathbb{K}^D)$ leads to a Jordan basis of the f-polynomials with the *spectral decomposition of the endomorphism*:

$$\begin{aligned} \text{basis of } \mathbb{K}[f] :\ & \{\mathcal{P}_j(f), \mathcal{N}_j(f)^{k_j} \mid j = 1, \ldots, m,\ k_j = 1, \ldots, N_j - 1\}, \\ & f = \sum_{j=1}^{m} [\alpha_j \mathcal{P}_j(f) + \mathcal{N}_j(f)]. \end{aligned}$$

An endomorphism $h \in \mathbf{AL}(\mathbb{K}^D)$ is triagonalizable iff it is split, i.e., the minimal polynomial is a product of polynomials $(X - \alpha)^N$. All complex endomorphisms are triagonalizable; all real endomorphisms are block-triagonalizable.

Each endomorphism $f \in \mathbf{AL}(\mathbb{K}^D)$ is a unique sum of a semisimple endomorphism $h(f)$, i.e., V is decomposable into $h(f)$-irreducible vector subspaces, and a nilpotent endomorphism $n(f)$. Both parts $h(f)$ and $n(f)$ are polynomials in f and therefore commute with each other:

$$f = h(f) + n(f), \quad h(f) \circ n(f) = n(f) \circ h(f).$$

An endomorphism $n \in \mathbf{AL}(\mathbb{K}^D)$ is nilpotent, $n^k = 0$, iff it is *niltriagonalizable* or iff the traces of all its powers vanish: $\mathrm{tr}\, n^m = 0$, $m \in \mathbb{N}$, and, for $n \neq 0$, iff the characteristic or minimal polynomial are pure powers X^k.

An endomorphism $h \in \mathbf{AL}(\mathbb{K}^D)$ is semisimple iff its minimal polynomial is a product of irreducible polynomials, all different ($p_i \neq p_j$ for $i \neq j$):

$$p_h(X) = \prod_i p_i(X), \quad p_i(X) = \begin{cases} X - \alpha, & \alpha \in \mathbb{C}, \mathbb{R}, \\ (X-\beta)^2 + \gamma^2, & \beta, \gamma \in \mathbb{R}, \ \gamma \neq 0. \end{cases}$$

An endomorphism $h \in \mathbf{AL}(\mathbb{K}^D)$ is diagonalizable iff it is semisimple and split, i.e., the minimal polynomial is a product of factors $(X - \alpha_j)$ with different $\{\alpha_j\}$:

$$\begin{array}{rcl} h \in \mathbf{AL}(\mathbb{K}^D) \text{ diagonalizable} & \iff & p_h(X) \text{ has irreducible factors} \\ & & \text{(first degree)} \\ & \iff & \operatorname{diag} h = g \circ h \circ g^{-1}, \\ h \in \mathbf{AL}(\mathbb{R}^D) \text{ block-diagonalizable} & \iff & p_h(X) \text{ has irreducible factors} \\ & & \text{(first and second degree).} \end{array}$$

A subset of the diagonalizable complex endomorphisms are even unitarily diagonalizable; this involves a nondegenerate sesquilinear form,

$$\begin{array}{rcl} f \in \mathbf{AL}(\mathbb{C}^D) \ \mathbf{U}\text{-diagonalizable} & \iff & f \circ f^* = f^* \circ f \ \mathbf{U}\text{-normal} \\ & \iff & f = f_+ + f_-, \ \ f_\pm^* = \pm f_\pm, \ \ [f_+, f_-] = 0 \\ & \iff & \operatorname{diag} f = u \circ f \circ u^* \text{ with } u^* = u^{-1}, \end{array}$$

and for orthogonal diagonalizability of real endomorphisms $f \in \mathbf{AL}(\mathbb{R}^D)$, this involves a nondegenerate symmetric bilinear form,

$$\begin{array}{rcl} f \in \mathbf{AL}(\mathbb{R}^D) \ \mathbf{O}\text{-diagonalizable} & \iff & f = f^T \\ & \iff & \operatorname{diag} f = o \circ f \circ o^T \text{ with } o^T = o^{-1}, \\ \Rightarrow \text{block-}\mathbf{O}\text{-diagonalizable} & \iff & f \circ f^T = f^T \circ f \\ & \iff & f = f_+ + f_-, \ \ f_\pm^T = \pm f_\pm, \ \ [f_+, f_-] = 0. \end{array}$$

Unitarity and orthogonality may be indefinite $\mathbf{U}(p,q)$ and $\mathbf{O}(p,q)$.

For any endomorphism one obtains

$$f \in \mathbf{AL}(\mathbb{K}^D) \Rightarrow \det e^f = e^{\operatorname{tr} f}.$$

For $\mathbb{K} = \mathbb{C}$ one transforms to the triagonal form; for $\mathbb{K} = \mathbb{R}$ one triagonalizes in the complex $\mathbb{C}^D$. Then one uses the determinant and trace property $\det g \circ f \circ g^{-1} = \det f$, $\operatorname{tr} g \circ f \circ g^{-1} = \operatorname{tr} f$.

2.22.2 Eigenspaces and Eigenvalues

An endomorphism $f \neq 0$ of a vector space $V \cong \mathbb{K}^D$ defines a vector subspace for any scalar $\alpha \in \mathbb{K}$:

$$V_\alpha(f) = \{v \in V \mid f(v) = \alpha v\} \in \underline{\mathbf{vec}}_{\mathbb{K}}.$$

A nontrivial $V_\alpha(f) \neq \{0\}$ is called the *eigenspace of f for the eigenvalue α* with the *eigenvectors* of f for eigenvalue α. If α is not an eigenvalue for f, one has $V_\alpha(f) = \{0\}$.

The eigenvalues of the endomorphism f are the zeros of the characteristic and the minimal polynomials. They constitute the *spectrum* of the endomorphism f, a subset of the scalars:

$$\begin{aligned}\operatorname{spec} f &= \{\alpha \in \mathbb{K} \mid \text{There exists } v \neq 0 \text{ with } f(v) = \alpha v\} \\ &= \{\alpha \in \mathbb{K} \mid c_f(\alpha) = 0\} = \{\alpha \in \mathbb{K} \mid p_f(\alpha) = 0\} \\ &= \{\alpha \in \mathbb{K} \mid f - \alpha \operatorname{id}_V \text{ not invertible}\} \\ &= \{\alpha_j \mid j = 1, \ldots, m\} \in \underline{\mathbf{set}}.\end{aligned}$$

On a *real* space $V \cong \mathbb{R}^D$ the spectrum may be empty, there may be no eigenvector for $f \in \mathbf{AL}(\mathbb{R}^D)$, e.g., for $f \sim \begin{pmatrix} 0 & 1 \\ -1 & 0 \end{pmatrix}$.

The dimension of $V_\alpha(f)$ is the *multiplicity* M_α of the eigenvalue α. It defines a spectral measure (counting measure):

$$\begin{aligned}&M : \operatorname{spec} f \longrightarrow \mathbb{N}_0, \quad \alpha \longmapsto M_\alpha = \dim_\mathbb{K} V_\alpha(f), \\ &\operatorname{card} \operatorname{spec} f \leq \dim_\mathbb{K} V.\end{aligned}$$

The spectrum of an algebra element $a \in A \in \underline{\mathbf{aag}}_\mathbb{K}$ is the spectrum of its action by left multiplication $\operatorname{spec} a = \operatorname{spec} L_a$.

Each eigenvalue α has at least one eigenvector $v \cong (v_j)_{j=1}^n \neq 0$. After the characteristic equation is solved for an $(n \times n)$ matrix $(f)^j_k$ to obtain the eigenvalues, the vectors for an α-eigenspace are found with the n equations $(f - \alpha \mathbf{1})^j_k v_j = 0$. For a base numbering with $v_n \neq 0$ the ratios $\frac{v_j}{v_n}$ are obtained from $(n-1)$ equations with $(f - \alpha\mathbf{1})(n-1)$ the $(n-1) \times (n-1)$ matrix of the first $(n-1)$ rows and colums

$$\frac{1}{v_n}\begin{pmatrix} v_1 \\ \cdots \\ v_{n-1} \end{pmatrix} = -[(f - \alpha\mathbf{1})(n-1)]^{-1}\begin{pmatrix} f^1_n \\ \cdots \\ f^{n-1}_n \end{pmatrix}.$$

For multiplicity $M_\alpha = 1$ the solution is unique; otherwise, one can find $M_\alpha > 1$ linear independent solutions.

All eigenvectors of f span the *eigenspace of* f

$$V \supseteq V(f) = \bigoplus_{\alpha \in \operatorname{spec} f} V_\alpha(f) = \{\sum_{j=1}^m \gamma_j v_j \mid v_i \in V_{\alpha_j}(f)\},$$

which can be a proper subspace of V, both in the real and complex cases. This occurs for a nontrivial nilpotent part $n(f)$ or, only for the real case, for a semisimple endomorphism with $\begin{pmatrix} 0 & 1 \\ -1 & 0 \end{pmatrix}$ contributions.

An endomorphism is diagonalizable iff its eigenvectors span the full space. The basis is unique if all eigenvalues are nondegenerate. With an eigenvector basis and the unique dual basis, the diagonal form, the *spectral projector decomposition*, displays the eigenvalues (spectral theorem for finite dimensions)

$$V(h) = V \iff h = g \circ \operatorname{diag} h \circ g^{-1} = \sum_{i=1}^D \alpha_i e^i \otimes \check{e}_i = \sum_{j=1}^m \alpha_j \mathcal{P}_j(h), \ \langle \check{e}_j, e^i \rangle = \delta^i_j.$$

The Jordan basis of the h-polynomials consists of projectors:

$$\text{basis of } \mathbb{K}[h]: \ \{\mathcal{P}_j(h) \mid j = 1, \ldots, m\}.$$

$\mathbf{U}(n)$- and $\mathbf{O}(n)$-diagonalizable endomorphisms define bases with even orthogonal eigenvectors, again unique if all eigenvalues are nondegenerate (in the bra-ket notation):

$$f = u \circ \operatorname{diag} f \circ u^\star = \sum_{i=1}^{D} \alpha_i |e^i\rangle\langle e^i|, \ \langle e^i| = \delta^{ij}\langle \check{e}_j|, \ \langle e^j|e^i\rangle = \delta^{ji}.$$

In the decomposition $f = h(f) + n(f)$, f and $h(f)$ have equal eigenvalues if they have eigenvalues at all:

$$\alpha \in \operatorname{spec} f = \operatorname{spec} h(f), \quad V_\alpha(f) = V_\alpha(h(f)).$$

An f-eigenvector is an $n(f)$-eigenvector with eigenvalue 0. An $n(f)$-eigenvector with eigenvalue 0 is a linear combination of f-eigenvectors

$$V(f) = V(h(f)) = \mathbf{INV}_{n(f)}V = \{v \in V \mid n(f)(v) = 0\}.$$

A triagonalizable endomorphism is nilpotent iff it has only the eigenvalue 0.

2.22.3 Principal Spaces

The *generalized eigenspace (principal space) of an endomorphism f on $V \cong \mathbb{K}^D$ for the eigenvalue $\alpha \in \mathbb{K}$* is defined by

$$V^\alpha(f) = \{v \in V \mid \text{There exists } N \geq 1 \text{ with } (f - \alpha\, \mathrm{id}_V)^N(v) = 0\} \in \underline{\mathbf{vec}}_{\mathbb{K}}$$

if $V^\alpha(f)$ does not contain only the trivial vector. The minimal N that can be used in the definition of $V^\alpha(f)$ is the *order N_α of the eigenvalue*:

$$(f - \alpha\, \mathrm{id}_V)^{N_\alpha}.V^\alpha(f) = \{0\}.$$

The elements of $V^\alpha(f)$ are called *principal vectors* of f for the eigenvalue α. A *nilvector* $v \in V$ for the eigenvalue $\alpha \in \mathbb{K}$ satisfies

$$(f - \alpha\, \mathrm{id}_V)^N(v) \neq 0, \quad (f - \alpha\, \mathrm{id}_V)^{1+N}(v) = 0, \quad N \geq 1.$$

A nilvector v is not an eigenvector, but $(f - \alpha\, \mathrm{id}_V)^N(v)$ is one. A principal space contains the eigenspace $V^\alpha(f) \supseteq V_\alpha(f)$ (sup- and subindex), whose dimension is the *multiplicity M_α of the eigenvalue.* The dimension D_α of $V^\alpha(f)$ is the *degeneracy* of α. N_α and D_α are the powers of $(X - \alpha)$ arising in the minimal polynomial and the characteristic polynomial respectively:

$$M_\alpha = \dim_{\mathbb{K}} V_\alpha(f), \quad D_\alpha = \dim_{\mathbb{K}} V^\alpha(f), \quad M_\alpha + N_\alpha = D_\alpha.$$

The following example of a triagonalizable endomorphism of $\mathbb{K}^{12}$ may be useful as an illustration

$$f \simeq \left(\begin{array}{c|c|cc|ccc||c|cc|cc}
\alpha_1 & 0 & 0 & 0 & 0 & 0 & 0 & 0 & 0 & 0 & 0 & 0 \\ \hline
0 & \alpha_1 & 0 & 0 & 0 & 0 & 0 & 0 & 0 & 0 & 0 & 0 \\ \hline
0 & 0 & \alpha_1 & 1 & 0 & 0 & 0 & 0 & 0 & 0 & 0 & 0 \\
0 & 0 & 0 & \alpha_1 & 0 & 0 & 0 & 0 & 0 & 0 & 0 & 0 \\ \hline
0 & 0 & 0 & 0 & \alpha_1 & 1 & 0 & 0 & 0 & 0 & 0 & 0 \\
0 & 0 & 0 & 0 & 0 & \alpha_1 & 1 & 0 & 0 & 0 & 0 & 0 \\
0 & 0 & 0 & 0 & 0 & 0 & \alpha_1 & 0 & 0 & 0 & 0 & 0 \\ \hline\hline
0 & 0 & 0 & 0 & 0 & 0 & 0 & \alpha_2 & 0 & 0 & 0 & 0 \\ \hline
0 & 0 & 0 & 0 & 0 & 0 & 0 & 0 & \alpha_2 & 1 & 0 & 0 \\
0 & 0 & 0 & 0 & 0 & 0 & 0 & 0 & 0 & \alpha_2 & 0 & 0 \\ \hline
0 & 0 & 0 & 0 & 0 & 0 & 0 & 0 & 0 & 0 & \alpha_2 & 1 \\
0 & 0 & 0 & 0 & 0 & 0 & 0 & 0 & 0 & 0 & 0 & \alpha_2
\end{array}\right), \quad \alpha_1 \neq \alpha_2,$$

$$p_f(X) = (X - \alpha_1)^3 (X - \alpha_2)^2, \quad c_f(X) = (X - \alpha_1)^7 (X - \alpha_2)^5.$$

For each triagonal box one gets one eigenvector only:

eigenvalue α_j	multiplicity M_{α_j}	order N_j	degeneracy D_j
α_1	$2+1+1=4$	3	7
α_2	$1+2=3$	2	5

The restriction of f to the principal space $V^\alpha(f)$,

$$\begin{array}{l} f_{V^\alpha(f)} = f_\alpha = h(f_\alpha) + n(f_\alpha), \quad \text{semisimple } h(f_\alpha), \quad \text{nilpotent } n(f_\alpha), \\ c_{f_\alpha}(x) = (X - \alpha)^{D_\alpha}, \quad p_{f_\alpha}(x) = (X - \alpha)^{N_\alpha}, \end{array}$$

is triagonalizable with nondecomposable *Jordan matrices for the eigenvalue α of order k* being the sum of a $(k \times k)$-unit matrix and a power k *nilcyclic* matrix $\mathbf{N}_k$:

$$\begin{array}{l} J_k(\alpha) \cong \alpha \mathbf{1}_k + \mathbf{N}_k \simeq \begin{pmatrix} \alpha & 1 & 0 & \dots & 0 & 0 \\ 0 & \alpha & 1 & \dots & 0 & 0 \\ & \dots & & & \dots & \\ 0 & 0 & 0 & \dots & \alpha & 1 \\ 0 & 0 & 0 & \dots & 0 & \alpha \end{pmatrix}, \\ c_{J_k(\alpha)}(X) = p_{J_k(\alpha)}(X) = (X - \alpha)^k. \end{array}$$

The orders k occurring for the principal space $V^\alpha(f)$ are between 1 and the maximal order N_α. A Jordan matrix can arise with multiplicity $M_k = 0, 1, \dots$:

$$\begin{array}{l} V^\alpha(f) = \bigoplus_{k=1}^{N_\alpha} M_k \cdot V_k^\alpha(f), \quad f_\alpha = \bigoplus_{k=1}^{N_\alpha} M_k \cdot J_k(\alpha), \\ \sum_{k=1}^{N_\alpha} M_k = M_\alpha, \quad \sum_{k=1}^{N_\alpha} M_k k = D_\alpha, \quad V_k^\alpha(f) \cong \mathbb{K}^k. \end{array}$$

A vector v in a nondecomposable space $V_k^\alpha(f)$ with $(f - \alpha \, \mathrm{id}_V)^{k-1}.v \neq 0$ (nilvector for $k \geq 2$, eigenvector for $k = 1$) is called a *cyclic principal vector* for $V_k^\alpha(f)$ since the principal vectors

$$\{(f - \alpha \, \mathrm{id}_V)^N.v \bigm| N = 0, \dots, k-1\}$$

are a basis for $V_k^\alpha(f) \cong \mathbb{K}^k$. The minimal polynomial $(X - \alpha)^{N_\alpha}$ of the restriction f_α of f on the principal space $V^\alpha(f)$ allows the Jordan basis of the f_α-polynomials:

$$\text{basis of } \mathbb{K}[f_\alpha]: \quad \{\mathcal{P}_\alpha(f), \mathcal{N}_\alpha(f)^k \bigm| k = 1, \dots, N_\alpha - 1\}.$$

The minimal polynomial of a split endomorphism f,

$$p_f(X) = \prod_{j=1}^{m}(X - \alpha_j)^{N_j}, \quad c_f(X) = \prod_{j=1}^{m}(X - \alpha_j)^{D_j}, \quad N_j \leq D_j,$$

leads to a Jordan basis of the f-polynomials:

$$\text{basis of } \mathbb{K}[f]: \ \{\mathcal{P}_j(f), \mathcal{N}_j(f)^{k_j} \bigm| j = 1, \dots, m, \ k_j = 1, \dots, N_j - 1\}.$$

Therefore the vector space V is the direct sum of f-invariant *principal spaces* $\{V^{\alpha_j}(f)\}_{j=1}^m$ (decomposition with respect to different eigenvalues):

$$V = \bigoplus_{j=1}^{m} V^{\alpha_j}(f), \quad V^{\alpha_j}(f) = \mathcal{P}_j(f).V \cong \mathbb{K}^{D_j}, \quad f.V^{\alpha_j}(f) \subseteq V^{\alpha_j}(f),$$
$$f = \bigoplus_{j=1}^{m} f_{\alpha_j} = \sum_{j=1}^{m}[\alpha_j \mathcal{P}_j(f) + \mathcal{N}_j(f)], \quad f_{\alpha_j} = \mathcal{P}_j(f) \circ f$$

(spectral decomposition of f).

f is triagonalizable with the direct sum of triagonal $(D_j \times D_j)$ matrices; f_{α_j} on $V^{\alpha_j}(f)$ may be decomposable into nondecomposable Jordan matrices as described above.

Principal vectors for a matrix f and an eigenvalue α can be found by first determining the eigenvectors. For degeneracy $D_\alpha > 1$ one has multiplicity $M_\alpha > 1$ or order $N_\alpha > 1$. After determining the M_α-dimensional eigenspace as sketched above, one solves with an eigenvector v the equation for a "neighboring" nilvector v_{-1}

$$(f - \alpha\mathbf{1})v_{-1} = v.$$

For multiplicity $M_\alpha = 1$ and order $N_\alpha > 1$ a nilvector space $\mathbb{K}v_{-1}$ is determined up to $\mathbb{K}v$. Then one considers $(f - \alpha\mathbf{1})^2 v_{-2} = v$, etc. For multiplicity $N_\alpha > 1$ one has to associate the appropriate eigenvectors and nilvectors, which may be a rather complicated problem.

Bibliography

[1] N. Bourbaki, *Algebra I,II, Chapters 1-7* (1989, 90), Springer, Berlin, Heidelberg, New York, London, Paris, Tokyo.

[2] N. Bourbaki, *Algèbre, Chapitre 8* (1958) (Modules et anneaux semi-simples), Hermann, Paris.

[3] N. Bourbaki, *Lie Groups and Lie Algebras, Chapters 1-3* (1989), Springer, Berlin, Heidelberg, New York, London, Paris, Tokyo.

[4] H. Boerner, *Darstellungen von Gruppen* (1955), Springer, Berlin, Göttingen, Heidelberg.

[5] W. Fulton, J. Harris, *Representation Theory* (1991), Springer.

[6] A.A. Kirillov, *Elements of the Theory of Representations* (1976), Springer-Verlag, Berlin, Heidelberg, New York.

[7] H. Weyl, *The Theory of Groups and Quantum Mechanics* (1931), Dover, New York.

3

SPIN, ROTATIONS, AND POSITION

In addition to time with its order structure, formalized by the real Lie group $\mathbf{D}(1)$ (causal group) and its Lie algebra $\mathbb{R}$ (time translations) with energy (frequency) characterizing its eigenvectors, a physical dynamics also represents position with the related operations: The real 3-dimensional position translations come with a scalar product, invariant under rotations. The spatial form of an object, a sphere, an ellipsoid, a cube, a tree, etc., is characterized by its possibly very complicated properties with respect to rotations (chapter "Harmonic Analysis"). There is only one Lie algebra[1] of real dimension three, which Cartan called A_1^c, that realizes the position operations. The exponent of this compact Lie algebra is $\mathbf{SU}(2) = \exp A_1^c$, the *spin group*, whose classes with respect to its discrete center $\mathbb{I}(2) \cong \{\pm \mathbf{1}_2\}$ are the rotations $\mathbf{SO}(3) \cong \mathbf{SU}(2)/\mathbb{I}(2)$. The invariant property for these operations is angular momentum or spin. The abstract $\mathbf{SU}(2)$-operations are realized in basic physical interactions and particles, e.g., by spin, by isospin, and in Lorentz transformations. In the following the physically suggestive position- and spin-oriented language will be used.

In this chapter all finite-dimensional representations of the complex 3-dimensional Lie algebra A_1 and all representations of its compact form A_1^c are considered. From the group-theoretical point of view, the "simplest simple" Lie algebra A_1 is fundamental for all nonabelian, especially semisimple, Lie operations that, in some sense can be considered as "lumped together" spin structures (chapter "Simple Lie Operations"). In the representations of the nonabelian $\mathbf{SU}(2)$, the integer winding numbers $Z \in \mathbb{Z}$ for the representations $e^{i\alpha} \longmapsto e^{iZ\alpha}$ of the abelian circle group $\mathbf{U}(1) = \exp i\mathbb{R}$ (1-torus) come in reflected winding number pairs $(Z, -Z)$ for the dual $\mathbf{U}(1)$-subrepresentations in $\mathbf{SO}(2)$. The representation characteristic maximal natural number $|Z_{\max}| = 2J$ defines the spin J with the $\mathbf{SU}(2)$-representation space dimension $1+2J$, e.g., the position dimension $s = 3$ for the adjoint representation $J = 1$.

[1] In this chapter a Lie algebra structure of a vector space L is defined up to linear equivalence, i.e., up to vector space automorphisms $\mathbf{GL}(L)$.

3.1 Linear Operations on the Alternative

The spin structure can be motivated by discrete noncommutative operations: Such operations are defined, minimally, as acting on a 2-element set $\{\circ,\bullet\}$, a *basic alternative*, by an exchange $\bullet\leftrightarrow\circ$ ("shut $\leftrightarrow$ open" or "up $\leftrightarrow$ down") in addition to the identity $(\bullet,\circ)\leftrightarrow(\bullet,\circ)$ ("nothing changes").

To algebraize this structure: A *numerical valuations* of the operations on the two-element set $\{\circ,\bullet\}$ is the free vector space $K^{\{\circ,\bullet\}}$ with the mappings of the basic alternative $\{\circ,\bullet\}\longrightarrow K$ into a number field, i.e., a 2-dimensional linear space $V\cong K^2$. The simplest field $\mathbb{Z}_2=\{0,1\}$ allows a valuation with the truth values "false" and "true" ("bit"), the embedding into real and then complex numbers $\mathbb{R}$ and $\mathbb{C}$ allows extended valuations (modalities) with probabilities and even "probability amplitudes" (chapter "Quantum Probability")

$$\begin{array}{lccccc}\text{field } K: & \mathbb{Z}_2 & \subset & \mathbb{R} & \subset & \mathbb{C},\\ \text{valuation:} & \{0,1\} & \subset & [0,1] & \subset & \mathbf{U}(1)\times[0,1].\end{array}$$

With the alternative formalized by two basic vectors of a free vector space K^2,

$$\left.\begin{array}{lll}V\text{-basis: } \circ= & e^1\simeq\begin{pmatrix}1\\0\end{pmatrix}, & \bullet= \ e^2\simeq\begin{pmatrix}0\\1\end{pmatrix},\\ V^T\text{-basis:} & \check e_1\simeq(1,0), & \check e_2\simeq(0,1)\end{array}\right., \quad \text{dual } \langle\check e_A,e^B\rangle=\delta_A^B.$$

it can be acted on by the transitions $\{\sigma_+,\sigma_-\}$

$$\begin{array}{llll}\sigma_+ & =e^1\otimes\check e_2=\begin{pmatrix}0&1\\0&0\end{pmatrix}, & \sigma_+(e^1)=0, & \sigma_+(e^2)=e^1,\\ \sigma_- & =e^2\otimes\check e_1=\begin{pmatrix}0&0\\1&0\end{pmatrix}, & \sigma_-(e^1)=e^2, & \sigma_-(e^2)=0.\end{array}$$

The sum $\sigma^1=\sigma_++\sigma_-=\begin{pmatrix}0&1\\1&0\end{pmatrix}$ represents the exchange operation $\bullet\leftrightarrow\circ$.

From now on the field K is chosen to be complex or real. The operations $\sigma_\pm$ together with their diagonal commutator σ_0 constitute a basis for the 3-dimensional operation Lie algebra of the alternative

$$\begin{array}{l}[l_-,l_+]=h,\quad [h,l_\pm]=\pm2l_\pm,\\ l_\pm\simeq\pm\sigma_\pm,\quad h\simeq\sigma_0=\begin{pmatrix}1&0\\0&-1\end{pmatrix}.\end{array}$$

The complex linear combinations of the three elements $\{l_\pm,h\}$ as *spherical Weyl (Cartan) basis* with the Lie brackets above define the Lie algebra $A_1\cong\mathbb{C}^3$. There exist *Cartesian (Euclidean) bases* $\{l^a\}_{a=1}^3$ with totally antisymmetric structure constants $\epsilon^{ab}_c=-\epsilon^{abc}$, $\epsilon^{123}=1$,

$$\begin{array}{l}l^1=\frac{l_--l_+}{2i},\quad l^2=\frac{l_-+l_+}{2},\quad l^3=i\frac{h}{2},\\ il_+=l^1+il^2,\quad -il_-=l^1-il^2,\quad ih=2l^3,\\ [l^a,l^b]=\epsilon^{ab}_cl^c=-\epsilon^{abc}l^c.\end{array}$$

The real span of a Cartesian basis defines the lowest-dimensional compact Lie algebra $A_1^c\cong\mathbb{R}^3$ (supindex c for compact) as compact form of A_1, isomorphic to the angular momentum Lie algebra for the rotation group $\mathbf{SO}(3)$. The real span of a Weyl basis defines the Lie algebra $A_1^n \cong \mathbb{R}^3$, isomorphic to the Lie algebra of the noncompact Lorentz group $\mathbf{SO}_0(1,2)$ for two position dimensions.

A_1 is the "smallest" simple Lie algebra. It has complex dimension 3. The adjoint representation is defined by the Lie bracket

$$\mathrm{ad} : A_1 \longrightarrow \mathbf{AL}(A_1), \quad \mathrm{ad}\, l(m) = [l, m].$$

The spherical basis $(l_\pm, h)$ consists of $\mathrm{ad}\, h$-eigenvectors with eigenvalues $(\pm 2, 0)$. The "double trace" of the adjoint representation defines the nondegenerate *Killing form* κ^2 as invariant Lie algebra inner product:

$$\kappa^2 : A_1 \times A_1 \longrightarrow \mathbb{C}, \quad \kappa^2(l, m) = \mathrm{tr}\ \mathrm{ad}\, l \circ \mathrm{ad}\, m.$$

The adjoint representation of the compact form A_1^c leads to the angular momenta

$$\mathrm{ad} : A_1^c \longrightarrow \log \mathbf{SO}(3), \quad \begin{cases} \mathrm{ad}\, l^a = \mathcal{O}^a = -\epsilon^{abc} l^c \otimes \check{l}_b, & \langle \check{l}_b, l^a \rangle = \delta_b^a, \\ \kappa^2(l^a, l^b) = \mathrm{tr}\, \mathcal{O}^a \circ \mathcal{O}^b = \epsilon^{acd}\epsilon^{bdc} = -2\delta^{ab}. \end{cases}$$

The maximal abelian Lie subalgebras (Cartan subalgebras) of A_1 and of its real normal and compact forms A_1^n and A_1^c are complex and real 1-dimensional. This defines the *rank* 1 (therefore the subindex 1). The adjoint action $\mathrm{ad}\, l(m) = [l, m]$ can be diagonalized for maximally one nontrivial Lie algebra operation: For example, $\mathrm{ad}\, h$ in the case of A_1 and A_1^n whereas the compact A_1^c has no nontrivial diagonalizable action.

The inverse Killing form is the *Casimir element* in the complex and real enveloping algebras $\mathbf{E}(A_1)$ and $\mathbf{E}(A_1^n), \mathbf{E}(A_1^c)$

$$I(A_1) = \kappa^{-2} = -\tfrac{\delta_{ab}}{2} l^a \otimes l^b \in \mathbf{E}(A_1), \ [A_1, I(A_1)] = \{0\}.$$

This one Casimir element generates the polynomial ring with all Lie algebra A_1 invariants; this also reflects rank 1.

The distinction of a compact Lie algebra in $A_1 \cong \mathbb{C}^3$ with antisymmetric elements by a Euclidean conjugation $l^\star = -l \in A_1^c \cong \mathbb{R}^3$ defines, together with the symmetric elements $b = b^\star \in iA_1^c$, a canonically complexified Lie algebra $A_1^c \oplus iA_1^c$ - the real 6-dimensional Lie algebra of the Lorentz group (chapter "Lorentz Symmetry"):

$$\text{basis of } A_1^c \oplus iA_1^c \cong \mathbb{R}^6 : \ \{l^a, b^a = il^a\}_{a=1,2,3}, \begin{cases} [l^a, l^b] &= -\epsilon^{abc} l^c, \\ [l^a, b^b] &= -\epsilon^{abc} b^c, \\ [b^a, b^b] &= +\epsilon^{abc} l^c. \end{cases}$$

3.2 Pauli Spinors

The Stern-Gerlach experiment splits an initial ray of silver atoms by an inhomogeneous magnetic field into two final rays. Such an even-fold split can be taken as the decisive hint that the symmetry operations, underlying position with odd dimension 3, are not completely described by the rotation group $\mathbf{SO}(3)$, which has only odd-dimensional irreducible representations spaces, but by its simply connected universal cover group $\mathbf{SU}(2)$, which also has even-dimensional irreducible representation spaces.

The *fundamental representation* of the position underlying Lie operations acts on *Pauli spinors*. Pauli spinors with $\mathbf{SU}(2)$-transformations play a fundamental role: They arise for external spacetime-related transformations, characterized by spin $\frac{1}{2}$, as well as for internal ones, characterized by isospin $\frac{1}{2}$.

All representations of the complex Lie algebra A_1 are obtainable, up to isomorphism, as tensor powers of the fundamental A_1-representation on Pauli spinors, i.e., on a complex 2-dimensional vector space $V \cong \mathbb{C}^2$. If the spinors and their forms carry a scalar product and therefore a *Euclidean conjugation*

$$\langle\ |\ \rangle : V \times V \longrightarrow \mathbb{C}, \quad \begin{cases} \langle v|v\rangle = \langle v^\star, v\rangle > 0 \iff v \neq 0, \\ \text{Euclidean basis: } \langle e^A|e^B\rangle = \delta^{AB}, \quad e^A \overset{\star}{\leftrightarrow} \delta^{AB}\check{e}_B, \end{cases}$$

V and V^T are Hilbert spaces. Also, the spinor endomorphisms, the unital *Pauli algebra* $\mathbf{AL}(V) \cong \mathbb{C}^4$, can be conjugated. The Pauli spinors feel the action of the complex Lie algebra A_1 via the representation by the *traceless* endomorphisms, and of the compact Lie algebra A_1^c by the *traceless Euclidean antisymmetric* endomorphisms

$$\begin{aligned} A_1 &\cong \{l \in \mathbf{AL}(\mathbb{C}^2) \,\big|\, \operatorname{tr} l = 0\} && \in \underline{\mathbf{lag}}_{\mathbb{C}}, \\ A_1^c &\cong \{l \in \mathbf{AL}(\mathbb{C}^2) \,\big|\, \operatorname{tr} l = 0,\ \ l^\star = -l\} && \in \underline{\mathbf{lag}}_{\mathbb{R}}. \end{aligned}$$

The fundamental Pauli representation allows a basis $\{\vec{l}\} = \{l^a\}_{a=1,2,3}$ using the three Euclidean symmetrical *Pauli matrices* $\{\vec{\sigma}\} = \{\sigma^a\}_{a=1,2,3}$,

$$\vec{l} = \tfrac{i}{2}\vec{\sigma}_A^B\ e^A \otimes \check{e}_B = -\vec{l}^\star, \begin{cases} \alpha_a\sigma^a &= \vec{\alpha} = \begin{pmatrix} \alpha_3 & \alpha_1 - i\alpha_2 \\ \alpha_1 + i\alpha_2 & -\alpha_3 \end{pmatrix}, \\ \sigma_+ &= \frac{\sigma^1 + i\sigma^2}{2} = \begin{pmatrix} 0 & 1 \\ 0 & 0 \end{pmatrix}, \quad \sigma_- = \frac{\sigma^1 - i\sigma^2}{2} = \begin{pmatrix} 0 & 0 \\ 1 & 0 \end{pmatrix}, \\ \vec{\sigma} &= \vec{\sigma}^\star, \quad \sigma_+^\star = \sigma_-, \\ \sigma^a\sigma^b &= \delta^{ab}\mathbf{1}_2 + i\epsilon^{abc}\sigma^c, \quad \det\sigma^a = -1. \end{cases}$$

In the following, vectors $\vec{\alpha} = \vec{\alpha}^\star \in \mathbb{R}^3$ are to be understood as traceless Hermitian 2×2 matrices.

The invariant nondegenerate inner product, associate to the Pauli representation

$$\kappa^1(\ ,\ ) : A_1 \times A_1 \longrightarrow \mathbb{C}, \quad \kappa^1(l^a, l^b) = \tfrac{i^2}{4}\operatorname{tr}\sigma^a\sigma^b = -\tfrac{1}{2}\delta^{ab},$$

gives the *Casimir element*, realized in the Pauli algebra by a multiple of the identity

$$I^1(A_1) = -\tfrac{1}{2}\,\vec{l} \circ \vec{l} = \tfrac{3}{8}\,\mathrm{id}_V.$$

The *dual Pauli spinor representation* is (with transposition sign $\epsilon = \pm 1$)

$$A_1, A_1^c \longrightarrow \mathbf{AL}(V^T), \quad \vec{l} \longmapsto -\vec{l}^T = -\tfrac{i}{2}\vec{\sigma}_A^B\ \epsilon\check{e}_B \otimes e^A.$$

3.3 Spin Group

By exponentiating the Lie algebra A_1 in the Pauli algebra, there arises the complex Lie group $\mathbf{SL}(\mathbb{C}^2)$, the special automorphisms in two complex dimensions:

$$\begin{array}{l} \exp A_1 \cong \mathbf{SL}(\mathbb{C}^2) = \{s \in \mathbf{GL}(\mathbb{C}^2) \,\big|\, \det s = 1\} \in \underline{\mathbf{lgrp}}_{\mathbb{C}}, \\ e^l = s = s_A^B\, e^A \otimes \check{e}_B \cong e^{\vec{z}} = \mathbf{1}_2 e_+(z) + \frac{\vec{z}}{z} e_-(z),\ z_a \in \mathbb{C}, \\ \text{since } \vec{z}^2 = z^2 \mathbf{1}_2, \end{array}$$

with the even and odd powers

$$\begin{array}{l} e_+(z) = \sum\limits_{k \geq 0} \frac{z^{2k}}{(2k)!}, \quad e_-(z) = \sum\limits_{k \geq 0} \frac{z^{1+2k}}{(1+2k)!}, \\ e_\pm(z) = \frac{e^z \pm e^{-z}}{2} = \left\{ \begin{array}{ll} (\cosh \beta, \sinh \beta), & z = \beta \in \mathbb{R}, \\ (\cos \alpha, i \sin \alpha), & z = i\alpha \in i\mathbb{R}. \end{array} \right. \end{array}$$

The real Lie group $\mathbf{SU}(2)$ with the special Euclidean unitary automorphisms can be locally (in an open set around the neutral element) parametrized with coefficients of a Lie algebra basis:

$$\begin{array}{rl} A_1^c \ni & i\vec{\alpha} \longmapsto e^{i\vec{\alpha}} \in \mathbf{SU}(2),\ \ l \longmapsto \exp l = e^{i\vec{\alpha}} = \mathbf{1}_2 \cos \alpha + i\frac{\vec{\alpha}}{\alpha} \sin \alpha, \\ \mathbf{SU}(2) & = \{u \in \mathbf{GL}(\mathbb{C}^2) \,\big|\, \det u = 1,\ \ u^\star = u^{-1}\} \in \underline{\mathbf{lgrp}}_{\mathbb{R}}. \end{array}$$

The logarithm of the abelian group

$$\mathbf{U}(1) \longrightarrow \log \mathbf{U}(1),\ \ u = e^{i\alpha z} \longmapsto \log u = \alpha(\tfrac{d}{d\alpha} u) u^{-1} = \alpha \tfrac{d}{d\alpha} \log u = i\alpha z,$$

is embedded into the nonabelian logarithm, locally ($\alpha < 2\pi$) defined by a sum of "directed" logarithms:

$$\mathbf{SU}(2) \ni e^{i\vec{\alpha}} \longmapsto i\vec{\alpha} \in A_1^c,\ \ u \longmapsto \log u = \alpha_a \log^a u.$$

The directed logarithms with derivatives with respect to the Lie parameters $\partial^a = \frac{\partial}{\partial \alpha_a}$ define a Lie algebra basis $\{i\sigma^a(u)\}$ at each group element, related to each other by the Lie-Jacobi isomorphisms $u_* : A_c^1 \longrightarrow A_c^1$:

$$\begin{array}{rl} \log^a e^{i\vec{\alpha}} = (\partial^a e^{i\vec{\alpha}}) \circ e^{-i\vec{\alpha}} & = i\sigma^a(u) = (u_*)_b^a i\sigma^b,\ \ i\sigma^a(\mathbf{1}_2) = i\sigma^a, \\ (u_*)_b^a & = \delta_{ab} \frac{\sin 2\alpha}{2\alpha} + \epsilon^{abc} \frac{\alpha_c}{\alpha} \frac{1 - \cos 2\alpha}{2\alpha} + \frac{\alpha_a \alpha_b}{\alpha^2}(1 - \frac{\sin 2\alpha}{2\alpha}), \\ \log u = \alpha_a \log^a u & = \alpha_a i\sigma^a(u) = i\vec{\alpha}. \end{array}$$

3.4 Spinor Reflections

By definition, the complex representation of the compact Lie algebra A_1^c and its unitary Lie group $\mathbf{SU}(2)$ is positive self-dual with a *Euclidean conjugation*, an antilinear reflection

$$\begin{array}{ccc} V & \xrightarrow{l,\ u} & V \\ \star \downarrow & & \downarrow \star \\ V^T & \xrightarrow[-l^T,\ u^{-1T}]{} & V^T \end{array}, \qquad \begin{array}{rcrl} u \in \mathbf{SU}(2) & \iff & u^{-1} & = u^\star, \\ l \in A_1^c & \iff & -l & = l^\star, \\ & \vec{\sigma} = \vec{\sigma}^\star, & \vec{\sigma}_B^A & = \delta^{AC} \overline{\vec{\sigma}_C^D} \delta_{DB}. \end{array}$$

The Pauli representations of $\mathbf{SL}(\mathbb{C}^2)$ and A_1 by special automorphisms and traceless endomorphisms respectively leaves invariant the volume elements of the spinor space. Only for dimension two, $V \cong \mathbb{C}^2$, are the volume elements bilinear forms (inner products), also called *spinor "metric"* (they do not define a topology):

$$\epsilon^1(\ ,\ ) : V \times V \longrightarrow \mathbb{C}, \quad \begin{cases} \epsilon^1(v,w) = -\epsilon^1(w,v),\ \epsilon^1(e^A,e^B) = \epsilon^{AB} = -\epsilon^{BA}, \\ \epsilon_{AB}\epsilon^{BC} = \delta_A^C,\quad \epsilon_{12} = 1,\quad \det \epsilon^1 = 1. \end{cases}$$

The related symplectic *spinor dual isomorphism* ϵ^1, a linear reflection, gives the self-duality of the Pauli representation:

$$\begin{array}{ccc} V & \xrightarrow{l,\ s} & V \\ \epsilon^1 \downarrow & & \downarrow \epsilon^1 \\ V^T & \xrightarrow[-l^T,\ s^{-1T}]{} & V^T \end{array}, \quad \begin{array}{rcl} s \in \mathbf{SL}(\mathbb{C}^2) \iff & s^{-1} & = \epsilon^{-1} \circ s^T \circ \epsilon^1, \\ l \in A_1 \iff & -l & = \epsilon^{-1} \circ l^T \circ \epsilon^1, \\ -\vec{\sigma} = \epsilon^{-1} \circ \vec{\sigma}^T \circ \epsilon^1, & -\vec{\sigma}_B^A & = \epsilon^{AC}\vec{\sigma}_C^D\epsilon_{DB}, \end{array}$$

$$\begin{pmatrix} 0 & -1 \\ 1 & 0 \end{pmatrix}\begin{pmatrix} \alpha_3 & \alpha_1 + i\alpha_2 \\ \alpha_1 - i\alpha_2 & -\alpha_3 \end{pmatrix}\begin{pmatrix} 0 & 1 \\ -1 & 0 \end{pmatrix} = -\begin{pmatrix} \alpha_3 & \alpha_1 - i\alpha_2 \\ \alpha_1 + i\alpha_2 & -\alpha_3 \end{pmatrix}.$$

The existence of an invariant inner product (bilinear form) for the defining representation and the related self-duality distinguishes the case $n = 2$ in the groups $\{\mathbf{SL}(\mathbb{C}^n) \mid n \geq 2\}$ and $\{\mathbf{SU}(n)) \mid n \geq 2\}$ with the $\mathbb{C}^n$-volume elements as invariant n-linear forms.

Since there is only one $\mathbf{SL}(\mathbb{C}^2)$-invariant (Lie algebra has rank 1), the invariant "double" trace for the Lie algebra is proportional to the invariant determinant:

$$l, m \in \mathbf{AL}(\mathbb{C}^2)_0 : \quad \begin{array}{rl} \kappa^1(l,l) & = \operatorname{tr} l \circ l = -2 \det l, \\ \operatorname{tr} \vec{\alpha}^2 & = -2 \det \vec{\alpha} = 2\vec{\alpha}^2, \\ \kappa^1(l,m) & = -\frac{\det(l+m) - \det(l-m)}{2}. \end{array}$$

The linear and antilinear spinor reflections define reflections of the Pauli algebra

$$\mathbf{AL}(V) \xleftrightarrow{\epsilon^1, \star} \mathbf{AL}(V), \quad \begin{array}{l} f \overset{\epsilon^1}{\leftrightarrow} \epsilon^{-1} \circ f^T \circ \epsilon^1, \\ f \overset{\star}{\leftrightarrow} f^\star, \end{array}$$

ϵ^1 realizes the *Lie algebra inversion* of A_1 and $A_1^{\mathbb{c}}$:

$$A_1 \overset{\epsilon^1}{\leftrightarrow} A_1, \quad l \leftrightarrow -l \text{ with } \begin{array}{rl} \epsilon^{-1} \circ (\mathbf{1}_2, \vec{\sigma})^T \circ \epsilon^1 & = (\mathbf{1}_2, -\vec{\sigma}), \\ (\mathbf{1}_2, \vec{\sigma})^\star & = (\mathbf{1}_2, \vec{\sigma}). \end{array}$$

3.5 Spin Representations

Each complex finite-dimensional representation of the simple Lie algebra A_1 and its Lie group $\mathbf{SL}(\mathbb{C}^2)$ and each complex representation of the compact spin Lie algebra A_1^c and of the compact spin group $\mathbf{SU}(2)$ is semisimple, i.e., decomposable into irreducible ones.

For complex representations of the real Lie structures A_1^c and $\mathbf{SU}(2)$ a positive conjugation $\star$ of the representation space W, i.e., a Hilbert space structure, is necessary. Therefore each irreducible finite-dimensional A_1- and $\mathbf{SL}(\mathbb{C}^2)$-representation gives an irreducible $\star$-antisymmetric and $\star$-unitary representation of its compact form; in such a way all irreducible A_1^c- and $\mathbf{SU}(2)$-representations are obtained. The self-dual complex A_1-representation space $\mathbf{W} \cong W \oplus W^T$ is decomposable into the $\star$-symmetric and $\star$-antisymmetric real subspaces $\mathbf{W}_+, \mathbf{W}_-$ with equivalent representations of A_1^c.

The complex powers $\zeta \in \mathbb{C}$, characterizing the irreducible number representations $\mathbb{C}^\diamond \ni e^z \longrightarrow (e^z)^\zeta \in \mathbb{C}^\diamond$ (chapter "Time Representations"), are used, for $\mathbf{SL}(\mathbb{C}^2)$, in the representation of a maximal abelian subgroup with dimension one (rank 1): $\mathbf{SL}(\mathbb{C}^2) \ni e^{z_3\sigma_3} \longrightarrow (e^{z_3\sigma_3})^\zeta \in \mathbf{SL}(\mathbb{C}^2)$. Finite-dimensional representations require integer powers $\zeta \in \mathbb{Z} \oplus i\mathbb{Z}$. Continuous powers are necessary for faithful Hilbert representations of $\mathbf{SL}(\mathbb{C}^2)$, which have to be infinite-dimensional (chapter "Harmonic Analysis").

Because of rank 1, the finite-dimensional irreducible representations are characterized by *one* natural number $n = 0, 1, \dots$:

$$\begin{array}{lll} A_1, A_1^c \longrightarrow \mathbf{AL}(W)_0, & l \longmapsto \mathcal{D}^n(l), & \\ \mathbf{SL}(\mathbb{C}^2), \mathbf{SU}(2) \longrightarrow \mathbf{SL}(W), & g \longmapsto D^n(g), & \end{array} \quad W \cong \mathbb{C}^{1+n};$$

$\mathcal{D}^0$ and D^0 are the trivial representations by $\{0\}$ (Lie algebra) and $\{1\}$ (Lie group) respectively; $n = 1$ characterizes the defining fundamental Pauli spinor representation

$$l = \mathcal{D}^1(l) \cong i\vec{\alpha}, \quad g = D^1(g) \cong e^{i\vec{\alpha}},$$

with complex and real parameters $(\alpha_a)_{a=1,2,3}$; $J = \frac{n}{2}$ is called *spin*, with $1+n = 1+2J$ the dimensionality of the representation.

The irreducible representations arise, up to equivalence, by the *totally symmetric tensor powers* of the fundamental one:

$$\begin{aligned} W \cong \bigvee^n V \subseteq \bigotimes^n V, &\quad V \cong \mathbb{C}^2, \ W \cong \mathbb{C}^{1+n}, \\ \mathcal{D}^n(l) \cong \sum_{k=1}^n \underbrace{\mathrm{id}_V \otimes \cdots \otimes l \otimes \cdots \otimes \mathrm{id}_V}_{n \text{ factors, } k\text{th place}}, &\quad l \in A_1 \text{ and } A_1^c, \\ D^n(g) \cong \bigvee^n g, &\quad g \in \mathbf{SL}(\mathbb{C}^2) \text{ and } \mathbf{SU}(2); \end{aligned}$$

l is extended, as derivation, to the tensor products.

Sometimes, representations $D^n, \mathcal{D}^n$ with even dimension (odd tensor power n, half-integer spin $J = \frac{n}{2}$) are called *spinor representations*, those with odd dimension (even tensor power n, integer spin $J = \frac{n}{2}$) *vector representations*. (Obviously, also spinors are vectors.) The vector representation spaces

$\bigvee^{2n} \mathbb{C}^2 \cong \mathbb{C}^{1+2n}$ are up to $n = 0, 1$ proper subspaces in the symmetric products of the Lie algebras A_1 and A_1^c respectively, $\bigvee^N \mathbb{K}^3 \cong \mathbb{K}^{\binom{2+N}{2}}$; i.e., those totally symmetric product representations of the adjoint representation on $\mathbb{C}^3$ or $\mathbb{R}^3$ are decomposable.

In the following the integer power notation D^{2J} is used in contrast to the perhaps more familiar spin notation D^J. Both notations have advantages and disadvantages.

$\mathbf{SU}(2)$ has a nontrivial discrete center

$$\text{centr}\, \mathbf{SU}(2) = \{\pm \mathbf{1}_2\} \cong \mathbb{I}(2).$$

The adjoint group of $\mathbf{SU}(2)$ is the locally isomorphic rotation group

$$\text{Int}\, \mathbf{SU}(2) = \mathbf{SU}(2)/\mathbb{I}(2) \cong \mathbf{SO}(3);$$

$\mathbf{SU}(2)$-vector representations realize $\mathbb{I}(2)$ trivially, and therefore faithfully, only the classes for the rotations $\mathbf{SO}(3)$:

$$\vec{e}^2 = 1 \Rightarrow e^{i\pi\vec{e}} = -\mathbf{1}_2, \quad D^n(e^{i\pi\vec{e}}) = (-1)^n \mathbf{1}_{1+n}.$$

The faithfulness of the center representation defines the *two-ality* of the irreducible $\mathbf{SU}(2)$-representations, trivial for the vector representations

$$2J \bmod 2 = \begin{cases} 0, & J = \frac{n}{2} = 0, 1, \ldots, \\ 1, & J = \frac{n}{2} = \frac{1}{2}, \frac{3}{2}, \ldots. \end{cases}$$

All irreducible representations have a nondegenerate, invariant inner product ("metric"). Therefore they are self-dual

$$\mathcal{D}^n(l) \cong \mathcal{D}^n(-l)^T, \quad D^n(g) \cong D^n(g^{-1})^T,$$

symplectically for even dimension (half-integer spin) and orthogonally for odd dimension (integer spin):

$$\begin{aligned} \epsilon^n : \bigvee^n V &\longrightarrow \bigvee^n V^T, \quad \epsilon^n = \bigvee^n \epsilon^1, \\ \epsilon^n \simeq \epsilon_{c,d} &= \epsilon_{C_1 \ldots C_n, D_1 \ldots D_n} = \delta^{A_1 \ldots A_n}_{C_1 \ldots C_n} \delta^{B_1 \ldots B_n}_{D_1 \ldots D_n} \epsilon_{A_1 B_1} \cdots \epsilon_{A_n B_n} \\ &= \begin{cases} -\epsilon_{d,c} & \text{for } n = 2J = 1, 3, \ldots, \\ +\epsilon_{d,c} & \text{for } n = 2J = 0, 2, \ldots. \end{cases} \end{aligned}$$

Therefore the representation images of the group and Lie algebra are subgroups of symplectic and orthogonal automorphism groups or corresponding subalgebras:

$$\begin{aligned} \mathcal{D}^n[A_1] &\subseteq \begin{cases} \log \mathbf{Sp}(\mathbb{C}^{1+n}) & \text{for } n = 2J = 1, 3, \ldots, \\ \log \mathbf{SO}(\mathbb{C}^{1+n}) & \text{for } n = 2J = 0, 2, \ldots, \end{cases} \\ \mathcal{D}^n[A_1^c] &\subseteq \begin{cases} \log \mathbf{SpU}(1+n) & \text{for } n = 2J = 1, 3, \ldots, \\ \log \mathbf{SO}(1+n) & \text{for } n = 2J = 0, 2, \ldots. \end{cases} \end{aligned}$$

The compact groups are defined by

$$\mathbf{SpU}(2N) = \mathbf{Sp}(\mathbb{C}^{2N}) \cap \mathbf{SU}(2N), \quad \mathbf{SO}(N) \cong \mathbf{SO}(\mathbb{C}^N) \cap \mathbf{SU}(N).$$

The Euclidean conjugation $\delta^1 = \star$, necessary for the fundamental $\mathbf{SU}(2)$-representation induces Euclidean conjugations $\delta^n = \bigvee^n \delta^1 = \star$ for the product representations. The odd-dimensional representation Hilbert spaces of A_1^c and $\mathbf{SO}(3)$ can be chosen to be real $\mathbb{R}^{1+2n}$ with coinciding conjugation and metric $\epsilon^{2n} = \delta^{2n}$.

The product of the "metrics" $\epsilon^n \otimes (\epsilon^n)^{-1}$ gives the associate inner product κ^n ("double trace") of the Lie algebra

$$\kappa^n : A_1 \times A_1 \longrightarrow \mathbb{C}, \quad \kappa^n(l^a, l^b) = \operatorname{tr} \mathcal{D}^n(l^a) \circ \mathcal{D}^n(l^b) = -\tfrac{1}{2}\tbinom{2+n}{3}\delta^{ab},$$

where the factor is computed with

$$\kappa^n(l^a, l^b) = \delta^{ab}\kappa^n(l^3, l^3) \text{ and } \sum_{j=-J}^{J} j^2 = \tfrac{1}{2}\tbinom{2+2J}{3}.$$

The Casimir element uses the inverse Killing form, in Euclidean bases $\kappa_{ab} = -\frac{\delta_{ab}}{2}$. It is the identity with a representation characteristic scalar factor

$$I^n(A_1) = -\tfrac{\delta_{ab}}{2}\mathcal{D}^n(l^a) \circ \mathcal{D}^n(l^b) = \tfrac{n(2+n)}{8}\,\mathrm{id}_W = \tbinom{1+J}{2}\,\mathrm{id}_W.$$

A tensor product of irreducible representations for $\mathbf{SL}(\mathbb{C}^2)$ and A_1 or $\mathbf{SU}(2)$ and A_1^c has as Clebsch-Gordan decomposition into irreducible representations:

$$\mathcal{D}^{2J_1} \otimes \mathcal{D}^{2J_2} \cong \bigoplus_{J=|J_1-J_2|}^{J_1+J_2} \mathcal{D}^{2J}, \quad D^{2J_1} \otimes D^{2J_2} \cong \bigoplus_{J=|J_1-J_2|}^{J_1+J_2} D^{2J}.$$

An example: The tensor product decomposition for two fundamental representations gives in addition to the trivial representation the adjoint one (fundamental vector representation)

$$D^1 \otimes D^1 \cong D^0 \oplus D^2, \quad e^A \otimes \check{e}_B = \tfrac{1}{2}\delta^A_B(e^C \otimes \check{e}_C) + \tfrac{1}{2}\vec{\sigma}^A_B(e^C \vec{\sigma}^D_C \otimes \check{e}_D),$$

with the involutory Fierz recouplings of the Pauli spinor representations

$$\begin{pmatrix} \mathbf{1}_2 \otimes \mathbf{1}_2 \\ \vec{\sigma} \otimes \vec{\sigma} \end{pmatrix}^{AD}_{CB} \overset{\text{Fierz}}{=} \begin{pmatrix} \frac{1}{2} & \frac{1}{2} \\ \frac{3}{2} & -\frac{1}{2} \end{pmatrix} \begin{pmatrix} \mathbf{1}_2 \otimes \mathbf{1}_2 \\ \vec{\sigma} \otimes \vec{\sigma} \end{pmatrix}^{DA}_{CB}, \quad \epsilon^{AD}\epsilon_{CB} = \tfrac{1}{2}\delta^A_B\delta^D_C - \tfrac{1}{2}\vec{\sigma}^A_B\vec{\sigma}^D_C.$$

Since the square of a recoupling matrix (with "$6J$-coefficients" as matrix elements) as a reflection has to be the unit matrix, one has the recoupling eigenvalues $\{\pm 1\}$ for the totally (anti)symmetric combinations as eigenvectors:

$$\begin{pmatrix} S \\ T \end{pmatrix} \cong \begin{pmatrix} \mathbf{1}_2 \otimes \mathbf{1}_2 \\ \vec{\sigma} \otimes \vec{\sigma} \end{pmatrix}, \quad \begin{pmatrix} S - T \\ S + \frac{T}{3} \end{pmatrix} \overset{\text{Fierz}}{\longleftrightarrow} \begin{pmatrix} -(S - T) \\ S + \frac{T}{3} \end{pmatrix}.$$

With the self-duality of the representations, the Pauli matrices $\vec{\sigma}_C^B$ lead also to symmetric Clebsch-Gordan coefficients: In the case $D^1 \otimes D^1 \cong D^0 \oplus D^2$ they project to the adjoint representation $D^2 \cong D^1 \vee D^1$ with $\vec{\sigma}_{AC} = \vec{\sigma}_C^B \epsilon_{BA} = \vec{\sigma}_{CA}$. The dual isomorphism ϵ_{AB} gives the antisymmetric Clebsch-Gordan coefficients for the projection to the trivial representation $D^0 \cong D^1 \wedge D^1$.

The Cartan subalgebras of the Lie algebras A_1 and A_1^n, A_1^c are 1-dimensional, e.g., $\mathbb{C}h$ and $i\mathbb{R}h, \mathbb{R}l^3$. They lead to Cartan subgroups, e.g.,

$$e^{i\mathbb{R}h} \cong \{e^{i\alpha_3\sigma^3} \mid \alpha_3 \in \mathbb{R}\} \cong \mathbf{SO}(2) \subset \mathbf{SU}(2).$$

Because of the "special" group and the "traceless" Lie algebra, the eigenvalues (third spin components) of the Cartan group representations come in reflected pairs. They are $\mathbf{U}(1)$-winding numbers $z = 2j$ with $|z| \leq n = 2J$:

$$\begin{array}{l} \text{Pauli representation: } \operatorname{spec} h = \operatorname{spec} \sigma^3 = \{2j \mid j = \pm\frac{1}{2}\}, \\ \operatorname{spec} \mathcal{D}^{2J}(h) = \{2j \mid j = -J, -J+1, \dots, J-1, J\} \subset \mathbb{Z}. \end{array}$$

All complex representation spaces for A_1, A_1^n, and A_1^c can be spanned by spin eigenvectors of a Cartan subalgebra (diagonalizable). The real representation spaces $\mathbb{R}^{1+2n}$, $n = 0, 2, 4, \dots$, of $\log \mathbf{SO}(3) \cong A_1^c$ do not have an angular momentum eigenvector basis; a Cartan subalgebra is only box-diagonalizable; there exist no real $\mathbf{SO}(2)$-eigenvectors with nontrivial eigenvalue. Two complex 1-dimensional nontrivial $\mathbf{U}(1)$-representations with reflected winding numbers $\pm 2j \in 2\mathbb{Z}$ come together in one real 2-dimensional nondiagonalizable $\mathbf{SO}(2)$-representation

$$\begin{pmatrix} e^{i\alpha_3 2j} & 0 \\ 0 & e^{-i\alpha_3 2j} \end{pmatrix} \cong \begin{pmatrix} \cos\alpha_3 2j & \sin\alpha_3 2j \\ -\sin\alpha_3 2j & \cos\alpha_3 2j \end{pmatrix}.$$

$\mathbf{SO}(2)$ has no eigenvectors in $\mathbb{R}^2$.

3.6 Position Translations from Adjoint Spin Structures

All structures of the real 3-dimensional position translations are "very close" - up to multiplication with imaginary i - to adjoint spin structures.

With $\mathbf{SU}(2)$ as underlying group, the position translations are, as a vector space with rotation group action, not as a Lie algebra, isomorphic to the Lie algebra $\log \mathbf{SU}(2) = A_1^c$ with adjoint $\mathbf{SU}(2)$-action. The position translations can be represented, with imaginary scalar multiplication of the Lie algebra, by the real vector space iA_1^c. Their *Pauli representation* is constituted by the Euclidean symmetric traceless endomorphisms in the Pauli algebra $\mathbf{AL}(\mathbb{C}^2)$, i.e., by (2×2) matrices:

$$\begin{aligned} \mathbb{S} \cong\ & \{\vec{x} : \mathbb{C}^2 \longrightarrow \mathbb{C}^2 \mid \vec{x} = \vec{x}^\star, \quad \operatorname{tr} \vec{x} = 0\} = iA_1^c \cong \mathbb{R}^3, \\ & \vec{x} = x_a \sigma^a = \begin{pmatrix} x_3 & x_1 - ix_2 \\ x_1 + ix_2 & -x_3 \end{pmatrix}, \end{aligned}$$

with a translation basis $\{\sigma^a\}_{a=1,2,3}$. The position translations can be interpreted as linear transformations of Pauli spinors.

In contrast to the time translations $\mathbb{T} \cong \mathbb{R}$ as Lie algebra of the noncompact causal group $\mathbf{D}(1)$, the position translations $\mathbb{S} \cong \mathbb{R}^3$ are not the Lie algebra of the compact spin $\mathbf{SU}(2)$ group; they have only an isomorphic vector space structure. Both time and position translations can be embedded into tangent structures of the extended noncompact Lorentz group (chapter "Lorentz Symmetry").

The semidirect product of the rotations acting on the position translations is isomorphic to the adjoint affine spin group:

$$\begin{array}{lcll} \mathbf{SO}(3) & \vec{\times} & \mathbb{R}^3 \cong \mathbf{SU}(2) & \vec{\times} \quad iA_1^c, \\ \log \mathbf{SO}(3) & \vec{\oplus} & \mathbb{R}^3 \cong A_1^c & \vec{\oplus} \quad iA_1^c. \end{array}$$

The rotation action on the position translations arises in the form of inner automorphisms with $\mathbf{SU}(2)$; the Lie algebra A_1^c acts adjointly:

$$\begin{array}{ll} u \in \mathbf{SU}(2): \quad \mathrm{Ad}\, u : iA_1^c \longrightarrow iA_1^c, & \left\{ \begin{array}{l} \mathrm{Ad}\, u(\vec{x}) = u \circ \vec{x} \circ u^\star, \quad u^\star = u^{-1}, \\ \sigma^a \longmapsto u \circ \sigma^a \circ u^{-1} = O_b^a(u)\sigma^b, \end{array} \right. \\ l \in A_1^c: \quad \mathrm{ad}\, l : iA_1^c \longrightarrow iA_1^c, & \left\{ \begin{array}{l} \mathrm{ad}\, l(\vec{x}) = [l, \vec{x}], \\ \sigma^a \longmapsto [l^a, \sigma^b] = -\epsilon^{abc}\sigma^c \text{ with } l^a \cong \frac{i}{2}\sigma^a. \end{array} \right. \end{array}$$

With an appropriate normalization $\sigma_{\mathbf{P}}^2$ the associate inner product (double trace on $iA_1^c \times iA_1^c$) is the Euclidean scalar product σ of the position translations:

$$\begin{array}{rcll} \frac{\sigma(\vec{x},\vec{y})}{\sigma_{\mathbf{P}}^2} & = & \frac{1}{2}\mathrm{tr}\, \vec{x} \circ \vec{y} & = \frac{1}{2}\kappa^1(\vec{x},\vec{y}) = \frac{1}{2}x_a y_b \,\mathrm{tr}\, \sigma^a\sigma^b, \quad \frac{1}{2}\mathrm{tr}\, \vec{x} \circ \vec{x} = -\det \vec{x}, \\ \kappa^1 & = & \epsilon^1 \otimes \epsilon^{-1} & \cong \epsilon^{AD}\epsilon_{BC}, \\ & & \frac{1}{2}\mathrm{tr}\, \sigma^a\sigma^b & = \frac{1}{2}\epsilon^{AD}\epsilon_{BC}(\sigma^a)_A^B(\sigma^b)_D^C = \delta^{ab}. \end{array}$$

With $\mathbf{SU}(2)$-parameters the normalized rotation axis is $\frac{\vec{\alpha}}{\alpha}$, the $\mathbf{SO}(3)$-rotation angle uses the doubled angle $\varphi = \sqrt{\vec{\varphi}^2} = 2\alpha$; explicitly,

$$\begin{array}{l} u(\vec{\alpha}) = e^{i\vec{\alpha}} \in \mathbf{SU}(2), \quad O(u) = O(\vec{\varphi}) = e^{\varphi_c \mathcal{O}^c} \in \mathbf{SO}(3), \quad \mathcal{O}^c = \mathrm{ad}\, l^c, \ \vec{\varphi} = 2\vec{\alpha}, \\ O_b^a = e^{-\varphi_c \epsilon^{abc}} = \frac{1}{2}\mathrm{tr}\, u\sigma^a u^\star \sigma^b = \delta^{ab}\cos\varphi - \epsilon^{abc}\varphi_c \frac{\sin\varphi}{\varphi} + \varphi_a\varphi_b \frac{1-\cos\varphi}{\varphi^2}, \end{array}$$

e.g., the matrix for a rotation around the third axis:

$$\vec{e}_3 = \begin{pmatrix} 0 \\ 0 \\ 1 \end{pmatrix} \Rightarrow u(\tfrac{\varphi\vec{e}_3}{2}) = \begin{pmatrix} e^{i\frac{\varphi}{2}} & 0 \\ 0 & e^{-i\frac{\varphi}{2}} \end{pmatrix}, \quad O(\varphi\vec{e}_3) = \begin{pmatrix} \cos\varphi & -\sin\varphi & 0 \\ \sin\varphi & \cos\varphi & 0 \\ 0 & 0 & 1 \end{pmatrix}.$$

The simply connected compact spin Lie group can be parametrized by the points of a full 3-sphere:

$$\mathbf{SU}(2) \cong \{\vec{\alpha} \in \mathbb{R}^3 \mid |\vec{\alpha}| \le 2\pi\} \cong \Omega^3 \cong \mathbf{SO}(4)/\mathbf{SO}(3).$$

The $\mathbf{SO}(3)$-manifold is the full sphere with an identification of antipodal points. Related to the nontrivially represented center, this identification is removed for the $\mathbf{SU}(2)$-manifold:

$$2\alpha = \varphi \in [-\pi, \pi] \Rightarrow \left\{ \begin{array}{lll} \cos\pi & = -1 = \cos(-\pi) & \text{for } \mathbf{SO}(3), \\ e^{i\frac{\pi}{2}} & = i = -e^{-i\frac{\pi}{2}} & \text{for } \mathbf{SU}(2). \end{array} \right.$$

3.7 Polynomials with Spin Group Action

For a compact Lie group, all irreducible Lie algebra representations can be obtained as derivatives acting on finite-degree polynomials. The polynomials in the vectors of any spin representation space carry the spin action via derivatives that extend the linear forms with dual bases

$$\langle \check{e}_B, e^A\rangle = \partial_B e^A = \delta_B^A \text{ from } \partial_B = \frac{\partial}{\partial e^B}.$$

The fundamental Pauli spinor and the adjoint angular momentum representations are of special interest.

3.7.1 Spinor Polynomials

All irreducible representations $\mathcal{D}^{2J}$ with $J = 0, \frac{1}{2}, 1, \ldots$ for the Lie algebra A_1 can be realized by derivations of the complex *Pauli spinor polynomials* in two indeterminates (polynomial algebra $\bigvee \mathbb{C}^2$) taken from a $V \cong \mathbb{C}^2$-basis $\{e^1, e^2\}$ for the fundamental representation

$$\mathbb{C}[e^1, e^2] \cong \bigvee \mathbb{C}^2$$

by the induced derivations

$$A_1 \longrightarrow \operatorname{der} \mathbb{C}[e^1, e^2],\ \ \vec{l} = \tfrac{i}{2}\vec{\sigma}_A^B e^A \otimes \check{e}_B \longmapsto \vec{l}_{\text{der}} = \tfrac{i}{2}\vec{\sigma}_A^B e^A \partial_B.$$

The totally symmetric tensor powers $\bigvee^{2J} \mathbb{C}^2$ of the Pauli spinors as irreducible $\mathbf{SU}(2)$-representation spaces are isomorphic to the spinor polynomials, *homogeneous* of degree $2J = 0, 1, 2, \ldots$:

$$\text{basis of } \mathbb{C}[e^1, e^2]^{2J} \cong \bigvee^{2J} \mathbb{C}^2 \cong \mathbb{C}^{1+2J} :\ \{(e^1)^{J+j}(e^2)^{J-j} \mid j = -J, \ldots, J\}.$$

The monomials in the given basis are Cartan subalgebra eigenvectors (third spin component):

$$h_{\text{der}} = e^1\partial_1 - e^2\partial_2 \Rightarrow \begin{cases} h \bullet e^1 = e^1,\ \ h \bullet e^2 = -e^2, \\ h \bullet (e^1)^{J+j}(e^2)^{J-j} = 2j(e^1)^{J+j}(e^2)^{J-j}. \end{cases}$$

3.7.2 Harmonic Polynomials and Spherical Harmonics

All irreducible representations $\mathcal{D}^{2L}$ with integer spin $L = 0, 1, 2, \ldots$ for the Lie algebra A_1 can be realized by derivations of the complex *position polynomials* $\bigvee \mathbb{C}^3$ in three indeterminates from a basis $\{l_\pm, h\}$ of an adjoint A_1-representation space

$$\mathbb{C}[l_\pm, h] \cong \bigvee \mathbb{C}^3.$$

The adjoint representation of the Lie algebra A_1 with an eigenvector basis (Weyl basis)

$$[l_-, l_+] = h, \quad [h, l_\pm] = \pm 2 l_\pm$$

induces the derivative action

$$A_1 \longrightarrow \operatorname{der} \mathbb{C}[l_\pm, h], \quad l \longmapsto \operatorname{ad} l_{\text{der}} = \mathcal{L}, \quad \begin{cases} \operatorname{ad} l^a_{\text{der}} & = \mathcal{L}^a = -\epsilon_{abc} l^c \frac{\partial}{\partial l^b}, \\ \partial_\pm & = \frac{\partial}{\partial l_\pm}, \ \partial_0 = \frac{\partial}{\partial h}, \\ \mathcal{L}_+ & = -h\partial_- - 2l_+\partial_0, \\ \mathcal{L}_- & = h\partial_+ + 2l_-\partial_0, \\ \mathcal{H} & = 2(l_+\partial_+ - l_-\partial_-). \end{cases}$$

The compact Lie algebra A_1^c is representable with a Cartesian basis with position translations

$$\begin{aligned} \{x^a \mid a = 1,2,3\} : & \begin{pmatrix} l_+ \\ l_- \\ h \end{pmatrix} = \begin{pmatrix} -i & 1 & 0 \\ i & 1 & 0 \\ 0 & 0 & -2i \end{pmatrix} \begin{pmatrix} x^1 \\ x^2 \\ x^3 \end{pmatrix}, \\ \partial_a = \frac{\partial}{\partial x^a} : & \begin{pmatrix} \partial_+ \\ \partial_- \\ \partial_0 \end{pmatrix} = \frac{1}{2} \begin{pmatrix} i & -i & 0 \\ 1 & 1 & 0 \\ 0 & 0 & i \end{pmatrix} \begin{pmatrix} \partial_1 \\ \partial_2 \\ \partial_3 \end{pmatrix}, \\ \mathcal{L}^a = \epsilon_{abc} x^b \partial^c. & \end{aligned}$$

Its action leaves invariant the real subspaces in the decomposition

$$\mathbb{C}[l_\pm, h] = \mathbb{C}[x^a] \cong \mathbb{R}[x^a] \oplus i\mathbb{R}[x^a];$$

$\mathbb{R}[x^a]$ are the real polynomials in the position translations.

The A_1-polynomials are decomposable into *homogeneous* polynomials $\bigvee^N A_1$ of degree $N = L_{\max} = 0, 1, \ldots$:

$$\bigvee^N \mathcal{D}^2 : A_1 \longrightarrow \mathbf{AL}(\mathbb{C}[l_\pm, h]^N), \quad \mathbb{C}[l_\pm, h]^N = \bigvee^N A_1 \cong \mathbb{C}^{\binom{2+N}{2}}.$$

In contrast to the Pauli spinor polynomials, homogeneity of position polynomials does not entail irreducibility. Using the rotation-invariant degree-2 polynomial (Casimir element)

$$l_+ l_- - \tfrac{1}{4} h^2 = (x^1)^2 + (x^2)^2 + (x^2)^2 = \delta_{ab} x^a x^b = r^2,$$

the degree-N position polynomials are decomposable

$$\bigvee^N \mathcal{D}^2 = \begin{cases} \bigoplus_{L=0,2,\ldots,N} \mathcal{D}^{2L}, & \binom{2+N}{2} = \sum_{L=0,2,\ldots,N} (1+2L), \quad N = 0, 2, \ldots, \\ \bigoplus_{L=1,3,\ldots,N} \mathcal{D}^{2L}, & \binom{2+N}{2} = \sum_{L=1,3,\ldots,N} (1+2L), \quad N = 1, 3, \ldots, \end{cases}$$

into irreducible $\mathbf{SO}(3)$-representation spaces with $\vec{x}^2$-powers as factors and even or odd integer spin L for even and odd degree N, respectively:

$$\begin{aligned} \mathbb{C}[l_\pm, h]^0 &= \mathbb{C}, \quad \mathbb{C}[l_\pm, h]^1 \cong \mathbb{C}^3, \\ \mathbb{C}[l_\pm, h]^N &= \vec{x}^2 \mathbb{C}[l_\pm, h]^{N-2} \oplus \mathbb{C}^{1+2N} \cong \mathbb{C}^{\binom{2+N}{2}}, \quad N = 2, 3, \ldots, \\ &\cong \begin{cases} r^N \mathbb{C} \oplus r^{N-2}\mathbb{C}^5 \oplus \ldots \oplus \mathbb{C}^{1+2N}, & N = 0, 2, \ldots, \\ r^{N-1}\mathbb{C}^3 \oplus r^{N-3}\mathbb{C}^7 \oplus \ldots \oplus \mathbb{C}^{1+2N}, & N = 1, 3, \ldots. \end{cases} \end{aligned}$$

With a Cartesian basis and the invariant inner product for the subtraction of the "traces" one has the decompositions into irreducible position polynomials $[x]^L$ of degree L with the bases

$$\begin{array}{lll} N=0: & \{1\}, & \\ N=1: & \{x^a \mid a=1,2,3\}, & \\ N=2: & \{x^a x^b\} & \cong \{\vec{x}^2\} \oplus \{x^a x^b - \frac{\vec{x}^2}{3}\delta^{ab}\}, \\ N=3: & \{x^a x^b x^c\} & \cong \{\vec{x}^2 x^a\} \oplus \{x^a x^b x^c - \frac{\vec{x}^2}{3}(\delta^{ab}x^c + \delta^{ac}x^b + \delta^{bc}x^a)\}, \\ & \ldots. & \end{array}$$

Only the complex polynomials have a basis with eigenvectors of a Cartan subalgebra.

The irreducible representation with the highest spin in the homogeneous degree-N polynomials $\mathbb{C}[l_\pm, h]^N$ are the *harmonic polynomials* with vanishing invariant derivative

$$\vec{\partial}^2 P^N(\vec{x}) = 0 \text{ for } \begin{cases} P^0(\vec{x}) & = 1, \\ P^1(\vec{x}) & \in \{x^a\}, \\ P^2(\vec{x}) & \in \{x^a x^b - \frac{\vec{x}^2}{3}\delta^{ab}\}, \\ \ldots. & \end{cases}$$

They can be obtained as follows: The eigenvalue $2m$ of a Cartan subalgebra eigenpolynomial y,

$$\mathrm{y} \in \mathbb{C}[l_\pm, h] \quad \text{with } \mathcal{H}(\mathrm{y}) = 2m\mathrm{y}, \quad 2m \in \mathbb{Z},$$

is raised (lowered) by the action with $\mathcal{L}_\pm$ by two windings:

$$\mathcal{H}(\mathcal{L}_\pm.\mathrm{y}) = (2m \pm 2)\mathcal{L}_\pm.\mathrm{y} \quad \text{since } [\mathcal{H}, \mathcal{L}_\pm] = \pm 2\mathcal{L}_\pm.$$

Descending with $\mathcal{L}_-$ from the highest spin monomial $(l_+)^N \in \mathbb{C}[l_\pm, h]^N$, one obtains an irreducible A_1-representation $\mathcal{D}^{2N}$ on a $\mathbb{C}^{1+2N}$-isomorphic vector subspace of the N-homogeneous harmonic polynomials with the h-eigenvector basis

$$\{\mathrm{y}^N_m = (\mathcal{L}_-)^{N-m}.(l_+)^N \mid m = -N, \ldots, N\}, \quad \mathcal{H}(\mathrm{y}^N_m) = 2m\, \mathrm{y}^N_m,$$

with the examples for spin (angular momentum) $0, 1, 2$:

$$\begin{array}{ll} N=0: & \{\mathrm{y}^0_0 = 1\}, \\ N=1: & \{\mathrm{y}^1_1 = l_+,\ \mathrm{y}^1_0 = h,\ \mathrm{y}^1_{-1} = 2l_-\}, \\ N=2: & \{\mathrm{y}^2_2 = l_+^2,\ \mathrm{y}^2_1 = 2hl_+,\ \mathrm{y}^2_0 = 2h^2 + 4l_-l_+,\ \mathrm{y}^2_{-1} = 12l_-h,\ \mathrm{y}^1_{-2} = 24l_-^2\}. \end{array}$$

The definite Killing form defines a scalar product

$$\kappa(l^a, l^b) = -2\delta^{ab} \Rightarrow \begin{cases} \langle l_+|l_+\rangle = \langle l_-|l_-\rangle = 1 = -\frac{\kappa(l_+,l_-)}{4}, \\ \langle h|h\rangle = 2 = \frac{\kappa(h,h)}{4}, \end{cases}$$

which can be naturally extended to the totally symmetric tensor products

$\bigvee^N A_1^c$. It is determined up to an N-dependent overall normalization.

From the harmonic polynomials, the *spherical harmonics* as homogeneous polynomials in the position space directions on the 2-sphere $\frac{\vec{x}}{r} \in \Omega^2 \cong \mathbf{SO}(3)/\mathbf{SO}(2)$ are obtained by normalizing with the invariant $r = |\vec{x}|$:

$$\mathrm{Y}_m^N \sim \frac{1}{r^N} \mathrm{y}_m^N \sim \bigvee^N \frac{\vec{x}}{r}\Big|_{\text{traceless}}.$$

They can be expressed with polar coordinates involving the Cartan group parameter $\varphi = 2\alpha_3$:

$$\vec{x} \cong \begin{pmatrix} x^3 & x^1 - ix^2 \\ x^1 + ix^2 & -x^3 \end{pmatrix} \cong r \begin{pmatrix} \cos\varphi \sin\theta \\ \sin\varphi \sin\theta \\ \cos\theta \end{pmatrix}, \quad \begin{cases} \int d^3x &= \int_0^\infty r^2 dr \int d^2\omega, \\ \int d^2\omega &= \int_0^{2\pi} d\varphi \int_0^\pi \sin\theta \, d\theta, \end{cases}$$

e.g., for spin 1 and 2

$$N = 1: \ \vec{x} \cong i \begin{pmatrix} \frac{h}{2} & -l_- \\ l_+ & -\frac{h}{2} \end{pmatrix} = r \sqrt{\frac{4\pi}{3}} \begin{pmatrix} \mathrm{Y}_0^1 & \sqrt{2}\mathrm{Y}_{-1}^1 \\ -\sqrt{2}\mathrm{Y}_1^1 & -\mathrm{Y}_0^1 \end{pmatrix} (\varphi, \theta),$$

$$\frac{1}{r} \begin{pmatrix} l_\pm \\ \frac{h}{\sqrt{2}} \end{pmatrix} = \begin{pmatrix} \mp i e^{\pm i\varphi} \sin\theta \\ -i\sqrt{2}\cos\theta \end{pmatrix} = i\sqrt{\frac{8\pi}{3}} \begin{pmatrix} \mathrm{Y}_{\pm 1}^1 \\ \mathrm{Y}_0^1 \end{pmatrix} (\varphi, \theta),$$

$$N = 2: \quad \frac{1}{r^2} \begin{pmatrix} \frac{l_\pm^2}{\sqrt{2}} \\ \frac{hl_\pm}{\sqrt{2}} \\ \frac{h^2 + 2l_+ l_-}{\sqrt{12}} \end{pmatrix} = \begin{pmatrix} -e^{\pm 2i\varphi} \frac{\sin^2\theta}{\sqrt{2}} \\ \pm e^{\pm i\varphi} \sqrt{2} \cos\theta \sin\theta \\ \frac{1 - 3\cos^2\theta}{\sqrt{3}} \end{pmatrix} = -\sqrt{\frac{16\pi}{15}} \begin{pmatrix} \mathrm{Y}_{\pm 2}^2 \\ \mathrm{Y}_{\pm 1}^2 \\ \mathrm{Y}_0^2 \end{pmatrix} (\varphi, \theta).$$

3.8 Spin Representation Matrix Elements

In addition to the local Lie algebra parametrization of $\mathbf{SU}(2)$ with half the angles $2\vec{\alpha}$ for $\mathbf{SO}(3)$-rotations, related to Cartesian bases

$$\begin{aligned} \mathbf{SU}(2) \ni u = & \begin{pmatrix} z_1 & -\overline{z_2} \\ z_2 & \overline{z_1} \end{pmatrix} = e^{i\vec{\alpha}} = \begin{pmatrix} \cos\alpha + i\frac{\alpha_3}{\alpha} \sin\alpha & \frac{\alpha_2 + i\alpha_1}{\alpha} \sin\alpha \\ -\frac{\alpha_2 - i\alpha_1}{\alpha} \sin\alpha & \cos\alpha - i\frac{\alpha_3}{\alpha} \sin\alpha \end{pmatrix} \\ & \text{with } |z_1|^2 + |z_2|^2 = 1, \qquad |\vec{\alpha}| = \alpha \leq \pi, \\ \mathbf{SO}(3) \ni \tfrac{1}{2} \operatorname{tr} u\sigma^a u^\star \sigma^b \cong & \begin{pmatrix} \mathrm{Re}(z_1^2 - z_2^2) & -\mathrm{Im}(z_1^2 + z_2^2) & -2\,\mathrm{Re}\, z_1 z_2 \\ \mathrm{Im}(z_1^2 - z_2^2) & \mathrm{Re}(z_1^2 + z_2^2) & -2\,\mathrm{Im}\, z_1 z_2 \\ 2\,\mathrm{Re}\, z_1 \overline{z_2} & -2\,\mathrm{Im}\, z_1 \overline{z_2} & z_1\overline{z_1} - z_2\overline{z_2} \end{pmatrix}, \end{aligned}$$

the *Euler angle* (χ, φ, θ)-parametrization is related to spherical bases

$$\begin{aligned} \mathbf{SU}(2) \ni u & \cong [1]_m^k(\chi, \varphi, \theta) \cong \begin{pmatrix} e^{i\frac{\varphi+\chi}{2}} \cos\frac{\theta}{2} & ie^{i\frac{\varphi-\chi}{2}} \sin\frac{\theta}{2} \\ ie^{-i\frac{\varphi-\chi}{2}} \sin\frac{\theta}{2} & e^{-i\frac{\varphi+\chi}{2}} \cos\frac{\theta}{2} \end{pmatrix} \\ & = e^{i\frac{\varphi}{2}\sigma^3} e^{i\frac{\theta}{2}\sigma^1} e^{i\frac{\chi}{2}\sigma^3} = \begin{pmatrix} e^{i\frac{\varphi}{2}} & 0 \\ 0 & e^{-i\frac{\varphi}{2}} \end{pmatrix} \begin{pmatrix} \cos\frac{\theta}{2} & i\sin\frac{\theta}{2} \\ i\sin\frac{\theta}{2} & \cos\frac{\theta}{2} \end{pmatrix} \begin{pmatrix} e^{i\frac{\chi}{2}} & 0 \\ 0 & e^{-i\frac{\chi}{2}} \end{pmatrix}, \\ & \text{with } m, k \in \{\pm\tfrac{1}{2}\} \text{ and } \begin{cases} -2\pi & \leq \chi < 2\pi, \\ 0 & \leq \theta < \pi, \\ 0 & \leq \varphi < 2\pi. \end{cases} \end{aligned}$$

Euler angles are appropriate for the decomposition with the 2-sphere $\mathbf{SU}(2) = \mathbf{SU}(2)/\mathbf{SO}(2)\circ\mathbf{SO}(2)$ and the Cartan decomposition $\mathbf{SU}(2) = \mathbf{SO}(2)\circ\mathbf{SO}(2)\circ\mathbf{SO}(2)$. They give no Lie algebra coefficients near the unit $\mathbf{1}_2 \in \mathbf{SU}(2)$:

$$\begin{aligned} d\, e^{i\vec{\alpha}}|_{\mathbf{1}_2} &= i d\vec{\alpha}|_{\mathbf{1}_2}, \\ d\, e^{i\frac{\varphi}{2}\sigma^3} e^{i\frac{\theta}{2}\sigma^1} e^{i\frac{\chi}{2}\sigma^3}|_{\mathbf{1}_2} &= \tfrac{i}{2}\sigma^3(d\varphi + d\chi)|_{\mathbf{1}_2} + \tfrac{i}{2}\sigma^1 d\theta|_{\mathbf{1}_2}. \end{aligned}$$

The *Rodriguez formula* gives the Euler angle parametrization of the matrix elements of any representation as a totally symmetrized product,

$$\begin{aligned} \bigvee^{2J} u &\cong [2J]^k_m(\chi,\varphi,\theta) = i^{m-k} e^{i(k\chi+m\varphi)}[2J]^k_m(z), \quad z = \cos\theta \\ &\text{for } m,k \in \{-J,\dots,J\}, \end{aligned}$$

with a polynomial in $(\cos\frac{\theta}{2}, \sin\frac{\theta}{2}) = \sqrt{\frac{1\pm z}{2}}$, homogeneous of degree $2J$:

$$[2J]^k_m(z) = \frac{(-1)^{J-m}}{2^J}\sqrt{\frac{(J+m)!}{(J-k)!(J+k)!(J-m)!}}\, \frac{(1-z)^{\frac{k-m}{2}}}{(1+z)^{\frac{k+m}{2}}} \left(\frac{d}{dz}\right)^{J-m} (1-z)^{J-k}(1+z)^{J+k}.$$

The matrix elements are orthonormalized with respect to the integration over the group volume with the normalized $\mathbf{SU}(2)$-Haar measure (Schur's orthogonality, chapter "Harmonic Analysis")

$$\begin{aligned} &d^3u \sim dx_1 dy_1 dx_2 dy_2 \delta(|z_1|^2 + |z_2|^2 - 1) \sim d\chi\, d^2\omega, \\ &\int_{\mathbf{SU}(2)} d^3u = \int_{-2\pi}^{2\pi} \frac{d\chi}{4\pi} \int_0^{2\pi} \frac{d\varphi}{2\pi} \int_{-1}^{1} \frac{d\cos\theta}{2}, \\ &\int_{\mathbf{SU}(2)} d^3u\; [2J]^k_m(u)\, \overline{[2J']^{k'}_{m'}(u)} = \frac{1}{1+2J}\delta_{JJ'}\delta_{mm'}\delta_{kk'}. \end{aligned}$$

The $\mathbf{SU}(2)$-representations are in $\mathbf{SU}(1+2J)$ matrices, e.g., $\mathbf{SO}(3)$ in a spherical basis in $\mathbf{SU}(3)$:

$$u \vee u \cong [2]^k_m(\chi,\varphi,\theta) \cong \begin{pmatrix} e^{i\chi}e^{i\varphi}\cos^2\frac{\theta}{2} & ie^{i\varphi}\frac{\sin\theta}{\sqrt{2}} & -e^{-i\chi}e^{i\varphi}\sin^2\frac{\theta}{2} \\ ie^{i\chi}\frac{\sin\theta}{\sqrt{2}} & \cos\theta & ie^{-i\chi}\frac{\sin\theta}{\sqrt{2}} \\ -e^{i\chi}e^{-i\varphi}\sin^2\frac{\theta}{2} & ie^{-i\varphi}\frac{\sin\theta}{\sqrt{2}} & e^{-i\chi}e^{-i\varphi}\cos^2\frac{\theta}{2} \end{pmatrix} \in \mathbf{SU}(3),$$

with the properties

$$\begin{aligned} &\overline{[2J]^k_m(u)} = [2J]^m_k(u^{-1}), \quad [2J]^k_m(1) = \delta^k_m, \\ &[2J]^k_m(z) = (-1)^{m+k}[2J]^m_k(z) = [2J]^{-m}_{-k}(z) = (-1)^{J+k}[2J]^k_{-m}(-z). \end{aligned}$$

Each column can be associated with a $(1+2J)$-dimensional representation space, e.g., the χ-independent 0th column for integer spin $L = 0, 1\dots$:

$$\begin{aligned} [2L]^0_m(\chi,\varphi,\theta) &= i^m e^{im\varphi}[2L]^0_m(z), \quad z = \cos\theta, \; m \in \{-L,\dots,L\}, \\ [2L]^0_m(z) &= \frac{(-1)^{L-m}}{2^L L!}\sqrt{\frac{(L+m)!}{(L-m)!}} \times (1-z^2)^{-\frac{m}{2}} \left(\frac{d}{dz}\right)^{L-m} (1-z^2)^L \\ &= \frac{(-1)^{L-m}}{2^L L!}\sqrt{\frac{(L+m)!}{(L-m)!}} \times \frac{1}{\sin^m\theta}\left(\frac{d}{d\cos\theta}\right)^{L-m} \sin^{2L}\theta. \end{aligned}$$

The orthonormalized spherical harmonics have as conventional renormalization (chapter "Quantum Probability"):

$$\begin{aligned}
\mathrm{Y}^1_m(\varphi,\theta) &= \sqrt{\tfrac{3}{4\pi}}\,[2]^0_m(\chi,\varphi,\theta) \cong -i\sqrt{\tfrac{3}{4\pi}\tfrac{1}{r}}\begin{pmatrix}\frac{l_\pm}{\sqrt{2}}\\ \frac{h}{2}\end{pmatrix},\\
\mathrm{Y}^L_m(\varphi,\theta) &= \sqrt{\tfrac{1+2L}{4\pi}}\,[2L]^0_m(\chi,\varphi,\theta)\\
&= \tfrac{(-1)^L}{2^L L!}\sqrt{\tfrac{(1+2L)(L+m)!}{4\pi(L-m)!}}\,\left(\tfrac{e^{i\varphi}}{\sin\theta}\right)^m\left(\tfrac{d}{d\cos\theta}\right)^{L-m}\sin^{2L}\theta,\\
\int d^2\omega\ \mathrm{Y}^L_m(\varphi,\theta)\,\overline{\mathrm{Y}^{L'}_{m'}(\varphi,\theta)} &= \delta^{LL'}\delta_{mm'}.
\end{aligned}$$

3.9 Spin Invariants and Weights

The equivalence classes $[n]=[2J]$, both of the finite-dimensional irreducible representations for the complex Lie group $\mathbf{SL}(\mathbb{C}^2)$ and its Lie algebra A_1 and of the irreducible complex representations for the compact Lie group $\mathbf{SU}(2)$ and its Lie algebra A_1^c, constitute *representation cones*, where the composition takes the highest spin in the product representation

$$\begin{aligned}
\mathbf{irrep}_{\,\mathrm{fin}}\mathbf{SL}(\mathbb{C}^2) = \mathbf{irrep}\,\mathbf{SU}(2) \cong\ & \{[2J]\ \big|\ J=0,\tfrac{1}{2},\dots\}\cong\mathbb{N}_0\\
\text{with}\ & \begin{cases}[2J_1]\vee[2J_2]=[2J_1+2J_2],\\ \text{neutral element: } [0].\end{cases}
\end{aligned}$$

The 2-dimensional Pauli representation [1] is *defining* and *fundamental*, i.e., it combines $\mathbb{N}_0$-linearily all irreducible representations

$$[2J]=\bigvee^{2J}[1],\quad 2J\in\mathbb{N}_0.$$

The totally symmetric natural tensor powers $\bigvee^{2J} e^{i\vec{\alpha}}$, $2J\in\mathbb{N}_0$, embed the integer powers $(e^{i\alpha})^Z$, $Z\in\mathbb{Z}$, for the abelian group $\mathbf{U}(1)$ in the self-dual $\mathbf{SO}(2)$-representations.

The *weights* of a representation $[2J]$ are given by the spectrum of a Cartan subalgebra $\log\mathbf{SO}(2)\subset\log\mathbf{SU}(2)$ (e.g., third spin or isospin component):

$$\mathbf{weights}\,[2J]=\{2j\ \big|\ j=-J,-J+1,\dots,J-1,J\}.$$

All weights for the irreducible representations above constitute the *weight modules*

$$\mathbf{weights}_{\,\mathrm{fin}}\mathbf{SL}(\mathbb{C}^2)=\mathbf{weights}\,\mathbf{SU}(2)=\{2j\ \big|\ j=0,\pm\tfrac{1}{2},\dots\}=\mathbb{Z}.$$

The weights are a discrete subgroup of the linear Lie algebra forms $\log\mathbf{SO}(\mathbb{C}^2)^T\cong\mathbb{C}$ and $\log\mathbf{SO}(2)^T\cong\mathbb{R}$ of a Cartan subalgebra and therefore of the dual Lie algebra $\log\mathbf{SL}(\mathbb{C}^2)^T\cong\mathbb{C}^3$ and $\log\mathbf{SU}(2)^T\cong\mathbb{R}^3$ respectively.

Weights, reflected with the spinor "metric" $2j \overset{\epsilon}{\leftrightarrow} -2j$ belong to the same representation. The representation cone is isomorphic to the classes $\{\pm 2j\}$, i.e., to the positive weights.

The locally isomorphic rotation group $\mathbf{SO}(3) \cong \mathbf{SU}(2)/\mathbb{I}(2)$ has as representation cone and weight module a subcone and submodule respectively with even eigenvalues $4j$ only:

$$\begin{aligned} \mathbf{irrep}\,\mathbf{SO}(3) &\cong \{[4J] \mid 2J = 0, 1, \dots\} &&\cong 2\mathbb{N}_0, \\ \mathbf{weights}\,\mathbf{SO}(3) &= \{[4j] \mid 2j = 0, \pm 1, \dots\} &&= 2\mathbb{Z}. \end{aligned}$$

The *fundamental* and *defining* representation is the adjoint one [2]. The real representation spaces have no eigenvector basis for a Cartan subalgebra or $\mathbf{SO}(2)$-isomorphic subgroup.

3.10 Summary

The compact Lie algebra $A_1^c \cong \mathbb{R}^3$ in the "smallest" simple complex Lie algebra $A_1 \cong \mathbb{C}^3$ is the operation structure for the Euclidean position translations $\mathbb{S} \cong \mathbb{R}^3$. All finite-dimensional representations of the Lie algebra A_1 with Lie group $\mathbf{SL}(\mathbb{C}^2) = \exp A_1$ and all representations of A_1^c with Lie group $\mathbf{SU}(2) = \exp A_1^c$ (spin group) are semisimple. The simple ones are equivalent to the totally symmetric tensor powers of Pauli spinors acted on by the fundamental Pauli representation. The complex 2-dimensional Pauli representation for A_1 is symplectic self-dual with the spinor volume elements, for A_1^c, by definition, also positive selfdual with a scalar product. The symplectic dual isomorphism implements the inversion of the Lie algebras A_1 and A_1^c and the Lie groups $\mathbf{SL}(\mathbb{C}^2)$ and $\mathbf{SU}(2)$. The affine position group $\mathbf{SO}(3) \vec{\times} \mathbb{R}^3$ is realized by the adjoint action of the spin group on the vector space structure iA_1^c of its Lie algebra, i.e., by $\mathbf{SU}(2) \vec{\times} iA_1^c$.

	$\mathbf{SL}(\mathbb{C}^2) = \exp A_1$	$\mathbf{SU}(2) = \exp A_1^c$	$\mathbf{SO}(3) \cong \mathbf{SU}(2)/\mathbb{I}(2)$
weight module	$\mathbf{weights}_{\,\mathrm{fin}}\mathbf{SL}(\mathbb{C}^2) = \mathbb{Z}$	$\mathbf{weights}\,\mathbf{SU}(2) = \mathbb{Z}$	$\mathbf{weights}\,\mathbf{SO}(3) = 2\mathbb{Z}$
representation cone	$\mathbf{irrep}_{\,\mathrm{fin}}\mathbf{SL}(\mathbb{C}^2) \cong \mathbb{N}_0$	$\mathbf{irrep}\,\mathbf{SU}(2) \cong \mathbb{N}_0$	$\mathbf{irrep}\,\mathbf{SO}(3) \cong 2\mathbb{N}_0$
representations	$D^{2J}(s) \in \begin{cases} \mathbf{Sp}(\mathbb{C}^{1+2J}) \\ J = \frac{1}{2}, \frac{3}{2}, \dots \\ \mathbf{SO}(\mathbb{C}^{1+2J}) \\ J = 0, 1, \dots \end{cases}$ $D^{2J}(s) \cong \bigvee^{2J} s$	$D^{2J}(u) \in \begin{cases} \mathbf{SpU}(1+2J) \\ J = \frac{1}{2}, \frac{3}{2}, \dots \\ \mathbf{SO}(1+2J) \\ J = 0, 1, \dots \end{cases}$ $D^{2J}(u) \cong \bigvee^{2J} u$	$D^{2J}(O) \in \mathbf{SO}(1+2J)$ $J = 0, 1, \dots$
Lie algebra	$A_1 = \{\vec{z} = \begin{pmatrix} z_3 & z_1 - iz_2 \\ z_1 + iz_2 & -z_3 \end{pmatrix} \mid \vec{z} \in \mathbb{C}^3\} \cong \mathbb{C}^3$	$A_1^c = \{i\vec{\alpha} \mid \vec{\alpha} \in \mathbb{R}^3\} \cong \mathbb{R}^3$	$\log \mathbf{SO}(3) \ni \varphi_a \,\mathrm{ad}\, l^a$ $\vec{\varphi} = 2\vec{\alpha}$ $(\mathrm{ad}\, l^a)^b_c = -\epsilon^{abc}$
fundamental representation	$s = e^{\vec{z}},\ J = \frac{1}{2}$ Pauli spinors	$u = e^{i\vec{\alpha}},\ J = \frac{1}{2}$ Pauli spinors	$O \cong u \vee u,\ J = 1$ rotations
fundamental bilinear form	$\epsilon = -\epsilon^T \cong \begin{pmatrix} 0 & 1 \\ -1 & 0 \end{pmatrix}$ volume form	$\epsilon = -\epsilon^T$ spinor "metric"	$\epsilon \vee \epsilon = -\delta = -\delta^T$ $\cong -\mathbf{1}_3$ Killing form
conjugation	–	$\star \cong \mathbf{1}_2$ scalar product	scalar product $\mathbf{1}_3$ $\cong$ –Killing form

$\mathbf{SL}(\mathbb{C}^2)$ and $\mathbf{SU}(2)$-representations and weights

MATHEMATICAL TOOLS

3.11 Derivations of Algebras

A *derivation of an algebra* $A \in \underline{\mathbf{ag}}_{\mathbb{K}}$ is a linear mapping with *Leibniz rule*

$$v : A \longrightarrow A, \quad \begin{cases} v(a+b) = v(a) + v(b), & v(\alpha a) = \alpha v(a), \\ v(a \diamond b) = v(a) \diamond b + a \diamond v(b). \end{cases}$$

For a unital algebra the scalars are "constant" $v(\mathbb{K}1_A) = \{0\}$.

In general, all derivations do not constitute an associative subalgebra of the vector space endomorphisms $\mathbf{AL}(A)$, however they are a Lie subalgebra with the commutator

$$\begin{aligned} &\operatorname{der} A = \{v \in \mathbf{AL}(A), \text{ derivation}\} \in \underline{\mathbf{lag}}_{\mathbb{K}} \\ &\text{with } [v, w](a) = (v \circ w - w \circ v)(a). \end{aligned}$$

The functor $\operatorname{der} : \underline{\mathbf{ag}}_{\mathbb{K}} \longrightarrow \underline{\mathbf{lag}}_{\mathbb{K}}$ is covariant.

A finite generating system for an unital algebra defines associate *generating derivations*, which have to be extended with the Leibniz rule

$$\begin{aligned} &\text{generating system } \{e^j\}_{j=1}^n \text{ for } A \in \underline{\mathbf{aag}}_{\mathbb{K}} \\ &\partial_j : A \longrightarrow A, \quad \partial_j(e^l) = \delta_j^l. \end{aligned}$$

The *invariance algebra for a subset* V *of derivations* is

$$V \subseteq \operatorname{der} A, \quad \mathrm{INV}_V A = \{a \in A \,\big|\, V(a) = \{0\}\} \in \underline{\mathbf{ag}}_{\mathbb{K}}.$$

With a stable ideal $v[I] \subseteq I$, a derivation v is well defined also on the quotient algebra A/I. If a minimal ideal $I(S)$ in a unital algebra arises from a set S, stable for a derivation v, then this derivation is well defined on $A/I(S)$ too:

$$v \in \operatorname{der} A, \quad v[S] \subseteq S \Rightarrow v \in \operatorname{der} A/I(S).$$

3.11.1 Inner Derivations

The natural Lie algebra structure of an associative algebra $A \in \underline{\mathbf{aag}}_{\mathbb{K}}$ defines the *inner derivations* of the algebra via the commutator

$$\operatorname{ad} a : A \longrightarrow A, \quad \operatorname{ad} a(b) = [a, b], \quad [a, bc] = [a, b]c + b[a, c].$$

The *inner derivations of a Lie algebra* $L \in \underline{\mathbf{lag}}_{\mathbb{K}}$ use the Lie bracket

$$\begin{aligned} &\operatorname{ad} l : L \longrightarrow L, \quad \operatorname{ad} l(m) = [l, m] \\ &\Rightarrow \operatorname{ad} l([m, n]) = [\operatorname{ad} l(m), n] + [m, \operatorname{ad} l(n)]. \end{aligned}$$

They are an ideal in the Lie algebra of all derivations

$$\begin{aligned} &[\operatorname{der} A, \operatorname{ad} A] \subseteq \operatorname{ad} A, \quad [\operatorname{der} L, \operatorname{ad} L] \subseteq \operatorname{ad} L, \\ &v \in \operatorname{der} A \Rightarrow [v, \operatorname{ad} a](b) = v([a, b]) - [a, v(b)] = [v(a), b] = \operatorname{ad} v(a)(b), \end{aligned}$$

with $\operatorname{der} A/\operatorname{ad} A$ and $\operatorname{der} L/\operatorname{ad} L$ respectively the classes of the *outer derivations.* For abelian algebras the inner derivations are trivial.

Since ideals I are invariant under inner derivations, inner derivations are well defined on the quotient algebras A/I and L/I. *Characteristic ideals* are invariant even with respect to all derivations.

The eigenspace $A_\alpha(\operatorname{ad} a)$ with the eigenvalue $\alpha \in \mathbb{K}$ is

$$\begin{array}{l} A_\alpha(\operatorname{ad} a) = \{b \in A \mid [a,b] = \alpha b\} \in \underline{\mathbf{vec}}_{\mathbb{K}}, \\ A \in *\underline{\mathbf{ag}}_{\mathbb{K}} \Rightarrow A_\alpha(\operatorname{ad} a^*) = A_{-\alpha^*}(\operatorname{ad} a). \end{array}$$

For an (anti-)symmetric element $a = \mp a^*$ the eigenvalues of $\operatorname{ad} a$ lie reflection symmetrically with respect to the (real) imaginary axis for a conjugate linear reflection ($\alpha^* = \overline{\alpha} \in \mathbb{C}$). The eigenspace $A_0(\operatorname{ad} a)$ for eigenvalue 0 is an algebra. It contains the elements that are constant under inner derivation with a:

$$\begin{array}{l} A_0(\operatorname{ad} a) = \{b \in A \mid [a,b] = 0\} = \operatorname{INV}_a A \in \underline{\mathbf{ag}}_{\mathbb{K}}, \\ A \in *\underline{\mathbf{ag}}_{\mathbb{K}} \Rightarrow A_0(\operatorname{ad} a) = A_0(\operatorname{ad} a^*) \in *\underline{\mathbf{ag}}_{\mathbb{K}}. \end{array}$$

3.11.2 Adjoint Affine Lie Algebra

The mapping of a Lie algebra L into its inner derivations defines its *adjoint representation* acting on its vector space structure:

$$\begin{array}{l} \operatorname{ad} : L \longrightarrow \mathbf{AL}(L), \quad l \longmapsto \operatorname{ad} l, \\ \operatorname{ad} l : L \longrightarrow L, \quad \operatorname{ad} l(k) = [l,k], \\ \operatorname{ad} [l,m] = [\operatorname{ad} l, \operatorname{ad} m]. \end{array}$$

The center of L (an ideal) is the kernel of the adjoint representation leading to the *adjoint Lie algebra*

$$\operatorname{ad} L = L/\operatorname{centr} L.$$

For an n-dimensional Lie algebra with dual bases the adjoint representation $\operatorname{ad} l^j$ can be given by $n \times n$ matrices with the structure constants $(\epsilon^j)^k_m = \epsilon^{jk}_m$ as matrix elements with dual bases

$$\{l^j, \check{l}_j\}_{j=1}^n : \left\{ \begin{array}{rl} \operatorname{ad} l^j(l^k) & = [l^j, l^k] = \epsilon^{jk}_m l^m, \\ \operatorname{ad} l & = [l, l^k] \otimes \check{l}_k, \\ \operatorname{ad} l^j & = \epsilon^{jk}_m l^m \otimes \check{l}_k. \end{array} \right.$$

For example, the adjoint representation of the Heisenberg Lie algebra $\log \mathbf{H}(1) \cong \mathbb{R}^3$ with basis $\{\mathbf{x}, \mathbf{p}, \mathbf{I}\}$ loses the only nontrivial bracket for position-momentum; the image is the classical position-momentum Lie algebra

$$[\mathbf{x}, \mathbf{p}] = \mathbf{I}, \quad \operatorname{ad} \mathbf{x} = \begin{pmatrix} 0 & 0 & 0 \\ 0 & 0 & 1 \\ 0 & 0 & 0 \end{pmatrix}, \quad \operatorname{ad} \mathbf{p} = \begin{pmatrix} 0 & 0 & -1 \\ 0 & 0 & 0 \\ 0 & 0 & 0 \end{pmatrix}, \quad \operatorname{ad} \mathbf{I} = 0.$$

Each Lie algebra representation on a finite-dimensional vector space is a tensor of power 3:

$$\begin{array}{ll} \mathcal{D}: \; L \longrightarrow \mathbf{AL}(V), & \mathcal{D} = \mathcal{D}(l^j) \otimes \check{l}_j = \mathcal{D}^{jB}_A \; e^A \otimes \check{e}_B \otimes \check{l}_j \in V \otimes V^T \otimes L^T, \\ \mathrm{ad} : \; L \longrightarrow \mathbf{AL}(L), & \mathrm{ad} = \mathrm{ad}\, l^j \otimes \check{l}_j = \epsilon^{jk}_m \; l^m \otimes \check{l}_k \otimes \check{l}_j \in L \otimes L^T \otimes L^T, \end{array}$$

invariant under the Lie algebra action

$$\begin{array}{rl} l^i \bullet \mathcal{D} & = [\mathcal{D}(l^i), \mathcal{D}(l^j)] \otimes \check{l}_j - \mathcal{D}(l^j) \otimes (\,\mathrm{ad}\, l^i)^T(\check{l}_j) \\ & = \epsilon^{ij}_k \mathcal{D}(l^k) \otimes \check{l}_j - \mathcal{D}(l^j) \otimes \epsilon^{ik}_j \check{l}_k = 0 \\ \Rightarrow l \bullet \mathcal{D} & = 0, \quad l \bullet \; \mathrm{ad} \, = 0. \end{array}$$

The *adjoint tensor* ad_L of a finite-dimensional Lie algebra coincides with the Lie bracket tensor

$$\begin{array}{l} [\;,\;]_L : L \otimes L \longrightarrow L, \quad l^j \otimes l^k \longrightarrow \epsilon^{jk}_m l^m, \\ [\;,\;]_L = \mathrm{ad}_L = \epsilon^{jk}_m \; l^m \otimes \check{l}_k \otimes \check{l}_j \in L \otimes L^T \otimes L^T. \end{array}$$

The semidirect product of adjoint Lie algebra and the Lie algebra as a vector space is the *adjoint affine Lie algebra*

$$\begin{array}{l} \mathrm{ad}\, L \;\vec{\oplus}\; \underline{L} \subseteq \mathbf{AL}(L) \;\vec{\oplus}\; \underline{L}, \\ [l_1 + k_1, l_2 + k_2] = [l_1, l_2] + \mathrm{ad}\, l_1(k_2) - \mathrm{ad}\, l_2(k_1). \end{array}$$

The translation factor keeps the vector space feature, but "forgets" the Lie bracket, indicated by underlining.

3.11.3 Adjoint Affine Lie Group

For a representation both of a Lie group G and its Lie algebra $L = \log G \cong \mathbb{K}^n$ on $V \in \underline{\mathbf{vec}}_{\mathbb{K}}$,

$$\begin{array}{ll} D : G \longrightarrow D[G] & \subseteq \mathbf{GL}(V), \\ \mathcal{D} : L \longrightarrow \mathcal{D}[L] & \subseteq \mathbf{AL}(V), \end{array}$$

the Lie algebra image is stable with respect to inner automorphisms with the group image, $\mathrm{Ad}\, D(g)[\mathcal{D}[L]] \subseteq \mathcal{D}[L]$. For a simpler notation in the following, $G \subseteq \mathbf{GL}(V) \subset \mathbf{AL}(V) \supseteq L$, i.e., one considers an endomorphism Lie group and Lie algebra.

The restriction of the inner automorphisms to the Lie algebra gives the *adjoint representation of the Lie group G on its Lie algebra L*:

$$\begin{array}{ll} \mathrm{Ad} : G \longrightarrow \mathbf{GL}(L), & g \longmapsto \mathrm{Ad}\, g, \\ \mathrm{Ad}\, g : L \longrightarrow L, & \mathrm{Ad}\, g(l) = g \circ l \circ g^{-1}, \\ & \mathrm{Ad}\, g(l^j) = (\,\mathrm{Ad}\, g)^j_k l^k. \end{array}$$

The adjoint action of a Lie group on its Lie algebra can also be defined without using a representation $(D, \mathcal{D})$ on a vector space V.

If the Lie algebra has a nondegenerate double trace for an invariant inner product κ, one obtains for the adjoint representation of a Lie group on its Lie algebra

$$\mathrm{tr}\, l^m \circ l^j = \kappa^{jm}, \quad \check{l}_k = \kappa_{km} l^m \in L^T \Rightarrow (\,\mathrm{Ad}\, g)^j_k = \mathrm{tr}\, g \circ l^j \circ g^{-1} \circ \check{l}_k.$$

The semidirect product of Lie group and its Lie algebra as a vector space defines the *adjoint affine Lie group*. The center acts trivially:

$$\begin{aligned} &G \;\vec{\times}\; \underline{\log G} = \{(g,l) \mid g \in G,\ l \in \log G\},\\ &(g_1,l_1) \circ (g_2,l_2) = (g_1 g_2, l_1 + \operatorname{Ad} g(l_2)). \end{aligned}$$

3.12 Differentiable Manifolds

Since Lie groups are analytic manifolds, some sketchy remarks for finite-dimensional manifolds $M \in \underline{\mathbf{dif}}_{\mathbb{R}}$ are given in this section. Differentiability details are not specified ("appropriately smooth").

The functions on an open real set valued in a Banach space, e.g., $\Phi : O \longrightarrow \mathbb{K}$, differentiable at a point $\alpha = (\alpha_j) \in O \subseteq \mathbb{R}^d$, can be collected in classes (*function germs*) $f \in \mathcal{C}_\alpha(O)$ that are characterized by "locally (at α) different" functions. Their nonstationary derivation classes, at least one derivative $\partial^j_\alpha.f = \partial^j f(\alpha) \neq 0$, give the *tangent vector space* $\mathbf{T}_\alpha(\mathbb{R}^d) \cong \mathbb{R}^d$ with a basis given by the partial derivatives $\{\partial^j_\alpha\}_{j=1}^d$. The dual *cotangent space* $\mathbf{T}^T_\alpha(\mathbb{R}^d)$ is isomorphic to the function germs modulo the stationary ones, i.e., $d_\alpha f \in \mathbf{T}^T_\alpha(\mathbb{R}^d) \cong \mathcal{C}_\alpha(O)/\mathcal{S}_\alpha(O)$ with $\langle d_\alpha f, \partial^j_\alpha\rangle = \partial^j_\alpha.f$. It has the ∂^j_α-dual basis $\langle d_\alpha \alpha_k, \partial^j_\alpha\rangle = \delta^j_k$.

Three contravariant functors are of interest: after the functor from the pointed open sets into the vector spaces follows the tangent functor into the Lie algebras and the dual functor on them. The morphisms are: A differentiable point-hitting mapping $\Phi : (O,\alpha) \longrightarrow (O',\alpha')$ gives, by composition $f' \circ \Phi \in \mathcal{C}_\alpha(O)$, a vector space morphism that leads to the tangent *Jacobi mapping* Φ_*, a Lie algebra morphism, with the transposed mapping for the linear forms

$$\underline{\mathbb{R}}^{\mathbb{N}}_{\bullet} \xrightarrow{\mathcal{C}_\bullet} \underline{\mathbf{vec}}_{\mathbb{R}} \xrightarrow{\mathbf{T}_\bullet} \underline{\mathbf{lag}}_{\mathbb{R}} \xrightarrow{T} \underline{\mathbf{vec}}_{\mathbb{R}},$$

$$\begin{array}{ccccccc} (\alpha_j) \in O & & \mathcal{C}_\alpha(O) & & \mathbf{T}_\alpha(\mathbb{R}^d) & & \mathbf{T}^T_\alpha(\mathbb{R}^d) \\ \downarrow \Phi & \longmapsto & \uparrow \circ\Phi & \longmapsto & \downarrow \Phi_*(\alpha) & \longmapsto & \uparrow \Phi_*(\alpha)^T \\ (\alpha'_a) \in O' & & \mathcal{C}_{\alpha'}(O') & & \mathbf{T}_{\alpha'}(\mathbb{R}^{d'}) & & \mathbf{T}^T_{\alpha'}(\mathbb{R}^{d'}) \end{array},$$

$$\Phi_*(\alpha) \cong \left(\frac{\partial \Phi_a}{\partial \alpha_j}\right)^{j=1,\dots,d}_{a=1,\dots,d'}(\alpha) \text{ since } \left\{ \begin{array}{ll} \Phi_*(\alpha)(\partial^j_\alpha).f' & = \partial^j_\alpha(f' \circ \Phi) \\ & = (\partial^j \Phi_a)(\alpha)\ \partial^a f'(\Phi(\alpha)). \end{array}\right.$$

Differentiable manifolds are covered by open charts $M(p) \cong O$ for each point homeomorphic to open real sets, $M \supseteq M(p) \ni p \longmapsto \alpha \in O \subseteq \mathbb{R}^d$,

$$\begin{array}{ccccccc} M(p) & & \mathcal{C}_p(M) & & \mathbf{T}_p(M) & & \mathbf{T}^T_p(M) \\ \downarrow \Phi & \longmapsto & \uparrow \circ\Phi & \longmapsto & \downarrow \Phi_*(p) & \longmapsto & \uparrow \Phi_*(p)^T \\ M'(p') & & \mathcal{C}_{p'}(M') & & \mathbf{T}_{p'}(M') & & \mathbf{T}^T_{p'}(M') \end{array}.$$

There are the tangent tensors

$$\mathbf{T}^{(k,l)}_p(M) = \bigotimes^{k} \mathbf{T}_p(M) \otimes \bigotimes^{l} \mathbf{T}^T_p(M) \in \underline{\mathbf{vec}}_{\mathbb{R}}.$$

Local structures are smoothly patched together with an atlas:

$$\begin{array}{ccccccc} M & & \mathcal{C}(M) & & \mathbf{T}(M) & & \mathbf{T}^T(M) \\ \downarrow \Phi & \longmapsto & \uparrow \circ\Phi & \longmapsto & \downarrow \Phi_* & \longmapsto & \uparrow \Phi_*^T \\ M' & & \mathcal{C}(M') & & \mathbf{T}(M') & & \mathbf{T}^T(M') \end{array} .$$

The functions $\mathcal{C}(M)$, the tangent bundle $\mathbf{T}(M)$, and cotangent bundle $\mathbf{T}^T(M)$ are unions of the point-related structures. The tensor bundles are modules over the function ring

$$\mathbf{T}^{(k,l)}(M) = \bigcup_{p \in M} (p, \mathbf{T}_p^{(k,l)}(M)) \in \underline{\mathbf{mod}}_{\mathcal{C}(M)}, \quad \begin{cases} \mathbf{T}^{(0,0)}(M) & = \mathcal{C}(M) \\ & = \bigcup_{p \in M} (p, \mathcal{C}_p(M)), \\ \mathbf{T}^{(1,0)}(M) & = \mathbf{T}(M), \\ \mathbf{T}^{(0,1)}(M) & = \mathbf{T}^T(M). \end{cases}$$

Tensor fields (vector fields, differential forms, tangent inner products like $v_j \partial^j$, $\omega^j d\alpha_j$, $\gamma^{jk} d\alpha_j \otimes d\alpha_k$) are differentiable mappings

$$M \ni p \longmapsto t(p) \in \mathbf{T}_p^{(k,l)}(M) \subseteq \mathbf{T}^{(k,l)}(M),$$

e.g., moving identities (frames) with dual bases $p \longmapsto e^j(p) \otimes \check{e}_j(p)$ like $\partial^j \otimes d\alpha_j$.

For manifolds with additional structures, e.g., vector spaces, Lie groups, the manifold morphisms Φ have to be corresponding morphisms. In the diagrams above one may use as entries

$$\left.\begin{array}{l} v \in V \in \underline{\mathbf{vec}}_{\mathbb{R}} \\ \text{e.g., } v = 0 \end{array}\right\}: \quad V(v) \longmapsto \mathcal{C}_v(V) \longmapsto \mathbf{T}_v(V) \longmapsto \mathbf{T}_v^T(V),$$

$$\left.\begin{array}{l} g \in G \in \underline{\mathbf{lgrp}}_{\mathbb{R}} \\ \text{e.g., } g = e \end{array}\right\}: \quad G(g) \longmapsto \mathcal{C}_g(G) \longmapsto \mathbf{T}_g(G) \longmapsto \mathbf{T}_g^T(G).$$

A vector space is isomorphic to its tangent space:

$$V(v) \longmapsto \mathcal{C}_v(V) \longmapsto \mathbf{T}_v(V) \cong V \longmapsto \mathbf{T}_v^T(V) \cong V^T.$$

It can be taken as a chart for all vectors, i.e., $V(v) = V$ and $\mathcal{C}_v(V) = V^T$.

3.13 Exponential and Logarithmic Mappings

A real Lie group G is considered as given with charts and mappings into a matrix algebra where the matrix elements are analytic functions. A *Lie group germ* is a Lie group with a chart around $\mathbf{1} \in G$, defining the local Lie group structures characteristic for the homogeneous group:

$$\begin{array}{ccccccc} L(\mathbf{0}) & & \mathcal{C}_{\mathbf{0}}(L) & & \mathbf{T}_{\mathbf{0}}(L) & & \mathbf{T}_{\mathbf{0}}^T(L) \\ \downarrow \exp & \longmapsto & \uparrow \circ \exp & \longmapsto & \downarrow \exp_*(\mathbf{0}) & \longmapsto & \uparrow \exp_*(\mathbf{0})^T \\ G(\mathbf{1}) & & \mathcal{C}_{\mathbf{1}}(G) & & \mathbf{T}_{\mathbf{1}}(G) & & \mathbf{T}_{\mathbf{1}}^T(G) \end{array} .$$

With Lie algebra coefficients for a basis as Lie group parameters, one has an *exponential homeomorphism* for a local *Lie algebra (canonical) parametrization of the group*:

$$\begin{aligned} \mathbb{R}^d \supseteq L \supseteq L(\mathbf{0}) \ni l &= \alpha_j l^j \longmapsto \exp l = g(l) \in G \subseteq \mathbf{GL}(\mathbb{C}^n), \\ g(\mathbf{0}) &= \mathbf{1}, \ \ g(-l) = g(l)^{-1}, \ \ g(l) \cong g(l)^A_B, \ \ A = 1, \dots, n, \end{aligned}$$

The *logarithmic mapping* as inverse uses the parameter derivatives $\partial^j = \frac{\partial}{\partial \alpha_j}$:

$$\begin{aligned} G \supseteq G(\mathbf{1}) \ni g(l) = e^l &\longmapsto \log g(l) = \alpha_j l^j(g) \in L \subseteq \mathbf{AL}(\mathbb{C}^n), \\ l^j(g) &= (\partial^j g(l)) \circ g(l)^{-1} = \sum_{k \geq 0} \tfrac{(\operatorname{ad} l)^k}{(1+k)!}(l^j) \\ &= l^j + \tfrac{[l,l^j]}{2} + \tfrac{[l,[l,l^j]]}{3!} + \tfrac{[l,[l,[l,l^j]]]}{4!} + \cdots, \\ \log e^l &= \alpha_j l^j(g) = l. \end{aligned}$$

The logarithm is a sum over the *directed logarithms* that define a Lie algebra basis $\{l^j(g)\}_j$ at each group element around the neutral element:

$$\begin{aligned} G \ni g(\alpha_j l^j) \longmapsto \log^j g(l) = l^j(g) &= (\partial^j g(l)) \circ g(l)^{-1} \in L, \\ l^j(\mathbf{1}) = l^j, \ \ l^j(g^{-1}) = -l^j(g), \ \ l^j(g) &\cong l^j(g)^A_B. \end{aligned}$$

The local tangent structures are isomorphic:

$$\begin{aligned} g = e^l = e^{\alpha_j l^j}: &\ \ L \cong \mathbf{T}_l(L) \cong \mathbf{T}_g(G) \cong \mathbb{R}^d, \\ \text{dual bases:} &\ \ \partial^j_\alpha \otimes d_\alpha \alpha_j \cong l^j(g) \otimes \check{l}_j(g). \end{aligned}$$

3.13.1 Lie Algebra-Lie Group Relations

A Lie group has its Lie algebra with the covariant *Lie algebra functor* - surjective, but not injective:

$$\log : \underline{\mathbf{lgrp}}_{\mathbb{K}} \longrightarrow \underline{\mathbf{lag}}_{\mathbb{K}}, \quad f \begin{array}{c} G_1 \\ \big\downarrow \\ G_2 \end{array} \longmapsto \begin{array}{c} \log G_1 \\ \big\downarrow \\ \log G_2 \end{array} \log f,$$

$$\log[G_1 \times G_2] = \log G_1 \oplus \log G_2.$$

The logarithmic mapping may be definable by a series in an algebra:

$$\begin{aligned} G \longrightarrow \log G, \ \ g \longmapsto \log g &= -\sum_{k \geq 1} \tfrac{(1-g)^k}{k}, \\ \log e^l &= l \text{ for } g - 1 = e^l - 1 = l + \tfrac{l \circ l}{2} + \cdots . \end{aligned}$$

The logarithmic additivity holds for the product of two elements iff they commute:

$$\begin{aligned} g(l_1) \circ g(l_2) \neq g(l_2) \circ g(l_1) \iff &\log g(l_1) \circ g(l_2) = l_1 \mathbin{+\!\!+} l_2 \\ &\neq \log g(l_1) + \log g(l_2) = l_1 + l_2. \end{aligned}$$

The exponent of a Lie algebra (all Lie structures finite-dimensional) comes with the injective covariant Lie algebra functor

$$\exp : \underline{\mathbf{lag}}_{\mathbb{K}} \longrightarrow \underline{\mathbf{lgrp}}_{\mathbb{K}},\quad f \begin{array}{c} L_1 \\ \downarrow \\ L_2 \end{array} \longmapsto \begin{array}{c} \overline{\exp L_1} \\ \downarrow \\ \overline{\exp L_2} \end{array} \exp f,$$

$$[L_1, L_2] = \{0\} \Rightarrow \overline{\exp(L_1 \oplus L_2)} = \overline{\exp L_1} \times \overline{\exp L_2},$$

An example is the *Heisenberg group* $\mathbf{H}(1)$ in a real 3-dimensional faithful representation,

$$\log \mathbf{H}(1) \ni q\mathbf{x} + y\mathbf{p} + t\mathbf{I} = \begin{pmatrix} 0 & q & t \\ 0 & 0 & y \\ 0 & 0 & 0 \end{pmatrix} \longmapsto e^{q\mathbf{x}+y\mathbf{p}+t\mathbf{I}} = \begin{pmatrix} 1 & q & t+qy \\ 0 & 1 & y \\ 0 & 0 & 1 \end{pmatrix} \in \mathbf{H}(1),$$

$$\textit{Weyl product: } e^{q\mathbf{x}}e^{y\mathbf{p}} = \begin{pmatrix} 1 & q & qy \\ 0 & 1 & y \\ 0 & 0 & 1 \end{pmatrix} = e^{qy\mathbf{I}}e^{y\mathbf{p}}e^{q\mathbf{x}}.$$

The Lie algebra exponent can be locally defined in an exponentially closed algebra:

$$L \ni l \longmapsto e^l \in \exp L,\quad e^l = \sum_{k \geq 0} \tfrac{l^k}{k!} = 1 + l + \tfrac{l \circ l}{2} + \cdots .$$

The exponential product property holds for the sum of two elements iff they commute:

$$[l_1, l_2] \neq 0 \iff e^{l_1+l_2} \neq e^{l_1} \circ e^{l_2} = e^{l_1 \mathbin{+\!\!+} l_2} = g(l_1) \circ g(l_2).$$

The exponent of a simple Lie algebra is simply connected and has a discrete center. Two Lie groups with isomorphic Lie algebras are called *locally isomorphic*. All locally isomorphic connected Lie groups $\{G_i\}$, $\log G_i \cong L$, arise as quotients from their *universal cover Lie group* $\overline{\exp L}$ by the discrete normal subgroups $\{\mathbb{I}_i\}$ in the center of $\overline{\exp L}$:

$$\begin{array}{rlll} G_i & \cong \overline{\exp L}/\mathbb{I}_i & \text{with discrete} & \mathbb{I}_i \subseteq \mathbb{I}(G) = \operatorname{centr} \overline{\exp L}, \\ \log G_i & \cong \log \overline{\exp L} & \text{with trivial} & \log \mathbb{I}_i = \log \mathbb{I}(G) = \{0\}. \end{array}$$

The logarithm of a Lie group representation gives a Lie algebra representation, the exponent of a Lie algebra representation a Lie group representation with the compatibility $e^{\mathcal{D}(l)} = D(e^l)$:

$$\mathcal{D} \begin{array}{c} L \\ \downarrow \\ \mathbf{AL}(V) \end{array} \longmapsto \begin{array}{c} \exp L \\ \downarrow \\ \mathbf{GL}(V) \end{array} D{=}\exp \mathcal{D},\qquad D \begin{array}{c} G \\ \downarrow \\ \mathbf{GL}(V) \end{array} \longmapsto \begin{array}{c} \log G \\ \downarrow \\ \mathbf{AL}(V) \end{array} \mathcal{D}{=}\log D.$$

For example, the inner automorphisms of a Lie group are related to the adjoint group representation on the Lie algebra:

$$\operatorname{Int} g \begin{array}{c} G \\ \downarrow \\ G \end{array} \longmapsto \begin{array}{c} \log G = L \\ \downarrow \\ \log G = L \end{array} \log(\operatorname{Int} g){=}\operatorname{Ad} g,$$

$$\operatorname{Int} g(h) = ghg^{-1} \longmapsto \operatorname{Ad} g(l) = glg^{-1}.$$

There are the exponential compatibilities

$$\begin{aligned} e^{\,\mathrm{Ad}\,g(l)} &= e^{glg^{-1}} = ge^{l}g^{-1} = \mathrm{Int}\,g(e^{l}),\\ \mathrm{Ad}\,e^{l} &= e^{\,\mathrm{ad}\,l},\text{ i.e., } \mathrm{Ad}\,e^{l}(m) = e^{l}me^{-l} &&= \sum_{k\geq 0}\tfrac{(\mathrm{ad}\,l)^k}{k!}(m)\\ & &&= 1 + [l,m] + \tfrac{[l,[l,m]]}{2} + \cdots . \end{aligned}$$

3.13.2 Lie-Jacobi Transformation

The vector space isomorphisms induced by left multiplication in the group $G(\mathbf{1}) \ni h \longmapsto L_g(h) = g \circ h \in G(\mathbf{1})$ (section "Exponential and Logarithmic Mappings"),

$$\begin{array}{ccccccc} G(\mathbf{1}) & & \mathcal{C}_{\mathbf{1}}(G) & & \mathbf{T}_{\mathbf{1}}(G) & & \mathbf{T}^T_{\mathbf{1}}(G)\\ \big\downarrow L_g & \longmapsto & \big\uparrow \circ L_g & \longmapsto & \big\downarrow L_{g*}(\mathbf{1}) = g_* & \longmapsto & \big\uparrow g_*^T\\ G(g) & & \mathcal{C}_g(G) & & \mathbf{T}_g(G) & & \mathbf{T}^T_g(G) \end{array},$$

is called a *Lie-Jacobi isomorphism*:

$$\begin{aligned} g_* &: L \longrightarrow L,\;\; l^j \longmapsto l^j(g) = (g_*)^j_k\, l^k,\\ g = e^l \Rightarrow g_* &= \sum_{k\geq 0}\tfrac{(\mathrm{ad}\,l)^k}{(1+k)!} = \tfrac{\exp \mathrm{ad}\,l - 1}{\mathrm{ad}\,l} \in \mathbf{GL}(\mathbf{AL}(\mathbb{C}^n)),\;\; (g_*)^j_k \cong [(g_*)^j_k]^A_B, \end{aligned}$$

with the transformed Lie bracket

$$\begin{aligned} [l^j(g), l^k(g)] &= \epsilon^{jk}_r(g) l^r(g),\\ F^{jk}(g) &= \partial^j l^k(g) - \partial^k l^j(g) - [l^j(g), l^k(g)] = 0. \end{aligned}$$

The canonical differential *Lie-Jacobi form* is valued in the Lie algebra:

$$\begin{aligned} g : L &\longmapsto G,\;\; l = \alpha_j l^j \longmapsto g(l),\\ (dg)\circ g^{-1} : L &\longrightarrow \mathbf{T}(G)\otimes \mathbf{T}^T(L),\\ l &\longmapsto (dg(l))\circ g(l)^{-1} = (\partial^j g(l))\circ g(l)^{-1}\otimes d\alpha_j\\ & \qquad = l^j(g)\otimes d\alpha_j = (g_*)^j_k\, l^k \otimes\; d\alpha_j. \end{aligned}$$

In general, a group-valued mapping on a manifold (chapter "Gauge Interactions") defines a corresponding Lie-Jacobi form,

$$\begin{aligned} U : M &\longrightarrow G,\;\; x \longmapsto U(x),\\ (dU)\circ U^{-1} : M &\longrightarrow \mathbf{T}(G)\otimes \mathbf{T}^T(M),\\ x &\longmapsto (dU(x))\circ U(x)^{-1} = (\partial^a U(x))\circ U(x)^{-1}\otimes dx_a, \end{aligned}$$

which for a Lie algebra parametrization involves the Lie-Jacobi transformation

$$\begin{aligned} M \longrightarrow L \longrightarrow G,\;\; x \longmapsto l(x) = \alpha_j(x) l^j &\longmapsto u(l(x)) = U(x),\\ x \longmapsto (dU(x))\circ U(x)^{-1} &= l^j(U(x))\otimes d\alpha_j(x) \in L\otimes \mathbf{T}^T_x(M),\\ d\alpha_j(x) &= (\partial^a \alpha_j(x)) dx_a. \end{aligned}$$

Forms for Lie group elements connected by left multiplication are related to each other by an action of the affine group $G \vec{\times} \underline{\log G}$:

$$\begin{aligned} (d\,U\circ V)\circ(U\circ V)^{-1} &= (dU)\circ U^{-1} + U\circ(dV)\circ V^{-1}\circ U^{-1}\\ &= (\,\mathrm{Ad}\,U, (dU)\circ U^{-1})\bullet (dV)\circ V^{-1}. \end{aligned}$$

3.14 (Semi)Simple Lie Algebras

The *commutator Lie algebra* $[L, L]$ of a Lie algebra L is the span of all Lie brackets. It is an ideal

$$[L, L] = \{[l_j, m_j] \mid l_j, m_j \in L\} \in \underline{\mathbf{lag}}_{\mathbb{K}}.$$

A Lie algebra L with $[L, L] = \{0\}$ is called *abelian* (nilquadratic). The quotient $L/[L, L]$ for any Lie algebra is abelian. A Lie algebra is called *perfect (idempotent)* for $L = [L, L]$.

A finite-dimensional Lie algebra L is called *simple* if it is *nonabelian* and without proper ideals. The trivial Lie algebra $\{0\}$ is not simple.

A finite-dimensional Lie algebra L is called *semisimple* with the following equivalent characterizations:

$$\begin{array}{l} L \in \underline{\mathbf{lag}}_{\mathbb{K}} \\ L \cong \mathbb{K}^n \\ \text{semisimple} \end{array} : \left\{ \begin{array}{ll} \iff & L = \bigoplus_{i=1}^{j} L_i \ \ L_i \text{ simple ideal} \\ \iff & \text{Each commutative ideal is trivial } \{0\} \\ \iff & \text{Each finite-dimensional representation} \\ & \text{is semisimple, i.e., decomposable into} \\ & \text{irreducible representations } (\textit{theorem of Weyl}). \end{array} \right.$$

The trivial Lie algebra $\{0\}$ is semisimple. A simple Lie algebra is semisimple.

A semisimple Lie algebra is perfect (idempotent). Its adjoint representation is injective, $L \cong \operatorname{ad} L$. Each derivation is inner, $\operatorname{der} L = \operatorname{ad} L$, i.e., each ideal is characteristic.

A finite-dimensional Lie algebra L is called *reductive* if its adjoint representation is semisimple, i.e., $[L, L]$ is semisimple.

One has the inclusions

$$\text{simple} \Rightarrow \text{semisimple} \Rightarrow \left\{ \begin{array}{l} \text{reductive} \\ \text{perfect (idempotent).} \end{array} \right.$$

The lowest-dimensional Lie algebras (chapter "Simple Lie Operations"): $L \cong \mathbb{K}$ is abelian, $L \cong \mathbb{K}^2$ is either abelian or semidirect, characterizable by the Lie bracket $[l^1, l^2] = l^2$ with $L \cong \log[\mathbf{GL}(\mathbb{K}) \vec{\times} \mathbb{K}]$. Simple Lie algebras start with $L \cong \mathbb{K}^3$: they are $\log \mathbf{SO}(\mathbb{C}^3) \cong \mathbb{C}^3$, and $\log \mathbf{SO}(3), \log \mathbf{SO}(1,2) \cong \mathbb{R}^3$. The other nondecomposable Lie algebras $L \cong \mathbb{R}^3$ are, up to isomorphism, - the contractions $\log[\mathbf{SO}(2) \times \mathbb{R}^2]$ and $\log[\mathbf{SO}(1,1) \times \mathbb{R}^2]$ and the Heisenberg Lie algebra as double contraction $\log \mathbf{H}(1)$ from the semidirect group $\mathbf{H}(1) \cong \mathbb{R} \vec{\times} \mathbb{R}^2$.

For $V \cong \mathbb{K}^n$ the endomorphisms Lie algebra $\mathbf{AL}(V) = \log \mathbf{GL}(V)$ is reductive, for $n \geq 2$ the traceless endomorphisms $[\mathbf{AL}(V), \mathbf{AL}(V)] = \mathbf{AL}(V)_0 = \log \mathbf{SL}(V)$ constitute a simple Lie algebra. The Lie algebra $\log(\mathbf{SL}(V) \vec{\times} V)$ of the semidirect affine group is perfect, in general, however, not semisimple.

3.15 Lie Algebra Inner Products

A Lie algebra form $\omega \in L^T$, trivial on its commutator $\omega([L,L]) = \{0\}$, is called *invariant*. Since $[L,L] = L$ for a semisimple Lie algebra, such L has no nontrivial invariant linear form. Each finite-dimensional representation of a semisimple Lie algebra is traceless, e.g., the adjoint one (structure constants):

$$\begin{array}{l} \mathcal{D} : L \longrightarrow \mathbf{AL}(V), \quad \mathrm{tr}_V : L \longrightarrow \mathbb{K}, \quad \mathrm{tr}_V l = \mathrm{tr}\, \mathcal{D}(l), \\ \text{semisimple } L \Rightarrow \left\{ \begin{array}{l} \mathrm{tr}\, \mathcal{D}(l) = 0, \\ \mathrm{tr}\ \mathrm{ad}\, l^j = \epsilon_k^{jk} = 0. \end{array} \right. \end{array}$$

For an (anti-) symmetric *invariant inner product* (bilinear form) of a Lie algebra

$$\kappa(\ ,\ ) : L \times L \longrightarrow \mathbb{K}, \quad \left\{ \begin{array}{l} \kappa(l,m) = \pm\kappa(m,l), \\ \kappa([n,l],m) + \kappa(l,[n,m]) = 0, \end{array} \right.$$

the orthogonal is an L-ideal

$$L_\kappa^\perp = \{l \mid \kappa(l,L) = \{0\}\}.$$

A nontrivial invariant inner product of a simple Lie algebra is nondegenerate.

A finite-dimensional representation of a Lie algebra has as *associate inner product* κ_V the symmetric "double trace"

$$\kappa_V(\ ,\ ) : L \times L \longrightarrow \mathbb{K}, \quad \left\{ \begin{array}{ll} \kappa_V(l,m) & = \mathrm{tr}\, \mathcal{D}(l) \circ \mathcal{D}(m) = \kappa_V(m,l), \\ \kappa_V(l^k, l^j) & = \kappa_V^{kj} = \mathcal{D}_A^{kB} \mathcal{D}_B^{jA}. \end{array} \right.$$

It is invariant under the adjoint action

$$\begin{array}{l} f,g,h \in \mathbf{AL}(V) \Rightarrow \mathrm{tr}\, [f,g] \circ h + \mathrm{tr}\, g \circ [f,h] = 0 \\ \Rightarrow \kappa_V([n,l],m) + \kappa_V(l,[n,m]) = 0, \quad \epsilon_k^{ij} \kappa_V^{kr} + \kappa_V^{jk} \epsilon_k^{ir} = 0. \end{array}$$

An associate inner product can be written in Sylvester bases of the Lie algebra L with $\kappa_V(l^a, l^b) = \pm\delta^{ab}, 0$.

Associated with the adjoint representation of a Lie algebra $L \cong \mathbb{K}^n$ is the *Killing form*

$$\begin{array}{l} \kappa(\ ,\ ) : L \times L \longrightarrow \mathbb{K}, \quad \left\{ \begin{array}{l} \kappa(l,m) = \ \mathrm{tr}\ \mathrm{ad}\, l \circ \mathrm{ad}\, m, \\ \kappa(l^i, l^j) = \kappa^{ij} = \epsilon_r^{ik} \epsilon_k^{jr}, \end{array} \right. \\ \kappa \circ \mathrm{ad}\, l = -(\kappa \circ \mathrm{ad}\, l)^T : \quad \epsilon_k^{ij} \kappa^{kr} + \epsilon_k^{ir} \kappa^{jk} = 0, \quad \epsilon_k^{mn} \kappa_{jn} + \epsilon_j^{mn} \kappa_{kn} = 0. \end{array}$$

An abelian Lie algebra has a trivial Killing form κ. However, the associate inner product κ_V is not necessarily trivial for all its representations.

Precisely for semisimple Lie algebras, the Killing form κ is nondegenerate. This can be seen immediately in the case of a diagonal matrix for the symmetric Killing form with ± 1 and 0, from which one can read off the (non)existence of a nontrivial abelian ideal.

Nonsemisimple Lie algebras, e.g., the abelian Lie algebra $\mathbb{K}$, can have representations with nondegenerate inner products κ_V. Such products define dual isomorphisms between L and L^T:

$$\begin{array}{ccc} L & \xrightarrow{\operatorname{ad} l} & L \\ \kappa_V \downarrow & & \downarrow \kappa_V, \\ L^T & \xrightarrow[-(\operatorname{ad} l)^T]{} & L^T \end{array} \qquad \begin{array}{l} \kappa_V \circ \operatorname{ad} l + (\operatorname{ad} l)^T \circ \kappa_V = 0, \\ \kappa_V(l^i) = \kappa_V^{ij} \check{l}_j, \quad \kappa_V^{-1}(\check{l}_i) = \kappa_{Vij} l^j. \end{array}$$

If a faithful irreducible complex representation of a simple Lie algebra L on $V \cong \mathbb{C}^n$ has a nondegenerate, invariant inner product γ, it is unique up to a scalar factor (theorem of Schur). Therefore, the invariant nondegenerate inner product $\Gamma = \gamma \otimes \gamma^{-1}$ on the endomorphisms $\mathbf{AL}(V)$ has to coincide on $\mathcal{D}[L] \subseteq \mathbf{AL}(V)$ with the associate inner product κ_V up to a nontrivial factor. In addition, γ has to be either *symmetric or antisymmetric*:

$$\begin{array}{l} \kappa_V(l,m) = \operatorname{tr} \mathcal{D}(l) \circ \mathcal{D}(m) = \alpha \Gamma(l,m), \quad \Gamma = \gamma \otimes \gamma^{-1}, \quad \alpha \in \mathbb{C}^\diamond, \\ \Gamma = \Gamma^T \iff \gamma = \gamma^T \text{ or } \gamma = -\gamma^T. \end{array}$$

For a complex, simple Lie algebra the Killing form is, up to a scalar factor, the unique invariant inner product (theorem of Schur), e.g., the "double trace" $\operatorname{tr} f_0 \circ g_0$ for the complex traceless endomorphisms $\mathbf{AL}(V)_0$.

The invariance of the Killing form for a semisimple *complex* Lie algebra allows bases with $\kappa^{ij} = \delta^{ij}$ and totally antisymmetric structure constants

$$\epsilon^{ijr} = \epsilon^{ij}_k \delta^{kr} = -\epsilon^{jir} = -\epsilon^{irj}.$$

Morover, there exists, up to linear equivalence, exactly one semisimple Lie algebra $A_1 \cong \mathbb{C}^3$; it allows a Cartesian basis with $[l^a, l^b] = -\epsilon^{abc} l^c$.

A real Lie algebra *representation* is *compact (Hilbert)* if the associate inner product is strictly negative. A semisimple real *Lie algebra* is *compact* with strictly negative Killing form. Compact Lie algebras have bases with totally antisymmetric structure constants. Also, noncompact Lie algebras, e.g., $\mathbb{R}$, can have Hilbert representations. All complex representations of compact Lie algebras are semisimple and decomposable into irreducible Hilbert representations, all of which are finite-dimensional (*theorem of Weyl*).

The adjoint representation of a semisimple Lie algebra $L \cong \mathbb{K}^d$ goes into the orthogonal Lie algebra of its endomorphisms, which defines the *signature* (d_+, d_-) *of the Lie algebra* L:

$$\operatorname{ad}: L \longrightarrow \begin{cases} \log \mathbf{SO}(\mathbb{C}^d) \subset \mathbf{AL}(L), & \mathbb{K} = \mathbb{C}, \\ \log \mathbf{SO}(d_+, d_-) \subset \mathbf{AL}(L), & \mathbb{K} = \mathbb{R},\ d_+ + d_- = d, \end{cases}$$

where $d_+ (d_-)$ *noncompact (compact) dimensions* have positive (negative) definite Killig form, $\gamma(l,l) > 0, < 0$ for $l \neq 0$. The orthogonal and unitary Lie groups have as maximal compact groups

$$\begin{array}{lll} \mathbf{SO}(p,q) \supseteq \mathbf{SO}(p) \times \mathbf{SO}(q): & d_- = \binom{p}{2} + \binom{q}{2}, & d_+ = pq, \\ \mathbf{SU}(p,q) \supseteq \mathbf{SO}(2) \times \mathbf{SU}(p) \times \mathbf{SU}(q): & d_- = p^2 + q^2 - 1, & d_+ = 2pq. \end{array}$$

3.16 Lie Algebra Decompositions

Each semisimple real Lie algebra has a *Cartan decomposition* into the direct sum of a maximal compact Lie subalgebra K and a vector space P with noncompact vectors on which the Killing form is strictly negative and positive respectively:

$$\begin{aligned}
&L \cong K \oplus P \cong \mathbb{R}^{d_-} \oplus \mathbb{R}^{d_+}, \quad K \in \underline{\mathbf{lag}}_{\mathbb{R}}, \quad P \in \underline{\mathbf{vec}}_{\mathbb{R}},\\
&[K,K] \subseteq K, \quad [K,P] \subseteq P, \quad [P,P] \subseteq K,\\
&\kappa \cong \begin{pmatrix} -\mathbf{1}_{d_-} & 0 \\ 0 & \mathbf{1}_{d_+} \end{pmatrix}.
\end{aligned}$$

A characteristic example is the decomposition of the Lorentz Lie algebra into angular momenta and boosts as tangent space of the 3-hyperboloid:

$$\begin{aligned}
\log \mathbf{SL}(\mathbb{C}^2) &\cong \log \mathbf{SU}(2) \oplus \log \mathcal{Y}^3,\\
\{\vec{l} \cong i\vec{\sigma}, \vec{b} \cong \vec{\sigma}\} &\cong \{\vec{l}\} \oplus \{\vec{b}\}.
\end{aligned}$$

An *Iwasawa decomposition* uses in addition to a maximal compact subalgebra a maximal abelian (diagonal as matrix) subalgebra of the noncompact subspace and a nilpotent (strictly triangular) Lie algebra:

$$L \cong K \oplus A \oplus N, \quad \begin{cases} A \subset P, \quad [A,A] = \{0\},\\ [N,[N,\dots[N,N]\dots]N] = \{0\},\\ [A \oplus N, A \oplus N] \subseteq N, \end{cases}$$

$$\begin{aligned}
\text{e.g., } \log \mathbf{SL}(\mathbb{C}^2) &\cong \log \mathbf{SU}(2) \oplus \log \mathbf{SO}_0(1,1) \oplus \mathbb{R}^2,\\
\{\vec{l}, \vec{b}\} &\cong \{\vec{l}\} \oplus \{b_3\} \oplus \{b_1 + l_2, b_2 - l_1\}.
\end{aligned}$$

The uniquely determined dimension of A is called the *real (noncompact) rank of* L. In general, N is constructed with a basis $\{l_{+\omega} \mid \omega \neq 0\}$ consisting of nilpotent raising operators (chapter "Simple Lie Algebras").

For semisimple connected real Lie groups, the corresponding factorizations (decompositions, parametrizations) with a maximal compact, maximal noncompact abelian and unipotent subgroup are obtainable by exponentiation $(\mathcal{K}, \mathcal{A}, \mathcal{N}) = \exp(K, A, N)$:

$$\begin{aligned}
&\textit{Cartan factorization (diagonalization)}: && G = \mathcal{K} \circ \mathcal{A} \circ \mathcal{K}, \quad g = u_1 \circ a \circ u_2,\\
&\textit{Iwasawa factorization (triagonalization)}: && G = \mathcal{K} \circ \mathcal{A} \circ \mathcal{N}, \quad g = u \circ a \circ n.
\end{aligned}$$

A Cartan factorization with diagonal $\mathcal{A}$ involves a polar decomposition $G = \mathcal{K} \circ \mathcal{D}$ and an orthogonal or unitary diagonalization of the noncompact part $\mathcal{D} = \mathcal{D}^* = \mathcal{K} \circ \mathcal{A} \circ \mathcal{K}$. An Iwasawa factorization has triagonal $\mathcal{A} \circ \mathcal{N}$.

3.17 Multilinearity and Tensor Algebra

The *tensor powers* $\bigotimes^k$ of a vector space V,

$$\bigotimes^0 V = \mathbb{K}, \quad \bigotimes^k V = \underbrace{V \otimes \cdots \otimes V}_{k \text{ times}} \in \underline{\mathbf{vec}}_{\mathbb{K}},$$

$$\dim_{\mathbb{K}} V = n \Rightarrow \dim_{\mathbb{K}} \bigotimes^k V = n^k,$$

are covariant functors

$$\bigotimes^{k}: \underline{\mathbf{vec}_{\mathbb{K}}} \longrightarrow \underline{\mathbf{vec}_{\mathbb{K}}}, \quad f \begin{array}{c} V \\ \downarrow \\ W \end{array} \longmapsto \begin{array}{c} \bigotimes^{k} V \\ \downarrow \\ \bigotimes^{k} W \end{array} \bigotimes^{k} f,$$

$$\bigotimes^{0} f(\alpha) = \alpha, \ \alpha \in \mathbb{K}, \quad \bigotimes^{1} f(v) = f(v), \ v \in V,$$

$$\bigotimes^{k} f(v_1 \otimes \cdots \otimes v_k) = f(v_1) \otimes \cdots \otimes f(v_k),$$

$$V \cong \mathbb{K}^n: \quad \operatorname{tr} \bigotimes^{k} f = (\operatorname{tr} f)^k, \quad k \geq 1: \ \det \bigotimes^{k} f = [(\det f)^{n^{k-1}}]^k.$$

The tensor power vector space $\bigotimes^{k} V$ realizes the concept of *k-linearity* with respect to the linear space V.

For finite-dimensional spaces, duality commutes with k-linearity:

$$V \cong \mathbb{K}^n: \quad \bigotimes^{k} V^T = (\bigotimes^{k} V)^T.$$

One can define a product by juxtaposition:

$$\begin{aligned} &\bigotimes^{k} V \times \bigotimes^{l} V \longrightarrow \bigotimes^{k+l} V, \\ &(v_1 \otimes \cdots \otimes v_k, w_1 \otimes \cdots \otimes w_l) \longmapsto v_1 \otimes \cdots \otimes v_k \otimes w_1 \otimes \cdots \otimes w_l. \end{aligned}$$

The associative *tensor algebra* $\bigotimes V$ *for* V with unit $1 \in \mathbb{K}$,

$$\bigotimes V = \bigoplus_{k \geq 0} \bigotimes^{k} V \in \underline{\mathbf{aag}_{\mathbb{K}}},$$

defines "multilinearity." $\mathbb{K} \cup V$ is a generating system of the tensor algebra $\bigotimes V$, already a vector space basis $\{e^i\}_{i \in I}$ together with $1 \in \mathbb{K}$.

The tensor algebra is *universal*: Each linear mapping F of a vector space V into any unital algebra A is factorizable into a canonical injection ι and a uniquely determined algebra morphism $\tilde{F}$:

$$\begin{array}{l} V, \iota, F \ \in \underline{\mathbf{vec}_{\mathbb{K}}} \\ \bigotimes V, A, \tilde{F} \ \in \underline{\mathbf{aag}_{\mathbb{K}}}, \end{array} \quad \begin{array}{ccc} V & \stackrel{\iota}{\longrightarrow} & \bigotimes V \\ {\scriptstyle F}\downarrow & & \downarrow{\scriptstyle \tilde{F}} \\ A & \underset{\mathrm{id}_A}{\longrightarrow} & A \end{array}, \quad \begin{array}{l} \tilde{F}(1) = 1_A, \ \ \tilde{F}(v) = F(v), \\ \tilde{F}(t_1 \otimes t_2) = \tilde{F}(t_1)\tilde{F}(t_2). \end{array}$$

With this factorization condition the tensor algebra $\bigotimes V$ is determined up to algebra isomorphisms.

This defines the covariant *tensor algebra functor* $\bigotimes$:

$$\bigotimes : \underline{\mathbf{vec}}_{\mathbb{K}} \longrightarrow \underline{\mathbf{aag}}_{\mathbb{K}}, \quad f \begin{array}{c} V \\ \downarrow \\ W \end{array} \longmapsto \begin{array}{c} \bigotimes V \\ \downarrow \\ \bigotimes W \end{array} \bigotimes f.$$

For a direct sum one has the *exponential property* of the functor

$$\bigotimes(V_1 \oplus V_2) \cong \bigotimes V_1 \otimes \bigotimes V_2, \quad v_1 + v_1 \cong v_1 \otimes 1 + 1 \otimes v_2.$$

If $V \in *\underline{\mathbf{vec}}_{\mathbb{K}}$ is a vector space with (conjugate) linear reflection, $\alpha v \longmapsto \alpha^* v^*$, then $\bigotimes V$ is an algebra with (conjugate) linear reflection

$$* : \bigotimes V \longrightarrow \bigotimes V, \quad \left\{ \begin{array}{l} \alpha \longmapsto \alpha^* \in \mathbb{K}, \quad v \longmapsto v^* \in V, \\ (a \otimes b)^* = b^* \otimes a^*, \quad a, b \in \bigotimes V. \end{array} \right.$$

The quotient algebra $\bigotimes V/I$ with an ideal I remains universal for all linear mappings that leave trivial the ideal I:

$$\begin{array}{ccc} V & \stackrel{\iota}{\longrightarrow} & \bigotimes V/I \\ F \downarrow & & \downarrow \tilde{F} \\ A & \underset{\mathrm{id}_A}{\longrightarrow} & A \end{array}, \qquad \tilde{F}[I] = 0.$$

3.17.1 Grassmann and Polynomial Algebra

The direct sum of all *Grassmann powers* $\bigwedge^k$ *(covariant functors)* of a vector space V:

$$\begin{aligned}
&\bigwedge^0 V = \mathbb{K}, \quad \bigwedge^k V = \underbrace{V \wedge \cdots \wedge V}_{k \text{ times}} \in \underline{\mathbf{vec}}_{\mathbb{K}}, \\
&v_{j_1} \wedge \cdots \wedge v_{j_k} = \frac{\epsilon^{m_1 \dots m_k}_{j_1 \dots j_k}}{k!} v_{m_1} \otimes \cdots \otimes v_{m_k}, \\
&\epsilon^{m_1 \dots m_k}_{j_1 \dots j_k} = \left\{ \begin{array}{l} 0, \quad \{j_1, \dots, j_k\} \neq \{m_1, \dots, m_k\}, \\ \pm 1, \quad \{j_1, \dots, j_k\} \text{ even (odd) permutation of } \{m_1, \dots, m_k\}, \end{array} \right. \\
&\epsilon^{j_1 \dots j_k}_{1 \dots k} = \epsilon^{j_1 \dots j_k}
\end{aligned}$$

has a product by juxtaposition

$$\begin{aligned}
&\bigwedge^k V \times \bigwedge^l V \longrightarrow \bigwedge^{k+l} V, \\
&(v_1 \wedge \cdots \wedge v_k, w_1 \wedge \cdots \wedge w_l) \longmapsto v_1 \wedge \cdots \wedge v_k \wedge w_1 \wedge \cdots \wedge w_l, \\
&\bigwedge V = \bigoplus_{k \geq 0} \bigwedge^k V \in \underline{\mathbf{aag}}_{\mathbb{K}} \text{ with } 1 \in \mathbb{K}.
\end{aligned}$$

The *Grassmann algebra* $\bigwedge V \subseteq \bigotimes V$ *with the covariant functor* $\bigwedge$ realizes the "totally antisymmetric multilinearity" with respect to V. It is a vector subspace of the tensor algebra $\bigotimes V$, however not a subalgebra. As an algebra it is isomorphic to the quotient algebra of the tensor algebra $\bigotimes V$ with the minimal ideal, generated by the squares of the basic vectors:

$$\bigwedge V = \bigotimes V/I^0_+(V) = \bigoplus_{k\geq 0} \bigwedge^k V, \quad \begin{cases} I^0_+(V) = \text{ideal } S^0_+(V), \\ S^0_+(V) = \{v \otimes v \mid v \in V\}, \end{cases}$$

$$\text{in } \bigwedge V: \begin{cases} \{v_1, v_2\} = v_1 \otimes v_2 + v_2 \otimes v_1 = 0, \quad v_j \in V, \\ v_1 \wedge \cdots \wedge v_k = (-1)^{\binom{k}{2}} v_k \wedge \cdots \wedge v_1, \quad k \geq 2, \end{cases}$$

$$\dim_{\mathbb{K}} V = n \Rightarrow \dim_{\mathbb{K}} \bigwedge^k V = \binom{n}{k}, \quad \dim_{\mathbb{K}} \bigwedge V = 2^n.$$

A Grassmann algebra is the linear extension of a power set in the following sense: If a set S has n elements, its power set 2^S contains the 2^n subsets of S. This is reflected by the dimensions for a vector space and its Grassmann algebra:

$$\begin{array}{lll} S \in \underline{\mathbf{set}}: & \text{card } S = n & \Rightarrow \text{card } 2^S = 2^n, \\ V \in \underline{\mathbf{vec}}_{\mathbb{K}}: & \dim_{\mathbb{K}} V = n & \Rightarrow \dim_{\mathbb{K}} \bigwedge V = 2^n. \end{array}$$

A linear mapping $f : V \longrightarrow W$ of vector spaces with equal finite dimension $V, W \cong \mathbb{K}^n$ has a *determinant*:

$$\begin{aligned} \bigwedge^n V, \bigwedge^n W \cong \mathbb{K}, \quad & \bigwedge^n f = \det f : \bigwedge^n V \longrightarrow \bigwedge^n W, \\ \det f(e^1 \wedge \cdots \wedge e^n) &= f^1_{j_1} \cdots f^n_{j_n} e^{j_1} \wedge \cdots \wedge e^{j_n}, \\ &= (\epsilon^{j_1 \ldots j_n} f^1_{j_1} \cdots f^n_{j_n}) e^1 \wedge \cdots \wedge e^n. \end{aligned}$$

With the functor properties of $\bigwedge^n$ the determinant leads to a monoid morphism from the endomorphism algebra to the scalars, involving a group morphism for the regular groups

$$\begin{aligned} \det : \mathbf{AL}(V) \longrightarrow \mathbb{K}, \quad & \begin{cases} \det(f \circ g) = \det f \det g, \\ \det \mathrm{id}_V = 1, \\ \text{kern } \det = \mathbf{SL}(V), \end{cases} \\ \det : \mathbf{GL}(V) \longrightarrow \mathbb{K}^\diamond, \quad & \det f^{-1} = (\det f)^{-1}. \end{aligned}$$

The *totally symmetric* quotient algebra of the tensor algebra (analogous functor properties as for the Grassmann algebra)

$$\bigvee V = \bigotimes V/I^0_-(V) = \bigoplus_{k\geq 0} \bigvee^k V, \begin{cases} I^0_-(V) = \text{ideal } S^0_-(V), \\ S^0_-(V) = \{v \otimes w - w \otimes v \mid v, w \in V\}, \\ \text{in } \bigvee V: [v, w] = v \otimes w - w \otimes v = 0, \end{cases}$$

$$\dim_{\mathbb{K}} V = n \Rightarrow \dim_{\mathbb{K}} \bigvee^k V = \binom{n-1+k}{k},$$

is a commutative unital ring. For a finite basis $\{e^i\}_{i=1}^n$ of V the symmetric algebra is isomorphic to the *polynomial ring over* $\mathbb{K}$ with n indeterminates:

$$\mathbb{K}[e^1, \ldots, e^n] \cong \bigotimes V \Big/ \text{ideal } \{e^i \otimes e^j - e^j \otimes e^i \mid i,j = 1, \ldots, n\} = \bigvee V.$$

The individual tensor powers $\bigvee^k V$ are the *homogeneous polynomials of degree* k with the *monomials* as basis:

$$\mathbb{K}[e^1, \ldots, e^n]^k \cong \bigvee^k V.$$

For finite-dimensional spaces, duality commutes with totally (anti)symmetric k-linearity

$$V \cong \mathbb{K}^n : \begin{cases} \bigwedge^k V^T = (\bigwedge V^k)^T, \ \bigwedge V^T = (\bigwedge V)^T, \\ \bigvee^k V^T = (\bigvee^k V)^T. \end{cases}$$

3.17.2 Volume Elements and Axial Vectors

For a vector space $V \cong \mathbb{K}^n$ the elements of the Grassmann powers are called as indicated in the following table:

k	0	1	2	3	...	$n-2$	$n-1$	n
	scalars $\mathbb{K}$	vectors V	areas $V \wedge V$	3-volumes $V \wedge V \wedge V$	...	$(n-2)-$ volumes	$(n-1)-$ volumes	$n-$ volumes
		...			...	or	or	or
		...			...	axial areas	axial vectors	axial scalars

$$\bigwedge^k V: \ k\text{-volumes or } (n-k) \text{ axial volumes}$$

They constitute the unital Grassmann algebra.

The isomorphisms between the highest Grassmann power $\bigwedge^n V$ with the *n-volumes* and the scalar field $\mathbb{K}$ are given by $e^1 \wedge \cdots \wedge e^n \longmapsto \alpha_B$ with a nontrivial constant α_B as the *volume unit* for a basis that characterizes one $\mathbf{SL}(\mathbb{K}^n)$-equivalent class of bases $\{e^j\}_{j=1}^n$ by its totally antisymmetric multilinear form

$$\epsilon_n : \underbrace{V \times \cdots \times V}_{n \text{ times}} \longrightarrow \mathbb{K}, \quad \epsilon_n(e^{j_1}, \ldots, e^{j_n}) = \alpha_B \epsilon^{j_1 \cdots j_n},$$

$$\alpha_B \in \mathbb{K}^\diamond \cong \mathbf{GL}(\mathbb{K}^n)/\mathbf{SL}(\mathbb{K}^n) \cong \overset{\circ}{\mathbf{vec}}_{\mathbb{K}} \left(\bigwedge^n V, \mathbb{K}\right),$$

$$\frac{\epsilon^{j_1 \cdots j_n}}{n!} \check{e}_{j_1} \otimes \cdots \otimes \check{e}_{j_n} = \check{e}_1 \wedge \cdots \wedge \check{e}_n \in \bigwedge^n V^T.$$

The volume elements are the $\mathbf{SL}(\mathbb{K}^n)$-invariant *natural isomorphisms* between the dual partner spaces $\bigwedge^k V \cong \mathbb{K}^{\binom{n}{k}}$ and $\bigwedge^{n-k} V^T$:

$$\begin{array}{ccc} \bigwedge^k V & \xrightarrow{\bigotimes^k g} & \bigwedge^k V \\ \epsilon_n \downarrow & & \downarrow \epsilon_n \\ \bigwedge^{n-k} V^T & \xrightarrow[\bigotimes^{n-k} g^{-1T}]{} & \bigwedge^{n-k} V^T \end{array}, \qquad \begin{array}{l} g \in \mathbf{SL}(V) \cong \mathbf{SL}(\mathbb{K}^n), \\ \epsilon_n(e^{j_1} \wedge \cdots \wedge e^{j_k}) = \alpha_B \epsilon^{j_1 \cdots j_n} \check{e}_{j_{k+1}} \wedge \cdots \wedge \check{e}_{j_n}, \end{array}$$

for $k = 1$ called the *Hodge dual isomorphisms* $\epsilon_n : V \longrightarrow \bigwedge^{n-1} V^T$.

Only for dimension 2, $V \cong \mathbb{K}^2 \cong V^T$, are the volume elements $\mathbf{SL}(\mathbb{K}^n)$-invariant, symplectic, nondegenerate inner products $\epsilon_2 : V \longrightarrow V^T$.

Given a dual isomorphism (nondegenerate inner product) $\gamma : V \longrightarrow V^T$ there exist isomorphisms between vector and axial vectors that are $(n-1)$-volumes, etc.:

$$\left. \begin{array}{rl} \epsilon_n : & \bigwedge^k V \longrightarrow \bigwedge^{n-k} V^T \\ \bigwedge^k \gamma : & \bigwedge^k V \longrightarrow \bigwedge^k V^T \end{array} \right\} \Rightarrow \bigwedge^k V \cong \bigwedge^{n-k} V.$$

3.17.3 Derivations of Tensor Algebras

A *derivation of a tensor algebra* is determined by its action on the scalars and the basic vectors $V \in \underline{\mathbf{vec}}_{\mathbb{K}}$ as generators and the extension by Leibniz rule:

$$D : \bigotimes V \longrightarrow \bigotimes V, \quad \begin{cases} D : \mathbb{K} \longrightarrow \bigotimes V, \ \ D : V \longrightarrow \bigotimes V, \\ D(t_1 \otimes t_2) = D(t_1) \otimes t_2 + t_1 \otimes D(t_2). \end{cases}$$

The derivative property on $\mathbb{K}$ gives $D\alpha = 0$.

The derivations considered in the following involve a vector space form $\omega : V \longrightarrow \mathbb{K}$ and the uniquely extended derivation $D = \omega_{\text{der}}$ of the tensor algebra:

$$\begin{array}{l} V^T \longrightarrow \operatorname{der} \bigotimes V, \ \ \omega \longmapsto \omega_{\text{der}}, \\ \omega_{\text{der}} : \bigotimes V \longrightarrow \bigotimes V, \quad \begin{cases} \omega_{\text{der}}(\alpha) = 0, & \alpha \in \mathbb{K}, \\ \omega_{\text{der}}(v) = \langle \omega, v \rangle, & v \in V, \end{cases} \\ \omega_{\text{der}} [\bigotimes^k V] \subseteq \bigotimes^{k-1} V. \end{array}$$

For example, the *derivations with respect to dual bases of* $V \cong \mathbb{K}^n$,

$$\partial_j : \bigotimes V \longrightarrow \bigotimes V, \ \ \partial_j(e^k) = \langle \check{e}_j, e^k \rangle = \delta_j^k.$$

Since the induced derivation is trivial on the ideal for the polynomial algebra (in general not for the Grassmann algebra)

$$\omega_{\text{der}}(v \otimes u - u \otimes v) = 0, \quad \omega_{\text{der}} : \bigvee V \longrightarrow \bigvee V,$$

it is well defined on the polynomial algebra.

Forms can be embedded into higher tensor powers. With $V \otimes V^T \subseteq \mathbf{AL}(V)$, each endomorphism $f : V \longrightarrow V$ defines a unique derivation f_{der} of the tensor algebra as a Lie algebra morphism, i.e., with trivial action on the scalars:

$$\mathbf{AL}(V) \longrightarrow \text{der} \bigotimes V, \quad f \longmapsto f_{\text{der}}, \quad [f, h]_{\text{der}} = [f_{\text{der}}, h_{\text{der}}],$$
$$f_{\text{der}} : \bigotimes V \longrightarrow \bigotimes V, \quad \begin{cases} f_{\text{der}}(\alpha) = 0, & \alpha \in \mathbb{K}, \\ f_{\text{der}}(v) = f(v), & v \in V, \text{ e.g., } (u \otimes \omega)(v) = \langle \omega, v \rangle u, \end{cases}$$
$$f_{\text{der}}[\bigotimes^k V] \subseteq \bigotimes^k V.$$

If f is an inner derivation of a (Lie) algebra V, then f_{der} is an inner derivation of the tensor algebra. Since the ideals for Grassmann and polynomial algebras are stable with respect to the induced derivation, f_{der} is well defined for both Grassmann and polynomial algebras. On the highest Grassmann power of $V \cong \mathbb{K}^n$ (n-volumes), f_{der} is the multiplication with trace

$$f_{\text{der}}[S^0_{\pm}] \subseteq S^0_{\pm}, \quad f_{\text{der}} \in \text{der} \bigwedge V, \text{ der} \bigvee V,$$
$$v^{(n)} \in \bigwedge^n V \Rightarrow f_{\text{der}}(v^{(n)}) = (\operatorname{tr} f)\, v^{(n)}.$$

Therefore the action of endomorphisms W on a vector space $V \cong \mathbb{K}^n$, e.g., of a represented Lie algebra, can be uniquely extended to derivations W_{der} on the tensor, Grassmann, and polynomial algebras.

The relation between extension of linear forms and endomorphisms for $V \cong \mathbb{K}^n$ is given by

$$\begin{aligned} f &= v^a \otimes \omega^a & &\Rightarrow f_{\text{der}} = v^a \otimes \omega^a_{\text{der}}, \\ f &= f^j_k e^k \otimes \check{e}_j & &\Rightarrow f_{\text{der}} = f^j_k e^k \otimes \partial_j, \end{aligned}$$

which can be also used for the tensor and polynomial algebras (in general not for the Grassmann algebra). One obtains for a represented Lie algebra the action by derivations of the polynomials for the representation space,

$$\begin{aligned} L &\longrightarrow \mathbf{AL}(V), & l &\longmapsto l^j_k e^k \otimes \check{e}_j, \\ L &\longrightarrow \text{der}\, \mathbb{K}[e^1, \ldots, e^n], & l &\longmapsto l_{\text{der}} = l^j_k e^k \partial_j, \end{aligned}$$

decomposable into *Lie algebra representations on the homogeneous polynomials of degree* k:

$$\mathcal{D}^k : L \longrightarrow \mathbf{AL}(\mathbb{K}[e^1, \ldots, e^n]^k), \quad l \longmapsto l_{\text{der}}.$$

Via the Leibniz rule for the dual product an endomorphism-induced derivation is defined on the dual space (negative transposition),

$$\begin{aligned} \theta \in V^T : &\quad \langle D(\theta), v \rangle + \langle \theta, D(v) \rangle = 0, \\ \text{e.g., } f = u \otimes \omega : &\quad Dv = f(v) = \langle \omega, v \rangle u, \quad D\theta = -f^T(\theta) = -\langle \theta, u \rangle \omega, \end{aligned}$$

its tensor algebra, and all tensors $\bigotimes(V \oplus V^T)$.

3.18 Enveloping Algebra

The *enveloping algebra* $\mathbf{E}(L) \in \underline{\mathbf{aag}}_{\mathbb{K}}$ with unit $1 \in \mathbb{K}$ for a Lie algebra $L \in \underline{\mathbf{lag}}_{\mathbb{K}}$ is defined by the tensor algebra $\bigotimes L$ of the vector space L, factorized by the minimal ideal of $\bigotimes L$, which enforces equality for Lie brackets and tensor commutators:

$$S^{\text{com}}(L) = \{l \otimes m - m \otimes l - [l,m] \bigm| l, m \in L\}, \quad I^{\text{com}}(L) = \text{ideal } S^{\text{com}}(L),$$
$$\text{in } \mathbf{E}(L) = \bigotimes L \Big/ I^{\text{com}}(L): \quad \bigotimes^2 L \ni l \otimes m - m \otimes l = [l,m] \in \bigotimes^1 L = L.$$

The ideal connects even and odd tensor powers. The Lie algebra is embedded by a Lie algebra morphism

$$\iota : L \longrightarrow \mathbf{E}(L), \quad \iota([l,m]) = [\iota(l), \iota(m)].$$

Using the equality of Lie bracket and tensor commutator an antisymmetric tensor from $\bigotimes L$ as an element of $\mathbf{E}(L)$ can be reduced pairwise with respect to its power. Therefore, there exists for each element in the enveloping algebra a totally symmetric representative. As vector space the enveloping algebra is isomorphic to the totally symmetric polynomial algebra:

$$\begin{aligned} \text{as vector space: } \mathbf{E}(L) &\cong \bigvee L \cong \mathbb{K}[e^1, \dots, e^n], \\ \bigotimes L &\cong \bigvee L \oplus I^{\text{com}}(L). \end{aligned}$$

The enveloping algebra for a 1-dimensional Lie algebra $L^{(1)} \cong \mathbb{K}$ is generated by two elements: the unit $1 \in \mathbb{K} = \bigotimes^0 L^{(1)}$ and a basis $I \in L^{(1)}$; $\mathbf{E}(L^{(1)})$ is the polynomial ring $\mathbb{K}[I]$ with one indeterminate.

The enveloping algebra solves a *universal mapping problem*: For each Lie algebra morphism $\mathcal{D}$ (representation of L) in a unital algebra $A \in \underline{\mathbf{aag}}_{\mathbb{K}}$ there is exactly one algebra morphism $\tilde{\mathcal{D}}$ of the enveloping algebra $\mathbf{E}(L)$, defined for the generating system, which gives rise to the following commutative diagram:

$$\begin{array}{rl} L, \iota, \mathcal{D} & \in \underline{\mathbf{lag}}_{\mathbb{K}} \\ \mathbf{E}(L), A, \tilde{\mathcal{D}} & \in \underline{\mathbf{aag}}_{\mathbb{K}} \end{array}, \quad \begin{array}{ccc} L & \xrightarrow{\iota} & \mathbf{E}(L) \\ \mathcal{D} \downarrow & & \downarrow \tilde{\mathcal{D}} \\ A & \xrightarrow[\text{id}_A]{} & A \end{array}, \quad \begin{array}{l} \tilde{\mathcal{D}}(1) = 1_A, \ \tilde{\mathcal{D}}(l) = \mathcal{D}(l), \\ \tilde{\mathcal{D}}(t_1 \otimes t_2) = \tilde{\mathcal{D}}(t_1)\tilde{\mathcal{D}}(t_2). \end{array}$$

With the universal enveloping algebra $\mathbf{E}(L)$ of a Lie algebra L all multilinear forms on the dual space L^T, e.g., volume elements, Casimir element (below), can be transferred into the endomorphism algebras $\mathbf{AL}(V)$ of its representation spaces. The tensor product is represented by the representation algebra composition product:

$$\begin{array}{ccc} L & \stackrel{\iota}{\longrightarrow} & \mathbf{E}(L) \\ {\scriptstyle \mathcal{D}}\downarrow & & \downarrow {\scriptstyle \tilde{\mathcal{D}}} \\ \mathbf{AL}(V) & \underset{\mathrm{id}_{\mathbf{AL}(V)}}{\longrightarrow} & \mathbf{AL}(V) \end{array},$$

$$\tilde{\mathcal{D}}(1) = \mathrm{id}_V, \quad \tilde{\mathcal{D}}(l \otimes \cdots \otimes m) = \mathcal{D}(l) \circ \cdots \circ \mathcal{D}(m), \quad l, m \in L.$$

Furthermore, the representation space is a module $V \in \underline{\mathbf{mod}}_{\mathbf{E}(L)}$ with the action of the full enveloping algebra:

$$\mathbf{E}(L) \times V \longrightarrow V, \quad (l \otimes \cdots \otimes m) \bullet v = \mathcal{D}(l) \circ \cdots \circ \mathcal{D}(m)(v).$$

The covariant *Lie algebra enveloping functor* $\mathbf{E}$ is given by

$$\mathbf{E} : \underline{\mathbf{lag}}_{\mathbb{K}} \longrightarrow \underline{\mathbf{aag}}_{\mathbb{K}}, \quad {\scriptstyle f} \begin{array}{c} L \\ \downarrow \\ M \end{array} \longmapsto \begin{array}{c} \mathbf{E}(L) \\ \downarrow \\ \mathbf{E}(M) \end{array} \otimes f,$$

For a direct sum with commuting summands one has the exponential property

$$[L_1, L_2] = \{0\} \Rightarrow \mathbf{E}(L_1 \oplus L_2) \cong \mathbf{E}(L_1) \otimes \mathbf{E}(L_2).$$

If $L \in *\underline{\mathbf{lag}}_{\mathbb{K}}$ has a (conjugate) linear reflection, then $\mathbf{E}(L) \in *\underline{\mathbf{aag}}_{\mathbb{K}}$ also has one.

The Lie algebra acts by inner derivations on its enveloping algebra $\mathbf{E}(L)$:

$$\begin{array}{ccc} L & \stackrel{\mathrm{ad}\, l}{\longrightarrow} & L \\ {\scriptstyle \iota}\downarrow & & \downarrow {\scriptstyle \iota} \\ \mathbf{E}(L) & \underset{\mathrm{ad}\, l}{\longrightarrow} & \mathbf{E}(L) \end{array}, \qquad \begin{array}{l} l \bullet 1 = [l, 1] = 0, \\ l \bullet (m_1 \otimes m_2) = [l, m_1 \otimes m_2] \\ = [l, m_1] \otimes m_2 + m_1 \otimes [l, m_2]. \end{array}$$

3.18.1 Lie Algebra Invariants

The center of the enveloping algebra $\mathbf{E}(L)$ is the abelian unital *ring of Lie algebra invariants*

$$\mathrm{INV}_L \mathbf{E}(L) = \{I \in \mathbf{E}(L) \mid [L, I] = \{0\}\} = \mathrm{centr}\, \mathbf{E}(L).$$

Invariants can be mapped into representations: For complex finite-dimensional irreducible representations $\mathcal{D}: L \longrightarrow \mathbf{AL}(V)$ the invariants are, up to a factor, the identity (theorem of Schur):

$$\begin{aligned} \mathrm{INV}_L\mathbf{E}(L) \ni I &= \gamma_{r_1 \ldots r_k}\, l^{r_1} \otimes \cdots \otimes l^{r_k} \\ \longmapsto \tilde{\mathcal{D}}(I) &= \gamma_{r_1 \ldots r_k}\, \mathcal{D}(l^{r_1}) \circ \cdots \circ \mathcal{D}(l^{r_k}) \\ &= \beta_V(I)\, \mathrm{id}_V \in \mathrm{INV}_L\mathbf{AL}(V), \quad \beta_V(I) \in \mathbb{C}. \end{aligned}$$

The abelian Lie algebra $\mathbb{K}l^0$ has one generator for all invariants:

$$\mathrm{centr}\,\mathbf{E}(\mathbb{K}l^0) = \mathbf{E}(\mathbb{K}l^0) \cong \mathbb{K}[l^0].$$

The invariant *trace forms* for a representation $\mathcal{D}$ on $V \cong \mathbb{K}^n$ (associate multilinear forms) are defined by the multilinear "multiple" traces

$$\kappa_V^k : \bigotimes^k L \longrightarrow \mathbb{K}, \quad \begin{cases} \kappa_V^k(l_1 \otimes \cdots \otimes l_k) = \mathrm{tr}\,\mathcal{D}(l_1) \circ \cdots \circ \mathcal{D}(l_k), \\ \kappa_V^k = \kappa_V^{i_1 \ldots i_k} \check{l}_{i_k} \otimes \cdots \otimes \check{l}_{i_1} \in \bigotimes^k L^T, \\ \kappa_V^0 = 1, \quad \kappa_V^2 = \kappa_V, \end{cases}$$

$$\begin{array}{ccc} \bigotimes^k L & \xrightarrow{\;l\bullet\;} & \bigotimes^k L \\ \kappa_V^k \big\downarrow & & \big\downarrow \kappa_V^k \\ \mathbb{K} & \xrightarrow[\;0\;]{} & \mathbb{K} \end{array}, \quad l \in L,$$

e.g., for the adjoint representation

$$\text{Killing multilinear forms: } \kappa^{i_1 \cdots i_k} = \epsilon^{i_1 m_1}_{m_k} \epsilon^{i_2 m_2}_{m_1} \cdots \epsilon^{i_k m_k}_{m_{k-1}}.$$

With a nondegenerate inner product κ_V^2, e.g., the Killing form for a semisimple Lie algebra, its totally symmetric *trace invariants* are given in $\mathbf{E}(L)$ as follows:

$$\begin{aligned} &I_V^k(L) = \kappa_{V r_1 \ldots r_k} l^{r_1} \otimes \cdots \otimes l^{r_k} \in \mathbf{E}(L) \cong \bigvee L, \\ &\kappa_{V r_1 \ldots r_k} = \kappa_V^{i_1 \ldots i_k} \kappa_{V i_1 r_1} \cdots \kappa_{V i_k r_k}, \\ &I_V^0(L) = 1 \in \mathbb{K}, \quad [L, I_V^k(L)] = \{0\}. \end{aligned}$$

The inverse associate inner product $I_V^2(L)$ (e.g., inverse Killing form $I^2(L)$ of a semisimple Lie algebra L) is called the associate *Casimir element*, a symmetric L-invariant power-2 tensor

$$I_V^2(L) = \kappa_{V jk}\, l^j \otimes l^k \in \bigvee^2 L \subset \mathbf{E}(L).$$

The adjoint representation of a semisimple Lie algebra defines a totally antisymmetric L-invariant tensor of power 3, in $\mathbf{E}(L)$ proportional to the Casimir element

$$\begin{aligned} &\mathrm{ad}\,(L) = \epsilon_{rim}\, l^r \otimes l^i \otimes l^m \in \bigwedge^3 L, \quad \epsilon_{rim} = \epsilon^{jk}_m \kappa_{ki} \kappa_{jr}, \\ &\text{in } \mathbf{E}(L): \quad \mathrm{ad}\,(L) = \tfrac{1}{2}\epsilon_{rim}[l^r, l^i] \otimes l^m = \tfrac{1}{2} I^2(L). \end{aligned}$$

Also, the *volume elements of a semisimple Lie algebra* $L \cong \mathbb{K}^n$ in $\mathbf{E}(L)$ are invariant ($\operatorname{ad} l$ is traceless):

$$\epsilon_n(L) = \tfrac{\epsilon_{j_1\dots j_n}}{n!} l^{j_1} \otimes \dots \otimes l^{j_n} \in \bigwedge^n L \subseteq \mathbf{E}(L), \quad [L, \epsilon^n(L)] = \{0\}.$$

They have a totally symmetric representative; for $n = 3$ they are proportional to the Casimir element.

The coefficients of the characteristic polynomial for a finite-dimensional space endomorphism are the traces of its Grassmann powers ("multiple traces"). Being invariant under inner automorphisms, they are called the *similarity invariants of the endomorphism* f:

$$\begin{aligned} f \in \mathbf{AL}(V),\ \ V \cong \mathbb{K}^n: \ \det[f - X \operatorname{id}_V] &= (-X)^n + \sum_{a=1}^{n} (-X)^{n-a} \kappa_V^a(f), \\ \kappa_V^a(f) = \operatorname{tr} \bigwedge^a f \text{ with } &\left\{ \begin{array}{ll} \kappa_V^1(f) &= \operatorname{tr} f, \dots, \\ \kappa_V^n(f) &= \det f, \end{array} \right. \\ g \in \mathbf{GL}(V) \Rightarrow \ \kappa_V^a(f) = \kappa_V^a(g \circ f \circ g^{-1}). \end{aligned}$$

For a triagonalizable matrix the similarity invariants give the sum over the eigenvalues, the sum over the products of proper pairs, products of proper triplets, etc.:

$$f \cong \begin{pmatrix} \alpha_1 & \beta_2^1 & \dots & \beta_n^1 \\ 0 & \alpha_2 & \dots & \beta_n^2 \\ \dots & & \dots & \\ 0 & \dots & \dots & \alpha_n \end{pmatrix} \Rightarrow \kappa_V^1 = \sum_{a=1}^{n} \alpha_a, \ \kappa_V^2 = \sum_{a \neq b} \alpha_a \alpha_b, \ \dots, \ \kappa_V^n = \alpha_1 \cdots \alpha_n.$$

For a matrix Lie algebra (Lie algebra representation) the similarity invariants give homogeneous polynomials in the coefficients of a basis $\{l^j\}_{j=1}^d$:

$$\det[\alpha_j l^j - X \operatorname{id}_V] = (-X)^n + \sum_{a=1}^{n} (-X)^{n-a} \kappa_V^a(\alpha_1, \dots \alpha_d), \ \ \deg \kappa_V^a(\alpha) = a.$$

For a semisimple Lie algebra, the representation is traceless, $\kappa_V^1(\alpha) = 0$; κ_L^2 for the adjoint representation is the Killing form.

An example for an abelian Lie algebra:

$$\begin{aligned} \log \mathbf{SO}(2) \longrightarrow \mathbf{AL}(\mathbb{R}^2) \ni \mathcal{O}(\alpha) &= \begin{pmatrix} 0 & \alpha \\ -\alpha & 0 \end{pmatrix} \cong \begin{pmatrix} i\alpha & 0 \\ 0 & -i\alpha \end{pmatrix} \text{ in } \mathbf{AL}(\mathbb{C}^2) \\ \Rightarrow \det[\mathcal{O}(\alpha) - X \mathbf{1}_2] &= X^2 + \alpha^2, \end{aligned}$$

and for compact Lie algebras

$$\begin{aligned} A_1^c = \log \mathbf{SU}(2) \longrightarrow \ & \mathbf{AL}(\mathbb{C}^2) \ni l(\alpha) = \begin{pmatrix} i\alpha_3 & i\alpha_1 + \alpha_2 \\ i\alpha_1 - \alpha_2 & -i\alpha_3 \end{pmatrix} \\ \Rightarrow \ & \det[l(\alpha) - X \mathbf{1}_2] = X^2 + \vec{\alpha}^2, \end{aligned}$$

$$\begin{aligned} A_2^c = \log \mathbf{SU}(3) \longrightarrow \ & \mathbf{AL}(\mathbb{C}^3) \ni \mathcal{U}(\alpha) = \begin{pmatrix} i\alpha_3 + i\frac{\alpha_8}{\sqrt{3}} & i\alpha_1 + \alpha_2 & i\alpha_4 + \alpha_5 \\ i\alpha_1 - \alpha_2 & -i\alpha_3 + i\frac{\alpha_8}{\sqrt{3}} & i\alpha_6 + \alpha_7 \\ i\alpha_4 - \alpha_5 & i\alpha_6 - \alpha_7 & -i\frac{2\alpha_8}{\sqrt{3}} \end{pmatrix} \\ \Rightarrow \ & \det[\mathcal{U}(\alpha) - X \mathbf{1}_3] = (-X)^3 - X \sum_{j=1}^{8} \alpha_j^2 + \det \mathcal{U}(\alpha), \end{aligned}$$

or the Lorentz Lie algebra acting on a real 4-dimensional Minkowski space,

$$\begin{aligned} \mathcal{L}(\varphi,\psi) &= \begin{pmatrix} 0 & \psi_1 & \psi_2 & \psi_3 \\ \psi_1 & 0 & \varphi_3 & -\varphi_2 \\ \psi_2 & -\varphi_3 & 0 & \varphi_1 \\ \psi_3 & \varphi_2 & -\varphi_1 & 0 \end{pmatrix} \in \mathbf{AL}(\mathbb{R}^4) \\ \Rightarrow \det[\mathcal{L}(\varphi,\psi) - X\mathbf{1}_4] &= X^4 + X^2(\vec{\varphi}^2 - \vec{\psi}^2) - (\vec{\varphi}\vec{\psi})^2, \end{aligned}$$

and $\log \mathbf{SL}(\mathbb{C}^2)$ on a complex 2-dimensional space giving one real and one imaginary degree-2 polynomial,

$$\begin{aligned} \mathcal{S}(\alpha,\beta) &= \begin{pmatrix} i\alpha_3+\beta_3 & i\alpha_1+\beta_1+\alpha_2-i\beta_2 \\ i\alpha_1+\beta_1-\alpha_2+i\beta_2 & -i\alpha_3-\beta_3 \end{pmatrix} \in \mathbf{AL}(\mathbb{C}^2) \\ \Rightarrow \det[\mathcal{S}(\alpha,\beta) - X\mathbf{1}_2] &= X^2 - (i\vec{\alpha}+\vec{\beta})^2 = X^2 + (\vec{\alpha}^2 - \vec{\beta}^2) - 2i\vec{\alpha}\vec{\beta}. \end{aligned}$$

The factorization of the characteristic polynomial into irreducible $\mathbb{K}$-polynomials (degree 1 and 2 for $\mathbb{R}$, degree 1 for $\mathbb{C}$) displays the eigenvalues arising in (maximally) triagonalized matrices for the representation $l(\alpha)$, e.g., $\pm i|\vec{\alpha}|$ for A_1^c.

The similarity invariants define invariant symmetric multilinear forms of the Lie algebra $\kappa_V(\alpha) = \kappa_V^{i_1\dots i_a}\alpha_{i_1}\cdots\alpha_{i_a}$ which, with an invariant nondegenerate bilinear form, define invariant tensors in the enveloping algebra $I^a(L) \in \operatorname{centr}\mathbf{E}(L)$. The number of functionally independent polynomials, i.e., of independent multilinear invariants, is given by the rank of the Jacobi matrix $\left(\frac{\partial\kappa_V^a}{\partial\alpha_j}\right)^{a=1,\dots,n}_{j=1,\dots,d}$.

With the representation of the Lie algebra by derivations on the polynomials of a representation space, the invariants come in the form of *Laplace-Beltrami operators*

$$\begin{aligned} L \longrightarrow \operatorname{der}\mathbb{K}[e^c],\ \ l^j &\longmapsto (l^j)^b_c e^c\partial_b, \\ \kappa_V^a(l) &\longmapsto \kappa_V^a(l^b_c e^c\partial_b),\ \ \text{e.g., } a=2\text{: } \kappa_{jk}(l^j)^b_c e^c\partial_b(l^k)^d_a e^a\partial_d. \end{aligned}$$

The independent number of invariants in the faithful adjoint representation of a semisimple Lie algebra $L \ni l \longmapsto \operatorname{ad} l = \operatorname{ad}\alpha_j l^j$ is its *rank*: $\operatorname{rank}_{\mathbb{K}} L = r$. The adjoint representation gives, via the multiple traces $\operatorname{tr}\operatorname{ad} l \circ \cdots \circ \operatorname{ad} l$ or the characteristic polynomial coefficients $\det[\operatorname{ad} l - X\operatorname{id}_L]$, the ring of all invariants:

$$\operatorname{centr}\mathbf{E}(L) \cong \mathbb{K}[I^2(L),\dots,I^{1+r}(L)].$$

Bibliography

[1] N. Bourbaki, *Algebra I, Chapters 1-3* (1989), Springer, Berlin, Heidelberg, New York, London, Paris, Tokyo.

[2] N. Bourbaki, *Lie Groups and Lie Algebras, Chapters 1-3* (1989), Springer, Berlin, Heidelberg, New York, London, Paris, Tokyo.

[3] S. Helgason, *Differential Geometry, Lie Groups and Symmetric Spaces* (1978), Academic Press, New York. London, Sydney, Tokyo, Toronto, etc.

[4] N. Jacobson, *Lie Algebras* (1961), Dover, New York.

[5] A. Knapp, *Representation Theory of Semisimple Groups* (1986), Princeton University Press, Princeton.

[6] N.Ja. Vilenkin, A.V. Klimyk, *Representations of Lie Groups and Special Functions* (1991), Kluwer Academic Publishers, Dordrecht, Boston, London.

4

ANTISTRUCTURES: The Real in the Complex

Quantum theory is a real theory, formulated with unitary operations, e.g. with $\mathbf{U}(1)$ or $\mathbf{SU}(n)$, which are real Lie groups acting on complex spaces with a conjugation. Complex structures with a conjugation have to be seen as doubled real structures, i.e., $\mathbb{C}_{\mathbb{R}} = \mathbb{R} + i\mathbb{R}$.

Complex numbers have two physically important properties: First, the involutive canonical conjugation implements nontrivially the *future-past reflection* $\overline{\alpha} \overset{\mathrm{T}}{\leftrightarrow} \alpha$, $\mathrm{T} = \star$, i.e., the reflection of the causal order, for complex vector spaces in the suggestive bra-ket notation $\langle v| \overset{\mathrm{T}}{\leftrightarrow} |v\rangle$. The real in the complex is established with a conjugation-induced sesquilinear form, e.g., with a scalar product (probability amplitudes) which in quantum theory leads to probabilities to describe experiments. The probabilities as products, for the numbers $\langle\alpha|\alpha\rangle = \overline{\alpha}\alpha$ or in the past-future connecting scalar products $\langle v|v\rangle$, are positive definite. Second, by the algebraic completeness of the complex numbers, there exist eigenvalues and eigenvectors for all complex linear transformations and eigenvector bases for semisimple finite-dimensional complex linear transformations. The irreducibility of real degree-2 polynomials, e.g., of X^2+1, is reflected in the real nondiagonalizability, e.g., of the harmonic oscillator time translation matrix $\begin{pmatrix} 0 & 1 \\ -1 & 0 \end{pmatrix}$. In the real, the Hamiltonian $H = \frac{p^2+x^2}{2}$ of a harmonic oscillator has no time translation eigenvalues and no eigenvectors. For the same reason, the rotation group $\mathbf{SO}(3)$ acting on a real 3-dimensional vector space, e.g., on the position translations, also has no nontrivial diagonalizable subgroup, e.g., there do not exist eigenvectors with nontrivial eigenvalues for a maximal abelian subgroup $\mathbf{SO}(2)$, e.g., for the generating angular momentum $\mathcal{O}^3 = \begin{pmatrix} 0 & 1 & 0 \\ -1 & 0 & 0 \\ 0 & 0 & 0 \end{pmatrix}$. If objects are defined as eigenvectors with respect to time and spin group action, both modeled by real Lie groups, a complex formulation is necessary.

Even the canonical conjugation of the numbers has its intricacies: It cannot be used for the definition of the reals as its invariants[1] since it is not uniquely

[1] Any involutive field automorphism keeps the rationals $\mathbb{Q}$ fixed. Since the natural $\mathbb{C}$-topology is determined from the natural $\mathbb{R}$-topology, topological arguments cannot be used to define $\mathbb{R}$ in $\mathbb{C}$.

determined by the property to be a nontrivial involutive automorphism of the complex field $\mathbb{C}$; there are infinitely many, e.g., one with

$$\sqrt[4]{2} \leftrightarrow i\sqrt[4]{2} \Rightarrow \sqrt{2} \leftrightarrow -\sqrt{2} \Rightarrow 2 \leftrightarrow 2,$$

whose existence can be proved by general arguments (involving the axiom of choice) but whose explicit form is unknown. The complex numbers with the canonical conjugation are a real 2-dimensional algebra $\mathbb{C}_\mathbb{R} = \mathbb{R} \oplus i\mathbb{R}$, e.g., as real Clifford algebra (chapter "Quantum Algebras").

Vector spaces over a field K have K-linear mappings as morphisms. Conjugate linear mappings of complex vector spaces, i.e., $f(\alpha v) = \overline{\alpha} f(v)$, are real linear. Therefore conjugate linear (antilinear) mappings play a role for complex representations of real structures, and only there.

Even with the canonical number conjugation, the conjugation for vector spaces is not unique; concepts like Hermitian and unitary require the specification of the conjugation they are defined with, in the mathematical literature called a *complex structure*. With the real isomorphism for the complex numbers $\mathbb{C}_\mathbb{R} \cong \mathbb{R}^2$, complex n-dimensional representations are real $2n$-dimensional. There is no natural isomorphism of a complex n-dimensional space with $n \geq 2$ to one of its $2n$-dimensional real forms, and therefore in general no natural realization of real structures in complex spaces. For a vector space $V \cong \mathbb{C}^n$, there are different conjugation types. They are characterized by different signatures in $\mathbf{U}(p,q)$, $p+q=n$, starting nontrivially for two dimensions with definite $\mathbf{U}(2)$ and indefinite $\mathbf{U}(1,1)$. All possible conjugations and all real forms are taken account of by using two complex vector spaces (space and antispace), canonically conjugated to each other. The induced canonical conjugation ("anticonjugation") doubles the complex scheme with duality to a quartet structure.

The mathematical discussion of antistructures, although rather formal, has important physical applications: The anticonjugation induced doubling arises most prominently in the "particle-antiparticle" dichotomy, which is extensively used in quantum fields (chapter "Massive Particle Quantum Fields"). The nontrivial action of the Lorentz group on finite-dimensional complex representation spaces, necessarily with dimension $n \geq 2$, is formulated with doubling anticonjugation. This is visible in the transition from one Pauli representation for spin $\mathbf{SU}(2)$ to the two left- and right-handed Weyl representations for Lorentz $\mathbf{SL}(\mathbb{C}^2)$ (chapter "Lorentz Operations"). More generally, the anticonjugation gives a natural doubling of representations of real Lie groups (chapter "Simple Lie Operations").

4.1 Anticonjugation

A vector space is an additive group with a compatible scalar multiplication. For each vector space V with complex scalar multiplication, there is its *antispace (canonical conjugated vector space)* $\overline{V} \in \underline{\mathbf{vec}}_\mathbb{C}$, defined by the same additive group, but equipped with a scalar multiplication, canonically conjugated in comparison to V:

$$\mathbb{C} \times \overline{V} \longrightarrow \overline{V}, \quad (\alpha, v) \longmapsto \alpha \bullet v = \overline{\alpha} v.$$

V and $\overline{V}$ are *two different complex vector spaces* with an antilinear vector space isomorphism co_V, an isomorphism for the additive group structure, called *anticonjugation (canonical conjugation).* It is useful to denote the group elements $v \in V$ as element of the vector spaces $\overline{V}$ with $\overline{v}$; then the different action of the scalar field can be written without the symbol $\bullet$:

$$\mathrm{co}_V : V \longrightarrow \overline{V}, \quad \mathrm{co}_V(v) = \overline{v}, \quad \begin{cases} \mathrm{co}_V(v+w) = \mathrm{co}_V(v) + \mathrm{co}_V(w) = \overline{v} + \overline{w}, \\ \mathrm{co}_V(\alpha v) = \alpha \bullet v = \overline{\alpha}\mathrm{co}_V(v) = \overline{\alpha}\, \overline{v}, \end{cases}$$
$$\mathrm{co}_V^{-1} = \mathrm{co}_{\overline{V}};$$

V and $\overline{V}$ are isomorphic as real vector spaces.

What is done in this chapter with vector spaces over $\mathbb{C}$ with canonical conjugation can be done in general with each involution $\alpha \leftrightarrow \alpha^{\mathtt{I}}$ of a ring R: There arise pairs of $\mathtt{I}$-involuted (conjugated) modules $V, V^{\mathtt{I}} \in \underline{\mathbf{mod}}_R$ with the $\mathtt{I}$-involuted (conjugated) action of the scalars R.

Antilinear mappings $\tilde{f}(\alpha v) = \overline{\alpha}\tilde{f}(v)$ of $V \in \underline{\mathbf{vec}}_{\mathbb{C}}$ and linear mappings $f(\alpha v) = \alpha f(v)$ of the antispace $\overline{V}$ are in a bijective correspondence:

$$\begin{array}{ccc} V & \xrightarrow{\mathrm{co}_V} & \overline{V} \\ {\scriptstyle \tilde{f}}\downarrow & & \downarrow{\scriptstyle f} \\ W & \xrightarrow[\mathrm{id}_W]{} & W \end{array}, \quad \begin{array}{l} f = \tilde{f} \circ \mathrm{co}_{\overline{V}}, \\ \tilde{f}(v) = f(\overline{v}). \end{array}$$

Therefore, the antispace $\overline{V}$ is the solution of a universal problem: Using both V and $\overline{V}$, only $\mathbb{C}$-linear mappings have to be considered up to the universal anticonjugation co_V. The related covariant and additive *antifunctor* for complex vector spaces is involutive:

$$- : \underline{\mathbf{vec}}_{\mathbb{C}} \longrightarrow \underline{\mathbf{vec}}_{\mathbb{C}}, \quad f \begin{array}{c} V \\ \downarrow \\ W \end{array} \longmapsto \begin{array}{c} \overline{V} \\ \downarrow \\ \overline{W} \end{array} {\scriptstyle \overline{f} = \mathrm{co}_W \circ f \circ \mathrm{co}_V^{-1}, \quad \overline{f}(\overline{v}) = \overline{f(v)},}$$
$$\overline{\mathrm{id}_V} = \mathrm{id}_{\overline{V}}, \quad \overline{f \circ g} = \overline{f} \circ \overline{g}, \quad \overline{f+g} = \overline{f} + \overline{g},$$
$$\overline{\overline{V}} = V, \quad \overline{\overline{f}} = f.$$

With f also $\overline{f}$ is a $\mathbb{C}$-linear mapping.

The algebra product in the antialgebra is defined by a contramorphism

$$\text{in } \overline{A}: \quad \overline{a} \diamond \overline{b} = \overline{b \diamond a}, \quad a, b \in A \in \underline{\mathbf{ag}}_{\mathbb{C}}.$$

The canonical conjugate anti-tensor algebra is the tensor algebra of the antispace

$$\overline{\bigotimes V} = \bigotimes \overline{V}.$$

4.2 The Complex Quartet

Now the relationship between duality (complex linear forms) and anticonjugation is discussed.

A complex vector space V comes with its dual space V^T, its antispace $\overline{V}$, and its dual antispace $\overline{V^T}$:

$$\begin{aligned} \mathrm{co}_V : \ & V \longrightarrow \overline{V}, && v \longmapsto \overline{v}, \\ \mathrm{co}_{V^T} : \ & V^T \longrightarrow \overline{V^T}, && \omega \longmapsto \overline{\omega}. \end{aligned}$$

They define the

$$\textit{complex quartet: } (V, \overline{V}, V^T, \overline{V^T})$$

isomorphic for finite dimension, however without natural $\mathbb{C}$-isomorphism. Antipartners are naturally $\mathbb{R}$-linear isomorphic.

The dual products will be related to each other *symmetrically*:

$$\left.\begin{array}{l} V^T \times V \longrightarrow \mathbb{C} \\ \overline{V} \times \overline{V^T} \longrightarrow \mathbb{C} \end{array}\right\} \text{ with } \langle \overline{v}, \overline{\omega} \rangle = \overline{\langle \omega, v \rangle}.$$

One thus obtains with equal transposition sign $\epsilon_V = \epsilon_{\overline{V}} = \epsilon = \pm 1$ the four dual products related to each other as follows:

$$\langle \omega, v \rangle = \epsilon \langle v, \omega \rangle = \epsilon \overline{\langle \overline{\omega}, \overline{v} \rangle} = \overline{\langle \overline{v}, \overline{\omega} \rangle}.$$

The dual and antifunctor commute with each other:

$$\overline{V}^T = \overline{V^T}, \quad \overline{f}^T = \overline{f^T}.$$

The composite *antiduality functor* $\times = \overline{} \circ T$ is contravariant and additive:

$$\times : \mathbf{vec}_{\mathbb{C}} \longrightarrow \mathbf{vec}_{\mathbb{C}}, \quad f \begin{array}{c} V \\ \downarrow \\ W \end{array} \longmapsto \begin{array}{c} \overline{V}^T \\ \uparrow f^\times = \overline{f}^T = \mathrm{co}_{V^T} \circ f^T \circ \mathrm{co}_{W^T}^{-1} \\ \overline{W}^T \end{array},$$

$$\mathrm{id}_V^\times = \mathrm{id}_{\overline{V}^T}, \quad (f \circ g)^\times = g^\times \circ f^\times, \quad (f+g)^\times = f^\times + g^\times,$$

and involutive $f^{\times\times} = f$ for finite dimension $V \cong \mathbb{C}^n$. The special star $\times$ is used also for the scalars and the basic space vectors and forms as follows $\alpha^\times = \overline{\alpha}$, $v^\times = \overline{v}$, $\omega^\times = \overline{\omega}$.

It is useful to fix a notation (final and initial roman letters) in the complex quartet that takes care of the anticonjugation:

$$\text{notation} : \begin{cases} \mathrm{u}, \mathrm{v}, \dots & \in V, \quad \mathrm{u}^\times, \mathrm{v}^\times, \dots \in \overline{V}, \\ \mathrm{a}^\times, \mathrm{b}^\times, \dots & \in V^T, \quad \mathrm{a}, \mathrm{b}, \dots \quad \in \overline{V}^T. \end{cases}$$

For finite-dimensional vector spaces there are dual bases and antibases

$$\begin{array}{llll} \text{for } V: & \{e^A\}_{A=1}^n = \{\mathrm{u}^A\}_{A=1}^n, & \text{for } \overline{V}: & \{\overline{e}^A\}_{A=1}^n = \{\mathrm{u}^{\times A}\}_{A=1}^n; \\ \text{for } V^T: & \{\check{e}_A\}_{A=1}^n = \{\mathrm{a}_A^\times\}_{A=1}^n, & \text{for } \overline{V}^T: & \{\overline{\check{e}}_A\}_{A=1}^n = \{\mathrm{a}_A\}_{A=1}^n, \end{array}$$

$$\langle \mathrm{a}_B^\times, \mathrm{u}^A \rangle = \epsilon \langle \mathrm{u}^A, \mathrm{a}_B^\times \rangle = \epsilon \langle \mathrm{a}_B, \mathrm{u}^{\times A} \rangle = \langle \mathrm{u}^{\times A}, \mathrm{a}_B \rangle = \delta_B^A.$$

There are other notations for involution-related doublings, e.g., Weyl's dotted and undotted indices $(e^A, \overline{e}^A) \to (e^A, e^{\dot{A}})$ for the antistructures as used for the Lorentz group (chapter "Lorentz Operations"), called an *index notation.*

4.2.1 Canonical Real Substructures

The direct sum spaces $V \oplus \overline{V}$ and $V^T \oplus \overline{V}^T$ with anticonjugation

$$V, \overline{V}, V^T, \overline{V}^T \ni \mathrm{u}, \mathrm{u}^\times, \mathrm{a}^\times, \mathrm{a},$$

can be decomposed into two *real subspaces*, antisymmetric and symmetric with respect to the anticonjugation and isomorphic as real spaces by multiplication by i (wherein vectors are denoted with uppercase boldface letters $\mathbf{U}_\pm$ and $\mathbf{A}_\pm$):

$$\begin{array}{lll} V \oplus \overline{V} \cong & \mathbb{C} \otimes [V_- \oplus V_+] & \cong \mathbb{C}^{2n} \\ & \text{with} \left\{ \begin{array}{l} V_- = \{i\mathbf{U}_- = \frac{\mathrm{u}-\mathrm{u}^\times}{\sqrt{2}}\} \cong \mathbb{R}^n, \\ V_+ = \{\mathbf{U}_+ = \frac{\mathrm{u}+\mathrm{u}^\times}{\sqrt{2}}\} \cong \mathbb{R}^n, \end{array} \right. & \\ V^T \oplus \overline{V}^T \cong & \mathbb{C} \otimes [(V^T)_- \oplus (V^T)_+] & \cong \mathbb{C}^{2n} \\ & \text{with} \left\{ \begin{array}{l} (V^T)_- = \{-i\mathbf{A}_- = \frac{\mathrm{a}^\times-\mathrm{a}}{\sqrt{2}}\} \cong \mathbb{R}^n, \\ (V^T)_+ = \{\mathbf{A}_+ = \frac{\mathrm{a}^\times+\mathrm{a}}{\sqrt{2}}\} \cong \mathbb{R}^n. \end{array} \right. & \end{array}$$

The dual products for the real subspaces are given by the real and imaginary parts of the dual product for the full space $V \oplus \overline{V}$:

$$\begin{array}{ll} \langle \mathrm{a}^\times, \mathrm{u} \rangle & = \epsilon \overline{\langle \mathrm{u}, \mathrm{a}^\times \rangle} \\ = \overline{\langle \mathrm{u}^\times, \mathrm{a} \rangle} & = \epsilon \langle \mathrm{a}, \mathrm{u}^\times \rangle \end{array} \Rightarrow \left\{ \begin{array}{lll} \langle \mathbf{A}_-, \mathbf{U}_- \rangle & = \langle \mathbf{A}_+, \mathbf{U}_+ \rangle & = \frac{\langle \mathrm{a}^\times,\mathrm{u}\rangle + \epsilon\overline{\langle \mathrm{a}^\times,\mathrm{u}\rangle}}{2}, \\ \langle \mathbf{A}_+, i\mathbf{U}_- \rangle & = \langle -i\mathbf{A}_-, \mathbf{U}_+ \rangle & = \frac{\langle \mathrm{a}^\times,\mathrm{u}\rangle - \epsilon\overline{\langle \mathrm{a}^\times,\mathrm{u}\rangle}}{2}, \end{array} \right.$$

which leads for bases of the real spaces to

$$\begin{array}{ll} i\mathbf{U}_-^A = \frac{\mathrm{u}^A-\mathrm{u}^{\times A}}{\sqrt{2}}, & \mathbf{U}_+^A = \frac{\mathrm{u}^A+\mathrm{u}^{\times A}}{\sqrt{2}} \\ -i\mathbf{A}_{A-} = \frac{\mathrm{a}_A^\times-\mathrm{a}_A}{\sqrt{2}}, & \mathbf{A}_{A+} = \frac{\mathrm{a}_A^\times+\mathrm{a}_A}{\sqrt{2}} \end{array} \Rightarrow \left\{ \begin{array}{lll} \langle \mathbf{A}_{B-}, \mathbf{U}_-^A \rangle & = \langle \mathbf{A}_{B+}, \mathbf{U}_+^A \rangle & = \frac{1+\epsilon}{2}\delta_B^A, \\ \langle \mathbf{A}_{B+}, i\mathbf{U}_-^A \rangle & = \langle -i\mathbf{A}_{B-}, \mathbf{U}_+^A \rangle & = \frac{1-\epsilon}{2}\delta_B^A. \end{array} \right.$$

With respect to the real subspaces the Fermi case ($\epsilon = +1$) has as dual pairing symmetric-symmetric and antisymmetric-antisymmetric or real-real and imaginary-imaginary, whereas there is a crossover dual pairing for the Bose case ($\epsilon = -1$), i.e., antisymmetric-symmetric or real-imaginary

$$\begin{array}{rl} \text{Fermi } \epsilon = +1: & (V^T)_\pm = (V_\pm)^T, \\ \text{Bose } \epsilon = -1: & (V^T)_\pm = (V_\mp)^T. \end{array}$$

That is visible, e.g., in the pairing of the real position with the imaginary multiplied momentum in the Heisenberg Bose commutator $[i\mathbf{p}, \mathbf{x}] = \mathbf{1}$ (chapter "Quantum Algebras") in contrast to the anticommutator for a Dirac field $\{\overline{\Psi}, \Psi\}(\vec{x}) = \gamma^0\delta(\vec{x})$ without the imaginary unit i (chapter "Massive Particle Quantum Fields").

4.2.2 Isomorphisms in the Complex Quartet

If the anticonjugation $\times$ comes together with an additional dual conjugation (antilinear dual isomorphism) $\tilde{\zeta} = *$; all four spaces of the complex quartet are isomorphic to each other as real vector spaces:

$$\begin{array}{ccc} V & \overset{\times}{\leftrightarrow} & \overline{V} \\ *\updownarrow & & \updownarrow * \\ V^T & \underset{\times}{\leftrightarrow} & \overline{V}^T \end{array}, \qquad \begin{array}{ccc} \mathrm{u} & \leftrightarrow & \mathrm{u}^\times = \mathrm{a}^* \\ \updownarrow & & \updownarrow \\ \mathrm{a}^\times = \mathrm{u}^* & \leftrightarrow & \mathrm{a} \end{array},$$

$$\tilde{\zeta}(\mathrm{u}, \mathrm{v}) = \langle \mathrm{u}^*, \mathrm{v} \rangle = \overline{\langle \mathrm{v}^*, \mathrm{u} \rangle}\ , \qquad \begin{array}{l} \mathrm{u}^{A*} = \tilde{\zeta}^{AB} \mathrm{u}_B \\ \tilde{\zeta}^{AB} = \overline{\tilde{\zeta}^{BA}} \end{array}.$$

A sesquilinear form $\tilde{\zeta} : V \times V \longrightarrow \mathbb{C}$ is not a tensor. Since the product of both conjugations defines $\mathbb{C}$-linear isomorphisms between antispace and dual space,

$$\zeta = \tilde{\zeta} \circ \times = * \circ \times : \quad \begin{cases} V \overset{\zeta}{\leftrightarrow} \overline{V}^T, & \mathrm{u} \leftrightarrow \mathrm{a}, \\ V^T \overset{\zeta}{\leftrightarrow} \overline{V}, & \mathrm{u}^\star \leftrightarrow \mathrm{a}^\star, \end{cases}$$

there is an equivalent tensor expression for the bilinear form ζ:

$$\begin{array}{ccc} V \times V & \overset{\tilde{\zeta}}{\longrightarrow} & \mathbb{C} \\ {\scriptstyle (\times, \mathrm{id}_V)} \downarrow & & \downarrow {\scriptstyle \mathrm{id}_{\mathbb{C}}}, \\ \overline{V} \times V & \underset{\zeta}{\longrightarrow} & \mathbb{C} \end{array} \qquad \zeta = \tilde{\zeta}^{AB} \mathrm{a}^\times_A \otimes \mathrm{a}_B.$$

4.3 Antidoubling

In analogy to the self-dual vector space sum $\mathbf{V} = V \oplus V^T$, there is the *antidoubling* V_{doub} of a complex vector space V. It is defined as the direct sum with its dual antispace $\overline{V}^T$:

$$V_{\text{doub}} = V \oplus \overline{V}^T, \quad V^T_{\text{doub}} = V^T \oplus \overline{V}.$$

For example, Dirac spinors are antidoubled Weyl spinors (chapter "Lorentz Operations").

The anticonjugation defines a dual conjugation $\times = \mathrm{co}_V \oplus \mathrm{co}_{\overline{V}^T}$:

$$\begin{array}{l} V_{\text{doub}} \overset{\times}{\leftrightarrow} V^T_{\text{doub}}, \quad \mathbf{v} = \mathrm{u} + \mathrm{a} \leftrightarrow \mathbf{v}^\times = \mathrm{u}^\times + \mathrm{a}^\times, \\ \langle \mathbf{v}^\times, \mathbf{v} \rangle = \langle \mathrm{a}^\times + \mathrm{u}^\times, \mathrm{u} + \mathrm{a} \rangle = \langle \mathrm{a}^\times, \mathrm{u} \rangle + \overline{\langle \mathrm{a}^\times, \mathrm{u} \rangle} = 2\,\mathrm{Re}\langle \mathrm{a}^\times, \mathrm{u} \rangle, \\ \text{with } \langle \mathrm{u}^\times, \mathrm{u} \rangle = 0 = \langle \mathrm{a}^\times, \mathrm{a} \rangle. \end{array}$$

An additional dual conjugation $\tilde{\zeta} : V \leftrightarrow V^T$ gives a linear involution for the antidoubling $V_{\text{doub}} = V \oplus \overline{V}^T$ and an isomorphism for the (anti-) symmetric real subspaces of $V \oplus \overline{V}$ and those of $V^T \oplus \overline{V}^T$:

$$\begin{aligned} V \oplus \overline{V}^T &\overset{\zeta}{\leftrightarrow} V \oplus \overline{V}^T, \\ V_\pm &\overset{\zeta}{\leftrightarrow} (V^T)_\pm = \begin{cases} (V_\pm)^T, & \epsilon = +1, \\ (V_\mp)^T, & \epsilon = -1, \end{cases} \quad \text{for bases: } (\mathbf{u}_\pm^A)^* = \tilde{\zeta}^{AB}\mathbf{a}_{B\pm}. \end{aligned}$$

The $\times$-symmetric basis vectors are not $*$-symmetric.

4.3.1 The Anticonjugation Invariance Group

The endomorphism algebra of the antidoubling $V_{\text{doub}} = \begin{pmatrix} V \\ V^T \end{pmatrix}$ is the direct sum of four vector subspaces; for $V \cong \mathbb{C}^n$

$$\begin{aligned} \mathbf{AL}(V_{\text{doub}}) = V_{\text{doub}} \otimes V_{\text{doub}}^T &= \begin{pmatrix} V \otimes V^T & V \otimes \overline{V} \\ \overline{V}^T \otimes V^T & \overline{V}^T \otimes \overline{V} \end{pmatrix} \\ &= \begin{pmatrix} \mathbf{AL}(V) & \mathbf{P}(V) \\ \mathbf{P}(V)^T & \mathbf{AL}(\overline{V}^T) \end{pmatrix} \cong \mathbb{C}^{4n^2}. \end{aligned}$$

The diagonal endomorphism algebras in $\mathbf{AL}(V_{\text{doub}})$ are self-dual, $\mathbf{AL}(V)^T = \mathbf{AL}(V)$, and canonically conjugate to each other (antialgebras):

$$\mathbf{AL}(V) \overset{\times}{\leftrightarrow} \mathbf{AL}(\overline{V}^T), \quad (\mathrm{u} \otimes \mathrm{a}^\times)^\times = \mathrm{a} \otimes \mathrm{u}^\times.$$

The skew diagonal conjugation-stable vector spaces (no algebras) are dual to each other, $\mathbf{P}(\overline{V}^T) = \mathbf{P}(V)^T$:

$$\mathbf{P}(V) \ni (\mathrm{u} \otimes \mathrm{v}^\times)^\times = \mathrm{v} \otimes \mathrm{u}^\times, \quad \mathbf{P}(V)^T \ni (\mathrm{a} \otimes \mathrm{b}^\times)^\times = \mathrm{b} \otimes \mathrm{a}^\times.$$

For finite dimension, $V \cong \mathbb{C}^n$, $V_{\text{doub}} \cong \mathbb{C}^{2n}$, the invariance group of the anticonjugation-induced sesquilinear form

$$\begin{aligned} \langle\ |\ \rangle_\times : V_{\text{doub}} \times V_{\text{doub}} \longrightarrow \mathbb{C}, &\quad \begin{cases} \langle \mathbf{v}|\mathbf{w}\rangle_\times = \langle \mathbf{v}^\times, \mathbf{w}\rangle, \\ \langle \mathbf{v}|\mathbf{v}\rangle_\times \in \mathbb{R}, \end{cases} \\ \text{with bases} &\quad \begin{pmatrix} \langle \mathrm{u}^A|\mathrm{u}^B\rangle_\times & \langle \mathrm{u}^A|\mathrm{a}_B\rangle_\times \\ \langle \mathrm{a}_A|\mathrm{u}^B\rangle_\times & \langle \mathrm{a}_A|\mathrm{a}_B\rangle_\times \end{pmatrix} = \begin{pmatrix} 0 & \delta^A_B \\ \delta^B_A & 0 \end{pmatrix}, \end{aligned}$$

is the *indefinite unitary group* $\mathbf{U}(n,n)$ with neutral signature (n,n):

$$\begin{aligned} \{g \in \mathbf{GL}(V_{\text{doub}}) \,\big|\, g^\times = g^{-1}\} &\cong \mathbf{U}(n,n), \\ \{l \in \mathbf{AL}(V_{\text{doub}}) \,\big|\, l^\times = \ \ l\} &\cong \log \mathbf{U}(n,n). \end{aligned}$$

For example, $\mathbf{U}(2,2)$ is the conjugation group for Dirac spinors.

The composition of Euclidean $\mathbf{U}(2n)$-conjugation $\star$ and $\mathbf{U}(n,n)$-conjugation $\times$ defines a $\mathbb{C}$-*linear involutive automorphism* P of the antidoubling

$$\begin{aligned} &(V, V^T) \overset{\star}{\leftrightarrow} (V^T, V) \overset{\times}{\leftrightarrow} (\overline{V}^T, \overline{V}), \\ &\mathsf{P} = \times \circ \star \cong \begin{pmatrix} 0 & \mathbf{1}_n \\ \mathbf{1}_n & 0 \end{pmatrix} \in \mathbf{GL}(V_{\text{doub}}), \quad \times = \mathsf{P} \circ \star, \\ &\mathsf{P} : V_{\text{doub}} \longrightarrow V_{\text{doub}}, \ \begin{pmatrix} \mathrm{u} \\ \mathrm{a} \end{pmatrix} \longmapsto \begin{pmatrix} \mathrm{a} \\ \mathrm{u} \end{pmatrix}. \end{aligned}$$

For Dirac spinors the automorphism is used as parity involution, exchanging the chiral partners in the doubling. On this basis, the anticonjugation exchanges the block-diagonal matrices, not the skew-diagonal ones, explicitly given with $(n \times n)$ matrices as entries:

$$\begin{pmatrix} f & \xi \\ \pi & g \end{pmatrix}^\times = \begin{pmatrix} 0 & \mathbf{1}_n \\ \mathbf{1}_n & 0 \end{pmatrix}\begin{pmatrix} f & \xi \\ \pi & g \end{pmatrix}^\star \begin{pmatrix} 0 & \mathbf{1}_n \\ \mathbf{1}_n & 0 \end{pmatrix} = \begin{pmatrix} g^\star & \xi^\star \\ \pi^\star & f^\star \end{pmatrix}, \quad \begin{pmatrix} f & \xi \\ \pi & g \end{pmatrix}^\star = \begin{pmatrix} f^\star & \pi^\star \\ \xi^\star & g^\star \end{pmatrix}.$$

4.4 Dual and Antirepresentations

With the complex quartet of vector spaces there is a fourfoldness of the related concepts: For example, all linear mappings of complex vector spaces come in quartets (with $\epsilon_V = \epsilon_W = \epsilon$):

$$\begin{array}{l} (f, \overline{f}, f^T, \overline{f^T}) : (V, \overline{V}, W^T, \overline{W^T}) \longrightarrow (W, \overline{W}, V^T, \overline{V^T}) \\ \text{with } \langle \omega, f(v) \rangle = \overline{\langle \overline{f}(\overline{v}), \overline{\omega} \rangle} = \langle f^T(\omega), v \rangle = \overline{\langle \overline{v}, \overline{f^T}(\overline{\omega}) \rangle}, \end{array}$$

with the tensor expressions for transposed endomorphisms and antipartners

$$\begin{array}{llll} f & = f_A^B \, \mathrm{u}^A \otimes \mathrm{a}_B^\times & \in \mathbf{AL}(V), & f^{\times T} = \overline{f} & = \overline{f_A^B} \, \epsilon \mathrm{u}^{\times A} \otimes \mathrm{a}_B & \in \mathbf{AL}(\overline{V}), \\ f^T & = f_A^B \, \epsilon \mathrm{a}_B^\times \otimes \mathrm{u}^A & \in \mathbf{AL}(V^T), & f^\times = \overline{f}^T & = \overline{f_A^B} \, \mathrm{a}_B \otimes \mathrm{u}^{\times A} & \in \mathbf{AL}(\overline{V}^T). \end{array}$$

This is used for quadrupling representations: With a representation D of a group G on a complex vector space V there comes the dual representation $\check{D}$ on V^T, the *antirepresentation* $\overline{D}$ on $\overline{V}$, and the dual antirepresentation $\hat{D}$ on $\overline{V}^T$:

$$\begin{array}{lll} D : G \longrightarrow \mathbf{GL}(V), & D(g) \cong g & = \begin{pmatrix} \alpha & \beta \\ \gamma & \delta \end{pmatrix}, \\ \check{D} : G \longrightarrow \mathbf{GL}(V^T), & \check{D}(g) \cong g^{-1T} & = \frac{1}{\det g}\begin{pmatrix} \delta & -\gamma \\ -\beta & \alpha \end{pmatrix}, \\ \overline{D} : G \longrightarrow \mathbf{GL}(\overline{V}), & \overline{D}(g) \cong \overline{g} & = \begin{pmatrix} \overline{\alpha} & \overline{\beta} \\ \overline{\gamma} & \overline{\delta} \end{pmatrix}, \\ \hat{D} : G \longrightarrow \mathbf{GL}(\overline{V}^T), & \hat{D}(g) \cong g^{-1\times} & = \frac{1}{\det \overline{g}}\begin{pmatrix} \overline{\delta} & -\overline{\gamma} \\ -\overline{\beta} & \overline{\alpha} \end{pmatrix}, \\ \text{with } \hat{D} = \check{\overline{D}} = \overline{\check{D}}, \ \overline{D} = D^{\times T} = D^{T\times}. \end{array}$$

In the simplest illustration with (2×2) matrices the anticonjugation $\times$ can be taken as the transposition with canonical conjugation. Starting from a representation, the antirepresentation takes the canonically conjugated matrix elements. One has the involutive properties

$$\check{\check{D}} = D, \quad \overline{\overline{D}} = D, \quad \hat{\hat{D}} = D.$$

Analogous structures arise for the quartet of complex Lie algebra representations

$$\begin{array}{lll} \mathcal{D} : L \longrightarrow \mathbf{AL}(V), & \mathcal{D}(l) \cong \ l & = \begin{pmatrix} \alpha & \beta \\ \gamma & \delta \end{pmatrix}, \\ \check{\mathcal{D}} : L \longrightarrow \mathbf{AL}(V^T), & \check{\mathcal{D}}(l) \cong \ -l^T & = -\begin{pmatrix} \alpha & \gamma \\ \beta & \delta \end{pmatrix}, \\ \overline{\mathcal{D}} : L \longrightarrow \mathbf{AL}(\overline{V}), & \overline{\mathcal{D}}(l) \cong \ \overline{l} & = \begin{pmatrix} \overline{\alpha} & \overline{\beta} \\ \overline{\gamma} & \overline{\delta} \end{pmatrix}, \\ \hat{\mathcal{D}} : L \longrightarrow \mathbf{AL}(\overline{V}^T), & \hat{\mathcal{D}}(l) \cong \ -l^\times & = -\begin{pmatrix} \overline{\alpha} & \overline{\gamma} \\ \overline{\beta} & \overline{\delta} \end{pmatrix}, \end{array}$$

$$\text{with} \quad \hat{\mathcal{D}} = \check{\overline{\mathcal{D}}} = \overline{\check{\mathcal{D}}}, \quad \overline{\mathcal{D}} = \mathcal{D}^{\times T} = \mathcal{D}^{T\times}, \\ \check{\check{\mathcal{D}}} = \mathcal{D}, \quad \overline{\overline{\mathcal{D}}} = \mathcal{D}, \quad \hat{\hat{\mathcal{D}}} = \mathcal{D}.$$

All four representations are real-linear isomorphic. In addition to the natural $\mathbb{C}$-linear isomorphisms $\cong$ for dual pairs by the inverse and negative transposition $(g, l) \leftrightarrow (g^{-1T}, -l^T)$, one has, with the anticonjugation, natural antilinear ($\mathbb{R}$-linear) isomorphisms $\cong_{\mathbb{R}}$ for antipairs $(g, l) \leftrightarrow (\overline{g}, \overline{l})$, where the equivalence is induced by the anticonjugation co_V.

As an example, take the special groups $\mathbf{SL}(\mathbb{C}^n)$, $n \geq 2$, as real $2(n^2-1)$-dimensional Lie groups with rank $2(n-1)$ in the complex quartet with the defining representation, e.g., for $n = 3$ with Gell-Mann matrices $\lambda_8^{a=1}$

$$\begin{array}{llll} \mathbf{SL}(\mathbb{C}^3) \ni s & = e^{(i\gamma_a+\delta_a)\lambda^a}, & \hat{s} \ = s^{\star -1} & = e^{(i\gamma_a-\delta_a)\lambda^a}, \\ \check{s} = s^{T-1} & = e^{-(i\gamma_a+\delta_a)(\lambda^a)^T}, & \overline{s} \ = s^{\star T} & = e^{-(i\gamma_a-\delta_a)(\lambda^a)^T}. \end{array}$$

The four representation types of $\mathbf{SL}(\mathbb{C}^n)$, $n \geq 2$, are inequivalent with the exception of the real 6-dimensional Lorentz group cover $\mathbf{SL}(\mathbb{C}^2)$ with rank 2, where with the equivalence of dual representations $s \cong \check{s}$ there are two inequivalent ones, with Pauli matrices $s = e^{(i\vec{\alpha}+\vec{\beta})\vec{\sigma}}$ and $\hat{s} = e^{(i\vec{\alpha}-\vec{\beta})\vec{\sigma}}$ (chapter "Lorentz Operations"). For the real 16-dimensional group $\mathbf{SL}(\mathbb{C}^3)$ with rank 4 all four (3×3)-representations are inequivalent.

For the representations of the unitary subgroups $\mathbf{SU}(n)$ with half the real dimension and rank the dual antirepresentations are equivalent with the isomorphism $\star \circ \times$: $u \cong \hat{u}$, $\check{u} \cong \overline{u}$. For example, the rank 2 color group $\mathbf{SU}(3)$ has two inequivalent representations $u = e^{i\gamma_a\lambda^a}$ ("quark") and $\check{u} = e^{-i\gamma_a(\lambda^a)^T}$ ("antiquark"). For the rank 1 spin group $\mathbf{SU}(2)$ all four representations are equivalent, $u \cong \check{u}$.

Direct sum representations on the antidoubling for a group G or a Lie algebra L,

$$\begin{array}{ll} D_{\text{doub}} : G \longrightarrow \mathbf{GL}(V_{\text{doub}}), & D_{\text{doub}}(g) = (D \oplus \hat{D})(g), \\ \mathcal{D}_{\text{doub}} : L \longrightarrow \mathbf{AL}(V_{\text{doub}}), & \mathcal{D}_{\text{doub}}(l) = (\mathcal{D} \oplus \hat{\mathcal{D}})(l), \end{array}$$

are conjugate linear self-dual with the anticonjugation

$$\begin{array}{ccc} V_{\text{doub}} & \xrightarrow{D_{\text{doub}}(g),\mathcal{D}_{\text{doub}}(l)} & V_{\text{doub}} \\ \times \updownarrow & & \updownarrow \times \\ V_{\text{doub}}^T & \xrightarrow[\check{D}_{\text{doub}}(g),\check{\mathcal{D}}_{\text{doub}}(l)]{} & V_{\text{doub}}^T \end{array},$$

i.e., the represented group is a $\mathbf{U}(n,n)$-subgroup ($\times$-unitary)

$$D_{\text{doub}}[G] \subseteq \mathbf{U}(n,n): \begin{cases} D_{\text{doub}}(g^{-1}) = D_{\text{doub}}(g)^{\times}, \\ \langle D_{\text{doub}}(g)(\mathbf{v}) | D_{\text{doub}}(g)(\mathbf{w})\rangle_{\times} = \langle \mathbf{v}|\mathbf{w}\rangle_{\times}, \end{cases}$$

and the represented Lie algebra a $\log \mathbf{U}(n,n)$-Lie subalgebra ($\times$-antisymmetrical)

$$\mathcal{D}_{\text{doub}}[L] \subseteq \log \mathbf{U}(n,n): \begin{cases} -\mathcal{D}_{\text{doub}}(l) = \mathcal{D}_{\text{doub}}(l)^{\times}, \\ \langle \mathcal{D}_{\text{doub}}(l)(\mathbf{v}) | \mathbf{w}\rangle_{\times} + \langle \mathbf{v}|\mathcal{D}_{\text{doub}}(l)(\mathbf{w})\rangle_{\times} = 0. \end{cases}$$

The inner automorphisms for the endomorphism algebra $\mathbf{AL}(V_{\text{doub}})$ effected by the represented group

$$\begin{pmatrix} D(g) & 0 \\ 0 & \hat{D}(g) \end{pmatrix} \begin{pmatrix} \mathbf{AL}(V) & \mathbf{P}(V) \\ \mathbf{P}(V)^T & \mathbf{AL}(\overline{V}^T) \end{pmatrix} \begin{pmatrix} D(g^{-1}) & 0 \\ 0 & \hat{D}(g^{-1}) \end{pmatrix}$$

contain four group representations: In addition to the G-representations by inner automorphisms of the endomorphism algebras

$$\begin{array}{llll} G \times \mathbf{AL}(V) & \longrightarrow \mathbf{AL}(V), & \mathrm{u} \otimes \mathrm{a}^{\times} & \longmapsto D(g) \circ (\mathrm{u} \otimes \mathrm{a}^{\times}) \circ D(g^{-1}), \\ G \times \mathbf{AL}(\overline{V}^T) & \longrightarrow \mathbf{AL}(\overline{V}^T), & \mathrm{a} \otimes \mathrm{u}^{\times} & \longmapsto D(g^{-1})^{\times} \circ (\mathrm{a} \otimes \mathrm{u}^{\times}) \circ D(g)^{\times}, \end{array}$$

the group G is represented on the anticonjugation stable vector spaces

$$\begin{array}{llll} G \times \mathbf{P}(V) & \longrightarrow \mathbf{P}(V), & \mathrm{u} \otimes \mathrm{v}^{\times} & \longmapsto D(g) \circ (\mathrm{u} \otimes \mathrm{v}^{\times}) \circ D(g)^{\times}, \\ G \times \mathbf{P}(V)^T & \longrightarrow \mathbf{P}(V)^T, & \mathrm{a} \otimes \mathrm{b}^{\times} & \longmapsto D(g^{-1})^{\times} \circ (\mathrm{a} \otimes \mathrm{b}^{\times}) \circ D(g^{-1}). \end{array}$$

If a representation $\mathcal{D}$ of a real Lie algebra on V is conjugate linear selfdual with $*$, i.e., if there exists an invariant sesquilinear form of V, then it is equivalent with ζ to the dual antirepresentation $\hat{\mathcal{D}}$:

$$\begin{array}{ccc} V & \xrightarrow{\mathcal{D}(l)} & V \\ \zeta \updownarrow & & \updownarrow \zeta, \\ \overline{V}^T & \xrightarrow[\hat{\mathcal{D}}(l)]{} & \overline{V}^T \end{array} \qquad \hat{\mathcal{D}}(l) = \zeta \circ \mathcal{D}(l) \circ \zeta^{-1}.$$

One has for a decomposable representation on the antidoubling $V_{\text{doub}} = V \oplus V^T$,

$$\begin{aligned} (\mathcal{D} \oplus \hat{\mathcal{D}})(l) &= \mathcal{D}(l)^B_A \mathrm{u}^A \otimes \mathrm{a}^{\times}_B - \overline{\mathcal{D}(l)^B_A} \mathrm{a}_B \otimes \mathrm{u}^{\times A} \\ &= \mathcal{D}(l)^B_A (\mathrm{u}^A \otimes \mathrm{a}^{\times}_B + \tilde{\zeta}^{AC} \mathrm{a}_C \otimes \mathrm{u}^{\times D} \tilde{\zeta}_{BD}). \end{aligned}$$

4.5 Particles and Antiparticles

If irreducible time representations with Euclidean conjugation $\star$ act on a complex quartet with anticonjugation $\times$, the representation space vectors are interpreted as particles and antiparticles (chapter "Massive Particle Quantum

Fields"). The particle-antiparticle doubling is a natural consequence of the anticonjugation doubling for complex vector spaces with more than one dimension, e.g., for the simplest 2-dimensional case with the Lorentz group $\mathbf{SL}(\mathbb{C}^2)$ and the two conjugations with Euclidean invariance group $\mathbf{U}(2)$ and anticonjugation invariance group $\mathbf{U}(1,1)$.

Irreducible time $\mathbf{D}(1)$-representations in $\mathbf{U}(1)$ on the 1-dimensional complex quartet (Fermi $\epsilon = +1$ and Bose $\epsilon = -1$),

$$\begin{array}{llllll}
D(t|m) & = e^{itH} & = e^{imt}\,\mathrm{id}_V, & H & = m\,\mathrm{id}_V & = m\mathrm{u}\otimes\mathrm{a}^\times, \\
 & & & d_t\mathrm{u} & = im\mathrm{u}, & \\
\check{D}(t|m) & = e^{-itH^T} & = e^{-imt}\,\mathrm{id}_{V^T}, & H^T & = m\,\mathrm{id}_{V^T} & = m\epsilon\mathrm{a}^\times\otimes\mathrm{u}, \\
 & & & d_t\mathrm{a}^\times & = -im\mathrm{a}^\times, & \\
\overline{D}(t|m) & = e^{-it\overline{H}} & = e^{-im}\,\mathrm{id}_{\overline{V}}, & \overline{H} & = m\,\mathrm{id}_{\overline{V}} & = m\epsilon\mathrm{u}^\times\otimes\mathrm{a}, \\
 & & & d_t\mathrm{u}^\times & = -im\mathrm{u}^\times, & \\
\hat{D}(t|m) & = e^{it\overline{H}^T} & = e^{imt}\,\mathrm{id}_{\overline{V}^T}, & H^\times & = m\,\mathrm{id}_{\overline{V}^T} & = m\mathrm{a}\otimes\mathrm{u}^\times, \\
 & & & d_t\mathrm{a} & = im\mathrm{a}, &
\end{array}$$

have conjugated imaginary eigenvalues

$$\pm im \in \mathbf{weights}\,\mathbf{D}(1) = i\mathbb{R}.$$

With the canonical $\mathbf{U}(1,1)$ and the dual $\mathbf{U}(1)$-conjugation all spaces in the quartet are isomorphic as real vector spaces.

The following names are used for the compact time representation eigenvectors in the complex quartet

particle creators $\mathrm{u}\in V$	$\overset{\times}{\leftrightarrow}$	*antiparticle annihilators* $\mathrm{u}^\times = \mathrm{a}^\star \in \overline{V}$
$\star\updownarrow$		$\updownarrow\star$
particle annihilators $\mathrm{a}^\times = \mathrm{u}^\star \in V^T$	$\overset{\times}{\leftrightarrow}$	*antiparticle creators* $\mathrm{a}\in\overline{V}^T$

"Creation-annihilation" is formalized with duality, i.e., *dual pairs* (V, V^T) and $(\overline{V}^T, \overline{V})$, whereas anticonjugation with *antipairs* $(V, \overline{V})$ and $(V^T, \overline{V}^T)$ leads to the *"particle-antiparticle"* property. The $\mathbf{U}(1,1)$-anticonjugation $\times$ connects particle (antiparticle) creation with antiparticle (particle) annihilation, the $\mathbf{U}(1)$-conjugation $\star$ particle (antiparticle) creation with particle (antiparticle) annihilation.

There is an action on the canonical doubling $V_{\mathrm{doub}} = V \oplus \overline{V}^T \cong \mathbb{C}^2$ by a decomposable time representation with the Hamiltonian both $\star$- and $\times$-symmetric:

$$\begin{aligned}
H_{\mathrm{doub}} &= H \oplus H^\times = m(\mathrm{u}\otimes\mathrm{a}^\times + \mathrm{a}\otimes\mathrm{u}^\times) = m(\mathrm{u}\otimes\mathrm{u}^\star + \mathrm{a}\otimes\mathrm{a}^\star), \quad m\in\mathbb{R}, \\
H_{\mathrm{doub}} &= H^\times_{\mathrm{doub}} = H^\star_{\mathrm{doub}}.
\end{aligned}$$

The anticonjugation $\times$-symmetric (canonical real) combinations combine particle creators and antiparticle annihilators:

$$\begin{array}{llllll}
\mathbf{U}_+ & = \frac{\mathrm{u}+\mathrm{u}^\times}{\sqrt{2}} & = \frac{\mathrm{u}+\mathrm{a}^\star}{\sqrt{2}}, & \mathbf{A}_+ & = \frac{\mathrm{a}^\times+\mathrm{a}}{\sqrt{2}} & = \frac{\mathrm{u}^\star+\mathrm{a}}{\sqrt{2}}, \\
i\mathbf{U}_- & = \frac{\mathrm{u}-\mathrm{u}^\times}{\sqrt{2}} & = \frac{\mathrm{u}-\mathrm{a}^\star}{\sqrt{2}}, & -i\mathbf{A}_- & = \frac{\mathrm{a}^\times-\mathrm{a}}{\sqrt{2}} & = \frac{\mathrm{u}^\star-\mathrm{a}}{\sqrt{2}}.
\end{array}$$

They are not symmetric (self-dual real) for Euclidean conjugation $\star$

$$\mathbf{U}_{\pm}^{\star} = \mathbf{A}_{\pm}.$$

The dual pairing is different for Bose and Fermi (chapter "Massive Particle Quantum Fields"):

$$\begin{array}{lllll} \langle \mathbf{A}_-, \mathbf{U}_- \rangle & = \langle \mathbf{U}_-^{\star}, \mathbf{U}_- \rangle & = \langle \mathbf{A}_+, \mathbf{U}_+ \rangle & = \langle \mathbf{U}_+^{\star}, \mathbf{U}_+ \rangle & = \begin{cases} 1, & \text{Fermi}, \\ 0, & \text{Bose}, \end{cases} \\ \langle \mathbf{A}_+, i\mathbf{U}_- \rangle & = \langle \mathbf{U}_+^{\star}, i\mathbf{U}_- \rangle & = \langle -i\mathbf{A}_-, \mathbf{U}_+ \rangle & = \langle -i\mathbf{U}_-^{\star}, \mathbf{U}_+ \rangle & = \begin{cases} 0, & \text{Fermi}, \\ 1, & \text{Bose}. \end{cases} \end{array}$$

The self-dual time development in $\mathbf{SO}(2)$ is generated by

$$\begin{array}{rl} H_{\text{doub}} - H_{\text{doub}}^T & = m([\mathrm{u}, \mathrm{u}^{\star}]_{-\epsilon} + [\mathrm{a}, \mathrm{a}^{\star}]_{-\epsilon}) \\ & = \begin{cases} -im([\mathbf{U}_+, \mathbf{U}_-^{\star}] - [\mathbf{U}_-, \mathbf{U}_+^{\star}]) & \text{(Fermi)}, \\ m(\{\mathbf{U}_+, \mathbf{U}_+^{\star}\} + \{\mathbf{U}_-, \mathbf{U}_-^{\star}\}) & \text{(Bose)}, \end{cases} \\ \multicolumn{2}{l}{(\mathbf{U}_+(t), \mathbf{U}_-(t)) = (\mathbf{U}_+, \mathbf{U}_-)e^{\mathbf{i}mt}, \quad (\mathbf{U}_+^{\star}(t), \mathbf{U}_-^{\star}(t)) = (\mathbf{U}_+^{\star}, \mathbf{U}_-^{\star})e^{\mathbf{i}mt}} \\ \multicolumn{2}{l}{\text{with } \mathbf{i} = \begin{pmatrix} 0 & 1 \\ -1 & 0 \end{pmatrix}, \quad \exp \mathbf{i}mt = \begin{pmatrix} \cos mt & \sin mt \\ -\sin mt & \cos mt \end{pmatrix}.} \end{array}$$

In addition to the $\times$-symmetric Hamiltonian there is the $\times$-antisymmetric generator in the indefinite unitary group $\mathbf{U}(1,1) = \mathbf{U}(\mathbf{1}_2) \circ \mathbf{SU}(1,1)$. It is used for the *particle-antiparticle charge* Q_{doub}:

$$\begin{array}{rl} Q_{\text{doub}} & = z(\mathrm{u} \otimes \mathrm{a}^{\times} - \mathrm{a} \otimes \mathrm{u}^{\times}) = z(\mathrm{u} \otimes \mathrm{u}^{\star} - \mathrm{a} \otimes \mathrm{a}^{\star}), \quad z \in \mathbb{Z}, \\ Q_{\text{doub}} & = -Q_{\text{doub}}^{\times} = Q_{\text{doub}}^{\star}, \\ iH_{\text{doub}} & \in \log \mathbf{U}(\mathbf{1}_2), \quad iQ_{\text{doub}} \in \log \mathbf{U}(1)_3 \subset \log \mathbf{SU}(1,1), \\ & [H_{\text{doub}}, Q_{\text{doub}}] = 0. \end{array}$$

Its action defines as eigenvalue an integer particle-antiparticle charge number $\pm z$:

$$\begin{array}{rlrl} Q_{\text{doub}}(\mathrm{u}) & = z\mathrm{u}, & -Q_{\text{doub}}^T(\mathrm{a}^{\star}) & = z\mathrm{a}^{\star}, \\ -Q_{\text{doub}}^T(\mathrm{u}^{\star}) & = -z\mathrm{u}^{\star}, & Q_{\text{doub}}(\mathrm{a}) & = -z\mathrm{a}. \end{array}$$

The self-dual combinations for the particle-antiparticle charge are:

$$\begin{array}{rl} Q_{\text{doub}} - Q_{\text{doub}}^T & = z[\mathrm{u}, \mathrm{u}^{\star}]_{-\epsilon} - z[\mathrm{a}, \mathrm{a}^{\star}]_{-\epsilon} \\ & = \begin{cases} z([\mathbf{U}_+, \mathbf{U}_+^{\star}] + [\mathbf{U}_-, \mathbf{U}_-^{\star}]) & \text{(Fermi)}, \\ -iz(\{\mathbf{U}_+, \mathbf{U}_-^{\star}\} - \{\mathbf{U}_-, \mathbf{U}_+^{\star}\}) & \text{(Bose)}, \end{cases} \\ \multicolumn{2}{l}{(Q_{\text{doub}} - Q_{\text{doub}}^T).\mathbf{U}_{\pm} = z\mathbf{U}_{\pm}, \quad (Q_{\text{doub}} - Q_{\text{doub}}^T).\mathbf{U}_{\pm}^{\star} = -z\mathbf{U}_{\pm}^{\star}.} \end{array}$$

The weights of the irreducible $\mathbf{D}(1) \times \mathbf{U}(1)$-representations on the four 1-dimensional eigenspaces of the complex quartet

$$\mathbf{D}(1) \times \mathbf{U}(1) \longrightarrow \mathbf{U}(1) \times \mathbf{U}(1), \quad e^{t+i\alpha} \longmapsto e^{\pm imt \pm i\alpha z},$$

involve the real continuous energy (frequency) m and the integer particle-antiparticle charge number z

$$\textbf{weights}\, \mathbf{D}(1) \times \mathbf{U}(1) = \{(\pm im, \pm z)\} = i\mathbb{R} \times \mathbb{Z}.$$

4.6 Summary

The canonical conjugation of the complex numbers defines the anticonjugation $\times$ for complex vector spaces. The complex vector space quartet $(V, \overline{V}, V^T, \overline{V}^T)$ contains two dual vector space pairs $(V, \overline{V}) \overset{T}{\leftrightarrow} (V^T, \overline{V}^T)$ and two antispace pairs $(V, V^T) \overset{\times}{\leftrightarrow} (\overline{V}, \overline{V}^T)$. Antispaces are naturally isomorphic as real vector spaces.

Antispace and dual space are isomorphic, if an additional dual conjugation $V \overset{*}{\leftrightarrow} V^T$ is given.

For finite dimension $V \cong \mathbb{C}^n$, the anticonjugation is connected with a sesquilinear form of the antidoubling $V_{\text{doub}} = V \oplus \overline{V}^T \cong \mathbb{C}^{2n}$ with neutral signature invariance group $\mathbf{U}(n,n)$. All complex representations of real Lie groups and Lie algebras come in quadruples - the proper, the dual, the anti- and the antidual representation. The direct sum representations on the antidoubling are $\mathbf{U}(n,n)$-subgroups and $\log \mathbf{U}(n,n)$-Lie subalgebras respectively.

$f \in \mathbf{AL}(V)$	$\overset{-}{\leftrightarrow}$	$\overline{f} \in \mathbf{AL}(\overline{V})$
$T \updownarrow$		$\updownarrow T$
$f^T \in \mathbf{AL}(V^T)$	$\overset{-}{\leftrightarrow}$	$f^\times \in \mathbf{AL}(\overline{V}^T)$

endomorphisms of a complex quartet

$D(g) = g$ $\mathcal{D}(l) = l$	anti $\leftrightarrow$	$\overline{D}(g) = \overline{g}$ $\overline{\mathcal{D}}(l) = \overline{l}$
dual $\updownarrow$		$\updownarrow$ dual
$\check{D}(g) = g^{-1T}$ $\check{\mathcal{D}}(l) = -l^T$	anti $\leftrightarrow$	$\hat{D}(g) = g^{-1\times}$ $\hat{\mathcal{D}}(l) = -l^\times$

representations with dual and antipartners

$V_{\text{doub}} = V \oplus \overline{V}^T$

antidoubling with $\mathbf{U}(n,n)$-conjugation

A complex quartet of 1-dimensional spaces with irreducible time action carries the weights $(\pm im, \pm z) \in i\mathbb{R} \times \mathbb{Z}$ as eigenvalues for a quartet of compact $\mathbf{D}(1) \times \mathbf{U}(1)$ (time and particle-antiparticle group) representations.

$V \ni \mathrm{u} : (im, z)$ **particle creation**	$\overset{\times}{\longleftrightarrow}$	$\overline{V} \ni \mathrm{a}^\star = \mathrm{u}^\times : (-im, z)$ **antiparticle annihilation**
$\star \updownarrow$		$\updownarrow \star$
$V^T \ni \mathrm{u}^\star = \mathrm{a}^\times : (-im, -z)$ **particle annihilation**	$\overset{\times}{\longleftrightarrow}$	$\overline{V}^T \ni \mathrm{a} : (im, -z)$ **antiparticle creation**

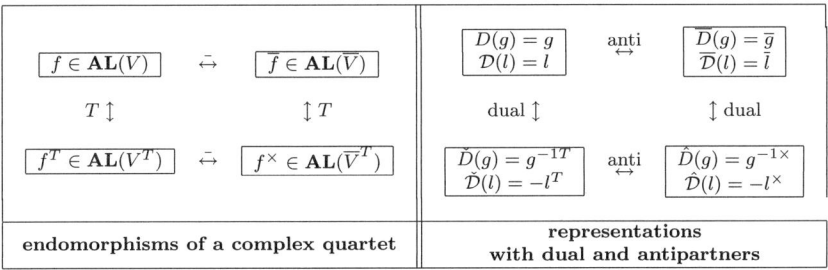

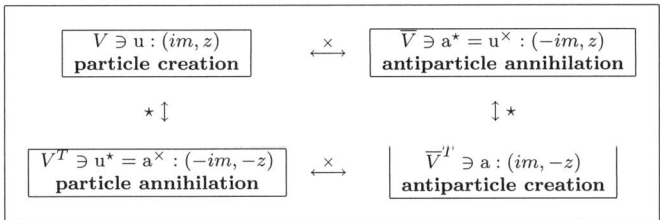

MATHEMATICAL TOOLS

4.7 Twin Vector Spaces

The direct sum $V_1 \oplus V_2 \cong \mathbb{K}^{2n}$ of two vector spaces $V_{1,2} \cong \mathbb{K}^n$ with a given *exchange isomorphism* $\mathrm{is}_{V_1V_2}$ is called a *twin vector space.* V_1 and V_2 carry isomorphic vector space structures, e.g., bases, bilinear forms, representations:

$$\begin{aligned} &\mathrm{is}_{V_1V_2} : V_2 \longrightarrow V_1, \quad \mathrm{is}_{V_1V_2}^{-1} = \mathrm{is}_{V_2V_1}, \\ &\text{dual \textit{twin bases}:} \quad \begin{cases} \langle \check{e}_B, e^A\rangle = \epsilon_{V_1}\langle e^A, \check{e}_B\rangle = \delta_B^A, & \epsilon_{V_1} = \pm 1, \\ \langle \check{c}_B, c^A\rangle = \epsilon_{V_2}\langle c^A, \check{c}_B\rangle = \delta_B^A, & \epsilon_{V_2} = \pm 1. \end{cases} \end{aligned}$$

For example, a dual isomorphism gives $V \oplus V^T$ a twin vector space structure

$$\gamma : V \longrightarrow V^T, \quad \gamma = \mathrm{is}_{V^TV}, \quad \gamma^{-1} = \mathrm{is}_{VV^T}.$$

As twin space mappings the identities are *idempotent* (projectors); the exchange isomorphism is *nilquadratic*

$$\begin{aligned} &\mathrm{id}_{V_1}, \mathrm{id}_{V_2}, \mathrm{is}_{V_1V_2}, \mathrm{is}_{V_2V_1} : V_1 \oplus V_2 \longrightarrow V_1 \oplus V_2, \\ &\mathrm{id}_{V_1} \circ \mathrm{id}_{V_1} = \mathrm{id}_{V_1}, \quad \mathrm{is}_{V_1V_2} \circ \mathrm{is}_{V_1V_2} = 0, \quad \mathrm{is}_{V_1V_2} \circ \mathrm{is}_{V_2V_1} = \mathrm{id}_{V_1}, \quad \text{etc.}, \\ &\mathrm{is}_{V_1 \oplus V_2} = \mathrm{is}_{V_1V_2} + \mathrm{is}_{V_2V_1}, \quad \mathrm{is}_{V_1 \oplus V_2} \circ \mathrm{is}_{V_1 \oplus V_2} = \mathrm{id}_{V_1 \oplus V_2} = \mathrm{id}_{V_1} + \mathrm{id}_{V_2}. \end{aligned}$$

With respect to the twin structure one can use a 2-component notation:

$$\begin{aligned} &e^A \simeq \begin{pmatrix} 1 \\ 0 \end{pmatrix}, \quad c^A \simeq \begin{pmatrix} 0 \\ 1 \end{pmatrix}, \quad \check{e}_A \simeq (1,0), \quad \check{c}_A \simeq (0,1), \\ &\mathrm{id}_{V_1} = e^A \otimes \check{e}_A \simeq \begin{pmatrix} 1 & 0 \\ 0 & 0 \end{pmatrix}, \quad \mathrm{id}_{V_2} = c^A \otimes \check{c}_A \simeq \begin{pmatrix} 0 & 0 \\ 0 & 1 \end{pmatrix}, \\ &\mathrm{is}_{V_1V_2} = e^A \otimes \check{c}_A \simeq \begin{pmatrix} 0 & 1 \\ 0 & 0 \end{pmatrix}, \quad \mathrm{is}_{V_2V_1} = c^A \otimes \check{e}_A \simeq \begin{pmatrix} 0 & 0 \\ 1 & 0 \end{pmatrix}. \end{aligned}$$

With equal transposition sign $\epsilon_{V_1}\epsilon_{V_2} = 1$, one calls the twin space *monovular*, for $\epsilon_{V_1}\epsilon_{V_2} = -1$ *binovular*. Monovular twins have a transposed exchange isomophism

$$\mathrm{is}_{V_1V_2}^T = \mathrm{is}_{V_2^TV_1^T}.$$

4.8 Complexification of Real Vector Spaces

A real vector space $V_\mathbb{R}$ is *complexified* by the extension of the scalars $\mathbb{R}$ to $\mathbb{C}$,

$$V = \mathbb{C} \otimes V_\mathbb{R} = \mathbb{R}^2 \otimes_\mathbb{R} V_\mathbb{R},$$

with the scalar multiplication defined as follows:

$$\mathbb{C} \times V \longrightarrow V, \quad (\alpha, \beta \otimes v) \longmapsto (\alpha\beta) \otimes v = (\alpha\beta)v.$$

Each real linear mapping of $V_{\mathbb{R}}$ in a complex vector space W can be uniquely extended to a complex linear mapping f on $\mathbb{C} \otimes V_{\mathbb{R}}$:

$$\begin{array}{ccc} V_{\mathbb{R}} & \overset{\sigma}{\longrightarrow} & \mathbb{C} \otimes V_{\mathbb{R}} \\ f_{\mathbb{R}} \downarrow & & \downarrow f \\ W & \underset{\mathrm{id}_W}{\longrightarrow} & W \end{array}, \qquad \begin{array}{l} f_{\mathbb{R}}(\alpha v) = \alpha f_{\mathbb{R}}(v), \quad \alpha \in \mathbb{R}, \\ f(\beta v) = \beta f_{\mathbb{R}}(v), \quad \beta \in \mathbb{C}, \end{array}$$

which defines the covariant *complexification functor*

$$\underline{\mathbf{vec}}_{\mathbb{R}} \longrightarrow \underline{\mathbf{vec}}_{\mathbb{C}}, \quad f_{\mathbb{R}} \begin{array}{c} V_{\mathbb{R}} \\ \downarrow \\ W_{\mathbb{R}} \end{array} \longmapsto \begin{array}{c} \mathbb{C} \otimes V_{\mathbb{R}} \\ \downarrow \\ \mathbb{C} \otimes W_{\mathbb{R}} \end{array} f_{\mathbb{C}},$$
$$(\mathrm{id}_{V_{\mathbb{R}}})_{\mathbb{C}} = \mathrm{id}_{\mathbb{C} \otimes V_{\mathbb{R}}}, \quad (f \circ g)_{\mathbb{C}} = f_{\mathbb{C}} \circ g_{\mathbb{C}}.$$

Complexification is compatible with duality and multilinearity (tensor algebra):

$$(\mathbb{C} \otimes V_{\mathbb{R}})^T = \mathbb{C} \otimes V_{\mathbb{R}}^T, \quad \mathbb{C} \otimes \bigotimes V_{\mathbb{R}} = \bigotimes (\mathbb{C} \otimes V_{\mathbb{R}}).$$

In analogy one defines a *complexified algebra* $\mathbb{C} \otimes A_{\mathbb{R}}$.

Bibliography

[1] N. Bourbaki, *Algèbre, Chapitre 2* (Algèbre linéaire), Hermann, Paris (1955).

[2] N. Bourbaki, *Algebra I, Chapters 1-3* (1989), Springer, Berlin, Heidelberg, New York, London, Paris, Tokyo.

[3] C. Segre, *Atti Torino* 25 (1889), 276.

5

SIMPLE LIE OPERATIONS

Operators in quantum theory come as linear transformations. They are characterized by invariants and eigenvectors with eigenvalues. The numerical results for a measurement of physical properties produce eigenvalues of operations. Given a set of operators, one will ask for a maximal subset that allows a simultaneous measurement for the eigenvalues of the operators therein. This involves the problem to find common eigenvectors or, for matrices, a common diagonal, i.e., a maximal common abelian structure, which for quantum operations may be called a classical projection. Such considerations for a Lie algebra of operators lead to the concept of a *Cartan subalgebra*, which is mathematically important for the classification and the representation of Lie algebras. An exhaustive physical measurement is mathematically formalized as a maximal diagonalization ("Cartanization").

Representations of Lie algebras and Lie groups are characterized by eigenvalues with the associated principal spaces (generalized eigenspaces with eigenvectors and, possibly, also nilvectors) of the endomorphisms involved. As seen in the triagonal Jordan form for principal vector bases, one endomorphism of a complex vector space is characterized by its eigenvalues, their degeneracy, order and multiplicity. A set of complex endomorphisms that constitutes a solvable Lie algebra can be brought simultaneously to a triagonal matrix form: there exist common eigenvectors with the eigenvalues constituing weights i.e., Lie algebra forms. For an even nilpotent complex Lie algebra of operations, the direct decomposition of the space acted on into weight-related principal spaces is possible, i.e., the endomorphisms have a basis of common principal vectors.

A Cartan subalgebra of a Lie algebra is a maximal nilpotent Lie subalgebra. A Lie algebra representation is characterizable by eigenforms (weights) of a Cartan subalgebra. The representation vector space allows a spectral decomposition into principal spaces with respect to a Cartan Lie subalgebra. For semisimple Lie algebras the Cartan subalgebras are even abelian; their representations are diagonalizable. In this case, the representation space can be spanned by simultaneous eigenvectors of a Cartan subalgebra; there are no nilvectors; the representation vector space allows a spectral projector decomposition without nilpotents. This leads - for finite dimensions - to the *Cartan classification of semisimple Lie algebras* (this chapter) und their representations (next chapter), which will be seen to be a beautiful theory after one has become familiar with the initially rather complicated-looking concepts involved.

Throughout this chapter all vector spaces are assumed to be finite $\mathbb{K}$-dimensional. With Ado's theorem finite-dimensional Lie algebras can be represented by finite matrices.

5.1 Diagonalization of Operations

5.1.1 Eigenspaces and Eigenforms (Weights)

An important concept for operations are weights as a generalization and collection of eigenvalues. To remember: Each endomorphism f of a vector space $V \cong \mathbb{K}^D$, $D \geq 1$, is a unique sum of a semisimple and a nilpotent endomorphism $f = h + n \sim \left(\begin{array}{c|c} h_1 & n \\ \hline 0 & h_2 \end{array}\right)$. For the algebraically closed complex numbers, it is *triagonalizable*, i.e., there exist appropriate bases for a Jordan matrix with the eigenvalues $\{\alpha_a\}_{a=1}^m$ on the diagonal. A complex endomorphism h is even *diagonalizable* iff semisimple, i.e., iff h has only degree-1 factors $(X - \alpha)$ in the minimal polynomial.

Generalizing from one endomorphism f and therefore from a 1-dimensional operation space $\mathbb{K}f$, one analyzes the *action of an operator vector space* $W \cong \mathbb{K}^r$, $r \geq 1$, consisting of endomorphisms $W \subseteq \mathbf{AL}(V)$ of $V \cong \mathbb{K}^D$. A common eigenvector v for two endomorphisms $f^{1,2}(v) = w^{1,2}v$ with eigenvalues $w^{1,2}$ (for more than one endomorphism α is replaced by w) is an eigenvector for all linear combinations $\beta_1 f^1 + \beta_2 f^2$ with the corresponding linear combinations $\beta_1 w^1 + \beta_2 w^2$ as eigenvalues, i.e., the eigenvalues arise from the action of a linear form $w \in W^T$ (eigenform) on the operator vector space:

$$\langle w, \beta_1 f^1 + \beta_2 f^2 \rangle = \beta_1 \langle w, f^1 \rangle + \beta_2 \langle w, f^2 \rangle = \beta_1 w^1 + \beta_2 w^2.$$

Now the general definition: If for a linear form $w : W \longrightarrow \mathbb{K}$ the *eigenvector space, common for all endomorphisms from* W,

$$V_w(W) = \{v \in V \mid f(v) = \langle w, f \rangle v \text{ for all } f \in W\} \in \underline{\mathbf{vec}}_{\mathbb{K}},$$
$$w \in W^T, \quad V_w(W) = \bigcap_{f \in W} V_{\langle w, f \rangle}(f),$$

does not consist only of the trivial vector $0 \in V$, the vector space $V_w(W)$ is called the *eigenspace for the weight (eigenform)* $w \in W^T$ *of the operators* W *on* V. The eigenspace $V_w(W)$ consists of the vectors $v \in V$ that are eigenvectors for *all* operators $f \in W$ with an operator dependent eigenvalue $\langle w, f \rangle \in \mathbb{K}$. Then v is a W-eigenvector iff v is an eigenvector for a W-basis $\{f^j\}^{j=1,\ldots,r}$. The eigenvalue $\alpha = \langle w, f \rangle$ for *one* operator f and thus for a 1-dimensional operator space $\mathbb{K}f$ is generalized to the weight (eigenform) w for the operator space W. The *multiplicity* of a weight w is the dimension of its eigenspace:

$$M_w = \dim_{\mathbb{K}} V_w(W) \leq \dim_{\mathbb{K}} V.$$

The generalization of the spectrum (eigenvalues) $\operatorname{spec} f = \{\alpha_a\}_{a=1}^m$ for one operator f is the *weight set* of W:

$$\begin{aligned}\operatorname{spec}_V W &= \{w : W \longrightarrow \mathbb{K} \mid V_w(W) \neq \{0\}\} \\ &= \{w_a \mid a = 1, \dots, m\} \in \underline{\mathbf{set}}.\end{aligned}$$

The weights constitute a subset of the linear forms W^T; in general they are not a vector subspace. The number of different weights is equal to or less than the dimension D of the space V:

$$\operatorname{spec}_V W \subset W^T, \quad \operatorname{Card} \operatorname{spec}_V W \leq \dim_{\mathbb{K}} V.$$

There may not exist a nontrivial common eigenvector and therefore no weight, e.g., for a real endomorphism space W containing $f \cong \begin{pmatrix} 0 & 1 \\ -1 & 0 \end{pmatrix}$. For a basis $\{f^j\}^{l=1,\dots,r}$ of the endomorphisms W the components of the weights give the $(m \times r)$*-eigenvalue matrix* (w_a^j)

$$\begin{array}{lcl}\text{for } f \in \mathbf{AL}(\mathbb{K}^D) & & \text{for } W \subseteq \mathbf{AL}(\mathbb{K}^D) \\ \begin{pmatrix} \alpha_1 \\ \dots \\ \alpha_m \end{pmatrix} & \hookrightarrow & \begin{pmatrix} w_1 \\ \dots \\ w_m \end{pmatrix} = \begin{pmatrix} \langle w_1, f^1 \rangle & \dots & \langle w_1, f^r \rangle \\ \dots & \dots & \dots \\ \langle w_m, f^1 \rangle & \dots & \langle w_m, f^r \rangle \end{pmatrix} = (w_a^j)_{a=1,\dots,m}^{j=1,\dots,r}.\end{array}$$

If the basic endomorphisms are triagonalizable, the components of a weight $(w_a^j)^{j=1,\dots,r}$ (rows in the matrix) occupy analogous diagonal positions.

For example, if an operator space W with a basis of diagonal matrices $\{f^j\}^{j=1,\dots,r}$, $f = \beta_j f^j \in W$, acts on the vector space V, the diagonal elements $\{w_a^j = \langle w_a, f^j \rangle\}_{a=1}^D$ of the matrices yield the weights $\langle w_a, f \rangle = \beta_j w_a^j$. In the example of a 2-dimensional operator space $W \cong \mathbb{K}^2$ acting on a 3-dimensional vector $V \cong \mathbb{K}^3$ one has three weights:

$$f^1 = \begin{pmatrix} w_1^1 & 0 & 0 \\ 0 & w_2^1 & 0 \\ 0 & 0 & w_3^1 \end{pmatrix}, \quad f^2 = \begin{pmatrix} w_1^2 & 0 & 0 \\ 0 & w_2^2 & 0 \\ 0 & 0 & w_3^2 \end{pmatrix}, \qquad \begin{aligned} w_1 &= (w_1^1, w_1^2), \\ w_2 &= (w_2^1, w_2^2), \\ w_3 &= (w_3^1, w_3^2), \end{aligned}$$

$$V_{w_1}(W) = \begin{pmatrix} \mathbb{K} \\ 0 \\ 0 \end{pmatrix}, \quad V_{w_2}(W) = \begin{pmatrix} 0 \\ \mathbb{K} \\ 0 \end{pmatrix}, \quad V_{w_3}(W) = \begin{pmatrix} 0 \\ 0 \\ \mathbb{K} \end{pmatrix}.$$

The *principal space (generalized eigenspace) for the weight (eigenform)* $w \in W^T$ *of the operator space* W *acting on the vector space* V consists of the nontrivial common *principal vectors*:

$$\begin{aligned} V^w(W) &= \{v \in V \mid \begin{array}{l}\text{There exists a minimal } N_w \geq 1 \text{ with} \\ [f - \langle w, f \rangle \operatorname{id}_V]^{N_w}(v) = 0 \text{ for all } f \in W\end{array}\} \in \underline{\mathbf{vec}}_{\mathbb{K}}, \\ V^w(W) &= \bigcap_{f \in W} V^{\langle w, f \rangle}(f). \end{aligned}$$

The eigenspace (subindex w) is a vector subspace of the principal space (supindex w):

$$V_w(W) \subseteq V^w(W), \quad V_{\langle w, f \rangle}(f) \subseteq V^{\langle w, f \rangle}(f).$$

The dimension of the principal space is the *degeneracy* D_w of the weight w, with N_w the *order* of the weight w

$$D_w = \dim_{\mathbb{K}} V^w(W), \quad M_w = \dim_{\mathbb{K}} V_w(W), \quad M_w, N_w \leq D_w.$$

In the following example with vector space $V \cong \mathbb{K}^3$ and endomorphisms $W \cong \mathbb{K}^2$ two diagonal elements are equal with a nontrivial nilpotent part. There are two weights:

$$f^1 = \left(\begin{array}{cc|c} w_1^1 & 1 & 0 \\ 0 & w_1^1 & 0 \\ \hline 0 & 0 & w_2^1 \end{array}\right), \quad f^2 = \left(\begin{array}{cc|c} w_1^2 & 0 & 0 \\ 0 & w_1^2 & 0 \\ \hline 0 & 0 & w_2^2 \end{array}\right), \qquad \begin{array}{l} w_1 = (w_1^1, w_1^2), \\ w_2 = (w_2^1, w_2^2), \end{array}$$

$$V_{w_1}(W) = \begin{pmatrix} \mathbb{K} \\ 0 \\ 0 \end{pmatrix} \cong \mathbb{K}, \quad V^{w_1}(W) = \begin{pmatrix} \mathbb{K} \\ \mathbb{K} \\ 0 \end{pmatrix} \cong \mathbb{K}^2,$$

$$V_{w_2}(W) = V^{w_2}(W) = \begin{pmatrix} 0 \\ 0 \\ \mathbb{K} \end{pmatrix},$$

$$\text{if } w_1 \neq w_2 : \quad (M_{w_1}, N_{w_1}, D_{w_1}) = (1, 2, 2), \quad (M_{w_2}, N_{w_2}, D_{w_2}) = (1, 1, 1).$$

5.1.2 Projectors and Nilpotents

To remember (chapter "Time Representations"): An endomorphism f of a *complex* space $V \cong \mathbb{C}^D$ is triagonalizable (box-diagonalizable) with Jordan matrices

$$J_N(\alpha) = \alpha \mathbf{1}_N + \mathbf{N}_N = \begin{pmatrix} \alpha & 1 & 0 & \dots & 0 & 0 \\ 0 & \alpha & 1 & \dots & 0 & 0 \\ & \dots & & & \dots & \\ 0 & 0 & 0 & \dots & \alpha & 1 \\ 0 & 0 & 0 & \dots & 0 & \alpha \end{pmatrix}.$$

It can be characterized by its complex eigenvalues $\{\alpha_a\}_{a=1}^m$ with degeneracies $D_a \in \mathbb{N}$. Associated with the eigenvalue α_a is one projector $\mathcal{P}_a(f)$ (diagonal) and, if not semisimple, one nontrivial $\mathcal{N}_a(f)$ nilpotent to the order N_a. Hence the action space V for the endomorphism f is decomposable into principal spaces for each eigenvalue (*spectral decomposition with respect to* f):

$$\begin{aligned} f \in \mathbf{AL}(V) \Rightarrow \quad & f = \sum_{a=1}^m [\alpha_a \mathcal{P}_a(f) + \mathcal{N}_a(f)], \\ & \sum_{a=1}^m \mathcal{P}_a(f) = \mathrm{id}_V, \quad \mathcal{N}_a(f)^{N_a} = 0, \\ & V = \bigoplus_{a=1}^m V^{\alpha_a}(f), \quad V^{\alpha_a}(f) = \mathcal{P}_a(f)[V] \cong \mathbb{C}^{D_a}, \end{aligned}$$

$$\text{e.g., } f = \left(\begin{array}{ccc|c} \alpha_1 & 1 & 0 & 0 \\ 0 & \alpha_1 & 0 & 0 \\ 0 & 0 & \alpha_1 & 0 \\ \hline 0 & 0 & 0 & \alpha_2 \end{array}\right), \quad \left\{ \begin{array}{l} \mathcal{P}_1(f) = \left(\begin{array}{ccc|c} 1 & 0 & 0 & 0 \\ 0 & 1 & 0 & 0 \\ 0 & 0 & 1 & 0 \\ \hline 0 & 0 & 0 & 0 \end{array}\right), \\ \mathcal{N}_1(f) = \left(\begin{array}{ccc|c} 0 & 1 & 0 & 0 \\ 0 & 0 & 0 & 0 \\ 0 & 0 & 0 & 0 \\ \hline 0 & 0 & 0 & 0 \end{array}\right), \\ \mathcal{P}_2(f) = \left(\begin{array}{ccc|c} 0 & 0 & 0 & 0 \\ 0 & 0 & 0 & 0 \\ 0 & 0 & 0 & 0 \\ \hline 0 & 0 & 0 & 1 \end{array}\right). \end{array} \right.$$

The f-polynomials have a Jordan basis that obviously consists of f-polynomials

$$\mathbb{C}[f]_{\mathbf{AL}(V)}\text{-basis}: \ \{\mathcal{P}_a(f),\ \mathcal{N}_a^{k_a}(f) \mid a = 1, \dots m,\ \ k_a = 1, \dots, N_a - 1\}.$$

To generalize from one endomorphism with $W = \mathbb{C}f$, one tries to characterize a *complex* operator space $W \cong \mathbb{C}^r$, represented on $V \cong \mathbb{C}^D$, by its *common* projectors $\mathcal{P}_a(W)$ and nilpotents $\mathcal{N}_a(W)$ associated with each complex-valued weight $w_a \in W^T$ with degeneracy D_a and order N_a and to decompose the vector space V into w_a-associated principal spaces for W (common principal spaces):

$$\begin{aligned} W \subseteq \mathbf{AL}(V) &\overset{?}{\Rightarrow} \sum_{a=1}^{m} \mathcal{P}_a(W) = \mathrm{id}_V,\ \ \mathcal{N}_a(W)^{N_a} = 0, \\ & V = \bigoplus_{a=1}^{m} V^{w_a}(W),\ V^{w_a}(W) = \mathcal{P}_a(W)[V] \cong \mathbb{C}^{D_a}. \end{aligned}$$

Obviously, for $r \geq 2$, it is not enough for a simultaneous triagonalization to have complex endomorphisms. A vector space W of diagonalizable endomorphisms is *simultaneously diagonalizable* if and only if it is *a commutative Lie algebra* $[W, W] = \{0\}$ with commutator. As will be discussed below in more detail, a *simultaneous triagonalization* and an associate decomposition of the representation space V (*spectral decomposition with respect to* W) proves possible if the operator space W with the commutator $[f, g]$ constitutes *a nilpotent complex Lie algebra.* In this case the W-polynomials have a Jordan basis that can be written as polynomials in a W-basis:

$$\begin{aligned} \mathbb{C}[W]_{\mathbf{AL}(V)}\text{-basis}: &\ \ \{\mathcal{P}_a(W),\ \mathcal{N}_a^{k_a}(W) \mid a = 1, \dots m,\ \ k_a = 1, \dots, N_a - 1\}, \\ W &= \sum_{a=1}^{m} [w_a \mathcal{P}_a(W) + \sum_{k_a=1}^{N_a} \kappa_{k_a} \mathcal{N}_a(W)^{k_a}], \\ f \in W \Rightarrow f &= \sum_{a=1}^{m} [\langle w_a, f\rangle \mathcal{P}_a(W) + \sum_{k_a=1}^{N_a} \langle \kappa_{k_a}, f\rangle \mathcal{N}_a(W)^{k_a}]. \end{aligned}$$

The nilpotent contributions come with coefficients from $(N_a - 1)$ linear forms $\kappa_{k_a} : W \longrightarrow \mathbb{C}$, $k_a = 1, \dots, N_a - 1$, for each weight w_a, $a = 1, \dots, m$. If the operator space W is even a semisimple complex Lie algebra the nilpotents are trivial $\mathcal{N}_a(W) = 0$.

5.2 Abelian, Nilpotent, and Solvable

The three Lie algebra product (bracket) related concepts "abelian," "nilpotent" and "solvable" are discussed in the following, in general and for matrices. The "smallest" examples are given which are characteristic for nonrelativistic, relativistic, and quantum structures (below).

A $\mathbb{K}$-Lie algebra L is *abelian* for a trivial *commutator ideal* $[L, L]$. For any Lie algebra the classes "up to noncommutativity" constitute an abelian Lie algebra

$$L \longrightarrow L/[L, L] \text{ abelian.}$$

To generalize the concept "abelian" one considers the iterated Lie bracket. There exist two extensions of "abelian"; they use ideals: With two ideals $I, J \subseteq L$, the intersection, sum and Lie bracket (finite linear combinations of individual Lie brackets) are also ideals:

$$\text{ideals } I, J \Rightarrow \text{ ideals } I \cap J,\ I + J,\ [I, J] = \{[i_\alpha, j_\alpha]\}.$$

5.2.1 Solvable and Nilpotent Algebras

The first generalization of "abelian" considers the series of the *derived Lie algebras* by the multiple commutator ideals, which form a descending series:

$$\begin{array}{l} \partial^0 L = L,\ \ \partial L = [L, L],\ \partial^2 L = [\partial L, \partial L],\ \ \ldots, \\ \partial^{n+1} L = [\partial^n L, \partial^n L],\ n = 0, 1, \ldots, \\ \text{ideals: } L \supseteq \partial L \supseteq \partial^2 L \supseteq \cdots; \end{array}$$

$\partial^n L$ for $n \geq k$ is an ideal in $\partial^k L$. If this series ends trivially, the Lie algebra is called *solvable*:

$$\partial^n L = \{0\},\quad n \in \mathbb{N}.$$

The unique maximal solvable ideal is called the *radical.* Each Lie algebra L is isomorphic to the direct vector space sum of its radical $\mathcal{R}$ and the classes "up to solvability," a semisimple Lie algebra. There always exists an isomorphic *Levi subalgebra* $\mathcal{S}$ with the direct vector space decomposition

$$\text{as vector space } L = \mathcal{S} \oplus \mathcal{R} \text{ with semisimple } \mathcal{S}.$$

A second generalization of "abelian" defines the *(bracket) powers of a Lie algebra* by the multiple Lie bracket, which leads again to a descending series of ideals:

$$\begin{array}{l} L^1 = L,\ \ L^2 = [L, L],\ \ L^3 = \Big[L, [L, L]\Big], \ldots, \\ L^{n+1} = [L, L^n],\ n = 1, 2, \ldots, \\ \text{ideals: } L \supseteq L^2 \supseteq L^3 \supseteq \cdots. \end{array}$$

If for a minimal natural number $n_0 \in \mathbb{N}$,

$$L^{n_0} = \{0\} \iff (\operatorname{ad} l)^{n_0} = 0 \text{ for all } l \in L,$$

the Lie algebra is called *nilpotent to the power n_0*.

The unique maximal nilpotent ideal of a Lie algebra is called its *nilradical* $\mathcal{N}$. It is also the nilradical of the radical. The nilradical contains the *nilpotent radical* $\mathcal{N}_{\mathcal{R}}$:

$$\mathcal{R} \supseteq [L, \mathcal{R}] = \mathcal{N} \supseteq [L, L] \cap \mathcal{R} = \mathcal{N}_{\mathcal{R}}.$$

The nilradical is a semidirect factor

$$L \cong L/\mathcal{N} \ \vec{\oplus}\ \mathcal{N} \text{ (semidirect Lie algebra)}.$$

Each finite-dimensional Lie algebra has an injective (faithful) representation on a finite-dimensional $\mathbb{K}$-vector space, where the nilradical is represented by nilpotent endomorphisms (*theorem of Ado*)

$$L \cong \left(\begin{array}{c|c} L/\mathcal{N} & \mathcal{N} \\ \hline 0 & L/\mathcal{N} \end{array}\right) \subset \mathbf{AL}(V).$$

A Lie algebra is semisimple if and only if the nilradical or the radical is trivial.

Lie subalgebras, quotient Lie algebras, and representations of solvable (nilpotent) Lie algebras keep those properties.

For a Lie algebra L with Killing form κ one has

$$\begin{array}{l} L \text{ nilpotent} \Rightarrow \kappa = 0 \Rightarrow L \text{ solvable}, \\ L \text{ solvable} \iff [L, L] \text{ nilpotent} \iff \kappa(L, [L, L]) = 0. \end{array}$$

For an associative algebra A the properties "abelian" and "solvable" are defined by its natural Lie algebra structure. However, A is *associative nilpotent* if the kth power A^k, defined by the linear combinations of the k-fold products $a_1 a_2 \cdots a_k$, is trivial, i.e., $A^k = \{0\}$ for a $k \in \mathbb{N}$. A^k is a two-sided ideal in A. An associative nilpotent algebra is also nilpotent with respect to its natural Lie algebra structure. Not the inverse: take diagonal matrices.

Each abelian Lie algebra is nilpotent; each nilpotent Lie algebra is solvable:

$$\begin{array}{lcccc} L \in \underline{\mathbf{lag}}_{\mathbb{K}}: & \text{abelian} & \Rightarrow & \text{nilpotent} & \Rightarrow \text{ solvable} \\ \Uparrow & & & \Uparrow & \\ A \in \underline{\mathbf{aag}}_{\mathbb{K}}: & & & \text{nilpotent} & \end{array}$$

5.2.2 Theorems of Engel and Lie

The endomorphisms of a vector space $V \cong \mathbb{K}^D$ with a fixed basis include the *diagonal matrices* $H(\mathbb{K}^D)$, the *niltriagonal matrices* $N(\mathbb{K}^D)$, and the *(upper) triagonal matrices (Jordan matrices)* $\mathcal{R}(\mathbb{K}^D)$:

$$\begin{array}{rll} \textit{diagonal} \ H(\mathbb{K}^D) & = \{ \sum\limits_{1 \le i = j \le D} \alpha_i^j e^i \otimes \check{e}_j \mid \alpha_i^j \in \mathbb{K} \} = \begin{pmatrix} \times & 0 \\ 0 & \times \end{pmatrix}, \\ \textit{niltriagonal} \ N(\mathbb{K}^D) & = \{ \sum\limits_{1 \le i < j \le D} \alpha_i^j e^i \otimes \check{e}_j \mid \alpha_i^j \in \mathbb{K} \} = \begin{pmatrix} 0 & \times \\ 0 & 0 \end{pmatrix}, \\ \textit{triagonal} \ \mathcal{R}(\mathbb{K}^D) & = \{ \sum\limits_{1 \le i \le j \le D} \alpha_i^j e^i \otimes \check{e}_j \mid \alpha_i^j \in \mathbb{K} \} = \begin{pmatrix} \times & \times \\ 0 & \times \end{pmatrix}, \end{array}$$

with the dimensions and substructures

$$\begin{array}{ll} \dim_{\mathbb{K}} H(\mathbb{K}^D) = D, \quad \dim_{\mathbb{K}} N(\mathbb{K}^D) = \binom{D}{2}, & \dim_{\mathbb{K}} \mathcal{R}(\mathbb{K}^D) = \binom{1+D}{2}, \\ H(\mathbb{K}^D) \oplus N(\mathbb{K}^D) = \mathcal{R}(\mathbb{K}^D) \subseteq \mathbf{AL}(\mathbb{K}^D), & \dim_{\mathbb{K}} \mathbf{AL}(\mathbb{K}^D) = D^2. \end{array}$$

These concepts are interesting only for dimension $D \geq 2$; therefore the illustration uses 2×2 matrices.

The diagonal matrices are abelian, $[H(\mathbb{K}^D), H(\mathbb{K}^D)] = \{0\}$, and nilpotent as a Lie algebra, in general not as an associative algebra. The niltriagonal matrices are closed with respect to matrix multiplication where the nontrivial elements wander, step by step, into the right upper corner; finally, one obtains the null matrix. $N(\mathbb{K}^D)$ is nilpotent, both as associative and as Lie algebra:

$$\begin{aligned} N(\mathbb{K}^D) \circ N(\mathbb{K}^D) &\cong N(\mathbb{K}^{D-1}) \subseteq N(\mathbb{K}^D), \\ N(\mathbb{K}^D)^D &= \underbrace{N(\mathbb{K}^D) \circ \cdots \circ N(\mathbb{K}^D)}_{D \text{ times}} = \{0\}, \\ [N(\mathbb{K}^D), N(\mathbb{K}^D)] &\subseteq N(\mathbb{K}^D), \quad [N(\mathbb{K}^D), H(\mathbb{K}^D)] \subseteq N(\mathbb{K}^D). \end{aligned}$$

Also the triagonal matrices are closed under matrix multiplication. The commutator is niltriagonal. $\mathcal{R}(\mathbb{K}^D)$ is a solvable, not necessarily a nilpotent Lie algebra:

$$\begin{aligned} &\mathcal{R}(\mathbb{K}^D) \circ \mathcal{R}(\mathbb{K}^D) \subseteq \mathcal{R}(\mathbb{K}^D), \\ &\partial\mathcal{R}(\mathbb{K}^D) = \left[\mathcal{R}(\mathbb{K}^D), \mathcal{R}(\mathbb{K}^D)\right] \subseteq N(\mathbb{K}^D), \\ &\partial^{D+1}\mathcal{R}(\mathbb{K}^D) = \partial^D N(\mathbb{K}^D) = \{0\}. \end{aligned}$$

One nilpotent endomorphism can be niltriagonalized, i.e., there exist appropriate bases

$$V \cong \mathbb{K}^D, \ n \in \mathbf{AL}(V): \ n^k = 0 \iff n \in \begin{pmatrix} 0 & \times \\ 0 & 0 \end{pmatrix}.$$

The *theorem of Engel* states that a associative nilpotent endomorphism algebra can be characterized by individually nilpotent endomorphisms and by their simultaneous niltriagonalization:

$$V \cong \mathbb{K}^D, \ \underline{\mathbf{aag}}_{\mathbb{K}} \ni A \subset \mathbf{AL}(V): \ \left\{ \begin{array}{l} A \text{ associative nilpotent}, \\ n^k = 0 \text{ for all } n \in A \\ \iff A \subseteq \begin{pmatrix} 0 & \times \\ 0 & 0 \end{pmatrix} \text{(niltriagonalizable)}. \end{array} \right.$$

Each complex endomorphism f can be triagonalized:

$$V \cong \mathbb{C}^D, \ f \in \mathbf{AL}(V) \iff f \in \begin{pmatrix} \times & \times \\ 0 & \times \end{pmatrix}.$$

The *theorem of Lie* states that the solvable complex endomorphism Lie algebras coincide with the simultaneously triagonalizable ones:

$$V \cong \mathbb{C}^D, \ \underline{\mathbf{lag}}_{\mathbb{C}} \ni L \subset \mathbf{AL}(V): \ \left\{ \begin{array}{l} L \text{ solvable} \\ \iff L \subseteq \begin{pmatrix} \times & \times \\ 0 & \times \end{pmatrix} \text{(triagonalizable)}. \end{array} \right.$$

Summarizing: For finite complex dimension, the solvable Lie algebras and the nilpotent associative algebras are uniquely characterizable by matrices:

$$\begin{array}{lccccc}
\underline{\mathbf{lag}_{\mathbb{C}}} \ni L \subseteq \mathbf{AL}(\mathbb{C}^D): & \text{abelian} & \Rightarrow & \text{nilpotent} & \Rightarrow & \text{solvable} \\
 & & & & & \Updownarrow \text{(Lie)} \\
 & \Uparrow & & \Uparrow & & L \subseteq \begin{pmatrix} \times & \times \\ 0 & \times \end{pmatrix} \\
\underline{\mathbf{aag}_{\mathbb{C}}} \ni A \subseteq \mathbf{AL}(\mathbb{C}^D): & & & \text{nilpotent} & & \\
 & & & \Updownarrow \text{(Engel)} & & \\
 & & & A \subseteq \begin{pmatrix} 0 & \times \\ 0 & 0 \end{pmatrix} & &
\end{array}$$

5.3 The Basic Lie Operations

The physically relevant Lie operations, translations, rotations, Lorentz boosts, Poincaré operations, and the Heisenberg Lie algebra for quantum mechanics, show up in the "basic Lie operations."

The abelian and simple real and complex Lie algebras are the building blocks for all real and complex Lie operations. The classification of the Lie algebras with dimensions one, two, and three gives the simplest nontrivial examples for the concepts "abelian $\Rightarrow$ nilpotent $\Rightarrow$ solvable" and "simple".

5.3.1 The Lie Algebras with Dimensions up to Three

It makes sense to call the trivial 0-dimensional Lie algebra $L = \{0\}$ semisimple, but not simple.

The abelian real translations originate from the complex 1-dimensional Lie algebra $L \cong \mathbb{C}$, it is abelian $\partial L = [L, L] = \{0\}$. It generates the linear group $\exp \mathbb{C} = \mathbf{GL}(\mathbb{C})$. Complex Lie operations have real forms: $\mathbf{D}(1) = \exp \mathbb{R}$ and $\mathbf{U}(1) = \exp i\mathbb{R}$ are the connected real 1-dimensional Lie groups.

There are two complex 2-dimensional Lie algebras $L \cong \mathbb{C}^2$, the abelian decomposable one and, new for two dimensions, the nonabelian, solvable one. The latter one is a semidirect product, isomorphic to the Lie algebra of the 1-dimensional affine group. It is given in the second column with the bracket in a basis $\{l^1, l^2\}$ and its faithful adjoint representation:

abelian	solvable
$\log \mathbf{GL}(\mathbb{C}) \cong \mathbb{C}$	$\log[\mathbf{GL}(\mathbb{C}) \vec{\times} \mathbb{C}] \cong \mathbb{C} \vec{\oplus} \mathbb{C}$
$[l^1, l^1] = 0$ $xl^1 \longmapsto x \in \mathbb{C}$	$[l^1, l^2] = l^2$ $\psi l^1 + xl^2 \longmapsto \begin{pmatrix} 0 & 0 \\ -x & \psi \end{pmatrix} \in \mathbf{AL}(\mathbb{C}^2)$
$\log \mathbf{D}(1) \cong \mathbb{R}$ $\log \mathbf{U}(1) \cong \mathbb{R}$	$\log[\mathbf{D}(1) \vec{\times} \mathbb{R}] \cong \mathbb{R} \vec{\oplus} \mathbb{R}$

There are three nondecomposable complex 3-dimensional Lie algebras $L \cong \mathbb{C}^3$: simple, solvable and nilpotent. Since there are no semisimple Lie algebras of dimension 1 and 2, the complex 3-dimensional Lie algebras have radical

$N \in \{\{0\}, \mathbb{C}^3\}$, i.e., they are either simple or solvable. They can be classified and constructed with the commutator ideal

$$L \cong \mathbb{C}^3 \Rightarrow \partial L = [L, L] \in \{\mathbb{C}^3, \mathbb{C}^2, \mathbb{C}, \{0\}\}.$$

The perfect case is the simple Lie algebra A_1 with a basis $\{l^1, l^2, l^3\}$ for totally antisymmetric structure constants

$$\mathbb{C}^3 \cong \partial L = A_1 \text{ with } \begin{cases} [l^1, l^2] &= l^3, \\ [l^2, l^3] &= l^1, \\ [l^3, l^1] &= l^2. \end{cases}$$

The ideals $\partial L \cong \mathbb{C}^2$ belong to semidirect product Lie algebras. An abelian ∂L gives a solvable, not nilpotent Lie algebra with possible basis:

$$\begin{matrix} \partial L = [L, \partial L] = \mathbb{C}l^1 \oplus \mathbb{C}l^2, \ \partial^2 L = \{0\}, \\ L = \mathbb{C}l^3 \ \vec{\oplus} \ [\mathbb{C}l^1 \oplus \mathbb{C}l^2] \end{matrix} \text{ with } \begin{cases} [l^1, l^2] &= 0, \\ [l^2, l^3] &= l^1, \\ [l^3, l^1] &= l^2. \end{cases}$$

A semidirect $\mathbb{C}^2$-ideal is not possible, since

$$\begin{array}{rl} \partial L = \mathbb{C}l^1 \ \vec{\oplus} \ \mathbb{C}l^2 : & \begin{cases} [l^1, l^2] &= l^2, \\ [l^2, l^3] &= \gamma l^1 + \delta l^2, \\ [l^3, l^1] &= \alpha l^1 + \beta l^2, \end{cases} \\ \text{Jacobi identity:} & \begin{cases} 0 &= [l^1, [l^2, l^3]] + [l^2, [l^3, l^1]] + [l^3, [l^1, l^2]] \\ &= \delta l^2 - \alpha l^2 - \gamma l^1 - \delta l^2, \end{cases} \\ \Rightarrow (\alpha, \gamma) = (0, 0) \Rightarrow & \left. \begin{matrix} [l^1, l^2] &= l^2, \\ [l^2, l^3] &= \delta l^2, \\ [l^3, l^1] &= \beta l^2, \end{matrix} \right\} \Rightarrow [L, L] = \mathbb{C}l^2. \end{array}$$

For $\partial L \cong \mathbb{C}$ there is a basis with

$$\partial L = \mathbb{C}l^2 : \begin{cases} [l^1, l^2] &= \beta_1 l^2, \\ [l^2, l^3] &= \beta_3 l^2, \\ [l^3, l^1] &= \alpha l^2; \end{cases}$$

l^1 and l^3 can be exchanged (first and second line). By renormalizations, a basis with three nontrivial brackets leads to a basis with two nontrivial brackets, and even to a basis with one nontrivial bracket,

$$\begin{cases} [l^1, l^2] &= l^2, \\ [l^2, l^3] &= l^2, \\ [l^3, l^1] &= l^2, \end{cases} \iff \begin{cases} [l^1, l^2] &= l^2, \\ [\underline{l}^3, l^2] &= 0, \\ [\underline{l}^3, l^1] &= l^2, \\ \text{with } \underline{l}^3 &= l^3 + l^1, \end{cases} \iff \begin{cases} [l^1, l^2] &= l^2, \\ [\underline{\underline{l}}^3, l^2] &= 0, \\ [\underline{\underline{l}}^3, l^1] &= 0 \\ \text{with } \underline{\underline{l}}^3 &= \underline{l}^3 + l^2, \end{cases}$$

which arises also from a basis with the following two nontrivial brackets:

$$\begin{cases} [l^1, l^2] &= l^2, \\ [l^3, l^2] &= l^2, \\ [l^3, l^1] &= 0, \end{cases} \iff \begin{cases} [l^1, l^2] &= l^2, \\ [\underline{l}^3, l^2] &= 0, \\ [\underline{l}^3, l^1] &= 0, \end{cases} \quad \underline{l}^3 = l^3 - l^1,$$

This characterizes the decomposable Lie algebra $\mathbb{C} \oplus [\mathbb{C} \,\vec{\oplus}\, \mathbb{C}]$.

Therefore, the only nondecomposable $\mathbb{C}^3$-Lie algebra with $\partial L \cong \mathbb{C}$ is the nilcubic Heisenberg Lie algebra, in a basis with two trivial brackets

$$L = \mathbb{C}l^3 \,\vec{\oplus}\, [\mathbb{C}l^1 \oplus \mathbb{C}l^2],\ \partial L = \mathbb{C}l^2,\ [L, \partial L] = \{0\} \text{ with } \begin{cases} [l^1, l^2] & = 0, \\ [l^2, l^3] & = 0, \\ [l^3, l^1] & = l^2. \end{cases}$$

The nondecomposable $\mathbb{C}^3$-Lie algebras have faithful representations for a basis $\{l^1, l^2, l^3\}$ by (3×3) matrices $\alpha_1 l^1 + \alpha_2 l^2 + \alpha_3 l^3 \longmapsto \mathbf{AL}(\mathbb{C}^3)$:

simple	solvable	nilpotent
$\log \mathbf{SO}(\mathbb{C}^3)$	$\log[\mathbf{SO}(\mathbb{C}^2) \,\vec{\times}\, \mathbb{C}^2]$ $\cong \log \mathbf{SO}(\mathbb{C}^2) \,\vec{\oplus}\, \mathbb{C}^2$	$\log \mathbf{H}(\mathbb{C})$ $\cong \mathbb{C} \,\vec{\oplus}\, \mathbb{C}^2$
$[l^1, l^2] = l^3$ $[l^2, l^3] = l^1$ $[l^3, l^1] = l^2$	$[l^1, l^2] = 0$ $[l^2, l^3] = l^1$ $[l^3, l^1] = l^2$	$[l^1, l^2] = 0$ $[l^2, l^3] = 0$ $[l^3, l^1] = l^2$
$\begin{pmatrix} 0 & \alpha_3 & \alpha_2 \\ -\alpha_3 & 0 & \alpha_1 \\ -\alpha_2 & -\alpha_1 & 0 \end{pmatrix}$	$\left(\begin{array}{cc\|c} 0 & \alpha_3 & \alpha_2 \\ -\alpha_3 & 0 & \alpha_1 \\ \hline 0 & 0 & 0 \end{array}\right)$	$\left(\begin{array}{cc\|c} 0 & \alpha_3 & \alpha_2 \\ 0 & 0 & \alpha_1 \\ \hline 0 & 0 & 0 \end{array}\right)$
$\log \mathbf{SO}(3)$ $\log \mathbf{SO}_0(1,2)$	$\log \mathbf{SO}(2) \,\vec{\oplus}\, \mathbb{R}^2$ $\log \mathbf{SO}_0(1,1) \,\vec{\oplus}\, \mathbb{R}^2$	$\log \mathbf{H}(1)$ $\cong \mathbb{R} \,\vec{\oplus}\, \mathbb{R}^2$
rotations flat Lorentz	flat Euclidean flat Poincaré	Heisenberg

Their real forms (second last line) are, up to $\log \mathbf{SO}(3)$, noncompact. The simple real Lie structures with dimension 3 add spherical and hyperbolic degrees of freedom to the abelian $\mathbb{R}$-structures:

$$\begin{array}{lll} \mathbf{SO}(2) & \cong \Omega^1, & \mathbf{SO}_0(1,1) \cong \mathcal{Y}^1, \\ \mathbf{SO}(3) & \text{rotation group}, & \mathbf{SO}(3)/\mathbf{SO}(2) \cong \Omega^2, \\ \mathbf{SO}_0(1,2) & \text{flat Lorentz group}, & \begin{cases} \mathbf{SO}_0(1,2)/\mathbf{SO}(2) & \cong \mathcal{Y}^2, \\ \mathbf{SO}_0(1,2)/\mathbf{SO}_0(1,1) & \cong \mathcal{Y}^1 \times \Omega^1. \end{cases} \end{array}$$

The twofold covering groups are (iso)spin $\mathbf{SU}(2)$ (simply connected) and $\mathbf{SU}(1,1)$ ($\mathbb{Z}$-connected) as real forms of the complex 3-dimensional special Lie group $\mathbf{SL}(\mathbb{C}^2)$ (considered as a 3-dimensional complex Lie group). The not simple nondecomposable real 3-dimensional Lie operations are all semidirect groups, i.e., affine subgroups in $\mathbf{GL}(\mathbb{R}^2) \,\vec{\times}\, \mathbb{R}^2$

$$\begin{array}{rl} \mathbf{SO}(2) \,\vec{\times}\, \mathbb{R}^2 & \text{Euclidean (flat Galileo) group}, \\ \mathbf{SO}_0(1,1) \,\vec{\times}\, \mathbb{R}^2 & \text{flat Poincaré group}, \\ \mathbf{H}(1) \cong \mathbb{R} \,\vec{\times}\, \mathbb{R}^2 & \text{Heisenberg group}. \end{array}$$

They are contractions of $\mathbf{SO}(3)$ and $\mathbf{SO}_0(1,2)$. $\mathbf{D}(1) \,\vec{\times}\, \mathbb{R}$ is a subgroup of the flat Poincaré group.

All nondecomposable Lie algebras with dimensions $1, 2$ and 3 have rank 1. There is one generating linear invariant for the abelian case, and one quadratic Casimir invariant (inverse Killing form) for the simple case and its contracted forms for the contractions. The 2-dimensional Lie algebra $\mathbb{R} \,\vec{\oplus}\, \mathbb{R}$ has no nontrivial invariant.

5.3.2 Heisenberg Lie Algebras

The Heisenberg Lie algebra with s-position momentum pairs, $s \geq 1$,

$$\log \mathbf{H}(s) = \mathbf{h}(s) \cong \mathbb{R}^{1+2s}: \quad [\mathbf{x}_a, \mathbf{p}^b] = \delta_a^b \mathbf{I}, \quad [\mathbf{x}_a, \mathbf{I}] = 0 = [\mathbf{I}, \mathbf{p}^b],$$

is nilcubic with its center as commutator ideal:

$$[\mathbf{h}(s), \mathbf{h}(s)] = \mathbb{R}\mathbf{I} = \operatorname{centr} \mathbf{h}(s), \quad [[\mathbf{h}(s), \mathbf{h}(s)], \mathbf{h}(s)] = \{0\}.$$

It has a faithful representation by $((2+s) \times (2+s))$ matrices:

$$\mathbf{h}(s) \ni q^a \mathbf{x}_a + y_b \mathbf{p}^b + t\mathbf{I} \longmapsto \left(\begin{array}{c|c|c} 0 & q^a & t \\ \hline 0 & 0 & y_b \\ \hline 0 & 0 & 0 \end{array}\right) \in \mathbf{AL}(\mathbb{R}^{2+s}).$$

The *extended Heisenberg Lie algebra* includes the linear transformations $f \in \mathbf{AL}(\mathbb{R}^s)$ of the position-momentum pairs, in the matrix representation

$$\mathbf{h}(s) \subset \mathbf{Ah}(s) \ni \left(\begin{array}{c|c|c} 0 & q^a & t \\ \hline 0 & f_b^a & y_b \\ \hline 0 & 0 & 0 \end{array}\right) \in \left(\begin{array}{c|c|c} 0 & \check{\mathbb{R}}^s & \mathbb{R} \\ \hline 0 & \mathbf{AL}(\mathbb{R}^s) & \mathbb{R}^s \\ \hline 0 & 0 & 0 \end{array}\right) \in \mathbf{AL}(\mathbb{R}^{2+s})$$

with the Lie bracket involving the linear transformations acting on position and momentum and their dual product $\langle q, y \rangle$:

$$[\left(\begin{array}{c|c|c} 0 & q_1 & t_1 \\ \hline 0 & f_1 & y_1 \\ \hline 0 & 0 & 0 \end{array}\right), \left(\begin{array}{c|c|c} 0 & q_2 & t_2 \\ \hline 0 & f_2 & y_2 \\ \hline 0 & 0 & 0 \end{array}\right)] = \left(\begin{array}{c|c|c} 0 & f_2^T(q_1) - f_1^T(q_2) & \langle q_1, y_2 \rangle - \langle q_2, y_1 \rangle \\ \hline 0 & [f_1, f_2] & f_1(y_2) - f_2(y_1) \\ \hline 0 & 0 & 0 \end{array}\right)$$

with $f_a^b y_b = f(y)_a, \quad f_a^b q^a = f^T(q)^b, \quad \langle q, y \rangle = q^q y_a.$

The Heisenberg Lie algebra is a semidirect product of the abelian position Lie algebra with the ideal build by the center $\mathbb{R}\mathbf{I}$ and the subspace $\mathbb{R}\mathbf{p}$ spanned by the momenta

$$\mathbf{h}(s) = \check{\mathbb{R}}^s \ \vec{\oplus} \ [\mathbb{R}^s \oplus \mathbb{R}]: \quad [\mathbb{R}\mathbf{x}, \mathbb{R}\mathbf{p} + \mathbb{R}\mathbf{I}] \subseteq \mathbb{R}\mathbf{I}.$$

It is the nilradical of the extended Heisenberg Lie algebra

$$[\left(\begin{array}{c|c|c} 0 & \check{\mathbb{R}}^s & \mathbb{R} \\ \hline 0 & \mathbf{AL}(\mathbb{R}^s) & \mathbb{R}^s \\ \hline 0 & 0 & 0 \end{array}\right), \left(\begin{array}{c|c|c} 0 & \check{\mathbb{R}}^s & \mathbb{R} \\ \hline 0 & 0 & \mathbb{R}^s \\ \hline 0 & 0 & 0 \end{array}\right)] \subseteq \left(\begin{array}{c|c|c} 0 & \check{\mathbb{R}}^s & \mathbb{R} \\ \hline 0 & 0 & \mathbb{R}^s \\ \hline 0 & 0 & 0 \end{array}\right).$$

Hence the extended Heisenberg Lie algebra is a double semidirect product:

$$\mathbf{Ah}(s) = \mathbf{AL}(\mathbb{R}^s) \ \vec{\oplus} \ \mathbf{h}(s) = \mathbf{AL}(\mathbb{R}^s) \ \vec{\oplus} \ [\check{\mathbb{R}}^s \ \vec{\oplus} \ [\mathbb{R}^s \oplus \mathbb{R}]].$$

5.4 Spectral Decompositions of Lie Algebras

Eigenspaces and eigenvalues for Lie algebra actions are described by Cartan algebras which are nilpotent subalgebras, even abelian for semisimple Lie algebras.

5.4.1 Spectral Decompositions for Nilpotent Lie Algebras

Since a complex representation of a solvable complex Lie algebra is solvable, i.e., triagonalizable, there exist, for the diagonal part, projectors with a vector space decomposition, associated to the weights (eigenforms) w_a. Those spaces have not to be principal spaces.

If the Lie algebra $\mathcal{N} \cong \mathbb{C}^r$ is even *nilpotent*

$$\mathcal{D} : \mathcal{N} \longrightarrow \mathbf{AL}(V), \quad h \longmapsto \mathcal{D}(h) \in \begin{pmatrix} \times & \times \\ 0 & \times \end{pmatrix},$$

there is a decomposition into $\mathcal{N}$-invariant principal spaces (spectral decomposition with respect to $\mathcal{D}[\mathcal{N}]$):

$$\begin{aligned} V^{w_a}(\mathcal{N}) &= \left\{ \begin{matrix} v \in V & \big| \text{There exists a minimal } N_a \geq 1 \text{ with} \\ & [\mathcal{D}(h) - \langle w_a, h \rangle \,\mathrm{id}_V]^{N_a}(v) = 0 \text{ for all } h \in \mathcal{N} \end{matrix} \right\}, \\ \mathrm{spec}\,_V\mathcal{N} &= \{w_a : \mathcal{D}[\mathcal{N}] \longrightarrow \mathbb{C} \mid a = 1, \ldots, m\}, \\ V &= \bigoplus_{j=1}^{m} V^{w_a}(\mathcal{N}), \quad w_a \in \mathcal{N}^T, \quad V^{w_a}(\mathcal{N}) \cong \mathbb{C}^{D_a}, \quad \mathcal{N} \bullet V^{w_a}(\mathcal{N}) \subseteq V^{w_a}(\mathcal{N}). \end{aligned}$$

The representation space has bases in which all elements of $\mathcal{N}$ are represented by the same triagonal matrix structure

$$\mathcal{D}[\mathcal{N}] = \bigoplus_{j=1}^{m} \mathcal{D}_a[\mathcal{N}], \quad \mathcal{D}_a[\mathcal{N}] = \mathcal{D}[\mathcal{N}] \circ \mathcal{P}_a(\mathcal{D}[\mathcal{N}]) \subseteq \begin{pmatrix} \times & \times \\ 0 & \times \end{pmatrix}.$$

The polynomials of the represented milpotent complex Lie algebra $\mathcal{D}[\mathcal{N}]$ have a Jordan basis

$$\mathbb{C}[\mathcal{D}[\mathcal{N}]]_{\mathbf{AL}(V)}\text{-basis} : \{\mathcal{P}_a(\mathcal{D}[\mathcal{N}]), \mathcal{N}_a^{k_a}(\mathcal{D}[\mathcal{N}]) \mid a{=}1, \ldots, m, \ k_a{=}1, \ldots, N_a{-}1\}.$$

An important special case comes for the adjoint representation of a complex Lie algebra $\mathcal{N} \subseteq L \cong \mathbb{C}^d$ restricted to a nilpotent Lie subalgebra

$$\mathrm{ad}\big|_{\mathcal{N}} : \mathcal{N} \longrightarrow \mathbf{AL}(L), \quad h \longmapsto \mathrm{ad}\, h.$$

Here one obtains an *$\mathcal{N}$-decomposition of the full Lie algebra* L with weights w_j. The principal space for the eigenform 0 contains the nilpotent Lie subalgebra

$$\begin{aligned} L^{w_j}(\mathcal{N}) &= \left\{ \begin{matrix} l \in L & \big| \text{There exists a minimal } N_j \geq 1 \text{ with} \\ & [\mathrm{ad}\, h - \langle w_j, h \rangle \,\mathrm{id}_L]^{N_j}(l) = 0 \text{ for all } h \in \mathcal{N} \end{matrix} \right\}, \\ \mathrm{spec}\,_L\mathcal{N} &= \{w_j : \mathrm{ad}\,\mathcal{N} \longrightarrow \mathbb{C} \mid j = 1, \ldots, r\}, \\ L &= \bigoplus_{j=1}^{r} L^{w_j}(\mathcal{N}), \quad w_j \in \mathcal{N}^T, \\ L^{w_j}(\mathcal{N}) &\cong \mathbb{C}^{d_j}, \quad L^0(\mathcal{N}) \supseteq \mathcal{N}, \quad \mathcal{N} \bullet L^{w_j}(\mathcal{N}) \subseteq L^{w_j}(\mathcal{N}). \end{aligned}$$

For a representation of a nilpotent Lie subalgebra $\mathcal{N} \subseteq L$ on a vector space V, e.g., adjoint on L, one has for linear Lie algebra forms $w, \theta : \mathcal{N} \longrightarrow \mathbb{C}$,

$$L^w(\mathcal{N}) \bullet V^\theta(\mathcal{N}) \subseteq V^{\theta+w}(\mathcal{N}), \quad [L^w(\mathcal{N}), L^\theta(\mathcal{N})] \subseteq L^{\theta+w}(\mathcal{N}).$$

Therefore the nontrivial principal space $L^0(\mathcal{N})$ for the trivial weight is a Lie subalgebra whose action leaves invariant the other spaces $V^\theta(\mathcal{N})$.

5.4.2 Cartan Subalgebras

Since nilpotent complex Lie algebras can be simultaneously triagonalized for each complex representation, one looks for maximal nilpotent Lie subalgebras to characterize a complex Lie algebra L and their representations. "Maximal" nilpotent, however, is not enough: "maximality" is replaced by "normalizer."

The *normalizer of a vector subspace* $W \subseteq L$ in a Lie algebra consist of all those Lie algebra elements whose adjoint action keep W stable:

$$N(W) = \{l \in L \mid [l, W] \subseteq W\} \in \underline{\mathbf{lag}}_{\mathbb{K}}.$$

A Lie subalgebra is an ideal $W \subseteq N(W)$ of its normalizer.

A nilpotent Lie subalgebra $\mathcal{N} \subseteq L \in \underline{\mathbf{lag}}_{\mathbb{K}}$, identical with its normalizer, is called a *Cartan subalgebra of* L:

$$\mathcal{N} \in \underline{\mathbf{lag}}_{\mathbb{K}} \text{ nilpotent}, \quad \mathcal{N} = N(\mathcal{N}) = \{h \in L \mid [h, \mathcal{N}] \subseteq \mathcal{N}\}.$$

A Lie subalgebra is always contained in its normalizer, $\mathcal{N} \subseteq N(\mathcal{N})$; therefore a Cartan subalgebra is a maximal nilpotent Lie subalgebra. However, there exist maximal nilpotent Lie subalgebras that are not Cartan algebras.

An equivalent definition characterizes a Cartan subalgebra $\mathcal{N} \subseteq L$ by its nilpotency and the coincidence with the *principal space for the weight* 0:

$$\begin{array}{l} \mathcal{N} \text{ nilpotent and} \\ \mathcal{N} = L^0(\mathcal{N}) = \{l \in L \mid \begin{array}{l} \text{There exists an } N_0 \geq 1 \text{ with} \\ \operatorname{ad} h^{N_0}(l) = 0 \text{ for all } h \in \mathcal{N}\}. \end{array} \end{array}$$

For semisimple Lie algebras there exists a simpler characterization of its Cartan subalgebras (below).

Each Lie algebra $L \in \underline{\mathbf{lag}}_{\mathbb{K}}$ *has nontrivial Cartan subalgebras* $\{\mathcal{N}\}$. All Cartan subalgebras have equal dimension; this defines the *rank of a Lie algebra*

$$\dim_{\mathbb{K}} \mathcal{N} = \operatorname{rank}_{\mathbb{K}} L.$$

Any two Cartan subalgebras of a complex Lie algebra L are even isomorphic and related to each other by an L-automorphism. This is not the case for real Lie algebras, e.g., for the noncompact $\log \mathbf{SO}_0(1,2)$, where there exist Cartan subalgebras of type $\log \mathbf{SO}(2)$ and $\log \mathbf{SO}(1,1)$.

For an endomorphism Lie algebra $\mathbf{AL}(V)$, $V \cong \mathbb{K}^D$, the diagonal matrices $\mathcal{N}(\mathbb{K}^D)$ with respect to a basis are a Cartan subalgebra.

The similarity invariants (chapter "Spin, Rotations, and Position") for a Lie algebra, represented on $V \cong \mathbb{K}^D$, are obtained from the characteristic polynomial. They define D invariant k-linear symmetric Lie algebra forms:

$$\begin{array}{ll} \mathcal{D} : L \longrightarrow \mathbf{AL}(V), & \det(\mathcal{D}(l) - X \operatorname{id}_V) = (-X)^D + \sum_{k=1}^{D} (-X)^{D-k} \kappa_V^k(l), \\ \kappa_V^k : \bigvee^k L \longrightarrow \mathbb{K}, & \kappa_V^k(l) = \operatorname{tr} \bigwedge^k \mathcal{D}(l) : \left\{ \begin{array}{ll} \kappa_V^1(l) & = \operatorname{tr} \mathcal{D}(l), \ldots, \\ \kappa_V^D(l) & = \det \mathcal{D}(l). \end{array} \right. \end{array}$$

For semisimple L, the tracelessness of all representations involves $\kappa_V^1(l) = 0$.

The characteristic polynomial of a Lie algebra element in the adjoint representation, with dual bases $\{l^k, \check{l}_k\}_{k=1}^d$ and structure constants $[l^j, l^k] = \epsilon_r^{jk} l^r$,

$$l = \alpha_j l^j \in L: \quad \mathrm{ad}\, l = [l, l^k] \otimes \check{l}_k = \alpha_j \epsilon_r^{jk}\, l^r \otimes \check{l}_k,$$

has always the eigenvalue 0 with degeneracy at least 1, since $\mathrm{ad}\, l(l) = 0$:

$$\det(\mathrm{ad}\, l - X\, \mathrm{id}_L) = (-X)^d + \sum_{k=1}^{d-1} (-X)^{d-k} \kappa^k(l), \quad \kappa^d(l) = \det \mathrm{ad}\, l = 0,$$

κ^2 is the Killing form.

A Lie algebra element $h \in L$ is called *regular* if the dimension of its principal space for the eigenvalue 0 with respect to adjoint action,

$$L^0(\mathrm{ad}\, h) = \{l \in L \mid \text{There exists } n_0 \geq 1 \text{ with } (\mathrm{ad}\, h)^{n_0}(l) = 0\},$$

is *minimal*, i.e., $\mathrm{ad}\, h$ acts trivially on as few Lie algebra elements as possible.

Since each Cartan subalgebra $\mathcal{N} \subseteq L$ is the principal space for the eigenvalue 0 of its elements h, its nontrivial elements are *regular*:

$$\mathcal{N} = L^0(\mathcal{N}) = L^0(\mathrm{ad}\, h), \quad \text{regular } 0 \neq h \in \mathcal{N}.$$

Therefore, the rank of L is the degeneracy of the eigenvalue 0 for the regular elements: the minimal degeneracy of the eigenvalue 0 for all elements in the adjoint representation

$$\begin{aligned} r = \mathrm{rank}_{\mathbb{K}} L = \dim_{\mathbb{K}} \mathcal{N} &= \min_{l \in L} \dim_{\mathbb{K}} L^0(\mathrm{ad}\, l) = \dim_{\mathbb{K}} L^0(\mathrm{ad}\, h), \\ \det(\mathrm{ad}\, h - X\, \mathrm{id}_L) &= (-X)^d + (-X)^{d-1}\kappa^1(h) + \cdots + (-X)^r \kappa^{d-r}(h) \\ \text{with } &\kappa^{d-r}(h) \neq 0. \end{aligned}$$

The smallest simple Lie algebra $\mathbb{K}^3$ with $[l^a, l^b] = -\epsilon^{abc} l^c$ has rank 1; each nontrivial element is regular and a basis for a Cartan subalgebra:

$$\begin{aligned} \mathrm{ad}\, \varphi_a l^a &= -\varphi_a \epsilon^{abc} l^c \otimes \check{l}_b \simeq \begin{pmatrix} 0 & -\varphi_3 & \varphi_2 \\ \varphi_3 & 0 & -\varphi_1 \\ -\varphi_2 & \varphi_1 & 0 \end{pmatrix}, \\ \det(\mathrm{ad}\, \varphi_a l^a - X\mathbf{1}_3) &= -X(X^2 + \vec{\varphi}^2). \end{aligned}$$

With a Cartan subalgebra $\mathcal{N}$ one has the *Cartan decomposition a complex Lie algebra* L with $\mathcal{N} = L^0(\mathcal{N})$:

$$L = \mathcal{N} \oplus \bigoplus_{\omega_j \neq 0} L^{\omega_j}(\mathcal{N}).$$

The *nontrivial* weights of a Cartan subalgebra in the adjoint representation $\omega_j : \mathrm{ad}\, \mathcal{N} \longrightarrow \mathbb{C}$ are called the *roots of* $\mathcal{N}$.

The Killing form, restricted to a Cartan subalgebra, can be diagonalized with the roots and the dimension d_j of the associated principal spaces:

$$\text{Killing form} \begin{cases} \kappa|_{\mathcal{N} \otimes \mathcal{N}} = \sum_{\omega_j} d_j \omega_j \otimes \omega_j, \\ \kappa(h, h') = \sum_{\omega_j} d_j \langle \omega_j, h \rangle \langle \omega_j, h' \rangle. \end{cases}$$

5.4.3 Spectral Decomposition of Simple Lie Algebras

Going from nilpotent Cartan subalgebras to even abelian ones (for $\mathbb{K} = \mathbb{C}$ from triagonal to diagonal) the structures become much simpler. For a semisimple Lie algebra $L \cong \mathbb{K}^d$ a Lie subalgebra $H \cong \mathbb{K}^r$ is a Cartan subalgebra if and only if it is *abelian*, $[H, H] = \{0\}$, and *maximal*. In the rest of this chapter only semisimple Lie algebras will be considered. For a semisimple $\mathbb{K}$-Lie algebra the rank, i.e., the dimension of a Cartan subalgebra, coincides with the number of independent invariants (chapter "Spin, Rotations, and Position"). For a complex semisimple Lie algebra the rank coincides with the maximal number of simultaneously diagonalizable linearly independent elements in a faithful representation. This is not the case for real semisimple Lie algebras. They are only block-diagonizable, i.e., also with 2×2 matrices. For example, in the adjoint representation of the real 3-dimensional $\log \mathbf{SO}(3)$ with rank 1 ($\log \mathbf{SO}(2)$ as Cartan Lie subalgebra), no nontrivial element (angular momentum) is diagonalizable.

The polynomials of the represented abelian complex Lie algebra $\mathcal{D}[H]$ have a basis that consists of projectors only (no nilpotents):

$$\mathbb{C}[\mathcal{D}[H]]_{\mathbf{AL}(V)}\text{-basis}: \ \{\mathcal{P}_a(\mathcal{D}[H]) \mid a = 1, \ldots, m\}.$$

A complex semisimple Lie algebra $L \cong \mathbb{C}^d$ is decomposable with respect to the diagonalizable adjoint action of a Cartan subalgebra $H \cong \mathbb{C}^r$ into H as eigenspace for the trivial weight $0 \in H^T$ and into *pairs of 1-dimensional H-eigenspaces*, $L_\omega(H) \oplus L_{-\omega}(H) \cong \mathbb{C}^2$, for *nontrivial reflected H*-weights (H-eigenforms), the roots (root vectors)

$$L = H \oplus \bigoplus_{\omega \in R^H} L_\omega(H) : \begin{cases} \text{spec}_L H = \{w \in H^T \mid L_w(H) \neq \{0\}\}, \\ \text{roots: } R^H = \{\omega \in H^T \mid \text{weight}, \ \omega \neq 0\}, \\ L_\omega(H) = \{l \in L \mid [h, l] = \langle \omega, h \rangle l \text{ for all } h \in H\} \cong \mathbb{C}, \\ \omega \in R^H \iff -\omega \in R^H, \\ \omega\text{-proportional weights: } \{\omega, 0, -\omega\}. \end{cases}$$

The Lie algebra has a basis of $\text{ad}\, H$-eigenvectors with $(d-r)$ different nontrivial H-weights (roots) and r vectors from H with trivial H-weight 0. The root set R^H contains $\frac{d-r}{2}$ different root pairs $\{\pm\omega_j\}_{j=1}^{\frac{d-r}{2}}$. With the exception of the smallest simple Lie algebra $A_1 \cong \mathbb{C}^3$ the $\frac{d-r}{2}$ different roots $\{\omega_j\}_{j=1}^{\frac{d-r}{2}}$ in $H^T \cong \mathbb{C}^r$ are linearly dependent, $\frac{d-r}{2} \geq r$,

$$d - r \in 2\mathbb{N}, \ \ \frac{d-3r}{2} \geq 0.$$

The natural number $\frac{d-3r}{2}$ is called *diagonal degeneracy* of the complex semisimple Lie algebra L.

The nontrivial diagonalizable Lie algebra elements of a Cartan subalgebra H are called *diagonal operators (Cartan operators)* $h \in H$. H is diagonalizable in any representation $\mathcal{D}[H] \in \begin{pmatrix} \times & 0 \\ 0 & \times \end{pmatrix}$. The nontrivial elements of the eigenspaces for the roots are called *eigenoperators* , they come in reflected pairs

$l_{\pm\omega} \in L_{\pm\omega}(H)$. $\mathrm{ad}\, h$ for the diagonal operators is semisimple; $\mathrm{ad}\, l_\pm$ for the eigenoperators is nilpotent.

The nondegenerate Killing form κ on $L \cong \mathbb{C}^d$ goes with a Cartan decomposition as follows: Eigenoperators to different, but not reflected, H-weights are orthogonal. κ is nondegenerate on reflected eigenoperators $L_\omega(H) \times L_{-\omega}(H)$ and on $H \times H$, where it is the sum of the products of the weights:

$$\begin{array}{l} \kappa(\ ,\ ) : L \times L \longrightarrow \mathbb{C}, \\ l_{\omega,\theta} \in L_{\omega,\theta}(H), \\ h, h' \in H, \end{array} : \left\{ \begin{array}{l} \omega + \theta \neq 0 : \quad \kappa(l_\omega, l_\theta) = 0, \\ \omega + \theta = 0, \quad l_\omega \neq 0 : \quad \kappa(l_\omega, l_{-\omega}) \neq 0, \\ \kappa(h, h') = \sum_{\omega \in R^H} \langle \omega, h \rangle \langle \omega, h' \rangle, \\ \kappa(h, l_\omega) = 0, \end{array} \right.$$

$$L = H \oplus L_+(H) \oplus L_-(H) : \quad \kappa \simeq \left(\begin{array}{c|cc} \times & 0 & 0 \\ \hline 0 & 0 & \times \\ 0 & \times & 0 \end{array} \right).$$

In addition one has for each L-invariant bilinear form, especially for the Killing form κ,

$$h \in H, \quad \omega \in R^H : \quad \kappa(h, [l_\omega, l_{-\omega}]) = \langle \omega, h \rangle \kappa(l_\omega, l_{-\omega}).$$

5.5 "Spin" Structure of Simple Lie Algebras

The complex 3-dimensional Lie algebra $A_1 \cong \mathbb{C}^3$ with Lie group $\exp A_1 \cong \mathbf{SL}(\mathbb{C}^2)$ is the smallest simple Lie algebra. Its structures are characteristic and constitutive for all semisimple Lie algebras.

Its fundamental physical importance shows up in the spin and isospin group $\mathbf{SU}(2) \cong \exp A_1^c$ and in the universal Lorentz cover group $\mathbf{SL}(\mathbb{C}^2_{\mathbb{R}}) \cong \exp(A_1^c \oplus iA_1^c)$.

From the compact form with the real Lie algebra $A_1^c \cong \mathbb{R}^3$ with Cartesian basis

$$[l^a, l^b] = -\epsilon^{abc} l^c, \quad \kappa(l^a, l^b) = -2\delta^{ab},$$

a $\mathbb{C}$-linear mapping leads to a *canonical (spherical) basis of* A_1 with diagonal operator h and eigenoperators $l_\pm$:

$$\begin{array}{l} l_\pm = l^2 \mp il^1, \quad h = -2il^3 \iff l^1 = i\frac{l_+ - l_-}{2}, \quad l^2 = \frac{l_+ + l_-}{2}, \quad l^3 = i\frac{h}{2}, \\ [l_-, l_+] = h, \quad [h, l_\pm] = \pm 2 l_\pm, \\ A_1 = \mathbb{C}h \oplus \mathbb{C}l_+ \oplus \mathbb{C}l_-, \quad \text{weights } \{\omega_\pm, 0\}, \quad \langle \omega_\pm, h \rangle = \pm 2, \\ \kappa(h, h) = 8, \quad \kappa(l_+, l_-) = -4, \quad \kappa \simeq 4 \begin{pmatrix} 2 & 0 & 0 \\ 0 & 0 & -1 \\ 0 & -1 & 0 \end{pmatrix}. \end{array}$$

The fundamental complex 2-dimensional defining representation is in the traceless endomorphisms

$$A_1 \longrightarrow \log \mathbf{SL}(\mathbb{C}^2), \quad \begin{cases} l_+ \longmapsto \sigma_+ & = \begin{pmatrix} 0 & 1 \\ 0 & 0 \end{pmatrix}, \\ l_- \longmapsto -\sigma_- & = \begin{pmatrix} 0 & 0 \\ -1 & 0 \end{pmatrix}, \\ h \longmapsto \sigma_0 & = \begin{pmatrix} 1 & 0 \\ 0 & -1 \end{pmatrix}, \end{cases}$$
$$\sigma^1 = \sigma_+ + \sigma_-, \quad \sigma^2 = -i(\sigma_+ - \sigma_-), \quad \sigma^3 = \sigma_0,$$
$$\operatorname{tr} \sigma_0\sigma_0 = 2 = 2 \operatorname{tr} \sigma_+\sigma_-, \quad \operatorname{tr} \sigma^a\sigma^b = 2\delta^{ab}.$$

Obviously there exists only one diagonal basis element σ_0. The A_1-*triplet* $(\sigma_0, \pm\sigma_\pm)$ is the minimal simple extension of the abelian 1-dimensional Lie algebra $\mathbb{C}\sigma_0$.

In a semisimple Lie algebra $L \cong \mathbb{C}^d$ with Cartan subalgebra $H \cong \mathbb{C}^r$ each nontrivial nilpotent element $l \in L$, $(\operatorname{ad} l)^n = 0$, $n \geq 1$, can be completed to an A_1-triplet $(h, l = l_+, l_-)$ (*theorem of Jacobson and Morozov*). For each root ω one gets an A_1-isomorphic structure: The vector space H_ω with the commutators of reflected eigenoperators has dimension 1,

$$H_\omega = [L_\omega(H), L_{-\omega}(H)] \cong \mathbb{C}, \quad H_\omega \subseteq H,$$

and completes an A_1-isomorphic Lie subalgebra of L:

$$A_1(\omega) = H_\omega \oplus L_\omega(H) \oplus L_{-\omega}(H) \cong \mathbb{C}^3 \in \underline{\mathbf{lag}}_{\mathbb{C}}.$$

For each eigenoperator $l_\omega \in L_\omega(H)$ there is a unique reflected eigenoperator $l_{-\omega} \in L_{-\omega}(H)$ and an appropriately normalized diagonal operator $h_\omega \in H_\omega$ to form an A_1-triplet $(l_\omega, l_{-\omega}, h_\omega)$. The ratio of the Killing form values is as for A_1:

$$[l_{-\omega}, l_\omega] = h_\omega, \quad [h_\omega, l_{\pm\omega}] = \pm\langle\omega, h_\omega\rangle l_{\pm\omega} = \pm 2 l_{\pm\omega}, \quad h_{-\omega} = -h_\omega,$$
$$\langle\omega, h_\omega\rangle = 2 = -\frac{\kappa(h_\omega, h_\omega)}{\kappa(l_\omega, l_{-\omega})} = -\frac{\kappa(h,h)}{\kappa(l_+, l_-)} = \frac{\operatorname{tr} \sigma_0\sigma_0}{\operatorname{tr} \sigma_+\sigma_-}.$$

Therefore, each root $\omega \in R^H$ induces a representation of the Lie algebra A_1 on the Lie algebra $L \cong \mathbb{C}^d$

$$\mathcal{D}_\omega : A_1 \longrightarrow \mathbf{AL}(L), \quad l_\pm \longmapsto \operatorname{ad} l_{\pm\omega}, \quad h \longmapsto \operatorname{ad} h_\omega,$$

where A_1 is a simple Lie algebra and the Lie algebra L as representation space for $A_1(\omega)$ with each root $\omega \in R^H$ is decomposable into irreducible $A_1(\omega)$-representation spaces characterized by integer eigenvalues $\langle\theta, h_\omega\rangle$ and with dimension $|\langle\theta, h_\omega\rangle| + 1$:

$$\omega \in R^H, \text{ decomposition for } A_1(\omega): \quad \begin{cases} L \cong \mathbb{C}^{r-1} \oplus \bigoplus_\theta V_\omega(\theta), \quad \theta \in R^H, \\ V_\omega(\theta) \cong \mathbb{C}^{|\langle\theta, h_\omega\rangle|+1}, \quad \begin{cases} \langle\theta, h_\omega\rangle \in \mathbb{Z}, \\ \langle\omega, h_\omega\rangle = 2; \end{cases} \end{cases}$$

$\mathbb{C}^{r-1}$ is the space with $r-1$ trivial representations, $V_\omega(\omega) \cong \mathbb{C}^3$ with basis $\{h_\omega, l_{\pm\omega}\}$ is acted on with the adjoint A_1-representation. The direct sum over roots θ does not need to include all roots of L, since an A_1-multiplet can contain eigenoperators for different roots. As integer eigenvalues there arise only those with

$$|\langle\theta, h_\omega\rangle| = 0, 1, 2, 3.$$

Therefore in the adjoint representation of a semisimple Lie algebra there can be *only A_1-singlets, doublets, triplets and quartets.* This natural number structure induced by the dimensions of A_1-representations (integer winding numbers) occurs everywhere in the theory of semisimple Lie algebras (more below).

5.5.1 Canonical Triplet Generators

Each semisimple Lie algebra $L \cong \mathbb{C}^d$ with a chosen Cartan subalgebra $H \cong \mathbb{C}^r$ allows a *canonical (spherical) generating system* for the Lie algebra with $\frac{d-r}{2}$ A_1-triplets associated to a root set:

$$\begin{aligned}
\text{roots: } & R^H = \{\pm\omega_j \mid j = 1, \dots, \tfrac{d-r}{2}\} \subset H^T, \\
\text{generating system of } L\text{: } & \{l_{\pm\omega_j}, h_{\omega_j} \mid j = 1, \dots, \tfrac{d-r}{2}\}, \\
\text{diagonal operators: } & \check{R}^H = \{\pm h_{\omega_j} \mid j = 1, \dots, \tfrac{d-r}{2}\} \subset H.
\end{aligned}$$

Examples for $\log \mathbf{SL}(\mathbb{C}^n)$ and $\log \mathbf{SU}(n)$ are given below.

In general, a canonical generating system is not a basis, i.e., $3\frac{d-r}{2} \geq d$. It consists of pairs with reflected *canonical eigenoperators* $l_{\pm\omega_j} \in L_{\pm\omega_j}(H)$ with a *canonical diagonal operator* (Cartan operator) $h_{\omega_j} = -h_{-\omega_j} \in H$ with Lie brackets

$$\begin{aligned}
&[h_\theta, l_{\pm\omega}] = \pm\langle\omega, h_\theta\rangle l_{\pm\omega}, \quad [h_\omega, h_\theta] = 0; \quad \omega, \theta \in R^H, \ \langle\omega, h_\theta\rangle \in \mathbb{Z}, \\
&[l_{-\omega}, l_\omega] = h_\omega, \quad \omega + \theta \neq 0 \Rightarrow [l_\omega, l_\theta] = N_{\omega,\theta} l_{\omega+\theta} \\
&\qquad \text{with } N_{\omega,\theta} \begin{cases} = 0 & \text{if } \omega + \theta \notin R^H, \\ \in \mathbb{Z} \setminus \{0\} & \text{if } \omega + \theta \in R^H. \end{cases}
\end{aligned}$$

All *canonical structure constants* $\langle\omega, h_\theta\rangle$ and $N_{\omega,\theta}$ are integers. The canonical generating system spans a $\mathbb{Z}$-module with a $\mathbb{Z}$-Lie algebra structure.

The roots $\{\omega_j\}_{j=1}^{\frac{d-r}{2}}$ with the associated canonical diagonal operators $\{h_{\omega_j}\}$ define the quadratic $(\frac{d-r}{2} \times \frac{d-r}{2})$ *Cartan eigenvalue matrix* (ω_i^j). It contains the integer-valued eigenvalues (root components) for the adjoint action

$$\begin{aligned}
R^H \times \check{R}^H \longrightarrow \mathbb{Z}, \ (\omega_i, h_{\omega_j}) &\longmapsto \langle\omega_i, h_{\omega_j}\rangle = \omega_i^j, \\
\langle\omega_j, h_{\omega_j}\rangle &= 2 \qquad \text{(no summation over } j\text{)}, \\
[h_{\omega_j}, l_{\omega_i}] &= \omega_i^j l_{\omega_i} \ \text{(no summation over } i\text{)}.
\end{aligned}$$

For the root ω, the vectors $l_{\pm\theta}$ with $[h_\omega, l_{\pm\theta}] = \pm\langle\theta, h_\omega\rangle l_{\pm\theta}$ belong to a $|\langle\theta, h_\omega\rangle|+1$-multiplet of $A_1(\omega)$. The columns of the Cartan eigenvalue matrix give the corresponding $A_1(\omega)$-decompositions since the natural numbers $|\omega_i^j|$ are the powers of the fundamental Pauli representation of $A_1(\omega)$.

As an example the Cartan eigenvalue matrix for the dimension $d = 8$ and rank $r = 2$ simple Lie algebra $A_2 \cong \log \mathbf{SL}(\mathbb{C}^3)$

$$\omega_i^j = \begin{pmatrix} 2 & -1 & 1 \\ -1 & 2 & 1 \\ 1 & 1 & 2 \end{pmatrix},$$

which shows the A_2-decomposition with respect to $A_1(\omega_2)$ in the second column: into doublet, triplet, and doublet from $|w_2^j|+1 = 2,3,2$.

The canonical structure constants $N_{\omega,\theta}$ can be computed with the integer limit numbers p,q of the θ *root chain in direction* ω (this uninterrupted chain exists):

$$\begin{aligned}
\theta,\omega \in R^H: &\quad \{\theta + k\omega \mid k = -q, -q+1, \dots, 0, \dots, p, \text{ maximal } p,q\} \subseteq R^H, \\
\text{with} &\quad p - q = -\langle\theta, h_\omega\rangle, \\
&\quad N_{\omega,\theta} = -N_{\theta,\omega} = N_{-\omega,-\theta}, \quad (N_{\omega,\theta})^2 = (q+1)^2, \\
\text{if } \theta+\omega \in R^H: &\quad N_{-\omega,\omega+\theta} = -N_{\omega,\theta}, \quad N_{\omega,\theta}N_{-\omega,\omega+\theta} = -p(q+1).
\end{aligned}$$

The Killing form $\kappa \simeq \begin{pmatrix} \times & 0 & 0 \\ 0 & 0 & \times \\ 0 & \times & 0 \end{pmatrix}$ for the canonical operators is integer valued with block diagonal form: The eigenoperators have "individual" (2×2) matrices

$$\kappa(l_\omega, l_\theta) = 0 \text{ for } \omega+\theta \notin R^H, \quad \begin{pmatrix} 0 & \kappa(l_\omega, l_{-\omega}) \\ \kappa(l_\omega, l_{-\omega}) & 0 \end{pmatrix} \text{ with } \kappa(l_\omega, l_{-\omega}) \in -\mathbb{N}.$$

The diagonal operators have positive square with the ratio -2 to their eigenoperator pair Killing product

$$\kappa(h_\omega, h_\theta) \in \mathbb{Z}, \quad \kappa(h_\omega, h_\omega) = -2\kappa(l_\omega, l_{-\omega}) \in \mathbb{N}.$$

The Killing form for the diagonal operators with the real span $H_\mathbb{R}$ defines a *scalar product*

$$\kappa|_{H_\mathbb{R}}(\ ,\ ): H_\mathbb{R} \times H_\mathbb{R} \longrightarrow \mathbb{R}: \quad \begin{cases} \kappa(h,h') = \sum_{j=1}^{d-r} \langle\omega_j, h\rangle\langle\omega_j, h'\rangle, \\ \kappa(h,h) > 0 \iff h \neq 0. \end{cases}$$

The canonical diagonal operators yield the canonical Killing matrix, in the A_2 example:

$$\kappa(h_{\omega_i}, h_{\omega_j}) = \kappa^{ij} = 6\begin{pmatrix} 2 & -1 & 1 \\ -1 & 2 & 1 \\ 1 & 1 & 2 \end{pmatrix}.$$

The dual space $H_\mathbb{R}^T \cong \mathbb{R}^r$ is the $\mathbb{R}$-span of the root set; it is called the *weight space* for the Lie algebra L. The Killing-form-induced isomorphisms between Cartan subalgebra H and its linear forms H^T associate to each root $\omega \in R^H \subseteq H^T$ a unique diagonal operator $h'_\omega \in H_\omega \subseteq H$. In general, this diagonal operator does *not* coincide with the canonical diagonal operator h_ω:

$$\langle\omega, h_\theta\rangle = \kappa(h'_\omega, h_\theta).$$

For each root $\omega \in R^H$ there exists an automorphism $\mathsf{S}_\omega \in \mathbf{GL}(H_\mathbb{R}^T)$ of the real weight space that keeps invariant an $\mathbb{R}^{r-1}$-hyperplane and maps the roots to the roots. It is an *"integer-valued" reflection* S_ω *of the roots* in the direction of ω:

$$\mathsf{S}_\omega : H_\mathbb{R}^T \longrightarrow H_\mathbb{R}^T: \quad \begin{cases} \mathsf{S}_\omega \circ \mathsf{S}_\omega = \mathrm{id}_{H_\mathbb{R}^T}, \\ \{w \in H_\mathbb{R}^T \mid \mathsf{S}_\omega(w) = w\} \cong \mathbb{R}^{r-1}, \\ \mathsf{S}_\omega(\theta) = \theta - \langle\theta, h_\omega\rangle\omega \in R^H, \quad \mathsf{S}_\omega(\omega) = -\omega, \\ \langle\omega, h_\theta\rangle \in \mathbb{Z}, \quad \langle\omega, h_\omega\rangle = 2, \quad \mathsf{S}_\omega[R^H] = R^H. \end{cases}$$

This integer-valued reflection structure of the root set in the weight space allows only a finite number of types for semisimple complex Lie algebras, which will be discussed below.

5.5.2 A_{n-1} and A^c_{n-1}: The Lie Algebras of $\mathbf{SL}(\mathbb{C}^n)$ and $\mathbf{SU}(n)$

The endomorphism Lie algebra of $V \cong \mathbb{C}^n$, $n \geq 2$, is decomposed into trace part and the simple Lie algebra with the traceless matrices, isomorphic to the abstract Lie algebra $A_{n-1} \cong \mathbb{C}^{n^2-1} \cong \log \mathbf{SL}(\mathbb{C}^n)$. With Ado's theorem each complex finite-dimensional semisimple Lie algebra has a faithful representation in a Lie algebra A_r.

With a scalar product of V the Euclidean Hermitian traceless endomorphisms are a compact simple Lie algebra, isomorphic to the abstract Lie algebra $A^c_{n-1} \cong \mathbb{R}^{n^2-1} \cong \log \mathbf{SU}(n)$, with a basis of traceless and Hermitian *generalized Pauli matrices*, constructed inductively from the proper Pauli matrices $\vec{\sigma} = \vec{\sigma}(2)$:

$$\begin{aligned} &\{\sigma(n)^a\}_{a=1}^{n^2-1}, \ \operatorname{tr}\sigma(n)^a = 0, \ \ \sigma(n)^a = (\sigma(n)^a)^\star, \\ &\sigma(1+n)^a = \left(\begin{array}{c|c} \sigma(n)^a & 0 \\ \hline 0 & 0 \end{array}\right), \ \ a = 1, \ldots, n^2-1. \end{aligned}$$

The new off-diagonal matrices for $a = n^2, \ldots, (1+n)^2 - 2$ come in pairs with unit column vectors (e)

$$\sigma(1+n)^a = \left(\begin{array}{c|c} \mathbf{0}_n & e \\ \hline e^T & 0 \end{array}\right), \ \ \sigma(1+n)^{1+a} = \left(\begin{array}{c|c} \mathbf{0}_n & -ie \\ \hline ie^T & 0 \end{array}\right),$$

as illustrated in the first step from $\sigma(2)$ to $\sigma(3)$:

$$\begin{aligned} \sigma(3)^4 &= \left(\begin{array}{cc|c} 0 & 0 & 1 \\ 0 & 0 & 0 \\ \hline 1 & 0 & 0 \end{array}\right), \quad \sigma(3)^5 = \left(\begin{array}{cc|c} 0 & 0 & -i \\ 0 & 0 & 0 \\ \hline i & 0 & 0 \end{array}\right), \\ \sigma(3)^6 &= \left(\begin{array}{cc|c} 0 & 0 & 0 \\ 0 & 0 & 1 \\ \hline 0 & 1 & 0 \end{array}\right), \quad \sigma(3)^7 = \left(\begin{array}{cc|c} 0 & 0 & 0 \\ 0 & 0 & -i \\ \hline 0 & i & 0 \end{array}\right). \end{aligned}$$

The diagonal matrices are defined by

$$\sigma(n)^{n^2-1} = \frac{1}{\sqrt{\binom{n}{2}}} \left(\begin{array}{c|c} \mathbf{1}_{n-1} & 0 \\ \hline 0 & -(n-1) \end{array}\right)$$

with the normalization as for the proper Pauli matrices

$$\operatorname{tr}\sigma(n)^a\sigma(n)^b = 2\delta^{ab}.$$

For $n = 3$, the generalized Pauli matrices are called *Gell-Mann matrices*,

$$\sigma(3)^a = \lambda^a, \ \ \beta_a\lambda^a = \begin{pmatrix} \beta_3 + \frac{\beta_8}{\sqrt{3}} & \beta_1 - i\beta_2 & \beta_4 - i\beta_5 \\ \beta_1 + i\beta_2 & -\beta_3 + \frac{\beta_8}{\sqrt{3}} & \beta_6 - i\beta_7 \\ \beta_4 + i\beta_5 & \beta_6 + i\beta_7 & -\frac{2\beta_8}{\sqrt{3}} \end{pmatrix}.$$

The Pauli matrices and the unit matrix $\mathbf{1}_n$ as $\mathbb{C}$-basis for the simple associative algebra $\mathbf{AL}(\mathbb{C}^n) \cong \mathbb{C}^{n^2}$ have the products with totally antisymmetric and symmetric structure constants

$$\begin{array}{ll} \sigma(n)^a\sigma(n)^b = \frac{2}{n}\delta^{ab}\mathbf{1}_n + \delta^{abc}\sigma(n)^c + i\epsilon^{abc}\sigma(n)^c, & \\ \text{with} & \left\{ \begin{array}{l} \text{totally antisymmetrical } \epsilon^{abc}, \\ \text{totally symmetrical } \delta^{abc}, \end{array} \right. \\ n=2: & \left\{ \begin{array}{l} \epsilon^{123} = 1, \\ \delta^{abc} = 0, \end{array} \right. \\ n=3: & \left\{ \begin{array}{l} \epsilon^{123} = 1, \quad \epsilon^{147,246,257,345,165,376} = \frac{1}{2}, \quad \epsilon^{458,678} = \frac{\sqrt{3}}{2}, \\ \delta^{118,228,338} = -\delta^{888} = \frac{1}{\sqrt{3}}, \quad \delta^{448,558,668,778} = -\frac{1}{2\sqrt{3}}, \\ \delta^{146,157,256,344,355} = -\delta^{247,366,377} = \frac{1}{2}, \end{array} \right. \end{array}$$

leading to the Lie bracket and the negative definite associated bilinear form for the compact Lie algebra A^c_{n-1}:

$$[\tfrac{i}{2}\sigma(n)^a, \tfrac{i}{2}\sigma(n)^b] = -\epsilon^{abc}\tfrac{i}{2}\sigma(n)^c, \quad \operatorname{tr} \tfrac{i}{2}\sigma(n)^a \tfrac{i}{2}\sigma(n)^b = -\tfrac{1}{2}\delta^{ab}.$$

A Cartan subalgebra is spanned by the diagonal matrices

$$A_{n-1}\text{-Cartan subalgebra basis: } \{\tfrac{i}{2}\sigma(n)^{m^2-1} \mid m = 2, 3, \ldots, n\}.$$

With dimension $n^2 - 1$ and rank $n - 1$ the Lie algebra A_{n-1} has diagonal degeneracy for

$$A_{n-1} \cong \log \mathbf{SL}(\mathbb{C}^n) : \tfrac{d-3r}{2} = \binom{n-1}{2}.$$

In abstract notation: The complex span of the compact Lie algebra A^c_{n-1} with basis $\{l^a\}_{a=1}^{n^2-1}$ and Lie bracket

$$[l^a, l^b] = -\epsilon^{abc} l^c$$

is the Lie algebra $A_{n-1} \cong \mathbb{C}^{n^2-1}$. A_1-triplets arise by combining two "nondiagonal" basis elements to one *raising and lowering element*:

$$\begin{array}{l} \text{for } A_1: \quad l_\pm = l^2 \mp i l^1, \\ \text{for } A_{n-1}: \quad \{l^{AB}_\pm \mid A < B, \quad A, B = 1, \ldots, n\}, \left\{ \begin{array}{ll} (l^{AB}_+)^j_k = \delta^{jA}\delta^B_k & \cong \begin{pmatrix} 0 & 1 \\ 0 & 0 \end{pmatrix}, \\ (l^{AB}_-)^j_k = -\delta^{jB}\delta^A_k & \cong \begin{pmatrix} 0 & 0 \\ -1 & 0 \end{pmatrix}. \end{array} \right. \end{array}$$

These $\binom{n}{2}$ pairs with their Lie bracket represent $\binom{n}{2}$ A_1-Lie algebras

$$h^{AB} = [l^{AB}_-, l^{AB}_+], \qquad [h^{AB}, l^{AB}_\pm] = \pm 2 l^{AB}_\pm.$$

For example, for $A_2 \cong \mathbb{C}^8$ with basis $\{l^a\}_{a=1}^8$ one obtains three A_1-triplets

$$\begin{array}{ll} l^{12}_\pm = l^2 \mp i l^1 = l_{\pm\omega_1}, & [l^{12}_-, l^{12}_+] = h^{12} = h_{\omega_1} = -2il^3, \\ l^{23}_\pm = l^7 \mp i l^6 = l_{\pm\omega_2}, & [l^{23}_-, l^{23}_+] = h^{23} = h_{\omega_2} = i(l^3 - \sqrt{3}l^8), \\ l^{13}_\pm = l^5 \mp i l^4 = l_{\pm\omega_3}, & [l^{13}_-, l^{13}_+] = h^{13} = h_{\omega_3} = -i(l^3 + \sqrt{3}l^8). \end{array}$$

The $\binom{n}{2}$ A_1-triplets $(l_\pm^{AB}, h^{AB})$ with raising and lowering operator and their diagonal partners constitute a spherical (canonical) generating system of the Lie algebra A_{n-1} with $3\binom{n}{2}$ elements. The diagonal elements generate a Cartan Lie subalgebra H of A_{n-1},

$$[h^{AB}, h^{CD}] = 0.$$

There exist $\binom{n-1}{2}$ independent nontrivial linear combinations for those diagonal elements:

$$\begin{array}{rl} \text{for } A_2: & h_{\omega_1} + h_{\omega_2} = h_{\omega_3}, \quad H \cong \mathbb{C}^2, \\ \text{for } A_{n-1}: & h^{AB} + h^{BC} = h^{AC}, \quad A < B < C, \quad H \cong \mathbb{C}^{n-1}. \end{array}$$

For each $A < B$, the adjoint representation of the full Lie algebra A_{n-1}, restricted to the associated A_1, is decomposable into irreducible A_1-representations

$$\begin{array}{l} \mathcal{D}_{AB} : A_1 \longrightarrow \mathbf{AL}(A_{n-1}), \quad l^{AB} \longmapsto \operatorname{ad} l^{AB}, \\ A_{n-1} \cong \bigoplus_N V_{AB}(N), \quad [A_1, V_{AB}(N)] \subseteq V_{AB}(N). \end{array}$$

Here are some more explicit details for $A_2 \cong \log \mathbf{SL}(\mathbb{C}^3) \cong \mathbb{C}^8$ and $A_2^c \cong \log \mathbf{SU}(3) \cong \mathbb{R}^8$: A_2 is decomposed with respect to the three A_1 Lie subalgebras with $\{l_{\pm\omega_i}, h_{\omega_i}\}_{i=1,2,3}$ into a trivial, an adjoint, and two Pauli representations, explicitly for $\mathcal{D}_{12}$,

$$A_2 \cong \mathbb{C} \oplus \mathbb{C}^3 \oplus \mathbb{C}^2 \oplus \mathbb{C}^2 : \begin{cases} [h_{\omega_1}, h_{\omega_2}] = 0 \text{ (singlet)}, \\ [h_{\omega_1}, h_{\omega_1}] = 0, \quad [h_{\omega_1}, l_{\pm\omega_1}] = \pm 2 l_{\pm\omega_1} \text{ (triplet)}, \\ [h_{\omega_1}, l_{\pm\omega_2}] = \mp l_{\pm\omega_2} \text{(doublet)}, \\ [h_{\omega_1}, l_{\pm\omega_3}] = \pm l_{\pm\omega_3} \text{(doublet)}. \end{cases}$$

A Cartan subalgebra $H \cong \mathbb{C}^2$ of A_2 can be spanned by two regular elements (Euclidean basis),

$$h_{\omega_1} = -2il^3, \quad h^8 = \sqrt{3}(h_{\omega_2} + h_{\omega_3}) = -6il^8,$$

with the adjoint action

$$\begin{array}{ll} [h_{\omega_1}, l_{\pm\omega_1}] = \pm\langle \omega_1, h_{\omega_1}\rangle l_{\pm\omega_1} = \pm 2 l_{\pm\omega_1}, & [h^8, l_{\pm\omega_1}] = \pm\langle \omega_1, h^8\rangle l_{\pm\omega_1} = 0, \\ [h_{\omega_1}, l_{\pm\omega_2}] = \pm\langle \omega_2, h_{\omega_1}\rangle l_{\pm\omega_2} = \mp l_{\pm\omega_2}, & [h^8, l_{\pm\omega_2}] = \pm\langle \omega_2, h^8\rangle l_{\pm\omega_2} = \pm\sqrt{3} l_{\pm\omega_2}, \\ [h_{\omega_1}, l_{\pm\omega_3}] = \pm\langle \omega_3, h_{\omega_1}\rangle l_{\pm\omega_2} = \pm l_{\pm\omega_3}, & [h^8, l_{\pm\omega_3}] = \pm\langle \omega_3, h^8\rangle l_{\pm\omega_3} = \pm\sqrt{3} l_{\pm\omega_3}. \end{array}$$

Hence one has six roots $\{\pm\omega_j\}_{j=1}^3 \in \mathbb{R}^2$ with two Euclidean components as (h_{ω_1}, h^8)-eigenvalues,

$$\begin{array}{rl} \text{in } H_{\mathbb{R}}^T: & \omega_1 = (2,0), \quad \omega_2 = (-1, \sqrt{3}), \quad \omega_3 = (1, \sqrt{3}), \quad \omega_1 + \omega_2 = \omega_3, \\ \text{in } H: & h_{\omega_1} = \begin{pmatrix} 1 \\ 0 \end{pmatrix}, \quad h_{\omega_2} = \begin{pmatrix} -\frac{1}{2} \\ \frac{\sqrt{3}}{2} \end{pmatrix}, \quad h_{\omega_3} = \begin{pmatrix} \frac{1}{2} \\ \frac{\sqrt{3}}{2} \end{pmatrix}, \quad h^8 = \begin{pmatrix} 0 \\ 1 \end{pmatrix}. \end{array}$$

The Cartan eigenvalue matrix is symmetrical:

$$\begin{array}{l} [h_{\omega_j}, l_{\omega_i}] = \omega_i^j l_{\omega_i} \text{ (no summation over } i), \\ \omega_i^j = \langle \omega_i, h_{\omega_j}\rangle = \begin{pmatrix} 2 & -1 & 1 \\ -1 & 2 & 1 \\ 1 & 1 & 2 \end{pmatrix}. \end{array}$$

The Killing form on the Cartan subalgebra H is

$$\begin{aligned}
&\kappa(h_{\omega_1}, h_{\omega_1}) = 2[\langle\omega_1, h_{\omega_1}\rangle^2 + \langle\omega_2, h_{\omega_1}\rangle^2 + \langle\omega_3, h_{\omega_1}\rangle^2] = 12,\\
&\kappa(h^8, h^8) = 108, \quad \kappa(h_{\omega_1}, h^8) = 0\\
&\Rightarrow \kappa(h_{\omega_i}, h_{\omega_j}) = \kappa^{ij} = 6\begin{pmatrix} 2 & -1 & 1\\ -1 & 2 & 1\\ 1 & 1 & 2\end{pmatrix}.
\end{aligned}$$

One obtains with the root chain $\omega_1 + \omega_2 = \omega_3$,

$$[l_{\omega_1}, l_{\omega_2}] = l_{\omega_3} \Rightarrow N_{\omega_1,\omega_2} = 1.$$

5.6 Roots and Weights

As mentioned above, roots $\{\omega\}$ as adjoint weights of diagonal elements $\{h_\omega\}$ of semisimple complex Lie algebras have an *"integer-valued" reflection symmetry*, induced by their spin structure from $A_1 \cong \log \mathbf{SO}(\mathbb{C}^3)$ with one root pair and "angular momenta" $\{\pm 1\}$. A simple complex Lie algebra has a compact form whose definite Killing form endows the weight space with an Euclidean structure.

Semisimple Lie algebras can be classified by their reflection symmetries and Euclidean structure for root systems which are investigated in their own right.

5.6.1 Root Systems with Reflections

An $(r-1)$-dimensional reflection $\mathbf{S}_\omega \in \mathbf{GL}(\Gamma)$ of a vector space $\Gamma \cong \mathbb{R}^r$ is an involutive linear mapping, $\mathbf{S}_\omega \circ \mathbf{S}_\omega = \mathrm{id}_\Gamma$ (order 2), which leaves invariant a hyperplane $\Gamma_{\mathrm{inv}}(\mathbf{S}_\omega) \cong \mathbb{R}^{r-1}$, its *mirror*. Each hyperplane is defined by a nontrivial form $h_\omega \in \Gamma^T$,

$$\Gamma_{\mathrm{inv}}(\mathbf{S}_\omega) = \{w \in \Gamma \,\big|\, \langle w, h_\omega\rangle = 0\} = \mathrm{kern}\, h_w.$$

Therefore each reflection is characterizable by a nontrivial vector $\omega \in \Gamma$,

$$\begin{aligned}
\mathbf{S}_\omega &= \mathrm{id}_\Gamma - \omega\otimes h_\omega, \quad \langle\omega, h_\omega\rangle = 2, \begin{cases} \text{projector: } \mathcal{P}_{||}(\omega) = \frac{\omega\otimes h_\omega}{2},\\ \mathrm{id}_\Gamma = (\mathrm{id}_\Gamma - \mathcal{P}_{||}(\omega)) + \mathcal{P}_{||}(\omega),\\ \mathbf{S}_\omega = (\mathrm{id}_\Gamma - \mathcal{P}_{||}(\omega)) - \mathcal{P}_{||}(\omega),\end{cases}\\
\Gamma &\ni w \leftrightarrow \mathbf{S}_\omega(w) = w - \langle w, h_\omega\rangle\omega;
\end{aligned}$$

here $\mathbf{S}_\omega$ maps ω in its negative, $\mathbf{S}_\omega(\omega) = -\omega$, and therefore is called the *reflection of the vector space* Γ *along* ω. Taking ω and h_ω as elements of dual bases, one has

$$\omega \cong (2, 0, \dots, 0), \quad h_\omega \cong \begin{pmatrix}1\\0\\\dots\\0\end{pmatrix}, \quad \mathbf{S}_\omega \cong \begin{pmatrix} -1 & 0 & \dots & 0\\ 0 & 1 & \dots & 0\\ & \dots & \dots & \\ 0 & 0 & \dots & 1\end{pmatrix}.$$

A *root system R of rank r* is a set of vectors of a real space $\Gamma \cong \mathbb{R}^r$ with an "integer-valued" reflection symmetry:

(1) finite generating system $R \subset \Gamma$, $0 \notin R$;
(2) for each root $\omega \in R$ there exist a linear form (diagonal operator) in the dual space $h_\omega \in \Gamma^T$ and hence a reflection $\mathsf{S}_\omega = \mathrm{id}_\Gamma - \omega \otimes h_\omega$:
$$\Gamma \overset{\mathsf{S}_\omega}{\leftrightarrow} \Gamma, \quad \begin{cases} \mathsf{S}_\omega(w) = w - \langle w, h_\omega \rangle \omega, \\ \mathsf{S}_\omega(\omega) = -\omega, \quad \langle \omega, h_\omega \rangle = 2, \end{cases}$$
$\mathsf{S}_\omega[R] \subseteq R$;
(3) with integer-valued reflection matrix (Cartan eigenvalue matrix)
$$R \times R \longrightarrow \mathbb{Z} : \ (\omega, \theta) \longmapsto \langle \omega, h_\theta \rangle, \quad \langle \omega, \omega \rangle = 2.$$

There exists in Γ at most one reflection along $\gamma \in \Gamma$ that leaves invariant a finite generating system.

The existing linear forms (diagonal operators) h_ω associated with the roots ω constitute a root system in Γ^T with the contragredient dual reflections $\mathsf{S}_{h_\omega} = \mathsf{S}_\omega^{-1T}$, called the *inverse root system* $\check{R}$. The vectors h_ω are linear in the roots:

$$\begin{aligned} &R \longrightarrow \check{R}, \quad \omega \longmapsto h_\omega, \\ &\text{if } \omega, \theta, \omega + \theta, \alpha\omega \in R \Rightarrow h_{\omega+\theta} = h_\omega + h_\theta, \quad h_{\alpha\omega} = \alpha h_\omega. \end{aligned}$$

With $R = -R$, a root system (with cardinality $d - r$) consists of pairs $\pm\omega$ with reflected roots
$$d - r \in 2\mathbb{N}.$$

In a *reduced root system* $\pm\omega$ are the only ω-proportional roots.

In a reduced root system of a semisimple Lie algebra $L \cong \mathbb{C}^d$ with Cartan subalgebra $H \cong \mathbb{C}^r$ and weight space $\Gamma = H_\mathbb{R}^T$, the $(d-r)$ roots $\pm\omega \in R^H \subseteq H^T$ represent the canonical eigenoperators $l_{\pm\omega}$; the $\frac{d-r}{2}$ vectors $h_\omega \in H$ as the canonical diagonal operators represent the inverse roots.

If it is possible to decompose with the weights $\Gamma = \bigoplus_n \Gamma_n$ the root system $R = \biguplus_n R_n$ nontrivially into subsystems, R is called *decomposable*, otherwise *nondecomposable*.

The kernels of the linear forms $h_\omega \in \Gamma^T$ (*mirrors*) decompose the weight space Γ into connection components, the *Weyl chambers* Γ_W:

$$\Gamma \setminus \bigcup_{\omega \in R} \mathrm{kern}\, h_\omega = \biguplus \Gamma_W.$$

They are open cones.

The *Weyl group of the root system* $\mathrm{Weyl}(R)$ is the reflection generated group
$$\mathrm{Weyl}(R) = \{\mathsf{S}_{\omega_1} \circ \cdots \circ \mathsf{S}_{\omega_k}\}.$$

With $\mathsf{S}_\omega[R] = R$ each reflection is a root permutation; therefore the Weyl group is a subgroup of the permutations group $\mathbf{G}(d-r)$ generated by $\frac{d-r}{2}$

permutations of order 2. In the Weyl group there exists exactly one *inversion of all roots*:

$$I_R \in \text{Weyl}(R), \quad I_R(\omega) = -\omega, \quad \omega \in R.$$

5.6.2 Fundamental Roots and Weights

A subset of the roots $B(R) \subseteq R \subseteq \Gamma$ that is linearly independent in the weight space Γ is called a *basis of the root system* R if all roots are linearly combinable with *either positive integers or negative integers*:

$$\begin{aligned} B(R) &= \{\omega_j \mid j = 1, \dots r\}, \quad R = R_+(B) \uplus R_-(B), \\ R_+(B) &= -R_-(B) = \{\sum_{j=1}^{r} n^j \omega_j \mid n^j \in \mathbb{N}_0\} \text{ contains } \tfrac{d-r}{2} \text{ roots.} \end{aligned}$$

Each root system of rank r has a basis with r *fundamental roots*. Different bases are related to each other by a Weyl transformation. The reflections of a basis $\{\mathrm{S}_{\omega_j}\}_{j=1}^{r}$ generate the Weyl group as permutation subgroup $\text{Weyl}(R) \subseteq \mathbf{G}(d-r)$.

The *diagonal operator basis* $B(\check{R}) = \{h_{\omega_j}\}_{j=1}^{r}$ associated to a root basis are the *fundamental diagonal operators*. Their kernels define the mirrors of the *fundamental Weyl chamber* $\Gamma_W(B)$ associated with the fundamental roots $B(R)$.

For an illustration the example $A_2 \cong \log \mathbf{SL}(\mathbb{C}^3)$ with six roots

$$\begin{aligned} R = \{\pm\omega_j \mid j = 1,2,3\}, &\quad B(R) = \{\omega_1 = (2,0), \omega_2 = (-1, \sqrt{3})\}, \\ &\quad \omega_1 + \omega_2 = \omega_3, \\ \check{R} = \{\pm h_{\omega_j} \mid j = 1,2,3\}, &\quad B(\check{R}) = \{h_{\omega_1} = \begin{pmatrix} 1 \\ 0 \end{pmatrix}, \ h_{\omega_2} = \begin{pmatrix} -\frac{1}{2} \\ \frac{\sqrt{3}}{2} \end{pmatrix}\}, \\ &\quad h_{\omega_1} + h_{\omega_2} = h_{\omega_3}. \end{aligned}$$

With respect to the definite combinations from a basis, the roots have a *partial order*

$$\omega_1 \succeq \omega_2 \iff \omega = \omega_1 - \omega_2 \succeq 0; \quad \omega \succeq 0 \iff \omega \in R_+(B)$$

with a *maximal root* $\omega_{\max}$.

Basis roots of a Lie algebra give a basis for the Cartan subalgebra H. From the A_1-triplet relations $[l_{-\omega_j}, l_{\omega_j}] = h_{\omega_j}$ for the fundamental roots, $j = 1, \dots, r$, one obtains triplet relations for all roots:

$$[l_{-\omega}, l_\omega] = h_\omega = \sum_{j=1}^{r} n^j h_{\omega_j} \text{ if } \omega = \sum_{j=1}^{r} n^j \omega_j \in R^H.$$

The eigenoperators $l_{\pm\omega}$ for a positive root $\omega \succeq 0$ are called the corresponding *raising and lowering operators*. The Lie algebra is the direct sum of diagonal, raising, and lowering Lie subalgebras:

$$\begin{aligned} L &= H \oplus L_+(H) \oplus L_-(H), \\ L_\pm(H) &= \bigoplus_{\omega \in R_\pm(B)} L_\omega(H) \cong \mathbb{C}^{\frac{d-r}{2}} \in \underline{\mathbf{lag}}_{\mathbb{C}}. \end{aligned}$$

The r fundamental diagonal operators $\{h_{\omega_j}\}_{j=1}^r \subseteq \Gamma^T$ as a basis $B(\check{R})$ of the inverse root system are not the dual basis for the fundamental roots $B(R)$. The Γ-basis dual to the basis $B(\check{R}) \subset \Gamma^T$ with the fundamental diagonal operators, for a Lie algebra a basis of the linear forms H^T of a Cartan subalgebra, is called the associated *weight basis* $B(\Gamma) \subseteq \Gamma$ with the r *fundamental weights.*

Altogether one has three important bases, two for the weight space Γ,

$$\begin{array}{rll} \text{fundamental roots:} & B(R) & = \{\omega_j \,\big|\, j = 1, \ldots, r\} \subset \Gamma, \\ \text{fundamental diagonal operators:} & B(\check{R}) & = \{h_{\omega_j} \,\big|\, j = 1, \ldots, r\} \subset \Gamma^T, \\ \text{fundamental weights:} & B(\Gamma) & = \{\gamma_j \,\big|\, j = 1, \ldots, r\} \subset \Gamma, \end{array}$$

with the properties

$$\begin{array}{rl} \text{mirror-inverse bases:} & \langle \omega_j, h_{\omega_j} \rangle = \omega_j^j = 2 \text{ (no summation over } j), \\ \text{dual bases:} & \langle \gamma_i, h_{\omega_j} \rangle = \delta_i^j. \end{array}$$

The fundamental weight γ_j is a vector in all mirrors (walls of the fundamental Weyl chamber $\Gamma_W(B)$) with the exeption of the ω_j-mirror

$$\gamma_j \in \bigcap_{k=1, k \neq j}^{r} \operatorname{kern} h_{\omega_k}.$$

The sum ρ of all r fundamental weights is half the sum of the $\frac{d-r}{2}$ positive roots in a reduced root system:

$$\begin{array}{c} \rho = \sum_{j=1}^{r} \gamma_j = \frac{1}{2} \sum_{\omega \in R_+(B)} \omega, \\ \text{for } A_2: \quad \rho = \gamma_1 + \gamma_2 = (1, \sqrt{3}), \quad \omega_1 + \omega_2 + \omega_3 = (2, 2\sqrt{3}). \end{array}$$

A reflection along a fundamental root changes the weight sum ρ by this root:

$$\langle \rho, h_{\omega_i} \rangle = 1 \Rightarrow \mathrm{S}_{\omega_i}(\rho) = \rho - \langle \rho, h_{\omega_i} \rangle \omega_i = \rho - \omega_i, \quad i = 1, \ldots, r.$$

Different fundamental roots of a semisimple Lie algebra give the following relations for the eigenoperators:

$$\omega_{i,j} \in B(R^H), \quad \omega_i \neq \omega_j \quad \left\{ \begin{array}{rl} \langle \omega_i, h_{\omega_j} \rangle = \omega_i^j & < 0, \\ [l_{\omega_i}, l_{-\omega_j}] & = 0, \\ (\operatorname{ad} l_{\pm\omega_j})^{1 - \langle \omega_i, h_{\omega_j} \rangle} l_{\pm\omega_i} & = 0. \end{array} \right.$$

Therefore the *basis restricted* $(r \times r)$ *Cartan eigenvalue matrix* has 2 as diagonal elements and negative integers off the diagonals:

$$(\omega_i^j)_{i,j=1}^r = \begin{pmatrix} 2 & -|\omega_1^2| & -|\omega_1^3| & \ldots \\ -|\omega_2^1| & 2 & -|\omega_2^3| & \ldots \\ -|\omega_3^1| & -|\omega_3^2| & 2 & \ldots \\ \ldots & & \ldots & \end{pmatrix}, \quad |\omega_i^j| \in \mathbb{N}_0.$$

With fundamental diagonal operators and weights as dual bases $\mathrm{id}_W = \sum_{j=1}^{r} \gamma_j \otimes h_{\omega_j}$ the $(r \times r)$ Cartan matrix gives a *bijection between fundamental roots and weights*:

$$B(\Gamma) \cong B(R), \quad \begin{cases} \omega_i = \mathrm{id}_W(\omega_i) = \langle \omega_i, h_{\omega_j} \rangle \gamma_j = \omega_i^j \gamma_j, \\ \gamma_i = (\omega^{-1})_j^i \omega_j, \\ (\omega_i^j)_{i,j=1}^r \in \mathbb{Z} \otimes \mathbb{Z}, \quad (\omega^{-1})_j^i \in \mathbb{Q} \otimes \mathbb{Q}. \end{cases}$$

The fundamental roots (weights) are integer (rational) linear combinations of the fundamental weights (roots).

5.6.3 Weight Modules and Weight Cones

In the real weight vector space $\Gamma \cong \mathbb{R}^r$ the positive linear combinations of the fundamental weights span the closed fundamental Weyl chamber with the mirrors for the fundamental roots as walls:

$$\text{fundamental Weyl chamber: } \overline{\Gamma_W(B)} = \{\sum_{j=1}^{r} \alpha^j \gamma_j \mid \alpha^j \geq 0\}.$$

The integer linear combinations of the fundamental weights constitute a $\mathbb{Z}$-module, the

$$\textit{weight module: } \Gamma_{\mathbb{Z}} = \{w = \sum_{j=1}^{r} z^j \gamma_j \mid z^j \in \mathbb{Z}\} \cong \mathbb{Z}^r.$$

The integer coefficients for a *weight* are called *canonical coordinates* - they are the *winding numbers* as eigenvalues of compact Lie algebra representations

$$w \in \Gamma_{\mathbb{Z}} \Rightarrow w = [z^1, \ldots, z^r].$$

The winding number coordinates of the roots are the rows of the $(r \times r)$ Cartan eigenvalue matrix, reduced to the fundamental root basis:

$$\begin{aligned} \omega_i = \omega_i^j \gamma_j \quad &= [\omega_i^1, \ldots, \omega_i^r], \\ \text{e.g., for } A_2 : \begin{pmatrix} \omega_1 \\ \omega_2 \end{pmatrix} &= \begin{pmatrix} [2, -1] \\ [-1, 2] \end{pmatrix}. \end{aligned}$$

The weights with natural numbers are called *dominant weights*, they constitute the

$$\textit{weight cone: } \Gamma_{\mathbb{N}_0} = \{\sum_{j=1}^{r} n^j \gamma_j \mid n^j \in \mathbb{N}_0\} \cong \mathbb{N}_0^r.$$

as subset of the weight module and the Weyl chamber.

The Weyl group transformations leave the weights stable and decompose the weight module into orbits with exactly one dominant weight:

$$\begin{aligned}
&\mathrm{Weyl}(R) \times \Gamma_{\mathbb{Z}} \longrightarrow \Gamma_{\mathbb{Z}}, \quad \mathtt{S}_\omega(w) = w - \langle w, h_\omega \rangle \omega, \\
&\Gamma_{\mathbb{Z}} \times R \longrightarrow \mathbb{Z}, \quad \langle w, h_\omega \rangle \in \mathbb{Z}, \\
&\Gamma_{\mathbb{Z}} = \biguplus_{w_{\mathrm{dom}}} \mathrm{Weyl}(R) \bullet w_{\mathrm{dom}}, \quad w_{\mathrm{dom}} \in \Gamma_{\mathbb{N}_0}.
\end{aligned}$$

The roots R are a weight orbit with the maximal root as dominant weight:

$$R = \mathrm{Weyl}(R) \bullet \omega_{\max} \subset \Gamma_{\mathbb{Z}}.$$

The weight structures for a reduced root system

$\Gamma \cong \mathbb{R}^r$	$\supset$	$\overline{\Gamma_W(B)}$		
$\cup$		$\cup$		
$\Gamma_{\mathbb{Z}} \cong \mathbb{Z}^r$ ($\mathbb{Z}$-module)	$\supset$	$\Gamma_{\mathbb{N}_0} \cong \mathbb{N}_0^r$ ($\mathbb{N}_0$-cone)	$\supset$	$B(\Gamma) = \{\gamma_j\}_{j=1}^r$ ($\mathbb{Z}$-basis)

reduced root system $R \subset \Gamma$

are translated for a semisimple complex Lie algebra into the weights of a Cartan subalgebra in all irreducible representations with the weight basis corresponding to the fundamental representations. The weight cone characterizes the irreducible representations (chapter "Rational Quantum Numbers")

$H_{\mathbb{R}}^T$ (weight space)	$\supset$	$\overline{\Gamma_W(B)}$ (fundamental Weyl chamber)		
$\cup$		$\cup$		
weights L (weight module)	$\supset$	**irrep** L (irreducible representations)	$\supset$	**funrep** L (fundamental representations)

Cartan subalgebra $H \subset L$ **(semisimple)**

5.6.4 Euclidean Structure for Weights

The diagonal operators of a root system give a *canonical scalar product* for the weight space $\Gamma \cong \mathbb{R}^r$, invariant under the root permutation group $\mathbf{G}(R)$:

$$\begin{aligned}
&\langle\ |\ \rangle : \quad \Gamma \times \Gamma \longrightarrow \mathbb{R}, \quad \langle w | w' \rangle = \sum_{\omega \in R} \langle w, h_\omega \rangle \langle w', h_\omega \rangle, \\
&\langle\ |\ \rangle = \sum_{\omega \in R} h_\omega \otimes h_\omega, \quad \langle w | w \rangle > 0 \iff w \neq 0, \\
&\langle g(w) | g(w') \rangle = \langle w | w' \rangle, \quad g \in \mathbf{G}(R), \\
&\omega, \theta \in R \Rightarrow \langle \omega | \theta \rangle \in \mathbb{Z}, \quad \langle \omega | \omega \rangle \in \mathbb{N}.
\end{aligned}$$

The summation goes over all roots, not only over the fundamental roots. An irreducible reduced root system has, up to a scalar, a unique invariant scalar product $\langle\ |\ \rangle$. For a Lie algebra with Cartan subalgebra H, it is given by the inverse Killing form κ^{-1} restricted to $H_{\mathbb{R}}^T \times H_{\mathbb{R}}^T$.

With a root-reflection-invariant scalar product $\langle\ |\ \rangle$ on the weight space Γ and the associated isomorphism between Γ and Γ^T, a diagonal operator $h_\omega \in \Gamma^T$ is paired with a unique vector from the weight space $\omega' \in \Gamma$:

$$\begin{aligned} \Gamma^T \longrightarrow \Gamma,\quad & h_\omega \longmapsto \omega' \text{ with } \langle w, h_\omega\rangle = \langle w|\omega'\rangle;\\ \mathrm{S}_\omega \in \mathbf{G}(R),\ \text{since } \langle w|\omega\rangle &= \langle \mathrm{S}_\omega(w)|\mathrm{S}_\omega(\omega)\rangle \text{ with } \begin{cases} \mathrm{S}_\omega(w) &= w - \langle w, h_\omega\rangle\omega \\ \mathrm{S}_\omega(\omega) &= -\omega \end{cases}\\ \Rightarrow \langle w|\omega'\rangle &= 2\tfrac{\langle w|\omega\rangle}{\langle\omega|\omega\rangle} \Rightarrow \omega' = 2\tfrac{\omega}{\langle\omega|\omega\rangle}. \end{aligned}$$

Hence one obtains the following: The fundamental weight γ_j is orthogonal to all fundamental roots ω_i for $i \neq j$; its projection on "its" root ω_j is half the rooot $\frac{1}{2}\omega_j$:

$$\begin{aligned} &\delta_i^j = \langle\gamma_j, h_{\omega_i}\rangle = \langle\gamma_j|\omega_i'\rangle = 2\tfrac{\langle\gamma_j|\omega_i\rangle}{\langle\omega_i|\omega_i\rangle},\quad i,j = 1,\dots,r;\\ &\omega \in R,\quad w \in \Gamma_{\mathbb{Z}}:\quad \langle w, h_\omega\rangle = 2\tfrac{\langle w|\omega\rangle}{\langle\omega|\omega\rangle} \in \mathbb{Z}. \end{aligned}$$

The connection of the integer-valued Cartan eigenvalue matrix for the roots with the definite scalar product restricts strongly the *length ratio of and the angle between two roots*:

$$\langle\theta, h_\omega\rangle = 2\tfrac{\langle\theta|\omega\rangle}{\langle\omega|\omega\rangle} \Rightarrow \begin{cases} \langle\omega, h_\theta\rangle = 0 \iff \langle\theta, h_\omega\rangle = 0 \iff \langle\omega|\theta\rangle = 0,\\ \langle\omega|\theta\rangle \neq 0: \quad \frac{\langle\theta, h_\omega\rangle}{\langle\omega, h_\theta\rangle} = \frac{\langle\theta|\theta\rangle}{\langle\omega|\omega\rangle} = \frac{\|\theta\|^2}{\|\omega\|^2},\\ \omega,\theta \neq 0: \quad \begin{cases} \langle\omega, h_\theta\rangle\langle\theta, h_\omega\rangle = 4\frac{\langle\omega|\theta\rangle^2}{\langle\omega|\omega\rangle\langle\theta|\theta\rangle} = 4\cos^2\Phi(\omega,\theta),\\ 4\cos^2\Phi(\omega,\theta) \in \mathbb{Z}. \end{cases} \end{cases}$$

The integer-valuedness allows only the following possibilities:

$4\cos^2\Phi(\omega,\theta)$	$= \langle\omega, h_\theta\rangle\langle\theta, h_\omega\rangle$	$\Rightarrow \Phi(\omega,\theta)$	$\frac{\|\theta\|}{\|\omega\|}$	$(\mathrm{S}_\omega \circ \mathrm{S}_\theta)^k = \mathrm{id}_W$
0	$= 0\cdot 0$	$\Rightarrow \frac{\pi}{2}$	—	$k = 2$
1	$= 1\cdot 1$ $= (-1)\cdot(-1)$	$\Rightarrow \frac{\pi}{3}$ $\Rightarrow \frac{2\pi}{3}$	1	$k = 3$
2	$= 1\cdot 2$ $= (-1)\cdot(-2)$	$\Rightarrow \frac{\pi}{4}$ $\Rightarrow \frac{3\pi}{4}$	$\sqrt{2}$	$k = 4$
3	$= 1\cdot 3$ $= (-1)\cdot(-3)$	$\Rightarrow \frac{\pi}{6}$ $\Rightarrow \frac{5\pi}{6}$	$\sqrt{3}$	$k = 6$
4	$= 1\cdot 4$ $= (-1)\cdot(-4)$ $= 2\cdot 2$ $= (-2)\cdot(-2)$	$\Rightarrow 0$ $\Rightarrow \pi$ $\Rightarrow 0$ $\Rightarrow \pi$	$\theta = \omega$ $\theta = -\omega$ $\theta = 2\omega$ $\theta = -2\omega$	–

For a cosine with modulus 1 the roots are proportional to each other, for a reduced root system this situation is uninteresting.

Between different fundamental roots of a reduced root system *all angles are obtuse*:

$$0 \geq \omega_i^j = \langle\omega_i, h_{\omega_j}\rangle = 2\tfrac{\langle\omega_i|\omega_j\rangle}{\|\omega_i\|^2},\quad i,j = 1,\dots,r.$$

Hence the nondiagonal elements in a reduced Cartan matrix ω_i^j have to be negative, $\{0,-1,-2,-3\}$, with an angle $\Phi = \frac{\pi}{2}, \frac{2\pi}{3}, \frac{3\pi}{4}, \frac{5\pi}{6}$ between the roots, i.e., $90, 120, 135, 150$ degrees. For those angles an $A_1(\omega_i)$-triplet is connected to a singlet, doublet, triplet and quartet. For the last three cases the length ratios for the two fundamental roots are given by $1, \sqrt{2}, \sqrt{3}$:

$$\begin{array}{l} B(R) = \{\omega_j \mid j = 1, \dots, r\}, \\ \langle \omega_i, h_{\omega_j} \rangle = \omega_i^j, \\ n_i^j = \sqrt{\frac{\omega_i^j}{\omega_j^i}} = \frac{\|\omega_i\|}{\|\omega_j\|}, \\ \Phi_{ij} = \Phi(\omega_i, \omega_j), \\ 4\cos^2 \Phi_{ij} = \omega_i^j \omega_j^i \\ \text{(no summations)} \end{array} \quad \left\{ \begin{array}{l} \omega_i^i = 2, \text{ triplet}, \\ i \neq j: \ \omega_i^j = 0 = \omega_j^i, \ \Phi_{ij} = \frac{\pi}{2}, \text{ singlet}, \\ i \neq j: \ \omega_j^i = -1 \ \left\{ \begin{array}{l} \omega_i^j = -1, -2, -3, \\ \Phi_{ij} = \frac{2\pi}{3}, \frac{3\pi}{4}, \frac{5\pi}{6}, \\ n_i^j = 1, \text{ doublet}, \\ n_i^j = \sqrt{2}, \text{ triplet}, \\ n_i^j = \sqrt{3}, \text{ quartet}. \end{array} \right. \end{array} \right.$$

These three possibilities are related to three isosceles triangles, the *Platonic triangles*, which have two angles of 60, 45, or 30 degrees, and the corresponding length ratio of basis to one of the equal sides:

$$4\cos^2 \Phi_{ij} = 1, 2, 3: \left\{ \begin{array}{rcl} \text{one angle:} & 2\Phi_{ij} - \pi & = \frac{\pi}{3}, \frac{\pi}{2}, \frac{2\pi}{3}, \\ \text{two angles:} & \pi - \Phi_{ij} & = \frac{\pi}{3}, \frac{\pi}{4}, \frac{\pi}{6}, \\ \text{length ratio:} & \frac{\|\omega_i\|^2}{\|\omega_j\|^2} & = 1, 2, 3. \end{array} \right.$$

An $(r \times r)$ Cartan eigenvalue matrix (ω_j^i) with the integer-valued canonical (winding number) coordinates contains all information to draw a *root diagram* in a Euclidean space $\mathbb{R}^r$ with *Euclidean coordinates* (examples below): One starts with the first fundamental root ω_1, which is normalized to 2. It defines the first axis. Then one uses the length ratio n_2^1 and the obtuse angle Φ_{21} to draw the second root in the $(1,2)$-plane. This determines its two Euclidean coordinates, etc. In general, the coordinates are chosen in such a way that for the kth fundamental root only the first k components can be nontrivial:

$$\begin{pmatrix} \omega_1 \\ \omega_2 \\ \dots \\ \omega_j \\ \dots \\ \omega_r \end{pmatrix} = \underbrace{\begin{pmatrix} [2 & \omega_1^2 & \dots & \dots & \dots & \omega_1^r] \\ [\omega_2^1 & 2 & \dots & \dots & \dots & \omega_2^r] \\ & \dots & \dots & & & \\ [\omega_j^1 & \dots & \omega_j^{j-1} & 2 & \dots & \omega_j^r] \\ & \dots & \dots & \dots & & \\ [\omega_r^1 & \dots & \dots & & \omega_r^{r-1} & 2] \end{pmatrix}}_{\text{winding numbers}} \cong \underbrace{\begin{pmatrix} 2 & 0 & \dots & & \dots & 0 \\ \omega_2^1 & \alpha_2^2 & 0 & & \dots & 0 \\ & \dots & \dots & & & \\ \omega_j^1 & \dots & \alpha_j^j & 0 & \dots & 0 \\ & \dots & \dots & \dots & & \\ \omega_r^1 & \alpha_r^2 & & \dots & & \alpha_r^r \end{pmatrix}}_{\text{Euclidean coordinates}},$$

$$\begin{array}{l} j \neq 1: \ \omega_j^1 \in \{0, -1, -2, -3\}, \ \alpha_j^j \neq 0, \\ \text{e.g., for } A_2: \ \begin{pmatrix} \omega_1 \\ \omega_2 \end{pmatrix} = \begin{pmatrix} [2, -1] \\ [-1, 2] \end{pmatrix} = \begin{pmatrix} 2 & 0 \\ -1 & \sqrt{3} \end{pmatrix}. \end{array}$$

Only the first Euclidean coordinate α_j^1 coincides with the integer h_{ω_1} eigenvalue. It gives the multiplicity for an $A_1(\omega_1)$-decomposition $l_{\pm\omega_j}$: singlet, doublet, triplet or quartet.

With the orthogonal mirrors for the r fundamental roots one obtains the remaining roots by reflections, e.g for A_2 the roots $-\omega_{1,2}$ and the root pair $\pm\omega_3 = \pm(\omega_1 + \omega_2)$. With all roots one has the canonical generating triplets with canonical Lie bracket and also the Killing form.

The fundamental weights $\{\gamma_j\}_{j=1}^r$ are obtained by linear combination of the fundamental roots with rational coeffiicents from the inverse Cartan eigenvalue matrix (examples below):

$$\omega_i = \sum_{j=1}^r \omega_i^j \gamma_j, \quad \gamma_j = \sum_{i=1}^r (\omega^{-1})_j^i \omega_i.$$

The invariant scalar product for the fundamental weights is up to a common factor the inverse Killing form of the linear forms H^T of the Cartan subalgebra. It can be obtained also by renormalization of the columns in the inverse Cartan matrix:

$$\kappa_{ij} \sim 2\langle\gamma_i|\gamma_j\rangle = 2\sum_{l=1}^{r}(\omega^{-1})^l_j\langle\gamma_i|\omega_l\rangle = (\omega^{-1})^i_j \parallel \omega_i \parallel^2$$

(no summation over i).

The free factor for κ_{ij} is obtained with the normalization of the diagonal elements (no i-summation) in the inverted matrix for the Killing form of the diagonal elements (j-summation goes over all roots)

$$\kappa^{ii} = \kappa(h_{\omega_i}, h_{\omega_i}) = \sum_{j=1}^{d-r}[\omega_j(h_{\omega_i})]^2 = 2\sum_{j=1}^{\frac{d-r}{2}}[\omega^i_j]^2.$$

5.7 Classification of Complex Simple Lie Algebras and Dynkin Diagrams

For each reduced root system R_r there is a exactly one rank r semisimple Lie algebra ${}_dL_r \cong \mathbb{C}^d$, which is simple for a nondecomposable reduced root system.

A simple complex Lie algebra ${}_dL_r$ is characterized by a (topologically) connected *Dynkin diagram* which is a picture for the $(r \times r)$ Cartan eigenvalue matrix for the basis $B(R)$ of a reduced root system: The diagram consists of r *vertices* which have *connections* by $0, 1, 2, 3$ lines. Over each vertex there is written a natural number $1, 2, 3$.

A Dynkin diagram with r vertices gives the construction prescription for a root diagram of a Lie algebra L_r with rank r in the Euclidean weight space $\mathbb{R}^r$. Each vertex i is associated to a fundamental root ω_i. The connection of two vertices (i, j) gives the angle Φ_{ij} between these roots via the number of the connecting lines with $0, 1, 2, 3 = 4\cos^2\Phi_{ij} = \omega^j_i\omega^i_j$ (no summation). The number over the vertex gives the squared root length $\parallel \omega_i \parallel^2$.

The classification of simple Lie algebras has much to do with elementary Euclidean geometry: For the rank two Lie algebras with the root diagrams in the Euclidean plane, take one of the three isosceles Platonic triangles with basic angle $\Phi = 60°, 45°, 30°$,

$$4\cos^2\Phi = 1, 2, 3 \Rightarrow \Phi = \begin{cases} \frac{\pi}{3}, & \text{regular triangle,} \\ \frac{\pi}{4}, & \text{half square triangle,} \\ \frac{\pi}{6}, & \text{centrally trisected regular triangle,} \end{cases}$$

fix one corner of the basis at the origin of a 2-dimensional Euclidean plane and reflect the triangle as long over its sides (Weyl reflections) until one has a closed polygon (here the plane filling property of square and regular triangle is relevant). With this method one obtains either a hexagon (6 points and 6 Platonic $\frac{\pi}{3}$-triangles) or a square, pointed in the middle of its four sides (8 points and 8 Platonic $\frac{\pi}{4}$-triangles and 4 squares), or a 12-cornered star of David (12 points and 12 Platonic $\frac{\pi}{6}$-triangles). The peripheral 6, 8, and 12 points as roots, supplemented by two points in the origin as trivial adjoint weights, represent the 8, 10 and 14 weights of the three simple Lie algebras of rank 2

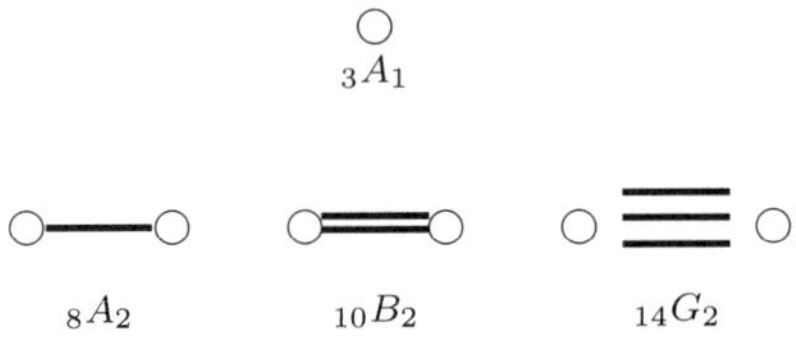

One vertex stands for the rank 1 Lie algebra $_3A_1$ with the 1-dimensional weight space. Its three weights are the two endpoints of a line as roots and the central point in the origin as trivial adjoint weight.

The root diagram at the corners of a star of David of is a peculiarity in two dimensions. With only the two Platonic $\frac{\pi}{3}$- and $\frac{\pi}{4}$-triangles the Dynkin diagrams of all simple Lie algebras of rank $r \geq 3$ can be obtained - the Dynkin diagrams of A_2 and B_2 have to be connected. The three-line connection of G_2 ($\frac{\pi}{6}$-triangle) does not occur for more than two vertices. Hence the equilateral triangle as building block for A_2 and the square (with its half, the Platonic $\frac{\pi}{4}$-triangle as building block for B_2) determine the roots of all simple Lie algebras. The subdiagrams of the simple $L_{r\geq 3}$ Dynkin diagrams with two connected vertices show the Lie subalgebra A_2 (simple line as connection) or B_2 (double line as connection). Unconnected vertices describe orthogonal roots, $\cos^2 \Phi = 0$.

The number over the vertices, giving the squared root lengths up to a common factor, is redundant, since the length ratio is given by the angle. Only for the Lie algebras B_r, C_r is the order relevant.

From the angle restrictions for the fundamental roots the following rules can be derived for the Dynkin diagrams of simple complex Lie algebras with rank r:

(1) There are not more than $(r-1)$ connections, a connection can be a line, a double line, or a triple line.

(2) There are no loops (closed connections).

(3) Maximally three lines leave one vertex. Therefore there is only one diagram G_2 with a triple line and hence only one simple Lie algebra with an A_1-quartet (spin $J = \frac{3}{2}$) in the adjoint representation.

(4) By shrinking in a Dynkin diagram a chain of simple line-connected vertices to one vertex one obtains a Dynkin diagram. Therefore a diagram can have not more than one double line or not more than one threefold branching; not, however, both.

(5) There is only one diagram F_4 with a double line not at the end.

(6) There are only three diagrams $E_{6,7,8}$ with a threefold branching not at the vertex one connection from the end.

In addition to the *five simple exceptional Lie algebras*

$$_{14}G_2,\ _{52}F_4,\ _{78}E_6,\ _{133}E_7,\ _{248}E_8$$

there exist the four *classical Lie algebra types*

$$_{r(2+r)}A_r, \quad _{r(1+2r)}C_r, \quad _{r(1+2r)}B_r, \quad _{r(2r-1)}D_r \text{ for } r \geq 1$$

with the inclusions

$$A_r \subset D_{1+r} \subset B_{1+r} \cap C_{1+r}.$$

All those Lie algebras are simple up to the semisimple $D_2 \cong A_1 \oplus A_1$ and $D_1 \cong \mathbb{C}$, which is used for the abelian Lie symmetry.

From the Dynkin diagrams one can read off all isomorphisms for simple root systems up to rank 3:

$r = 1$		$_1D_1$		$_3A_1$	$\cong$	$_3B_1$	$\cong$	$_3C_1$	
$r = 2$	$_3A_1 \oplus {_3A_1} \cong$	$_6D_2$		$_8A_2$		$_{10}B_2$	$\cong$	$_{10}C_2$	$_{14}G_2$
$r = 3$		$_{15}D_3$	$\cong$	$_{15}A_3$		$_{21}B_3$		$_{21}C_3$	
$r = 4$		$_{28}D_4$		$_{24}A_4$		$_{36}B_4$		$_{36}C_4$	$_{52}F_4$

For rank $r \geq 4$ the four main series have no isomorphisms with each other.

5.8 Simple Complex and Compact Lie Groups

Two Lie groups with isomorphic Lie algebras are called *locally isomorphic.* All connected Lie groups $\{G_i\}$, locally isomorphic to its universal cover group $\overline{\exp L}$, arise as classes with respect to the discrete normal subgroups $\{\mathbb{I}_i\}$ of the discrete center of $\overline{\exp L}$:

$$G_i \cong \overline{\exp L}/\mathbb{I}_i \text{ with discrete } \mathbb{I}_i \subseteq \mathbb{I}(G) = \text{centr}\, \overline{\exp L}.$$

For the simple Lie algebras L_r from the four main series and their compact forms L_r^c (more below) one has the Weyl groups and the universal cover groups with their centers, equal for complex groups and their real forms (below):

Lie algebra L_r L_r^c	Weyl group (permutations)	universal cover group $G = \overline{\exp L_r}$ compact form $G^c = \overline{\exp L_r^c}$	center $\mathbb{I}(G) = \mathbb{I}(G^c)$ (cyclic group)
A_r, A_r^c, $r \geq 1$	$\mathbf{G}(1+r)$	$\mathbf{SL}(\mathbb{C}^{1+r})$ $\mathbf{SU}(1+r)$	$\mathbb{I}(1+r)$
C_r, C_r^c, $r \geq 1$	$\mathbf{G}(r) \vec{\times} \mathbf{G}(2)^r$	$\mathbf{Sp}(\mathbb{C}^{2r})$ $\mathbf{SpU}(2r)$	$\mathbb{I}(2)$
B_r, B_r^c, $r \geq 1$	$\mathbf{G}(r) \vec{\times} \mathbf{G}(2)^r$	$\overline{\mathbf{SO}(\mathbb{C}^{1+2r})}$ $\overline{\mathbf{SO}(1=2r)}$	$\mathbb{I}(2)$
D_r, D_r^c, $r \geq 3$	$\mathbf{G}(r) \vec{\times} \mathbf{G}(2)^{r-1}$	$\overline{\mathbf{SO}(\mathbb{C}^{2r})}$ $\overline{\mathbf{SO}(2r)}$	$\mathbb{I}(4)$ if r odd; $\mathbb{I}(2) \times \mathbb{I}(2)$ if r even

with the inclusions

$$\mathbf{SL}(\mathbb{C}^{1+r}) \subset \overline{\mathbf{SO}(\mathbb{C}^{2+2r})} \subset \overline{\mathbf{SO}(\mathbb{C}^{2r+3})} \cap \mathbf{Sp}(\mathbb{C}^{2+2r}).$$

The universal covers of the orthogonal groups are also called *spin groups* because of their faithful spinor representations (chapter "Rational Quantum

Numbers"). The compact form of the universal cover groups are isomorphic to the definite unitary and orthogonal groups with the definitions

$$\mathbf{SU}(1+r) = \mathbf{SL}(\mathbb{C}^{1+r}) \cap \mathbf{U}(1+r), \quad \mathbf{SpU}(2r) = \mathbf{Sp}(\mathbb{C}^{2r}) \cap \mathbf{U}(2r)$$

and the corresponding inclusions and isomorphisms.

The Lie algebra isomorphisms for $r = 1, 2, 3$ have corresponding group isomorphisms that lead to the isomorphisms of the complex and compact spin groups to matrix groups only for the cases

Lie algebra	complex		group			compact		group		
$A_1 \cong B_1 \cong C_1$	$\mathbf{SL}(\mathbb{C}^2)$	$\cong$	$\overline{\mathbf{SO}(\mathbb{C}^3)}$	$\cong$	$\mathbf{Sp}(\mathbb{C}^2)$	$\mathbf{SU}(2)$	$\cong$	$\overline{\mathbf{SO}(3)}$	$\cong$	$\mathbf{SpU}(2)$
$D_2 \cong A_1 \oplus A_1$	$\overline{\mathbf{SO}(\mathbb{C}^4)}$	$\cong$	$\mathbf{SL}(\mathbb{C}^2)$	$\times$	$\mathbf{SL}(\mathbb{C}^2)$	$\overline{\mathbf{SO}(4)}$	$\cong$	$\mathbf{SU}(2)$	$\times$	$\mathbf{SU}(2)$
$B_2 \cong C_2$	$\overline{\mathbf{SO}(\mathbb{C}^5)}$	$\cong$	$\mathbf{Sp}(\mathbb{C}^4)$			$\overline{\mathbf{SO}(5)}$	$\cong$	$\mathbf{SpU}(4)$		
$D_3 \cong A_3$	$\overline{\mathbf{SO}(\mathbb{C}^6)}$	$\cong$	$\mathbf{SL}(\mathbb{C}^4)$			$\overline{\mathbf{SO}(6)}$	$\cong$	$\mathbf{SU}(4)$		

The centers in $\overline{\mathbf{SO}(\mathbb{C}^{2r})}$ are either the Klein group $\mathbb{I}(2) \times \mathbb{I}(2)$ for even r, inherited from the starting point $\overline{\mathbf{SO}(\mathbb{C}^4)} \cong \mathbf{SL}(\mathbb{C}^2) \times \mathbf{SL}(\mathbb{C}^2)$, or the cyclic $\mathbb{I}(4)$ for odd r, inherited from $\overline{\mathbf{SO}(\mathbb{C}^6)} \cong \mathbf{SL}(\mathbb{C}^4)$:

$$n \geq 3: \ \overline{\mathbf{SO}(\mathbb{C}^n)}/\mathbb{I}(2) \cong \mathbf{SO}(\mathbb{C}^n), \begin{cases} \operatorname{centr} \mathbf{SO}(\mathbb{C}^{1+2r}) & = \{\mathbf{1}_{1+2r}\} \cong \{1\}, \\ \operatorname{centr} \mathbf{SO}(\mathbb{C}^{2r})|_{r\geq 2} & = \{\pm \mathbf{1}_{2r}\} \cong \mathbb{I}(2). \end{cases}$$

The orthogonal abelian Lie symmetry with dimension 1 and rank 1 (complex, compact, noncompact) is

$$D_1 = \log \mathbf{SO}(\mathbb{C}^2), \quad D_1^c = \log \mathbf{SO}(2), \quad D_1^n = \log \mathbf{SO}_0(1,1).$$

5.9 Simple Root Systems

5.9.1 The Root Systems A_r

One has for the *root systems* A_r $(r > 1)$ associated with the Lie algebra of the special group $\mathbf{SL}(\mathbb{C}^{1+r})$ as volume-invariance group:

$$\text{Cartan matrix } (r \times r)\text{: } \omega_i^j = \begin{pmatrix} 2 & -1 & 0 & 0 & \dots & 0 & 0 \\ -1 & 2 & -1 & 0 & \dots & 0 & 0 \\ 0 & -1 & 2 & -1 & \dots & 0 & 0 \\ \dots & & & \dots & & & \\ 0 & 0 & 0 & 0 & \dots & 2 & -1 \\ 0 & 0 & 0 & 0 & \dots & -1 & 2 \end{pmatrix},$$

Dynkin diagram: 

dimension: $d = r(2+r)$, diagonal degeneracy: $\frac{d-3r}{2} = \frac{r(r-1)}{2}$,

$$\text{weight space (barycentric):} \Gamma = \{w \in \mathbb{R}^{1+r} \,\big|\, w = \sum_{j=1}^{1+r} \alpha^j e_j, \ \sum_{j=1}^{1+r} \alpha^j = 0\} \cong \mathbb{R}^r,$$

$$\text{fundamental weights: } B(\Gamma) = \{\gamma_i = e_1 + \cdots + e_i - \tfrac{i}{1+r} \sum_{j=1}^{1+r} e_j \,\big|\, i = 1, \dots, r\},$$

$$\begin{array}{rl}
\text{roots:} & R = \{e_i - e_j \,\big|\, i \neq j, \ 1 \leq i, j \leq 1 + r\}, \\
\text{fundamental roots:} & B(R) = \{\omega_1 = e_1 - e_2, \dots, \omega_r = e_r - e_{1+r}\}, \\
\text{dominant root:} & \omega_{\max} = e_1 - e_{1+r} = \gamma_1 + \gamma_r, \\
\text{Weyl group:} & \mathrm{Weyl}(R) \cong \mathbf{G}(1 + r), \\
\text{root inversion:} & r = 1: \ I_R = -\mathrm{id}_W, \ \ r \geq 2: \ I_R(\omega_i) = -\omega_{1+r-i}.
\end{array}$$

The simplest case is A_1:

$$\text{in } \Gamma \cong \mathbb{R}: \ d = 3, \ \ \omega_1^1 = 2, \ \ \omega_1 = 2, \ \ \gamma_1 = 1 = \rho;$$

the next one A_2:

$$\text{in } \Gamma \cong \mathbb{R}^2: \ \left\{ \begin{array}{l}
d = 8, \\
\omega_i^j = \begin{pmatrix} 2 & -1 \\ -1 & 2 \end{pmatrix}, \ \ (\omega^{-1})_j^i = \frac{1}{3}\begin{pmatrix} 2 & 1 \\ 1 & 2 \end{pmatrix}, \\
\Phi_{12} = \frac{2}{3}\pi, \ \ \left\{ \begin{array}{l} \| \omega_1 \| = \| \omega_2 \|, \\ \omega_1 = (2, 0), \ \omega_2 = (-1, \sqrt{3}), \end{array} \right. \\
\omega_{\max} = (1, \sqrt{3}), \\
\frac{1}{3}\begin{pmatrix} 2 & 1 \\ 1 & 2 \end{pmatrix}\begin{pmatrix} 2 & 0 \\ -1 & \sqrt{3} \end{pmatrix} = \begin{pmatrix} 1 & \frac{1}{\sqrt{3}} \\ 0 & \frac{2}{\sqrt{3}} \end{pmatrix} \Rightarrow \left\{ \begin{array}{l} \gamma_1 = (1, \frac{1}{\sqrt{3}}), \\ \gamma_2 = (0, \frac{2}{\sqrt{3}}), \\ \rho = (1, \sqrt{3}), \end{array} \right. \\
\text{inverse Killing form in } H^T\text{: } \kappa_{ij} \sim \frac{2}{3}\begin{pmatrix} 2 & 1 \\ 1 & 2 \end{pmatrix}, \\
A_1\text{-decomposition} : A_1(\omega_1) \text{ or } A_1(\omega_2) : 1 \ \oplus \ 3 \ \oplus \ 2 \times 2.
\end{array} \right.$$

5.9.2 The Root Systems C_r

One has for the *root systems* C_r ($r \geq 1$) associated to the Lie algebra of the symplectic group $\mathbf{Sp}(\mathbb{C}^{2r})$ as invariance group of antisymmetric nondegenerate bilinear forms

$$\text{Cartan matrix } (r \times r)\text{: } \omega_i^j = \begin{pmatrix}
2 & -1 & 0 & 0 & \dots & 0 & 0 \\
-1 & 2 & -1 & 0 & \dots & 0 & 0 \\
0 & -1 & 2 & -1 & \dots & 0 & 0 \\
\dots & & & \dots & & & \\
0 & 0 & 0 & 0 & \dots & 2 & -1 \\
0 & 0 & 0 & 0 & \dots & -2 & 2
\end{pmatrix},$$

Dynkin diagram: 

,

$$\text{dimension: } d = r(1 + 2r), \ \ \text{diagonal degeneracy: } \tfrac{d-3r}{2} = r(r - 1),$$

$$\begin{aligned} \text{weight space: } & \Gamma \cong \mathbb{R}^r, \\ \text{fundamental weights: } & B(\Gamma) = \{\gamma_i = e_1 + \cdots + e_i; 1 \le i \le r\}, \end{aligned}$$

$$\begin{aligned} \text{roots: } & R = \{\pm 2e_i, 1 \le i \le r; \pm e_i \pm e_j, 1 \le i < j \le r, \\ & \qquad \pm \text{ independent } \}, \\ \text{fundamental roots: } & B(R) = \{\omega_1 = e_1 - e_2, \ldots, \omega_{r-1} = e_{r-1} - e_r, \\ & \qquad \omega_r = 2e_r\}, \\ \text{dominant root: } & \omega_{\max} = 2e_1 = 2\gamma_1, \\ \text{Weyl group: } & \mathrm{Weyl}(R) \cong \mathbf{G}(r) \vec{\times} \mathbf{G}(2)^r, \\ \text{root inversion: } & I_R = -\,\mathrm{id}_W. \end{aligned}$$

5.9.3 The Root Systems B_r

One has for the *root systems* B_r ($r \ge 1$) associated to the Lie algebra of the spin group $\overline{\mathbf{SO}(\mathbb{C}^{1+2r})}$ (universal cover) as invariance group of symmetric nondegenerate bilinear forms

$$\text{Cartan matrix } (r \times r)\text{: } \omega_i^j = \begin{pmatrix} 2 & -1 & 0 & 0 & \ldots & 0 & 0 \\ -1 & 2 & -1 & 0 & \ldots & 0 & 0 \\ 0 & -1 & 2 & -1 & \ldots & 0 & 0 \\ \ldots & & & \ldots & & & \\ 0 & 0 & 0 & 0 & \ldots & 2 & -2 \\ 0 & 0 & 0 & 0 & \ldots & -1 & 2 \end{pmatrix},$$

$$\text{Dynkin diagram: } \begin{array}{ccccccc} 2 & & 2 & & 2 & & 1 \\ \circ & \text{---} & \circ & \text{---}\cdots\cdots\text{---} & \circ & = & \circ \\ \omega_1 & & \omega_2 & & \omega_{r-1} & & \omega_r \end{array},$$

$$\text{dimension: } d = r(1+2r), \quad \text{diagonal degeneracy: } \tfrac{d-3r}{2} = r(r-1),$$

$$\begin{aligned} \text{weight space: } & \Gamma \cong \mathbb{R}^r, \\ \text{fundamental weights: } & B(\Gamma) = \{\gamma_i = e_1 + \cdots + e_i; 1 \le i \le r-1; \\ & \qquad \gamma_r = \tfrac{1}{2}(e_1 + \cdots + e_r)\}, \end{aligned}$$

$$\begin{aligned} \text{roots: } & R = \{\pm e_i, \ 1 \le i \le r; \quad \pm e_i \pm e_j, 1 \le i < j \le r; \\ & \qquad \pm \text{ independent}\}, \\ \text{fundamental roots: } & B(R) = \{\omega_1 = e_1 - e_2, \ldots, \omega_{r-1} = e_{r-1} - e_r, \omega_r = e_r\}, \\ \text{dominant root: } & \omega_{\max} = e_1 + e_2 = \begin{cases} 2\gamma_2, & r = 2, \\ \gamma_2, & r \ge 3, \end{cases} \\ \text{Weyl group: } & \mathrm{Weyl}(R) \cong \mathbf{G}(r) \vec{\times} \mathbf{G}(2)^r, \\ \text{root inversion: } & I_R = -\,\mathrm{id}_W. \end{aligned}$$

After $B_1 \cong A_1$ the simplest case is B_2:

$$\text{in } \Gamma \cong \mathbb{R}^2: \begin{cases} d = 10, \\ \omega_i^j = \begin{pmatrix} 2 & -2 \\ -1 & 2 \end{pmatrix}, \quad (\omega^{-1})_j^i = \frac{1}{2}\begin{pmatrix} 2 & 2 \\ 1 & 2 \end{pmatrix}, \\ \Phi_{12} = \frac{3\pi}{4}, \quad \begin{cases} \| \omega_1 \| = \sqrt{2} \| \omega_2 \|, \\ \omega_1 = (2,0), \quad \omega_2 = (-1,1), \end{cases} \\ \omega_{\max} = (0,2), \\ \frac{1}{2}\begin{pmatrix} 2 & 2 \\ 1 & 2 \end{pmatrix}\begin{pmatrix} 2 & 0 \\ -1 & 1 \end{pmatrix} = \begin{pmatrix} 1 & 1 \\ 0 & 1 \end{pmatrix} \Rightarrow \begin{cases} \gamma_1 = (1,1), \\ \gamma_2 = (0,1), \\ \rho = (1,2), \end{cases} \\ \text{inverse Killing form in } H^T\text{: } \kappa_{ij} \sim \begin{pmatrix} 2 & 1 \\ 1 & 1 \end{pmatrix}, \\ A_1\text{-decomposition: } \begin{cases} A_1(\omega_1): & 1 \oplus 3 \oplus 2 \times (2 \oplus 1), \\ A_1(\omega_2): & 1 \oplus 3 \oplus 2 \times 3. \end{cases} \end{cases}$$

5.9.4 The Root Systems D_r

One has for the *root systems* D_r ($r \geq 3$) associated to the Lie algebra of the spin group $\overline{\mathbf{SO}(\mathbb{C}^{2r})}$ (universal cover) as invariance group of symmetric nondegenerate bilinear forms

$$\text{Cartan matrix } (r \times r)\text{: } \omega_i^j = \begin{pmatrix} 2 & -1 & 0 & 0 & \dots & 0 & 0 & 0 & 0 \\ -1 & 2 & -1 & 0 & \dots & 0 & 0 & 0 & 0 \\ 0 & -1 & 2 & -1 & \dots & 0 & 0 & 0 & 0 \\ \dots & & & \dots & & & & & \\ 0 & 0 & 0 & 0 & \dots & -1 & 2 & -1 & -1 \\ 0 & 0 & 0 & 0 & \dots & 0 & -1 & 2 & 0 \\ 0 & 0 & 0 & 0 & \dots & 0 & -1 & 0 & 2 \end{pmatrix},$$

Dynkin diagram: 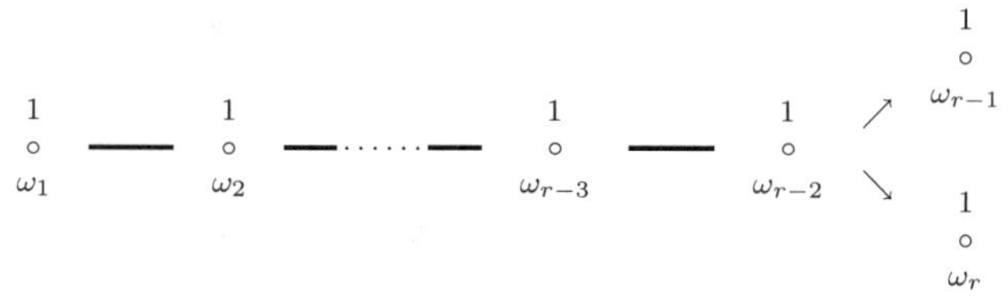

,

$$\text{dimension: } d = r(2r-1), \quad \text{diagonal degeneracy: } \frac{d-3r}{2} = r(r-2),$$

$$\begin{aligned} \text{weight space: } & \Gamma \cong \mathbb{R}^r, \\ \text{fundamental weights: } & B(\Gamma) = \{\gamma_i = e_1 + \ldots + e_i, 1 \leq i \leq r-2; \\ & \gamma_{r-1} = \tfrac{1}{2}(e_1 + \cdots + e_{r-1} - e_r) \\ & \gamma_r = \tfrac{1}{2}(e_1 + \cdots + e_{r-1} + e_r)\}, \end{aligned}$$

$$
\begin{aligned}
\text{roots: } & R = \{\pm e_i \pm e_j \,\big|\, 1 \le i < j \le r,\ \pm \text{ independent }\},\\
\text{fundamental roots: } & B(R) = \{\omega_1 = e_1 - e_2, \cdots, \omega_{r-1} = e_{r-1} - e_r,\\
& \qquad \omega_r = e_{r-1} + e_r\},\\
\text{dominant root: } & \omega_{\max} = e_1 + e_2 = \begin{cases} \gamma_2 + \gamma_3, & r = 3,\\ \gamma_2, & r \ge 4, \end{cases}\\
\text{Weyl group: } & \mathrm{Weyl}(R) \cong \mathbf{G}(r) \vec{\times} \mathbf{G}(2)^{r-1},\\
\text{root inversion: } & \begin{cases} r \in 2\mathbb{N}: & I_R = -\mathrm{id}_W,\\ r \in 2\mathbb{N}+1: & \begin{cases} I_R(\omega_j) = -\omega_j, & j = 1, \dots, r-2,\\ I_R(\omega_{r-1}) = -\omega_r,\\ I_R(\omega_r) = -\omega_{r-1}. \end{cases} \end{cases}
\end{aligned}
$$

In the isomorpism $D_3 \cong A_3$ the roots have a different numbering:

$$A_3: \ (\omega_1, \omega_2, \omega_3) \cong D_3: \ (\omega_2, \omega_1, \omega_3).$$

5.9.5 The Root System G_2

One has for the *root system* G_2

$$\text{Cartan matrix: } \omega_i^j = \begin{pmatrix} 2 & -1 \\ -3 & 2 \end{pmatrix}, \quad (\omega^{-1})_j^i = \begin{pmatrix} 2 & 1 \\ 3 & 2 \end{pmatrix},$$

Dynkin diagram: 

,

$$\text{dimension: } d = 14, \quad \text{diagonal degeneracy: } \tfrac{d-3r}{2} = 4,$$

$$
\begin{aligned}
\text{weight space (barycentric): } & \Gamma = \{w \in \mathbb{R}^3 \,\big|\, w = \sum_{j=1}^{3} \alpha^j e_j,\ \sum_{j=1}^{3} \alpha^j = 0\} \cong \mathbb{R}^2,\\
\text{fundamental weights: } & B(\Gamma) = \{\gamma_1 = e_3 - e_2,\ \gamma_2 = 2e_3 - e_1 - e_2\},
\end{aligned}
$$

$$
\begin{aligned}
\text{roots: } & R = \{\pm(e_1 - e_2), \pm(e_2 - e_3), \pm(e_3 - e_1),\\
& \qquad \pm(2e_1 - e_2 - e_3),\\
& \qquad \pm(2e_3 - e_1 - e_2), \pm(2e_2 - e_3 - e_1)\},\\
\text{fundamental roots: } & B(R) = \{\omega_1 = e_1 - e_2,\ \omega_2 = -2e_1 + e_2 + e_3\},\\
\text{dominant root: } & \omega_{\max} = 2e_3 - e_1 - e_2 = \gamma_2,\\
\text{Weyl group: } & \mathrm{Weyl}(R) \cong \mathbb{I}(6) \vec{\times} \mathbf{G}(2) \text{ (Dieder group)},\\
\text{root inversion: } & I_R = -\mathrm{id}_W,
\end{aligned}
$$

$$\text{in } \Gamma \cong \mathbb{R}^2: \quad \begin{cases} \Phi_{12} = \frac{5}{6}\pi, & \begin{cases} \| \omega_1 \| = \frac{1}{\sqrt{3}} \| \omega_2 \|, \\ \omega_1 = (2,0), \quad \omega_2 = (-3, \sqrt{3}), \end{cases} \\ \begin{pmatrix} 2 & 1 \\ 3 & 2 \end{pmatrix} \begin{pmatrix} 2 & 0 \\ -3 & \sqrt{3} \end{pmatrix} = \begin{pmatrix} 1 & \sqrt{3} \\ 0 & 2\sqrt{3} \end{pmatrix} \Rightarrow & \begin{cases} \gamma_1 = (1, \sqrt{3}), \\ \gamma_2 = (0, 2\sqrt{3}) = \omega_{\max}, \\ \rho = (1, 3\sqrt{3}), \end{cases} \\ \text{inverse Killing form in } H^T\text{: } \kappa_{ij} \sim 2\begin{pmatrix} 2 & 3 \\ 3 & 6 \end{pmatrix}, \\ A_1\text{- decomposition: } & \begin{cases} A_1(\omega_1): \ 1 \oplus 3 \oplus 2 \times (4 \oplus 1), \\ A_1(\omega_2): \ 1 \oplus 3 \oplus 2 \times (2 \times 2 \oplus 1). \end{cases} \end{cases}$$

5.9.6 The Root System F_4

One has for the *root system* F_4

$$\text{Cartan matrix: } \omega_i^j = \begin{pmatrix} 2 & -1 & 0 & 0 \\ -1 & 2 & -2 & 0 \\ 0 & -1 & 2 & -1 \\ 0 & 0 & -1 & 2 \end{pmatrix},$$

$$\text{Dynkin diagram: } \quad \overset{2}{\underset{\omega_1}{\circ}} \text{ --- } \overset{2}{\underset{\omega_2}{\circ}} = \overset{1}{\underset{\omega_3}{\circ}} \text{ --- } \overset{1}{\underset{\omega_4}{\circ}},$$

$$\text{dimension: } d = 52, \quad \text{diagonal degeneracy: } \frac{d-3r}{2} = 20,$$

$$\begin{array}{rl} \text{weight space:} & \Gamma \cong \mathbb{R}^4, \\ \text{fundamental weights:} & B(\Gamma) = \{\gamma_1 = e_1 + e_2, \ \ \gamma_2 = 2e_1 + e_2 + e_3 \\ & \qquad \gamma_3 = \frac{1}{2}(3e_1 + e_2 + e_3 + e_4), \ \ \gamma_4 = e_1\}, \end{array}$$

$$\begin{array}{rl} \text{roots:} & R = \{\pm e_i, \ 1 \le i \le 4; \ \ \pm (e_i - e_j), \ 1 \le i < j \le 4; \\ & \qquad \frac{1}{2}(\pm e_1 \pm e_2 \pm e_3 \pm e_4), \pm \text{ independent}\}, \\ \text{fundamental roots:} & B(R) = \{\omega_1 = e_2 - e_3, \ \omega_2 = e_3 - e_4, \ \omega_3 = e_4, \\ & \qquad \omega_4 = \frac{1}{2}(e_1 - e_2 - e_3 - e_4)\}, \\ \text{dominant root:} & \omega_{\max} = e_1 + e_2 = \gamma_1, \\ \text{Weyl group:} & \mathrm{Weyl}(R) \cong \mathbf{G}(3) \ \vec{\times} \left(\mathbf{G}(4) \ \vec{\times}\, \mathbf{G}(2)^3\right), \\ \text{root inversion:} & I_R = -\,\mathrm{id}_W. \end{array}$$

5.9.7 The Exceptional Lie Algebras $E_{6,7,8}$

One has for the three *exceptional algebras* $E_{6,7,8}$:

$$E_{p+q},\ \frac{1}{p}+\frac{1}{q}>\frac{1}{2},\quad p,q\in\mathbb{N},\quad p,q\geq 3.$$

This *Diophantine inequality* has a connection with the angle relations for Platonic polyhedra. For those regular convex polyhedra at each corner there meet p regular q-cornered polygons, each with an angle $\frac{q-2}{q}\pi$:

$$\left.\begin{array}{l}\frac{p(q-2)}{q}\pi<2\pi\\ p,q\in\mathbb{N},\quad p,q\geq 3\end{array}\right\}\Rightarrow\left\{\begin{array}{ll}(p,q)=(3,3): & \text{tetrahedron},\\ (p,q)=(3,4): & \text{hexahedron (cube)},\\ (p,q)=(4,3): & \text{octahedron},\\ (p,q)=(3,5): & \text{dodecahedron},\\ (p,q)=(5,3): & \text{icosahedron}.\end{array}\right.$$

For the Dynkin diagrams the pair (p,q) gives the number of vertices on both sides of the branching point:

Dynkin diagrams:
$E_6:\ (p,q)=(3,3);\quad E_7:\ (p,q)=(3,4);\quad E_8:\ (p,q)=(3,5)$

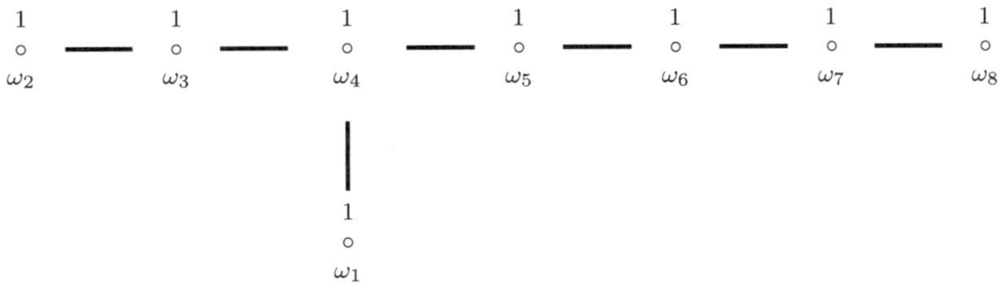

where the diagram for E_7 comes without ω_8, that for E_6 without $\omega_{7,8}$:

$$\text{Cartan matrix for } E_8:\ \omega_i^j=\begin{pmatrix}2&0&-1&0&0&0&0&0\\0&2&0&-1&0&0&0&0\\-1&0&2&-1&0&0&0&0\\0&-1&-1&2&-1&0&0&0\\0&0&0&-1&2&-1&0&0\\0&0&0&0&-1&2&-1&0\\0&0&0&0&0&-1&2&-1\\0&0&0&0&0&0&-1&2\end{pmatrix}.$$

For E_7 and E_6 the last and the two last row(s) and column(s) have to be omitted:

$$\begin{array}{l}\text{dimension: } d_6=78,\quad d_7=133,\quad d_8=248,\\ \text{diagonal degeneracy: } \frac{d_6-3r_6}{2}=30,\quad \frac{d_7-3r_7}{2}=56,\quad \frac{d_8-3r_8}{2}=112.\end{array}$$

5.10 Real Simple Lie Algebras

The faithful adjoint representation of a semisimple real Lie algebra gives the nondegenerate symmetric Killing form, which is characterized by a signature

(d_+, d_-). $d_-(d_+)$ denotes the number of compact (noncompact) dimensions with negative (positive) Killing square:

$$\text{semisimple } L^{\mathbb{R}} \cong \mathbb{R}^d \Rightarrow \text{ad } L^{\mathbb{R}} \subseteq \log \mathbf{SO}(d_+, d_-).$$

The complex span $\mathbb{C} \otimes L^{\mathbb{R}} \cong \mathbb{C}^d$ of a real semisimple Lie algebra $L^{\mathbb{R}} \cong \mathbb{R}^d$ is semisimple. Hence all real semisimple Lie algebras $L^{\mathbb{R}} \cong \mathbb{R}^d$ are real forms of complex semisimple Lie algebras $L \cong \mathbb{C}^d$.

5.10.1 The Normal and Compact Forms of a Simple Complex Lie Algebra

By restricting to real scalars, a complex Lie algebra $L \cong \mathbb{C}^d$ leads to different real Lie algebras. The normal and compact forms are easy to construct.

From a canonical generating triplet system of a complex semisimple Lie algebra $L \cong \mathbb{C}^d$ with integer structure constants,

$$L = \mathbb{C}\{h_{\omega_j}, l_{\pm\omega_j} \mid j = 1, \ldots, \tfrac{d-r}{2}\}$$
$$\text{with } \begin{cases} [h_{\omega_k}, l_{\pm\omega_j}] = \pm\omega_j^k l_{\pm\omega_j}, \ \text{Cartan matrix } \omega_j^k \in \mathbb{Z}, \\ [l_{-\omega_j}, l_{\omega_j}] = h_{\omega_j}, \ j \neq k: \ [l_{\omega_k}, l_{\omega_j}] = N_j^k l_{\omega_k+\omega_j}, \quad N_j^k \in \mathbb{Z}, \\ N_j^k = 0 \iff \omega_j + \omega_k \text{ no root}, \end{cases}$$

and Killing form

$$\kappa(h_{\omega_j}, h_{\omega_j}) = 4\kappa_j^2 > 0, \quad \kappa(l_{\pm\omega_j}, l_{\pm\omega_k}) = 2\kappa_j^2 \begin{pmatrix} 0 & -1 \\ -1 & 0 \end{pmatrix} \delta^{jk},$$

the *normal form* $L^n \cong \mathbb{R}^d$ is obtained by the real span:

$$\begin{aligned} \mathcal{L}^n : \underline{\mathbf{lag}}_{\mathbb{C}}^{\text{semisimple}} &\longrightarrow \underline{\mathbf{lag}}_{\mathbb{R}}^{\text{semisimple}}, \\ L = \mathbb{C}\{h_{\omega_j}, l_{\pm\omega_j}\}_{j=1}^{\frac{d-r}{2}} &\longmapsto L^n = \mathbb{R}\{h_{\omega_j}, l_{\pm\omega_j}\}_{j=1}^{\frac{d-r}{2}}. \end{aligned}$$

To obtain the *compact form* $L^c \cong \mathbb{R}^d$ one generalizes the transition from $A_1 \cong \mathbb{C}^3$ to $A_1^c \cong \mathbb{R}^3$: All diagonal operators H (Killing form is positive definite) are multiplied by $\frac{i}{2}$ as $l^3 = \frac{i}{2}h$, the eigenoperator pairs are transformed from a spherical to a Euclidean generating system:

$$l^{j3} = \tfrac{i}{2}h_{\omega_j}, \quad l^{j1} = i\frac{l_{\omega_j} - l_{-\omega_j}}{2}, \quad l^{j2} = \frac{l_{\omega_j} + l_{-\omega_j}}{2},$$
$$[l^{ja}, l^{jb}] = -\epsilon^{abc} l^{jc}, \quad a, b, c \in \{1, 2, 3\},$$
$$\text{with } \begin{cases} \kappa(l^{j3}, l^{j3}) &= -\kappa_j^2, \quad \kappa(ih, ih) \leq 0 \text{ for } ih \in iH, \\ \kappa(l^{jA}, l^{kB}) &= -\kappa_j^2 \begin{pmatrix} 1 & 0 \\ 0 & 1 \end{pmatrix} \delta^{AB}\delta^{jk}, \quad A, B \in \{1, 2\}, \end{cases}$$

leading to the functor from complex to compact:

$$\begin{aligned} \mathcal{L}^c : \underline{\mathbf{lag}}_{\mathbb{C}}^{\text{semisimple}} &\longrightarrow \underline{\mathbf{lag}}_{\mathbb{R}}^{\text{semisimple}}, \\ L = \mathbb{C}\{h_{\omega_j}, l_{\pm\omega_j}\}_{j=1}^{\frac{d-r}{2}} &\longmapsto L^c = \mathbb{R}\{l^{ja} \mid a = 1, 2, 3\}_{j=1}^{\frac{d-r}{2}}. \end{aligned}$$

5.10.2 Reflections of Compact Lie Algebras

Any semisimple real Lie algebra can be obtained from the compact form of a semisimple complex Lie algebra $L \cong \mathbb{C}^d \overset{\mathcal{L}^c}{\longmapsto} L^c \cong \mathbb{R}^d$ by replacing "real" l by "imaginary" $il = b$ as exemplified for A_1:

$$\begin{array}{llll} A_1^c \cong \log \mathbf{SO}(3): & [l^1, l^2] = -l^3, & [l^2, l^3] = -l^1, & [l^3, l^1] = -l^2, \\ A_1^n \cong \log \mathbf{SO}(2,1): & [il^1, l^2] = -il^3, & [l^2, il^3] = -il^1, & [il^3, il^1] = l^2, \\ & [b^1, l^2] = -b^3, & [l^2, b^3] = -b^1, & [b^3, b^1] = l^2. \end{array}$$

An appropriate subspace for the real-imaginary transition is obtained by an involutive automorphism of a real compact Lie algebra

$$\mathtt{R} : L^c \longrightarrow L^c, \quad [\mathtt{R}(l), \mathtt{R}(m)] = \mathtt{R}([l, m]), \quad \mathtt{R} \circ \mathtt{R} = \mathrm{id}_{L^c}.$$

L^c has the reflection associated *Cartan decomposition* into symmetric Lie subalgebra and antisymmetric vector subspace

$$\begin{array}{l} L_\pm^{\mathtt{R}} = \{l \in L^c \,\big|\, \mathtt{R}(l) = \pm l\} \cong \mathbb{R}^{d\pm}, \quad L^c = L_+^{\mathtt{R}} \oplus L_-^{\mathtt{R}}, \\ [L_\pm^{\mathtt{R}}, L_\pm^{\mathtt{R}}] \subseteq L_+^{\mathtt{R}}, \quad [L_+^{\mathtt{R}}, L_-^{\mathtt{R}}] \subseteq L_-^{\mathtt{R}}, \end{array}$$

orthogonal with respect to the Killing form:

$$\kappa(L_+^{\mathtt{R}}, L_-^{\mathtt{R}}) = \kappa(\mathtt{R}.L_+^{\mathtt{R}}, \mathtt{R}.L_-^{\mathtt{R}}) = -\kappa(L_+^{\mathtt{R}}, L_-^{\mathtt{R}}) = \{0\}.$$

By multiplication of the antisymmetric subspace with the imaginary unit i (*Weyl's unitary trick*) one obtains the R-*reflected Lie algebra form*, another semisimple Lie algebra as real form of L:

$$\begin{array}{lll} \mathcal{L}^{\mathtt{R}} : \underline{\mathbf{lag}}_{\mathbb{R}}^{\text{compact}} \longrightarrow \underline{\mathbf{lag}}_{\mathbb{R}}, & L^c = L_+^{\mathtt{R}} \oplus L_-^{\mathtt{R}} \longmapsto L^{\mathtt{R}} = L_+^{\mathtt{R}} \oplus iL_-^{\mathtt{R}} \\ \text{with} & [L_\pm^{\mathtt{R}}, L_\pm^{\mathtt{R}}] \subseteq L_+^{\mathtt{R}}, \; [L_+^{\mathtt{R}}, iL_-^{\mathtt{R}}] \subseteq iL_-^{\mathtt{R}}, \\ & \mathrm{ad}\, L^{\mathtt{R}} \subset \log \mathbf{SO}(d_+, d_-). \end{array}$$

$L_+^{\mathtt{R}} \cong \mathbb{R}^{d-}$ is a maximal compact Lie subalgebra in $L^{\mathtt{R}}$. The structure constants of the noncompact subspace $iL_-^{\mathtt{R}} \cong \mathbb{R}^{d+}$ bracket get a negative sign. The nondegenerate Killing form, diagonal in $L_\pm^c$, has a positive signature in $iL_-^{\mathtt{R}}$.

An example is the transition to the normal form, now starting from the compact form: With a Euclidean generating system for L^c,

$$L^c = \mathbb{R}\{l^{ja} \,\big|\, j = 1, \dots, \tfrac{d-r}{2}, \; a = 1, 2, 3\}, \quad \mathrm{ad}\, L^c \subseteq \log \mathbf{SO}(0, d),$$

the anticonjugation is an involutive automorphism:

$$\begin{array}{l} (l^{j3}, l^{j1}, l^{j2}) \overset{\times}{\leftrightarrow} (-l^{j3}, -l^{j1}, l^{j2}), \\ L^c = L_+^\times \oplus L_-^\times : \quad L_+^\times = \mathbb{R}\{l^{j2}\}, \quad L_-^\times = \mathbb{R}\{l^{j3}, l^{j1}\}. \end{array}$$

Weyl's unitary trick gives the normal real form with the functor from compact to normal:

$$\begin{array}{l} \mathcal{L}^\times : \underline{\mathbf{lag}}_{\mathbb{R}}^{\text{compact}} \longrightarrow \underline{\mathbf{lag}}_{\mathbb{R}}, \quad L^c \longmapsto L^\times = L^n, \\ L^n = L_+^n \oplus iL_-^n : \left\{ \begin{array}{l} L_+^n = \mathbb{R}\{l^{j2} = \frac{l_{\omega_j} + l_{-\omega_j}}{2}\}, \\ iL_-^n = \mathbb{R}\{-il^{j3} = \frac{1}{2} h_{\omega_j}, \; -il^{j1} = \frac{l_{\omega_j} - l_{-\omega_j}}{2}\}, \end{array} \right. \\ \mathrm{ad}\, L^n \subseteq \log \mathbf{SO}_0(\frac{d+r}{2}, \frac{d-r}{2}), \end{array}$$

e.g.,

$$A_1^c \cong \log \mathbf{SO}(3) \overset{\times}{\leftrightarrow} A_1^n \cong \log \mathbf{SO}_0(2,1).$$

With respect to the signature character $J({}_dL_r^{\mathtt{R}})$ of a real form, the compact and the normal forms are the extremes

$$J({}_dL_r^c) = -d \leq J({}_dL_r^{\mathtt{R}}) = d_+ - d_- \leq r = J({}_dL_r^n).$$

In the faithful adjoint representation the compact Lie algebra L^c comes with real antisymmetric matrices, block-diagonal and block-skew-diagonal for $\mathtt{R}$-eigenvalues ± 1

$$\begin{array}{ll} L^c = L_+^{\mathtt{R}} \oplus L_-^{\mathtt{R}} \cong \begin{pmatrix} L_+^{\mathtt{R}} \\ L_-^{\mathtt{R}} \end{pmatrix}, & \operatorname{ad} L^c \subset \log \mathbf{SO}(0,d), \\ \operatorname{ad} l_+ = -(\operatorname{ad} l_+)^T \cong \begin{pmatrix} A & 0 \\ 0 & B \end{pmatrix}, & \begin{cases} A = \operatorname{ad} l_+|_{L_+^{\mathtt{R}}} = -A^T, \\ B = \operatorname{ad} l_+|_{L_-^{\mathtt{R}}} = -B^T, \end{cases} \\ \operatorname{ad} l_- = -(\operatorname{ad} l_-)^T \cong \begin{pmatrix} 0 & C \\ -C^T & 0 \end{pmatrix}, & C = \operatorname{ad} l_-|_{L_+^{\mathtt{R}}} = -C^T = \operatorname{ad} l_-|_{L_-^{\mathtt{R}}}, \end{array}$$

This gives a symmetric matrix for the noncompact vector subspace $iL_-^{\mathtt{R}}$ of the $\mathtt{R}$-reflected Lie algebra:

$$\begin{array}{ll} L^{\mathtt{R}} = L_+^{\mathtt{R}} \oplus iL_-^{\mathtt{R}} \cong \begin{pmatrix} L_+^{\mathtt{R}} \\ iL_-^{\mathtt{R}} \end{pmatrix}, & \operatorname{ad} L^{\mathtt{R}} \subset \log \mathbf{SO}(d_+, d_-), \\ \operatorname{ad} l_+ = -(\operatorname{ad} l_+)^T \cong \begin{pmatrix} A & 0 \\ 0 & B \end{pmatrix}, & \begin{cases} A = \operatorname{ad} l_+|_{L_+^{\mathtt{R}}} = -A^T, \\ B = \operatorname{ad} l_+|_{iL_-^{\mathtt{R}}} = -B^T, \end{cases} \\ \operatorname{ad} il_- = +(\operatorname{ad} il_-)^T \cong \begin{pmatrix} 0 & C \\ +C^T & 0 \end{pmatrix}, & C = \operatorname{ad} il_-|_{L_+^{\mathtt{R}}} = +C^T = \operatorname{ad} il_-|_{iL_-^{\mathtt{R}}}. \end{array}$$

5.10.3 Classification of Real Simple Lie Algebras

The classification of the simple real Lie algebras has two parts: the real forms of simple complex Lie algebras and the antidoubling of their compact forms.

The classification of the real forms is equivalent to the classification of the involutive automorphisms (reflections) of the compact form. For endomorphism Lie algebras (matrix Lie algebras) the reflections come as inner automorphisms of $L^c \subset \mathbf{AL}(\mathbb{C}^n)$:

$$\mathtt{I} \in \mathbf{GL}(\mathbb{C}^n), \quad \mathtt{R} = \operatorname{Int} \mathtt{I} : L^c \longrightarrow L^c, \quad \operatorname{Int} \mathtt{I}(l) = \mathtt{I} \circ l \circ \mathtt{I}^{-1}, \quad \mathtt{I}^2 = \mathbf{1}_n.$$

The reflections are of three types, called *anticonjugation, orthogonal reflections (with signature), and symplectic reflection*:

$$\mathbf{GL}(\mathbb{C}^n) \ni \mathtt{I} = \begin{cases} \times \circ T = - & \text{(anticonjugation)}, \\ \delta_{p,q} = \begin{pmatrix} \mathbf{1}_p & 0 \\ 0 & -\mathbf{1}_q \end{pmatrix}, \quad p+q=n & \text{(orthogonal)}, \\ \epsilon_n = \begin{pmatrix} 0 & \mathbf{1}_{\frac{n}{2}} \\ -\mathbf{1}_{\frac{n}{2}} & 0 \end{pmatrix} \quad \text{for even } n & \text{(symplectic)}, \end{cases}$$

with the corresponding functors

$$\begin{array}{ccccc} \underline{\mathbf{lag}}_{\mathbb{C}}^{\text{semisimple}} & \longrightarrow & \underline{\mathbf{lag}}_{\mathbb{R}}^{\text{compact}} & \longrightarrow & \underline{\mathbf{lag}}_{\mathbb{R}}^{\text{semisimple}}, \\ L \cong \mathbb{C}^d & \longmapsto & L^c \cong \mathbb{R}^d & \longmapsto & (L^{\mathbb{R}}, L^{\mathbb{H}}, L^{(p,q)}). \end{array}$$

Also, the normal forms arise by the reflections above, however, not with only one type.

The anticonjugation (chapter "Antistructures: The Real in the Complex") is the canonical number conjugation of the matrix elements, e.g., for A_1^c,

$$\alpha_j \in \mathbb{R}, \quad \overline{\begin{pmatrix} i\alpha_3 & i\alpha_1 + \alpha_2 \\ i\alpha_1 - \alpha_2 & -i\alpha_3 \end{pmatrix}} = \begin{pmatrix} -i\alpha_3 & -i\alpha_1 + \alpha_2 \\ -i\alpha_1 - \alpha_2 & i\alpha_3 \end{pmatrix}.$$

It leads from the complex matrix Lie algebra to the corresponding real matrix Lie algebra, e.g., $\log \mathbf{SL}(\mathbb{C}^n) \overset{\mathcal{L}^c}{\longmapsto} \log \mathbf{SU}(n) \overset{\mathcal{L}^\times}{\longmapsto} \log \mathbf{SL}(\mathbb{R}^n)$.

Sometimes a reflection or a product of two reflections is trivial, leading to a double entry in the summarizing table:

L	complex L_r $\dim_{\mathbb{C}} L = d$	compact L_r^c	anti $L_r^{\mathbb{R}}$ in $\log \mathbf{SL}(\mathbb{R}^n)$	symplectic $L_r^{\mathbb{H}}$ in $\log \mathbf{SL}(\mathbb{H}^n)$	orthogonal $L_r^{p,q}$
A	$\log \mathbf{SL}(\mathbb{C}^{1+r})$ $r(2+r)$	$\log \mathbf{SU}(1+r)$ $r(2+r)$	$\log \mathbf{SL}(\mathbb{R}^{1+r})$ $\supset \log \mathbf{SO}(1+r)$ $\binom{1+r}{2}$	for $1+r$ even $\log \mathbf{SU}^\star(1+r)$ $\supset \log \mathbf{SpU}(1+r)$ $\binom{2+r}{2}$	$p+q=1+r$ $\log \mathbf{SU}(p,q)$ $\supset \log \mathbf{SU}(p) \times \mathbf{U}(1) \times \mathbf{SU}(q)$ p^2+q^2-1
C	$\log \mathbf{Sp}(\mathbb{C}^{2r})$ $\binom{1+2r}{2}$	$\log \mathbf{SpU}(2r)$ $\binom{1+2r}{2}$	$\log \mathbf{Sp}(2r)$ $\supset \log \mathbf{U}(r)$ r^2	$= \log \mathbf{Sp}(2r)$	$p+q=r$ $\log \mathbf{SpU}(2p,2q)$ $\supset \log \mathbf{SpU}(2p) \times \mathbf{SpU}(2q)$ $\binom{2p+1}{2} + \binom{2q+1}{2}$
D	$\log \mathbf{SO}(\mathbb{C}^{2r})$ $\binom{2r}{2}$	$\log \mathbf{SO}(2r)$ $\binom{2r}{2}$	$= \log \mathbf{SO}(2r)$	$\log \mathbf{SO}^\star(2r)$ $\supset \log \mathbf{U}(r)$ r^2	$p+q=2r$ $\log \mathbf{SO}(p,q)$ $\supset \log \mathbf{SO}(p) \times \mathbf{SO}(q)$ $\binom{p}{2} + \binom{q}{2}$
B	$\log \mathbf{SO}(\mathbb{C}^{1+2r})$ $\binom{1+2r}{2}$	$\log \mathbf{SO}(1+2r)$ $\binom{1+2r}{2}$	$= \log \mathbf{SO}(1+2r)$	–	$p+q=1+2r$ $\log \mathbf{SO}(p,q)$ $\supset \log \mathbf{SO}(p) \times \mathbf{SO}(q)$ $\binom{p}{2} + \binom{q}{2}$

simple complex and real Lie algebras
of real form type L_r^c, $L_r^{\mathbb{R}}$, $L_r^{\mathbb{H}}$, $L_r^{p,q}$
($\supset$ maximal compact Lie subalgebra with dimension)
(without the exceptional Lie algebras)

The real forms from the symplectic reflection ϵ_{2n} can be related to quaternionic Lie algebras in endomorphisms of vector spaces $V \cong \mathbb{H}^n$ with $\dim_{\mathbb{R}} \mathbf{SL}(\mathbb{H}^n) = 4n^2 - 1$. The nonabelian *quaternionic field* $\mathbb{H} \cong \mathbb{R}^4$ with three imaginaries is isomorphic to $\{\alpha_0 \mathbf{1}_2 + i\vec{\alpha}\vec{\sigma} \mid \alpha \in \mathbb{R}\}$ with the Pauli matrices (chapter "Quantum Algebras"):

	$\mathbf{SU}^\star(2n) \cong \mathbf{SL}(\mathbb{H}^n)$	
$\mathbf{SpU}(2r) \cong \mathbf{U}(\mathbb{H}^r)$		$\mathbf{SpU}(2p,2q) \cong \mathbf{U}(\mathbb{H}^r, p, q)$
	$\mathbf{SO}^\star(2r) \cong \mathbf{SO}(\mathbb{H}^r)$	

The groups G_{d_c} (compact dimension as right subindex) for the rank $1, 2, 3$ real simple Lie algebras with the (local) isomorphies and for the abelian D_1 are the following ones:

D_1
$\mathbf{SO}(2)_1 \cong \mathbf{SO}^\star(2)_1 \cong \mathbf{SO}(\mathbb{H})_1$
$\mathbf{SO}_0(1,1)_0$

A_1	$\cong$	B_1	$\cong$	C_1
$\mathbf{SU}(2)_3$	$\sim$	$\mathbf{SO}(3)_3$	$\sim$	$\mathbf{SpU}(2)_3 \sim \mathbf{SU}^\star(2)_3 \sim \mathbf{SL}(\mathbb{H})_3$
$\mathbf{SU}(1,1)_1 \sim \mathbf{SL}(\mathbb{R}^2)_1$	$\sim$	$\mathbf{SO}(1,2)_1$	$\sim$	$\mathbf{Sp}(2)_1$

D_2	$\cong$	A_1	$\oplus$	A_1
$\mathbf{SO}(4)_6$	$\sim$	$\mathbf{SU}(2)_3$	$\times$	$\mathbf{SU}(2)_3$
$\mathbf{SO}^\star(4)_4 \sim \mathbf{SO}(\mathbb{H}^2)_4$	$\sim$	$\mathbf{SU}(1,1)_1$	$\times$	$\mathbf{SU}(2)_3$
$\mathbf{SO}(1,3)_3$	$\sim$	$\mathbf{SL}(\mathbb{C}^2_\mathbb{R})$		
$\mathbf{SO}(2,2)_2$	$\sim$	$\mathbf{SU}(1,1)_1$	$\times$	$\mathbf{SU}(1,1)_1$

A_2
$\mathbf{SU}(3)_8$
$\mathbf{SU}(1,2)_4$
$\mathbf{SL}(\mathbb{R}^3)_3$

B_2	$\cong$	C_2
$\mathbf{SO}(5)_{10}$	$\sim$	$\mathbf{SpU}(4)_{10}$
$\mathbf{SO}(1,4)_6$	$\sim$	$\mathbf{SpU}(2,2)_6$
$\mathbf{SO}(2,3)_4$	$\sim$	$\mathbf{Sp}(4)_4$

D_3	$\cong$	A_3
$\mathbf{SO}(6)_{15}$	$\sim$	$\mathbf{SU}(4)_{15}$
$\mathbf{SO}(1,5)_{10}$	$\sim$	$\mathbf{SU}^\star(4)_{10} \sim \mathbf{SL}(\mathbb{H}^2)_{10}$
$\mathbf{SO}^\star(6)_9 \sim \mathbf{SO}(\mathbb{H}^3)_9$	$\sim$	$\mathbf{SU}(1,3)_9$
$\mathbf{SO}(2,4)_7$	$\sim$	$\mathbf{SU}(2,2)_7$
$\mathbf{SO}(3,3)_6$	$\sim$	$\mathbf{SL}(\mathbb{R}^4)_6$

B_3
$\mathbf{SO}(7)_{21}$
$\mathbf{SO}(1,6)_{15}$
$\mathbf{SO}(2,5)_{11}$
$\mathbf{SO}(3,4)_9$

C_3
$\mathbf{SpU}(6)_{21}$
$\mathbf{SpU}(2,4)_{13}$
$\mathbf{Sp}(6)_9$

For D_4 there is the local isomorphism $\mathbf{SO}^\star(8)_{16} \sim \mathbf{SO}(2,6)_{16}$.

In addition to the real forms of simple complex Lie algebras there are simple real Lie algebras which are the *canonically complexified* (complex doubled) compact Lie algebras

$$\underline{\mathbf{lag}}_\mathbb{R}^{\text{compact}} \longrightarrow \underline{\mathbf{lag}}_\mathbb{R}^{\text{semisimple}}, \quad L^c \longmapsto L^c \oplus iL^c.$$

They have doubled dimension $d_\pm = d$ and equal real and imaginary rank $r_\pm = r$ and the Lie brackets, suggested by $i^2 = -1$,

$$\begin{aligned}
&[l_1 + il_2, k_1 + ik_2] = \Big([l_1, k_1] - [l_2, k_2]\Big) + i\Big([l_2, k_1] + [l_1, k_2]\Big);\\
&[L^c, L^c] \subseteq L^c, \quad [iL^c, iL^c] \subseteq L^c, \quad [L^c, iL^c] \subseteq iL^c;
\end{aligned}$$

L^c is maximal compact. The canonical complexifications are in Lie algebras of complex linear groups, where $\mathbb{C}$ is used with the canonical conjugation as $\mathbb{C}_\mathbb{R} = \mathbb{R} \oplus i\mathbb{R} \cong \mathbb{R}^2$:

Lie algebra		Lie group	Dimension $d = 2d_c$
$A_{(r,r)}$	$\cong$	$\log \mathbf{SL}(\mathbb{C}^{1+r}_\mathbb{R})$	$2r(2+r)$
$C_{(r,r)}$	$\cong$	$\log \mathbf{Sp}(\mathbb{C}^{2r}_\mathbb{R})$	$2\binom{1+2r}{2}$
$D_{(r,r)}$	$\cong$	$\log \mathbf{SO}(\mathbb{C}^{2r}_\mathbb{R})$	$2\binom{2r}{2}$
$B_{(r,r)}$	$\cong$	$\log \mathbf{SO}(\mathbb{C}^{2r}_\mathbb{R})$	$2\binom{1+2r}{2}$

simple real Lie algebras of type
canonical complexification $L_{(r,r)} = L^c_r \oplus iL^c_r$
(without the exceptional Lie algebras)

$D_{(1,1)} \cong \log \mathbf{GL}(\mathbb{C}_\mathbb{R})$ is abelian (not semisimple)

$D_{(1,1)}$	$A_{(1,1)}$	$\cong$	$B_{(1,1)}$	$\cong$	$C_{(1,1)}$
$\mathbf{SO}(\mathbb{C}^2_\mathbb{R}) \sim \mathbf{SO}(2) \times \mathbf{SO}(1,1)$	$\mathbf{SL}(\mathbb{C}^2_\mathbb{R})$	$\sim$	$\mathbf{SO}(\mathbb{C}^3_\mathbb{R}) \sim \mathbf{SO}(1,3)$	$\sim$	$\mathbf{Sp}(\mathbb{C}^2_\mathbb{R})$

Obviously, the canonical complexification of all real forms above, if they exist, coincide

$$L_{(r,r)} = L^c_r \oplus iL^c_r \cong L^\mathbb{R}_r \oplus iL^\mathbb{R}_r \cong L^\mathbb{H}_r \oplus iL^\mathbb{H}_r \cong L^{(p,q)}_r \oplus iL^{(p,q)}_r.$$

Bibliography

[1] N. Bourbaki, *Lie Groups and Lie Algebras, Chapters 1-3* (1989), Springer, Berlin, Heidelberg, New York, London, Paris, Tokyo.

[2] N. Bourbaki, *Groupes et Algèbres de Lie, Chapitre 4* (Groupes de Coxeter et systémes de Tits) (1968), Hermann, Paris.

[3] N. Bourbaki, *Groupes et Algèbres de Lie, Chapitre 5* (Groupes engendrés par des réflexions) (1968), Hermann, Paris.

[4] N. Bourbaki, *Groupes et Algèbres de Lie, Chapitre 6* (Systèmes de racines) (1968), Hermann, Paris.

[5] N. Bourbaki, *Groupes et Algèbres de Lie, Chapitre 7* (Sous-algèbres de Cartan, éléments réguliers) (1975), Hermann, Paris.

[6] H.S.M. Coxeter, *Regular Polytopes* (1973), Dover, New York.

[7] W. Fulton, J. Harris, *Representation Theory* (1991), Springer, Berlin, Heidelberg, New York, London, Paris, Tokyo.

[8] R. Gilmore, *Lie Groups, Lie Algebras and Some of Their Applications* (1974), John Wiley & Sons, New York, London, Sidney, Toronto.

[9] S. Helgason, *Differential Geometry, Lie Groups and Symmetric Spaces* (1978), Academic Press, New York. London, Toronto, Sidney, San Francisco.

[10] N. Jacobson, *Lie Algebras* (1961), Dover, New York.

[11] A. Knapp, *Representation Theory of Semisimple Groups* (1986), Princeton University Press, Princeton.

6

RATIONAL QUANTUM NUMBERS

Physical objects, basically bound and scattering waves or elementary particles, are "collections of properties" of quantum operations, i.e., vectors with invariants and eigenvalues. That all the real numbers resulting from physical measurements can be interpreted by discrete or continuous spectra of real Lie operations is, perhaps, the strongest argument for the relevance of the Lie group approach to physics. Obviously, it is principally undecidable, because of the error bars, if the experimental numbers yielding spectra are rational or continuous. On the one hand, the electric charge number and the spin numbers seem to come from a discrete rational spectrum. On the other hand, energy and momenta for scattering states seem to come from a continuous real spectrum and probably also the masses of elementary particles.

All discrete *rational* invariants and eigenvalues, quantum numbers in the original sense ("natura facit saltus"), can be related to representations of *compact* groups, sometimes as subgroups of noncompact groups, e.g., $\mathbf{SO}(2) \subset \mathbf{SO}(1,2)$. Continuous quantum numbers come only from noncompact operations with their faithful infinite-dimensional Hilbert space representations (chapter "Harmonic Analysis").

There occur mixed situtations: Energy levels for nonrelativistic bound states start from irreducible, not faithful representations of time $\mathbb{R} \ni t \longmapsto e^{iEt} \in \mathbf{U}(1)$ with a continuous invariant $E \in \mathbb{R}$. The discrete energy levels, e.g., in harmonic oscillators (chapter "Quantum Algebras") or atomic bound states (chapter "The Kepler Factor"), are related to tensor representations as integer powers of the defining one, e.g., $\mathbf{U}(1) \ni e^{iEt} \longmapsto (e^{iEt})^z \in \mathbf{U}(1)$, $z \in \mathbb{Z}$.

Ultimately, the representation theory of compact groups is simple (not trivial); it is connected, via the Cartan subgroups, to the representations of the compact 1-dimensional Lie group $\mathbf{U}(1) \cong \exp i\mathbb{R}$ (circle, 1-dimensional torus) whose irreducible ones are characterized by integer winding numbers (e.g., charge numbers) as powers of the defining representation $\mathbf{U}(1) \ni e^{i\alpha} \longmapsto (e^{i\alpha})^z \in \mathbf{U}(1)$

$$z \in \mathbb{Z} = \mathbf{weights}\,\mathbf{U}(1) \cong \mathbf{irrep}\,\mathbf{U}(1).$$

Any $\mathbf{U}(1)$-representation space is decomposable into irreducible complex 1-dimensional spaces (Fourier series). For a compact Lie algebra[1] the $\mathbf{U}(1)$-structure comes in self-dual representations via $\mathbf{SO}(2)$-Lie algebras that constitute the Cartan subalgebras of a compact Lie algebra. In the integer-power representations $\mathbf{SO}(2) \ni e^{i\sigma_3\alpha} \longmapsto (e^{i\sigma_3\alpha})^n \in \mathbf{SO}(2)$ the winding numbers come in pairs $\{\pm n\}$

$$n \in \mathbb{N}_0 \cong \mathbf{irrep}\,\mathbf{SO}(2).$$

The dimensions of representation spaces go with the rank r integer winding numbers as familiar from the spin group $\mathbf{SU}(2)$ (chapter "Spin, Rotations, and Position"). All representation spaces have a polynomial structure with r indeterminates (totally symmetric tensor powers) as familiar from the spherical polynomials and harmonics.

The simplest simple Lie operations of $\mathbf{SU}(2)$ and their representations are characteristic for all semisimple ones (chapter "Simple Lie Operations"). Compact Lie groups have an Euclidean structure, e.g., definite Killing forms. Therefore, it does not come as surprise that the regular Platonic polytopes play a prominent role in the weight and root diagrams of the simplest simple Lie algebras with rank 1, 2 and 3, e.g., for $\mathbf{SO}(3)$, $\mathbf{SO}(5)$ or $\mathbf{SU}(3)$ and $\mathbf{SU}(4)$, and, via the hierarchical buildup structure, for all semisimple Lie algebra representations.

After the general structures of representations of simple Lie algebras, they are listed explicitly for the four main series (A_r, C_r, B_r, D_r).

6.1 Simple Representations of Simple Lie Symmetries

All compact Lie algebra representations and hence all *finite dimensional* representations of a semisimple complex Lie algebra $L \in \underline{\mathbf{lag}}_{\mathbb{C}}$ are semisimple, i.e., decomposable into simple ones:

$$\mathcal{D} : L \longrightarrow \mathbf{AL}(V), \quad V = \bigoplus_i V_i, \quad \mathcal{D} = \bigoplus_i \mathcal{D}|_{V_i}.$$

A representation of a direct sum-product of two Lie algebras is equivalent to a representation on a tensor product of vector spaces:

$$\begin{array}{l} [L_1, L_2] = \{0\}, \quad \mathcal{D} : L_1 \oplus L_2 \longrightarrow \mathbf{AL}(V), \\ V \cong V_1 \otimes V_2, \quad \mathcal{D} \cong \mathcal{D}^{V_1 \otimes V_2}, \quad \left\{ \begin{array}{l} \mathcal{D}(l_1) \cong \mathcal{D}_1(l_1) \otimes \mathrm{id}_{V_2}, \\ \mathcal{D}(l_2) \cong \mathrm{id}_{V_1} \otimes \mathcal{D}_2(l_2). \end{array} \right. \end{array}$$

Therefore the representation structure of semisimple Lie algebras is determined by the simple representations of simple Lie algebras.

All finite dimensional (simple) representations of (semi)simple Lie algebras are characterized by weights with real or imaginary integer components,

[1] All Lie algebras considered in this chapter are assumed to be finite-dimensional.

$\mathbb{Z}^{r_{nc}} \oplus (i\mathbb{Z})^{r_c}$ for noncompact (real) rank r_{nc} and compact (imaginary) rank r_c.

To obtain a representation of a real form $L^{\mathtt{R}}$ of the complex Lie algebra L, an R-corresponding reflection (conjugation) has to exist in the representation space of the complex Lie algebra L, e.g., a Euclidean conjugation $\star$ for the compact form L^c:

$$\begin{array}{ccccc} L^c & & L & & L^{\mathtt{R}} \\ \mathcal{D}^c \downarrow & \stackrel{\mathbb{C}}{\longmapsto} & \mathcal{D} \downarrow & \stackrel{\mathtt{R}}{\longmapsto} & \downarrow \mathcal{D}^{\mathtt{R}} \\ \mathbf{AL}(V, \star) & & \mathbf{AL}(V) & & \mathbf{AL}(V, \mathtt{R}) \end{array}$$

$$\text{e.g., } e^{i\alpha_a\sigma^a} \in \mathbf{SU}(2) \qquad e^{z_a\sigma^a} \in \mathbf{SL}(\mathbb{C}^2) \qquad e^{i\alpha_3\sigma^3+\alpha_1\sigma^1+\alpha_2\sigma^2} \in \mathbf{SU}(1,1).$$

6.2 Representation Invariants and Weights of Simple Lie Algebras

An irreducible complex representation of a *simple* Lie algebra $L \cong \mathbb{C}^d$ with a fixed Cartan subalgebra $H \cong \mathbb{C}^r$ (rank r) is given by traceless endomorphisms

$$\mathcal{D} : L \longrightarrow \mathbf{AL}(V), \quad V \cong \mathbb{C}^D, \quad \operatorname{tr} \mathcal{D}[H] = \{0\}.$$

It is either trivial or injective (faithful). For the faithful case, the dimension fulfills, with the tracelessness, $D \geq 1+r$ and $\binom{D}{2} \geq d-r$ because of the injective representation for the $(d-r)$ raising-lowering pairs outside the diagonal.

A generating system (including bases) is given by canonical triplets, consisting each of a raising, a lowering, and a Cartan operator:

canonical A_1-triplets:	$\{(l_{\pm\omega_j}, h_{\omega_j}) \mid j = 1, \ldots, \frac{d-r}{2}\} \subset L$,
fundamental roots:	$\{\omega_j \mid j = 1, \ldots, r\} \subset H^T \subset L^T$,
fundamental diagonal operators:	$\{h_{\omega_j} \mid j = 1, \ldots, r\} \subset H \subset L$,
Cartan eigenvalue matrix:	$\langle \omega_i, h_{\omega_j} \rangle = \omega_i^j \in \mathbb{Z}, \quad \omega_i^j = 2$ (no sum),
fundamental weights:	$\{\gamma_j \mid j = 1, \ldots, r\} \subset H^T$,
	$\langle \gamma_i, h_{\omega_j} \rangle = \delta_i^j, \quad \omega_i = \omega_i^j \gamma_j$.

For an irreducible finite dimensional representation of a simple Lie algebra there exists H-weights (vectors with eigenvalues) $\{w_a\}_{a=1}^m$ with degeneracy D_a. The minimal polynomial of any Cartan operator $\mathcal{D}(h)$, $H \in H$, is simple, i.e., $\mathcal{D}(h)$ is diagonalizable. The polynomials in the operators from the represented Cartan algebra $\mathcal{D}[H]$ have a basis of projectors (no nilpotents) expressible as polynomials in an H-basis. The representation space $V \cong \mathbb{C}^D$ is decomposable into Cartan algebra H-eigenspaces and can be spanned by H-eigenvectors (no nilvectors):

$$\begin{aligned}
&\operatorname{spec}_V H = \{w_a : H \longrightarrow \mathbb{C} \mid a = 1, \dots, m\},\\
&V_{w_a}(H) = \{v \in V \mid \mathcal{D}(h)(v) = w_a(h)(v) \text{ for all } h \in H\},\\
&\mathbb{C}[\mathcal{D}[H]]_{\mathbf{AL}(V)}\text{-basis}: \ \{\mathcal{P}_a(\mathcal{D}[H]) \mid a = 1, \dots m\},\ \mathcal{D}[H] = \sum_{a=1}^{m} w_a \mathcal{P}_a(\mathcal{D}[H]),\\
&V = \bigoplus_{a=1}^{m} V_{w_a}(H),\ \ V_{w_a}(H) = \mathcal{P}_a(\mathcal{D}[H]).V \cong \mathbb{C}^{D_a},\ \ D = \sum_{a=1}^{m} D_a.
\end{aligned}$$

For the adjoint representation $\operatorname{ad} h \in \mathbf{AL}(L)$ the roots (nontrivial weights) are nondegenerate $D_{\text{ad}} = 1$.

Explicit examples are given below for the Lie algebras A_r and A_r^{c}. The compact Lie algebra $A_2^{\text{c}} = \log \mathbf{SU}(3)$ is a good illustration: the Cartan subalgebras are planes $H \cong \mathbb{R}^2$ which contain as weight diagrams singlet, (anti)triplet, octet (root diagram), (anti)decuplet, 27-plet, etc.

The eigenvalue matrix with the r components of the m different weights for the diagonal operators

$$v \in V_{w_a}(H) \Rightarrow \mathcal{D}(h_{\omega_j})(v) = w_a^j v,\ \ (w_a^j)_{a=1,\dots,m}^{j=1,\dots,r} \cong \begin{pmatrix} w_1^1 & \dots & w_1^r \\ & \dots & \\ & \dots & \\ w_m^1 & \dots & w_m^r \end{pmatrix}$$

can be extended with the weight-multiplicities D_a. Then the sums of the columns vanish since semisimple Lie algebra representations are traceless:

$$(w_{\underline{a}}^j)_{\underline{a}=1,\dots,D}^{j=1,\dots,r} \cong \left(\begin{array}{ccc|c} w_1^1 & \dots & w_1^r & \\ & \dots & & D_1 \\ w_1^1 & \dots & w_1^r & \\ \hline & \dots & & \dots \\ \hline & \dots & & \dots \\ \hline w_m^1 & \dots & w_m^r & \\ & \dots & & D_m \\ w_m^1 & \dots & w_m^r & \end{array}\right),\ \ \sum_{a=1}^{m} D_a w_a^j = 0 \text{ for } j = 1, \dots, r.$$

Each weight w_a is an integer combination of the fundamental weights, i.e., an element from the weight module with the *canonical coordinates*

$$w_a = \sum_{j=1}^{r} z_a^j \gamma_j,\ \ w_a = [z_a^1, \dots, z_a^r] \in \mathbf{weights}_{\text{fin}} L \cong \mathbb{Z}^r.$$

The integers are *winding number coordinates* for the representations of the $\mathbf{U}(1)$-Cartan subgroups $\{\exp i\alpha^j h_{\omega^j}\}$ for the 1-dimensional Cartan subalgebras.

Any r *natural numbers* characterize uniquely an equivalence class of irreducible representations of a simple rank r Lie algebra; the trivial representation is characterized by $[0, \dots, 0]$. An eigenvector of a Cartan algebra in this representation can be labeled with these natural numbers for the representation invariant (before the semicolon) and (after the semicolon) its weight (r eigenvalues, either Cartan subalgebra eigenvalues or integers for the combination from fundamental weights)

$$|n^a; w_a\rangle = |n^1, \dots, n^r; w_a^1, \dots, w_a^r\rangle \cong |n^1, \dots, n^r; z_a^1, \dots, z_a^r\rangle \in V.$$

The, with respect to the lexicographic order, *largest weight* $w_{\max} = (n^a)$ of a representation is nondegenerate $D_{w_{\max}} = 1$ with eigenvector $|w_{\max}\rangle$ and characterizes an irreducible representation

$$\begin{aligned} w_{\max} &= [n^1, \dots, n^r] \in \mathbf{irrep}_{\,\mathrm{fin}} L \cong \mathbb{N}_0^r, \\ \mathcal{D}(h_{\omega_j})|w_{\max}\rangle &= \langle w_{\max}, h_{\omega_j}\rangle |w_{\max}\rangle = n^j |w_{\max}\rangle, \quad j = 1, \dots, r. \end{aligned}$$

The weights of a representation can be decomposed into Weyl group orbits. Each orbit has a dominant (largest) weight. All weights of one orbit have equal degeneracy. The weights of the orbit with the largest weight $w_{\max}$ are nondegenerate:

$$\begin{aligned} \mathbf{weights}\,[n^1, \dots, n^r] &= \biguplus_{w_{\mathrm{dom}}} \mathrm{Weyl}(R) \bullet w_{\mathrm{dom}}, \\ w_a \in \mathrm{Weyl}(R) \bullet w_b &\Rightarrow D_a = D_b, \\ w_a \in \mathrm{Weyl}(R) \bullet w_{\max} &\Rightarrow D_{\max} = 1. \end{aligned}$$

The degeneracy of a weight can be computed by the number of possibilities to reach it from the largest weight of the representation by lowering operators.

The raising and lowering operators $l_{\pm\omega}$ associated with the root ω are strictly triagonal matrices, $\begin{pmatrix} 0 & \times \\ 0 & 0 \end{pmatrix}$ and $\begin{pmatrix} 0 & 0 \\ \times & 0 \end{pmatrix}$, therefore nilpotent. They act on the eigenspace with weight w by adding the root

$$L_{\pm\omega}(H) \bullet V_w(H) \subseteq V_{\pm\omega+w}(H).$$

The action gives $\{0\}$ if $\omega + w$ is not a weight of the representation. Each raising operator acts trivially on the vectors with maximal weight

$$\omega \in R_+(B), \quad L_\omega(H) \bullet V_{w_{\max}}(H) = \{0\}.$$

Each weight of a representation can be obtained from the largest weight by descending with positive integer linear combinations of the fundamental roots

$$w_m = w_{\max} \iff \text{each weight } w_a = w_m - \sum_{\omega_j, k_a^j \in \mathbb{N}} k_a^j \omega_j.$$

The dimension of the representation $[n^1, \dots, n^r]$ is given by the *Weyl formula*, which uses the sum ρ of the fundamental weights

$$\rho = \tfrac{1}{2} \sum_{\omega \in R_+(B)} \omega = \sum_{j=1}^r \gamma_j$$

and can be written with an invariant nondegenerate scalar product $\langle\ |\ \rangle$ of the weights (e.g., with the inverse Killing form):

$$\dim_{\mathbb{C}}[n^1, \dots, n^r] = \prod_{\omega \in R_+(B)} \tfrac{\langle w_{\max} + \rho, h_\omega\rangle}{\langle \rho, h_\omega\rangle} = \prod_{\omega \in R_+(B)} \left(1 + \tfrac{\langle w_{\max}|\omega\rangle}{\langle \rho|\omega\rangle}\right).$$

The quadratic *Casimir element* of an irreducible representation is a power-two polynomial in raising, lowering, and Cartan operators. One has with trivial value $l_\omega|w_{\max}\rangle = 0$ for the raising operators

$$I^2(L) \bullet |w_{\max}\rangle = \left(\sum_{i,j=1}^{r} \kappa^{-1}(w_i, \gamma_j) h_{\omega_i} \otimes h_{\omega_j} + \sum_{\omega \in R_+(B)} \frac{[l_\omega, l_{-\omega}]}{\kappa(l_\omega, l_{-\omega})} \right) |w_{\max}\rangle .$$

The Killing form can be replaced by any invariant symmetric bilinear form. Choosing for this bilinear form on $L \times L$ and its "inverse" on $L^T \times L^T$ (both denoted by $\langle\ |\ \rangle$) the associated diagonal operator $h'_w \in H$ for a weight w,

$$\text{for all } h \in H: \quad \langle h'_w | h \rangle = \langle w, h \rangle \iff \text{for all } \omega \in H^T: \quad \langle w | \omega \rangle = \langle \omega, h'_w \rangle,$$

one obtains with the invariance of the bilinear form

$$\frac{[l_\omega, l_{-\omega}]}{\langle l_\omega | l_{-\omega} \rangle} |w\rangle = \frac{\langle h'_w | [l_\omega, l_{-\omega}] \rangle}{\langle l_\omega | l_{-\omega} \rangle} |w\rangle = \omega(h'_w)|w\rangle = \langle w|\omega\rangle |w\rangle .$$

Therefore the quadratic Casimir element in the irreducible representation $[n^1, \ldots, n^r]$ gives the invariant

$$I^2(L)^{[n^1,\ldots,n^r]} = \Big(\langle w_{\max} | w_{\max} \rangle + 2 \langle \rho | w_{\max} \rangle \Big) \, \mathrm{id}_V .$$

The values of the r invariant Casimir elements $I_1(L), \ldots, I_r(L)$ for a representation can replace the r natural numbers as familiar from spin $\mathbf{SU}(2)$ with Casimir value $4j(1+j) = n(2+n)$ and third direction (magnetic) quantum number $2m = z$:

$$\begin{aligned} |n^1, \ldots, n^r; z^1_a, \ldots, z^r_a\rangle &\cong |I_1^{[n^a]}, \ldots, I_r^{[n^a]}; w^1_a, \ldots, w^r_a\rangle, \\ \text{e.g., for } \log \mathbf{SU}(2): \quad |n; z\rangle &\cong |j; m\rangle . \end{aligned}$$

6.2.1 Weight Module and Representation Cone

For a simple complex Lie algebra L of rank r (or its compact form L^c) with a fixed Cartan subalgebra the weights for all complex finite dimensional representations form a $\mathbb{Z}$-module

$$\begin{aligned} \mathbf{weights}_{\,\text{fin}} L &= \mathbf{weights}\, L^c \\ &= \{[z^1, \ldots, z^r] \,\big|\, z^j = 0, \pm 1, \ldots\} = \mathbb{Z}^r \in \underline{\mathbf{mod}}_{\mathbb{Z}} . \end{aligned}$$

The weight module is ordered lexicographically. The equivalence classes of the irreducible complex finite-dimensional representations are characterized by r nonnegative numbers that define the *representation!cone* of L or L^c as positive cone of the weight module

$$\begin{aligned} \mathbf{irrep}_{\,\text{fin}} L &= \mathbf{irrep}\, L^c \\ &\cong \{[n^1, \ldots, n^r] \,\big|\, n^i = 0, 1, \ldots\} \cong \mathbb{N}_0^r . \end{aligned}$$

The weight module and the representation cone for the direct sum-product of two simple complex or two simple compact Lie algebras are the direct sum of the individual structures:

$$[L_1, L_2] = \{0\} : \quad \mathbf{weights}\,(L_1^c \oplus L_2^c) = \mathbf{weights}\,L_1^c \oplus \mathbf{weights}\,L_2^c.$$

The r *fundamental weights* $\{\gamma_j\}_{j=1}^r$ characterize r *fundamental representations* as $\mathbb{N}_0$-basis for the representation cone and $\mathbb{Z}$-basis for the weight module

$$\underbrace{[0, \ldots, 0, 1, 0, \ldots, 0]}_{j\text{th place}} \text{ with } w_{\max} = \gamma_j, \quad j = 1, \ldots, r.$$

Each fundamental root ω_j has a fundamental weight γ_j, uniquely associated with the Cartan eigenvalue matrix as bijection $\omega_j = \omega_j^i \gamma_i$. With respect to the Killing form induced metric, γ_j is the only fundamental weight not orthogonal to ω_j; its ω_j-projection is half the fundamental root:

$$\langle \omega_j | \gamma_k \rangle = \tfrac{\delta_{jk}}{2} \parallel \omega_j \parallel^2 .$$

With the root-weight correspondence $\omega_j \leftrightarrow \gamma_j$, a vertex of a *Dynkin diagram* can be related either to a fundamental root or to the associated fundamental representation. The full Dynkin diagram can be labeled with all fundamental representations. The *dominant root* is the largest weight of the adjoint representation

$$\omega_{\max} = N^j \gamma_j, \quad \mathrm{ad}\, L = [N^1, \ldots, N^r], \quad \dim_{\mathbb{C}} \mathrm{ad}\, L = d.$$

6.2.2 Dual Representations and Invariant Forms

Dual irreducible representations of a simple Lie algebra $L_r \cong \mathbb{C}^d$ have centrally reflected weights

$$\mathbf{weights}\,[n^1, \ldots, n^r] \ni [z^1, \ldots, z^r] \leftrightarrow [-z^1, \ldots, -z^r] \in \mathbf{weights}\,[n^1, \ldots, n^r]^{\mathrm{dual}}.$$

Therefore, an irreducible complex representation of L_r on V is self-dual, i.e., V has an L_r-invariant nondegenerate bilinear form iff the weights are stable under central reflection. To this end it is enough that the largest weight goes to its negative under weight reflection. With Schur's theorem the bilinear form is unique up to a factor.

If there exists an invariant bilinear form, it is orthogonal or symplectic, depending on even or odd sum respectively for the coefficients of the largest weight:

$$\begin{aligned} & [z^1, \ldots, z^r], [-z^1, \ldots, -z^r] \in \mathbf{weights}\,[n^1, \ldots, n^r] \\ \Rightarrow\ & \text{There exists } \langle\ |\ \rangle : V \times V \longrightarrow \mathbb{C} \text{ with } \langle v|w \rangle = \eta \langle w|v \rangle, \quad \eta = (-1)^N \\ & N = \sum_{j=1}^r n^j \in \begin{cases} 2\mathbb{N}, & \text{orthogonal bilinear } V\text{-form}, \\ 1 + 2\mathbb{N}, & \text{symplectic bilinear } V\text{-form}. \end{cases} \end{aligned}$$

The adjoint representation with a symmetric root diagram has the invariant Killing form. Not all representations have necessarily an invariant bilinear form, e.g., not the inequivalent $\mathbf{SU}(3)$-triplet and antitriplet with centrally reflected triangles as weight diagrams (quark and antiquark).

6.2.3 Permutation Symmetry of Product Representations

The nth tensor power of a vector space $\bigotimes^n V$ is acted on by the permutation group $\mathbf{G}(n) \in \underline{\mathbf{grp}}$ (permuting the tensor factors) and the complex permutation algebra

$$\mathbb{C}^{\mathbf{G}(n)} = \{\alpha_j p_j \mid \alpha_j \in \mathbb{C},\ p_j \in \mathbf{G}(n)\} \in \underline{\mathbf{aag}}_{\mathbb{C}}.$$

With the decomposition of the simple permutation algebra into *Young ideals*, characterizing the irreducible algebra representations (chapter "Time Representations")

$$\mathbb{C}^{\mathbf{G}(n)} = \bigoplus_z A_z^{\min}, \quad \text{simple ideals } A_z^{\min},$$

the tensor power is also Young-decomposable into symmetry classes, one totally symmetric, one totally antisymmetric, and, for $n \geq 3$, also mixed symmetric classes

$$\bigotimes^n V = \bigoplus_i V_i = \bigvee^n V \oplus \cdots \oplus \bigwedge^n V.$$

These vector subspaces are invariant under the Lie algebra action given by product representations of $\mathcal{D} : L \longrightarrow \mathbf{AL}(V)$, since permutations and L-actions commute:

$$\begin{array}{ccc} \bigotimes^n V & \xrightarrow{\bigotimes^n f} & \bigotimes^n V \\ \pi \downarrow & & \downarrow \pi \\ \bigotimes^n V & \xrightarrow[\bigotimes^n f]{} & \bigotimes^n V \end{array}, \qquad \begin{array}{l} f \in \mathbf{AL}(V), \quad \pi \in \mathbb{C}^{\mathbf{G}(n)}, \\ p \in \mathbf{G}(n) \Rightarrow p(v_1 \otimes \cdots \otimes v_n) = v_{p(1)} \otimes \cdots \otimes v_{p(n)}. \end{array}$$

The invariant subspaces for a Young decomposition of $\bigotimes^n V$ do not have to be irreducible.

The irreducible representation $[0, \ldots, 0, n^j, 0, \ldots, 0]$ of a simple Lie algebra is equivalent to the *totally symmetric* product of the generating fundamental representation

$$\begin{aligned} [0, \ldots, 0, n^j, 0, \ldots, 0] &= \bigvee^{n^j} [0, \ldots, 0, 1, 0, \ldots, 0], \\ \dim_{\mathbb{C}}[0, \ldots, 0, n^j, 0, \ldots, 0] &= \binom{n^j + d_j - 1}{n_j}. \end{aligned}$$

Those totally symmetric products reflect the $\mathbb{N}$-linear property of the representation cone.

Also, the *totally antisymmetric* product representations define representations. Using totally antisymmetric products of the fundamental representations of simple Lie algebras, one can distinguish maximally three fundamental representations, in the following called *cyclic representations*, which generate, by linear combinations of their totally antisymmetric products, all fundamental representations and hence, using the totally symmetric products above, all irreducible representations. Those cyclic representations are situated at the maximally three "loose" ends of the Dynkin diagram of the simple Lie algebra under consideration (examples below).

6.3 Representations of Simple Lie Algebras

6.3.1 Representations of the Lie Algebras A_r

The traceless endomorphisms of $V \cong \mathbb{C}^{1+r}$ constitute the *defining representation* of the Lie algebra A_r. It is the Lie algebra of the special invariance group of the nontrivial volume elements $e^1 \wedge \cdots \wedge e^{1+r}$ (totally antisymmetric):

$$r \geq 1: \quad {}_{r(2+r)}A_r \cong \log \mathbf{SL}(\mathbb{C}^{1+r}), \quad {}_{r(2+r)}A_r^c \cong \log \mathbf{SU}(1+r).$$

With Ado's theorem all finite-dimensional Lie algebras are isomorphic to Lie subalgebras of A_r.

Dual bases $\{E^A, \check{E}_A\}_{j=1}^{1+r}$ for V and V^T give canonical eigenoperators and diagonal operators

$$\begin{aligned}
&l_+^{AB} = E^A \otimes \check{E}_B \cong \begin{pmatrix} 0 & 1 \\ 0 & 0 \end{pmatrix}, \quad l_-^{AB} = -E^A \otimes \check{E}_B \cong \begin{pmatrix} 0 & 0 \\ -1 & 0 \end{pmatrix}, \\
&h^{AB} = [l_-^{AB}, l_+^{AB}] = E^A \otimes \check{E}_B - E^B \otimes \check{E}_A \cong \begin{pmatrix} 1 & 0 \\ 0 & -1 \end{pmatrix} \text{ (no summation)}, \\
&1 \leq A < B \leq 1+r.
\end{aligned}$$

Hence one obtains for the roots

$$[h^{AB}, l_+^{CD}] = \omega_{CD}(h^{AB})\, l_+^{CD}, \quad \omega_{CD}(h^{AB}) = \delta_C^A + \delta_D^B - \delta_C^B - \delta_D^A.$$

Writing for the $(1+r)$-dimensional space of the diagonal matrices with dual bases,

$$h^{AB} = e^A - e^B, \quad \omega_{CD} = \check{e}_C - \check{e}_D \Rightarrow \omega_{CD}(h^{AB}) = \langle \omega_{CD}, h^{AB} \rangle,$$

one has the connection with the root system A_r.

The equivalence classes of the irreducible $\mathbf{SU}(1+r)$-representations are characterized, according to rank r and a Cartan subgroup $\mathbf{U}(1)^r$, by r natural numbers $n^j = 2J^j$ (half-integer spin numbers J^j) in the $\mathbf{SU}(1+r)$-*representation cone*

$$\mathbf{irrep}\,\mathbf{SU}(1+r) = \{[2J^1, \ldots, 2J^r]\} \cong \mathbb{N}_0^r.$$

The dimensions are given by the formula

$$\begin{array}{lll}
\text{for } \mathbf{SU}(2): & \dim_{\mathbb{C}}[n] & = 1+n, \\
\text{for } \mathbf{SU}(3): & \dim_{\mathbb{C}}[n^1,n^2] & = \frac{(1+n^1)(1+n^2)(2+n^1+n^2)}{2}, \\
\text{for } \mathbf{SU}(4): & \dim_{\mathbb{C}}[n^1,n^2,n^3] & = \frac{(1+n^1)(1+n^2)(2+n^1+n^2)(1+n^3)(2+n^3+n^2)(3+n^3+n^2+n^1)}{2!3!}, \\
 & \dots & \\
\text{for } \mathbf{SU}(1+r): & \dim_{\mathbb{C}}[n^1,\dots,n^r] & = \dim_{\mathbb{C}}[n^1,\dots,n^{r-1}]\frac{(1+n^1)(2+n^r+n^{r-1})\cdots(r+n^r+\cdots+n^1)}{r!}, \\
 & & = \frac{(1+n^1)\cdots(1+n^r)(2+n^1+n^2)\cdots(2+n^{r-1}+n^r)\cdots(r+n^1+\cdots+n^r)}{2!3!\cdots r!}.
\end{array}$$

The integer components in the $\mathbf{SU}(1+r)$-*weight module*

$$\begin{aligned}\mathbf{weights}\,\mathbf{SU}(1+r) &= \{[2j^1,\dots,2j^r]\} = \mathbb{Z}^r \\ &= \mathbf{weights}\,\mathbf{U}(1)^r\end{aligned}$$

are winding numbers for the representation of a Cartan subgroup $\mathbf{U}(1)^r$. Since $\mathbf{SU}(1+r)$ is a special group, $\det u = 1$, the $\mathbf{U}(1)$-representations come in self-dual $\mathbf{SO}(2)$-representations.

There are r *fundamental* representations with the dimensions in the Dynkin diagram

$$\begin{array}{ccccccc}
\gamma_1 & & \gamma_2 & & \gamma_{r-1} & & \gamma_r \\
\binom{1+r}{1} & — & \binom{1+r}{2} & —\cdots\cdots— & \binom{1+r}{r-1} & — & \binom{1+r}{r} \\
[1,0,\dots,0] & & [0,1,0,\dots,0] & & [0,\dots,0,1,0] & & [0,\dots,0,1]
\end{array}$$

fundamental representations for A_r
with cyclic representation: $\{\gamma_1 \cong \binom{1+r}{1}\}$

e.g., the Pauli spinor representation $[1]$ for $\mathbf{SU}(2)$ or the quark and antiquark representations $[1,0]$ and $[0,1]$ for $\mathbf{SU}(3)$. All weights of the fundamental representations are nondegenerate.

The defining representation $[1,0,\dots,0]$ is *cyclic* fundamental. It gives the other $(r-1)$ fundamental representations as product representations on the totally antisymmetric Grassmann powers of $V \cong \mathbb{C}^{1+r}$:

$$\underbrace{[0,\dots,0,1,0,\dots 0]}_{j\text{th place}} = \bigwedge^{j}[1,0,\dots,0], \quad \dim_{\mathbb{C}} \bigwedge^{j} V = \binom{1+r}{j},$$

$$\mathcal{D}^{(0,\dots,1,\dots,0)}(l) = \bigoplus_{n=1}^{j} \underbrace{\mathrm{id}_V \otimes \cdots \otimes \mathcal{D}(l) \otimes \cdots \otimes \mathrm{id}_V}_{n\text{th place}}\Big|_{\bigwedge^{j} V}.$$

The dual to the fundamental representation on $\bigwedge^{j} V^T$ is equivalent, with the volume isomorphims, to the fundamental representation on $\bigwedge^{r-j} V$:

$$\text{dual: } \binom{1+r}{j} \leftrightarrow \binom{1+r}{1+r-j}, \quad j = 1,\dots,r.$$

Hence there is maximally one fundamental representation with an invariant nondegenerate bilinear form with symmetry (orthogonal $+1$, symplectic -1, no form 0):

$$\text{invariant bilinear form of } A_r \text{ for } \gamma_j: \quad \begin{cases} 0, & j \neq \frac{1+r}{2}, \\ (-1)^j, & j = \frac{1+r}{2}. \end{cases}$$

$A_{\rm odd}$ has an invariant form, that is symplectic self-dual for A_{4n-3} with γ_{2n-1} and orthogonal self-dual for A_{4n-1} with γ_{2n}, $n = 1, \ldots$.

The defining representation on a complex $(1+r)$-dimensional space can be written with generalized Pauli matrices $\{\sigma(1+r)^d\}_{d=1,\ldots,r(2+r)}$:

$$\textit{Pauli } \mathbf{SU}(1+r)\textit{-representation: } \exp[1,0,\ldots,0] = \exp i \sum_{d=1}^{r(2+r)} \alpha_d \sigma(1+r)^d.$$

The r diagonal Pauli matrices represent a Cartan subgroup $\mathbf{U}(1)^r$:

$$\mathbf{U}(1)^r = \{\exp i \sum_{j=1}^{r} \alpha_{j(2+j)} \sigma(1+r)^{j(2+j)} \mid \alpha_{j(2+j)} \in \mathbb{R}\}.$$

Together with the unit matrix they combine the projectors for the eigenspace decomposition of the defining representation space, e.g., the Pauli $\mathbf{SU}(2)$ and $\mathbf{SU}(3)$ representations:

$$\begin{array}{lll} \log \mathbf{SU}(2): & \mathcal{P}_{1,2}(\mathbf{U}(1)) & \in \{\frac{\mathbf{1}_2 \pm \sigma^3}{2}\}, \\ \log \mathbf{SU}(3): & \mathcal{P}_{1,2,3}(\mathbf{U}(1)^2) & \in \{\frac{2\mathbf{1}_3 + \sqrt{3}\lambda^8 \pm 3\lambda^3}{3}, \ \frac{\mathbf{1}_3 - \sqrt{3}\lambda^8}{3}\}. \end{array}$$

If one writes the diagonals of the r matrices $\{\sigma(1+r)^{j(2+j)}\}_{j=1,\ldots,r}$ as columns of a matrix, one obtains in the horizontal rows the $(1+r)$ weights $\{w_a\}_{a=1,\ldots,1+r}$ of the defining $\mathbf{SU}(1+r)$-representation in the real r-dimensional weight space. In the normalization with the Pauli matrices the defining weights occupy the corners of a regular fundamental simplex (distance, triangle, tetrahedron, etc. for $\mathbf{SU}(2)$, $\mathbf{SU}(3)$, $\mathbf{SU}(4)$ etc.), centered at the origin, as seen in the $[(1+r)\times r]$ matrix for the weights in Euclidean coordinates (the sums of the columns component vanish):

$$(w_a^j)_{a=1,\ldots,1+r}^{j=1,\ldots,r} = \begin{pmatrix} [1,0,0,\ldots,0,0] \\ [-1,1,0,\ldots,0,0] \\ [0,\ \ 1,1,\ldots,0,0] \\ \cdots \\ \cdots \\ [0,0,0,\ldots,-1,1] \\ [0,0,0,\ldots,0,-1] \end{pmatrix} \cong \underbrace{\begin{pmatrix} 1 & \frac{1}{\sqrt{3}} & \frac{1}{\sqrt{6}} & \cdots & \frac{1}{\sqrt{\binom{1+r}{2}}} \\ -1 & \frac{1}{\sqrt{3}} & \frac{1}{\sqrt{6}} & \cdots & \frac{1}{\sqrt{\binom{1+r}{2}}} \\ 0 & -\frac{2}{\sqrt{3}} & \frac{1}{\sqrt{6}} & \cdots & \frac{1}{\sqrt{\binom{1+r}{2}}} \\ 0 & 0 & -\frac{3}{\sqrt{6}} & \cdots & \frac{1}{\sqrt{\binom{1+r}{2}}} \\ & \cdots & & \cdots & \cdots \\ & \cdots & & \cdots & \cdots \\ 0 & 0 & 0 & \cdots & -\frac{r}{\sqrt{\binom{1+r}{2}}} \end{pmatrix}}_{\text{regular } \mathbb{R}^r\text{-simplex}},$$

$$\mathbf{weights}\,[1,0,\ldots,0],$$
$$\sum_{a=1}^{1+r} w_a = 0, \quad \| w_a - w_b \| = 2\delta_{ab}, \quad \| w_a \| = \sqrt{\tfrac{2r}{1+r}}.$$

The first row in the matrix gives the r components of the fundamental weight $w_1 = \gamma_1$, the other weights of the fundamental representation $[1,0,\ldots,0]$ are not fundamental weights. The cyclic property of the representations with γ_1

as largest weight allows one to calculate the weights for the other fundamental representations given by the $\binom{1+r}{j}$ different sums of j different weights:

$$\text{for } \gamma_j : \quad \sum_{k=1}^{j} w_{a_k}, \quad \{a_1, \dots, a_j\} \subseteq \{1, \dots, r\}.$$

Hence one obtains a bijection between the fundamental weights and the weights of the defining representation:

$$\begin{pmatrix} \gamma_1 \\ \gamma_2 \\ \dots \\ \gamma_r \end{pmatrix} = \begin{pmatrix} w_1 \\ w_1 + w_2 \\ \dots \\ w_1 + \dots + w_{r-1} + w_r \end{pmatrix}, \quad \begin{pmatrix} w_1 \\ w_2 \\ \dots \\ w_r \\ w_{1+r} \end{pmatrix} = \begin{pmatrix} \gamma_1 \\ -\gamma_1 + \gamma_2 \\ \dots \\ -\gamma_{r-1} + \gamma_r \\ -\gamma_r \end{pmatrix}.$$

The largest weight of the resulting fundamental representations is the unique one with only positive coefficients.

For $\mathbf{SU}(2)$, there are two weights in $\mathbb{R}$, the endpoints of a line:

$$\text{for } A_1 : \quad \mathbf{weights}\,[1] = \begin{pmatrix} [1] \\ [-1] \end{pmatrix} \cong \begin{pmatrix} 1 \\ -1 \end{pmatrix} \text{ (distance)}.$$

The weights for triplet and antitriplet of $\mathbf{SU}(3)$ are the corners of a regular triangle and centrally reflected antitriangle in the Euclidean plane $\mathbb{R}^2$; the winding number and Euclidean coordinates are different:

$$\text{for } A_2 : \begin{cases} \mathbf{weights}\,[1,0] = \begin{pmatrix} [1,0] \\ [-1,1] \\ [0,-1] \end{pmatrix} \cong \begin{pmatrix} 1 & \frac{1}{\sqrt{3}} \\ -1 & \frac{1}{\sqrt{3}} \\ 0 & -\frac{2}{\sqrt{3}} \end{pmatrix} \text{ (triangle)}, \\ \\ \mathbf{weights}\,[0,1] = \begin{pmatrix} [0,1] \\ [1,-1] \\ [-1,0] \end{pmatrix} \cong \begin{pmatrix} 0 & \frac{2}{\sqrt{3}} \\ 1 & -\frac{1}{\sqrt{3}} \\ -1 & -\frac{1}{\sqrt{3}} \end{pmatrix} \text{ (antitriangle)}. \end{cases}$$

The weights for the three fundamental $\mathbf{SU}(4)$-representations, quartet and antiquartet, dual to each other, and orthogonal self-dual sextet (with the isomorphism $\log \mathbf{SU}(4) \cong \log \mathbf{SO}(6)$), are, in the Euclidean space $\mathbb{R}^3$, the corners of a regular *Platonic tetrahedron*, the centrally reflected antitetrahedron and a regular octahedron with triplet and antitriplet in parallel hyperplanes:

$$\text{for } A_3 : \begin{cases} \mathbf{weights}\,[1,0,0] = \begin{pmatrix} [1,0,0] \\ [-1,1,0] \\ [0,-1,1] \\ [0,0,-1] \end{pmatrix} \cong \begin{pmatrix} 1 & \frac{1}{\sqrt{3}} & \frac{1}{\sqrt{6}} \\ -1 & \frac{1}{\sqrt{3}} & \frac{1}{\sqrt{6}} \\ 0 & -\frac{2}{\sqrt{3}} & \frac{1}{\sqrt{6}} \\ 0 & 0 & -\frac{3}{\sqrt{6}} \end{pmatrix}, \\ \qquad\qquad \text{(tetrahedron)} \\ \mathbf{weights}\,[0,0,1] = \begin{pmatrix} [0,0,1] \\ [0,1,-1] \\ [1,-1,0] \\ [-1,0,0] \end{pmatrix} \cong \begin{pmatrix} 0 & 0 & \frac{3}{\sqrt{6}} \\ 0 & \frac{2}{\sqrt{3}} & -\frac{1}{\sqrt{6}} \\ 1 & -\frac{1}{\sqrt{3}} & -\frac{1}{\sqrt{6}} \\ -1 & -\frac{1}{\sqrt{3}} & -\frac{1}{\sqrt{6}} \end{pmatrix}, \\ \qquad\qquad \text{(antitetrahedron)} \\ \mathbf{weights}\,[0,1,0] = \begin{pmatrix} [0,1,0] \\ [0,-1,1] \\ [-1,0,1] \\ [1,0,-1] \\ [0,1,-1] \\ [0,-1,0] \end{pmatrix} \cong \begin{pmatrix} 0 & \frac{2}{\sqrt{3}} & \frac{2}{\sqrt{6}} \\ 1 & -\frac{1}{\sqrt{3}} & \frac{2}{\sqrt{6}} \\ -1 & -\frac{1}{\sqrt{3}} & \frac{2}{\sqrt{6}} \\ 1 & \frac{1}{\sqrt{3}} & -\frac{2}{\sqrt{6}} \\ -1 & \frac{1}{\sqrt{3}} & -\frac{2}{\sqrt{6}} \\ 0 & -\frac{2}{\sqrt{3}} & -\frac{2}{\sqrt{6}} \end{pmatrix}. \\ \qquad\qquad \text{(octahedron)} \end{cases}$$

The *adjoint representation* with maximal root

$$\omega_{\max} = \gamma_1 + \gamma_r \Rightarrow \text{ad}\, A_r = \begin{cases} [2] & \text{for } r = 1, \\ [1, 0, \dots, 0, 1] & \text{for } r \geq 2, \end{cases}$$

is faithful only for the adjoint group $\mathbf{SU}(1 + r)/\mathbb{I}(1 + r)$, e.g., the adjoint $\mathbf{SU}(2)$-representation [2] for $\mathbf{SO}(3)$ or the $\mathbf{SU}(3)$-octet representation [1, 1] for $\mathbf{SU}(3)/\mathbb{I}(3)$. The Lie algebra A_r has r trivial adjoint weights and $\binom{1+r}{2}$ root pairs $\pm\omega_j$ whose lengths can be normalized to 2 (winding number). The weights of the adjoint representation for A_1 are the two endpoints of a line (roots) and the center point (trivial weight)

$$\text{for } A_1: \quad \mathbf{weights}\,[2] = \begin{pmatrix} \omega \\ \hline 0 \\ \hline -\omega \end{pmatrix} = \begin{pmatrix} [2] \\ \hline [0] \\ \hline [-2] \end{pmatrix} \cong \begin{pmatrix} 2 \\ \hline 0 \\ \hline -2 \end{pmatrix} \text{(distance)},$$
$$\text{Cartan matrix: } (\omega) = 2.$$

For A_2 the roots are the vertices of a regular *Platonic hexagon* with the two trivial weights as center points:

$$\text{for } A_2: \quad \mathbf{weights}\,[1, 1] = \begin{pmatrix} \omega_3 \\ \omega_2 \\ \omega_1 \\ \hline 0 \\ 0 \\ \hline -\omega_1 \\ -\omega_2 \\ -\omega_3 \end{pmatrix} = \begin{pmatrix} [1, 1] \\ [-1, 2] \\ [2, -1] \\ \hline [0, 0] \\ [0, 0] \\ \hline [-2, 1] \\ [1, -2] \\ [-1, -1] \end{pmatrix} \cong \begin{pmatrix} 1 & \sqrt{3} \\ -1 & \sqrt{3} \\ 2 & 0 \\ \hline 0 & 0 \\ 0 & 0 \\ \hline -2 & 0 \\ 1 & -\sqrt{3} \\ -1 & -\sqrt{3} \end{pmatrix}_{\text{(hexagon)}},$$
$$\text{Cartan matrix: } \begin{pmatrix} \omega_1 \\ \omega_2 \end{pmatrix} = \begin{pmatrix} 2 & -1 \\ -1 & 2 \end{pmatrix}.$$

In the general case the weights of the adjoint A_r-representation are in three $\mathbb{R}^{r-1}$-hyperplanes in a Euclidean $\mathbb{R}^r$, symmetric under central reflection. The decomposition with respect to simple A_{r-1}-representations gives the adjoint and the trivial one and the two dual defining representations:

$$\begin{aligned} A_r &\cong \bigoplus A_{r-1}, \\ [1, 0, \dots, 0, 1] &\cong [1, 0, \dots, 0, 1] \oplus [0, \dots, 0] \oplus [1, 0, \dots, 0] \oplus [0, \dots, 0, 1], \\ r(2 + r) &= (r - 1)(1 + r) + 1 + r + r \text{ for the dimensions.} \end{aligned}$$

Hence one obtains in one hyperplane the weights for the adjoint A_{r-1}-representation and the trivial one, whereas the two other hyperplanes contain the centrally reflected weights of the two defining A_{r-1}-representations, dual to each other. All $r(1 + r)$ roots have length 2 with the components in the eigenvalue matrix with r columns and $r(2+r)$ rows

$$\text{for } A_r: \quad \begin{pmatrix} \omega_{\binom{1+r}{2}} \\ \dots \\ \omega_2 \\ \omega_1 \\ \hline r \text{ zeros} \\ \hline -\omega_1 \\ -\omega_2 \\ \dots \\ -\omega_{\binom{1+r}{2}} \end{pmatrix} \cong \left(\begin{array}{c|c} \textbf{weights}\,[1,0,\dots,0] & \sqrt{\frac{2(1+r)}{r}} \\ \textbf{weights}\,[1,0,\dots,0,1] & 0 \\ \hline (r-1) \text{ zeros} & 0 \\ \hline -\textbf{weights}\,[1,0,\dots,0] & -\sqrt{\frac{2(1+r)}{r}} \end{array}\right) \subset \mathbb{R}^r,$$

$$\textbf{weights}\,[1,0,\dots,0,1],$$

$$\sum_{k=1}^{r(1+r)} \omega_k = 0, \quad \| \omega_k \| = 2.$$

The weight matrix contains the Cartan matrix for the fundamental roots:

$$\begin{pmatrix} \omega_1 \\ \omega_2 \\ \omega_3 \\ \dots \\ \omega_{r-1} \\ \omega_r \end{pmatrix} = \begin{pmatrix} 2 & -1 & 0 & 0 & \dots & 0 & 0 \\ -1 & 2 & -1 & 0 & \dots & 0 & 0 \\ 0 & -1 & 2 & -1 & \dots & 0 & 0 \\ \dots & & & \dots & & & \\ 0 & 0 & 0 & 0 & \dots & 2 & -1 \\ 0 & 0 & 0 & 0 & \dots & -1 & 2 \end{pmatrix}.$$

Starting with the line for A_1 in $\mathbb{R}^1$, the root diagrams give a nested series of convex A_r-polytopes in the Euclidean space $\mathbb{R}^r$, called *adjutopes*. The A_r-adjutope has r central points (trivial weights) and $r(1+r)$ vertices (roots) on a sphere with radius 2: For A_1 a distance, for A_2 in the plane a regular hexagon, and for A_3 in 3-space the 12 vertices (roots) of an *Archimedean hexoctahedron* with triplet, sextet and antitriplet in three parallel hyperplanes

$$\text{for } A_3: \quad \textbf{weights}\,[1,0,1] = \begin{pmatrix} \omega_6 \\ \omega_5 \\ \omega_3 \\ \omega_4 \\ \omega_2 \\ \omega_1 \\ \hline 3 \text{ zeros} \\ \hline -\omega_1 \\ -\omega_2 \\ -\omega_4 \\ -\omega_3 \\ -\omega_5 \\ -\omega_6 \end{pmatrix} = \begin{pmatrix} [1,0,1] \\ [-1,1,1] \\ [0,-1,2] \\ [1,1,-1] \\ [-1,2,-1] \\ [2,-1,0] \\ \hline 3 \text{ zeros} \\ \hline [-2,1,0] \\ [1,-2,1] \\ [-1,-1,1] \\ [0,1,-2] \\ [1,-1,-1] \\ [-1,0-1] \end{pmatrix} \cong \begin{pmatrix} 1 & \frac{1}{\sqrt{3}} & \frac{4}{\sqrt{6}} \\ -1 & \frac{1}{\sqrt{3}} & \frac{4}{\sqrt{6}} \\ 0 & -\frac{2}{\sqrt{3}} & \frac{4}{\sqrt{6}} \\ 1 & \sqrt{3} & 0 \\ -1 & \sqrt{3} & 0 \\ 2 & 0 & 0 \\ \hline & 3 \text{ zeros} & \\ \hline -2 & 0 & 0 \\ 1 & -\sqrt{3} & 0 \\ -1 & -\sqrt{3} & 0 \\ 0 & \frac{2}{\sqrt{3}} & -\frac{4}{\sqrt{6}} \\ 1 & -\frac{1}{\sqrt{3}} & -\frac{4}{\sqrt{6}} \\ -1 & -\frac{1}{\sqrt{3}} & -\frac{4}{\sqrt{6}} \end{pmatrix},$$

hexoctahedron

$$\text{Cartan matrix:} \quad \begin{pmatrix} \omega_1 \\ \omega_2 \\ \omega_3 \end{pmatrix} = \begin{pmatrix} 2 & -1 & 0 \\ -1 & 2 & -1 \\ 0 & -1 & 2 \end{pmatrix}.$$

The hexoctahedron as intersection of the Platonic hexahedron and octahedron has six squares and eight regular triangles as sides.

6.3.2 Representations of the Lie Algebras C_r

The simple Lie algebra C_r is definable as the Lie algebra of the symplectic invariance group of the nondegenerate antisymmetric bilinear forms:

$$r \geq 1: \quad {}_{r(1+2r)}C_r \cong \log \mathbf{Sp}(\mathbb{C}^{2r}), \quad {}_{r(1+2r)}C_r^c \cong \log \mathbf{SpU}(2r).$$

C_1 is isomorphic to A_1. The r fundamental representations γ_j with dimensions $\binom{2r}{j} - \binom{2r}{j-2}$ act on the Grassmann powers $\bigwedge^j V$ for $V \cong \mathbb{C}^{2r}$,

$$\tbinom{2r}{1}_- \text{---} [\tbinom{2r}{2}-1]_+ \text{---} [\tbinom{2r}{3}-\tbinom{2r}{1}]_- \text{---} \cdots\cdots \text{---} [\tbinom{2r}{r-1}-\tbinom{2r}{r-3}]_{(-1)^{r-1}} = [\tbinom{2r}{r}-\tbinom{2r}{r-2}]_{(-1)^r},$$

fundamental representations for C_r
with cyclic representation: $\{\gamma_1 \cong \binom{2r}{1}\}$.

All fundamental representations have an invariant nondegenerate bilinear form with the symmetry (orthogonal or symplectic) given as subindex:

$$\text{symmetry for bilinear forms of } C_r, \text{ for } \gamma_j: \quad (-1)^j.$$

6.3.3 Representations of the Lie Algebras B_r

The simple Lie algebra B_r is definable as the Lie algebra of the orthogonal invariance group of nondegenerate symmetric bilinear forms of odd-dimensional spaces:

$$r \geq 1: \quad {}_{r(1+2r)}B_r \cong \log \mathbf{SO}(\mathbb{C}^{1+2r}), \quad {}_{r(1+2r)}B_r^c \cong \log \mathbf{SO}(1+2r).$$

B_1 is isomorphic to A_1. For $r \geq 2$, the defining representation on $V \cong \mathbb{C}^{1+2r}$ is cyclic for $(r-1)$ fundamental representations on the Grassmann powers $\bigwedge^j V$ for $j = 1, \ldots, r-1$ with dimension $\binom{1+2r}{j}$. The remaining *spinorial* fundamental representation for the fundamental weight γ_r, obtainable from the Clifford algebra CLIFF$(1+2r)$ (chapter "Quantum Algebras"), has dimension 2^r:

$$\tbinom{1+2r}{1}_+ \text{---} \tbinom{1+2r}{2}_+ \text{---} \cdots\cdots \text{---} \tbinom{1+2r}{r-2}_+ \text{---} \tbinom{1+2r}{r-1}_+ = (2^r)_\eta,$$

fundamental representations for B_r, $r \geq 2$,
with cyclic representations: $\{\gamma_1 \cong \binom{1+2r}{1}, \; \gamma_r \cong 2^r\}$.

There are two cyclic representations from the two loose ends in the Dynkin diagram.

All fundamental representations have an invariant nondegenerate bilinear form:

$$\text{symmetry for bilinear forms of } B_r, \text{ for } \gamma_j: \quad \begin{cases} 1, & j \neq r, \\ (-1)^{\frac{r(1+r)}{2}}, & j = r, \end{cases}$$

with the subindex $\pm$ showing the symmetry property of the form. Up to the spinorial representation with $\eta = (-1)^{\frac{r(1+r)}{2}}$, all fundamental representation spaces inherit an orthogonal product.

The *adjoint representation* is fundamental for $r \geq 3$:

$$\omega_{\max} = \begin{cases} 2\gamma_2, & r = 2, \\ \gamma_2, & r \geq 3 \end{cases} \Rightarrow \operatorname{ad} B_r = \begin{cases} [0, 2], & r = 2, \\ [0, 1, 0, \ldots, 0], & r \geq 3. \end{cases}$$

The structure in the Euclidean plane $\mathbb{R}^2$ for ${}_{10}B_2 \cong \log \mathbf{SO}(5)$, isomorphic to $C_2 \cong \log \mathbf{Sp}(\mathbb{C}^4)$, shows up in the weight diagrams. The two fundamental B_2-representations have their weights on a centrally dotted square for 5 and on a $\frac{\pi}{4}$-rotated square for 4. The weights of the adjoint 10-dimensional representation constitute a square, dotted centrally on all 4 sides, with two central weights.

The Lie algebra ${}_{21}B_3 \cong \log \mathbf{SO}(7)$ has their weight diagrams in Euclidean 3-space. The fundamental representations have dimensions $(7, 21, 8)$. The adjoint 21-dimensional representation is decomposable into the B_2-representations as $21 = 5 + (1 + 10) + 5$. The number of roots at the convex vertices of the root diagram is $12 = 4 + 4 + 4$, i.e., the diagram has three squares in three parallel planes, the middle one larger and rotated by $\frac{\pi}{4}$ with respect to the upper and lower square. The roots constitute the 12 vertices of an Archimedean hexoctaedron.

More: All simple rank-3 Lie algebras $A_3 \cong D_3, B_3, C_3$ are characterizable in Euclidean 3-space with the Archimedean hexoctaedron whose surface has 8 triangles and 6 squares. The roots of ${}_{15}A_3 \cong {}_{15}D_3$ are the 12 vertices. The 18 roots of ${}_{21}B_3$ and ${}_{21}C_3$ are these 12 vertices and, in addition, the 6 central points in the 6 squares. The Lie algebra C_3 has the same $C_2 \cong B_2$-decomposition as B_3.

6.3.4 Representations of the Lie Algebras D_r

The Lie algebra D_r is definable as Lie algebra of the orthogonal invariance group of the nondegenerate symmetric bilinear forms of even-dimensional spaces

$$r \geq 1: \quad {}_{r(2r-1)}D_r \cong \log \mathbf{SO}(\mathbb{C}^{2r}), \quad {}_{r(2r-1)}D_r^c \cong \log \mathbf{SO}(2r),$$

$D_1 \cong \log \mathbf{SO}(\mathbb{C}^2)$ is abelian and $D_2 \cong A_1 \oplus A_1$ semisimple. For simple $D_{r\geq 3}$ one obtains the $(r-2)$ fundamental representations with dimension $\binom{2r}{j}$ for $j = 1, \dots, r-2$ on the Grassmann powers $\bigwedge^j V$ of $V \cong \mathbb{C}^{2r}$. The two additional *spinorial* fundamental representations from the Clifford algebra CLIFF$(2r)$ (chapter "Quantum Algebras") for the fundamental weights γ_{r-1} and γ_r both have dimension 2^{r-1}:

$$\begin{array}{lcll} & \nearrow & 2^{r-1} & \text{for } r \in 2\mathbb{N}: \ \eta = (-1)^{\frac{r}{2}} \\ \binom{2r}{1}_+ - \binom{2r}{2}_+ - \cdots\cdots - \binom{2r}{r-3}_+ - \binom{2r}{r-2}_+ & & & \nwarrow \ \text{dual for } r \in 1 + 2\mathbb{N} \ \swarrow \\ & \searrow & 2^{r-1} & \text{for } r \in 2\mathbb{N}: \ \eta = (-1)^{\frac{r}{2}} \end{array},$$

fundamental representations for D_r, $r \geq 3$,
with cyclic representations: $\{\gamma_1 \cong \binom{1+r}{1}, \ \gamma_{r-1} \cong 2^{r-1}, \ \gamma_r \cong 2^{r-1}\}$.

The three cyclic representations, the $2r$-dimensional defining representation and the two 2^{r-1}-dimensional spinor representations, come from the three loose ends of the Dynkin diagram.

All nonspinorial fundamental representations inherit an invariant symmetric bilinear form:

$$\text{symmetry for bilinear forms of } D_r$$
$$\text{for } \gamma_j : \quad \left\{ \begin{array}{ll} \text{for } j = 1, \ldots, r-2: & 1, \\ \text{for } j = r-1, r: & \left\{ \begin{array}{ll} 0, & r \in 1 + 2\mathbb{N}, \\ (-1)^{\frac{r}{2}}, & r \in 2\mathbb{N}. \end{array} \right. \end{array} \right.$$

Depending on the rank r the spinor representations are either dual to each other (for odd rank) or they are both self-dual (for even rank).

The *adjoint representation* is fundamental for $r \geq 4$:

$$\omega_{\max} = \left\{ \begin{array}{ll} \gamma_2 + \gamma_3, & r = 3 \\ \gamma_2, & r \geq 4 \end{array} \right. \Rightarrow \operatorname{ad} D_r = \left\{ \begin{array}{ll} [0, 1, 1], & r = 3, \\ [0, 1, 0, \ldots, 0], & r \geq 4. \end{array} \right.$$

With respect to the isomorpism $A_3 \cong D_3$ one has to take care of the different numbering of the roots and weights for $A_3 : \ (1, 2, 3) \leftrightarrow (2, 1, 3) \ :$ for D_3.

6.3.5 Representations of the Exceptional Lie Algebras

The adjoint representations of the exceptional Lie algebras (underlined in the Dynkin diagrams) are fundamental. The existence of an invariant bilinear form is denoted with a subindex ± 1 for orthogonally and symplectically invariant bilinear form and with 0 if there does not exist an invariant bilinear form.

For the Lie algebra ${}_{14}G_2$ the 7-dimensional fundamental representation is cyclic:

$$7_+ \equiv [\tbinom{7}{2} - 7]_+ = \underline{14}_+,$$

fundamental representations for G_2
with cyclic representation: $\{7\}$.

The Lie algebra ${}_{52}F_4$ has two cyclic representations with dimension 26 and 52:

$$26_+ — [\tbinom{26}{2} - 2 \cdot 26]_+ = [\tbinom{52}{2} - 52]_+ — \underline{52}_+,$$

$$26_+ — 273_+ = 1{,}274_+ — \underline{52}_+,$$

fundamental representations for F_4
with cyclic representations: $\{26, \ 52\}$.

For the series $E_{6,7,8}$ there are three cyclic representations at the three loose ends, first for ${}_{78}E_6$:

$$\begin{array}{ccccccccc} & & & & \underline{78}_+ & & & & \\ & & & & | & & & & \\ 27 & — & \binom{27}{2} & — & \binom{27}{3}_+ & — & \binom{27}{2} & — & 27, \end{array}$$

$$\text{dual}: 27 \leftrightarrow 27,\quad \binom{27}{2} \leftrightarrow \binom{27}{2}$$

$$\begin{array}{ccccccccc} & & & & \underline{78}_+ & & & & \\ & & & & | & & & & \\ 27_0 & — & 351_0 & — & 2{,}925_+ & — & 351_0 & — & 27_0, \end{array}$$

fundamental representations for E_6
with cyclic representations: $\{78,\ 27,\ 27\}$

In addition, $2,925 = \binom{78}{2} - 78$. Furthermore, for ${}_{133}E_7$,

$$\begin{array}{ccccccccccc} & & & & & & 912_- & & & & \\ & & & & & & | & & & & \\ 56_- & — & [\binom{56}{2}-1]_+ & — & [\binom{56}{3}-\binom{56}{1}]_- & — & [\binom{56}{4}-\binom{56}{2}]_+ & — & [\binom{133}{2}-\binom{133}{1}]_+ & — & \underline{133}_+, \end{array}$$

$$\begin{array}{ccccccccccc} & & & & & & 912_- & & & & \\ & & & & & & | & & & & \\ 56_- & — & 1{,}539_+ & — & 27{,}664_- & — & 365{,}750_+ & — & 8{,}645_+ & — & \underline{133}_+ \end{array}$$

fundamental representations for E_7
with cyclic representations: $\{912,\ 56,\ 133\}$

Finally, for ${}_{248}E_8$ the dimensions for the fundamental representations are astronomical numbers; first explicitly

$$\begin{array}{ccccccccccccc} & & & & & & & & 147{,}250_+ & & & & \\ & & & & & & & & | & & & & \\ \underline{248}_+ & — & 30{,}380_+ & — & 2{,}450{,}240_+ & — & 146{,}325{,}270_+ & — & 6{,}899{,}079{,}264_+ & — & 6{,}696{,}000_+ & — & 3{,}875_+ \end{array},$$

fundamental representations for E_8
with cyclic representations: $\{147250,\ 248,\ 3875\}$

The totally antisymmetric products of the three cyclic representations have the vertex-related fundamental representation as the highest-dimensional one in the following decompositions into irreducible representations:

$$\begin{aligned}
\binom{248}{2} &= 30{,}380 \oplus 248\\
\binom{248}{3} &= 2{,}450{,}240 \oplus 2{\bullet}30{,}380 \oplus 2{\bullet}248\\
\binom{248}{4} &= 146{,}325{,}270 \oplus (4{,}096{,}000) \oplus 2{,}450{,}240 \oplus (779{,}247) \oplus 147{,}250 \oplus (27{,}000) \oplus 3{,}875 \oplus 248\\
\binom{248}{5} &= 6{,}899{,}079{,}264 \oplus (344{,}452{,}500) \oplus 146{,}325{,}270 \oplus (76{,}271{,}625) \oplus (26{,}411{,}008)\\
&\quad \oplus 6{,}696{,}000 \oplus (4{,}096{,}000) \oplus (1{,}763{,}125) \oplus 2{\bullet}(779{,}247) \oplus 147{,}250 \oplus 2{\bullet}30{,}380 \oplus 248\\
\binom{147{,}250}{2} &= 6{,}899{,}079{,}264 \oplus (2{,}275{,}896{,}000) \oplus \ldots \oplus 248\\
\binom{3{,}875}{2} &= 6{,}696{,}000 \oplus (779{,}247) \oplus 30{,}380 \oplus 248\\
\binom{3{,}875}{3} &= 6{,}899{,}079{,}264 \oplus \ldots
\end{aligned}$$

There arise also (in parentheses) other not fundamental irreducible representations in the decompositions.

6.4 Centrality of Representations

The representations of locally isomorphic connected Lie groups are distinguished by the representations of their *discrete centers*. For the simply connected universal cover Lie groups $\overline{\exp L}$ and $\overline{\exp L^c}$ of a simple complex Lie algebra L of rank r with compact form L^c the equivalence classes of the irreducible complex finite-dimensional representations are characterized as for the Lie algebras, i.e., by the representation cone of L or L^c:

$$\begin{aligned}
&\mathbf{irrep}_{\text{fin}} L \cong \mathbf{irrep}\, L^c \cong \{[n^1, \ldots, n^r] \mid n^i = 0, 1, \ldots\} \cong \mathbb{N}_0^r,\\
&[n^1, \ldots, n^r] \in \mathbf{irrep}\, L^c \Rightarrow \exp[n^1, \ldots, n^r] \in \mathbf{irrep}\, \overline{\exp L^c},
\end{aligned}$$

and the associated weight modules. Locally isomorphic connected Lie groups arise from the universal cover group as classes $\overline{\exp L}/\mathbb{I}$ with respect to *discrete cyclic subgroups* $\mathbb{I}$ *of the center*, for the four main series

cover group $G = \overline{\exp L}$	$\mathbf{SL}(\mathbb{C}^{1+r})$	$\mathbf{Sp}(\mathbb{C}^{2r})$	$\mathbf{SO}(\mathbb{C}^{1+2r})$	$\mathbf{SO}(\mathbb{C}^{2r})$
center $\mathbb{I}(G)$	$\mathbb{I}(1+r)$	$\mathbb{I}(2)$	$\mathbb{I}(2)$	$\mathbb{I}(4)$ if r odd $\mathbb{I}(2) \times \mathbb{I}(2)$ if r even

The center of the universal cover group acts on the weight module. The irreducible representations and their weights of quotients $\overline{\exp L}/\mathbb{I}$ are subcones and submodules $\mathbf{weights}\, \frac{\overline{\exp L}}{\mathbb{I}} \cong \frac{\mathbf{weights}\, \overline{\exp L}}{\mathbb{I}}$. For example, the even weights $2z \in 2\mathbb{Z}$ of the rotations $\mathbf{SO}(3) \cong \mathbf{SU}(2)/\mathbb{I}(2)$ constitute a submodule of the integer spin weights $z \in \mathbb{Z}$.

Each representation projects the discrete center of the universal cover group to a subgroup, another cyclic group, called the *centrality of the representation*:

$$\exp[n^1, \ldots, n^r] : \text{centr}\, \overline{\exp L_r} = \mathbb{I}(G) \longrightarrow \mathbb{I}(D) \subseteq \mathbb{I}(G).$$

The center representation is trivial for the classes $\mathbb{I}(G)/\mathbb{I}(D)$.

Some remarks on the endomorphisms of cyclic groups: Any cyclic group is the direct product of the cyclic groups associated with its prime power factorization

$$\begin{aligned}
&\mathbb{I}(n) = \{\exp \tfrac{2i\pi}{n} z \mid z \in \mathbb{Z}\}, \quad n = 1, 2, \ldots,\\
&n = p_1^{k_1} \cdots p_l^{k_l}, \text{ relative prime } p_1, \ldots, p_l \Rightarrow \mathbb{I}(n) \cong \mathbb{I}(p_1^{k_1}) \times \cdots \times \mathbb{I}(p_l^{k_l}),\\
&m \text{ divisor of } n \Rightarrow \mathbb{I}(\tfrac{n}{m}) \cong \mathbb{I}(n)/\mathbb{I}(m).
\end{aligned}$$

The irreducible $\mathbb{I}(n)$-representations are determined by the value for the cyclic element $\exp \frac{2i\pi}{n}$:

$$\iota^C : \mathbb{I}(n) \longrightarrow \mathbb{I}(n), \quad \exp \tfrac{2i\pi}{n} \longmapsto \exp \tfrac{2i\pi}{n} C, \quad 0 \neq C \in \mathbb{Z}.$$

Obvioulsy only the class number $C \bmod n \in \mathbb{Z}_n \cong \{1, \dots, n\}$ is relevant. In the case of $\mathbf{SU}(n)$, the number $C \bmod n$ is called *n-ality*, *twoality* for $\mathbf{SU}(2)$, *triality* for $\mathbf{SU}(3)$, etc. The ι^C-image is a cyclic group which involves the greatest common divisor $\gcd(n, C)$:

$$\iota^C[\mathbb{I}(n)] = \mathbb{I}(\tfrac{n}{\gcd(n,C)}), \quad \mathbb{I}(\tfrac{n}{\gcd(n,n)}) = \mathbb{I}(1).$$

The centrality of the defining matrix representations with largest weight γ_1 for the four main series is the full center:

group	$\mathbf{SL}(\mathbb{C}^{1+r})$ $\exp A_r$	$\mathbf{Sp}(\mathbb{C}^{2r})$ $\exp C_r$	$\mathbf{SO}(\mathbb{C}^{1+2r})$ $\exp B_r/\mathbb{I}(2)$	$\mathbf{SO}(\mathbb{C}^{2r})$ $\exp D_r/\mathbb{I}(2)$
center	$\mathbb{I}(1+r)$	$\mathbb{I}(2)$	$\mathbb{I}(1)$	$\mathbb{I}(2)$

centrality of the cyclic representation $\gamma_1 \cong \exp[1, 0, \dots, 0]$

The centrality of a fundamental representation is obtained from the corresponding tensor product of the cyclic representations and the centrality of $\exp[n^1, \dots, n^r]$ by the centrality of the fundamental representations. The centrality of $\exp[n^1, \dots, n^r]$ for $\exp A_r$ and $\exp C_r$ are

$$\begin{array}{llll} \text{for } \mathbf{SU}(1+r): & \mathbb{I}(1+r) \ni \exp \frac{2i\pi}{1+r} \longmapsto & \exp \frac{2\pi}{1+r} C, & C = \sum_{j=1}^{r} j n^j, \\ \text{for } \mathbf{SpU}(2r): & \mathbb{I}(2) \ni \exp i\pi \longmapsto & \exp i\pi C, & C = \sum_{j=1}^{r} n^j, \end{array}$$

with the examples for the n-ality $\sum_{j=1}^{r} j n^j \bmod (1+r) \in \mathbb{Z}_{1+r}$:

$$\begin{array}{lrl} \mathbf{SU}(2), \text{ twoality for } [n]: & n \bmod 2 & \in \mathbb{Z}_2, \\ \mathbf{SU}(3), \text{ triality for } [n^1, n^2]: & (n^1 + 2n^2) \bmod 3 & \in \mathbb{Z}_3, \\ \mathbf{SU}(4), \text{ quadrality for } [n^1, n^2, n^3]: & (n^1 + 2n^2 + 3n^3) \bmod 4 & \in \mathbb{Z}_4. \end{array}$$

The possible centralities for $\mathbf{SU}(1+r)$-representations are given by all subgroups of $\mathbb{I}(1+r)$, i.e., by $\mathbb{I}(d)$ with d a divisor of $(1+r)$, e.g., $\mathbb{I}(1)$ and $\mathbb{I}(2)$ for $\mathbf{SU}(2)$, $\mathbb{I}(1)$, and $\mathbb{I}(3)$ for $\mathbf{SU}(3)$ and $\mathbb{I}(1)$, $\mathbb{I}(2)$, and $\mathbb{I}(4)$ for $\mathbf{SU}(4)$.

The centralities determine the weight modules of the corresponding quotient groups, e.g., in the case of $\mathbf{SU}(1+r)$ with d the divisors of $(1+r)$:

$$\mathbf{weights}\, \mathbf{SU}(1+r)/\mathbb{I}(d) = \{[z^1, \dots, z^r] \mid z^j \in \mathbb{Z}, \sum_{j=1}^{r} j z^j = d\mathbb{Z}\},$$

with the examples

$$\begin{array}{ll}
\mathbf{weights\,SU}(2) & = \{[z] \mid z \in \mathbb{Z}\}, \\
\mathbf{weights\,SU}(2)/\mathbb{I}(2) & = \{[z] \mid z \in 2\mathbb{Z}\}, \\
\mathbf{weights\,SU}(3) & = \{[z^1, z^2] \mid z^j \in \mathbb{Z}\}, \\
\mathbf{weights\,SU}(3)/\mathbb{I}(3) & = \{[z^1, z^2] \mid z^1 - z^2 \in 3\mathbb{Z}\}, \\
\mathbf{weights\,SU}(4) & = \{[z^1, z^2, z^3] \mid z^j \in \mathbb{Z}\}, \\
\mathbf{weights\,SU}(4)/\mathbb{I}(2) & = \{[z^1, z^2, z^3] \mid z^1 + 2z^2 - z^3 \in 2\mathbb{Z}\}, \\
\mathbf{weights\,SU}(4)/\mathbb{I}(4) & = \{[z^1, z^2, z^3] \mid z^1 + 2z^2 - z^3 \in 4\mathbb{Z}\}.
\end{array}$$

For the orthogonal groups, only the spinor representations are faithful for the full center:

$$\begin{array}{ll}
\overline{\mathbf{SO}(1+2r)}, \text{ centrality for} & \begin{cases} \gamma_j,\ j = 1, \dots, r-1: & \mathbb{I}(1), \\ \text{spinorial } \gamma_r: & \mathbb{I}(2), \end{cases} \\
\overline{\mathbf{SO}(2r)}, \text{ centrality for} & \begin{cases} \gamma_j,\ j = 1, \dots, r-2: & \mathbb{I}((\tfrac{2}{j})) \in \{\mathbb{I}(2), \mathbb{I}(1)\}, \\ \text{spinorial } \gamma_{r-1}, \gamma_r: & \begin{cases} \mathbb{I}(2) \times \mathbb{I}(2) & \text{for } r \text{ even}, \\ \mathbb{I}(4) & \text{for } r \text{ odd}. \end{cases} \end{cases}
\end{array}$$

6.4.1 Broken Winding Numbers as Lepton and Quark Hypercharges

With the transition from a representation of a Lie algebra direct sum-product $L_1 \oplus L_2$, $[L_1, L_2] = \{0\}$ to an associated Lie group $G_1 \circ G_2$, it may be possible that both normal factors have a common center subgroup $\mathbb{I} \subseteq \operatorname{centr} G_1 \cap \operatorname{centr} G_2$. Representations of centrally correlated groups play a role in the nonrelativistic hydrogen atom (chapter "The Kepler Factor") where they explain the squares $2 \times 1^2, 2 \times 2^2, 2 \times 3^2$ etc. in the shell ocupation numbers (energy degeneracy), and in the standard model of elementary particles (chapter "Gauge Theories") where they explain why isospin doublets have a hypercharge factor $\frac{1}{2}$, color triplets a factor $\frac{1}{3}$, and isospin doublet-color triplets a hypercharge factor $\frac{1}{2 \cdot 3} = \frac{1}{6}$.

A full unitary group $\mathbf{U}(p,q)$, $p+q = n \geq 2$, is not a direct product group: it has a central connection of its two normal Lie subgroups by the cyclic group $\mathbb{I}(n)$, the $\mathbf{SU}(p,q)$-center

$$\begin{array}{l}
\mathbf{U}(p,q) = \mathbf{U}(\mathbf{1}_n) \circ \mathbf{SU}(p,q) \cong \frac{\mathbf{U}(1) \times \mathbf{SU}(p,q)}{\mathbb{I}(n)}, \\
\text{with } \mathbf{U}(\mathbf{1}_n) \cap \mathbf{SU}(p,q) = \operatorname{centr} \mathbf{SU}(p,q) \cong \mathbb{I}(n).
\end{array}$$

The equivalence-classes-inducing denominator $\mathbb{I}(n)$ can be generated by a diagonal Cartan subalgebra element of $\log \mathbf{SU}(p,q)$:

$$\begin{array}{l}
\mathbf{U}(\mathbf{1}_n) \supset \mathbb{I}(n) \ni \exp \frac{2i\pi}{n} \mathbf{1}_n = \exp \frac{2i\pi}{n} \mathbf{w}_n \in \mathbb{I}(n) \subset \mathbf{SU}(p,q), \\
\mathbf{w}_n = \begin{pmatrix} \mathbf{1}_{n-1} & 0 \\ 0 & -(n-1) \end{pmatrix} = \sqrt{\binom{n}{2}} \sigma(n)^{n^2-1} \in A_{n-1},\ n \geq 2, \\
\text{e.g., } \mathbf{w}_2 = \sigma^3 \in A_1, \quad \mathbf{w}_3 = \sqrt{3}\lambda^8 \in A_2.
\end{array}$$

The weight module of the direct product group $\mathbf{U}(1)\times\mathbf{SU}(p,q)$ is the direct sum of the weight modules

$$\begin{aligned}\mathbf{weights}\,[\mathbf{U}(1)\times\mathbf{SU}(1+r)] &= \mathbf{weights}\,\mathbf{U}(1)\ \oplus\ \mathbf{weights}\,\mathbf{SU}(1+r)\\ &\cong \mathbb{Z}\ \oplus\ \mathbb{Z}^r \cong \mathbb{Z}^{1+r}.\end{aligned}$$

For $\mathbf{U}(1+r)$-representations, it has to be factorized with respect to the $\mathbb{I}(1+r)$ central correlation.

The *two defining representations of* $\mathbf{U}(1+r)$, dual to each other, are (with generalized Pauli matrices)

$$\begin{aligned}\exp[+\tfrac{1}{1+r}||\underbrace{1,0,\dots,0}_{r\text{ places}}] &= \exp[+i\alpha_0\tfrac{\mathbf{1}_{1+r}}{1+r}+i\vec{\alpha}\vec{\sigma}(1+r)],\\ \exp[-\tfrac{1}{1+r}||\underbrace{0,\dots,0,1}_{r\text{ places}}] &= \exp[-i\alpha_0\tfrac{\mathbf{1}_{1+r}}{1+r}-i\vec{\alpha}\vec{\sigma}(1+r)^T],\end{aligned}$$

with the examples for hyperisospin $\mathbf{U}(2)$ and hypercolor $\mathbf{U}(3)$

$$\mathbf{U}(2):\ \left\{\begin{aligned}\exp[+\tfrac{1}{2}||1] &= \exp[+i\alpha_0\tfrac{\mathbf{1}_2}{2}+i\vec{\alpha}\vec{\sigma}],\\ \exp[-\tfrac{1}{2}||1] &= \exp[-i\alpha_0\tfrac{\mathbf{1}_2}{2}-i\vec{\alpha}\vec{\sigma}^T],\end{aligned}\right.$$

$$\mathbf{U}(3):\ \left\{\begin{aligned}\exp[+\tfrac{1}{3}||1,0] &= \exp[+i\alpha_0\tfrac{\mathbf{1}_3}{3}+i\vec{\alpha}\vec{\lambda}],\\ \exp[-\tfrac{1}{3}||0,1] &= \exp[-i\alpha_0\tfrac{\mathbf{1}_3}{3}-i\vec{\alpha}\vec{\lambda}^T].\end{aligned}\right.$$

It is possible to represent $\mathbf{U}(1)$ by a broken winding number $\pm\frac{1}{1+r}$ in a maximal abelian subgroup of $\mathbf{U}(1+r)$ since uniqueness of the $\mathbf{U}(1)$-representation with $\alpha_0\longmapsto\alpha_0+2\pi$ is satified up to the centrally correlated $\mathbf{SU}(1+r)$ part

$$\begin{aligned}\exp[i(\alpha_0+2\pi)\tfrac{\mathbf{1}_{1+r}}{1+r}] &= \exp[i\alpha_0\tfrac{\mathbf{1}_{1+r}}{1+r}+i2\pi\tfrac{\mathbf{w}_{1+r}}{1+r}],\\ \text{e.g., for }\mathbf{U}(2):\ \exp[i(\alpha_0+2\pi)\tfrac{\mathbf{1}_2}{2}] &= \exp[i\alpha_0\tfrac{\mathbf{1}_2}{2}+i2\pi\tfrac{\sigma^3}{2}].\end{aligned}$$

The totally antisymmetric products of the two defining representations will be defined as $2(1+r)$ *fundamental* $\mathbf{U}(1+r)$*-representations.* They come in dual pairs:

$$\begin{aligned}&[+\tfrac{j}{1+r}||\underbrace{0,\dots,0,1,0\dots,0}_{j\text{th place}}],\ j=1,\dots,r \quad\text{and}\quad [+1||0,\dots,0],\\ &[-\tfrac{j}{1+r}||\underbrace{0,\dots,0,1,0\dots,0}_{(1+r-j)\text{th place}}],\ j=1,\dots,r \quad\text{and}\quad [-1||0,\dots,0].\end{aligned}$$

e.g., for hyperisospin and hypercolor

$$\mathbf{U}(2):\left\{\begin{matrix}[+\frac{1}{2}||1], & [+1||0]\\ [-\frac{1}{2}||1], & [-1||0]\end{matrix}\right.,\quad \mathbf{U}(3):\left\{\begin{matrix}[+\frac{1}{3}||1,0], & [+\frac{2}{3}||0,1], & [+1||0,0]\\ [-\frac{1}{3}||0,1], & [-\frac{2}{3}||1,0], & [-1||0,0]\end{matrix}\right..$$

The irreducible $\mathbf{U}(1+r)$-representations have the following $(1+r)$-ality correlation for $\mathbf{U}(1)$ and $\mathbf{SU}(1+r)$-representations

$$\begin{aligned}\mathbf{irrep}\,\mathbf{U}(1+r)) &= \{[\tfrac{\sum_{j=1}^r jn^j}{1+r}+z||n^1,\dots,n^r]\ \big|\ n^j\in\mathbb{N}_0,\ z\in\mathbb{Z}\},\\ \mathbf{weights}\,\mathbf{U}(1+r)) &= \{[\tfrac{\sum_{j=1}^r jz^j}{1+r}+z||z^1,\dots,z^r]\ \big|\ z^j\in\mathbb{Z},\ z\in\mathbb{Z}\}.\end{aligned}$$

Such central connections between the abelian compact group $\mathbf{U}(1)$ and simple compact groups $\mathbf{SU}(1+r)$ with broken winding numbers are used in the standard model of electroweak and strong interactions to describe the half, third, and sixth integer hypercharge numbers of lepton and quark fields: The 2·2 fundamental representations of the centrally correlated hypercharge-isospin group $\mathbf{U}(2)$ is used for the lepton isodoublet and lepton isosinglet fields $\mathbf{l}$ and $\mathbf{e}$ respectively with their antifields:

$$\mathbf{irrep}\,\mathbf{U}(2) = \{[[y||2T] = \tfrac{n}{2} + z||n] \mid n \in \mathbb{N}_0,\ z \in \mathbb{Z}\},$$

$[\frac{1}{2}\|\|1]$	$[1\|\|0]$
$\mathbf{l}^\star$	$\mathbf{e}^\star$
$[-\frac{1}{2}\|\|1]$	$[-1\|\|0]$
$\mathbf{l}$	$\mathbf{e}$

$\mathbf{U}(2)$-hypercharge-isospin representations.

The $2\cdot 3$ fundamental representations of the centrally correlated hypercharge-color group $\mathbf{U}(3)$ is used for the quark up and down isosinglet fields $\mathbf{u}$ and $\mathbf{d}$ respectively and the lepton isosinglet field $\mathbf{e}$:

$$\mathbf{irrep}\,\mathbf{U}(3) = \{[y||n^1, n^2] = [\tfrac{n^1-n^2}{3} + z||n^1, n^2] \mid n^j \in \mathbb{N}_0,\ z \in \mathbb{Z}\},$$

$[\frac{1}{3}\|\|1,0]$	$[\frac{2}{3}\|\|0,1]$	$[1\|\|0,0]$
$\mathbf{d}^\star$	$\mathbf{u}$	$\mathbf{e}^\star$
$[-\frac{1}{3}\|\|0,1]$	$[-\frac{2}{3}\|\|1,0]$	$[-1\|\|0,0]$
$\mathbf{d}$	$\mathbf{u}^\star$	$\mathbf{e}$

$\mathbf{U}(3)$-hypercharge-color representations.

The centers of the group factors in a direct product define a direct product center, e.g.,

$$\begin{aligned}\text{centr}\,[\mathbf{SU}(n)\times\mathbf{SU}(m)] &= \mathbb{I}(n)\times\mathbb{I}(m),\\ \text{centr}\,[\mathbf{U}(1)\times\mathbf{SU}(n)\times\mathbf{SU}(m)] &= \mathbf{U}(1)\times\mathbb{I}(n)\times\mathbb{I}(m).\end{aligned}$$

Subgroups of such product centers may be used to correlate the individual centers and hence the factor groups, e.g.,

$$\begin{array}{lll}\mathbb{I}(2)\cong\{1,I\} & \text{with } I=(-1,-1) & \in \mathbb{I}(2)\times\mathbb{I}(2),\\ & & \subset \mathbf{SU}(2)\times\mathbf{SU}(2),\\ \mathbb{I}(6)\cong\{1,I,\ldots,I^5\} & \text{with } I=(e^{\frac{2i\pi}{6}},e^{\frac{2i\pi}{2}},e^{\frac{2i\pi}{3}}) & \in \mathbb{I}(6)\times\mathbb{I}(2)\times\mathbb{I}(3),\\ & & \subset \mathbf{U}(1)\times\mathbf{SU}(2)\times\mathbf{SU}(3).\end{array}$$

The weight modules and representation cones of such product groups with central correlations are correspondingly modified.

The centrally correlated double spin group with dimension 6 and rank 2 arises in the nonrelativistic Kepler problem with rotation invariance and perihelion conservation

$$\frac{\mathbf{SU}(2)\times\mathbf{SU}(2)}{\mathbb{I}(2)} \cong \mathbf{SO}(4).$$

It gives an integer correlation condition for the half-integer spin weights

$$\begin{aligned}
\textbf{weights}\,\mathbf{SU}(2)\times\mathbf{SU}(2) &= \{[z^1;z^2] \mid z^{1,2}\in\mathbb{Z}\},\\
\text{fundamental representations:}\ & \dim_{\mathbb{C}}[1;0]=2,\quad \dim_{\mathbb{C}}([0;1]=2,\\
\textbf{weights}\,\mathbf{SO}(4) &= \{[z^1;z^2] \mid z^{1,2}\in\mathbb{Z},\quad z^1+z^2\in 2\mathbb{Z}\},\\
\text{fundamental representations:}\ & \dim_{\mathbb{R}}[1;1]=4,\quad \dim_{\mathbb{R}}([2;0]\ \oplus\ [0;2])=6.
\end{aligned}$$

The representations $[2J;2J]$ with the squared dimensions $(1+2J)^2$ characterize the bound states of the hydrogen atom.

The central correlation of the isospin-color group with the hypercharge group by the product center $\mathbb{I}(6)$ (star of David in the unit circle)

$$\mathbf{U}(2\times 3) = \frac{\mathbf{U}(1)\times\mathbf{SU}(2)\times\mathbf{SU}(3)}{\mathbb{I}(2)\times\mathbb{I}(3)}$$

has weights with possibly broken hypercharge number y

$$\textbf{weights}\,\mathbf{U}(2\times 3) = \{[y||z;z^1,z^2] \mid z,z^{1,2}\in\mathbb{Z},\quad y\in\tfrac{z}{2}-\tfrac{z^1-z^2}{3}+\mathbb{Z}\}.$$

The dimension-12, rank-4 group $\mathbf{U}(2\times 3)$ gives rise to $2\cdot(2+3)=10$ fundamental representations. They involve two representations with hypercharge $y=\pm\frac{1}{6}$ used for the quark fields $\mathbf{q}$ and $\mathbf{q}^\star$

$[+\frac{1}{6}\|\|1;1,0]$	$[+\frac{2}{6}\|\|0;0,1]$	$[+\frac{3}{6}\|\|1;0,0]$	$[+\frac{4}{6}\|\|0;1,0]$	$[+1\|\|0;0,0]$
$\mathbf{q}$	$\mathbf{d}^\star$	$\mathbf{l}^\star$	$\mathbf{u}$	$\mathbf{e}^\star$
$[-\frac{1}{6}\|\|1;0,1]$	$[-\frac{2}{6}\|\|0;1,0]$	$[-\frac{3}{6}\|\|1;0,0]$	$[-\frac{4}{6}\|\|0;0,1]$	$[-1\|\|0;0,0]$
$\mathbf{q}^\star$	$\mathbf{d}$	$\mathbf{l}$	$\mathbf{u}^\star$	$\mathbf{e}$

$\mathbf{U}(2\times 3)$-hypercharge-isospin-color representations.

The representations with hypercharge $\pm\frac{5}{6}$ are not fundamental:

$$[\tfrac{5}{6}||1;0,1] = [\tfrac{2}{6}||0;0,1]\ \oplus\ [\tfrac{3}{6}||1;0,0].$$

Bibliography

[1] H. Boerner, *Darstellungen von Gruppen* (1955), Springer, Berlin, Göttingen, Heidelberg.

[2] N. Bourbaki, *Groupes et Algèbres de Lie, Chapitre 8* (Algèbres de Lie semi-simples déployées) (1975), Hermann, Paris.

[3] H.S.M. Coxeter, *Regular Polytopes* (1973), Dover, New York.

[4] W. Fulton, J. Harris, *Representation Theory* (1991), Springer, Berlin, Heidelberg, New York, London, Paris, Tokyo.

[5] R. Gilmore, *Lie Groups, Lie Algebras and Some of Their Applications* (1974), John Wiley & Sons, New York, London, Sidney, Toronto.

[6] S. Helgason, *Differential Geometry, Lie Groups and Symmetric Spaces* (1978), Academic Press, New York, London, Toronto, Sidney, San Francisco.

7

QUANTUM ALGEBRAS

Quantum theory is a theory of really parametrizable operations acting on complex vector spaces with a conjugation. Quantum structures are unifying: The operations themselves are elements of their action spaces. Hence the classical distinction between operations and observables, e.g. between infinitesimal rotations and angular momenta, vanishes, in quantum theory the infinitesimal position rotations are identical with the angular momentum action.

Basically, the operational structure of quantum theory is algebraic, it does not start with states or with a Hilbert space or with "probability amplitudes," those concepts are needed for an ontology with particles and objects, necessary for a classical interpretation that arises by an experiment-induced projection of the operational algebraic structure. Hilbert spaces come with the definite unitarity of representations (chapter "Quantum Probability").

Linearity and concatenation of quantum operations are formalized with multilinearity (in distinction to "nonlinearity"). The multilinear structure for a vector space is its tensor algebra. The implementation of the basic vector space endomorphism Lie algebra in the form of inner algebra derivations leads to the *Fermi and Bose quantum algebra* of a vector space. Quantum elements are dual-product-induced equivalence classes in the tensor algebra, i.e., quantum algebras are tensor algebra quotient structures. The dual product of the quantum algebra underlying vector space leads to the *characteristic anticommutation and commutation relations of Fermi or Bose type* including the famous Born-Heisenberg position-momentum relation $[i\mathbf{p}, \mathbf{x}] = \hbar$.

The rich structure of quantum algebras, a mathematically basic and rather simple canonical factorization of the tensor algebra, will be considered in this chapter with respect to the implemented operations and the related invariants, with respect to their gradings and to the induced conjugations with the position-momentum formulation for the Bose quantum algebras.

The quantum algebras in this chapter are appropriate and enough for actions on finite-dimensional vector spaces, e.g., for irreducible representations of compact groups or of the time group $\mathbf{D}(1) \cong \mathbb{R}$. For faithful Hilbert representations of noncompact groups with infinite-dimensional spaces which occur in spacetime quantum theories, e.g., of the Poincaré group, they have to be generalized and play a role as value spaces (chapters "Massive Particle Quantum Fields" and "Harmonic Analysis").

The simplest examples are the Fermi and Bose oscillator quantum algebras for the abelian operation groups $\mathbf{U}(1)$ and $\mathbf{D}(1)$. They can be interpreted as the free algebras for two real 3-dimensional bracket algebras, for the Heisenberg Lie algebra $\log \mathbf{H}(1)$ with one position-momentum pair $[\mathbf{x}, \mathbf{p}] = \mathbf{I}$ in the Bose case and in the Fermi case for the hybrid Pauli Lie algebra $\log \mathbf{P}(1)$ with one creation-annihilation pair $\{\mathrm{u}^\star, \mathrm{u}\} = \mathbf{I}$.

The smallest simple compact group $\mathbf{SU}(2)$ gives rise to the spin quantum algebra of Fermi type which is used for spin $\frac{1}{2}$ particles, e.g., for electrons (chapter "Massive Particle Quantum Fields"). Adjoint Lie algebra representations with associated adjoint quantum algebras are exemplified by the position quantum algebra of Bose type as used in quantum mechanics with 3-dimensional position translations and the rotation group $\mathbf{SO}(3)$. In quantum field theories, adjoint quantum algebras are used for quantum gauge structures (chapter "Gauge Interactions").

7.1 Quantization

Quantization connects *multilinearity with "canonical pairs."* It is based on a pair of dual vector spaces (V, V^T), containing, e.g., position-momentum or creation-annihilation operators.

A quantum algebra is generated by the self-dual direct sum $\mathbf{V} = V \oplus V^T \cong \mathbb{K}^{2n}$ of a finite-dimensional vector space and its linear forms with the transposition sign $\epsilon = \pm 1$ (Fermi and Bose) defined by the symmetry property of the extended dual product

$$\mathbf{V} \times \mathbf{V} \longrightarrow \mathbb{K}, \begin{cases} \langle \mathbf{w}, \mathbf{v} \rangle = \langle u + \omega, v + \theta \rangle = \langle \omega, v \rangle + \langle u, \theta \rangle = \epsilon \langle \mathbf{v}, \mathbf{w} \rangle \\ \text{with } \langle \omega, v \rangle = \epsilon \langle v, \omega \rangle \text{ for } v \in V, \ \omega \in V^T. \end{cases}$$

All algebras for a vector space arise from its tensor algebra, e.g., Grassmann, polynomial, Clifford, or enveloping algebra. The tensor algebra $\bigotimes \mathbf{V}$ has two bilinear basic vector products: The ϵ-commutators $[\mathbf{w}, \mathbf{v}]_\epsilon \in \mathbf{V} \otimes \mathbf{V}$ which are power-2 tensors and the canonical products $\langle \mathbf{w}, \mathbf{v} \rangle \in \mathbb{K}$ which are scalars. In a *quantum algebra* $\mathbf{Q}_\epsilon(\mathbf{V})$ those two products are identified by working with equivalence classes with respect to the appropriate minimal, such an identity enforcing ideal $I(S_\epsilon^{\text{quant}})$:

$$S_\epsilon^{\text{quant}} = \{\mathbf{w} \otimes \mathbf{v} + \epsilon \mathbf{v} \otimes \mathbf{w} - \langle \mathbf{w}, \mathbf{v} \rangle \bigm| \mathbf{v}, \mathbf{w} \in \mathbf{V}\} \subset \bigotimes \mathbf{V},$$
$$\text{in } \mathbf{Q}_\epsilon(\mathbf{V}) = \bigotimes \mathbf{V} / I(S_\epsilon^{\text{quant}}) : \begin{cases} [\mathbf{w}, \mathbf{v}]_\epsilon = \langle \mathbf{w}, \mathbf{v} \rangle, \\ [\omega, v]_\epsilon = \langle \omega, v \rangle, \\ [u, v]_\epsilon = 0, \quad [\omega, \theta]_\epsilon = 0. \end{cases}$$

The embedding of a dual vector space pair into the tensor algebra with the basis independent factorization will be called *quantization.* For each vector space there exist both a *Fermi and a Bose quantum algebra* distinguished by the involutive transposition sign with $\epsilon^2 = 1$, i.e., by $\epsilon = \pm 1$ as *statistical sign.*

The main property of and motivation for such the tensor algebra factorization for a dual product-commutator identification is the implementation of derivations in the form of inner derivations (below), e.g., of the time derivation by an adjoint action with the Hamiltonian, $\frac{d}{dt} = [iH, \ ]$.

The factorization is basis independent. The dual products of dual bases of the basic vector space and its linear forms give (anti-) commutators in the quantum algebra with 1 or 0:

$$\text{in } \mathbf{Q}_\epsilon(\mathbf{V}) : \quad \begin{cases} [\check{e}_A, e^B]_\epsilon = \langle \check{e}_A, e^B \rangle = \delta^B_A, \\ [e^A, e^B]_\epsilon = \langle e^A, e^B \rangle = 0, \\ [\check{e}_A, \check{e}_B]_\epsilon = \langle \check{e}_A, \check{e}_B \rangle = 0. \end{cases}$$

New symbols for the classes in the quotient algebra will not be used e.g., $[a] = a + I(S_\epsilon^{\text{quant}})$ is simply written as a; one easily gets used to working with representatives modulo the quantum ideal. Later on, the tensor multiplication sign will also be omitted, i.e., $ab = a \otimes b$.

Quantum algebras of Bose type have only trivial divisors of zero:

$$a, b \in \mathbf{Q}_-(\mathbf{V}), \quad a \otimes b = 0 \Rightarrow a = 0 \text{ or } b = 0.$$

In contrast, Fermi quantum algebras have nontrivial nilpotent vectors; especially, all basic vectors and forms are nilquadratic. This can be used as structural origin of the *Pauli principle*:

$$v \in V, \ \omega \in V^T \Rightarrow v \otimes v = 0 = \omega \otimes \omega \text{ in } \mathbf{Q}_+(\mathbf{V}).$$

The quantum algebra generating scalars and basic vectors $\mathbb{K} \oplus \mathbf{V} \cong \mathbb{K}^{1+2n}$ constitute a bracket algebra. Quantum algebras are the respective enveloping algebras

$$\begin{aligned} \mathbb{K} \oplus \mathbf{V} : & \quad \begin{cases} [\alpha, \beta] = 0, \quad [\alpha, \mathbf{v}] = 0, \ [\mathbf{v}, \mathbf{w}]_\epsilon = \langle \mathbf{v}, \mathbf{w} \rangle, \\ \alpha, \beta \in \mathbb{K}, \ \mathbf{v}, \mathbf{w} \in \mathbf{V}, \end{cases} \\ \mathbf{Q}_\epsilon(\mathbf{V}) = \mathbf{E}(\mathbb{K} \oplus \mathbf{V}). & \end{aligned}$$

For the reals, the generating bracket algebras are representations of the *hybrid Pauli algebra* for the Fermi case and the *Heisenberg Lie algebra* for the Bose case with the central element represented by the complexified enveloping algebra as imaginary unit $\mathbf{I} \longmapsto i\mathbf{1}$:

$$\begin{aligned} \log \mathbf{P}(1) : & \quad \{\check{e}, e\} = \mathbf{I}, \quad [\check{e}, \mathbf{I}] = 0, \quad [e, \mathbf{I}] = 0, \\ \log \mathbf{H}(1) : & \quad [\check{e}, e] = \mathbf{I}, \quad [\check{e}, \mathbf{I}] = 0, \quad [e, \mathbf{I}] = 0. \end{aligned}$$

With a basic space $V \cong \mathbb{R}^n$ one obtains the Pauli and Heisenberg algebras with n-dual pairs $(\check{e}_A, e^A)$:

$$\mathbf{Q}_+(\mathbb{C}^{2n}) = \mathbb{C} \otimes \mathbf{E}(\log \mathbf{P}(n)), \quad \mathbf{Q}_-(\mathbb{C}^{2n}) = \mathbb{C} \otimes \mathbf{E}(\log \mathbf{H}(n)).$$

Fermi quantum algebras are Clifford algebras of the self-dual even-dimensional basic space with respect to the symmetric dual product

$$\mathbf{Q}_+(\mathbb{K}^{2n}) = \text{CLIFF}(\mathbb{K}^{2n}, \langle \ , \ \rangle) \text{ with } \{\mathbf{w}, \mathbf{v}\} = \langle \mathbf{w}, \mathbf{v} \rangle.$$

For real basic spaces the dual product has an orthogonal invariance group with neutral signature:

$$\begin{array}{l}\langle\ ,\ \rangle \cong \begin{pmatrix} 0 & \mathbf{1}_n \\ \mathbf{1}_n & 0 \end{pmatrix} \Rightarrow \text{invariance group: } \left\{ \begin{array}{ll} \mathbf{O}(\mathbb{C}^{2n}) & \text{for } V \cong \mathbb{C}^n, \\ \mathbf{O}(n,n) & \text{for } V \cong \mathbb{R}^n, \end{array} \right. \\ \mathbf{Q}_+(\mathbb{C}^{2n}) \cong \text{CLIFF}(2n), \quad \mathbf{Q}_+(\mathbb{R}^{2n}) \cong \text{CLIFF}(n,n).\end{array}$$

As quotient algebras of $\bigotimes \mathbf{V}$, quantum algebras are characterizable by a universal property: All dual-product-compatible linear mappings f of the self-dual space $\mathbf{V} \cong \mathbb{K}^{2n}$ into unital algebras A are factorizable via the quantum algebras

$$[f(\mathbf{v}),f(\mathbf{w})]_\epsilon = \langle \mathbf{v},\mathbf{w}\rangle 1_A,\ f \quad \begin{array}{ccc} (\mathbf{V},\epsilon) & \xrightarrow{\sigma} & \mathbf{Q}_\epsilon(\mathbf{V}) \\ \downarrow & & \downarrow \\ A & \xrightarrow[\text{id}_A]{} & A \end{array} \quad \tilde{f}\,.$$

For the related covariant quantum algebra functor $\mathbf{Q}_\epsilon$ from vector spaces with a fixed transposition sign ϵ into associative algebras only *vector space isomorphisms* $\text{is}_{WV} : V \longrightarrow W$ are admitted, since also the transposition of the inverse isomorphism $\text{is}_{VW} : W \longrightarrow V$ is involved for the mapping of the direct sum $\mathbf{is}_{\mathbf{VW}} : \mathbf{V} \longrightarrow \mathbf{W},\ \ \mathbf{is}_{\mathbf{VW}} = \text{is}_{WV} \oplus \text{is}_{WV}^{-1T}$,

$$\mathbf{Q}_\epsilon : \underline{\overset{\circ}{\mathbf{vec}}{}^{\epsilon}_{\mathbb{K}}} \longrightarrow \underline{\overset{\circ}{\mathbf{aag}}_{\mathbb{K}}}, \quad \text{is}_{WV} \begin{array}{c} V \\ \downarrow \\ W \end{array} \longmapsto \begin{array}{c} \mathbf{Q}_\epsilon(\mathbf{V}) \\ \downarrow \\ \mathbf{Q}_\epsilon(\mathbf{W}) \end{array} \bigotimes \mathbf{is}_{\mathbf{VW}}.$$

Hence, up to unital algebra isomorphisms, there is only one Fermi and one Bose quantum algebra $\mathbf{Q}_\pm(\mathbb{K}^{2n})$ for each basic vector space dimension. Below, the “smallest” nontrivial $n = 1$ quantum algebras $\mathbf{Q}_\epsilon(\mathbb{C}^2)$ (oscillator quantum algebras) are considered as well as the $n = 2$ spin Fermi quantum algebra $\mathbf{Q}_+(\mathbb{C}^4)$ and the $n = 3$ Bose quantum algebra $\mathbf{Q}_-(\mathbb{C}^6)$ of 3-dimensional quantum mechanics.

By arranging the basic vectors and forms in a quantum algebra element as follows,

$$(v_1 \otimes \cdots \otimes v_k) \otimes (\omega_1 \otimes \cdots \otimes \omega_l) \in \bigotimes^k V \otimes \bigotimes^l V^T,$$

one realizes the isomorphism of the quantum algebra to the endomorphism algebra of a vector space that is isomorphic either to the Grassmann algebra $\bigwedge V$ in the Fermi or to the polynomial algebra $\bigvee V$ in the Bose case:

$$\mathbf{Q}_\epsilon(\mathbf{V}) \cong \left\{ \begin{array}{ll} \mathbf{AL}(\bigwedge V) = \bigwedge V \otimes \bigwedge V^T, & \epsilon = +1, \\ \mathbf{AL}(\bigvee V) = \bigvee V \otimes \bigvee V^T, & \epsilon = -1. \end{array} \right.$$

The vector spaces $\bigwedge V$ and $\bigvee V$ can be related to Fock spaces (chapter “Quantum Probability”), i.e., quantum algebras can be considered as endomorphism algebras of Fock spaces for Bose and Fermi structures.

7.1.1 Oscillator Quantum Algebras

A 1-dimensional vector space $V \cong \mathbb{K}$ with dual bases $\langle \check{e}, e \rangle = 1$ leads to the smallest nontrivial quantum algebras $\mathbf{Q}_\epsilon(\mathbb{K}^2)$ with dimension 4 for Fermi and countably infinite dimension for Bose. In the complex case they are called *oscillator quantum algebras* $\mathbf{Q}_\pm(\mathbb{C}^2)$ (more below):

$$\text{Fermi } (\epsilon = +1): \begin{cases} \{\check{e}, e\} = 1, \quad \check{e} \otimes \check{e} = 0 = e \otimes e, \quad I = \frac{[e,\check{e}]}{2}, \quad I \otimes I = \frac{1}{4}, \\ \text{basis of } \mathbf{Q}_+(\mathbb{K}^2) \cong \mathbb{K}^4: \quad \{1, \ e, \ \check{e}, \ I\}, \\ \mathbf{Q}_+(\mathbb{C}^2) = \text{CLIFF}(2), \quad \mathbf{Q}_+(\mathbb{R}^2) = \text{CLIFF}(1,1), \end{cases}$$

$$\text{Bose } (\epsilon = -1): \begin{cases} [\check{e}, e] = 1, \quad I = \frac{\{e,\check{e}\}}{2}, \\ \text{basis of } \mathbf{Q}_-(\mathbb{K}^2) \cong \mathbb{K}^{\aleph_0}: \ \{e^m \otimes \check{e}^k \mid m, k \geq 0\}. \end{cases}$$

The action of the identity defines the $\mathbb{Z}_3 = \{-1, 0, 1\}$ grading for Fermi and $\mathbb{Z}$-grading for Bose (more on grading and I-action below):

$$[I, e^m \otimes \check{e}^k] = (m - k) e^m \otimes \check{e}^k.$$

The I-polynomials constitute the subalgebra with trivial grade

$$\text{INV}_I \mathbf{Q}_\epsilon(\mathbb{K}^2) = \begin{cases} \{\alpha + \beta I \mid \alpha, \beta \in \mathbb{K}\} \cong \mathbb{K}^2, & \epsilon = +1, \\ \mathbb{K}[I] \cong \mathbb{K}^{\aleph_0}, & \epsilon = -1. \end{cases}$$

7.1.2 Quantum Algebras and Classical Algebras

The tensor algebra as the free unital algebra of a vector space is the common origin and starting point for quantum and classical algebras: In the transition from the tensor algebra $\bigotimes \mathbf{V}$ to a quotient algebra, with the equivalence classes defined with the dual product (canonical bilinear forms), either symmetric or antisymmetric ($\epsilon = \pm 1$), it is possible to include a scalar $\hbar \in \mathbb{K}$ as duality normalization, i.e., to work in the tensor algebra modulo the identities

$$[\mathbf{w}, \mathbf{v}]_\epsilon = \hbar \langle \mathbf{w}, \mathbf{v} \rangle.$$

An identity enforcing corresponding factorization, either with trivial dual product $\hbar = 0$ or with a nontrivial one $\hbar \neq 0$,

$$A_\epsilon^\hbar(\mathbf{V}) = \bigotimes \mathbf{V} / \text{ideal}\{[\mathbf{v}, \mathbf{w}]_\epsilon - \hbar \langle \mathbf{v}, \mathbf{w} \rangle \mid \mathbf{v}, \mathbf{w} \in \mathbf{V}\}, \quad \epsilon = \pm 1, \quad \hbar \in \mathbb{K},$$
$$\text{in } A_\epsilon^\hbar(\mathbf{V}): \quad [\mathbf{v}, \mathbf{w}]_\epsilon = \hbar \langle \mathbf{v}, \mathbf{w} \rangle, \text{ for dual bases}: \begin{cases} [\check{e}_A, e^B]_\epsilon &= \hbar \delta_A^B, \\ [e^A, e^B]_\epsilon &= 0 = [\check{e}_A, \check{e}_B]_\epsilon, \end{cases}$$

gives rise to four types of *statistical algebras*: Two abelian "classical" algebras and two quantum algebras, both pairs with positive and negative statistical sign ϵ:

	$\epsilon = +1$	$\epsilon = -1$
$\hbar = 0$ (classical)	$\bigwedge \mathbf{V}$ Graßmann algebra (antiabelian)	$\bigvee \mathbf{V}$ polynomial algebra (abelian)
$\hbar \neq 0$ (quantal)	$\mathbf{Q}_+(\mathbf{V})$ Fermi quantum algebra	$\mathbf{Q}_-(\mathbf{V})$ Bose quantum algebra

four statistical algebras $A^\hbar_\epsilon(\mathbf{V})$

A nontrivial $\hbar \in \exp \mathbb{K}$ in quantum algebras can be absorbed in a basis renormalization, i.e., with $\hbar = 1$.

The *canonical representatives* of the tensors are the totally antisymmetrized tensors for $\epsilon = 1$ and the totally symmetrized tensors for $\epsilon = -1$. With the (anti-) commutators for the basic space vectors, each tensor can be written as linear combination of canonical representatives:

$$\begin{array}{ll} \text{in } A^\hbar_+(\mathbf{V}): & \mathbf{v}_1 \wedge \cdots \wedge \mathbf{v}_k = \frac{\epsilon^{j_1 \cdots j_k}}{k!} \mathbf{v}_{j_1} \otimes \cdots \otimes \mathbf{v}_{j_k}, \quad \mathbf{v}_j \in \mathbf{V}, \\ \text{in } A^\hbar_-(\mathbf{V}): & \mathbf{v}_1 \vee \cdots \vee \mathbf{v}_k = \frac{\delta^{j_1 \cdots j_k}}{k!} \mathbf{v}_{j_1} \otimes \cdots \otimes \mathbf{v}_{j_k}. \end{array}$$

As vector spaces, not as unital algebras, the quantum algebras and classical algebras of corresponding type $\epsilon = \pm 1$ are isomorphic via the canonical representatives $\iota_\epsilon : A^1_\epsilon(\mathbf{V}) \longrightarrow A^0_\epsilon(\mathbf{V})$.

Algebras with basic commutators have countably infinite dimension in contrast to the finite-dimensional algebras with basic anticommutators

$$\dim_{\mathbb{K}} V = n \Rightarrow \dim_{\mathbb{K}} A^\hbar_\epsilon(\mathbf{V}) = \begin{cases} 2^{2n}, & \epsilon = +1, \\ \aleph_0, & \epsilon = -1. \end{cases}$$

Bose and Fermi quantum algebras with their nontrivial (anti-) commutators $[\![a, b]\!] = [a, b]_{\pm 1}$ have a nontrivial natural Lie algebra and natural hybrid bracket algebra structure respectively, whereas Grassmann and polynomial algebras have trivial natural (anti-) commutators. With the aid of the canonical vector space isomorphism ι_ϵ the classical algebras $A^0_\epsilon(\mathbf{V})$ inherit from the quantum algebras $A^1_\epsilon(\mathbf{V})$ the structure of a nontrivial bracket algebra

$$\begin{array}{ccc} A^1_\epsilon(\mathbf{V}) \times A^1_\epsilon(\mathbf{V}) & \stackrel{[\![\ ,\ ]\!]}{\longrightarrow} & A^1_\epsilon(\mathbf{V}) \\ {\scriptstyle \iota_\epsilon \times \iota_\epsilon} \downarrow & & \downarrow {\scriptstyle \iota_\epsilon}, \\ A^0_\epsilon(\mathbf{V}) \times A^0_\epsilon(\mathbf{V}) & \underset{[\![\ ,\ ]\!]_{\text{Poisson}}}{\longrightarrow} & A^0_\epsilon(\mathbf{V}) \end{array}$$

$$[\![a, b]\!]_{\text{Poisson}} = \iota_\epsilon\Big([\![a, b]\!]\Big).$$

To transport the brackets with the vector space isomorphism, the quantum algebra bracket $[\![a, b]\!]$ for two canonical representatives is expanded in canonical representatives, which, then, is taken as an element of the corresponding classical algebra.

Only for the polynomial algebra, the commutator-induced *Poisson bracket* as classical Lie algebra structure can be written by derivatives (chapter "Spin, Rotations, and Position")

$$\begin{aligned} \text{in } A^1_-(\mathbb{K}^2): \quad & [\check{e}, e] = 1 \longmapsto [\check{e}, e]_{\text{Poisson}} = 1 \text{ in } A^0_-(\mathbb{K}^2), \\ \text{for } g, f \in A^0_-(\mathbb{K}^2): \quad & [\check{e}, g(e, \check{e})]_{\text{Poisson}} = \tfrac{\partial g}{\partial e}, \quad [f(e, \check{e}), e]_{\text{Poisson}} = -\tfrac{\partial f}{\partial \check{e}}, \\ & [f(e, \check{e}), g(e, \check{e})]_{\text{Poisson}} = \tfrac{\partial f}{\partial e}\tfrac{\partial g}{\partial \check{e}} - \tfrac{\partial g}{\partial e}\tfrac{\partial f}{\partial \check{e}}. \end{aligned}$$

This is familiar from the real quantum algebra $\mathbf{Q}_-(\mathbb{R}^2)$ with basic vector space $\mathbf{V} = \mathbb{R}\mathbf{x} \oplus \mathbb{R}i\mathbf{p}$ for position and momentum:

$$\begin{aligned} \text{in } A^1_-(\mathbb{R}^2): \quad & [i\mathbf{p}, \mathbf{x}] = \hbar \longmapsto [\mathbf{p}, \mathbf{x}]_{\text{Poisson}} = 1 \text{ in } A^0_-(\mathbb{R}^2), \\ \text{for } f, g \in A^0_-(\mathbb{R}^2): \quad & [\mathbf{p}, g(\mathbf{p}, \mathbf{x})]_{\text{Poisson}} = \tfrac{\partial g}{\partial \mathbf{x}}, \quad [f(\mathbf{p}, \mathbf{x}), \mathbf{x}]_{\text{Poisson}} = -\tfrac{\partial f}{\partial \mathbf{p}}, \\ & [f(\mathbf{p}, \mathbf{x}), g(\mathbf{p}, \mathbf{x})]_{\text{Poisson}} = \tfrac{\partial f}{\partial \mathbf{p}}\tfrac{\partial g}{\partial \mathbf{x}} - \tfrac{\partial g}{\partial \mathbf{p}}\tfrac{\partial f}{\partial \mathbf{x}}. \end{aligned}$$

7.1.3 Products of Quantum Algebras

The quantum algebra of a direct vector space sum $V \oplus W$ with equal transposition sign $\epsilon_V \epsilon_W = 1$ is isomorphic to the product of the individual quantum algebras. The quantum algebra functor is exponential:

$$\mathbf{Q}_\epsilon(\mathbf{V} \oplus \mathbf{W}) \cong \mathbf{Q}_\epsilon(\mathbf{V}) \otimes \mathbf{Q}_\epsilon(\mathbf{W}), \quad v + w \longmapsto v \otimes 1 + 1 \otimes w.$$

Hence any basis of the basic vector space $V \cong \mathbb{K}^n$ induces an isomorphism

$$\mathbf{Q}_\epsilon(\mathbb{K}^{2n}) \cong \mathbf{Q}_\epsilon(\mathbb{K}^2)^n = \bigotimes^n \mathbf{Q}_\epsilon(\mathbb{K}^2)$$

with the n-tensor power of the smallest nontrivial quantum algebras $\mathbf{Q}_\epsilon(\mathbb{K}^2)$ (oscillator algebras).

For a different transposition sign $\epsilon_V \epsilon_W = -1$ the quantum algebra of a direct vector space sum $V \oplus W$ will be defined as the product of both quantum algebras, where the bracket has to take care of the Fermi and Bose property

$$\begin{aligned} \mathbf{Q}_{\epsilon_V, \epsilon_W}(\mathbf{V} \oplus \mathbf{W}) \; & = \mathbf{Q}_{\epsilon_V}(\mathbf{V}) \otimes \mathbf{Q}_{\epsilon_W}(\mathbf{W}), \\ [\![a_V, a_W]\!] = 0 \; & = \begin{cases} [a_V, a_W] & \text{iff } a_V \text{ or } a_W \text{ are Bose,} \\ \{a_V, a_W\} & \text{iff } a_V \text{ and } a_W \text{ are Fermi,} \end{cases} \\ a_V \in \mathbf{Q}_{\epsilon_V}(\mathbf{V}), \quad & a_W \in \mathbf{Q}_{\epsilon_W}(\mathbf{W}). \end{aligned}$$

7.2 Actions in Quantum Algebras

The main property of quantum algebras, which can also be used to motivate the product identification in their definition above, is the multilinear extension of basic space operations, e.g., of rotations acting on basic positions-momenta.

7.2.1 Lie Algebra Actions on Quantum Algebras

A linear operation $f : V \longrightarrow V$ defines, via Leibniz rule extension, a unique derivation of its tensor algebra by requiring identical action on the basic vector space and trivial action on the scalars (chapter "Spin, Rotations, and Position"):

$$f_{\text{der}} : \bigotimes V \longrightarrow \bigotimes V, \quad \begin{cases} f_{\text{der}}(\alpha) = & 0, \quad \alpha \in \mathbb{K}, \\ f_{\text{der}}(v) = & f(v), \quad v \in V, \\ f_{\text{der}}(a \otimes b) = & f_{\text{der}}(a) \otimes b + a \otimes f_{\text{der}}(b), \\ & a, b \in \bigotimes V. \end{cases}$$

An analogous extension holds for the transposed endomorphism. A quantum algebra is factorized in such a way that the basic space *endomorphisms act as inner derivations* (adjoint action):

$$\mathbf{AL}(V) \times \mathbf{Q}_\epsilon(\mathbf{V}) \longrightarrow \mathbf{Q}_\epsilon(\mathbf{V}), \quad \text{ad}\, \tilde{f}(a) = [\tilde{f}, a] = [f, a] = [-f^T, a],$$
$$\begin{cases} \alpha \in \mathbb{C} : & [\tilde{f}, \alpha] & = 0, \\ v \in V : & [f, v] & = f \otimes v - v \otimes f = f(v), \\ \text{e.g.,} & [u \otimes \theta, v] & = [\theta, v]_\epsilon u = \langle \theta, v \rangle u, \\ \omega \in V^T : & [f^T, \omega] & = f^T \otimes \omega - \omega \otimes f^T = f^T(\omega), \\ \text{e.g.,} & [\epsilon\theta \otimes u, \omega] & = \epsilon [u, \omega]_\epsilon \theta = \langle \omega, u \rangle \theta. \end{cases}$$

The implementation of the basic space endomorphisms as inner derivations can be formulated for Bose quantum algebras with a given basis by expressing the duality-induced commutators with derivatives (chapter "Spin, Rotations, and Position"):

$$\begin{array}{rl} \text{for } \mathbf{Q}_-(\mathbb{K}^{2n}) : & \{e^a, \check{e}_a \mid a = 1, \dots, n\} \cong \{e^a, \partial_a \mid a = 1, \dots, n\} \\ \text{with} & [\check{e}_a, e^b] \cong [\partial_a, e^b] = \delta^b_a, \quad \partial_a = \frac{\partial}{\partial e^a}, \\ & [e^a, e^b] = 0 = [\check{e}_a, \check{e}_b] \cong [\partial_a, \partial_b]. \end{array}$$

Hence a Bose quantum algebra is representable by derivations of the unital ring of the polynomials in a basis $\mathbb{K}[e^1, \dots, e^n]$, extendable to derivations of the unital ring of the infinitely often differentiable functions

$$\begin{array}{l} \mathbf{Q}_-(\mathbb{K}^{2n}) \times \mathcal{C}(\mathbb{K}^n, \mathbb{K}) \longrightarrow \mathcal{C}(\mathbb{K}^n, \mathbb{K}), \\ \text{e.g., } \partial_a e^b f(e) = \delta^b_a f(e) + e^b \partial_a f(e). \end{array}$$

In quantum algebras, negative transposed endomorphisms $(f, -f^T)$ of the basic vector space pair (V, V^T) coincide up to the trace:

$$\text{in } \mathbf{Q}_\epsilon(\mathbf{V}) : \quad \begin{cases} v \otimes \omega + \epsilon \omega \otimes v & = \epsilon \langle \omega, v \rangle, \quad v \otimes \omega \in V \otimes V^T = \mathbf{AL}(V) \\ \Rightarrow f + f^T & = \epsilon \,\text{tr}\, f, \quad f \in \mathbf{AL}(V). \end{cases}$$

The canonical representative for an endomorphism is given by the *quantization-opposite commutator*

$$f \in \mathbf{AL}(V) \Rightarrow \tilde{f} = \tfrac{f - f^T}{2} = f^B_A \tfrac{[e^A, \check{e}_B]_{-\epsilon}}{2} \in \mathbf{Q}_\epsilon(\mathbf{V}),$$

e.g., for the identities of the basic vector spaces

$$\begin{array}{rl} I = \frac{\mathrm{id}_V - \mathrm{id}_{V^T}}{2} & = \frac{[e^A, \check{e}_A]_{-\epsilon}}{2} = \mathrm{id}_V - \frac{\epsilon n}{2} = -\,\mathrm{id}_{V^T} + \frac{\epsilon n}{2}, \\ \text{since } \mathrm{id}_V + \mathrm{id}_V^T & = \epsilon \dim_{\mathbb{K}} V = \epsilon n. \end{array}$$

Hence Lie algebra representations on a vector space and its dual,

$$\begin{array}{lll} \mathcal{D} : L \longrightarrow \mathbf{AL}(V), & \mathcal{D}(l) & = \mathcal{D}(l)_A^B \; e^A \otimes \check{e}_B, \\ \check{\mathcal{D}} : L \longrightarrow \mathbf{AL}(V^T), & \check{\mathcal{D}}(l) & = -\mathcal{D}(l)_A^B \; \epsilon \check{e}_B \otimes e^A, \end{array}$$

differ in quantum algebras only by a scalar:

$$\begin{array}{ll} L \longrightarrow \mathbf{Q}_\epsilon(\mathbf{V}), & \tilde{\mathcal{D}}(l) = \frac{\mathcal{D}(l) + \check{\mathcal{D}}(l)}{2} = \mathcal{D}(l)_A^B \frac{[e^A, \check{e}_B]_{-\epsilon}}{2}. \\ & \mathcal{D}(l) - \check{\mathcal{D}}(l) = \epsilon \operatorname{tr} \mathcal{D}(l). \end{array}$$

The naturally isomorphic basic space Lie algebras are represented by the same quantum algebra elements via the Lie algebra morphism

$$\begin{array}{rl} \mathbf{AL}(V) \ni f & \cong -f^T \in \mathbf{AL}(V^T) \cong \log \mathbf{GL}(\mathbb{K}^n) \\ \log \mathbf{GL}(\mathbb{K}^n) \longrightarrow \mathbf{Q}_\epsilon(\mathbf{V}), & f \cong -f^T \longmapsto \tilde{f} = \frac{f - f^T}{2} = f_A^B \frac{[e^A, \check{e}_B]_{-\epsilon}}{2}, \\ f_{1,2} \in \mathbf{AL}(V) : & \left\{ \begin{array}{ll} [f_1, f_2] & = f_1 \circ f_2 - f_2 \circ f_1, \\ [f_1, f_2]_\otimes & = f_1 \otimes f_2 - f_2 \otimes f_1, \end{array} \right. \\ \text{in } \mathbf{Q}_\epsilon(\mathbf{V}) : & [f_1, f_2] = [f_1, f_2]_\otimes = [-f_1^T, -f_2^T] = [-f_1^T, -f_2^T]_\otimes = [\tilde{f}_1, \tilde{f}_2]_\otimes. \end{array}$$

In the case of a nontrivial dual normalization factor $\hbar$ for the quantization, the endomorphism representatives have to be normalized correspondingly:

$$[\omega, v]_\epsilon = \hbar \langle \omega, v \rangle \Rightarrow \mathbf{AL}(V) \in f = v_i \otimes \omega_i \longmapsto \frac{1}{\hbar} \tilde{f} = \frac{[v_i, \omega_i]_{-\epsilon}}{2\hbar} \in \mathbf{Q}_\epsilon(\mathbf{V}).$$

Without renormalization the factor $\hbar$ arises in the endomorphism quantum commutator $[f, g]_\otimes = \hbar [f, g]$.

For real position-momentum quantum algebras with basis elements $i\mathbf{p}$, the imaginary i is often taken out to obtain the Hermitian part of a generator, e.g., for angular momenta $\hbar \mathcal{O}^a = \epsilon^{abc} \mathbf{x}_b i \mathbf{p}_c = i \mathcal{L}^a$, which gives rise to additional i-proportional renormalization factors, e.g., $[\mathcal{L}^a, \mathcal{L}^b] = i \hbar \epsilon^{abc} \mathcal{L}^c$ instead of $[\mathcal{O}^a, \mathcal{O}^b] = -\epsilon^{abc} \mathcal{O}^c$.

7.2.2 Quantum Enveloping Algebra and Invariants

With the endomorphism Lie algebra $\mathbf{AL}(V) \cong V \otimes V^T$ of the basic vector space also the corresponding enveloping algebra $\mathbf{E}(\mathbf{AL}(V))$ (chapter "Spin, Rotations, and Position") is represented in the quantum algebra

$$\begin{array}{ccc} \mathbf{AL}(V) & \stackrel{\sigma}{\longrightarrow} & \mathbf{E}(\mathbf{AL}(V)) \\ \downarrow & & \downarrow \\ \mathbf{Q}_\epsilon(\mathbf{V}) & \underset{\mathrm{id}_{\mathbf{Q}_\epsilon(\mathbf{V})}}{\longrightarrow} & \mathbf{Q}_\epsilon(\mathbf{V}) \end{array}$$

The quantum image of the enveloping algebra contains the quantum classes for the endomorphisms of the higher tensor powers. It is the invariance subalgebra of the the vector space identities $\{\mathrm{id}_V, \mathrm{id}_V^T\}$, $I = \frac{\mathrm{id}_V - \mathrm{id}_V^T}{2}$, representable by the quantum classes[1] of the tensors with equal powers of basic space vectors and forms:

$$\mathbf{E}(\mathbf{AL}(V))_{\text{quant}} = \mathrm{INV}_I \mathbf{Q}_\epsilon(\mathbf{V}) = \{a \in \mathbf{Q}_\epsilon(\mathbf{V}) \mid [I, a] = 0\}$$
$$\cong \begin{cases} \bigoplus_{k=0}^{n} \mathbf{AL}(\bigwedge^k V) \text{ with } \mathbb{K}\text{-dimension } \sum_{k=0}^{n} \binom{n}{k}^2 = \binom{2n}{n}, & \epsilon = +1, \\ \bigoplus_{k\geq 0} \mathbf{AL}(\bigvee^k V) \text{ with } \mathbb{K}\text{-dimension } \sum_{k\geq 0} \binom{n+k-1}{k}^2 = \aleph_0, & \epsilon = -1. \end{cases}$$

The complementary vector space is constituted by the tensors with different powers of basic vectors and forms

$$\mathbf{Q}_\epsilon(\mathbf{V}) = \mathrm{INV}_I \mathbf{Q}_\epsilon(\mathbf{V}) \oplus (VV^T)^{\neq}.$$

The rank-n Lie algebra $\mathbf{AL}(V) = \log \mathbf{GL}(\mathbb{K}^n)$ of the basic space has n invariants, which generate the center of its enveloping algebra. In the quantum algebra the identity I alone generates all the related *quantum invariants*

$$\begin{aligned} \mathrm{INV}_{\mathbf{AL}(V)} \mathbf{Q}_\epsilon(\mathbf{V}) &= \{a \in \mathbf{Q}_\epsilon(\mathbf{V}) \mid [f, a] = 0 \text{ for all } f \in \mathbf{AL}(V)\} \\ &= \mathbb{K}[I]_{\text{quant}} \cong \begin{cases} \mathbb{K}^{1+n}, & \epsilon = +1, \\ \mathbb{K}^{\aleph_0}, & \epsilon = -1. \end{cases} \end{aligned}$$

The center of the enveloping algebra is embedded into the I-polynomials of the quantum algebra. It is not the center of the full quantum algebra, which consists of the scalars

$$\mathrm{centr}\, \mathbf{E}(\mathbf{AL}(V))_{\text{quant}} = \mathbb{K}[I]_{\text{quant}}, \quad \mathrm{centr}\, \mathbf{Q}_\epsilon(\mathbf{V}) \cong \mathbb{K}.$$

In a complex Fermi quantum algebra, the minimal polynomial $p_I(X)$ of the semisimple identity has degree $1 + n$ and integer or half-integer roots, symmetrically distributed around 0 (the following sums and products go in integer steps):

$$\begin{aligned} \mathrm{INV}_{\mathbf{AL}(V)} \mathbf{Q}_+(\mathbb{C}^{2n}) &= \mathbb{C}[I]_{\text{quant}} \cong \mathbb{C}^{1+n}, \\ p_I(X) &= \prod_{k=-\frac{n}{2}}^{\frac{n}{2}} (X - k) = \begin{cases} (X - \frac{1}{2})(X + \frac{1}{2}), & n = 1, \\ (X - 1)X(X + 1), & n = 2, \\ \dots \end{cases} \\ p_{\mathrm{id}_V}(X) &= \prod_{k=0}^{n} (X - k), \quad p_{\mathrm{id}_V^T}(X) = \prod_{k=-n}^{0} (X - k). \end{aligned}$$

[1]For a vector subspace $U \subseteq \bigotimes \mathbf{V}$ the quantum classes are $U_{\text{quant}} = U / U \cap I(S_\epsilon^{\text{quant}})$.

The powers of the identity $\{I^m\}_{m=0}^n$ or the I-associated projectors

$$\mathcal{P}_l(I) = \prod_{k\neq l} \frac{I-k}{l-k}, \quad l = -\frac{n}{2}, \dots, \frac{n}{2},$$
$$I^m = \sum_{l=-\frac{n}{2}}^{\frac{n}{2}} l^m \mathcal{P}_l(I), \quad m = 0, 1, \dots,$$

are a a basis for the invariants $\mathbb{C}[I]_{\text{quant}}$. They decompose the Fermi algebra into invariant subspaces $\mathcal{P}_l(I)\mathbf{Q}_+(\mathbb{C}^{2n})\mathcal{P}_k(I)$.

For $n = 1, 2$ the I-associated projectors are

$$\begin{array}{lll} n=1: & I\otimes I = \frac{1}{4} & \Rightarrow \mathcal{P}_{\pm\frac{1}{2}}(I) = \frac{1}{2} \pm I, \\ n=2: & I\otimes I\otimes I = I & \Rightarrow \mathcal{P}_0(I) = 1 - I\otimes I, \quad \mathcal{P}_{\pm 1}(I) = \frac{I\otimes I \pm I}{2}. \end{array}$$

7.2.3 Quantum Algebras and Product Representations

With a Lie algebra representation on a vector space V, the totally antisymmetric powers of the basic space for Fermi or the totally symmetric ones for Bose in the quantum algebras $\mathbf{Q}_\epsilon(\mathbf{V})$ carry product representations (chapter "Spin, Rotations, and Position").

The maximal Lie algebra of the basic space with all endomorphisms $\mathbf{AL}(V) = \log \mathbf{GL}(\mathbb{K}^n) \cong \mathbb{K}^{n^2}$ is decomposable into the abelian trace and the traceless $\mathbf{SL}(\mathbb{K}^n)$-Lie algebra, in the complex case given by $\mathbb{C}$ and $A_{n-1} \cong \mathbb{C}^{n^2-1}$ with the compact form A_{n-1}^c, the Lie algebra of $\mathbf{SU}(n)$. Thus a quantum algebra for basic space $V \cong \mathbb{C}^n$ carries representations for the main series

$$\begin{array}{ll} A_r = \log \mathbf{SL}(\mathbb{C}^{1+r}), & C_r = \log \mathbf{Sp}(\mathbb{C}^{2r}), \\ D_r = \log \mathbf{SO}(\mathbb{C}^{2r}), & B_r = \log \mathbf{SO}(\mathbb{C}^{1+2r}), \quad r \geq 1. \end{array}$$

The $(n-1)$ fundamental $\mathbf{SL}(\mathbb{C}^n)$-representations

$$\begin{array}{l} [1,0,\dots,0], \quad \dots \quad, [0,\dots,0,1] \\ \text{with dimension } \binom{n}{k}, \ k = 1, \dots n-1, \end{array}$$

are isomorphic to the product representations acting on the totally antisymmetric powers $\bigwedge^k V$ of the defining representation $[1,0,\dots,0]$ on $V \cong \mathbb{C}^n$.

The Fermi quantum algebra

$$\mathbb{C}^{4^n} \cong \mathbf{Q}_+(\mathbf{V}) = \text{CLIFF}(2n) \supset \text{INV}_I \mathbf{Q}_+(\mathbf{V}) \cong \bigoplus_{k=0}^{n} \mathbf{AL}(\bigwedge^k V) \cong \mathbb{C}^{\binom{2n}{n}}$$

contains in the invariance algebra $\text{INV}_I\mathbf{Q}_+(\mathbf{V})$ of the basic space identity the quantum classes of *all fundamental* $\mathbf{SL}(\mathbb{C}^n)$*-representations*

$$\bigwedge^k V \cong \bigwedge^{n-k} V^T \cong \mathbb{C}^{\binom{n}{k}} \text{ with } \underbrace{[0,\dots,0,1,0,\dots,0]}_{k\text{th place}} \in \mathbf{irrep}\,\mathbf{SL}(\mathbb{C}^n)$$
$$\text{for } k = 1, \dots, n-1.$$

In contrast, the Bose quantum algebra

$$\mathbb{C}^{\aleph_0} \cong \mathbf{Q}_-(\mathbf{V}) \supset \mathrm{INV}_I\mathbf{Q}_-(\mathbf{V}) \cong \bigoplus_{k\geq 0} \mathbf{AL}(\bigvee^k V) \cong \mathbb{C}^{\aleph_0}$$

contains all the totally symmetric powers of the defining representation of the $\mathbf{SL}(\mathbb{C}^n)$-Lie algebra A_{n-1} on the basic vector space and its dual:

$$\begin{array}{ll} \bigvee^k V \cong \mathbb{C}^{\binom{n+k-1}{k}} & \text{with } [k,0,\dots,0] \in \mathbf{irrep}\,\mathbf{SL}(\mathbb{C}^n), \\ \bigvee^k V^T \cong \mathbb{C}^{\binom{n+k-1}{k}} & \text{with } [0,\dots,0,k] \in \mathbf{irrep}\,\mathbf{SL}(\mathbb{C}^n). \end{array}$$

7.2.4 Group Actions on Quantum Algebras

A quantum algebra implements the basic space endomorphism Lie algebra $\mathbf{AL}(V)$ by inner derivations. It is acted on also by the corresponding group $\mathbf{GL}(V)$.

A tensor algebra $\bigotimes V$ inherits the representation of a group in the automorphisms of the basic space V,

$$\begin{array}{ll} D : G \longrightarrow \mathbf{GL}(V), & D(g) = D(g)_A^B\, e^A \otimes \check{e}_B, \\ \check{D} : G \longrightarrow \mathbf{GL}(V^T), & \check{D}(g) = D(g^{-1})_A^B\, \epsilon\check{e}_B \otimes e^A, \end{array}$$

in the form of the induced algebra automorphisms $\bigotimes D(g)$. Since the dual product is invariant under dual group representations, the group action is defined also on the quantum algebras in the form of algebra automorphisms

$$G \times \mathbf{Q}_\epsilon(\mathbf{V}) \longrightarrow \mathbf{Q}_\epsilon(\mathbf{V}), \quad \begin{cases} g \bullet \alpha = \alpha \in \mathbb{K} \\ g \bullet v = v(g) = D(g)(v), \quad v \in V \\ g \bullet \theta = \theta(g) = \check{D}(g)(\theta), \quad \theta \in V^T \\ g \bullet (a \otimes b) = a(g) \otimes b(g), \quad a, b \in \mathbf{Q}_\epsilon(\mathbf{V}) \end{cases}$$

The group action defines for each quantum algebra element a the *quantum orbit* of the group G, e.g., the orbits of the causal group $\mathbf{D}(1)$ as quantum time orbits (below)

$$G \longrightarrow \mathbf{Q}_\epsilon(\mathbf{V}), \quad g \longmapsto a(g) = g \bullet a.$$

All orbits are generated by the orbits $g \longmapsto v(g), \theta(g)$ in the basic dual vector spaces

$$\mathbf{Q}_\epsilon(\mathbf{V})^G = \{G \longrightarrow \mathbf{Q}_\epsilon(\mathbf{V})\} \text{ generated by } \{\mathbb{K}, V^G, V^{TG}\}.$$

The quantization (anti-) commutators of the basic vector orbits depend only on the quotient of the groups elements and involve *matrix elements of the group representation*:

$$\begin{array}{ll} [\theta(g_2), v(g_1)]_\epsilon & = \langle \theta(g_2), v(g_1)\rangle = \langle \theta, v(g)\rangle = D_\theta^v(g), \\ [\check{e}_B(g_2), e^A(g_1)]_\epsilon & = D_B^A(g), \quad g = g_2^{-1} g_1, \\ [\check{e}_B(g_2), e^A(g_1)]_{-\epsilon} & = D_B^D(g_2^{-1})[\check{e}_D, e^C]_{-\epsilon} D_C^A(g_1). \end{array}$$

For a represented Lie group the corresponding Lie algebra at the unit element $e \in G$ acts by inner derivations

$$\begin{aligned} &\log G \times \mathbf{Q}_\epsilon(\mathbf{V}) \longrightarrow \mathbf{Q}_\epsilon(\mathbf{V}), \\ &l^a \bullet a = \partial^a|_{g=e} a(g) = [\tilde{\mathcal{D}}(l^a), a], \ \ \tilde{\mathcal{D}}(l^a) = \tfrac{\mathcal{D}(l^a)+\check{\mathcal{D}}(l^a)}{2}. \end{aligned}$$

With the endomorphism Lie algebra $\mathbf{AL}(V)$ represented in $\mathbf{Q}_\epsilon(\mathbf{V})$ via the tensor commutator $[f,g]_\circ \longmapsto [f,g]_\otimes$, also the adjoint group action is embedded as adjoint action

$$\begin{aligned} \mathbf{GL}(V) \times \mathbf{Q}_\epsilon(\mathbf{V}) &\longrightarrow \mathbf{Q}_\epsilon(\mathbf{V}) \text{ (if defined)}, \\ e^f \bullet a &= e^f \otimes a \otimes e^{-f} \text{ with } e^f = 1 + f + \tfrac{f\otimes f}{2} + \dots . \end{aligned}$$

In general, the exponent e^f with quantum tensor product $\otimes$ differs from the quantum representative of the exponent $e^f_\circ = \mathrm{id}_V + f + \frac{f \circ f}{2} + \dots \in \mathbf{AL}(\mathbb{V})$ with endomorphism composition product $\circ$. In the quantum algebra, the action is equal for basic space vectors and forms and basic endomorphisms:

$$\begin{aligned} v \in V: \ \ & e^f \otimes v \otimes e^{-f} &= e^{-f^T} \otimes v \otimes e^{f^T} &= e^f_\circ(v), \\ \omega \in V^T: \ \ & e^f \otimes \omega \otimes e^{-f} &= e^{-f^T} \otimes \omega \otimes e^{f^T} &= e^{-f^T}_\circ(\omega), \\ h \in \mathbf{AL}(V): \ \ & e^f \otimes h \otimes e^{-f} &= e^{-f^T} \otimes h \otimes e^{f^T} &= e^f_\circ \circ h \circ e^{-f}_\circ. \end{aligned}$$

Only quantum algebra elements with a well defined exponent give rise to an adjoint group action

$$\begin{aligned} e^a &= \sum_{k\ge 0} \tfrac{a^k}{k!} \in \mathbf{Q}_\epsilon(\mathbf{V}), \ \ \mathrm{Ad}\, e^a : \mathbf{Q}_\epsilon(\mathbf{V}) \longrightarrow \mathbf{Q}_\epsilon(\mathbf{V}), \\ \mathrm{Ad}\, e^a(b) &= e^a \otimes b \otimes e^{-a} = e^{\mathrm{ad}\, a}(b) = \sum_{k\ge 0} \tfrac{(\mathrm{ad}\, a)^k(b)}{k!} = b + [a,b] + \tfrac{[a,[a,b]]}{2} + \dots \\ & \text{for eigenvector } \mathrm{ad}\, a(b) = \beta b, \ \ \beta \in \mathbb{K} \Rightarrow \mathrm{Ad}\, e^a(b) = e^\beta b. \end{aligned}$$

In a complex Fermi quantum algebra, all elements have an exponent

$$\exp \mathbf{Q}_+(\mathbf{V}) = \mathbf{Q}_+(\mathbf{V})^\diamond.$$

7.3 Quantum Algebras with Conjugation

In addition to the dual product, defining the quantum structure, a quantum algebra basic space can come with a reflection, e.g., a time reflection implementing conjugation or the anticonjugation for particle-antiparticles. The associated product, e.g., a scalar product for a Euclidean conjugation, is different from the dual product.

A (conjugate) linear symmetric reflection of the basic space and hence of the self-dual sum

$$* : \mathbf{V} \longrightarrow \mathbf{V}, \ \ \mathbf{v} = u + \theta \longmapsto \mathbf{v}^* = u^* + \theta^*, \ \ \langle u^*, v \rangle = \overline{\langle v^*, u \rangle}$$

also endows the quantum algebra with a (conjugate) linear reflection (for a linear reflection without conjugation of the scalars)

$$* : \mathbf{Q}_\epsilon(\mathbf{V}) \longrightarrow \mathbf{Q}_\epsilon(\mathbf{V}), \quad \alpha^* = \overline{\alpha} \in \mathbb{K}, \quad (a \otimes b)^* = b^* \otimes a^*,$$

since the quantum ideal is stable under reflection:

$$I(S_\epsilon^{\text{quant}}) = I(S_\epsilon^{\text{quant}})^* : \quad \begin{cases} \Big([\mathbf{v}, \mathbf{w}]_\epsilon - \langle \mathbf{v}, \mathbf{w}\rangle\Big)^* = [\mathbf{w}^*, \mathbf{v}^*]_\epsilon - \overline{\langle \mathbf{v}, \mathbf{w}\rangle} = 0, \\ \text{since } \overline{\langle \mathbf{v}, \mathbf{w}\rangle} = \langle \mathbf{w}^*, \mathbf{v}^*\rangle, \quad \mathbf{v}, \mathbf{w} \in \mathbf{V}. \end{cases}$$

For a complex basic space with conjugation the self-dual sum is decomposable into the complexification of the symmetric and antisymmetric real vector subspaces,

$$V \cong \mathbb{C}^n, \quad V \overset{*}{\leftrightarrow} V^T \Rightarrow \begin{cases} \mathbf{V} = V \oplus V^T = \mathbb{C} \otimes_\mathbb{R} (V_+ \oplus V_-), \\ V_+ \cong \mathbb{R}^n \cong V_-, \end{cases}$$

leading to real generators $\{V_+, V_-\}$ for the complex quantum algebra

$$\mathbf{Q}_\epsilon(\mathbb{C}^{2n}) \cong \mathbb{C} \otimes_\mathbb{R} \mathbf{Q}_\epsilon(\mathbb{R}^{2n}).$$

7.3.1 Quantum Algebras with Euclidean Conjugation

For the oscillator quantum algebras with Euclidean conjugation for a dual basis $\mathrm{u} \overset{\star}{\leftrightarrow} \mathrm{u}^\star$ one has real and imaginary generators. They are denoted in the Fermi case by $(\mathbf{r}, \mathbf{l})$ ("right" and "left"),

$$\mathbf{Q}_+(\mathbb{C}^2) = \mathbb{C} \otimes_\mathbb{R} \text{CLIFF}(1,1) : \quad \begin{cases} \{\mathrm{u}^\star, \mathrm{u}\} = 1, \quad \{\mathrm{u}, \mathrm{u}\} = 0 = \{\mathrm{u}^\star, \mathrm{u}^\star\}, \\ \text{with } \mathbf{r} = \frac{\mathrm{u}+\mathrm{u}^\star}{\sqrt{2}} = \mathbf{r}^\star, \quad \mathbf{l} = \frac{\mathrm{u}-\mathrm{u}^\star}{\sqrt{2}} = -\mathbf{l}^\star, \\ \Rightarrow \mathbb{C}\mathrm{u} \oplus \mathbb{C}\mathrm{u}^\star = \mathbb{C} \otimes (\mathbb{R}\mathbf{r} \oplus \mathbb{R}\mathbf{l}), \\ \{\mathbf{r}, \mathbf{r}\} = 1 = -\{\mathbf{l}, \mathbf{l}\}, \quad \{\mathbf{r}, \mathbf{l}\} = 0, \\ 2 = \mathbf{r}^2 = -\mathbf{l}^2 = \frac{(\mathbf{rl})^2}{2}, \end{cases}$$

and by position-momentum $(\mathbf{x}, \mathbf{p})$ in the Bose case:

$$\mathbf{Q}_-(\mathbb{C}^2) = \mathbb{C} \otimes_\mathbb{R} \mathbf{Q}_-(\mathbb{R}^2) : \quad \begin{cases} [\mathrm{u}^\star, \mathrm{u}] = 1 \\ \text{with } \mathbf{x} = \frac{\mathrm{u}+\mathrm{u}^\star}{\sqrt{2}} = \mathbf{x}^\star, \quad -i\mathbf{p} = \frac{\mathrm{u}-\mathrm{u}^\star}{\sqrt{2}} = -i\mathbf{p}^\star \\ \Rightarrow \mathbb{C}\mathrm{u} \oplus \mathbb{C}\mathrm{u}^\star = \mathbb{C} \otimes (\mathbb{R}\mathbf{x} \oplus \mathbb{R}i\mathbf{p}), \\ [i\mathbf{p}, \mathbf{x}] = 1. \end{cases}$$

As seen in this example, the Fermi and the Bose case are different with respect to the duality of the real basic vector spaces. For Fermi, the symmetric and antisymmetric subspaces $V_\pm$ are self-dual - "right" with "right" and "left" with "left," whereas for Bose V_+ and V_- are dual to each other - position with momentum :

$$\begin{aligned} \epsilon = +1 : &\quad (V_+)^T = V_+, \quad (V_-)^T = V_-, \quad \langle V_+, V_-\rangle = \{0\}, \\ \epsilon = -1 : &\quad (V_+)^T = V_-, \quad \langle V_+, V_+\rangle = \{0\} = \langle V_+, V_+\rangle. \end{aligned}$$

7.3.2 Quantum Algebras with Anticonjugation

With complex antispaces $V, \overline{V} \cong \mathbb{C}^n$ (chapter "Antistructures: The Real in the Complex") the anti-quantum algebras are also related to each other by the anticonjugation $\times$. Their product gives the quantum algebras for the complex quartet used, e.g., in the Fermi case for electrons and positrons with creation and annihilation operators (chapter "Massive Quantum Particle Fields"):

$$\mathbf{Q}_\epsilon(\mathbf{V}_{\text{doub}}) \cong \mathbf{Q}_\epsilon(\mathbf{V}) \otimes \mathbf{Q}_\epsilon(\overline{\mathbf{V}}) : \begin{cases} \mathbf{V}_{\text{doub}} & = V_{\text{doub}} \oplus V_{\text{doub}}^T \\ & = \mathbf{V} \oplus \overline{\mathbf{V}} \quad \cong \mathbb{C}^{4n}, \\ V_{\text{doub}} & = V \oplus \overline{V}^T \quad \cong \mathbb{C}^{2n}, \\ \mathbf{V} & = V \oplus V^T \quad \cong \mathbb{C}^{2n}, \end{cases}$$

$$\times : \mathbf{Q}_\epsilon(\mathbf{V}_{\text{doub}}) \longrightarrow \mathbf{Q}_\epsilon(\mathbf{V}_{\text{doub}}), \quad \begin{cases} \mathbb{C} \ni \alpha & \leftrightarrow \overline{\alpha}, \\ V \ni \mathrm{u} & \leftrightarrow \mathrm{u}^\times \in \overline{V}, \\ V^T \ni \mathrm{a}^\times & \leftrightarrow \mathrm{a} \in \overline{V}^T. \end{cases}$$

7.4 Grading of Quantum Algebras

As quotient tensor algebras, quantum algebras inherit integer graduations from the tensor powers of the basic vector space and its linear forms.

The action of the basic vector space identity $I = \frac{\mathrm{id}_V - \mathrm{id}_V^T}{2}$ on a quantum algebra defines the *power or duality grading*, a $\mathbb{Z}_{1+2n}$-grading for Fermi and a $\mathbb{Z}$-grading for Bose quantum algebras:

$$z \in \mathbb{Z} : \quad V^{(z)} = \{a \in \mathbf{Q}_\epsilon(\mathbf{V}) \mid [I, a] = za\}, \quad V^{(0)} = \mathrm{INV}_I \mathbf{Q}_\epsilon(\mathbf{V}),$$

$$\mathbf{Q}_\epsilon(\mathbf{V}) = \begin{cases} \bigoplus_{z=-n}^{n} V^{(z)}, & \epsilon = +1, \quad V^{(z_1)} \otimes V^{(z_2)} \subseteq V^{((z_1+z_2) \bmod (1+2n))}, \\ \bigoplus_{z \in \mathbb{Z}} V^{(z)}, & \epsilon = -1, \quad V^{(z_1)} \otimes V^{(z_2)} \subseteq V^{(z_1+z_2)}. \end{cases}$$

The grades as eigenvalues of the identity action are used in the equidistant energy spectrum of the harmonic oscillator (below). The vector subspaces $V^{(z)}$ with fixed grade are the product of the 0-grade subspace $V^{(0)}$, the quantized enveloping algebra, with the powers of the basic vector space:

$$V^{(z)} = \begin{cases} (V^{(0)} \otimes \bigwedge^z V)_{\text{quant}} \cong \mathbb{K}^{\binom{2n}{n-z}}, & \epsilon = +1, \\ (V^{(0)} \otimes \bigvee^z V)_{\text{quant}}, & \epsilon = -1. \end{cases}$$

The negative powers are defined with the forms, e.g., $\bigwedge^{-n} V = \bigwedge^{n} V^T$ for $n \in \mathbb{N}$.

The quotient vector space with respect to the trivial grade subalgebra is isomorphic to the direct sum of the Grassmann algebras for basic vector space and its dual in the Fermi case and the corresponding direct sum of the polynomial algebras in the Bose case

$$\mathbf{Q}_\epsilon(\mathbf{V})/V^{(0)} \cong \begin{cases} \bigwedge V \oplus \bigwedge V^T, & \epsilon = +1, \\ \bigvee V \oplus \bigvee V^T, & \epsilon = -1. \end{cases}$$

The summands with even and odd grades in the coarser $\mathbb{Z}_2$-grading define the *hybrid structure* of quantum algebras:

$$\mathbf{Q}_\epsilon(\mathbf{V}) = \mathbf{Q}_\epsilon(\mathbf{V})_0 \oplus \mathbf{Q}_\epsilon(\mathbf{V})_1, \quad \begin{cases} \mathbf{Q}_\epsilon(\mathbf{V})_0 = \bigoplus_{z\in 2\mathbb{Z}} V^{(z)} \in \underline{\mathbf{aag}}_{\mathbb{K}}, \\ \mathbf{Q}_\epsilon(\mathbf{V})_1 = \bigoplus_{z\in 1+2\mathbb{Z}} V^{(z)}. \end{cases}$$

In Fermi algebras, the odd elements $\mathbf{Q}_+(\mathbf{V})_1$ are called *Fermi elements*, the even ones $\mathbf{Q}_+(\mathbf{V})_0$ *Bose elements.* Bose subalgebra and Fermi vector subspace have equal dimension:

$$\dim_{\mathbb{K}} \mathbf{Q}_+(\mathbf{V})_0 = \dim_{\mathbb{K}} \mathbf{Q}_+(\mathbf{V})_1 = 2^{n-1}.$$

A Bose algebra $\mathbf{Q}_-(\mathbf{V})$ has only Bose elements.

Quantum algebras have an adjoint self-action, for Bose quantum algebras a natural Lie algebra structure

$$\mathbf{Q}_-(\mathbf{V}) \times \mathbf{Q}_-(\mathbf{V}) \longrightarrow \mathbf{Q}_-(\mathbf{V}), \quad (a,b) \longmapsto [a,b].$$

The Bose-Fermi characterization of a Fermi quantum algebra defines its natural hybrid bracket (graded Lie algebra) structure

$$\begin{aligned} &\mathbf{Q}_+(\mathbf{V}) \times \mathbf{Q}_+(\mathbf{V}) \longrightarrow \mathbf{Q}_+(\mathbf{V}), \\ &(a,b) \longmapsto [\![a,b]\!] = \begin{cases} [a,b] & \text{iff } a \text{ or } b \text{ are Bose}, \\ \{a,b\} & \text{iff } a \text{ and } b \text{ are Fermi}. \end{cases} \end{aligned}$$

7.4.1 Phase and Dilation Grading of Bose Quantum Algebras

A complex Bose quantum algebra $\mathbf{Q}_-(\mathbb{C}^{2n})$ of a basic vector space $V \cong \mathbb{C}^n$ with Euclidean conjugation and adapted dual bases

$$[\check{\mathrm{u}}_b, \mathrm{u}^a] = \delta^a_b, \quad \mathrm{u}^a \overset{\star}{\leftrightarrow} \check{\mathrm{u}}_a = \mathrm{u}^\star_a,$$

allows the *definition of position-momentum pairs* $\{\mathbf{x}_a, i\mathbf{p}_a\}$,

$$\mathbf{x}_a = \tfrac{\mathrm{u}^a+\mathrm{u}^\star_a}{\sqrt{2}} \in V_+ = W, \quad -i\mathbf{p}_a = \tfrac{\mathrm{u}^a-\mathrm{u}^\star_a}{\sqrt{2}} \in V_- = W^T, \quad [\mathrm{u}^\star_b, \mathrm{u}^a] = \delta^a_b = [i\mathbf{p}_b, \mathbf{x}_a],$$

as dual bases for the real spaces in the decomposition of the basic space $\mathbf{V} = \mathbb{C} \otimes_{\mathbb{R}} (V_+ \oplus V_-)$. $\mathbf{Q}_-(\mathbb{C}^{2n})$ is the complexification of a real Bose quantum algebra with basic space $\mathbf{W} = W \oplus W^T \cong \mathbb{R}^{2n}$:

$$\begin{aligned} &\mathbf{Q}_-(\mathbb{C}^{2n}) = \mathbb{C} \otimes_{\mathbb{R}} \mathbf{Q}_-(\mathbb{R}^{2n}), \\ &\mathbf{Q}_-(\mathbb{R}^{2n}) = \bigotimes \mathbf{W}/\text{minimal ideal } \{[i\mathbf{p}_b, \mathbf{x}_a] - \delta^a_b, \ [\mathbf{x}_b, \mathbf{x}_a], \ [i\mathbf{p}_b, i\mathbf{p}_a]\}. \end{aligned}$$

In addition to the $\mathbb{Z}$-grading above with the tensor powers of the complex basic space $V \cong \mathbb{C}^n$

$$\mathbf{Q}_-(\mathbb{C}^{2n}) = \bigoplus_{z\in\mathbb{Z}} V^{(z)}, \quad V^{(z)} = \{a \in \mathbf{Q}_-(\mathbb{C}^{2n}) \,\big|\, [I,a] = za\},$$
$$V^{(0)} = \mathrm{INV}_I \mathbf{Q}_-(\mathbb{C}^{2n}), \quad V^{(z)} = (V^{(0)} \otimes \bigvee^{z} V)_{\mathrm{quant}},$$

as eigenvalues of the adjoint action with the basic space identity, in an example with $n = 1$,

$$\begin{array}{l} I = I^\star = \frac{\{\mathrm{u},\mathrm{u}^\star\}}{2} = \frac{\mathbf{x}^2+\mathbf{p}^2}{2} \Rightarrow \left\{ \begin{array}{l} [I,\mathrm{u}] = \mathrm{u}, \quad [I,\mathrm{u}^\star] = -\mathrm{u}^\star, \\ [I,\mathrm{u}^k\mathrm{u}^{\star l}] = (k-l)\mathrm{u}^k\mathrm{u}^{\star l}, \end{array}\right. \\ V^{(0)} = \mathbb{C}[I], \end{array}$$

there is the $\mathbb{Z}$-grading with the tensor powers of the real basic space $W \cong \mathbb{R}^n$ as eigenvalues for the corresponding identity, in the example with $n = 1$,

$$\begin{array}{l} J = -J^\star = \frac{\{\mathbf{x},i\mathbf{p}\}}{2} = \frac{\mathrm{u}^{\star 2}-\mathrm{u}^2}{2} \Rightarrow \left\{ \begin{array}{l} [J,\mathbf{x}] = \mathbf{x}, \quad [J,i\mathbf{p}] = -i\mathbf{p}, \\ [J,\mathbf{x}^k(i\mathbf{p})^l] = (k-l)\mathbf{x}^k(i\mathbf{p})^l, \end{array}\right. \\ W^{(0)} = \mathbb{R}[J]. \end{array}$$

The real space grading counts the relative powers of position $\mathbf{x}$ and momentum $\mathbf{p}$:

$$\mathbf{Q}_-(\mathbb{R}^{2n}) = \bigoplus_{z\in\mathbb{Z}} W^{(z)}, \quad W^{(z)} = \{a \in \mathbf{Q}_-(\mathbb{R}^{2n}) \,\big|\, [J,a] = za\},$$
$$W^{(0)} = \mathrm{INV}_J \mathbf{Q}_-(\mathbb{R}^{2n}), \quad W^{(z)} = (W^{(0)} \otimes \bigvee^{z} W)_{\mathrm{quant}}.$$

The quotient with respect to the trivial grade subalgebra is isomorphic to the direct sum of the position and momentum polynomials:

$$\mathbf{Q}_-(\mathbb{R}^{2n})/W^{(0)} \cong \bigvee W \,\oplus\, \bigvee W^T \cong \mathbb{R}[\mathbf{x}] \,\oplus\, \mathbb{R}[i\mathbf{p}].$$

The complex space identity generates the compact group $\mathbf{U}(1) \cong \mathbf{SO}(2)$,

$$I = \frac{\mathbf{x}^2+\mathbf{p}^2}{2}, \quad \mathrm{ad}\, iI(a) = [iI,a],$$
$$\begin{array}{llll} \mathrm{ad}\, iI\begin{pmatrix}\mathrm{u}\\ \mathrm{u}^\star\end{pmatrix} = \begin{pmatrix} i & 0\\ 0 & -i\end{pmatrix}\begin{pmatrix}\mathrm{u}\\ \mathrm{u}^\star\end{pmatrix}, & \exp\begin{pmatrix} i & 0\\ 0 & -i\end{pmatrix}\alpha & = \begin{pmatrix} e^{i\alpha} & 0\\ 0 & e^{-i\alpha}\end{pmatrix} & \cong \mathbf{U}(1), \\ \mathrm{ad}\, iI\begin{pmatrix}\mathbf{x}\\ \mathbf{p}\end{pmatrix} = \begin{pmatrix} 0 & 1\\ -1 & 0\end{pmatrix}\begin{pmatrix}\mathbf{x}\\ \mathbf{p}\end{pmatrix}, & \exp\begin{pmatrix} 0 & 1\\ -1 & 0\end{pmatrix}\alpha & = \begin{pmatrix} \cos\alpha & \sin\alpha\\ -\sin\alpha & \cos\alpha\end{pmatrix} & \cong \mathbf{SO}(2), \end{array}$$

whereas the real space identity generates the noncompact group $\mathbf{D}(1) \cong \mathbf{SO}_0(1,1)$,

$$J = \frac{\{\mathbf{x},i\mathbf{p}\}}{2}, \quad \mathrm{ad}\, J(a) = [J,a],$$
$$\begin{array}{llll} \mathrm{ad}\, J\begin{pmatrix}\mathbf{x}\\ \mathbf{p}\end{pmatrix} = \begin{pmatrix} 1 & 0\\ 0 & -1\end{pmatrix}\begin{pmatrix}\mathbf{x}\\ \mathbf{p}\end{pmatrix}, & \exp\begin{pmatrix} 1 & 0\\ 0 & -1\end{pmatrix}\beta & = \begin{pmatrix} e^{\beta} & 0\\ 0 & e^{-\beta}\end{pmatrix} & \cong \mathbf{D}(1), \\ \mathrm{ad}\, J\begin{pmatrix}\mathrm{u}\\ \mathrm{u}^\star\end{pmatrix} = \begin{pmatrix} 0 & 1\\ 1 & 0\end{pmatrix}\begin{pmatrix}\mathrm{u}\\ \mathrm{u}^\star\end{pmatrix}, & \exp\begin{pmatrix} 0 & 1\\ 1 & 0\end{pmatrix}\beta & = \begin{pmatrix} \cosh\beta & \sinh\beta\\ \sinh\beta & \cosh\beta\end{pmatrix} & \cong \mathbf{SO}_0(1,1). \end{array}$$

This characterizes the complex and real space gradings (eigenvalues) of the oscillator or position-momentum quantum algebra $\mathbf{Q}_-(\mathbb{C}^2)$ as the grading for the phase $\mathbf{U}(1)$ and for the dimension $\mathbf{D}(1)$ respectively.

Both gradings are incompatible, i.e., there are no common eigenvectors for dilation $\mathbf{D}(1)$ and phase $\mathbf{U}(1)$:

$$[I, J] = -(\mathrm{u}^2 + \mathrm{u}^{\star 2}) = \mathbf{p}^2 - \mathbf{x}^2$$

The eigenspace for a $\mathbf{U}(1)$-grade z contains the sum of eigenspaces with $\mathbf{D}(1)$-grades $|d| \leq |z|$, i.e., a degree-$|z|$ monomial in u or $\mathrm{u}^\star$ is a polynomial in position-momentum $(\mathbf{x}, i\mathbf{p})$ with maximal degree $|z|$:

$$\overset{z}{\bigvee} V = \mathbb{C} \otimes \bigoplus_{d=-|z|}^{|z|} \overset{d}{\bigvee} W,$$

$$\text{e.g., } z = 2: \quad \mathbb{C}\mathrm{u} \vee \mathrm{u} = \mathbb{C} \otimes [\mathbf{x} \vee \mathbf{x} \ \oplus \ \mathbf{x} \vee i\mathbf{p} \ \oplus \ i\mathbf{p} \vee i\mathbf{p}].$$

7.5 Symmetry and Statistics

A group or Lie algebra action on a vector space can be used to define the quantum algebra statistics, Bose or Fermi.

If there exists a nontrivial bilinear form γ of the basic space $V \cong \mathbb{K}^n$ which is either symmetric or antisymmetric,

$$\gamma : V \times V \longrightarrow \mathbb{K}, \quad \begin{cases} \gamma = \gamma^{BA} \check{e}_A \otimes \check{e}_B \in V^T \otimes V^T, \\ \gamma^{BA} = \epsilon_\gamma \gamma^{AB}, \quad \epsilon_\gamma = \pm 1, \end{cases}$$

its power-2 tensor for the bilinear form has a nontrivial class in the quantum algebra iff it is of Fermi type $\mathbf{Q}_+(\mathbf{V})$ for an antisymmetric form and of Bose type $\mathbf{Q}_-(\mathbf{V})$ for a symmetric one: The symmetry sign ϵ_γ for the bilinear form has to be opposite to the statistical sign ϵ :

$$\text{in } \mathbf{Q}_\epsilon(\mathbf{V}) : \quad \begin{cases} \gamma = \frac{\gamma^{BA} - \epsilon \gamma^{AB}}{2} \check{e}_A \otimes \check{e}_B = \frac{1 - \epsilon\epsilon_\gamma}{2} \gamma, \\ \gamma \neq 0 \iff \epsilon = -\epsilon_\gamma. \end{cases}$$

The dual isomorphism, associate to a nondegenerate bilinear form γ, implements the inversion of its invariance Lie algebra $L(\gamma)$:

$$\begin{array}{ccc} V & \overset{\mathcal{D}(l)}{\longrightarrow} & V \\ \gamma \downarrow & & \downarrow \gamma \\ V^T & \underset{\check{\mathcal{D}}(l)}{\longrightarrow} & V^T \end{array}, \qquad \begin{array}{l} l \in L(\gamma) \iff -\mathcal{D}(l) = \gamma^{-1} \circ \mathcal{D}(l)^T \circ \gamma, \\ \text{in } \mathbf{Q}_\epsilon(\mathbf{V}) : \ [\mathcal{D}(l), \gamma] = 0. \end{array}$$

A prominent example for the relation between the symmetry of a bilinear form and the quantum algebra statistics is the *spin statistics connection*: An irreducible representation of the spin group $\mathbf{SU}(2)$ on a vector space $V \cong \mathbb{C}^{1+2J}$ has, via the product of the antisymmetric Pauli spinor "metric" $\epsilon^1 \cong \epsilon_{AB} =$

$-\epsilon_{BA}$, an invariant bilinear form that is (anti)symmetric, i.e., (symplectic) orthogonal, for (half)integer spin J (chapter "Spin, Rotations, and Position")

$$\epsilon^{2J} = \bigvee^{2J} \epsilon^1 : V \times V \longrightarrow \mathbb{C}, \quad \epsilon^{2J}(v,w) = (-1)^{2J}\epsilon^{2J}(w,v).$$

In order to implement nontrivially the rotation invariant form and hence also the reflection of the spin Lie algebra $\log \mathbf{SU}(2) \cong A_1^c \cong \mathbb{R}^3$ (equally of $A_1 \cong \mathbb{C}^3$ for $\mathbf{SL}(\mathbb{C}^2)$) in the quantum algebra of the representation space, one needs a *Fermi algebra for half-integer spin* and a *Bose algebra for integer spin*:

$$\epsilon^{2J} \neq 0 \text{ in } \mathbf{Q}_\epsilon(\mathbb{C}^{2(1+2J)}) \iff \epsilon = (-1)^{1+2J} = \begin{cases} +1, & \text{i.e., Fermi for } J = \frac{1}{2}, \frac{3}{2}, \ldots, \\ -1, & \text{i.e., Bose for } J = 0, 1, \ldots. \end{cases}$$

A nontrivially implemented nondegenerate bilinear form $\gamma : V \longrightarrow V^T$ on a complex space and its inverse together with the basic space identity represent the $\mathbf{SL}(\mathbb{C}^2)$-Lie algebra $A_1(\gamma) \cong \mathbb{C}^3$, called a *self-duality Lie algebra* or Gürsey-Pauli Lie algebra, in spherical or Cartesian bases:

$$\left.\begin{array}{c} \gamma_- = \frac{i}{2}\gamma^{BA}\check{e}_A \otimes \check{e}_B, \quad \gamma_+ = -\frac{i}{2}\gamma_{AB}e^B \otimes e^A, \\ I = \frac{[e^A, \check{e}_A]_{-\epsilon}}{2} \end{array}\right\} \Rightarrow \left\{\begin{array}{ll} [\gamma_-, \gamma_+] & = I, \\ [I, \gamma_\pm] & = \pm 2\gamma_\pm, \end{array}\right.$$

$$\begin{array}{rl} \mathcal{T}^{1,2,3} = (i\frac{\gamma_+ - \gamma_-}{2}, \frac{\gamma_+ + \gamma_-}{2}, \frac{i}{2}I) & \Rightarrow \quad [\mathcal{T}^j, \mathcal{T}^k] = -\epsilon^{jkl}\mathcal{T}^l, \\ \text{Casimir element:} \quad -\frac{1}{2}\mathcal{T}^j \otimes \mathcal{T}^j = & 4\{\gamma_-, \gamma_+\} + 2I \otimes I. \end{array}$$

Such a self-duality Lie algebra $A_1(\gamma)$ commutes in the quantum algebra with the represented invariance Lie algebra $L(\gamma)$ for the bilinear form γ:

$$L(\gamma) \oplus A_1(\gamma) \longrightarrow \mathbf{Q}_\epsilon(\mathbf{V}), \quad [\mathcal{D}(l), \mathcal{T}] = 0, \quad l \in L(\gamma), \quad \mathcal{T} \in A_1(\gamma).$$

With the action of both Lie algebras, a complex quantum algebra is decomposable into irreducible representation spaces of the self-duality Lie algebra $A_1(\gamma)$, invariant under action with $L(\gamma)$. An example is given by the fundamental spin quantum algebra.

7.6 Fundamental Spin Quantum Algebra

The fundamental spin quantum algebra of Fermi type is used, e.g., for massive spin $\frac{1}{2}$-particles (chapter "Massive Quantum Particle Fields").

In contrast to the oscillator quantum algebras for basic space $V \cong \mathbb{C}$ with abelian Lie algebra $\log \mathbf{GL}(\mathbb{C})$, the quantum algebra for a complex 2-dimensional space has much structure. The basic space carries the fundamental Pauli representation of the $\mathbf{SL}(\mathbb{C}^2)$-Lie algebra A_1, the smallest simple Lie structure:

$$\begin{array}{l} A_1 \cong \mathbb{C}^3 : \quad [l^a, l^b] = -\epsilon^{abc}l^c, \; a,b,c = 1,2,3, \\ V \cong \mathbb{C}^2, \; A_1 \longrightarrow \mathbf{AL}(V), \; \vec{l} \longmapsto \vec{\Sigma} = i\frac{\vec{\sigma}_A^B}{2}e^A \otimes \check{e}_B, \quad A, B = 1, 2. \end{array}$$

The nontrivial implementation of the antisymmetric spinor "metric" (volume form)

$$\epsilon_- = \tfrac{i}{2}\epsilon^{BA}\check{e}_A \otimes \check{e}_B, \quad \epsilon_+ = -\tfrac{i}{2}\epsilon_{AB}e^B \otimes e^A$$

requires a Fermi quantum algebra (spin statistics connection). It is isomorphic to the Clifford algebra of the self-dual basic space $\mathbf{V} \cong \mathbb{C}^4$:

$$\begin{aligned}\mathbf{Q}_+(\mathbb{C}^4) &= \mathrm{CLIFF}(4) \cong \mathbb{C}^{16} \text{ with } \{\check{e}_A, e^B\} = \delta_A^B, \ \{e^A, e^B\} = 0 = \{\check{e}_A, \check{e}_B\} \\ &\cong \mathbb{C} \otimes \mathrm{CLIFF}(2,2) \text{ with } \mathbf{U}(2)\text{-conjugation } e^A \overset{\star}{\leftrightarrow} \check{e}_A.\end{aligned}$$

The identity, the bilinear form, and its inverse represent the selfduality Lie algebra $\mathcal{T}^j \in A_1(\epsilon)$, different from the spin Lie algebra $\Sigma^a \in A_1$:

$$\begin{array}{l}[\epsilon_-, \epsilon_+] = I = \frac{[e^A, \check{e}_A]}{2}, \quad [I, \epsilon_\pm] = \pm 2\epsilon_\pm, \\ \mathcal{T}^{1,2,3} = (i\frac{\epsilon_+ - \epsilon_-}{2}, \frac{\epsilon_+ + \epsilon_-}{2}, \frac{i}{2}I), \quad [\mathcal{T}^j, \mathcal{T}^k] = -\epsilon^{jkl}\mathcal{T}^l, \quad [\mathcal{T}^j, \vec{\Sigma}] = 0, \\ \{\epsilon_-, \epsilon_+\} = I \otimes I, \quad \{I, \epsilon_\pm\} = 0, \quad \{\vec{\Sigma}, \epsilon_\pm\}.\end{array}$$

The Casimir elements of spin and self-duality Lie algebras differ by a constant:

$$-\tfrac{1}{2}\vec{\Sigma} \otimes \vec{\Sigma} = \tfrac{3}{8} + \tfrac{1}{2}\mathcal{T}^j \otimes \mathcal{T}^j = \tfrac{3}{8}(1 - I \otimes I).$$

The spin quantum algebra is graded with $\mathbb{Z}_5 = \{0, \pm 1, \pm 2\}$. The $d(z) = \binom{4}{2-|z|}$-dimensional spaces with grade z are decomposable into irreducible spin A_1-representations $[2J]$ on spaces $\mathbb{C}^{1+2J}$ with the bases in the following table:

$z\downarrow / J\rightarrow$	0	$\frac{1}{2}$	1
$+2$	ϵ_+	$-$	$-$
-2	ϵ_-	$-$	$-$
$+1$	$-$	$e^A, \check{e}_A \otimes \epsilon_+$	$-$
-1	$-$	$\check{e}_A, e^A \otimes \epsilon_-$	$-$
0	$1, I, I \otimes I$	$-$	$\vec{\Sigma}$

The spin decomposition gives 5 trivial representations (scalars 1, dual isomorphisms $\epsilon_\pm$, identity I, Casimir element $\vec{\Sigma} \otimes \vec{\Sigma}$), two pairs of self-dual Pauli spinor spaces und one adjoint representation $\vec{\Sigma}$. The self-duality $A_1(\epsilon)$- decomposition is analogous.

In the coarser grading with even subalgebra (Bose elements) and odd vector subspace (Fermi elements),

$$\mathbf{Q}_+(\mathbb{C}^4) = \mathbf{Q}_+(\mathbb{C}^4)_0 \ \oplus \ \mathbf{Q}_+(\mathbb{C}^4)_1 \cong \mathbb{C}^8 \ \oplus \ \mathbb{C}^8,$$

the Fermi subspace with grades $\{\pm 1\}$ contains the quantum classes of the basic space $\mathbf{V} = V \oplus V^T \cong \mathbb{C}^4$ and those of third power $\bigwedge^3 \mathbf{V} \cong \mathbb{C}^4$. The Bose subalgebra with grades $\{0, \pm 2\}$,

$$\mathbf{Q}_+(\mathbb{C}^4)_0 = \mathrm{INV}_I \mathbf{Q}_+(\mathbb{C}^4) \ \oplus \ [V, V] \ \oplus \ [V^T, V^T],$$

is decomposable with respect to its natural Lie algebra structure into two abelian Lie algebras $\mathbb{C}$ with 1 and the Casimir element as bases and the two Lie algebras with bases $\{\vec{\Sigma}, \mathcal{T}^j\}$,

$$\text{as Lie algebra: } \mathbf{Q}_+(\mathbb{C}^4)_0 \cong \mathbb{C} \ \oplus \ \mathbb{C} \ \oplus \ A_1 \ \oplus \ A_1(\epsilon).$$

The spin quantum algebra is the square of the Fermi oscillator quantum algebra

$$\mathbf{Q}_+(\mathbb{C}^4) \cong \mathbf{Q}_+(\mathbb{C}^2)^2.$$

It is also acted on with the abelian group $\mathbf{GL}(\mathbb{C}) \cong \mathbf{GL}(\mathbb{C}^2)/\mathbf{SL}(\mathbb{C}^2)$. $\mathbf{GL}(\mathbb{C})$ is represented with the $\mathbf{SL}(\mathbb{C}^2)$-invariant identity I which can implement the time translations (more below).

The 6-dimensional invariance algebra of the identity I with grade 0 contains the quantum classes for the enveloping algebra $\mathbf{E}(\mathbf{AL}(V))$ of the spin Lie algebra representation with the Casimir invariant:

$$\mathrm{INV}_I\mathbf{Q}_+(\mathbb{C}^4) \cong \mathbb{C}^6 : \quad \begin{cases} \text{basis}: \ \{1, I, \vec{\Sigma}, I \otimes I\}, \\ [\Sigma^a, \Sigma^b] = -\epsilon^{abc}\Sigma^c, \ [I, \vec{\Sigma}] = 0, \\ \{\Sigma^a, \Sigma^b\} = -\frac{1}{8}\delta^{ab}(1 - I \otimes I), \ \{I, \vec{\Sigma}\} = 0, \\ I \otimes I \otimes I = I. \end{cases}$$

The quantum invariants of the basic space Lie algebra $\log \mathbf{GL}(\mathbb{C}^2) \cong \mathbb{C} \oplus A_1$ are the polynomials in the identity

$$\mathrm{INV}_{\mathbb{C} \,\oplus\, A_1}\mathbf{Q}_+(\mathbb{C}^4) = \{\alpha + \beta I + \gamma I \otimes I \,\big|\, \alpha, \beta, \gamma \in \mathbb{C}\} \cong \mathbb{C}^3.$$

The I-associated projectors

$$I \otimes I \otimes I = I \Rightarrow \mathcal{P}_0(I) = 1 - I \otimes I, \quad \mathcal{P}_{\pm 1}(I) = \tfrac{I \otimes I \pm I}{2}$$

lead to an involutor $\mathcal{E}_0$, commuting with the Bose subalgebra and anticommuting with the Fermi subspace:

$$\mathcal{E}_0 = 1 - 2\mathcal{P}_0(I), \quad \mathcal{E}_0 \otimes \mathcal{E}_0 = 1, \quad \begin{cases} [\mathcal{E}_0, \mathbf{Q}_+(\mathbb{C}^4)_0] = 0, \\ \{\mathcal{E}_0, \mathbf{Q}_+(\mathbb{C}^4)_1\} = 0. \end{cases}$$

The projectors have as (anti-) commutator property

$$\begin{aligned} &a_0 \in \mathbf{Q}_+(\mathbb{C}^4)_0 \Rightarrow [\mathcal{P}_0(I), a_0] = 0, \\ &a_1 \in \mathbf{Q}_+(\mathbb{C}^4)_1 \Rightarrow \mathcal{P}_0(I) \otimes a_1 = a_1 \otimes (1 - \mathcal{P}_0(I)). \end{aligned}$$

The invariant $I \otimes I$ has a "nonlinear" adjoint action

$$\begin{aligned} &a_0 \in \mathbf{Q}_+(\mathbb{C}^4)_0 \Rightarrow [I \otimes I, a_0] = 0, \\ &a_1 \in \mathbf{Q}_+(\mathbb{C}^4)_1 \Rightarrow \begin{cases} [I \otimes I, a_1 \otimes \mathcal{P}_0(I)] = a_1 \otimes \mathcal{P}_0(I), \\ [I \otimes I, a_1 \otimes (1 - \mathcal{P}_0(I))] = -a_1 \otimes (1 - \mathcal{P}_0(I)). \end{cases} \end{aligned}$$

7.7 Adjoint Quantum Algebras

For a finite-dimensional Lie algebra L, there exist many associative algebras - the Lie algebra endomorphism $L \otimes L^T$ with the adjoint L-representation, the enveloping algebra $\mathbf{E}(L)$ and the adjoint enveloping algebra $\mathbf{E}(L \otimes L^T)$. These algebras have to be distinguished from the *adjoint quantum algebras* $\mathbf{Q}_\epsilon(\mathbf{L})$ with $\mathbf{L} = L \oplus L^T$ as basic vector space. $\mathbf{Q}_\epsilon(\mathbf{L})$ contain the quantum classes of the adjoint enveloping algebra $\mathbf{E}(L \otimes L^T)$. Adjoint quantum algebras are used in quantum gauge theories (chapter "Gauge Interactions").

The Lie-bracket-induced commutator in the enveloping algebra has to be kept apart from the duality-induced (anti-) commutators in the adjoint quantum algebras:

$$\begin{array}{rl} \text{in } \mathbf{E}(L): & [l^a, l^b] = \epsilon^{ab}_c l^c, \\ \text{in } \mathbf{E}(L \otimes L^T): & [\mathcal{L}^a, \mathcal{L}^b] = \epsilon^{ab}_c \mathcal{L}^c \text{ with } \mathcal{L}^a = \operatorname{ad} l^a = \epsilon^{ab}_c l^c \otimes \check{l}_b, \\ \text{in } \mathbf{Q}_\epsilon(\mathbf{L}): & \left\{ \begin{array}{l} [\check{l}_a, l^b]_\epsilon = \delta^b_a, \quad [l^a, l^b]_\epsilon = 0 = [\check{l}_a, \check{l}_b], \\ [\mathcal{L}^a, \mathcal{L}^b] = \epsilon^{ab}_c \mathcal{L}^c, \\ [\mathcal{L}^a, l^b] = \epsilon^{ab}_c l^c, \quad [\mathcal{L}^a, \check{l}_c] = -\epsilon^{ab}_c \check{l}_b. \end{array} \right. \end{array}$$

A symmetric invariant nondegenerate bilinear Lie algebra form κ, especially the Killing form for a semisimple Lie algebra, has a nontrivial quantum representative only in the adjoint Bose quantum algebra $\mathbf{Q}_-(\mathbf{L})$. There, it defines a self-duality Lie algebra $A_1(\kappa)$:

$$\begin{array}{l} \kappa = \kappa^{ab} \check{l}_a \otimes \check{l}_b, \quad \kappa^{-1} = \kappa_{ab} l^a \otimes l^b, \quad [\mathcal{L}^a, \kappa] = 0 = [\mathcal{L}^a, \kappa^{-1}], \\ \quad \text{in } \mathbf{Q}_+(\mathbf{L}): \quad \kappa = 0 = \kappa^{-1}, \\ \quad \text{in } \mathbf{Q}_-(\mathbf{L}): \quad [\kappa, \kappa^{-1}] = 4I = 2\{l^a, \check{l}_a\}. \end{array}$$

The adjoint L-representation gives rise to an invariant power-3 tensor that is trivial in the adjoint Bose quantum algebra and a nilquadratic Fermi element in the adjoint Fermi quantum algebra

$$\operatorname{ad} = \mathcal{L}^a \otimes \check{l}_a = \epsilon^{ab}_c l^c \otimes \check{l}_b \otimes \check{l}_a, \quad [\mathcal{L}^a, \operatorname{ad}] = 0, \quad \left\{ \begin{array}{ll} \text{in } \mathbf{Q}_-(\mathbf{L}): & \operatorname{ad} = 0, \\ \text{in } \mathbf{Q}_+(\mathbf{L}): & \operatorname{ad} \otimes \operatorname{ad} = 0. \end{array} \right.$$

To prove the nilpotency in $\mathbf{Q}_+(\mathbf{L})$, the Jacobi Leibniz identity of the Lie algebra has to be used:

$$\begin{array}{rl} \text{in } \mathbf{Q}_+(\mathbf{V}): \ \operatorname{ad} \otimes \operatorname{ad} & = \frac{1}{2}\epsilon^{bc}_a \mathcal{L}^a \otimes \check{l}_c \otimes \check{l}_b - \frac{1}{2}\epsilon^{cd}_a \mathcal{L}^a \otimes \check{l}_d \otimes \check{l}_c + \frac{1}{2}\epsilon^{dc}_a \mathcal{L}^a \otimes \check{l}_d \otimes \check{l}_c \\ & = -\epsilon^{bc}_a \mathcal{L}^a \otimes \check{l}_c \otimes \check{l}_b = 0. \end{array}$$

The power-3 tensor plays a role as self-interaction vertex for Bose gauge fields and in nilpotent Becchi-Rouet-Stora transformations for Fermi Fadeev-Popov fields (chapter "Gauge Interactions").

7.8 The Quantum Algebra for Position Translations

An example for an adjoint quantum algebra is the Bose quantum algebra $\mathbf{Q}_-(\mathbb{C}^6)$ for the $\mathbf{SL}(\mathbb{C}^2)$-Lie algebra $A_1 \cong \mathbb{C}^3$ with the spin $\mathbf{SU}(2)$-Lie algebra $A_1^c \cong \mathbb{R}^3$ as compact form. It is the arena of *3-dimensional quantum mechanics.*

A spherical or a Cartesian basis for A_1,

$$l^1 = i\frac{l_+ - l_-}{2}, \quad l^2 = \frac{l_+ + l_-}{2}, \quad l^3 = i\frac{l_0}{2},$$

has the Lie brackets in the enveloping algebra:

$$\text{in } \mathbf{E}(A_1): \quad [l^a, l^b] = -\epsilon^{abc} l^c \text{ and } [l_-, l_+] = l_0, \quad [l_0, l_\pm] = \pm 2 l_\pm.$$

The adjoint A_1-representation

$$l^a \longmapsto \mathcal{O}^a = \epsilon^{abc} l^b \otimes \check{l}_c, \quad [\mathcal{O}^a, \mathcal{O}^b] = -\epsilon^{abc}\mathcal{O}^c$$

arises in the quantum algebra of $A_1 \oplus A_1^T \cong \mathbb{C}^6$:

$$\text{in } \mathbf{Q}_-(\mathbb{C}^6): \left\{ \begin{array}{llll} [\check{l}_a, l^b] & = \delta_a^b, & [l^a, l^b] = 0 & = [\check{l}_a, \check{l}_b], \\ [\mathcal{O}_0, \mathcal{O}_\pm] & = \pm 2\mathcal{O}_\pm, & [\mathcal{O}_+, \mathcal{O}_-] & = -\mathcal{O}_0, \ [\mathcal{O}^a, \mathcal{O}^b] = -\epsilon^{abc}\mathcal{O}^c, \\ [\mathcal{O}_0, l_\pm] & = \pm 2 l_\pm, & [\mathcal{O}_\pm, l_\mp] & = \mp l_0, \quad [\mathcal{O}_\pm, l_\pm] = 0, \\ [\mathcal{O}_0, l_0] & = 0, & [\mathcal{O}_\pm, l_0] & = \mp 2 l_\pm. \end{array} \right.$$

With a Euclidean conjugation

$$\star : \mathbf{Q}_-(\mathbb{C}^6) \longrightarrow \mathbf{Q}_-(\mathbb{C}^6), \quad l^a \leftrightarrow (l^a)^\star = \delta^{ab}\check{l}_b = \check{l}^a$$

positions and momenta $\{\mathbf{x}_a, \mathbf{p}_a\}_{a=1,2,3}$ are definable as conjugation-symmetric Cartesian basis

$$\mathbf{Q}_-(\mathbb{C}^6) \cong \mathbb{C} \otimes \mathbf{Q}_-(\mathbb{R}^6), \quad \left\{ \begin{array}{ll} \mathbf{x}_a & = \frac{l^a + \check{l}^a}{\sqrt{2}} = \mathbf{x}_a^\star, \\ \mathbf{p}_a & = i\frac{l^a - \check{l}^a}{\sqrt{2}} = \mathbf{p}_a^\star, \end{array} \right. \quad [i\mathbf{p}_a, \mathbf{x}_b] = \delta_{ab}.$$

They can be represented with derivations of the complex position polynomials

$$\{\mathbf{x}_a, i\mathbf{p}_a = \partial_a = \frac{\partial}{\partial \mathbf{x}_a}\} \text{ on } \mathbb{C}[\mathbf{x}_a].$$

The reflection antisymmetric angular momenta $\{\mathcal{O}^a\}_{a=1,2,3}$ are a Cartesian basis of the adjoint representation of the angular momenta $A_1^c \simeq \log \mathbf{SO}(3)$:

$$\mathcal{O}^a = -(\mathcal{O}^a)^\star = -i\epsilon^{abc}\mathbf{p}_b \otimes \mathbf{x}_c; \quad \mathcal{O}_\pm^\star = -\mathcal{O}_\mp, \quad \mathcal{O}_0^\star = \mathcal{O}_0.$$

Positions, momenta, and angular momenta are examples for vector subspaces $\Lambda \cong \mathbb{C}^3$ of the quantum algebra with adjoint A_1-representation and basis $\{\lambda^a\}_{a=1}^3$,

$$[\mathcal{O}^a, \lambda^b] = -\epsilon^{abc}\lambda^c \text{ ,e.g., } \lambda^b = \mathcal{O}^b, l^b, \check{l}_b, \mathbf{x}_b, \mathbf{p}_b,$$

$\lambda_+ = \lambda^2 - i\lambda^1$ is the $\mathcal{O}_0$-eigenvector in Λ with highest weight, $[\mathcal{O}_0, \lambda_+] = 2\lambda_+$. The products λ_+^L in the quantum algebra with power $L = 0, 1, 2, \ldots$ are the highest-weight eigenvectors for an A_1-representation $[2L]$ with dimension $1+2L$ (angular momentum L). Hence all other eigenvectors $\{\lambda_m^L \mid |m| \leq L\}$ of this representation can be reached with the lowering operator $\mathcal{O}_-$:

$$\begin{array}{ll} (\operatorname{ad}\mathcal{O}_-)^{L-m}(\lambda_+^L) & = \lambda_m^L \\ (\operatorname{ad}\mathcal{O}_-)^{1+2L}(\lambda_+^L) & = 0 \end{array} \Rightarrow \left\{ \begin{array}{l} m = L, L-1, \ldots, -L+1, -L, \\ [\mathcal{O}_0, \lambda_m^L] = 2m\lambda_m^L. \end{array} \right.$$

The angular momenta Casimir element, realized by

$$\begin{aligned} -\tfrac{1}{2}\vec{\mathcal{O}}\otimes\vec{\mathcal{O}} &= -\tfrac{\{\mathcal{O}_+,\mathcal{O}_-\}}{4} + \tfrac{\mathcal{O}_0^2}{8} \\ &= -\tfrac{\mathcal{O}_+\otimes\mathcal{O}_-}{2} - \tfrac{\mathcal{O}_0}{4} + \tfrac{\mathcal{O}_0^2}{8} = \tfrac{\mathcal{O}_-^\star\otimes\mathcal{O}_-}{2} + \tfrac{\mathcal{O}_0}{4}(\tfrac{\mathcal{O}_0}{2} - 1), \end{aligned}$$

produces the invariant by double adjoint action

$$[\vec{\mathcal{O}}, [\vec{\mathcal{O}}, \lambda^L_m]] = -L(1+L)\lambda^L_m.$$

The angular momentum decomposition of the Bose quantum algebra is used for the polar decomposition in Schrödinger's wave functions (chapter "Quantum Probability").

Because of the complex basic space $A_1 \cong \mathbb{C}^3$, the $\mathbf{SO}(3)$-structure (angular momentum) in the position translation quantum algebra $\mathbf{Q}_-(\mathbb{C}^6)$ is embedded into an $\mathbf{SU}(3)$-Lie algebra ("color") in the endomorphisms $\mathbf{AL}(\mathbb{C}^3)$: The traceless basic space endomorphisms are the two fundamental triplet representations $[1,0]$, $[0,1]$ of $A_2 \cong \log \mathbf{SL}(\mathbb{C}^3)$ with $A_2^c \cong \log \mathbf{SU}(3)$ (Gell-Mann matrices $\{\vec{\lambda}\}$):

$$A_2 \longrightarrow \mathbf{AL}(\mathbb{C}^3)_0,\quad l^j \longmapsto \left\{ \begin{array}{ll} \frac{i}{2}\lambda^{jb}_a l^a \otimes \check{l}_b & \text{for } [1,0], \\ -\frac{i}{2}\lambda^{jb}_a \check{l}_b \otimes l^a & \text{for } [0,1], \end{array} \right. \quad j = 1,\dots,8.$$

The symmetric tensor power $\bigvee^L A_1 \cong \mathbb{C}^{\binom{2+L}{2}}$ in the Bose quantum algebra is acted upon with the irreducible A_2-representation $[L,0]$. The $\mathbf{SU}(3)$-representation $[L,0]$ is decomposable into irreducible angular momentum $\mathbf{SO}(3)$-representations with the following dimensionalities of the representation spaces:

$$\mathbf{SU}(3) \cong \bigoplus \mathbf{SO}(3):\left\{ \begin{array}{l} L \text{ even:} \left\{ \begin{array}{l} [L,0] \cong [2L] \oplus [2(L-2)] \oplus \cdots \oplus [0], \\ \mathbb{C}^{\binom{2+L}{2}} \cong \mathbb{C}^{1+2L} \oplus \mathbb{C}^{2L-3} \oplus \cdots \oplus \mathbb{C}, \end{array} \right. \\ \\ L \text{ odd:} \left\{ \begin{array}{l} [L,0] \cong [2L] \oplus [2(L-2] \oplus \cdots \oplus [2], \\ \mathbb{C}^{\binom{2+L}{2}} \cong \mathbb{C}^{1+2L} \oplus \mathbb{C}^{2L-3} \oplus \cdots \oplus \mathbb{C}^3, \end{array} \right. \end{array} \right.$$

e.g., an $\mathbf{SU}(3)$-triplet $[1,0]$ into an $\mathbf{SO}(3)$-triplet, an $\mathbf{SU}(3)$-sextet $[2,0]$ into an $\mathbf{SO}(3)$-quintet and singlet, etc.

7.9 Quantum Implemented Time Action

A complex hybrid algebra A with unit 1_A and conjugation $*$, e.g., a quantum algebra $\mathbf{Q}_\epsilon(\mathbb{C}^{2n})$, contains its $*$-*antisymmetric real Lie algebra* with even (Bose) elements and its $*$-*unitary group*

$$\begin{aligned} L(A) &= \{il \in A \mid l = l^*,\ \deg l = 0\} &\in \underline{\mathbf{lag}}_{\mathbb{R}}, \\ U(A) &= \{u \in A \mid uu^* = u^*u = 1_A\} &\in \underline{\mathbf{grp}}. \end{aligned}$$

Lie algebra and unitary group act on the full algebra by inner derivations and adjoint representation, respectively:

$$\begin{aligned} L(A) \times A \longrightarrow A, &\quad (il, a) \longmapsto i\,\mathrm{ad}\,l(a) = [il, a], \\ U(A) \times A \longrightarrow A, &\quad (u, a) \longmapsto \mathrm{Ad}\,u(a) = uau^*. \end{aligned}$$

A representation of the abelian Lie algebra $\log \mathbf{D}(1) \cong \mathbb{R}$ (time translations) in the complex algebra with a symmetric Bose element $H = H^*$ defines a *dynamics (time development) of the algebra* A:

$$\begin{array}{l} \log \mathbf{D}(1) \longrightarrow L(A),\ \ 1 \longmapsto iH,\ H = H^* \ (\textit{Hamiltonian}), \\ D(t) = e^{iHt} \in U(A),\ \ D(-t) = D(t)^* \ (\textit{time operator}, \text{ if defined}). \end{array}$$

A Hamiltonian is called *compact* for a definite unitary conjugation group, and *noncompact* for an indefinite unitary one. Examples for both cases are given below.

The inner derivations with the Hamiltonian are the equations of motion, whose solutions are the time orbits, given by the adjoint action (if defined)

$$\tfrac{d}{dt}a(t) = [iH, a(t)],\ \ D(t)aD(-t) = a(t).$$

A dynamics is solved by the decomposition of the algebra A into nondecomposable time-development-invariant subspaces with eigenvectors and eigenvalues

$$[H, a] = E(a)a.$$

The eigenvectors of the time translations $\operatorname{ad} H$ span the *time translation eigenalgebra*

$$\begin{array}{rl} A(H) & = \{\alpha_k a_k \in A \mid [H, a_k] = E(a_k)a_k,\ \ \alpha_k, E(a_k) \in \mathbb{C}\} = A(H)^*, \\ \operatorname{spec} \operatorname{ad} H & = \overline{\operatorname{spec} \operatorname{ad} H}, \end{array}$$

in the simplest case the full algebra $A(H) = A$. If there occur nondecomposable reducible time representations, the H-eigenalgebra is a proper subalgebra. With the decomposition of the Hamiltonian into semisimple and nilpotent parts, the H-eigenalgebra $A(H)$ is characterized by its invariance under the adjoint action of the nil-Hamiltonian

$$H = H_S + N \Rightarrow a \in A(H) \iff [N, a] = 0.$$

The invariance algebra of a Hamiltonian H contains the - under time translations - *conserved elements.* It is conjugation stable:

$$\mathrm{INV}_H A = \{Q \in A \mid [H, Q] = 0\} = (\mathrm{INV}_H A)^*.$$

If $Q \in A$ represents a symmetry operation, e.g., in a representation of a Lie algebra $iQ \in \log G$ with Lie group G and Lie parameters φ, a trivial commutator with the Hamiltonian H can be read in two ways (*Noether's theorem*): The invariance of a dynamics under Q operations is equivalent to the corresponding conservation law

$$\begin{array}{ccc} Q \in \mathrm{INV}_H A & \iff & H \in \mathrm{INV}_Q A, \\ \tfrac{d}{dt}Q = [iH, Q] & = 0 = & -[iQ, H] = -\tfrac{d}{d\varphi}H, \\ \tfrac{d}{dt} \cong \operatorname{ad} iH, & & \tfrac{d}{d\varphi} \cong \operatorname{ad} iQ. \end{array}$$

For example, a rotation-invariant dynamics, $G = \mathbf{SO}(3)$, has conserved angular momenta $iQ = i\vec{\mathcal{L}} = \vec{\mathbf{x}} \times i\vec{\mathbf{p}} \in \log \mathbf{SO}(3)$.

7.9.1 Compact Time Representations in Quantum Algebras

Time developments for a quantum algebra $\mathbf{Q}_\epsilon(\mathbb{C}^{2n}) \cong \mathbf{Q}_\epsilon(\mathbb{C}^2)^n$, invariant for all basic space endomorphisms $\mathbf{AL}(V)$, are implemented by Hamiltonians which are polynomials in the basic space identity

$$H_0 = H_0^\star \in \mathrm{INV}_{\mathbf{AL}(V)}\mathbf{Q}_\epsilon(\mathbb{C}^{2n}) \cong \mathbb{C}[I]_{\text{quant}}, \quad I = \tfrac{[\mathrm{u}^A, \mathrm{u}_A^\star]_{-\epsilon}}{2} = I^\star.$$

With a positive conjugation the time representations with Hermitian Hamiltonians are positive unitary (compact).

For the finite-dimensional Fermi quantum algebras $\mathbf{Q}_+(\mathbb{C}^{2n}) \cong \mathbb{C}^{4^n}$ with invariants $\mathbb{C}[I]_{\text{quant}} \cong \mathbb{C}^{1+n}$, the invariant compact Hamiltonians involve $1+n$ real numbers

$$H_0(\mu) = \mu_0 + \mu_1 I + \cdots + \mu_n I^n, \quad \mu_j \in \mathbb{R}.$$

For inner derivations and adjoint actions the constant contribution μ_0 is irrelevant.

The quantum dynamics of the Fermi quantum algebra are solved as follows: Using the I-projector decomposition, constructed with the minimal polynomial $p_I(X)$ of degree $1+n$,

$$p_I(X) = \prod_{k=-\frac{n}{2}}^{\frac{n}{2}} (X-k) \Rightarrow \mathcal{P}_l(I) = \prod_{k\neq l} \tfrac{I-k}{l-k}, \; l = -\tfrac{n}{2}, \ldots, \tfrac{n}{2},$$

$$\mathcal{P}_l(I) \otimes \mathcal{P}_k(I) = \delta_{lk}\mathcal{P}_k(I), \quad I^m = \sum_{l=-\frac{n}{2}}^{\frac{n}{2}} l^m \mathcal{P}_l(I), \quad m = 0, 1, \ldots, n,$$

the time development has the spectral projector decomposition:

$$H_0(\mu) = \sum_{l=-\frac{n}{2}}^{\frac{n}{2}} E_l \mathcal{P}_l(I), \quad e^{iH_0(\mu)t} = \sum_{l=-\frac{n}{2}}^{\frac{n}{2}} e^{iE_l t}\mathcal{P}_l(I), \quad E_l = \sum_{m=0}^{n} \mu_m l^m.$$

The quantum algebras are $H_0(\mu)$-semisimple. They have bases of time translation eigenvectors.

The time development is trivial for the $\binom{2n}{n}$-dimensional I-invariance algebra

$$[H_0(\mu), \mathrm{INV}_I \mathbf{Q}_+(\mathbf{V})] = \{0\}.$$

For the subspace $(VV^T)^{\neq}$ with the elements having unequal powers of basic space vectors and forms, the eigenvalues for the time development are given by the differences of the projector coefficients; the invariant subspaces can be obtained with the projectors

$$\mathbf{Q}_+(\mathbb{C}^{2n}) = \bigoplus_{k,l} \mathcal{P}_l(I)\mathbf{Q}_+(\mathbb{C}^{2n})\mathcal{P}_k(I) \cong \mathbb{C}^{4^n},$$
$$[H_0(\mu), a] = (E_l - E_k)a \text{ for } a \in \mathcal{P}_l(I)\mathbf{Q}_+(\mathbb{C}^{2n})\mathcal{P}_k(I).$$

Two examples for $n = 1, 2$: The compact Fermi oscillator dynamics in $\mathbf{Q}_+(\mathbb{C}^2) \cong \mathbb{C}^4$:

$$H_0(\mu) = \mu I = \mu \frac{[\mathrm{u}, \mathrm{u}^\star]}{2} \Rightarrow [H_0(\mu), \mathrm{u}] = \mu \mathrm{u}, \quad [H_0(\mu), \mathrm{u}^\star] = -\mu \mathrm{u}^\star$$

is solved as follows:

$$n = 1: \left\{ \begin{array}{ll} p_I(X) & = X^2 - \frac{1}{4} \Rightarrow \mathcal{P}_{\pm\frac{1}{2}}(I) = \frac{1}{2} \pm I, \\ H_0(\mu) & = \mu I = \frac{\mu}{2}\mathcal{P}_{\frac{1}{2}}(I) - \frac{\mu}{2}\mathcal{P}_{-\frac{1}{2}}(I), \\ e^{iH_0(\mu)t} & = e^{i\frac{\mu}{2}t}\mathcal{P}_{\frac{1}{2}}(I) + e^{-i\frac{\mu}{2}t}\mathcal{P}_{-\frac{1}{2}}(I), \\ & = \cos\frac{\mu}{2}t + 2iI\sin\frac{\mu}{2}t, \\ \text{eigenspaces and -values}: & \left\{ \begin{array}{ll} \mathcal{P}_{\frac{1}{2}}(I)\mathbf{Q}_+(\mathbb{C}^2)\mathcal{P}_{-\frac{1}{2}}(I), & \mu, \\ \mathcal{P}_{-\frac{1}{2}}(I)\mathbf{Q}_+(\mathbb{C}^2)\mathcal{P}_{\frac{1}{2}}(I), & -\mu. \end{array} \right. \end{array} \right.$$

A compact dynamics in the Fermi spin quantum algebra $\mathbf{Q}_+(\mathbb{C}^4) \cong \mathbb{C}^{16}$ can have a nonlinear interaction

$$H_0(\mu) = \mu_1 I + \mu_2 I \otimes I = \mu_1 \frac{[\mathrm{u}^A, \mathrm{u}^\star_A]}{2} + \mu_2 \frac{[\mathrm{u}^A, \mathrm{u}^\star_A] \otimes [\mathrm{u}^B, \mathrm{u}^\star_B]}{4}.$$

It is solved as follows:

$$n = 2: \left\{ \begin{array}{ll} p_I(X) & = X(X^2 - 1) \Rightarrow \left\{ \begin{array}{l} \mathcal{P}_{\pm 1}(I) = \frac{I \otimes (I \pm 1)}{2}, \\ \mathcal{P}_0(I) = 1 - I \otimes I, \end{array} \right. \\ H_0(\mu) & = \mu_1 I + \mu_2 I \otimes I = (\mu_2 + \mu_1)\mathcal{P}_1(I) + (\mu_2 - \mu_1)\mathcal{P}_{-1}(I), \\ e^{tiH_0(\mu)} & = \mathcal{P}_0(I) + e^{ti(\mu_2+\mu_1)}\mathcal{P}_1(I) + e^{ti(\mu_2-\mu_1)}\mathcal{P}_{-1}(I), \\ \text{eigenspaces and -values}: & \left\{ \begin{array}{ll} \mathcal{P}_{\pm 1}(I)\mathbf{Q}_+(\mathbb{C}^4)\mathcal{P}_0(I), & \mu_2 \pm \mu_1, \\ \mathcal{P}_0(I)\mathbf{Q}_+(\mathbb{C}^4)\mathcal{P}_{\pm 1}(I), & -\mu_2 \mp \mu_1, \\ \mathcal{P}_{\pm 1}(I)\mathbf{Q}_+(\mathbb{C}^4)\mathcal{P}_{\mp 1}(I), & \pm 2\mu_1. \end{array} \right. \end{array} \right.$$

7.9.2 Harmonic Fermi and Bose Oscillators

The irreducible time representation $\mathbf{D}(1) \ni e^t \longmapsto e^{i\mu t} \in \mathbf{U}(1)$ on a complex 1-dimensional space $V \cong \mathbb{C}$ with dual bases $\langle \check{\mathrm{u}}, \mathrm{u} \rangle = 1$ and $\mathbf{U}(1)$-conjugation $\check{\mathrm{u}} = \mathrm{u}^\star$ leads to the quantum Hamiltonian for the implementation of the time translations

$$H_0 = \mu I = \mu \frac{[\mathrm{u}, \mathrm{u}^\star]_{-\epsilon}}{2}, \quad \mu \in \mathbb{R}.$$

It acts in the oscillator quantum algebras $\mathbf{Q}_\epsilon(\mathbb{C}^2)$ via the basic space identity

$$d_t a = [iH_0, a]: \quad [I, \mathrm{u}^m \otimes (\mathrm{u}^\star)^k] = (m - k)\mathrm{u}^m \otimes (\mathrm{u}^\star)^k \Rightarrow E = (m - k)\mu.$$

The energies as time translation eigenvalues reflect the integer $\mathbf{U}(1)$-winding numbers $\frac{E}{\mu} \in$ **weights** $\mathbf{U}(1) = \mathbb{Z}$.

For general dimension $V \cong \mathbb{C}^n$, the identity $I = \frac{\mathrm{id}_V - \mathrm{id}_V^T}{2}$ as compact Hamiltonian for the time representation $\mathbf{D}(1) \longrightarrow \mathbf{U}(\mathbf{1}_n) \subseteq \mathbf{U}(n)$ characterizes an isotropic *n-dimensional Fermi or Bose oscillator*

$$H_0 = \mu I = \mu \frac{[\mathrm{u}^A, \mathrm{u}^\star_A]_{-\epsilon}}{2} = \mu I^\star = H_0^\star \in \mathbf{Q}_\epsilon(\mathbb{C}^{2n}).$$

In the Bose case, a position-momentum basis can be used:

$$\begin{array}{l} \mathbf{p}_a = i\frac{\mathrm{u}^a-\mathrm{u}^\star_a}{\sqrt{2}}, \quad \mathbf{x}_a = \frac{\mathrm{u}^a+\mathrm{u}^\star_a}{\sqrt{2}}, \quad [\mathrm{u}^\star_a, \mathrm{u}^b] = \delta^b_a = [i\mathbf{p}_a, \mathbf{x}_b], \\ H_0 = \mu\frac{\{\mathrm{u}^a,\mathrm{u}^\star_a\}}{2} = \mu\frac{\mathbf{p}_a^2+\mathbf{x}_a^2}{2}, \quad [iH_0, \mathbf{x}_a] = \mu\mathbf{p}_a, \quad [iH_0, \mathbf{p}_a] = -\mu\mathbf{x}_a. \end{array}$$

The I-defined quantum algebra grading gives, up to the unit μ, the energy eigenvalues

$$\mathbf{Q}_\epsilon(\mathbb{C}^{2n}) = \left\{ \begin{array}{c} \bigoplus_{z=-n}^{n} V^{(z)} \\ \bigoplus_{z\in\mathbb{Z}} V^{(z)} \end{array} \right\}, \quad \mathrm{spec\,ad}\, H_0 \cong \left\{ \begin{array}{ll} \mathbb{Z}_{1+2n}, & \epsilon = +1, \\ \mathbb{Z}, & \epsilon = -1, \end{array} \right.$$
$$\text{with } [H_0, v^z] = z\mu v^z \text{ for } v^z \in V^{(z)}.$$

With $\mathbf{U}(n) \subset \mathbf{GL}(\mathbb{C}^n)$ the maximal compact subgroup in the invariance group of the basic space identity (Hamiltonian), the time translation eigenvalues (energies) have the degeneracy of $\mathbf{SU}(n)$-multiplets. In the Fermi quantum algebras, the degeneracy is given on the fundamental $\binom{n}{k}$-dimensional $\mathbf{SU}(n)$-representations, $k = 1, \dots, n-1$,

$$\mathbf{Q}_+(\mathbb{C}^{2n}) : \left\{ \begin{array}{ll} \mathbf{irrep}\,\mathbf{SU}(n) \ni \underbrace{[0,\dots,0,1,0,\dots,0]}_{k\text{th place}} \text{ on} & \left\{ \begin{array}{ll} \bigwedge^k V, & \frac{E}{\mu} = k, \\ \bigwedge^{n-k} V^T, & \frac{E}{\mu} = k-n, \end{array} \right. \\ \mathbf{irrep}\,\mathbf{SU}(n) \ni [0,\dots,0] \text{ (trivial)} \quad \text{on} & \left\{ \begin{array}{ll} \bigwedge^n V, \bigwedge^n V^T, & \frac{E}{\mu} = \pm n, \\ \mathbb{C}, & \frac{E}{\mu} = 0. \end{array} \right. \end{array} \right.$$

For example, the 3-dimensional oscillators show the degeneracy of $\mathbf{SU}(3)$-representations $[n_1, n_2]_d$ (with the representation space dimension d), for the Fermi oscillator:

$$\mathbf{Q}_+(\mathbb{C}^6) : \left\{ \begin{array}{ll} \mathrm{spec\,ad}\, H_0 & = \{\pm 3\mu, \pm 2\mu, \pm\mu, 0\} \cong \mathbb{Z}_7, \\ \frac{E}{\mu} = k : & \left\{ \begin{array}{lll} [1,0]_3, & [0,1]_3, & [0,0]_1, \\ \mathrm{u}^A, & \mathrm{u}^A\mathrm{u}^B, & \mathrm{u}^1\mathrm{u}^2\mathrm{u}^3, \end{array} \right. \\ \frac{E}{\mu} = -k : & \left\{ \begin{array}{lll} [0,1]_3, & [1,0]_3, & [0,0]_1, \\ \mathrm{u}^\star_A, & \mathrm{u}^\star_A\mathrm{u}^\star_B, & \mathrm{u}^\star_1\mathrm{u}^\star_2\mathrm{u}^\star_3. \end{array} \right. \end{array} \right.$$

A basis is written under each representation.

In the Bose quantum algebras, the degeneracy is given on the totally symmetric $\binom{n+k-1}{k}$-dimensional powers of the defining $\mathbf{SU}(n)$-representation and its dual:

$$\mathbf{Q}_-(\mathbb{C}^{2n}) : \left\{ \begin{array}{lll} \mathbf{irrep}\,\mathbf{SU}(n) \ni [k,0,\dots.0] & \text{on } \bigvee^k V, & \frac{E}{\mu} = k = 1,2,\dots, \\ \mathbf{irrep}\,\mathbf{SU}(n) \ni [0,\dots,0,k] & \text{on } \bigvee^k V^T, & \frac{E}{\mu} = -k, \\ \mathbf{irrep}\,\mathbf{SU}(n) \ni [0,\dots,0] & \text{on } \mathbb{C}, & \frac{E}{\mu} = 0, \end{array} \right.$$

e.g., for the Bose oscillator in position space:

$$\mathbf{Q}_-(\mathbb{C}^6): \begin{cases} \text{spec ad } H_0 & \cong \mathbb{Z}, \\ \frac{E}{\mu} = k: & \begin{cases} [k,0]_{\binom{2+k}{k}} = & [1,0]_3, \quad [2,0]_6, \quad \dots, \\ & \mathrm{u}^A, \quad \mathrm{u}^A\mathrm{u}^B, \quad \dots, \end{cases} \\ \frac{E}{\mu} = -k: & \begin{cases} [0,k]_{\binom{2+k}{k}} = & [0,1]_3, \quad [0,2]_6, \quad \dots, \\ & \mathrm{u}^\star_A, \quad \mathrm{u}^\star_A\mathrm{u}^\star_B, \quad \dots . \end{cases} \end{cases}$$

7.9.3 Noncompact Time Representations in Quantum Algebras

Nondecomposable time representations with Hamiltonian matrix $\begin{pmatrix} \mu & \nu \\ 0 & \mu \end{pmatrix}$ act upon the quantum algebra $\mathbf{Q}_\epsilon(\mathbf{V}_{\text{doub}})$ of the antidoubling $V_{\text{doub}} = V \oplus \overline{V}^T \cong \mathbb{C}^2$ with $\mathbf{U}(1,1)$-anticonjugation $\times$ (chapter "Anticonjugation: The Real in the Complex") by the Hamiltonian:

$$\begin{aligned} &[\mathrm{g}^\times, \mathrm{b}]_\epsilon = [\mathrm{b}^\times, \mathrm{g}]_\epsilon = 1, \ [\mathrm{g}^\times, \mathrm{g}]_\epsilon = [\mathrm{b}^\times, \mathrm{b}]_\epsilon = 0, \\ &H_1 = \mu I + \nu N = \mu\tfrac{[\mathrm{g},\mathrm{b}^\times]_{-\epsilon}+[\mathrm{b},\mathrm{g}^\times]_{-\epsilon}}{2} + \nu \mathrm{g} \otimes \mathrm{g}^\times = H_1^\times, \end{aligned}$$

The Hamiltonion is the sum of a semisimple part, given by the basic space identity $I = I^\times = I^\star$ and a nil-Hamiltonian $N = N^\times$. The quantum algebra $\mathbf{Q}_\epsilon(\mathbb{C}^4)$ is not H_1-semisimple. There occur reducible, but nondecomposable, time-invariant subspaces. The quantum algebra has no basis of energy eigenvectors only, there exist nontrivial nilvectors, e.g., the nilvectors $\mathrm{b}, \mathrm{b}^\times$ in the basic space

$$\begin{aligned} &[H_1, \mathrm{b}] = \mu\mathrm{b} + \nu\mathrm{g}, \quad &&[H_1, \mathrm{b}^\times] = -\mu\mathrm{b}^\times - \nu\mathrm{g}^\times, \\ &[H_1, \mathrm{g}] = \mu\mathrm{g}, &&[H_1, \mathrm{g}^\times] = -\mu\mathrm{g}^\times. \end{aligned}$$

The anticonjugation (anti-)symmetric combinations of basic space vectors and forms

$$\left.\begin{aligned} \mathbf{g}_+ &= \tfrac{\mathrm{g}+\mathrm{g}^\times}{\sqrt{2}}, \quad &i\mathbf{g}_- &= \tfrac{\mathrm{g}-\mathrm{g}^\times}{\sqrt{2}} \\ \mathbf{b}_+ &= \tfrac{\mathrm{b}+\mathrm{b}^\times}{\sqrt{2}}, \quad &i\mathbf{b}_- &= \tfrac{\mathrm{b}-\mathrm{b}^\times}{\sqrt{2}} \end{aligned}\right\} \Rightarrow \begin{cases} \{\mathbf{b}_-, \mathbf{g}_-\} = \{\mathbf{b}_+, \mathbf{g}_+\} = 1 & \text{for Fermi,} \\ [\mathbf{b}_+, i\mathbf{g}_-] = [-i\mathbf{b}_-, \mathbf{g}_+] = 1 & \text{for Bose,} \end{cases}$$

give the Hamiltonian in the form

$$H_1 = \begin{cases} i\mu\frac{[\mathbf{g}_-,\mathbf{b}_+]+[\mathbf{b}_-,\mathbf{g}_+]}{2} + i\nu\mathbf{g}_- \otimes \mathbf{g}_+ & \text{for Fermi,} \\ \mu\frac{\{\mathbf{g}_-,\mathbf{b}_-\}+\{\mathbf{g}_+,\mathbf{b}_+\}}{2} + \nu\frac{\mathbf{g}_-\otimes\mathbf{g}_-+\mathbf{g}_+\otimes\mathbf{g}_+}{2} & \text{for Bose.} \end{cases}$$

Only for the *Bose case* $\mathbf{Q}_-(\mathbb{C}^4)$ is there a reduced framework for $\mu = 0$ (quantized free Newtonian mass point):

$$\text{Bose with } \mu = 0: \quad \begin{cases} \mathbf{p} = -\mathbf{g}_-, \quad \mathbf{x} = \mathbf{b}_+ \Rightarrow [i\mathbf{p}, \mathbf{x}] = 1, \\ H_1^{\text{red}} = \nu\frac{\mathbf{p}\otimes\mathbf{p}}{2}. \end{cases}$$

Only in the *Fermi quantum algebra* $\mathbf{Q}_+(\mathbb{C}^4)$ is the nil-Hamiltonian N nil-quadratic both for the composition and the tensor product:

$$\begin{array}{ll} \text{in } \mathbf{AL}(\mathbb{C}^2): & N = \begin{pmatrix} 0 & 1 \\ 0 & 0 \end{pmatrix} \quad \Rightarrow N \circ N = 0 \\ \text{in } \mathbf{Q}_+(\mathbb{C}^4): & N = \mathrm{g} \otimes \mathrm{g}^\times \quad \Rightarrow N \otimes N = 0 \end{array}$$

$\mathbf{Q}_+(\mathbb{C}^4)$ is decomposable into ten vector spaces $W \cong \mathbb{C}^{1+n}$ with nondecomposable time representations D_n^{im}:

$$\mathbf{Q}_+(\mathbb{C}^4) \cong \mathbb{C}^{16} \text{ acted on with } \left\{ \begin{array}{l} 3 \times D_0^0(t) \oplus D_0^{2i\mu}(t) \oplus D_0^{-2i\mu}(t) \\ \oplus\ 2 \times D_1^{i\mu}(t) \oplus 2 \times D_1^{-i\mu}(t) \\ \oplus\ D_2^0(t). \end{array} \right.$$

Five spaces with dimension 1 are irreducible:

$$[H_1, I^k] = 0, \quad k = 0, 1, 2, \quad \left\{ \begin{array}{ll} [H_1, \mathrm{g} \otimes \mathrm{b}] & = 2\mu \mathrm{g} \otimes \mathrm{b}, \\ [H_1, \mathrm{b}^\times \otimes \mathrm{g}^\times] & = -2\mu \mathrm{b}^\times \otimes \mathrm{g}^\times. \end{array} \right.$$

Four spaces with dimension 2 are reducible, but nondecomposable. They have one eigenvector in the given basis,

$$\begin{array}{l} \text{bases: } \{\mathrm{b}, \mathrm{g}\}, \ \{\mathrm{b}^\times, \mathrm{g}^\times\}, \ \{\mathrm{b} \otimes I, \mathrm{g} \otimes I\}, \ \{I \otimes \mathrm{b}^\times, I \otimes \mathrm{g}^\times\}, \\ \left\{ \begin{array}{ll} [H_1, \mathrm{g}] & = \mu \mathrm{g}, \\ [H_1, \mathrm{g}^\times] & = -\mu \mathrm{g}^\times, \end{array} \right. \quad \left\{ \begin{array}{ll} [H_1, \mathrm{g} \otimes I] & = \mu \mathrm{g} \otimes I, \\ [H_1, I \otimes \mathrm{g}^\times] & = -\mu I \otimes \mathrm{g}^\times, \end{array} \right. \end{array}$$

as well as one space with dimension 3:

$$\text{basis: } \{\mathrm{b} \otimes \mathrm{b}^\times, \mathrm{g} \otimes \mathrm{b}^\times - \mathrm{b} \otimes \mathrm{g}^\times, \mathrm{g} \otimes \mathrm{g}^\times\}, \ [H_1, \mathrm{g} \otimes \mathrm{g}^\times] = 0.$$

The nil-Hamiltonian N has nilcubic adjoint action in the Fermi quantum algebra, $(\operatorname{ad} N)^3 = 0$. The decomposition of the quantum algebra with respect to the nildimension $(0, 1, 2)$,

$$\mathbf{Q}_+(\mathbb{C}^4) \cong \mathbb{C}^{10} \oplus \mathbb{C}^5 \oplus \mathbb{C},$$

contains the 10-dimensional subalgebra with trivial nildimension, i.e., with all energy eigenvector combinations

$$\begin{array}{l} \mathrm{INV}_N \mathbf{Q}_+(\mathbb{C}^4) = \{a \in \mathbf{Q}_+(\mathbb{C}^4) \,\big|\, [N, a] = 0\} \cong \mathbb{C}^{10} \\ \text{with basis } \{1, \ I, \ I \otimes I, \ \mathrm{g} \otimes \mathrm{b}, \ \mathrm{b}^\times \otimes \mathrm{g}^\times, \mathrm{g}, \ \mathrm{g}^\times, \ \mathrm{g} \otimes I, \ I \otimes \mathrm{g}^\times, \ N\}. \end{array}$$

7.10 Classical Lagrangians

"Quantizing" classical Lagrangians was the historical way. One can go in the other direction, starting from the quantum structure: The operational structure of the time translations in quantum algebras, implemented by the adjoint Hamiltonian action with $d_t = \frac{d}{dt} = i\,\mathrm{ad}\,H$,

$$\begin{array}{rl}\mathbf{Q}_\epsilon(\mathbf{V}): & [\check{e}_A, e^B]_\epsilon = \delta_A^B,\ [\check{e}_A, \check{e}_B]_\epsilon = 0 = [e^A, e^B]_\epsilon,\\ & d_t a = [iH, a],\ d_t e^A = [iH, e^A],\ \ d_t \check{e}_A = [iH, \check{e}_A],\end{array}$$

can be formulated for the corresponding classical polynomial and Grassmann algebras with Lagrangians.

A classical Lagrangian L, associated to a quantum time development, is the difference of a *kinetic term* and the Hamiltonian, assumed without explicit time dependence:

$$L(e, \check{e}, d_t e) = i\check{e}_A d_t e^A - H(e, \check{e}).$$

The kinetic term serves the purpose to establish, in correspondence with the quantum interpretation, the concept of *dual basic vector pairs* via derivations also in the classical algebra, here the duality of $\{e^A, \check{e}_A\}$:

$$\text{dual pairs: } (e^A, \check{e}_A = \frac{\partial L}{\partial i d_t e^A}).$$

A Lagrangian is determined up to time derivatives. In the case of a dual normalization factor $[\check{e}^A, e_B]_\epsilon = \hbar\delta_A^B$ the Hamiltonian and the Lagrangian come with the corresponding normalization $\frac{1}{\hbar}H$, $\frac{1}{\hbar}L$.

Lagrangians for the time development of Grassmann algebras with anticommuting basic vectors have to be used with some care.

Starting from a Lagrangian (without explicit time dependence) the Hamiltonian as time translation generator is constructed as

$$H(e, \check{e}) = \frac{\partial L}{\partial d_t e^A} d_t e^A - L(e, \check{e}, d_t e).$$

The Lagrangian is considered as dependent on the dual basic vectors and their first order time derivatives for the time Lie algebra (translations) action. The quantum implementation of the time translations by the adjoint Hamiltonian action is represented classically by the Euler-Lagrange equations

$$\begin{array}{rlll} 0 = & d_t \frac{\partial L}{\partial d_t \check{e}_A} - \frac{\partial L}{\partial \check{e}_A} & = & -i d_t e^A + \frac{\partial H}{\partial \check{e}_A},\\ 0 = & d_t \frac{\partial L}{\partial d_t e^A} - \frac{\partial L}{\partial e^A} & = & i d_t \check{e}_A + \frac{\partial H}{\partial e^A}.\end{array}$$

To obtain the time development for a Bose quantum algebra $\mathbf{Q}_-(\mathbf{V})$ in a commutative algebra of functions, the commutator is formulated as Poisson bracket $[\check{e}_A, e^B]_{\text{Poisson}} = \delta_A^B$ with the duality-induced derivative representation $\check{e}_A \cong \frac{\partial}{\partial e^A}$ and $e^A \cong -\frac{\partial}{\partial \check{e}_A}$. This gives the equations of motion

$$\begin{array}{rll} d_t e^A = & -i\frac{\partial H}{\partial \check{e}_A} & = [iH, e^A]_{\text{Poisson}},\\ d_t \check{e}_A = & i\frac{\partial H}{\partial e^A} & = [iH, \check{e}_A]_{\text{Poisson}},\end{array}$$

e.g., for a position-momentum formulation $[i\mathbf{p}, \mathbf{x}] = 1 = [\mathbf{p}, \mathbf{x}]_{\text{Poisson}}$,

$$L = \mathbf{p} d_t \mathbf{x} - H(\mathbf{x}, \mathbf{p}) \Rightarrow \left\{ \begin{array}{lll} d_t \mathbf{x} & = \frac{\partial H}{\partial \mathbf{p}} & = [H, \mathbf{x}]_{\text{Poisson}},\\ d_t \mathbf{p} & = -\frac{\partial H}{\partial \mathbf{x}} & = [H, \mathbf{p}]_{\text{Poisson}},\end{array}\right.$$

$$H(\mathbf{x}, \mathbf{p}) = \frac{\partial L}{\partial d_t \mathbf{x}} d_t \mathbf{x} - L.$$

7.11 Summary

Each finite-dimensional vector space has its Fermi and Bose quantum algebra $\mathbf{Q}_\pm(\mathbb{K}^{2n})$ with finite and countably infinite dimension respectively. The quantum algebras realize all multilinear structures, e.g., the enveloping algebra of the basic vector space endomorphisms, that act by inner derivations and their invariants. Fermi quantum algebras are isomorphic to Clifford algebras of even-dimensional spaces, for the real case with neutral signature. The quantization expresses the duality of the embedded basic vector space with its linear forms. Quantum algebras inherit from their basic space a reflection, e.g., a conjugation $\mathbf{Q}_\pm(\mathbb{C}^{2n}) \cong \mathbb{C} \otimes \mathbf{Q}_\pm(\mathbb{R}^{2n})$. The smallest nontrivial quantum algebras with a complex 1-dimensional basic space are the Fermi and Bose oscillator algebras $\mathbf{Q}_\pm(\mathbb{C}^2)$, with a definite conjugation for the irreducible time representations $\mathbf{D}(1) \longrightarrow \mathbf{U}(1)$. They build all quantum algebras $\mathbf{Q}_\pm(\mathbb{C}^{2n}) \cong \bigotimes^n \mathbf{Q}_\pm(\mathbb{C}^2)$. A symplectic and an orthogonal dual isomorphism, e.g., connected with a self-dual representation of a Lie algebra, requires for its nontrivial implementation a Fermi and a Bose quantum algebra respectively. The quantum algebra $\mathbf{Q}_+(\mathbb{C}^4)$ for the fundamental spin $\mathbf{SU}(2)$ representation by Pauli spinors is of Fermi type and complex 16-dimensional. The $\mathbf{SL}(\mathbb{C}^2)$-adjoint quantum algebra $\mathbf{Q}_-(\mathbb{C}^6)$ of Bose type with an $\mathbf{SU}(2)$-defining conjugation is the arena of quantum mechanics for 3-dimensional position space with the action of the angular momenta $\log \mathbf{SO}(3)$.

MATHEMATICAL TOOLS

7.12 Graded Algebras

A unital associative algebra A is *graded with an additive monoid* M, e.g., with the positive integers $\mathbb{Z}_+$, the integers $\mathbb{Z}$, or with a cyclic group $\mathbb{Z}_n = \mathbb{Z}/n\mathbb{Z}$, if it is a direct sum, indexed with M where the algebra multiplication is compatible with the monoid addition. The numbers $\mathbb{K}1_A \subseteq A$ have to be indexed with the neutral monoid element $0 \in M$,

$$A = \bigoplus_{h \in M} V_h, \quad V_h V_l \subseteq V_{h+l}, \quad \mathbb{K}1_A \subseteq V_0.$$

The grading can be expressed by a surjective monoid morphism, defining the *grade (degree, dimension)*:

$$\deg : A \longrightarrow M, \quad \deg(ab) = \deg(a) + \deg(b), \quad \deg^{-1}[h] = V_h.$$

By equivalence classes with respect to a submonoid $N \subseteq M$ one obtains a *coarser* M/N *grading*, i.e., projecting once more $M \longrightarrow M/N$:

$$V_{[h]} = \bigoplus_{k \in N} V_{h+k}.$$

Tensor algebras, Grassmann, and polynomial algebras are $\mathbb{Z}_+$-graded unital algebras. In general, the enveloping algebra of a Lie algebra is not graded.

A familiar grading in physics is a *dimensional analysis* with respect to the powers of "basic" units, e.g., the powers with respect to the three human units meter (length), kilogram (mass), and second (time), or with three intrinsic units, e.g., the speed of light c, Planck's unit (constant) $\hbar$, and Newton's unit (constant) G. All physical quantities constitute an associative real algebra $\mathbf{P} \in \underline{\mathbf{aag}}_{\mathbb{R}}$ with three rational powers, the dimensions of the units, as rational $\mathbb{Q}^3$ grading (the actual additive grading group may be coarser, e.g., $(\frac{\mathbb{Z}}{2})^3$). $[P]$ denotes a unit for P:

$$\dim : \mathbf{P} \longrightarrow \mathbb{Q}^3, \quad \dim(P) = \log[P] \cong \begin{pmatrix} q_1 \\ q_2 \\ q_3 \end{pmatrix},$$
$$\mathbf{P} = \bigoplus_{d \in \mathbb{Q}^3} V_d, \quad V_0 \supseteq \mathbb{R}, \quad V_{d_1} V_{d_2} \subseteq V_{d_1+d_2}.$$

The three rational dimensions depend on the basic units chosen in $\mathbb{Q}^3$, e.g.,

$$\dim_{\mathrm{m,s,kg}}(m) = \begin{pmatrix} 1 \\ 0 \\ 0 \end{pmatrix}, \quad \dim_{\hbar,c,G}(m) = \begin{pmatrix} \frac{1}{2} \\ -\frac{3}{2} \\ 1 \end{pmatrix},$$
$$\dim_{\mathrm{m,s,kg}}(\hbar) = \begin{pmatrix} 2 \\ -1 \\ 1 \end{pmatrix}, \quad \dim_{\hbar,c,G}(\hbar) = \begin{pmatrix} 1 \\ 0 \\ 0 \end{pmatrix}.$$

Different bases - different equivalent unit systems - are related to each other by $\mathbb{Q}^3$-automorphisms, e.g.,

$$\dim_{\hbar,c,G}(P) = \mathcal{E}_{\hbar,c,G}^{\mathrm{m,s,kg}} \dim_{\mathrm{m,s,kg}}(P),$$

$$\mathcal{E}_{\hbar,c,G}^{\mathrm{m,s,kg}} = \begin{pmatrix} \frac{1}{2} & \frac{1}{2} & \frac{1}{2} \\ -\frac{3}{2} & -\frac{5}{2} & \frac{1}{2} \\ \frac{1}{2} & \frac{1}{2} & -\frac{1}{2} \end{pmatrix}, \quad (\mathcal{E}_{\hbar,c,G}^{\mathrm{m,s,kg}})^{-1} = \begin{pmatrix} 2 & 1 & 3 \\ -1 & -1 & -2 \\ 1 & 0 & -1 \end{pmatrix}.$$

A grading can be defined with any number of units, e.g., with four units (m, s, kg, A) including Ampere A.

7.12.1 Hybrid Algebras

An even-odd grading with $\mathbb{Z}_2 = \mathbb{Z}/2\mathbb{Z} = \{0, 1\}$ is called *hybrid*. A hybrid algebra A is decomposable into an *even unital subalgebra* V_0 and an *odd vector subspace* V_1, compatible with the algebra product

$$A = V_0 \oplus V_1, \quad V_k V_l \subseteq V_{(k+l) \bmod 2}.$$

Only the trivial vector 0 is both even and odd. Each unital algebra has a *trivial hybrid structure* with $V_1 = \{0\}$. Each $\mathbb{Z}$- or $\mathbb{Z}_+$-graded algebra is hybrid with the even and odd summands, respectively.

Two even (odd) elements of a hybrid algebra define an *opposition sign*

$$\epsilon(v_k, v_l) = (-1)^{kl}, \quad v_{k,l} \in V_{k,l}, \quad k, l \in \{0, 1\}.$$

Two hybrid algebras $A^{1,2}$ define the hybrid product algebra $A^1 \otimes A^2$ with the multiplication for even or odd a^2, b^1:

$$\begin{array}{l} A^1 \otimes A^2: \quad (a^1 \otimes a^2)(b^1 \otimes b^2) = \epsilon(a^2, b^1)\ (a^1b^1) \otimes (a^2b^2), \\ (A^1 \otimes A^2)_0 = V_0^1 \otimes V_0^2 \ \oplus\ V_1^1 \otimes V_1^2, \\ (A^1 \otimes A^2)_1 = V_0^1 \otimes V_1^2 \ \oplus\ V_0^1 \otimes V_1^2. \end{array}$$

For a *(hybrid) bracket algebra* $H = V_0 \oplus V_1$ ($\mathbb{Z}_2$-graded Lie algebra) the hybrid bracket is defined as inner bilinear composition compatible with the $\mathbb{Z}_2$-grading

$$[\![\ ,\ ]\!]: H \times H \longrightarrow H, \quad \left\{ \begin{array}{l} [\![V_0, V_0]\!] \subseteq V_0, \\ [\![V_0, V_1]\!] \subseteq V_1, \\ [\![V_1, V_1]\!] \subseteq V_0. \end{array} \right.$$

In addition, the bracket has to have a *hybrid symmetry* with the opposition sign and a *hybrid Jacobi-Leibniz property*:

$$\begin{array}{l} [\![v_l, v_m]\!] + \epsilon(v_l, v_m)[\![v_m, v_l]\!] = 0, \\ \Big[\!\!\Big[v_l, [\![v_m, v_n]\!]\Big]\!\!\Big] = \Big[\!\!\Big[[\![v_l, v_m]\!], v_n\Big]\!\!\Big] + \epsilon(v_l, v_m)\Big[\!\!\Big[v_m, [\![v_l, v_n]\!]\Big]\!\!\Big]. \end{array}$$

The even subspace V_0 is a Lie algebra. All elements are nilpotent with respect to the hybrid bracket, even elements are nilquadratic, odd elements are nilcubic

$$a_0 \in V_0 \Rightarrow [\![a_0, a_0]\!] = 0, \quad a_1 \in V_1 \Rightarrow \Big[\!\!\Big[a_1, [\![a_1, a_1]\!]\Big]\!\!\Big] = 0.$$

Nilquadratic odd elements have a nilquadratic adjoint action

$$\begin{array}{l} a_1 \in V_1 \Rightarrow \Big[\!\!\Big[a_1, [\![a_1, b]\!]\Big]\!\!\Big] = \frac{1}{2}\Big[\!\!\Big[[\![a_1, a_1]\!], b\Big]\!\!\Big] \text{ for all } b \in H, \\ a_1 \in V_1, \ \ [\![a_1, a_1]\!] = 0 \Rightarrow \Big[\!\!\Big[a_1, [\![a_1, b]\!]\Big]\!\!\Big] = 0 \text{ for all } b \in H. \end{array}$$

Each Lie algebra has a bracket algebra structure with a trivial odd space. A bracket algebra with nilquadratic odd space $[\![V_1, V_1]\!] = \{0\}$ carries a Lie algebra structure with $[l, m] = [\![l, m]\!]$. A nilquadratic ideal $[F, F] = \{0\}$ of a Lie algebra L allows the structure of a bracket algebra $L \cong L/F \oplus F$ with $[\![l, m]\!] = [l, m]$.

Each hybrid unital algebra $A = V_0 \oplus V_1$ carries a natural hybrid bracket by commutator and anticommutator *(graded inner derivation)*,

$$\begin{array}{l} \mathrm{ad}\, a: A \longrightarrow A, \\ \mathrm{ad}\, a(b) = [\![a, b]\!] = [a, b]_{-\epsilon(a,b)} = \left\{ \begin{array}{ll} [a, b] = ab - ba & \text{if } a \text{ or } b \text{ even}, \\ \{a, b\} = ab + ba & \text{if } a \text{ and } b \text{ odd}, \end{array} \right. \end{array}$$

with the derivation properties

$$[\![a, bc]\!] = [\![a, b]\!]c + \epsilon(a, b)b[\![a, c]\!].$$

A product of (anti)commuting even and odd elements b_j and f_j is nilquadratic:

$$\begin{array}{l} m = b_j f_j \Rightarrow 2m^2 = \{m, m\} = b_j b_k \{f_j, f_k\} + [b_j, b_k] f_j f_k - 2b_j [b_k, f_j] f_k, \\ [b_j, b_k] = 0, \quad [b_j, f_k] = 0, \quad \{f_j, f_k\} = 0 \Rightarrow m^2 = 0. \end{array}$$

A *nilpotent even element* has a nilpotent adjoint action if applied sufficiently often:

$$\begin{array}{lll} a_0 \in V_0 & \Rightarrow (\operatorname{ad} a_0)^n(b) & = \sum_{k=0}^{n} \binom{n}{k} a_0^{n-k} b a_0^k, \\ a_0^r = 0 & \Rightarrow (\operatorname{ad} a_0)^n(b) & = 0 \text{ for } n \geq 2r - 1. \end{array}$$

A *representation of a bracket algebra* $H = L \oplus F$ in a hybrid unital algebra $A = V_0 \oplus V_1$ uses the natural hybrid bracket

$$\mathcal{D} : H \longrightarrow A, \quad \left\{ \begin{array}{l} \mathcal{D}[L] \subseteq V_0, \quad \mathcal{D}[F] \subseteq V_1, \\ \mathcal{D}([\![l, m]\!]) = \left[\!\!\left[\mathcal{D}(l), \mathcal{D}(m) \right]\!\!\right]. \end{array} \right.$$

Bracket nilquadratic odd elements are represented by product nilquadratic ones:

$$[\![l_1, l_1]\!] = 0, \quad l_1 \in F \Rightarrow \mathcal{D}(l_1)\mathcal{D}(l_1) = 0.$$

It is obvious how to construct the enveloping algebra $\mathbf{E}(H)$ for a bracket algebra: identification of the hybrid bracket with the corresponding (anti-) commutators.

7.13 Algebras with Bilinear Forms

An (anti-)symmetric bilinear form of a vector space V,

$$\gamma : V \times V \longrightarrow \mathbb{K}, \quad \gamma(w, v) = \epsilon\gamma(v, w), \quad \epsilon = \pm 1,$$

defines a unique minimal ideal $\mathcal{I}(S_\gamma)$ in the tensor algebra $\bigotimes V$ that gives rise to equivalence classes where form values (power 0 tensors, scalars) are identified with ϵ-commutators (power 2 tensors)

$$\begin{array}{l} S_\gamma = \{w \otimes v + \epsilon v \otimes w - 2\gamma(w, v) \mid v, w \in V\} \subseteq \bigotimes V \\ \text{in } \mathrm{AAG}(V, \gamma) = \bigotimes V / \mathcal{I}(S_\gamma) \in \underline{\mathbf{aag}}_{\mathbb{K}}, \\ \text{one has } \bigotimes^2 V \ni w \otimes v + \epsilon v \otimes w = [w, v]_\epsilon = 2\gamma(w, v) \in \mathbb{K} \\ \text{for } V\text{-basis } \{e^j\}_{j \in J} : [e^j, e^k]_\epsilon = \gamma^{jk}. \end{array}$$

The factorized tensor algebra is called an *algebra for V with bilinear form γ*.

For a symmetric bilinear form one obtains a *Clifford algebra*; there the following set is enough to define the ideal

$$\epsilon = +1 : \quad \mathrm{AAG}(V, \gamma) = \mathrm{CLIFF}(V, \gamma) \text{ with } S_\gamma = \{v \otimes v - \gamma(v, v) \mid v \in V\}.$$

The antisymmetric Grassmann algebra $\bigwedge V$ and the symmetric polynomial algebra $\bigvee V$ are the algebras with respect to the trivial forms $\gamma = 0$ and $\epsilon = \pm 1$ respectively.

Via the tensor powers $\bigotimes^k V$ the algebra $\mathrm{AAG}(V,\gamma)$ is graded with $\mathbb{N}_0$ for $\epsilon = -1$ and with $\mathbb{Z}_{1+n}$ for $V \cong \mathbb{K}^n$ and $\epsilon = +1$. A coarser grading is the hybrid structure ($\mathbb{Z}_2$-grading) with the odd and even tensor powers[2] for the even subalgebra and the odd vector subspace

$$\mathrm{AAG}(V,\gamma) = \mathrm{AAG}(V,\gamma)_0 \oplus \mathrm{AAG}(V,\gamma)_1.$$

As a vector space, $\mathrm{AAG}(V,\gamma)$ is isomorphic to the Grassmann algebra for $\epsilon = +1$ and to the polynomial algebra for $\epsilon = -1$. The totally (anti-)symmetric tensors are the canonical representatives for $\epsilon = \pm 1$ respectively.

$\mathrm{AAG}(V,\gamma)$ is the enveloping algebra of the bracket algebra $\mathbb{K} \oplus V$ with

$$[\![\alpha,\beta]\!] = 0, \quad [\![\alpha,v]\!] = 0, \quad [\![w,v]\!] = \gamma(w,v), \quad \alpha,\beta \in \mathbb{K}, \quad v,w \in V.$$

It is nilpotent with the power 3, $[\![[\![a,b]\!],c]\!] = 0$ for $a,b,c \in \mathbb{K} \oplus V$. In the case $\epsilon = +1$ the odd subspace is V; for $\epsilon = -1$, $\mathbb{K} \oplus V$ is a Lie algebra.

The functor AAG goes from the vector spaces (V,γ_V,ϵ) with an ϵ-symmetric bilinear form γ_V and form compatible mappings in the unital algebras

$$\mathrm{AAG} : \underline{\mathbf{vec}}_{\mathbb{K}}^{(\epsilon,\gamma)} \longrightarrow \underline{\mathbf{aag}}_{\mathbb{K}},$$

$$\gamma_W(f(v_1),f(v_2)) = \gamma_V(v_1,v_2), \quad f \; \begin{array}{c} (V,\gamma_V,\epsilon) \\ \downarrow \\ (W,\gamma_W,\epsilon) \end{array} \longmapsto \begin{array}{c} \mathrm{AAG}(V,\gamma_V) \\ \downarrow \\ \mathrm{AAG}(W,\gamma_W) \end{array} \; \bigotimes f.$$

The functor is exponential:

$$\mathrm{AAG}(V \oplus W, \gamma_V \oplus \gamma_W) \cong \mathrm{AAG}(V,\gamma_V) \otimes \mathrm{AAG}(W,\gamma_W).$$

For vector spaces with bilinear forms of different symmetry, $\epsilon_V = +1$ and $\epsilon_W = -1$, the algebra of the direct sum $V \oplus W$ is graded by defining the algebra of the vector space W with antisymmetric form as even:

$$\begin{array}{c} (V,\gamma_V,+1) \\ (W,\gamma_W,-1) \end{array} : \begin{cases} \mathrm{AAG}(V \oplus W, \gamma_V \oplus \gamma_W) = \mathrm{AAG}(V,\gamma_V) \otimes \mathrm{AAG}(W,\gamma_W), \\ \mathrm{AAG}(V \oplus W, \gamma_V \oplus \gamma_W)_+ = \mathrm{AAG}(V,\gamma_V)_+ \oplus \mathrm{AAG}(W,\gamma_W), \\ \mathrm{AAG}(V \oplus W, \gamma_V \oplus \gamma_W)_- = \mathrm{AAG}(V,\gamma_V)_-. \end{cases}$$

In an algebra $\mathrm{AAG}(V,\gamma)$ the "opposite" symmetric commutators of the basic vectors define the associate Lie algebra, *orthogonal for symmetric* and

[2] One has for a vector subspace $U \subseteq V \oplus W$ the isomorphism $(V \oplus W)/U \cong (V/V \cap U) \oplus (W/W \cap U)$.

symplectic for antisymmetric bilinear forms:

$$\begin{aligned}
\mathrm{LAG}(V,\gamma) &= [V,V]_{-\epsilon} = \{\sum_{\text{finite}}[v_\alpha, w_\alpha]_{-\epsilon} \mid v_\alpha, w_\alpha \in V\} \in \underline{\mathbf{lag}}_{\mathbb{K}} \\
&= \begin{cases} [V,V] = V \wedge V \cong \mathbb{K}^{\binom{n}{2}}, & \epsilon = +1, \\ \{V,V\} = V \vee V \cong \mathbb{K}^{\binom{1+n}{2}}, & \epsilon = -1, \end{cases}
\end{aligned}$$

with $l^{vw} = \frac{[v,w]_{-\epsilon}}{4}$:

$$[l^{vw}, l^{ur}] = \gamma(w,u)l^{vr} - \gamma(u,v)l^{wr} + \gamma(w,r)l^{uv} - \gamma(r,v)l^{uw}$$

with basis $\{l^{jk} = \frac{[e^j,e^k]_{-\epsilon}}{4} \mid j,k \in J \times J\}$

$$[l^{ij}, l^{km}] = \gamma^{jk}l^{im} - \gamma^{ki}l^{jm} + \gamma^{jm}l^{ki} - \gamma^{mi}l^{kj}.$$

A *representation of an algebra* $\mathrm{AAG}(V,\gamma)$ in a unital algebra $A \in \underline{\mathbf{aag}}_{\mathbb{K}}$ as an algebra morphism has to be compatible with the bilinear form

$$D : \mathrm{AAG}(V,\gamma) \longrightarrow A, \quad \begin{cases} 1 \longmapsto 1_A, \quad v \longmapsto D(v), \text{ algebra morphism} \\ \Rightarrow [D(v), D(w)]_\epsilon = 2\gamma(v,w)1_A. \end{cases}$$

Thus one also obtains a representation of the associate Lie algebra.

The action of the associate Lie algebra on the algebra leaves the basic vector space V stable:

$$\begin{aligned}
&\mathrm{LAG}(V,\gamma) \times \mathrm{AAG}(V,\gamma) \longrightarrow \mathrm{AAG}(V,\gamma), \quad (l,a) \longmapsto [l,a], \\
&[l^{vw}, u] = \gamma(w,u)v - \gamma(u,v)w, \quad [l^{ij}, e^k] = \gamma^{jk}e^i - \gamma^{ki}e^j.
\end{aligned}$$

Hence one has a representation of the associate Lie algebra on the basic space endomorphisms that leaves invariant the bilinear form

$$\begin{aligned}
&\mathcal{D} : \mathrm{LAG}(V,\gamma) \longrightarrow \mathbf{AL}(V), \quad l \longmapsto \mathcal{D}(l), \\
&\mathcal{D}(l) : V \longrightarrow V, \quad \mathcal{D}(l)(u) = [l,u], \\
&\gamma(\mathcal{D}(l)(v), w) + \gamma(v, \mathcal{D}(l)(w)) = 0.
\end{aligned}$$

In the following, Clifford algebras as a subclass of algebras with a bilinear form are considered in more detail.

7.14 Clifford Algebras

Clifford algebras can be motivated as generalizations of the concept of real and imaginary numbers. The first extension of the reals are the complex numbers with one real unit $1^2 = 1$ and one imaginary unit $i^2 = -1$.

The Clifford algebra of a nontrivial vector space $W \cong \mathbb{K}^n$ with symmetric bilinear form γ, e.g., of a Lie algebra with Killing form, for $\mathbb{K} = \mathbb{R}$ with signature (p,q), $p+q=n$ (then denoted by $\mathrm{CLIFF}(p,q)$), is graded with $\mathbb{Z}_{1+n} = \{0,1,\ldots,n\}$ as direct sum of vector spaces with the action of the orthogonal γ-invariance group $\mathbf{O}(\mathbb{C}^n)$ and $\mathbf{O}(p,q)$ respectively:

$$\mathrm{CLIFF}(W,\gamma) = \bigoplus_{k=0}^{n} \bigwedge^{k} W \cong \mathbb{K}^{2^n} \in \underline{\mathbf{aag}}_{\mathbb{R}}.$$

Its coarser graduation with $\mathbb{I}(2) = \{\pm 1\} \cong \mathbb{Z}_2$ defines the even subalgebra and the odd vector subspace with equal dimension

$$\mathrm{CLIFF}(W,\gamma) = \mathrm{CLIFF}(W,\gamma)_0 \oplus \mathrm{CLIFF}(W,\gamma)_1 \cong \mathbb{K}^{2^{n-1}} \oplus \mathbb{K}^{2^{n-1}}.$$

The tensor-algebra-induced associative unital algebra product is written as $a \otimes b = ab$.

The grade-n subspace, isomorphic to the scalars, contains the *coscalars* (volume or top elements, axial scalars)

$$e \in \overset{n}{\bigwedge} W \cong \overset{0}{\bigwedge} W = \mathbb{K}.$$

The dual isomorphism $\gamma : W \longrightarrow W^T$ defines vector space isomorphisms, especially between basic vectors and covectors (axial vectors):

$$\overset{k}{\bigwedge} W \cong \overset{n-k}{\bigwedge} W, \text{ e.g., } W \cong \overset{n-1}{\bigwedge} W, \quad v \cong ve.$$

With an orthonormal basic space basis

$$W\text{-basis: } \{e^j \mid j = 1, \ldots, n\}, \quad \{e^j, e^k\} = 2\eta^{jk}, \quad \text{for } \mathbb{K} = \mathbb{C} : \ \eta^{jk} = \delta^{jk},$$

the coscalars can be spanned by the product e as volume basis (basic top element)

$$\overset{n}{\bigwedge} W = \mathbb{K}e \text{ with } e = e^1 \cdots e^n.$$

It leads to a covector basis

$$\overset{n-1}{\bigwedge} W\text{-basis: } \{e^j_e \mid j = 1, \ldots, n\}, \quad e^j_e = e^j e = \eta^{jj}(-1)^{j-1} e^1 \cdots e^{j-1} e^{j+1} \cdots e^n.$$

Both all basic space vectors and all covectors either commute $[a,b] = [a,b]_{-1}$ or anticommute $\{a,b\} = [a,b]_{+1}$ with the coscalars, depending on the vector space dimension

$$[W, \overset{n}{\bigwedge} W]_\epsilon = \{0\} = [\overset{n-1}{\bigwedge} W, \overset{n}{\bigwedge} W]_\epsilon \text{ with } \epsilon = (-1)^n,$$

leading to the coscalar-valued opposite commutation for vectors and covectors

$$[W, \overset{n-1}{\bigwedge} W]_{-\epsilon} \subseteq \overset{n}{\bigwedge} W.$$

With a given basis, vectors, covectors, and coscalars have the (anti) commutators

	$e^j \in W$ (vectors)	$e^j_e = e^j e \in \overset{n-1}{\bigwedge} W$ (covectors)	$e = e^1 \cdots e^n \in \overset{n}{\bigwedge} W$ (coscalars)
$e^k \in W$	$\{e^j, e^k\} = 2\eta^{jk} 1$	•	•
$e^k_e \in \overset{n-1}{\bigwedge} W$	$[e^j, e^k_e]_{-\epsilon} = 2\eta^{jk} e$	$\{e^j_e, e^k_e\} = -2\epsilon\eta^{jk}(e)^2$	•
$e \in \overset{n}{\bigwedge} W$	$[e^j, e]_\epsilon = 0$	$[e^j_e, e]_\epsilon = 0$	$(e)^2 \in \mathbb{K}$

$\epsilon = (-1)^n$

The *center of a Clifford algebra* consists of the scalars and, for odd dimension, of the top elements having odd degree. For even basic space dimension the scalars and top elements constitute the center of the even subalgebra

$$\begin{array}{llll} p+q=2,4,\ldots & \Rightarrow e \in \mathrm{CLIFF}(p,q)_0, & \mathrm{centr}\ \mathrm{CLIFF}(p,q) & = \mathbb{K}, \\ & & \mathrm{centr}\ \mathrm{CLIFF}(p,q)_0 & = \mathbb{K} \oplus \mathbb{K}e, \\ p+q=1,3,\ldots & \Rightarrow e \in \mathrm{CLIFF}(p,q)_1, & \mathrm{centr}\ \mathrm{CLIFF}(p,q) & = \mathbb{K} \oplus \mathbb{K}e, \\ & \multicolumn{3}{r}{\mathrm{CLIFF}(p,q)/\ \mathrm{centr}\ \mathrm{CLIFF}(p,q) \cong \mathrm{CLIFF}(p,q)_0.} \end{array}$$

With the scalar-valued square of a coscalar basis

$$(e)^2 = \begin{cases} (-1)^{\binom{n}{2}}, & \mathbb{K} = \mathbb{C}, \\ (-1)^{\binom{n}{2}+q}, & \mathbb{K} = \mathbb{R}, \end{cases}$$

the square of the coscalars is definite, i.e., one obtains *either imaginary or real coscalars*

$$\bigwedge^n W = \mathbb{R}e \cong \begin{cases} i\mathbb{R} & \text{with } (e)^2 = -1 \text{ for } \binom{n}{2} + q = 1,3,\ldots, \\ \mathbb{R} & \text{with } (e)^2 = +1 \text{ for } \binom{n}{2} + q = 0,2,4,\ldots. \end{cases}$$

Both Clifford algebras for the Lorentz symmetry

$$\begin{array}{ccccccccccc} \mathbb{R}^{16} \cong \left\{ \begin{array}{c} \mathrm{CLIFF}(1,3) \\ \mathrm{CLIFF}(3,1) \end{array} \right\} \cong & \mathbb{R} & \oplus & \mathbb{R}^4 & \oplus & \mathbb{R}^6 & \oplus & \mathbb{R}^4 & \oplus & \mathbb{R} \\ & 1 & & e^j & & [e^j, e^k] & & e_5^j = e^j e & & e = e_5 \end{array}$$

have imaginary coscalars

$$\begin{array}{l} e_5 = \frac{\epsilon_{jklm}}{4!} e^j e^k e^l e^m, \ (e_5)^2 = -1, \ \{e^j, e^k\} = \{e_5^j, e_5^k\} = \pm 2\begin{pmatrix} 1 & 0 \\ 0 & -\mathbf{1}_3 \end{pmatrix}, \\ \left\{[e^j, e^k], [e^l, e^m]\right\} = 4\begin{pmatrix} \mathbf{1}_3 & 0 \\ 0 & -\mathbf{1}_3 \end{pmatrix}, \\ \{e^j, e_5\} = \{e_5^j, e_5\} = 0, \ [e^j, e_5^k] = \pm 2\begin{pmatrix} 1 & 0 \\ 0 & -\mathbf{1}_3 \end{pmatrix} e_5. \end{array}$$

CLIFF(1,3) has 6 positive squared basic elements and 10 negative ones, whereas for CLIFF(3,1) these numbers are interchanged.

7.14.1 The Abelian Clifford Algebras

The trivial Clifford algebras are defined by the complex and real numbers

$$\mathrm{CLIFF}(0) = \mathbb{C}, \quad \mathrm{CLIFF}(0,0) = \mathbb{R}.$$

The complex 2-dimensional algebra with the *complex Study numbers*

$$\mathrm{CLIFF}(1) = {}^2\mathbb{C} = \mathbb{C} \oplus \mathbb{C} = \{\alpha + \beta e \mid \alpha, \beta \in \mathbb{C}\} \in \underline{\mathbf{aag}}_{\mathbb{C}}$$

is a quadratic commutative unital ring (not a field). It has the two projectors

$$e^2 = 1, \quad P_\pm = \tfrac{1 \pm e}{2}.$$

The *complex numbers as a real 2-dimensional algebra* are a Clifford algebra for an $\mathbf{O}(0,1)$-form,

$$\mathrm{CLIFF}(0,1) = \mathbb{C}_{\mathbb{R}} = \mathbb{R} \oplus i\mathbb{R} = \{\alpha + \beta e \bigm| \alpha, \beta \in \mathbb{R}\} \in \underline{\mathbf{aag}}_{\mathbb{R}},$$

with the imaginary unit $e^2 = -1$. This field is different from the ring with the *real Study numbers* with two real units $e^2 = 1$ arising from an $\mathbf{O}(1,0)$-form:

$$\mathrm{CLIFF}(1,0) = {}^2\mathbb{R} = \mathbb{R} \oplus \mathbb{R} = \{\alpha + \beta e \bigm| \alpha, \beta \in \mathbb{R}\} \in \underline{\mathbf{aag}}_{\mathbb{R}}.$$

7.14.2 Clifford Algebras as Endomorphisms

All complex and real Clifford algebras are isomorphic, as unital algebras, to endomorphism algebras of free *spinor modules* $M \in \underline{\mathbf{mod}}_R$ with a finite number of unital algebras R - two complex and five real ones.

From the exponential property of the Clifford functor from the $\mathbb{K}$-vector spaces with symmetric bilinear form γ into the unital $\mathbb{K}$-algebras

$$\mathrm{CLIFF}(W_1 \oplus W_2, \gamma_1 \oplus \gamma_2) \cong \mathrm{CLIFF}(W_1, \gamma_1) \otimes \mathrm{CLIFF}(W_2, \gamma_2),$$

one will expect a buildup structure with "prime Clifford algebras."

For $n = p + q = 0, 1$, as seen above, the complex and real Clifford algebras are the abelian fields and rings $\mathbb{C}, {}^2\mathbb{C}$ and $\mathbb{R}, \mathbb{C}_{\mathbb{R}}, {}^2\mathbb{R}$ respectively.

For $n = p + q = 2$ the real orthogonal invariance Lie algebras of the basic space bilinear forms are abelian, not, however, the related Clifford algebras

$$p + q = 2: \quad \mathrm{CLIFF}(p,q) = \mathbb{R} \oplus (\mathbb{R}e^1 \oplus \mathbb{R}e^2) \oplus \mathbb{R}e \cong \mathbb{R}^4$$

	1^2	$(e^1)^2$	$(e^2)^2$	$(e)^2$
CLIFF(2,0)	1	1	1	−1
CLIFF(1,1)	1	1	−1	1
CLIFF(0,2)	1	−1	−1	−1

$e = e^1 e^2$

The $\mathbb{R}^4$-Clifford algebra with *three negative square* basic elements is Hamilton's *quaternionic field*, a nonabelian real unital algebra

$$\mathbb{R}^4 \cong \mathrm{CLIFF}(0,2) = \mathbb{H} \in \underline{\mathbf{aag}}_{\mathbb{R}}$$

representable by the Pauli matrices

$$\begin{aligned} (1, -e^1, -e^2, -e) = (1, \vec{e}) &\longmapsto (\mathbf{1}_2, i\vec{\sigma}) \\ q = \alpha_0 + \vec{\alpha}\vec{e} &\longmapsto \mathbf{q} = \alpha_0 \mathbf{1}_2 + i\vec{\alpha}\vec{\sigma} = \begin{pmatrix} \alpha_0 + \alpha_3 & i\alpha_1 + \alpha_2 \\ i\alpha_1 - \alpha_2 & \alpha_0 + \alpha_3 \end{pmatrix}. \end{aligned}$$

They have product, conjugation, and norm,

$$\begin{aligned} q(\alpha)q(\beta) &= (\alpha_0\beta_0 - \vec{\alpha}\vec{\beta}) + (\alpha_0\vec{\beta} + \beta_0\vec{\alpha} - \vec{\alpha} \times \vec{\beta})\vec{e}, \\ q \leftrightarrow \overline{q} &= \alpha_0 - \vec{\alpha}\vec{e}, \quad \mathbf{q} \leftrightarrow \mathbf{q}^\star, \\ |q|^2 = q\overline{q} &= \alpha_0^2 + \vec{\alpha}^2 = \det \mathbf{q} = \operatorname{tr} \mathbf{q}\mathbf{q}^\star, \end{aligned}$$

with the quaternion group and special subgroup, Lie groups of real dimension 4 and 3:

$$\mathbf{GL}(\mathbb{H}) = \{q \in \mathbb{H} \,\big|\, q \neq 0\}, \quad \mathbf{SL}(\mathbb{H}) = \{q \in \mathbb{H} \,\big|\, |q| = 1\} \cong \mathbf{SU}(2).$$

The two $\mathbb{R}^4$-Clifford algebras with only one negative square basic element are isomorphic to each other:

$$\mathrm{CLIFF}(2,0) \ni (1, e^1, e^2, e) \cong (1, e^1, e, e^2) \in \mathrm{CLIFF}(1,1).$$

This is the simplest case of the Cartan *signature swap isomorphisms* for the unital algebra structure of $\mathrm{CLIFF}(p,q)$, written with orthonormal basic space bases

$$\begin{aligned} \mathrm{CLIFF}(p+1, q-1) &\cong \mathrm{CLIFF}(q,p) \text{ for } p+q \geq 2 \\ \text{with } e^1 &\cong e^1, \\ e^j &\cong e^1 e^j, \; j = 2, \ldots, p+q. \end{aligned}$$

The neutral signature real Clifford algebra as well as the corresponding complex Clifford algebra are isomorphic to the endomorphism algebra of (2×2) matrices[3]

$$\begin{array}{llll} \mathbb{R}^4 & \cong \mathrm{CLIFF}(1,1) & \cong \mathbb{R}(2\times 2) & = \begin{pmatrix} \mathbb{R} & \mathbb{R} \\ \mathbb{R} & \mathbb{R} \end{pmatrix}, \\ \mathbb{C}^4 & \cong \mathrm{CLIFF}(2) & \cong \mathbb{C}(2\times 2) & = \begin{pmatrix} \mathbb{C} & \mathbb{C} \\ \mathbb{C} & \mathbb{C} \end{pmatrix}, \end{array}$$

with the (2×2)-*buildup isomorphism* using the Pauli matrices

$$\begin{aligned} & e^1 \cong \begin{pmatrix} 0 & 1 \\ 1 & 0 \end{pmatrix} = \sigma^1, \quad e^2 \cong \begin{pmatrix} 0 & 1 \\ -1 & 0 \end{pmatrix} = i\sigma^2 \\ \Rightarrow \; & 1 \cong \mathbf{1}_2 = \begin{pmatrix} 1 & 0 \\ 0 & 1 \end{pmatrix}, \quad e^1 e^2 = e \cong \begin{pmatrix} -1 & 0 \\ 0 & 1 \end{pmatrix} = -\sigma^3. \end{aligned}$$

The isomorphism to the Fermi quantum algebras is given by

$$\begin{aligned} & \mathrm{CLIFF}(1,1) = \mathbf{Q}_+(\mathbb{R}^2) \cong \mathbb{R}(2\times 2), \quad \mathrm{CLIFF}(2) = \mathbf{Q}_+(\mathbb{C}^2) \cong \mathbb{C}(2\times 2), \\ & e^1 = \mathrm{u} + \check{\mathrm{u}} \cong \begin{pmatrix} 0 & 1 \\ 1 & 0 \end{pmatrix} = \sigma^1, \qquad e^2 = \mathrm{u} - \check{\mathrm{u}} \cong \begin{pmatrix} 0 & 1 \\ -1 & 0 \end{pmatrix} = i\sigma^2, \\ & \tfrac{e^1+e^2}{2} = \mathrm{u} \cong \begin{pmatrix} 0 & 1 \\ 0 & 0 \end{pmatrix} = \sigma^+, \qquad \tfrac{e^1-e^2}{2} = \check{\mathrm{u}} \cong \begin{pmatrix} 0 & 0 \\ 1 & 0 \end{pmatrix} = \sigma^-, \\ \Rightarrow \; & \tfrac{1-e}{2} = \mathrm{u}\check{\mathrm{u}} \cong \begin{pmatrix} 1 & 0 \\ 0 & 0 \end{pmatrix} = \tfrac{\mathbf{1}_2+\sigma^3}{2}, \quad \tfrac{1+e}{2} = \check{\mathrm{u}}\mathrm{u} \cong \begin{pmatrix} 0 & 0 \\ 0 & 1 \end{pmatrix} = \tfrac{\mathbf{1}_2-\sigma^3}{2}, \\ & e = [\check{\mathrm{u}}, \mathrm{u}] \cong \begin{pmatrix} -1 & 0 \\ 0 & 1 \end{pmatrix} = -\sigma^3, \quad \{\check{\mathrm{u}}, \mathrm{u}\} = 1. \end{aligned}$$

The general (2×2) *matrix buildup isomorphism* is for the real case

$$\begin{aligned} \mathrm{CLIFF}(p+1, q+1) &\cong \mathrm{CLIFF}(p,q) \otimes \mathrm{CLIFF}(1,1) \\ \mathrm{CLIFF}(1,1) &\cong \mathbb{R}(2\times 2) \cong \mathbb{R}^4, \end{aligned}$$

[3]The endomorphism algebra of a free module $M \cong R^n$ over a unital ring R (also nonabelian like the quaternions $\mathbb{H}$) is denoted by $R(n\times n)$, the $(n\times n)$ matrices with R-entries.

i.e., the bigger Clifford algebra can be written as a (2×2) matrix with entries in the smaller one,

$$\mathrm{CLIFF}(p+1, q+1) \cong \begin{pmatrix} \mathrm{CLIFF}(p,q) & \mathrm{CLIFF}(p,q) \\ \mathrm{CLIFF}(p,q) & \mathrm{CLIFF}(p,q) \end{pmatrix},$$

and equally for the complex one,

$$\begin{aligned} \mathrm{CLIFF}(n+2) &\cong \mathrm{CLIFF}(n) \otimes \mathrm{CLIFF}(2) \\ \mathrm{CLIFF}(2) &\cong \mathbb{C}(2 \times 2) \cong \mathbb{C}^4, \end{aligned}$$

as given by orthonormal basic space bases

$$\begin{aligned} \underline{e}^j &\cong \begin{pmatrix} e^j & 0 \\ 0 & -e^j \end{pmatrix} &= \sigma^3 \otimes e^j, \quad j = 1, \dots, p+q, \\ \underline{e}^{p+q+1} &\cong \begin{pmatrix} 0 & 1 \\ 1 & 0 \end{pmatrix} &= \sigma^1 \otimes 1, \quad \underline{e}^{p+q+2} \cong \begin{pmatrix} 0 & 1 \\ -1 & 0 \end{pmatrix} = i\sigma^2 \otimes 1. \end{aligned}$$

With the buildup isomorphisms the *complex Clifford algebras* are isomorphic to endomorphism algebras over the field $\mathbb{C}$ and the ring ${}^2\mathbb{C}$:

$$\begin{aligned} \mathrm{CLIFF}(2r) &\cong \mathbb{C}(2^r \times 2^r), & \mathrm{CLIFF}(2r+1) &\cong {}^2\mathbb{C}(2^r \times 2^r), \\ \log \mathbf{SO}(\mathbb{C}^{2r}) &\cong D_r, & \log \mathbf{SO}(\mathbb{C}^{2r+1}) &\cong B_r. \end{aligned}$$

Also, all real Clifford algebras are isomorphic to endomorphism algebras, starting from the simplest cases

$p\downarrow$ $q\rightarrow$	0	1	2
0	$\mathbb{R}$	$\mathbb{C}_\mathbb{R}$	$\mathbb{H}$
1	${}^2\mathbb{R}$	$\mathbb{R}(2 \times 2)$	$\mathbb{C}_\mathbb{R}(2 \times 2)$
2	$\mathbb{R}(2 \times 2)$	${}^2\mathbb{R}(2 \times 2)$	$\mathbb{R}(4 \times 4)$

$\mathrm{CLIFF}(p,q)$ as endomorphisms

with

$$\mathbb{R}^8 \cong \mathrm{CLIFF}(1,2) \cong \mathbb{C}_\mathbb{R} \otimes \mathbb{R}(2 \times 2) \cong \mathbb{C}_\mathbb{R}(2 \times 2) = \begin{pmatrix} \mathbb{C}_\mathbb{R} & \mathbb{C}_\mathbb{R} \\ \mathbb{C}_\mathbb{R} & \mathbb{C}_\mathbb{R} \end{pmatrix},$$

$$\mathbb{R}^8 \cong \mathrm{CLIFF}(2,1) \cong {}^2\mathbb{R} \otimes \mathbb{R}(2 \times 2) \cong {}^2\mathbb{R}(2 \times 2) = \begin{pmatrix} {}^2\mathbb{R} & {}^2\mathbb{R} \\ {}^2\mathbb{R} & {}^2\mathbb{R} \end{pmatrix},$$

and using the signature swap isomorphism

$$\begin{aligned} \mathbb{R}^{16} \cong \mathrm{CLIFF}(2,2) &\cong \mathrm{CLIFF}(3,1) \\ &\cong \mathbb{R}(2 \times 2) \otimes \mathbb{R}(2 \times 2) \cong \mathbb{R}(4 \times 4) = \begin{pmatrix} \mathbb{R}(2 \times 2) & \mathbb{R}(2 \times 2) \\ \mathbb{R}(2 \times 2) & \mathbb{R}(2 \times 2) \end{pmatrix}. \end{aligned}$$

To obtain the isomorphic endomorphism algebras for all real Clifford algebras using the buildup isomorphisms one has still to look for the negative definite structures $\mathrm{CLIFF}(0,q)$. One obtains by direct inspection the real 8-dimensional quadratic quaternion algebra, a unital ring:

$$\mathrm{CLIFF}(0,3) = {}^2\mathbb{H} = \mathbb{H} \oplus \mathbb{H} = \{(\alpha h^1, \beta h^2) \,\big|\, \alpha, \beta \in \mathbb{R},\ h^{1,2} \in \mathbb{H}\} \in \underline{\mathbf{aag}}_\mathbb{R}.$$

One finally needs the Cartan isomorphisms for the *swap of four negative basic vectors with four positive ones*, written with orthonormal basic space bases

$$\begin{aligned} \mathrm{CLIFF}(p+4,q) &\cong \mathrm{CLIFF}(p,q+4) \\ \text{with } e^j &\cong e^j\, e^1e^2e^3e^4, && j=1,2,3,4, \\ e^j &\cong e^j, && j=4,\dots,p+q+4, \end{aligned}$$

leading in the smallest case to

$$\begin{aligned} \mathrm{CLIFF}(0,4) &\cong \mathrm{CLIFF}(4,0) \cong \mathrm{CLIFF}(1,3) \\ &\cong \mathrm{CLIFF}(0,2) \otimes \mathrm{CLIFF}(1,1) \cong \mathbb{H} \otimes \mathbb{R}(2\times 2) \cong \mathbb{H}(2\times 2). \end{aligned}$$

Thus the unital algebra isomorphisms for the real Clifford algebras are completely determined with the five basic unital rings

$$R \in \{\mathbb{R}, \mathbb{C}_{\mathbb{R}}, \mathbb{H} \mid \text{fields}\} \cup \{{}^2\mathbb{R}, {}^2\mathbb{H} \mid \text{rings}\},$$

the two signature swap isomorphisms (where applicable), and the (2×2)-buildup isomorphisms

$$\begin{aligned} \mathrm{CLIFF}(p,q) &\cong \mathrm{CLIFF}(q+1,p-1) \cong \mathrm{CLIFF}(p-4,q+4) \\ &\cong \mathrm{CLIFF}(p-1,q-1) \otimes \mathrm{CLIFF}(1,1), \\ \mathrm{CLIFF}(1,1) &\cong \mathbb{R}(2\times 2). \end{aligned}$$

Any real Clifford algebra different from the basic unital rings is the $\mathbb{R}$-tensor product of a basic ring with a real $(2^m \times 2^m)$ matrix algebra for appropriate m:

$$\mathrm{CLIFF}(p,q) \cong R \otimes \bigotimes^m \mathbb{R}(2\times 2) \cong R \otimes \mathbb{R}(2^m\times 2^m) \cong R(2^m \times 2^m), \quad m \geq 1.$$

The dimensionality of (2×2) matrices $R(2\times 2)$ with a unital ring R is given by

$$\begin{aligned} R(2\times 2) &= \begin{pmatrix} R & R \\ R & R \end{pmatrix}, & R(2^m\times 2^m) &\cong \bigotimes^m R(2\times 2), \quad m \geq 1, \\ \dim_R R(2\times 2) &= 4, & \dim_R R(2^m \times 2^m) &= [\dim_R R(2\times 2)]^m = 4^m. \end{aligned}$$

If the ring R is a vector space over a field K, e.g., over $\mathbb{R}$ or $\mathbb{C}$, the K-dimensions are

$$\begin{aligned} \dim_K R = d \Rightarrow \quad \dim_K R(N\times N) &= \dim_K R \cdot \dim_R R(N\times N) &&= d\cdot N^2, \\ \dim_{\mathbb{R}} R^N &= \dim_{\mathbb{R}} R \cdot \dim_R R^N &&= d\cdot N. \end{aligned}$$

For the real case there arise the dimensions

R	$\mathbb{R}$	$\mathbb{C}_{\mathbb{R}}$	${}^2\mathbb{R}$	$\mathbb{H}$	${}^2\mathbb{H}$
$\dim_{\mathbb{R}} R$	1	2	2	4	8
$(q-p) \bmod 8$	0, 6	1, 5	7	2, 4	3
$\dim_{\mathbb{R}} R^{2^m}$	2^m	2^{m+1}	2^{m+1}	2^{m+2}	2^{m+3}
$\dim_{\mathbb{R}} R(2^m \times 2^m)$	2^{2m}	2^{2m+1}	2^{2m+1}	2^{2m+2}	2^{2m+3}

$\mathrm{CLIFF}(p,q) \cong R(2^m \times 2^m)$

The periodicity 8 in the difference $q - p$ for modules over the ring R

$(q-p) \bmod 8$	0	1	2	3	4	5	6	7
$(p-q+6) \bmod 8$	6	5	4	3	2	1	0	7
R	$\mathbb{R}$	$\mathbb{C}_\mathbb{R}$	$\mathbb{H}$	${}^2\mathbb{H}$	$\mathbb{H}$	$\mathbb{C}_\mathbb{R}$	$\mathbb{R}$	${}^2\mathbb{R}$

$$\mathbb{Z}_8 = \{0, 1, \dots, 7\}$$

is a consequence of the isomorphism

$$\mathrm{CLIFF}(p, q+8) \cong \mathrm{CLIFF}(p, q) \otimes \mathbb{R}(16 \times 16)$$

obviously with the isomorphisms above. The rings that arise twice reflect the isomorphism $\mathrm{CLIFF}(p+1, q) \cong \mathrm{CLIFF}(q+1, p)$ and the five basic rings related to the five different decompositions

$$\begin{array}{rcccccl} 6 \bmod 8 = & (0+6, & 1+5, & 2+4, & 3+3, & 7+7) & \bmod 8, \\ R = & \mathbb{R} & \mathbb{C}_\mathbb{R} & \mathbb{H} & {}^2\mathbb{H} & {}^2\mathbb{R}. & \end{array}$$

The five basic rings give rise to basic nonabelian endomorphisms

$$\begin{array}{l} R(2^m \times 2^m) \cong \bigotimes^m R(2 \times 2), \\ R(2 \times 2) \in \{\mathbb{R}(2 \times 2),\ \mathbb{C}_\mathbb{R}(2 \times 2),\ \mathbb{H}(2 \times 2),\ {}^2\mathbb{H}(2 \times 2),\ {}^2\mathbb{R}(2 \times 2)\}, \end{array}$$

used in the isomorphies for the *real Clifford algebras*. The first (8×8)-isomorphies with equal rings R on the diagonal and its parallels give a chessboard where the dimensions 2^m can be computed from (p, q) and the real dimension of the rings in the table, e.g., for $(p, q) = (3, 5)$ with $R = \mathbb{H}$, there follows $2^m = 8$

$p{\downarrow}q{\rightarrow}$	0	1	2	3	4	5	6	7
0	$\mathbb{R}$	$\mathbb{C}_\mathbb{R}$	$\mathbb{H}$	${}^2\mathbb{H}$	$\mathbb{H}$	$\mathbb{C}_\mathbb{R}$	$\mathbb{R}$	${}^2\mathbb{R}$
1	${}^2\mathbb{R}$	$\mathbb{R}$	$\mathbb{C}_\mathbb{R}$	$\mathbb{H}$	${}^2\mathbb{H}$	$\mathbb{H}$	$\mathbb{C}_\mathbb{R}$	$\mathbb{R}$
2	$\mathbb{R}$	${}^2\mathbb{R}$	$\mathbb{R}$	$\mathbb{C}_\mathbb{R}$	$\mathbb{H}$	${}^2\mathbb{H}$	$\mathbb{H}$	$\mathbb{C}_\mathbb{R}$
3	$\mathbb{C}_\mathbb{R}$	$\mathbb{R}$	${}^2\mathbb{R}$	$\mathbb{R}$	$\mathbb{C}_\mathbb{R}$	$\mathbb{H}$	${}^2\mathbb{H}$	$\mathbb{H}$
4	$\mathbb{H}$	$\mathbb{C}_\mathbb{R}$	$\mathbb{R}$	${}^2\mathbb{R}$	$\mathbb{R}$	$\mathbb{C}_\mathbb{R}$	$\mathbb{H}$	${}^2\mathbb{H}$
5	${}^2\mathbb{H}$	$\mathbb{H}$	$\mathbb{C}_\mathbb{R}$	$\mathbb{R}$	${}^2\mathbb{R}$	$\mathbb{R}$	$\mathbb{C}_\mathbb{R}$	$\mathbb{H}$
6	$\mathbb{H}$	${}^2\mathbb{H}$	$\mathbb{H}$	$\mathbb{C}_\mathbb{R}$	$\mathbb{R}$	${}^2\mathbb{R}$	$\mathbb{R}$	$\mathbb{C}_\mathbb{R}$
7	$\mathbb{C}_\mathbb{R}$	$\mathbb{H}$	${}^2\mathbb{H}$	$\mathbb{H}$	$\mathbb{C}_\mathbb{R}$	$\mathbb{R}$	${}^2\mathbb{R}$	$\mathbb{R}$

$$\mathrm{CLIFF}(p, q) = R(2^m \times 2^m) \text{ for } 0 \le p, q < 8$$
$$\text{with } 2^m = \sqrt{\frac{2^{p+q}}{\dim_\mathbb{R} R}}$$

spinorial chessboard with six rings

7.14.3 Spinor Representations

The exponent of the orthogonal Lie algebra in the Clifford algebra $\mathrm{CLIFF}(W, \gamma)$ is the *spin group* (covering group for the orthogonal group)

$$\exp \bigwedge^2 W \cong \exp \log \mathbf{SO}(\gamma) = \overline{\mathbf{SO}(\gamma)} = \begin{cases} \overline{\mathbf{SO}(\mathbb{C}^n)}, & W \cong \mathbb{C}^n, \\ \overline{\mathbf{SO}_0(p, q)}, & W \cong \mathbb{R}^{p+q}, \end{cases}$$

with the centers

$\overline{\mathbf{SO}(\mathbb{C}^{2r+1})}$	$\mathbf{SO}(\mathbb{C}^{2r+1})$	$\overline{\mathbf{SO}(\mathbb{C}^{2r})}$ r odd	$\mathbf{SO}(\mathbb{C}^{2r})$ r odd	$\overline{\mathbf{SO}(\mathbb{C}^{2r})}$ r even	$\mathbf{SO}(\mathbb{C}^{2r})$ r even
$\mathbb{I}(2)$	$\{1\}$	$\mathbb{I}(4)$	$\mathbb{I}(2)$	$\mathbb{I}(2)\times\mathbb{I}(2)$	$\mathbb{I}(2)$

The Clifford subspaces with a definite grade are representation spaces with respect to the adjoint action

$$\begin{aligned}
&l\in\log\mathbf{SO}(\gamma): && [l,\bigwedge^{k}W]\subseteq\bigwedge^{k}W,\ \ [l,v^k]=\mathcal{D}^k(l)(v^k),\ \ k=0,\dots,n=p+q,\\
&g\in\overline{\mathbf{SO}(\gamma)}: && g(\bigwedge^{k}W)g^{-1}=\bigwedge^{k}(gWg^{-1})\subseteq\bigwedge^{k}W,\ \ gv^kg^{-1}=D^k(g)(v^k),
\end{aligned}$$

where *fundamental representations* $\mathcal{D}^k$, D^k arise for the rank r Lie algebras and groups for the first and last grades as given by

$$\begin{array}{llll}
n=2r+1: & B_{r\geq 2}\cong\log\mathbf{SO}(\mathbb{C}^{2r+1}) & \text{for } k=(1,\dots,r-1) & \cong(2r,\dots,r+2),\\
 & & \text{not fundamental} \quad r & \cong r+1,\\
n=2r: & D_{r\geq 3}\cong\log\mathbf{SO}(\mathbb{C}^{2r}) & \text{for } k=(1,\dots,r-2) & \cong(2r-1,\dots,r+2),\\
 & \text{for } r\geq 3: & \text{not fundamental } r-1 & \cong r+1 \text{ and } r,
\end{array}$$

and analogously for the real case. They represent $\mathbf{SO}(n)$ and their classes with respect to center subgroups.

The missing fundamental representations are the *spinor representations*; they represent faithfully the cover group. $B_1=\log\mathbf{SO}(\mathbb{C}^3)$ has one and $D_2\cong B_1\oplus B_1$ two spinor representations. In general, there is one Pauli spinor representation for odd dimensions (B_r with $r\geq 1$) and two Weyl spinor representations for even dimensions (D_r with $r\geq 2$). They are given via the free spinor modules M (over fields or rings), where the Clifford algebra is realized by endomorphisms.

The adjoint action of the spin group on the Clifford algebra gives the left multiplication action on the modules

$$\begin{aligned}
&\mathrm{CLIFF}(p,q)\cong M\otimes M^T,\\
&\overline{\mathbf{SO}(\gamma)}\times\mathrm{CLIFF}(p,q)\longrightarrow\mathrm{CLIFF}(p,q),\ \ ga\longmapsto gag^{-1}\\
\Rightarrow\ &\overline{\mathbf{SO}(\gamma)}\times M\longrightarrow M,\ \ m\longrightarrow gm=D(g)(m).
\end{aligned}$$

For odd-dimensional basic spaces $W\cong\mathbb{R}^{2r+1}$ the signature $(r+1,r)$ as the simplest case comes with the ring ${}^2\mathbb{R}$:

$$\mathrm{CLIFF}(r+1,r)\cong{}^2\mathbb{R}(2^r\times 2^r).$$

In this case the even Clifford subalgebra is isomorphic to the endomorphism algebra of a spinor space S with the spinor representation of $\overline{\mathbf{SO}_0(r+1,r)}$

$$\begin{aligned}
\mathrm{CLIFF}(r+1,r)_0 &\cong \mathrm{CLIFF}(r+1,r)/\operatorname{centr}\mathrm{CLIFF}(r+1,r)\\
&\cong S\otimes S^T,\ \ \dim_{\mathbb{R}}S=2^r.
\end{aligned}$$

For even complex dimension $W\cong\mathbb{C}^{2r}$ one has a simple algebra, i.e., a full matrix algebra

$$\mathrm{CLIFF}(2r)\cong\mathbb{C}(2^r\times 2^r).$$

For $W \cong \mathbb{R}^{2r}$ with a neutral signature

$$\mathrm{CLIFF}(r,r) = \mathbf{Q}_+(\mathbb{R}^{2r}) \cong \mathbb{R}(2^r \times 2^r),$$

the basic space W allows Witt decompositions with the two isotropic spaces as the dual basic spaces of the isomorphic Fermi quantum algebra

$$\begin{array}{l} W \cong W_+ \oplus W_- = V \oplus V^T = \mathbf{V} \cong \mathbb{R}^{2r}, \\ \text{basis of } \left\{ \begin{array}{ll} W = \mathbf{V}: & \{e^j \mid j = 1, \dots, 2r\}, \\ W_+ = V,\ W_- = V^T: & \{e^j_\pm = \frac{e^j \pm e^{r+j}}{2} \mid j = 1, \dots, r\}, \\ & e^j_+ = \mathrm{u}^j,\ \ e^j_- = \check{\mathrm{u}}_j, \end{array} \right. \\ \left(\{e^j, e^k\}\right) = \begin{pmatrix} 2\delta^{jk} & 0 \\ 0 & -2\delta^{jk} \end{pmatrix} \cong \begin{pmatrix} 0 & \delta^j_k \\ \delta^j_k & 0 \end{pmatrix} = \begin{pmatrix} \{\mathrm{u}^j, \mathrm{u}^k\} & \{\check{\mathrm{u}}_j, \mathrm{u}^k\} \\ \{\mathrm{u}^j, \check{\mathrm{u}}_k\} & \{\check{\mathrm{u}}_j, \check{\mathrm{u}}_k\} \end{pmatrix}. \end{array}$$

The Clifford algebra is isomorphic to the endomorphism algebra of a vector space that is isomorphic to, but different from, the Grassmann algebra of the quantum algebra basic space

$$\mathrm{CLIFF}(r,r) = \mathbf{Q}_+(\mathbf{V}) \cong \mathbf{AL}(\bigwedge V), \quad V \cong \mathbb{R}^r.$$

This vector space (Fock space) is given (chapter "Quantum Probability") by the classes of an annihilator left ideal and the dual creator right ideal, generated by dual top elements, a top creator u, and a top annihilator ŭ:

$$\begin{array}{l} e = e^1 \cdots e^{2r}, \quad e_\pm = e^1_\pm \cdots e^r_\pm, \quad e_+ = \mathrm{u}, \ e_- = \check{\mathrm{u}}, \quad e = \mathrm{u}\check{\mathrm{u}}, \\ \mathrm{CLIFF}(r,r)/\,\mathrm{CLIFF}(r,r)\check{\mathrm{u}} \ \cong \bigwedge V, \\ \mathrm{CLIFF}(r,r)/\mathrm{u}\,\mathrm{CLIFF}(r,r) \ \cong \bigwedge V^T. \end{array}$$

The intersection of this vector space with the even subalgebra and the odd subspace defines the 2^{r-1}-dimensional vector spaces for the two $\overline{\mathbf{SO}_0(r,r)}$-spinor representations

$$\begin{array}{l} S_{0,1} = \mathrm{CLIFF}(r,r)_{0,1}/\,\mathrm{CLIFF}(r,r)\check{\mathrm{u}} \cap \mathrm{CLIFF}(r,r)_{0,1}, \\ S_0 \cong \bigoplus\limits_{k=0,2,\dots} \bigwedge^k V, \quad S_1 \cong \bigoplus\limits_{k=1,3,\dots} \bigwedge^k V, \quad S_0 \oplus S_1 \cong \bigwedge V, \\ \mathrm{CLIFF}(r,r)_0 \cong S_0 \otimes S_0^T \ \oplus \ S_1 \otimes S_1^T. \end{array}$$

The dimensions for the used spaces are

U	W	V	$\mathrm{CLIFF}(r,r)$	$\mathrm{CLIFF}(r,r)_{0,1}$	$\bigwedge V$	$S_{0,1}$
$\dim_{\mathbb{R}} U$	$2r$	r	2^{2r}	2^{2r-1}	2^r	2^{r-1}

Bibliography

[1] N. Bourbaki, *Algèbre, Chapitre 9* (Formes sesquilineaires et formes quadratiques) (1959), Hermann, Paris.

[2] I.R. Porteous, *Clifford Algebras and the Classical Groups* (1995), Cambridge University Press.

[3] P. Lounesto, *Clifford Algebras and Spinors* (1997), Cambridge University Press.

8

QUANTUM PROBABILITY

Experiments measure numeric-valued properties of physical operations, especially invariants and eigenvalues like spin and its third direction or mass and energy-momenta. In quantum theory, the operations are formalized by endomorphisms of complex vector spaces with conjugation. The association of numbers to operators requires the study of the linear forms of an algebra, i.e., of the operation algebra dual, e.g., for a quantum algebra $\omega : \mathbf{Q}_\epsilon(\mathbb{K}^{2n}) \longrightarrow \mathbb{K}$. The endomorphisms of the quantum algebra underlying vector space $V \cong \mathbb{K}^n$ distinguish trace forms, which induce inner products, i.e., symmetric bilinear forms for real and sesquilinear forms for complex quantum algebras with a conjugation. The latter ones are the origin of the quantum characteristic "probability amplitudes" used in Hilbert spaces for the interpretation of experiments.

A quantum algebra with a conjugation determines "its" Hilbert spaces it is acting on. The unitary invariance group of an inner product of a complex basic vector space $V \cong \mathbb{C}^n$, definite or indefinite, and hence the associated conjugation can be determined by a representation of the real time group acting on it, e.g., given by a Hamiltonian in a complex quantum algebra $\mathbf{Q}_\epsilon(\mathbb{C}^{2n})$ with the conjugation implementing the time reflection. Abelian and nonabelian endomorphism algebras have two characteristic linear forms, induced by the trace and the "double trace," the abelian and nonabelian form. A quantum subalgebra with a positive inner product (prescalar product), sometimes, but not necessarily the full quantum algebra, is a pre-Hilbert space that in the associated Hilbert space allows a probability valuation of physical operations. Therefore, a complex representation of the really parametrized time transformations can be interpreted with the quantum characteristic "probability amplitude" structure, as established by Born.

In general, a real Lie group determines "its" Hilbert spaces (chapter "Harmonic Analysis"). The properties of Hilbert space vectors are given by their behavior with respect to the acting operations: physical objects are formalized by eigenvectors with respect to translations.

In addition to time translations, objects are acted on also with position translations. They lead to position orbits which, in nonrelativistic quantum theory, are described by Schrödinger's wave functions. They allow an interpretation of the quantum operations with a classical position space ontology and give a position spread of the time translation eigenvectors and, for normalizable functions, position densities of probabilities.

First, general structures of algebra forms are considered - the transition from linear forms of an algebra to inner products, especially to scalar products for the related Hilbert spaces. Then concrete forms are looked for: Trace forms of endomorphism algebras have canonical extensions to quantum algebras. Finally, the familiar representations of the Heisenberg groups on the Hilbert spaces of square integrable position functions are given.

8.1 From Operator Algebra to Hilbert Spaces

Physical objects are formalized, in quantum theory, by Hilbert space vectors which are constructed as linear-form-induced equivalence classes of operations.

Some general properties and concepts for linear forms of vector spaces: A linear form $\omega : V \longrightarrow \mathbb{K}$ is injective on the classes $V/\operatorname{kern}\omega$ with the kernel. Forms with (conjugate) linear reflection can be conjugated too:

$$V \in *\underline{\mathbf{vec}}_{\mathbb{K}} \Rightarrow V^T \in *\underline{\mathbf{vec}}_{\mathbb{K}},\ \omega \overset{*}{\leftrightarrow} \omega^*;\ \text{with } \omega^*(v) = \overline{\omega(v^*)} \text{ for all } v \in V.$$

ω is *symmetric* (conjugation compatible) for $\omega = \omega^*$.

Two forms compose a *product form of the tensor product* inheriting properties of the factors:

$$\begin{aligned} &\omega_1 \otimes \omega_2 : V_1 \otimes V_2 \longrightarrow \mathbb{K},\quad (\omega_1 \otimes \omega_2)(v_1 \otimes v_2) = \omega_1(v_1)\omega_2(v_2),\\ &V_1^T \otimes V_2^T \subseteq (V_1 \otimes V_2)^T, \text{ for finite dimensions } V_1^T \otimes V_2^T \cong (V_1 \otimes V_2)^T. \end{aligned}$$

A linear form is nontrivially *factorizable* (decomposable) if the vector space is nontrivially factorizable, $V = V_1 \otimes V_2$ with $V_{1,2} \neq \mathbb{K}$, with a corresponding form factorization $\omega = \omega_1 \otimes \omega_2$. Otherwise it is called a *nonfactorizable (irreducible)* form.

An inner product $\zeta : V \times V \longrightarrow \mathbb{K}$ of a vector space can be extended to its tensor products and the corresponding totally symmetric or antisymmetric subspaces, e.g., for power-2 tensors

$$\zeta^2 : \bigotimes^2 V \times \bigotimes^2 V \longrightarrow \mathbb{K},\quad \zeta^2(v_1 \otimes v_2, w_1 \otimes w_2) = \zeta(v_1, w_1)\zeta(v_2, w_2) \text{ etc.}$$

With a group G or Lie algebra L acting on a vector space, the action is given also on the forms by the corresponding dual representation. This defines G and L-invariant forms. $*$-symmetry is interpretable as time reflection invariance for a time-reflection-implementing conjugation.

The linear forms $A^T \in \underline{\mathbf{vec}}_{\mathbb{K}}$ of an associative algebra $A \in \underline{\mathbf{aag}}_{\mathbb{K}}$ (operator algebra), do not have to be compatible with the product structure, i.e., they do not have to be algebra morphisms for A in the abelian algebra $\mathbb{K}$ with the numbers, especially not in the case of the quantum characteristic nonabelian algebras. With respect to the *multiplication compatibility*, an algebra form can have the following properties, in decreasing strength:

$$\begin{array}{rl} \omega : A \longrightarrow \mathbb{K} \text{ is } \textit{algebra representation:} & \omega(ab) = \omega(a)\omega(b) \\ & \text{for all } a, b \in A \\ \Rightarrow \omega \text{ with } \omega(1_A) \neq 0 \text{ is } \textit{eigenform for } b \in A\text{:} & \omega(ab) = \omega(ba) = \frac{\omega(a)\omega(b)}{\omega(1_A)} \\ & \text{for all } a \in A \\ \Rightarrow \omega \text{ is } \textit{adjoint invariant for } b \in A\text{:} & \omega([b, a]) = 0 \\ & \text{for all } a \in A. \end{array}$$

For a form, nontrivial for the unit $\omega(1_A) \neq 0$, each element has its shifted *ω-trivial element*

$$A \ni b \longmapsto b_\omega \in A, \quad b_\omega = b - \frac{\omega(b)}{\omega(1_A)} 1_A \Rightarrow \omega(b_\omega) = 0.$$

An algebra form with $\omega(1_A) = 1$ is called *unital.* It is an algebra morphism for the numbers $\mathbb{K}1_A$, i.e., $\omega(\alpha\beta) = \alpha\beta$.

8.1.1 Inner Algebra Products

In the following, the algebra A ("operator algebra") is unital and equipped with a (conjugate) linear reflection, $1_A \in A \in *\underline{\mathbf{aag}}_{\mathbb{K}}$, e.g., a quantum algebra $\mathbf{Q}_\epsilon(\mathbb{K}^{2n})$. If the algebra has topological properties, the topological dual A' with the continuous linear A-forms has to be considered (chapter "Harmonic Analysis").

A symmetric linear form defines an *inner product,* it uses the algebra product (bilinear) and the (conjugate) linear reflection

$$\begin{array}{rl} \langle\ |\ \rangle_\omega : A \times A \longrightarrow \mathbb{K}, & \langle a|b\rangle_\omega = \omega(a^* b), \\ \omega = \omega^* \Rightarrow & \langle a|b\rangle_\omega = \overline{\langle b|a\rangle_\omega}. \end{array}$$

The *ω-orthogonal* $A_\omega^\perp$ is a left ideal, in general not reflection stable:

$$\begin{array}{lll} \text{left ideal} & A_\omega^\perp & = \{n \in A \,\big|\, \omega(An) = \langle A|n\rangle_\omega = \{0\}\}, \\ \text{right ideal} & (A_\omega^\perp)^* & = \{n \in A \,\big|\, \omega(nA) = \langle n^*|A\rangle = \{0\}\}. \end{array}$$

If ω is an eigenform of $b \in A$, the ω-trivial element is an element of the left and right ideals $b_\omega \in A_\omega^\perp \cap (A_\omega^\perp)^*$.

The *quotient vector space* $|A\rangle_\omega$ is constituted by operator classes $|a\rangle_\omega$ up to the left trivial ones $A_\omega^\perp$. In general, a form-induced projection goes from the operator algebra to a state vector space:

$$*\underline{\mathbf{aag}}_{\mathbb{K}} \ni A \stackrel{\omega}{\longrightarrow} |A\rangle_\omega \in \underline{\mathbf{vec}}_{\mathbb{K}}.$$

The conjugation relates left and right ideal space (written as Dirac ket and bra)

$$\begin{array}{rll} |A\rangle_\omega & = A/A_\omega^\perp & = \{|a\rangle_\omega = a + A_\omega^\perp \,\big|\, a \in A\}, \\ {}_\omega\langle A| = |A\rangle_\omega^* & = A/(A_\omega^\perp)^* & = \{{}_\omega\langle a| = a^* + (A_\omega^\perp)^* \,\big|\, a \in A\}. \end{array}$$

The induced inner product for the classes is nondegenerate:

$$\begin{aligned}&\langle\ ||\ \rangle_\omega : |A\rangle_\omega \times |A\rangle_\omega \longrightarrow \mathbb{K},\ \ \langle a||b\rangle_\omega = \omega(a^*b),\\ &\langle A||n\rangle_\omega = \{0\} \iff n \in A^\perp_\omega \iff |n\rangle_\omega = |0\rangle_\omega.\end{aligned}$$

The quotient space of $A_1 \otimes A_2$ with a product form $\omega_1 \otimes \omega_2$ comes as a product:

$$|A_1 \otimes A_2\rangle_{\omega_1\otimes\omega_2} \cong |A_1\rangle_{\omega_1} \otimes |A_2\rangle_{\omega_2}.$$

In addition to the properties, connected with the vector space structure, there are algebra structure-related features. The classes A/L of an algebra with a left ideal, here $L = A^\perp_\omega$, constitute an A-representation space (A-module). The class of the algebra unit (identity operator) 1_A is a *cyclic vector*, here $|1_A\rangle_\omega$. It generates, by left multiplication, all vectors, here

$$A \times |A\rangle_\omega \longrightarrow |A\rangle_\omega,\ \ b|a\rangle_\omega = |ba\rangle_\omega,\ \ |A\rangle_\omega = A|1_A\rangle_\omega.$$

The representation of the algebra A in the A/L-endomorphisms is irreducible iff A/L is isomorphic to a minimal left ideal $L_{\min} \subseteq A$. The group operational structure behind is the left group action on right subgroup orbits $G \times G/H \longrightarrow G/H$.

The *matrix elements of an algebra operator* $b \in A$ are denoted as follows

$$a, c \in A:\ \langle c|b|a\rangle_\omega = \langle c||ba\rangle_\omega = \langle b^*c||a\rangle_\omega = \omega(c^*ba).$$

With a sesquilinear form, the matrix elements are sesquilinear too.

For a unital eigenform for $b \in A$, the operator b acts on the cyclic vector by scalar multiplication:

$$\omega \text{ eigenform for } b \Rightarrow \left\{\begin{array}{ll} \langle a|b\rangle_\omega & = \omega(b)\langle a|1_A\rangle_\omega = \omega(a^*)\omega(b) \text{ for all } a \in A,\\ b|1_A\rangle_\omega & = \omega(b)|1_A\rangle_\omega.\end{array}\right.$$

The left multiplicative action with an orthogonal element can be replaced by the action of the (anti-)commutator

$$h \in A^\perp_\omega \iff |h\rangle_\omega = 0 \iff h|a\rangle_\omega = ha|1_A\rangle_\omega = [h,a]_\epsilon|1_A\rangle_\omega \text{ for all } a \in A.$$

8.1.2 Hilbert Spaces, States, and State Vectors

An inner product induced by a symmetric algebra form $\omega = \omega^* \in A^T$ allows "metric" (order) concepts: An algebra element a is called

$$\begin{aligned}\omega(a^*a) &= \langle a|a\rangle_\omega \left\{\begin{array}{l} \geq 0,\ \ \omega\text{-}\textit{positive},\\ \leq 0,\ \ \omega\text{-}\textit{negative},\\ = 0,\ \ a \neq 0,\ \ \omega\text{-}\textit{singular},\end{array}\right.\\ \omega(a^*a) &= 0 = \omega(aa^*),\ a \neq 0,\ \ \omega\text{-}\textit{ghost}.\end{aligned}$$

Without strictly negative elements the form itself is *positive*:

$$\omega \succeq 0 \iff \omega(a^*a) \geq 0 \text{ for all } a \in A.$$

A symmetric algebra morphism is necessarily positive:

$$\omega = \omega^*, \ \omega(ab) = \omega(a)\omega(b) \Rightarrow \omega(a^*a) = \omega(a)\overline{\omega(a)} \geq 0.$$

A unital algebra form is always positive on the subalgebra with the scalars $\mathbb{K}1_A \subseteq A$.

A positive form gives the algebra the structure of a *pre-Hilbert space* with prescalar product, prenorm, and topology (not necessarily Hausdorff):

$$\| \ \|_\omega : A \longrightarrow \mathbb{R}_+, \quad \| a \|_\omega^2 = \langle a|a\rangle_\omega = \omega(a^*a) \geq 0.$$

With the Cauchy-Schwarz inequality a positive form has to be symmetric. It is trivial if and only if it is trivial for the unit

$$\omega \succeq 0 \Rightarrow \begin{cases} \omega^* = \omega, \\ \omega = 0 \iff \omega(1_A) = 0. \end{cases}$$

A positive unital form, $\omega \succeq 0$ and $\omega(1_A) = 1$, is called a *state of the algebra*. For a positive form, the ω-trivial elements constitute the orthogonal $A_\omega^\perp$, with Cauchy-Schwarz

$$\omega \succeq 0 \Rightarrow A_\omega^\perp = \{n \in A \,\big|\, \omega(n^*n) = 0\}.$$

The classes $|A\rangle_\omega = A/A_\omega^\perp$ give a scalar product space

$$\begin{aligned} \langle \ || \ \rangle_\omega : \ |A\rangle_\omega \times |A\rangle_\omega \longrightarrow \mathbb{K}, \ \langle a||b\rangle_\omega = \omega(a^*b), \ \langle a||a\rangle_\omega \geq 0, \\ \langle n||n\rangle_\omega = 0 \iff n \in A_\omega^\perp \iff |n\rangle_\omega = 0, \end{aligned}$$

with the orthogonal or unitary invariance group $\mathbf{O}(D)$ or $\mathbf{U}(D)$ for possibly infinite dimension $D = \dim_\mathbb{K} |A\rangle_\omega$. The equivalence classes $|a\rangle_\omega$ are called *state vectors*. ω-strictly positive operators $a \in A$ define state vectors

The Cauchy completion of the scalar product space is the ω-*associate Hilbert space* $A \stackrel{\omega}{\longrightarrow} \overline{|A\rangle}_\omega$. The algebra is represented on $|A\rangle_\omega$, dense in $\overline{|A\rangle}_\omega$. The norm-bounded left multiplications are extendable to the Hilbert space:

$$\begin{aligned} &\text{if } \ \| ba \|_\omega \leq k \| a \|_\omega, \ k \in \mathbb{R}, \text{ for all } a \in |A\rangle_\omega, \\ &b : \overline{|A\rangle}_\omega \longrightarrow \overline{|A\rangle}_\omega, \ b|a\rangle_\omega = |ba\rangle_\omega. \end{aligned}$$

The dual vector space A^T with the linear algebra forms carries, in addition to the reflection $\omega \leftrightarrow \omega^*$, an order $\omega \succeq 0$ that defines the positive cone A_+^T, in which the states, i.e., positive and unital, are a convex "polytope." An algebra state is called *pure* (nondecomposable, irreducible) if there do not exist states for its nontrivial combination $\omega = \alpha_1\omega_1 + \alpha_2\omega_2$ with positive scalars. The pure states are the "corners" of the convex "polytope." They are related to the irreducible algebra representations (chapter "Harmonic Analysis").

Given a (symmetric and positive) form $\omega \in A^T$, *each algebra element* gives rise to a (symmetric and positive) form ω_a:

$$A \longrightarrow A^T, \quad a \longmapsto \omega_a,$$

$$\omega_a : A \longrightarrow \mathbb{K}, \quad \omega_a(b) = \omega(a^* b a), \quad \begin{cases} \omega_{1_A} = \omega, \\ \omega = \omega^* \Rightarrow \omega_a = \omega_a^*, \\ \omega \succeq 0 \Rightarrow \omega_a \succeq 0, \end{cases}$$

normalizable for positive elements $\omega(a^* a) > 0$.

8.2 Probability Amplitudes

Probability as such is not the distinguishing characteristics of quantum structures. Probability is used also in classical thermostatistics where it arises from measures on phase space subsets, i.e., from Boolean lattices with a classical logic. Probability in quantum theories, introduced by Born, arises from the quantum characteristic concept of probability *amplitudes*. As structurally analyzed by Birkhoff and von Neumann, the complex Hilbert subspaces constitute nondistributive linear lattices which can be used to define a "quantum logic" (chapter "The Kepler Factor"): "God plays complex dice".

8.2.1 Projectors and Probability Structure

In this section a fixed state ω is chosen for a complex algebra A and, although all depends on the state ω, the subindex ω is omitted, $|A\rangle_\omega = |A\rangle$, etc. The conjugation as antilinear isomorphism relates to each other vectors in the Hilbert space and its topological dual, on which Dirac's bra-ket notation relies:

$$\overline{|A\rangle} \overset{*}{\leftrightarrow} \overline{|A\rangle}' = \overline{\langle A|}, \quad |a\rangle \overset{*}{\leftrightarrow} \langle a|.$$

If for a dynamics in A the conjugation implements the time reflection $\mathtt{T} = *$, the scalar product connects future and past in the possibly nonreal "amplitudes" $\langle a||b\rangle$ with time-reflection-invariant probabilities $\langle a||a\rangle \geq 0$.

For a Hilbert space basis B, one has Parseval's equation and the decomposition of the Hilbert space identity that represents the algebra unit 1_A:

$$\sum_{|e\rangle \in B} \langle a||e\rangle\langle e||a\rangle = \langle a||a\rangle, \quad \sum_{|e\rangle \in B} |e\rangle \ \langle e| \cong \mathbf{1}_{\overline{|A\rangle}}.$$

For complex Hilbert spaces, the decomposition of the bilinear identity is sesquilinear, therefore the isomorphism notation $\cong$. There is a unique correspondence between Hilbert subspaces K and projectors $\mathcal{P}_K$ which can be written as sum of primitive (nondecomposable) projectors for an orthonormal basis. The subspace dimension, if finite, is given by the trace of the projector (summation over κ)

$$|A\rangle \supseteq K \leftrightarrow \mathcal{P}_K = \mathcal{P}_K^* = \mathcal{P}_K^2, \quad \begin{cases} \mathcal{P}_K & \cong |e^\kappa\rangle\langle e^\kappa|, \\ \operatorname{tr} \mathcal{P}_K & = \dim_{\mathbb{K}} K = d(K). \end{cases}$$

Each nontrivial finite-dimensional subspace K with the coarsest unital ring gives rise to the *yes-no K-probability measure* normalized with the discriminant of the scalar product (chapter "The Kepler Factor")

$$\mu : \{\emptyset, K\} \longrightarrow \mathbb{R}_+,\ \mu(\emptyset) = 0,\quad \mu(K) = \det\langle e^\kappa | e^{\kappa'}\rangle = 1.$$

Each nontrivial state vector $|a\rangle \neq 0$ defines the primitive projector for the ray $\mathbb{K}|a\rangle$:

$$|e^a\rangle = \frac{|a\rangle}{\|a\|},\quad \mathcal{P}_a \cong |e^a\rangle\langle e^a|,\quad \langle e^a||e^a\rangle = 1,\quad \text{e.g., } \mathcal{P}_{1_A} \cong |1_A\rangle\langle 1_A|.$$

The *expectation value* of an algebra operator in a subspace K (nontrivial, finite-dimensional) is given by the trace

$$b \in A:\ \mathcal{E}_K(b) = \operatorname{tr} \mathcal{P}_K \circ b,\ \text{e.g.,} \left\{ \begin{array}{ll} \mathcal{E}_a(b) & = \operatorname{tr} \mathcal{P}_a \circ b = \langle e^a|b|e^a\rangle = \frac{\langle a|b|a\rangle}{\langle a|a\rangle}, \\ \mathcal{E}_{1_A}(b) & = \operatorname{tr} \mathcal{P}_{1_A} \circ b = \langle 1_A|b|1_A\rangle = \langle b\rangle. \end{array} \right.$$

Characteristic for quantum theory are the *transition probabilities* between two subspaces, computable with the corresponding projectors

$$\begin{array}{lllll} & & \mathcal{E}_K(\mathcal{P}_L) & = \operatorname{tr} \mathcal{P}_K \circ \mathcal{P}_L & = \langle e^\kappa|e^\lambda\rangle\langle e^\lambda|e^\kappa\rangle \in [0, d(K)d(L)], \\ p_{a\to b} & = p_{b\to a} = & \mathcal{E}_a(\mathcal{P}_b) & = \operatorname{tr} \mathcal{P}_a \circ \mathcal{P}_b & = \frac{\langle a|b\rangle\langle b|a\rangle}{\langle a|a\rangle\langle b|b\rangle} \in [0, 1]. \end{array}$$

The transition probabilities involve the *transition amplitude* between two state vectors as scalar product of the normalized vectors with values in the complex unit disk

$$\frac{\langle a|b\rangle}{\|a\|\ \|b\|} = \langle e^a||e^b\rangle \in \mathbf{U}(1) \times [0, 1].$$

It is the complex generalization of the cosinus of the angle between two Euclidean vectors.

8.2.2 Uncertainty Relations

The *Cauchy-Schwarz inequality* for a positive form ω applied to the algebra product is relevant for Heisenberg's uncertainty relation: It gives a lower bound for the prenorm product of two operators $a, b \in A$:

$$\begin{array}{ll} & \omega((a^* + e^{-i\beta}b^*)(a + e^{i\beta}b)) \geq 0 \\ \Rightarrow & \omega(a^*a)\omega(b^*b) = \| a \|_\omega^2 \| b \|_\omega^2 \geq [\omega(\frac{e^{i\beta}a^*b + e^{-i\beta}b^*a}{2})]^2 = [\operatorname{Re}\, \omega(e^{i\beta}a^*b)]^2 \\ & \text{for all } e^{i\beta} \in \mathbf{U}(1) \cap \mathbb{K} = \left\{ \begin{array}{ll} \{\pm 1\}, & \mathbb{K} = \mathbb{R}, \\ \mathbf{U}(1), & \mathbb{K} = \mathbb{C}. \end{array} \right. \end{array}$$

If the algebra elements are both symmetric or both antisymmetric, the product of their prenorms is bounded with the absolute ω-value of a combination of their anticommutator and anticommutator

$$(a, b) = \pm(a^*, b^*) \Rightarrow \| a \|_\omega \| b \|_\omega \geq |\omega(\frac{\{a,b\}}{2} \cos\beta + \frac{[ia,b]}{2} \sin\beta)|.$$

For special values of the phase, there arises only the commutator or only the anticommutator

$$\begin{array}{ll} \beta = 0 & \Rightarrow \parallel a \parallel_\omega \parallel b \parallel_\omega \geq |\omega(\frac{\{a,b\}}{2})|, \\ \mathbb{K} = \mathbb{C}: \ \beta = \frac{\pi}{2} & \Rightarrow \parallel a \parallel_\omega \parallel b \parallel_\omega \geq |\omega(\frac{[ia,b]}{2})|. \end{array}$$

Here, the commutator, e.g., $[i\mathbf{p}, \mathbf{x}] = \hbar$, occurs only for complex algebras.

The *standard deviation* Δa of a symmetric algebra element is the square root of the variance in the cyclic state vector $|1_A\rangle$:

$$\begin{array}{rl} a = a^*, \ a_\omega = a - \langle a \rangle 1_A & = a_\omega^* \Rightarrow \langle a_\omega \rangle = 0, \\ (\Delta a)^2 = \mathcal{V}_{1_A}(a) & = \langle a_\omega^2 \rangle = \langle a^2 \rangle - \langle a \rangle^2 \geq 0. \end{array}$$

It is trivial for an a-eigenform, $\langle a^2 \rangle = \langle a \rangle^2$. For any pair of (anti)symmetric elements in a complex algebra (with an eventual i-multiplication both can assumed to be symmetric) the *uncertainty relation* - arising from the Cauchy-Schwarz inequality above - bounds the product of their standard deviations from below by half the absolute value of their commutator:

$$(a, b) = (a^*, b^*) \Rightarrow \Delta a \ \Delta b = \parallel a_\omega \parallel \parallel b_\omega \parallel \geq |\langle \frac{[a_\omega, b_\omega]}{2} \rangle| = |\langle \frac{[a,b]}{2} \rangle|.$$

8.3 Time Translation Eigenalgebras with Probability Interpretation

With a positive form $\omega \succeq 0$ a symmetric element $h = h^* \in A$, e.g., a Hamiltonian, acts *ω-monotonically* if $\mathrm{ad}\, h$ is *ω-positive*:

$$\mathrm{ad}\, h \succeq 0 \iff \omega(a^*[h, a]) \geq 0 \text{ for all } a \in A.$$

A *minimal state* ω for a symmetric operator h, especially a *ground state* for a Hamiltonian, is defined by an h-eigenform with ω-monotonic h-action

$$\omega \succ 0, \ h = h^*, \ \mathrm{ad}\, h \succeq 0: \ \left\{ \begin{array}{rl} \omega(ah) & = \omega(a)\omega(h) \text{ for all } a \in A, \\ \omega(a^*ha) & \geq \omega(h)\omega(a^*a), \\ \omega(a^*h_\omega a) & \geq 0, \ h_\omega = h - \omega(h). \end{array} \right.$$

If ω is a minimal (ground) state for h, then each strictly ω-positive h-conserving operator l keeping h invariant gives another minimal (ground) state $\frac{\omega_l}{\omega(l^*l)}$ for h:

$$[l, h] = 0: \left\{ \begin{array}{rl} \omega(ah) = \omega(a)\omega(h) & \Rightarrow \omega_l(ah) = \omega_l(a)\frac{\omega_l(h)}{\omega(l^*l)}, \\ \omega(a^*h_\omega a) \geq 0 & \Rightarrow \omega_l(a^*h_\omega a) \geq 0. \end{array} \right.$$

A minimal (ground) state for h is *degenerate* if there exists an l with $\frac{\omega_l}{\omega(l^*l)} \neq \omega$.

A representation of the time translations $\mathbb{R}$ in a complex unital algebra A via a Bose *Hamiltonian* $H = H^*$ (time reflection $\mathtt{T} = *$) defines a dynamics via its adjoint action. The eigenvectors of the time translations $\operatorname{ad} H$ span the *H-eigenalgebra*

$$\begin{aligned} A(H) &= \{\alpha_k a_k \in A \bigm| [H, a_k] = E(a_k)a_k, \quad \alpha_k, E(a_k) \in \mathbb{C}\} \\ &= A(H)^* = \mathrm{INV}_N A, \quad \operatorname{spec} H = \overline{\operatorname{spec} H}. \end{aligned}$$

It is the full algebra for a semisimple time translation action $H = H_s + N$ with $N = 0$. Time orbits are given by $t \longmapsto a(t) = D(t)aD(-t)$, if the time representation $D(t) = e^{iHt}$ exists.

The algebra is assumed to be equipped with a symmetric unital H-eigenform (therefore H-invariant), not necessarily positive:

$$\omega : A \longrightarrow \mathbb{C}, \quad \begin{cases} \omega(a^*) &= \overline{\omega(a)}, \ \ \omega(1_A) = 1, \\ \omega(Ha) &= \omega(H)\omega(a) = \omega(aH) \\ \Rightarrow \omega([H,a]) &= 0, \ \ \omega(a(t)) = \omega(a) \text{ for all } a \in A. \end{cases}$$

On the quotient space $|A(H)\rangle_\omega$ with inner product, the adjoint action of the ω-shifted Hamiltonian $H_\omega = H - \omega(H)$ is identifiable with the left action

$$a \in A(H), \ \ [H_\omega, a] = E(a)a \Rightarrow \begin{cases} E(a)|a\rangle_\omega = [H_\omega, a]|1\rangle_\omega = H_\omega|a\rangle_\omega, \\ H|a\rangle_\omega = (E(a) + \omega(H))|a\rangle_\omega. \end{cases}$$

This is the algebraic formulation of the Schrödinger equation for the energy eigenvalues.

A conjugation stable subalgebra $A_\omega(H)$ of the H-eigenalgebra on which ω is an H-ground state allows a probability interpretation; it is a pre-Hilbert space:

$$\begin{aligned} &A \text{ (full)} \supseteq A(H) \text{ (eigenalgebra)} \supseteq A_\omega(H) = A_\omega(H)^* \ (\omega\text{-positive}), \\ &a \in A_\omega(H) \Rightarrow \omega(a^*a) = \| a \|_\omega^2 \geq 0, \quad \omega(a^*Ha) \geq \omega(H)\omega(a^*a). \end{aligned}$$

The time operators $D(t)$ define *isometries*

$$\operatorname{Int} D(t) : A_\omega(H) \longrightarrow A_\omega(H), \quad \| D(t)aD(-t) \|_\omega = \| a(t) \|_\omega = \| a \|_\omega .$$

An ω-strictly positive time translation eigenvector $\mathrm{u} \in A_\omega(H)$ is called a

$$\textit{creation operator } [H, \mathrm{u}] = E(\mathrm{u})\mathrm{u}, \quad \omega(\mathrm{u}^*\mathrm{u}) = \| \mathrm{u} \|_\omega^2 > 0$$

with the time translation eigenstate $|\mathrm{u}\rangle_\omega \in |A\rangle_\omega$. There may be additional eigenvector properties with respect to other operations (position translations, spin, isospin, etc.). A creation operator u has positive energy:

$$\begin{aligned} [H, \mathrm{u}] = E(\mathrm{u})\mathrm{u} \ \ &\Rightarrow \begin{cases} [H, \mathrm{u}^*] = E(\mathrm{u}^*)\mathrm{u}^*, \quad E(\mathrm{u}^*) = -\overline{E(\mathrm{u})}, \\ E(\mathrm{u})\omega(\mathrm{u}) = 0 = E(\mathrm{u}^*)\omega(\mathrm{u}^*), \\ E(\mathrm{u})\omega(\mathrm{u}^*\mathrm{u}) \geq 0, \quad \overline{E(\mathrm{u})}\omega(\mathrm{u}\mathrm{u}^*) \leq 0, \end{cases} \\ \omega(\mathrm{u}^*\mathrm{u}) > 0 \ \ &\Rightarrow E(\mathrm{u}) \geq 0, \quad E(\mathrm{u}^*) \leq 0. \end{aligned}$$

In the case of even strictly positive energy, the time-reflected partner $\mathrm{u}^* \in A$ is an *annihilation operator* (ω-trivial element),

$$E(\mathrm{u}) > 0 \Rightarrow \omega(\mathrm{u}\mathrm{u}^*) = \| \mathrm{u}^* \|_\omega^2 = 0, \quad \omega(\mathrm{u}) = 0 = \omega(\mathrm{u}^*).$$

8.4 Tensor Algebra Forms

After the general considerations for algebra forms above, the canonical forms of quantum algebras $\mathbf{Q}_\pm(\mathbb{K}^{2n})$ will be constructed explicitly together with the associated canonical Hilbert spaces. In general, continuous linear forms for algebras with Lie group functions are used (chapter "Harmonic Analysis").

Quotient algebras of tensor algebras $\bigotimes \mathbf{AL}(V)$ with basic vector space endomorphisms $\mathbf{AL}(V) = V \otimes V^T \cong \mathbb{K}^{n^2}$ can be equipped with forms of $\mathbf{AL}(V) = \log \mathbf{GL}(\mathbb{K}^n)$. The endomorphism Lie algebra is a direct sum of an abelian and, for $n \geq 2$, a simple Lie subalgebra $\mathbb{K} \oplus \log \mathbf{SL}(\mathbb{K}^n)$. It has two types of linear forms: abelian and nonabelian.

8.4.1 Abelian and Nonabelian Forms

A form of a unital algebra $\omega \in M^T$, e.g., $M = \mathbf{AL}(V)$, can be extended to its tensor algebra $\bigotimes M$ (with product $f_1 \otimes f_2$),

$$\begin{array}{ccc} M & \xrightarrow{\iota} & \bigotimes M \\ {\scriptstyle\omega}\downarrow & & \downarrow{\scriptstyle\langle\ \rangle_\omega}, \\ \mathbb{K} & \xrightarrow[\mathrm{id}_\mathbb{K}]{} & \mathbb{K} \end{array} \qquad \begin{array}{l} \langle 1\rangle_\omega = 1, \\ \langle f\rangle_\omega = \omega(f),\ f \in M, \end{array}$$

in two ways: If it is compatible with the product of the scalars $\mathbb{K}$ it will be called an *abelian or a Fock form.* If it is compatible with the algebra product $f_1 \circ f_2$, it will be called a *nonabelian form* :

$$f_j \in M,\ \ \langle f_1 \otimes \cdots \otimes f_k\rangle_\omega = \left\{ \begin{array}{ll} \omega(f_1)\cdots\omega(f_k), & \text{abelian extension} \\ & \text{for } M \text{ abelian or nonabelian}, \\ \omega(f_1 \circ \cdots \circ f_k), & \text{nonabelian extension} \\ & \text{only for } M \text{ nonabelian}. \end{array} \right.$$

Symmetry of the M-form ω entails symmetry for the tensor algebra form

$$\omega = \omega^* \Rightarrow \langle a^*\rangle_\omega = \overline{\langle a\rangle_\omega},\quad a \in \bigotimes M.$$

A Fock form is necessarily positive:

$$\text{Fock form} \Rightarrow \| a \|_\omega^2 = \langle a^* \otimes a\rangle_\omega = \langle a^*\rangle_\omega \langle a\rangle_\omega = |\langle a\rangle_\omega|^2 \geq 0,\quad a \in \bigotimes M.$$

A nonabelian form is positive if the basic M-form is positive:

$$\begin{array}{l}\text{nonabelian form} \\ \text{with } \omega \succeq 0\end{array} \Rightarrow \left\{ \begin{array}{l} \langle (f_1 \otimes \cdots \otimes f_k)^* \otimes (f_1 \otimes \cdots \otimes f_k)\rangle_\omega = \omega(g^* \circ g) \geq 0, \\ f_j \in M,\ \ g = f_1 \circ \cdots \circ f_k \in M. \end{array} \right.$$

A sum form of the sum of two (anti)commuting unital algebras $[\![M_1, M_2]\!] = \{0\}$ induces the product form

$$\langle\ \rangle_{\omega_1+\omega_2} = \langle\ \rangle_{\omega_1} \otimes \langle\ \rangle_{\omega_2} \text{ of } \bigotimes(M_1 \oplus M_2) \cong \bigotimes M_1 \otimes \bigotimes M_2.$$

If M is unital[1] with 1_M, the extension of an M-form ω is called *trivial or normalized* with respect to its value for the squared unit $1_M \in M$,

$$0,1 = \langle 1_M \otimes 1_M \rangle_\omega = \begin{cases} \langle 1_M \rangle^2 \Rightarrow \langle 1_M \rangle = 0, \pm 1, & \text{abelian extension,} \\ \langle 1_M \rangle, & \text{nonabelian extension.} \end{cases}$$

An extension may be neither trivial nor normalized.

8.4.2 Trace Forms

The invariant linear forms of endomorphism algebras $\mathbf{AL}(V) = V \otimes V^T \cong \mathbb{K}^{n^2}$ use the trace, normalized with a real factor β,

$$\tfrac{\beta}{n}\,\mathrm{tr} : V \otimes V^T \longrightarrow \mathbb{K},\quad \tfrac{\beta}{n}\,\mathrm{tr}\, f = \langle f \rangle_\beta,\quad \langle \mathrm{id}_V \rangle_\beta = \beta \in \mathbb{R},$$

and conjugated with a dual symmetric isomorphism $\zeta = *$:

$$\zeta : V \longrightarrow V^T,\quad f^* = \zeta^{-1} \circ f^T \circ \zeta,\quad \langle f^* \rangle_\beta = \overline{\langle f \rangle}_\beta.$$

The extension of the trace forms to the tensor algebra $\bigotimes \mathbf{AL}(V)$ gives the always positive Fock forms and the symmetric nonabelian forms, the latter ones being positive only for a positive $\zeta \succeq 0$ (prescalar product). Trace forms are invariant under the action of the basic space endomorphisms

$$\langle [f, a] \rangle_\beta = 0,\quad f \in \mathbf{AL}(V),\quad a \in \bigotimes \mathbf{AL}(V).$$

They have been used for invariant forms of Lie algebra representations, e.g., $\mathrm{tr}\,\mathcal{D}(l) \circ \mathcal{D}(m)$ and the Killing form $\mathrm{tr}\ \mathrm{ad}\, l \circ \mathrm{ad}\, m$ (chapter "Spin, Rotations, and Position").

Fock forms give trivial values for all powers of a *traceless endomorphism*. In the case of nonabelian forms only *nilpotent endomorphisms* have this property:

$$\langle \underbrace{f \otimes \cdots \otimes f}_{k-\text{times}} \rangle = \begin{cases} \langle f \rangle \cdots \langle f \rangle = 0 \text{ for all } k \geq 1 \Longleftrightarrow \langle f \rangle = \mathrm{tr}\, f = 0, \text{ Fock,} \\ \mathrm{tr}\, f \circ \cdots \circ f = 0 \text{ for all } k \geq 1 \Longleftrightarrow f \text{ nilpotent, nonabelian.} \end{cases}$$

A trivial or normalized trace form is related to the normalization of the endomorphism identity on $V \cong \mathbb{K}^n$:

$$\langle \mathrm{id}_V \rangle_\beta = \beta = \begin{cases} 0, \pm 1, & \text{abelian extension,} \\ 0, 1, & \text{nonabelian extension.} \end{cases}$$

[1] The tensor algebra unit $\mathbb{K} \ni 1 \in \bigotimes M$ is different from the algebra M-unit as embedded in the tensor algebra $M \ni 1_M \in \bigotimes M$.

8.4.3 Quantum Algebra Forms

The trace forms of the endomorphism algebra $\mathbf{AL}(V)$ with normalization β and dual isomorphism ζ and, analogously, for the algebra with the transposed endomorphisms $\mathbf{AL}(V^T)$ with $(\check\beta, \zeta^{-1})$ are extended to the quantum algebra of the self-dual space $\mathbf{V} = V \oplus V^T$:

$$\langle\ \rangle_{\beta,\check\beta} : \mathbf{Q}_\epsilon(\mathbf{V}) \longrightarrow \mathbb{K}, \quad \begin{cases} \beta, \check\beta \in \mathbb{R}, \\ \text{dual isomorphism } \zeta = \zeta^*, \\ \langle a^* \rangle_{\beta,\check\beta} = \overline{\langle a \rangle}_{\beta,\check\beta}. \end{cases}$$

as folllows: A quantum algebra is the direct sum of the power grade trivial subalgebra and the vector space $(VV^T)^{\neq}$ with nontrivial power grades, i.e., with different powers of basic space vectors

$$\mathbf{Q}_\epsilon(\mathbf{V}) = \mathrm{INV}_I \mathbf{Q}_\epsilon(\mathbf{V}) \oplus (VV^T)^{\neq}, \quad I = \tfrac{\mathrm{id}_V - \mathrm{id}_V^T}{2}.$$

The quantum algebra form is required to be invariant under the basic space identity. This leads to trivial values on the nontrivial power grade subspace

$$\langle [I, a] \rangle_{\beta,\check\beta} = 0 \Rightarrow \langle (VV^T)^{\neq} \rangle_{\beta,\check\beta} = \{0\}.$$

Using the (anti-) commutators of basic vectors and forms, quantum algebra elements with power grade 0 can be written as linear combinations of products of basis space endomorphisms

$$\mathrm{INV}_I \mathbf{Q}_\epsilon(\mathbf{V}) \cong \bigotimes \mathbf{AL}(V) \cong \bigotimes \mathbf{AL}(V^T).$$

One has for a basic space endomorphism f and its transposed f^T,

$$f \in V \otimes V^T: \quad f + f^T = \epsilon \operatorname{tr} f, \quad \langle f - f^T \rangle_{\beta,\check\beta} = \tfrac{\beta - \check\beta}{n} \operatorname{tr} f.$$

Therefore traceless endomorphisms have trivial forms:

$$f \in V \otimes V^T, \ \operatorname{tr} f = 0 \Rightarrow \langle f \rangle_{\beta,\check\beta} = 0, \ \langle f^T \rangle_{\beta,\check\beta} = 0.$$

All this reduces the quantum algebra form definition to the definition on the quantum classes of the basic space identities. To be well defined there, the form has to be trivial on the quantum ideal, leading to the condition

$$\begin{aligned} \langle \mathrm{id}_V + \mathrm{id}_{V^T} \rangle_{\beta,\check\beta} &= \beta + \check\beta = \epsilon \dim_{\mathbb{K}} V = \epsilon n, \\ \langle \mathrm{id}_V - \mathrm{id}_{V^T} \rangle_{\beta,\check\beta} &= \beta - \check\beta \end{aligned}$$

Therefore the quantum algebra form is completely determined by the difference of the trace normalizations $\beta - \check\beta$ and its conjugation by a dual isomorphism $\zeta = *$.

The sum condition $\beta + \check\beta = \epsilon n$ for the quantum algebra form normalization factors of the dual identities and the additional requirement of a *normalized or trivial form*, i.e., with

$$\begin{aligned} &\beta + \check\beta \in \{\pm 2, \pm 1, 0\} && \text{for abelian extension with } \beta, \check\beta \in \{\pm 1, 0\} \\ &\beta + \check\beta \in \{2, 1, 0\} && \text{for nonabelian extension with } \beta, \check\beta \in \{1, 0\} \end{aligned}$$

leads to basic vector spaces $V \cong \mathbb{K}^n$ with dimension $n = 1, 2$ and normalization factors

$$\text{for } \mathbf{Q}_\epsilon(\mathbb{K}^{2n}) : \quad \begin{cases} n = 1 : & (\beta, \check{\beta}) = (\epsilon, 0) \text{ or } (0, \epsilon), \\ n = 2 : & (\beta, \check{\beta}) = (\epsilon, \epsilon), \\ n > 2 : & \text{no normalized trace forms.} \end{cases}$$

The normalized forms for $n = 1$ are called *irreducible Fock forms* of $\mathbf{Q}_\epsilon(\mathbb{C}^2)$. For $n = 2$, an *irreducible nonabelian form* exists only for the Fermi case $\epsilon = +1$, $(\beta, \check{\beta}) = (1, 1)$, i.e. on $\mathbf{Q}_+(\mathbb{C}^4)$. The irreducible quantum algebra forms and their product forms will be considered in more detail below.

Trace forms of quantum algebras, not necessarily normalized, will be built as product forms of the irreducible forms for $n = 1$ and $n = 2$:

$$\langle\ \rangle : \mathbf{Q}_{\epsilon_V}(\mathbf{V}) \otimes \mathbf{Q}_{\epsilon_W}(\mathbf{W}) \longrightarrow \mathbb{K}, \quad \langle a_V \otimes a_W \rangle = \langle a_V \rangle_V \langle a_W \rangle_W .$$

8.5 Fock States and Fock Spaces

Fock states with definite unitary time representations $\mathbb{R} \ni t \longmapsto e^{i\mu t} \in \mathbf{U}(1)$ are appropriate for normalizable translation eigenstates, e.g., for particles.

8.5.1 Irreducible Fock States

A 1-dimensional complex space has the $\mathbf{U}(1)$-conjugation

$$V \ni e = \mathrm{u} \overset{\star}{\leftrightarrow} \mathrm{u}^\star = \check{e} \in V^T .$$

The irreducible Fock forms act on the smallest quantum algebras $\mathbf{Q}_\epsilon(\mathbb{C}^2)$ with basic space $\mathbf{V} = \mathbb{C}\mathrm{u} \oplus \mathbb{C}\mathrm{u}^\star$ and quantization (anti)commutators and basic space identities

$$\text{in } \mathbf{Q}_\epsilon(\mathbb{C}^2) : \quad [\mathrm{u}^\star, \mathrm{u}]_\epsilon = [\mathrm{u}, \epsilon \mathrm{u}^\star]_\epsilon = 1, \quad \mathrm{id}_V = \mathrm{u} \otimes \mathrm{u}^\star, \quad \mathrm{id}_{V^T} = \epsilon \mathrm{u}^\star \otimes \mathrm{u}.$$

The Fock forms are determined by the trace normalization factors

$$(\beta, \check{\beta}) = \Big(\langle \mathrm{u} \otimes \mathrm{u}^\star \rangle, \langle \epsilon \mathrm{u}^\star \otimes \mathrm{u} \rangle\Big) = (\epsilon, 0) \text{ or } (0, \epsilon).$$

Both possibilities are equivalent by exchanging dual basic vectors

$$(\mathrm{u}^\star, \mathrm{u}) \leftrightarrow (\mathrm{u}, \epsilon \mathrm{u}^\star).$$

The irreducible Fock form with $(\beta, \check{\beta}) = (0, \epsilon)$ is chosen, denoted by $\langle\ \rangle_{\mathrm{F}}$. It gives the values on the endomorphisms $\bigotimes \mathbf{AL}(V)$ and $\bigotimes \mathbf{AL}(V^T)$. It is positive and normalized, i.e., a *state* of the quantum algebra:

$$\begin{array}{l} \langle\ \rangle_{\rm F} : \mathbf{Q}_\epsilon(\mathbb{C}^2) \longrightarrow \mathbb{C} \\ \langle {\rm u}^\star \otimes {\rm u}\rangle_{\rm F} = 1 \end{array} \Rightarrow \left\{ \begin{array}{rl} \langle({\rm u}^\star \otimes {\rm u})^k\rangle_{\rm F} & = 1, \quad \langle({\rm u} \otimes {\rm u}^\star)^k\rangle_{\rm F} = 0, \\ \langle {\rm u}^{\star k} \otimes {\rm u}^l\rangle_{\rm F} & = k!\delta_{kl}, \\ & k,l = \left\{ \begin{array}{ll} 0,1, & \epsilon = +1, \\ 0,1,2,\ldots, & \epsilon = -1, \end{array}\right. \\ \| {\rm u}^k \| & = \sqrt{k!}, \quad \| ({\rm u}^\star)^k \| = 0, \\ \langle a \otimes {\rm u}^\star\rangle_{\rm F} & = 0 = \langle {\rm u} \otimes a\rangle_{\rm F}, \quad a \in \mathbf{Q}_\epsilon(\mathbb{C}^2), \\ a & = \sum\limits_{k,l} \alpha_{kl} {\rm u}^k \otimes {\rm u}^{\star l} \\ \Rightarrow \langle a^\star \otimes a\rangle_{\rm F} & = \sum\limits_{k} |\alpha_{k0}|^2 k! \geq 0 \text{ (positive).} \end{array}\right.$$

The associate Fock state trivial left ideal $\mathbf{Q}_\epsilon(\mathbb{C}^2)^\perp$ is characterized by the *annihilation operator* $\mathrm{u}^\star$ as cyclic element

$$\langle a^\star \otimes a\rangle_{\rm F} = 0 \iff a \in \mathbf{Q}_\epsilon(\mathbb{C}^2){\rm u}^\star, \quad {\rm u}^\star|1\rangle_{\rm F} = 0.$$

The associate scalar product space $\mathrm{FOCK}_\epsilon(V)$ and Hilbert space with the operator classes can be generated with the creation operator u acting on the cyclic state vector $|1\rangle_{\rm F}$:

$$\begin{array}{c} \text{orthonormal basis of } \mathrm{FOCK}_\epsilon(\mathbb{C}) \cong \mathbf{Q}_\epsilon(\mathbb{C}^2)/\mathbf{Q}_\epsilon(\mathbb{C}^2){\rm u}^\star : \\ \{|k\rangle = \frac{{\rm u}^k}{\sqrt{k!}}|1\rangle_{\rm F} = \frac{{\rm u}^k}{\sqrt{k!}}|0\rangle \;\Big|\; \left\{ \begin{array}{ll} k = 0,1\}, & \epsilon = +1, \\ k = 0,1,\ldots\}, & \epsilon = -1. \end{array}\right. \end{array}$$

Here, one has to pay attention to two different notations: The notation without Fock form denoting subindex $|1\rangle = {\rm u}|1\rangle_{\rm F}$ designates not the class of the algebra unit $1 \in \mathbf{Q}_\epsilon(\mathbb{C}^2)$, but the "one-quantum" state vector. Analogously, the notation $|0\rangle = |1\rangle_{\rm F}$ (the class of the algebra unit) designates the "zero-quantum" state vector. The state vector $|k\rangle$ *with* k *quanta* – k is the invariant eigenvalue for the time translations – spans the classes of the basic vector space power $\bigotimes^k V$.

The invariance group of the Euclidean conjugation is $\mathbf{U}(1)$ with the basic vector space identity id_V as generator. Therefore the Fock state of $\mathbf{Q}_\epsilon(\mathbb{C}^2)$ with the associate Hilbert space is the form for an irreducible compact time representation $\mathbb{R} \longrightarrow \mathbf{U}(1)$, where the basic vector space identity comes as Hamiltonian for the harmonic Fermi or Bose oscillator

$$H_0 = \mu I = \mu\frac{[{\rm u},{\rm u}^\star]_{-\epsilon}}{2} = \mu({\rm u} \otimes {\rm u}^\star - \tfrac{\epsilon}{2}), \quad \mu \in \mathbb{R}.$$

The smallest quantum algebras $\mathbf{Q}_\epsilon(\mathbb{C}^2)$ are probability algebras for irreducible time representations with a basis of time translation eigenvectors

$$[H_0, {\rm u}^{\star k} \otimes {\rm u}^l] = \mu(l-k){\rm u}^{\star k} \otimes {\rm u}^l, \quad l - k \in \left\{ \begin{array}{ll} \{0,\pm 1\}, & \epsilon = +1, \\ \mathbb{Z}, & \epsilon = -1. \end{array}\right.$$

With a positive energy scale factor $\mu \geq 0$ this Hamiltonian (harmonic oscillator) has the Fock form as ground state

$$\text{for } \mu \geq 0: \quad \left\{ \begin{array}{l} \langle H_0^2\rangle_{\rm F} = \langle H_0\rangle_{\rm F}^2, \quad \langle H_0\rangle_{\rm F} = -\mu\frac{\epsilon}{2}, \\ \langle a^\star \otimes H_0 \otimes a\rangle_{\rm F} \geq \langle H_0\rangle_{\rm F}\langle a^\star \otimes a\rangle_{\rm F}, \quad a \in \mathbf{Q}_\epsilon(\mathbb{C}^2), \end{array}\right.$$

with the equations for eigenstates

$$[H_0, a] = E(a)a \Rightarrow H_0|a\rangle_F = E(a)|a\rangle_F + aH_0|1\rangle_F = (E(a) - \mu\tfrac{\epsilon}{2})|a\rangle_F.$$

Summarizing this section: Associated to the Fock state of the *Fermi quantum algebra* $\mathbf{Q}_+(\mathbb{C}^2) \cong \mathbb{C}^4$ with basis $\{1, \mathrm{u}, \mathrm{u}^\star, \mathrm{u}\otimes\mathrm{u}^\star\}$ is a complex 2-dimensional Hilbert space with the "zero-quantum" and "one-quantum" state vectors as a basis. As a vector space it is isomorphic to the Grassmann algebra of the basic vector space

$$\begin{array}{l} \mathrm{FOCK}_+(V) \cong \bigwedge V \cong \mathbb{C} \oplus V \cong \mathbb{C}^2, \quad V = \mathbb{C}\mathrm{u}, \\ \text{basis}: \ \{|0\rangle, |1\rangle\} = \{|1\rangle_F, |\mathrm{u}\rangle_F\}, \\ H_0 = \mu\frac{[\mathrm{u},\mathrm{u}^\star]}{2}, \quad H_0|k\rangle = \mu(k - \frac{1}{2})|k\rangle, \quad k \in \{0, 1\}. \end{array}$$

For the *Bose quantum algebra* $\mathbf{Q}_-(\mathbb{C}^2)$ the associate scalar product space is, as a vector space, isomorphic to the polynomial algebra of the basic vector space $V = \mathbb{C}\mathrm{u}$:

$$\begin{array}{l} \mathrm{FOCK}_-(V) \cong \bigvee V \cong \mathbb{C}^{\aleph_0}, \quad V = \mathbb{C}\mathrm{u}, \\ \text{basis}: \ \{|k\rangle \bigm| k = 0, 1, \dots\} = \{\frac{1}{\sqrt{k!}}|\mathrm{u}^k\rangle_F\}, \\ H_0 = \mu\frac{\{\mathrm{u},\mathrm{u}^\star\}}{2}, \quad H_0|k\rangle = \mu(k + \frac{1}{2})|k\rangle, \quad k \in \mathbb{N}_0. \end{array}$$

with the Hilbert space $\overline{\mathrm{FOCK}_-(V)}$ as Cauchy completion.

The scalar product on the Fock spaces is the scalar product $V \times V \longrightarrow \mathbb{C}$, $\langle \mathrm{u}|\mathrm{u}\rangle = 1$, induced to the Grassmann and polynomial algebra.

8.5.2 Factorizable Fock Forms

A Fock form for a basic vector space $V \cong \mathbb{C}^n$ with dimension $n \geq 2$ is factorizable in accordance with a quantum algebra factorization $\mathbf{Q}_\epsilon(\mathbf{V}) \cong \bigotimes^n \mathbf{Q}_\epsilon(\mathbb{C}^2)$. Since there are many decompositions $\mathbf{V} \cong \bigoplus_1^n \mathbb{C}^2$, one for each basis, there are correspondingly many isomorphisms for the quantum algebras and Fock form factorizations.

The Fock form is abelian and nontrivial either on the basic space identity or its transpose, e.g.,

$$f_j \in V \otimes V^T: \quad \left\{ \begin{array}{l} \langle f\rangle_F = 0, \ \langle f^T\rangle_F = \epsilon \operatorname{tr} f, \\ \langle f_1 \otimes \cdots \otimes f_k\rangle_F = \langle f_1\rangle_F \cdots \langle f_k\rangle_F = 0, \\ \langle f_1^T \otimes \cdots \otimes f_k^T\rangle_F = \langle f_1^T\rangle_F \cdots \langle f_k^T\rangle_F. \end{array} \right.$$

The transposed case with $\langle f\rangle_F = \epsilon \operatorname{tr} f$ and $\langle f^T\rangle_F = 0$ requires a redefinition with the factor ϵ in order to have a positive Euclidean conjugation.

For a direct sum basic space there arise products of scalar products and products of Hilbert spaces:

$$\mathrm{FOCK}_\epsilon(V_1 \oplus V_2) \cong \mathrm{FOCK}_\epsilon(V_1) \otimes \mathrm{FOCK}_\epsilon(V_2),$$
$$V \cong \mathbb{C}^n \Rightarrow \begin{cases} \mathrm{FOCK}_+(V) & \cong \bigwedge V \cong \mathbb{C}^{2^n}, \\ \mathrm{FOCK}_-(V) & \cong \bigvee V \cong \mathbb{C}^{\aleph_0}. \end{cases}$$

The Fock form is the ground state for the n-dimensional harmonic oscillator with the identity as Hamiltonian

$$H_0 = \mu I, \quad I = \frac{[\mathrm{u}^A, \mathrm{u}^\star_A]_{-\epsilon}}{2}, \quad \langle I \rangle_{\mathrm{F}} = -\epsilon \tfrac{n}{2}.$$

8.5.3 Particle-Antiparticle Spaces

In quantum algebras with Fock state and Hilbert space, a doubling with respect to the canonical conjugation (chapter "Anticonjugation: The Real in the Complex") is connected with the particle-antiparticle doubling. A quantum algebra with Euclidean conjugation $\star$ and anticonjugation $\times$ has a fourfold basic vector space $\mathbf{V}_{\mathrm{doub}} \cong \mathbb{C}^{4n}$ with two dual and two conjugated pairs:

$$\mathbf{V}_{\mathrm{doub}} = V_{\mathrm{doub}} \oplus V^T_{\mathrm{doub}} = \mathbf{V} \oplus \overline{\mathbf{V}}, \quad \begin{cases} V_{\mathrm{doub}} = V \oplus \overline{V}^T, \\ V^T_{\mathrm{doub}} = V^T \oplus \overline{V}, \\ \mathbf{V} = V \oplus V^T, \quad \overline{\mathbf{V}} = \overline{V} \oplus \overline{V}^T, \end{cases}$$
$$(\mathrm{u}^A, \mathrm{a}^{\star A}, \mathrm{u}^\star_A, \mathrm{a}_A)_{A=1}^n \text{ bases of } V, \overline{V}, V^T, \overline{V}^T.$$

The quantum algebra is the product of two conjugated factor algebras

$$\mathbf{Q}_\epsilon(\mathbf{V}_{\mathrm{doub}}) \cong \mathbf{Q}_\epsilon(\mathbf{V}) \otimes \mathbf{Q}_\epsilon(\overline{\mathbf{V}}), \quad \mathbf{Q}_\epsilon(\overline{\mathbf{V}}) = \overline{\mathbf{Q}_\epsilon(\mathbf{V})},$$
$$\text{basic vector space (anti) commutators: } [\mathrm{u}^\star_A, \mathrm{u}^B]_\epsilon = \delta^B_A = [\mathrm{a}^{B\star}, \mathrm{a}_A]_\epsilon.$$

With the Fock form of $\mathbf{Q}_\epsilon(\mathbf{V})$ and $\mathbf{Q}_\epsilon(\overline{\mathbf{V}})$ the quantum algebra $\mathbf{Q}_\epsilon(\mathbf{V}_{\mathrm{doub}})$ carries the product form, compatible with both conjugations,

$$\langle\ \rangle_{\mathrm{F}} : \mathbf{Q}_\epsilon(\mathbf{V}_{\mathrm{doub}}) \longrightarrow \mathbb{C}, \quad \begin{cases} \langle a \otimes b^\times \rangle_{\mathrm{F}} = \langle a \rangle_{\mathrm{F}} \langle b^\times \rangle_{\mathrm{F}}, \\ \langle a^\times \rangle_{\mathrm{F}} = \overline{\langle a \rangle_{\mathrm{F}}} = \langle a^\star \rangle_{\mathrm{F}}, \\ a, a^\star, b \in \mathbf{Q}_\epsilon(\mathbf{V}), \ a^\times, \ b^\times \in \mathbf{Q}_\epsilon(\overline{\mathbf{V}}). \end{cases}$$

The Fock form can be nontrivial only for elements $a \in \mathbf{Q}_\epsilon(\mathbf{V}_{\mathrm{doub}})$ with trivial power grade

$$\langle [\mathrm{id}_{V_{\mathrm{doub}}}, a] \rangle_{\mathrm{F}} = 0, \quad \mathrm{id}_{V_{\mathrm{doub}}} = \mathrm{u}^A \otimes \mathrm{u}^\star_A + \mathrm{a}_A \otimes \mathrm{a}^{\star A}.$$

As in the general factorizable case one has for the endomorphisms

$$f_j \in V \otimes V^T \text{ or } \overline{V}^T \otimes \overline{V} : \quad \begin{cases} \langle f \rangle_{\mathrm{F}} = 0, \ \langle f^T \rangle_{\mathrm{F}} = \epsilon \operatorname{tr} f, \\ \langle f_1 \otimes \cdots \otimes f_k \rangle_{\mathrm{F}} = \langle f_1 \rangle_{\mathrm{F}} \cdots \langle f_k \rangle_{\mathrm{F}} = 0, \\ \langle f_1^T \otimes \cdots \otimes f_k^T \rangle_{\mathrm{F}} = \langle f_1^T \rangle_{\mathrm{F}} \cdots \langle f_k^T \rangle_{\mathrm{F}}. \end{cases}$$

Therefore, one has the *creation operators for particles* $\{\mathrm{u}^A\}_{A=1}^n$ *and antiparticles* $\{\mathrm{a}_A\}_{A=1}^n$:

$$\begin{array}{lll} \langle \mathrm{u}_B^\star \otimes \mathrm{u}^A\rangle_{\mathrm{F}} & = \langle \mathrm{a}^{\star A} \otimes \mathrm{a}_B\rangle_{\mathrm{F}} & = \delta_B^A, \quad \langle \mathrm{u}^A \otimes \mathrm{u}_B^\star\rangle_{\mathrm{F}} = \langle \mathrm{a}_B \otimes \mathrm{a}^{\star A}\rangle_{\mathrm{F}} = 0, \\ \langle [\mathrm{u}_B^\star, \mathrm{u}^A]_{-\epsilon}\rangle_{\mathrm{F}} & = \langle [\mathrm{a}^{A\star}, \mathrm{a}_B]_{-\epsilon}\rangle_{\mathrm{F}} & = \delta_B^A. \end{array}$$

The dual vectors $\mathrm{u}_A^\star$, $\mathrm{a}^{A\star}$ are the corresponding annihilation operators.

There arise products of scalar product and Hilbert spaces

$$\mathrm{FOCK}_\epsilon(V_{\mathrm{doub}}) \cong \mathrm{FOCK}_\epsilon(V) \otimes \mathrm{FOCK}_\epsilon(\overline{V}^T).$$

The minimal case with particles and antiparticles and a 4-dimensional space $\mathbf{V} \cong \mathbb{C}^4$ is characteristic:

$$\begin{array}{l} \mathbf{Q}_\epsilon(\mathbb{C}^{4n}) \cong \bigotimes^n \mathbf{Q}_\epsilon(\mathbb{C}^4), \\ \text{basis of } \mathbf{Q}_\epsilon(\mathbb{C}^4): \quad \left\{ \begin{array}{ll} \{1,\ \mathrm{u},\ \mathrm{u}^\star,\ \mathrm{u}\otimes\mathrm{u}^\star,\ \mathrm{a},\ \mathrm{a}^\star,\ \mathrm{a}\otimes\mathrm{a}^\star\}, & \epsilon = +1, \\ \{\mathrm{u}^k \otimes \mathrm{u}^{\star l} \otimes \mathrm{a}^m \otimes \mathrm{a}^{\star n} \mid k,l,m,n \geq 0\}, & \epsilon = -1. \end{array}\right. \end{array}$$

The Fock state has the values

$$\langle\ \rangle_{\mathrm{F}} : \mathbf{Q}_\epsilon(\mathbb{C}^4) \longrightarrow \mathbb{C}, \quad \begin{array}{l} \langle (\mathrm{u}^\star \otimes \mathrm{u})^k\rangle_{\mathrm{F}} = \langle (\mathrm{a}^\star \otimes \mathrm{a})^k\rangle_{\mathrm{F}} = 1, \\ \langle (\mathrm{u} \otimes \mathrm{u}^\star)^k\rangle_{\mathrm{F}} = \langle (\mathrm{a} \otimes \mathrm{a}^\star)^k\rangle_{\mathrm{F}} = 0. \end{array}$$

The particle and antiparticle creation operators, $\mathrm{u} \in \mathbf{Q}_\epsilon(\mathbf{V})$ and $\mathrm{a} \in \mathbf{Q}_\epsilon(\overline{\mathbf{V}})$, have nontrivial state vectors $|\mathrm{u}\rangle_{\mathrm{F}}$ and $|\mathrm{a}\rangle_{\mathrm{F}}$. The associated Hilbert space with the Fock-form-relevant elements is the product of the individual Hilbert spaces, spanned by the algebra unit 1 with state vector $|0\rangle = |1\rangle_{\mathrm{F}}$ und the state vectors $|k,l\rangle = \frac{1}{\sqrt{k!l!}}|\mathrm{u}^k\mathrm{a}^l\rangle_{\mathrm{F}}$ for k particles and l antiparticles.

The Fock form is positive definite with the Euclidean $\mathbf{U}(2)$-conjugation $\star$, indefinite, however, with the $\mathbf{U}(1,1)$-conjugation $\times$:

$$\begin{array}{l} \text{e.g., } \mathrm{u}^\star - \mathrm{a}^\star = \mathrm{a}^\times - \mathrm{u}^\times \\ \Rightarrow \langle (\mathrm{u}^\star - \mathrm{a}^\star) \otimes (\mathrm{u} - \mathrm{a})\rangle_{\mathrm{F}} = 2 = -\langle (\mathrm{u}^\times - \mathrm{a}^\times) \otimes (\mathrm{u} - \mathrm{a})\rangle_{\mathrm{F}}. \end{array}$$

8.6 Position Representation

The Fock-Hilbert space of Bose quantum algebras can be used also for the Heisenberg group $\mathbf{H}(s)$ with s position-momentum pairs. The quantum algebra is the complexified enveloping algebra $\mathbf{Q}_-(\mathbb{C}^{2s}) = \mathbb{C} \otimes \mathbf{E}(\log \mathbf{H}(s))$, $\log \mathbf{H}(s) \cong \mathbb{R}^{1+2s}$.

8.6.1 Wave Functions

A symmetric basis in the Bose quantum algebra is interpretable as position and momentum:

$$\mathbf{Q}_-(\mathbb{C}^2) = \mathbb{C} \otimes \mathbf{Q}_-(\mathbb{R}^2), \quad \mathbf{x} = \tfrac{\mathrm{u}+\mathrm{u}^\star}{\sqrt{2}}, \quad -i\mathbf{p} = \tfrac{\mathrm{u}-\mathrm{u}^\star}{\sqrt{2}}, \quad [\mathrm{u}^\star, \mathrm{u}] = 1 = [i\mathbf{p}, \mathbf{x}].$$

Heisenberg's uncertainty relation for position and momentum arises with a scalar product (Fock form). It involves the Born-Heisenberg commutator

$$\left.\begin{array}{lll} (\Delta\mathbf{x})^2 & = \langle(\mathbf{x}-\langle\mathbf{x}\rangle_{\mathrm{F}})^2\rangle_{\mathrm{F}} & = \langle\mathbf{x}^2\rangle_{\mathrm{F}} - \langle\mathbf{x}\rangle_{\mathrm{F}}^2 \\ (\Delta\mathbf{p})^2 & = \langle(\mathbf{p}-\langle\mathbf{p}\rangle_{\mathrm{F}})^2\rangle_{\mathrm{F}} & = \langle\mathbf{p}^2\rangle_{\mathrm{F}} - \langle\mathbf{p}\rangle_{\mathrm{F}}^2 \end{array}\right\} \Rightarrow \Delta\mathbf{x}\Delta\mathbf{p} \geq |\langle\tfrac{[i\mathbf{p},\mathbf{x}]}{2}\rangle_{\mathrm{F}}| = \tfrac{\hbar}{2}.$$

The Fock space is isomorphic (below) to the complex $\mathcal{C}^\infty$-functions of the position translations $\mathcal{S}(\mathbb{R})$, whose absolute value decreases more rapidly than all position $\mathbf{x} \cong x$ and momentum $i\mathbf{p} \cong \frac{d}{dx}$ polynomials (chapter "Propagators"):

$$\begin{array}{rl} \mathbf{Q}_-(\mathbb{C}^2)/\mathbf{Q}_-(\mathbb{C}^2)\mathrm{u}^\star & \cong \mathrm{FOCK}_-(\mathbb{C}) \\ \cong \mathcal{S}(\mathbb{R}) & = \{\psi : \mathbb{R} \longrightarrow \mathbb{C} \mid \sup_{x\in\mathbb{R}} |\mathbf{a}\psi(x)| < \infty,\ \mathbf{a} \in \mathbf{Q}_-(\mathbb{C}^2)\}, \\ \mathbf{a} & = P_1(\mathbf{x})P_2(i\mathbf{p}) \cong P_1(x)P_2(\frac{d}{dx}) \text{ (polynomials).} \end{array}$$

Its Cauchy completion in the scalar product norm is constituted by the Lebesque almost everywhere defined square integrable complex functions with the elements called position *wave functions (Schrödinger functions)*

$$\overline{\mathcal{S}(\mathbb{R})} \cong L^2_{dx}(\mathbb{R},\mathbb{C}) = \{\psi : \mathbb{R} \longrightarrow \mathbb{C} \mid \langle\psi|\psi\rangle = \int dx |\psi(x)|^2 < \infty\}.$$

The wave functions $x \longmapsto \psi(x)$ are *position orbits*, e.g., ψ^k for the oscillator state vectors $|k\rangle$. Their values, obviously not in position space, are used for the position spread of probability amplitudes and for position densities of probabilities (below). They can be analyzed with respect to the representations of the position translations (chapter "The Kepler Factor").

In contrast to the Hilbert space, the rapidly decreasing functions do not coincide with their continuous linear forms, the *tempered distributions* (chapter "Propagators"):

$$\begin{array}{rlll} \mathcal{S}(\mathbb{R}) \subset & L^2_{dx}(\mathbb{R},\mathbb{C}) \cong [L^2_{dx}(\mathbb{R},\mathbb{C})]' & \subset \mathcal{S}'(\mathbb{R}), \\ \text{e.g., } \mathcal{S}(\mathbb{R}) \not\ni & e^{-|x|} \in L^2_{dx}(\mathbb{R},\mathbb{C}) \not\ni e^{ix} & \in \mathcal{S}'(\mathbb{R}). \end{array}$$

The scalar product space isomorphism $\mathrm{FOCK}_-(\mathbb{C}) \cong \mathcal{S}(\mathbb{R})$ can be shown in the *position representation*, where the position operator is represented by the spectrum $\mathrm{spec}\,\mathbf{x} \cong \mathbb{R}$. Creation and annihilation operator can be written in a Rodriguez form with derivatives:

$$\begin{array}{l} \mathbf{x} \cong x,\ i\mathbf{p} \cong \frac{d}{dx} \Rightarrow \left\{\begin{array}{lll} \sqrt{2}\mathrm{u} & \cong x - \frac{d}{dx} & = -e^{\frac{x^2}{2}}\frac{d}{dx}e^{-\frac{x^2}{2}}, \\ \sqrt{2}\mathrm{u}^\star & \cong x + \frac{d}{dx} & = e^{-\frac{x^2}{2}}\frac{d}{dx}e^{\frac{x^2}{2}}, \end{array}\right. \\ H_0 = \frac{\mu}{2}\{\mathrm{u},\mathrm{u}^\star\} = \mu\frac{\mathbf{p}^2+\mathbf{x}^2}{2} \cong \frac{\mu}{2}(-\frac{d^2}{dx^2} + x^2). \end{array}$$

The cyclic state vector $|0\rangle = |1\rangle_{\mathrm{F}}$ as the class of the algebra unit is represented by a positive definite rapidly decreasing function:

$$|0\rangle \cong \psi^0 : \left\{\begin{array}{lll} \mathrm{u}^\star|0\rangle = 0 & \iff & \frac{d}{dx}\left(e^{\frac{x^2}{2}}\psi^0(x)\right) = 0, \\ \langle 0|0\rangle = 1 & \iff & \int dx|\psi^0(x)|^2 = 1, \end{array}\right\} \Rightarrow \psi^0(x) = \frac{1}{\sqrt{\sqrt{\pi}}}e^{-\frac{x^2}{2}}.$$

From the cyclic ground state vector a scalar product space basis $\{\psi^k \mid k = 0, 1, \dots\}$ can be constructed by the action of the creation operator powers:

$$\frac{\mathrm{u}^k}{\sqrt{k!}}|0\rangle = |k\rangle \cong \psi^k : \begin{cases} \psi^k(x) & = \frac{1}{\sqrt{k!}}\left(\frac{x-\frac{d}{dx}}{\sqrt{2}}\right)^k \psi^0(x) = \frac{1}{\sqrt{2^k k!\sqrt{\pi}}} e^{\frac{x^2}{2}}\left(-\frac{d}{dx}\right)^k e^{-x^2} \\ & = \frac{1}{\sqrt{2^k k!\sqrt{\pi}}} e^{-\frac{x^2}{2}} \mathrm{H}^k(x). \end{cases}$$

It involves the Rodriguez formula for the *Hermite polynomials* with $[\frac{2n}{2}] = [\frac{1+2n}{2}] = n$:

$$\begin{aligned} \mathrm{H}^k(x) &= e^{x^2}\left(-\frac{d}{dx}\right)^k e^{-x^2} = k!\sum_{n=0}^{[\frac{k}{2}]} \frac{(-1)^n}{n!(k-2n)!}(2x)^{k-2n} = (-1)^k \mathrm{H}^k(-x) \\ \deg \mathrm{H}^k &= k, \ \left(\frac{d^2}{dx^2} - 2x\frac{d}{dx} + 2k\right)\mathrm{H}^k(x) = 0 \end{aligned}$$

Up to a factor from $\{1, x\}$, the Hermite polynomials are Laguerre polynomials (chapter "The Kepler Factor") with $\deg \mathrm{L}^N_\lambda(x^2) = N$, depending on the squared length x^2:

$$\mathrm{H}^{2N}(x) = N!(-4)^N \mathrm{L}^N_{-\frac{1}{2}}(x^2), \quad \text{e.g.,} \begin{cases} \mathrm{H}^0(x) & = 1, \\ \mathrm{H}^2(x) & = -2(1-2x^2), \\ \mathrm{H}^4(x) & = 4(3-12x^2+4x^4), \end{cases}$$

$$\mathrm{H}^{1+2N}(x) = N!(-4)^N 2x\, \mathrm{L}^N_{\frac{1}{2}}(x^2), \quad \text{e.g.,} \begin{cases} \mathrm{H}^1(x) & = 2x, \\ \mathrm{H}^3(x) & = -4x(3-2x^2), \\ \mathrm{H}^5(x) & = 8x(15-20x^2+4x^4). \end{cases}$$

Orthonormality and completeness of the harmonic oscillator eigenfunctions as Hilbert basis are expressed by

$$\begin{aligned} &\{x \longmapsto e^{-\frac{x^2}{2}}\mathrm{H}^k(x) \mid k = 0, 1, \dots\} \text{ basis of } L^2_{dx}(\mathbb{R}, \mathbb{R}), \\ &\text{with} \begin{cases} \int dx\ \mathrm{H}^k(x)\, e^{-x^2}\, \mathrm{H}^{k'}(x) & = 2^k k!\sqrt{\pi}\, \delta_{kk'}, \\ \sum_{k=0}^{\infty} \frac{1}{2^k k!\sqrt{\pi}}\, e^{-\frac{x^2}{2}}\mathrm{H}^k(x)\, e^{-\frac{x'^2}{2}}\mathrm{H}^k(x') & = \delta(x-x'). \end{cases} \end{aligned}$$

The Fourier transformation is a scalar product space isomorphism for rapidly decreasing functions and their completions:

$$\mathcal{S}(\mathbb{R}) \cong \mathcal{S}(\check{\mathbb{R}}) \subset L^2_{dx}(\mathbb{R}, \mathbb{C}) \cong L^2_{dp}(\check{\mathbb{R}}, \mathbb{C}).$$

The space $L^2(\mathbb{R}, \mathbb{C})$ is determined by the irreducible faithful Hilbert representations of the Heisenberg Lie algebra and group $\mathbf{H}(1)$ (chapter "Harmonic Analysis"). With the position-momentum symmetry of the oscillator Hamiltonian the position representation ground state function is essentially its own Fourier transform:

$$\begin{aligned} \sqrt{2^k k!}\psi^k(x) &= \mathrm{H}^k(x)\psi^0(x) = \int \frac{dp}{2\pi} e^{ipx}\mathrm{H}^k\left(i\frac{d}{dp}\right)\tilde{\psi}^0(p), \\ \mathcal{S}(\mathbb{R}) \ni \psi^0(x) &= \frac{1}{\sqrt{\sqrt{\pi}}} e^{-\frac{x^2}{2}} \overset{\mathbf{F}}{\leftrightarrow} \frac{1}{\sqrt{2\pi\sqrt{\pi}}} e^{-\frac{p^2}{2}} = \tilde{\psi}^0(p) \in \mathcal{S}(\check{\mathbb{R}}). \end{aligned}$$

The position wave functions ψ^k solve the Schrödinger equation with the energy eigenvalues for the harmonic oscillator; $\frac{E(k)}{\mu}$ are the degrees of the Hermite polynomials

$$\tfrac{1}{2}(-\tfrac{d^2}{dx^2} + x^2)\psi^k(x) = (k + \tfrac{1}{2})\psi^k(x).$$

By position spread of the yes-no probability $\langle\psi|\psi\rangle = 1$, probability densities $|\psi(x)|^2$ can be defined. Each normalized state vector ψ defines a positive measure $d^\psi x = |\psi(x)|^2 dx$ on position with Lebesgue measure basis. The Hilbert space has the structure of an orthogonal direct integral with the positive function for the ground state (chapter "The Kepler Factor") and an $\mathbb{C}$-isomorphic Hilbert space at each position:

$$L^2_{dx}(\mathbb{R}, \mathbb{C}) \cong {}^\perp\!\!\int_{\mathbb{R}} d^{\psi^0}x\ \mathbb{C}(x), \quad d^{\psi^0}x = |\psi^0(x)|^2 dx = \tfrac{e^{-x^2}}{\sqrt{\pi}}dx, \quad \mathbb{C}(x) \cong \mathbb{C}.$$

The Hilbert space basis orthocompleteness is reformulated with the positive ground state function:

$$\begin{aligned} \int \tfrac{e^{-x^2}}{\sqrt{\pi}} dx\ \mathrm{H}^k(x)\ \mathrm{H}^{k'}(x) &= 2^k k!\ \delta_{kk'}, \\ \sum_{k=0}^{\infty} \tfrac{1}{2^k k!}\ \mathrm{H}^k(x)\ \mathrm{H}^k(x') &= e^{x^2}\sqrt{\pi}\delta(x - x'), \\ \sum_{k=0}^{\infty} \tfrac{1}{k!}\mathrm{H}^k(x) y^k &= \tfrac{e^{x^2}}{e^{(x-y)^2}} = e^{2yx - y^2}. \end{aligned}$$

The degree-k polynomials multiplying the ground state exponential span a unital algebra. The degree reflects the tensor powers in the quantum algebra grading.

8.6.2 Position Space Quantum Mechanics

The Bose quantum algebra $\mathbf{Q}_-(\mathbb{C}^6)$ with a Euclidean conjugation is the arena for 3-dimensional position quantum mechanics. The angular momentum Lie algebra $\log \mathbf{SO}(3) \cong \mathbb{R}^3$ is implemented by position-momentum products

$$[\mathrm{u}^\star_a, \mathrm{u}^b] = [i\mathbf{p}_a, \mathbf{x}_b] = \delta^b_a, \quad \begin{cases} \mathcal{O}^a = i\mathcal{L}^a = \epsilon^{abc}\mathrm{u}_b \otimes \mathrm{u}^\star_c = -i\epsilon^{abc}\mathbf{p}_b \otimes \mathbf{x}_c, \\ \vec{\mathcal{O}}^\star = -\vec{\mathcal{O}}, \ \ \vec{\mathcal{L}}^\star = \vec{\mathcal{L}}, \\ \mathcal{O}_\pm = \mp i\mathcal{O}^1 + \mathcal{O}^2, \ \ \mathcal{O}_0 = -2i\mathcal{O}^3. \end{cases}$$

The factorizable Fock state of $\mathbf{Q}_-(\mathbb{C}^6) \cong \bigotimes^3 \mathbf{Q}_-(\mathbb{C}^2)$ is rotation invariant, since the angular momenta $\mathcal{O}$ on the basic space are traceless and hence belong to the Fock state trivial left ideal

$$\langle\ \rangle_{\mathrm{F}} : \mathbf{Q}_-(\mathbb{C}^6) \longrightarrow \mathbb{C}, \quad \langle \mathcal{O} \otimes a\rangle_{\mathrm{F}} = \langle\mathcal{O}\rangle_{\mathrm{F}}\langle a\rangle_{\mathrm{F}} = 0, \quad \mathcal{O} \in \log \mathbf{SO}(3).$$

Therefore, the angular momenta annihilate the cyclic state vector $|0\rangle$, the class of the algebra unit $1 \in \mathbf{Q}_-(\mathbb{C}^6)$

$$\mathrm{u}_a^\star|0\rangle = 0 \Rightarrow \mathcal{O}|0\rangle = 0, \quad [\mathcal{O}, a]|0\rangle = \mathcal{O}a|0\rangle.$$

The rotation action distinguishes invariant Hilbert subspaces (chapter "Quantum Algebra"): Taking from a vector subspace $\Lambda \cong \mathbb{C}^3$ of the quantum algebra with adjoint angular momentum representation and basis $\{\lambda^a\}_{a=1}^3$ an $\mathrm{ad}\,\mathcal{O}_0$-eigenvector λ_+, one obtains a $(1+2L)$-dimensional representation with highest-weight eigenvector as product $(\lambda_+)^L$:

$$\lambda_+ \in \mathbf{Q}_-(\mathbb{C}^6), \quad [\mathcal{O}_0, \lambda_+^L] = 2L\lambda_+^L, \quad L = 0, 1, \dots .$$

The $(1+2L)$ eigenstates of the Cartan operator $\mathcal{O}_0$ (third spin component) and of the Casimir element $\vec{\mathcal{O}}^2$ in the Fock space are

$$\begin{matrix} (\mathcal{O}_-)^{L-m}\lambda_+^L|0\rangle = |\lambda, L; m\rangle, \\ m = L, \dots, -L \end{matrix} \Rightarrow \left\{ \begin{array}{ll} \mathcal{O}_0|\lambda, L; m\rangle & = 2m|\lambda, L; m\rangle, \\ \vec{\mathcal{O}}^2|\lambda, L; m\rangle & = -L(1+L)|\lambda, L; m\rangle. \end{array} \right.$$

Eigenvectors from different angular momentum multiplets ($L \neq L'$) or with different Cartan eigenvalue ($m \neq m'$) are orthogonal

$$\langle \lambda, L; m||\lambda, L'; m'\rangle = |N_\lambda(L, m)|^2 \delta^{LL'}\delta_{mm'}.$$

The Fock norms are related to each other with the Casimir element and proportional to the norm of the eigenvector with highest weight:

$$\begin{aligned} \mathcal{O}_-^\star \otimes \mathcal{O}_- &= \vec{\mathcal{O}}^2 - \tfrac{\mathcal{O}_0}{2}(\tfrac{\mathcal{O}_0}{2} - 1) \\ \Rightarrow |N_\lambda(L, m-1)|^2 &= \Big(L(1+L) - m(m-1)\Big)|N_\lambda(L, m)|^2 \\ \Rightarrow |N_\lambda(L, m)|^2 &= (2L)!\tfrac{(L-m)!}{(L+m)!}|N_\lambda(L, L)|^2. \end{aligned}$$

Polar coordinates with the 2-sphere $\Omega^2 \cong \mathbf{SO}(3)/\mathbf{SO}(2)$,

$$\mathbb{R}^3 \ni \begin{pmatrix} x^1 \\ x^2 \\ x^3 \end{pmatrix} = r \begin{pmatrix} \cos\varphi\sin\theta \\ \sin\varphi\sin\theta \\ \cos\theta \end{pmatrix} \in \mathbb{R}_+ \times \Omega^2, \quad r \neq 0,$$

give the representation

$$\begin{aligned} \mathcal{O}^a &\cong \epsilon^{abc}x^b\tfrac{\partial}{\partial x^c}, \quad \mathcal{O}_\pm \cong i\sqrt{2}e^{\pm i\varphi}(\mp\tfrac{\partial}{\partial\theta} - i\cot\theta\tfrac{\partial}{\partial\varphi}), \quad \mathcal{O}_0 \cong -2i\tfrac{\partial}{\partial\varphi}, \\ -\vec{\mathcal{O}}^2 &= \tfrac{1}{1-\cos^2\theta}[(\tfrac{\partial}{\partial\cos\theta})^2 + (\tfrac{\partial}{\partial\varphi})^2] = \tfrac{1}{\sin^2\theta}[(\sin\theta\tfrac{\partial}{\partial\theta})^2 + (\tfrac{\partial}{\partial\varphi})^2] \\ &= \tfrac{1}{\sin\theta}\tfrac{\partial}{\partial\theta}\sin\theta\tfrac{\partial}{\partial\theta} + (\tfrac{1}{\sin\theta}\tfrac{\partial}{\partial\varphi})^2 = \tfrac{\partial^2}{\partial\theta^2} + \tfrac{1}{\tan\theta}\tfrac{\partial}{\partial\theta} + \tfrac{1}{\sin^2\theta}\tfrac{\partial^2}{\partial\varphi^2}. \end{aligned}$$

With $\lambda_+ \cong \frac{x_+}{r} = \frac{x^1+ix^2}{r} = e^{i\varphi}\sin\theta$ one obtains the spherical harmonics as eigenvectors

$$|L; m\rangle \cong \mathrm{Y}_m^L(\varphi, \theta) \sim \Big(i\sqrt{2}e^{-i\varphi}(\tfrac{\partial}{\partial\theta} - i\cot\theta\tfrac{\partial}{\partial\varphi})\Big)^{L-m}(e^{i\varphi}\sin\theta)^L.$$

The Cartesian product Hilbert space with, e.g., Hermite polynomials involving eigenstates for the 3-dimensional oscillator as a basis,

$$\begin{array}{rl} \overline{\text{FOCK}(\mathbb{C}^3)} \cong L_{d^3x}(\mathbb{R}^3,\mathbb{C}) \cong & \bigotimes_{a=1}^{3} L_{dx^a}(\mathbb{R},\mathbb{C}), \\ \text{Hilbert basis: } \{|k_1,k_2,k_3\rangle\} \cong & \{e^{-\frac{r^2}{2}}\mathrm{H}^{k_1}(x^1)\mathrm{H}^{k_2}(x^2)\mathrm{H}^{k_3}(x^3) \in \mathcal{S}(\mathbb{R}^3)\}, \end{array}$$

has a product form reflecting the polar decomposition

$$L_{d^3x}(\mathbb{R}^3,\mathbb{C}) \quad \cong L^2_{d^2\omega}(\Omega^2,\mathbb{C}) \otimes L^2_{dr}(\mathbb{R}_+,\mathbb{C}).$$

The completeness-relevant Dirac $\mathbb{R}^3$-distribution and scalar-product-relevant volume element are factorized correspondingly:

$$\delta(\vec{x}) = \delta(\vec{\omega})\tfrac{1}{r^2}\delta(r), \quad d^3x = d^2\omega \;\; r^2 dr.$$

The Hilbert space with the square integrable functions on the radial translation cone $\mathbb{R}_+$ (chapter "The Kepler Factor")

$$L^2_{dr}(\mathbb{R}_+,\mathbb{C}) = \{\psi \mid \textstyle\int_0^\infty r^2 dr \; |\psi(r)|^2 < \infty\}$$

is multiplied by the angular momentum Hilbert space with the square integrable functions on the 2-sphere surface, which is the orthogonal direct sum of the finite-dimensional scalar product spaces for each angular momentum L:

$$L^2_{d^2\omega}(\Omega^2,\mathbb{C}) = \{f \mid \textstyle\int d^2\omega \; |f(\varphi,\theta)|^2 < \infty\} = \bigoplus_{L=0,1,\dots} \mathbb{C}^{1+2L}(\Omega^2).$$

Its scalar product uses integration with the rotation-invariant measure $d^2\omega$ (not normalized):

$$\begin{array}{rl} \delta(\vec{\omega}) & = \delta(\varphi)\delta(\cos\theta) = \delta(\varphi)\frac{1}{\sin\theta}\delta(\theta), \\ \int d^2\omega & = \int_0^{2\pi} d\varphi \int_{-1}^{1} d(\cos\theta) = \int_0^{2\pi} d\varphi \int_0^{\pi} \sin\theta d\theta = 4\pi. \end{array}$$

The $\mathbb{C}^{1+2L}$-isomorphic subspace for the corresponding irreducible $\mathbf{SO}(3)$-representation has spherical harmonics, the traceless powers of the position space directions, as orthonormalized basic vectors

$$\mathrm{Y}^L_m : \Omega^2 \longrightarrow \mathbb{C}, \quad \tfrac{\vec{x}}{r} \cong (\varphi,\theta) \longmapsto \mathrm{Y}^L_m(\varphi,\theta), \quad r \neq 0.$$

The normalization of the highest-weight vector $|L,L\rangle \sim \left(\frac{x_+}{r}\right)^L$ of an irreducible representation

$$|N(L,L)|^2 = \textstyle\int d^2\omega \; |\frac{x_+}{r}|^{2L} = \int d^2\omega \; \sin^{2L}\theta = \frac{4\pi}{1+2L}\frac{(2^L L!)^2}{(2L)!}$$

gives the Rodrigues form for $\frac{|L;m\rangle}{N(L,m)}$,

$$\begin{array}{rl} \mathrm{Y}^L_m(\varphi,\theta) & = (-1)^m \overline{\mathrm{Y}^L_{-m}(\varphi,\theta)} \\ & = \sqrt{\frac{1+2L}{4\pi}\frac{(-1)^L}{2^L L!}}\sqrt{\frac{(L+m)!}{(L-m)!}}\left(\frac{e^{i\varphi}}{\sin\theta}\right)^m \left(\frac{\partial}{\partial\cos\theta}\right)^{L-m} \sin^{2L}\theta, \end{array}$$

with the lowest-dimensional examples

$$\mathrm{Y}^0_0(\varphi,\theta) = \sqrt{\tfrac{1}{4\pi}}, \qquad \begin{pmatrix}\mathrm{Y}^2_{\pm 2}\\ \mathrm{Y}^2_{\pm 1}\\ \mathrm{Y}^2_0\end{pmatrix}(\varphi,\theta) = \sqrt{\tfrac{5}{4\pi}}\begin{pmatrix}\sqrt{\frac{3}{8}}e^{\pm 2i\varphi}\sin^2\theta\\ \mp\sqrt{\frac{3}{2}}e^{\pm i\varphi}\sin\theta\cos\theta\\ \frac{3\cos^2\theta-1}{2}\end{pmatrix}.$$

$$\begin{pmatrix}\mathrm{Y}^1_{\pm 1}\\ \mathrm{Y}^1_0\end{pmatrix}(\varphi,\theta) = \sqrt{\tfrac{3}{4\pi}}\begin{pmatrix}\mp\frac{1}{\sqrt{2}}e^{\pm i\varphi}\sin\theta\\ \cos\theta\end{pmatrix},$$

The Hilbert basis properties of the spherical harmonics are

$$\begin{aligned}
&\{\mathrm{Y}^L_m \mid L=0,1,\dots;|m|\le L\} \text{ basis of } L^2_{d^2\omega}(\Omega^2,\mathbb{C}),\\
&\text{with}\left\{\begin{array}{rl}
\int d^2\omega\, \overline{\mathrm{Y}^{L'}_{m'}(\varphi,\theta)}\mathrm{Y}^L_m(\varphi,\theta) &= \delta^{LL'}\delta_{mm'},\\
\sum_{L=0}^{\infty}\sum_{m=-L}^{L}\overline{\mathrm{Y}^L_m(\varphi,\theta)}\mathrm{Y}^L_m(\varphi',\theta') &= \delta(\vec\omega-\vec\omega')\\
&= \delta(\varphi-\varphi')\frac{1}{\sin\theta}\delta(\theta-\theta')\\
\frac{1}{\sin^2\theta}[(\sin\theta\frac{\partial}{\partial\theta})^2+(\frac{\partial}{\partial\varphi})^2]\mathrm{Y}^L_m(\varphi,\theta) &= L(1+L)\mathrm{Y}^L_m(\varphi,\theta),\\
-i\frac{\partial}{\partial\varphi}\mathrm{Y}^L_m(\varphi,\theta) &= m\mathrm{Y}^L_m(\varphi,\theta).
\end{array}\right.
\end{aligned}$$

In the Schur orthonormality relations of the spherical harmonics (chapter "Harmonic Analysis") a normalization with the representation dimension is used:

$$[L]_m = \sqrt{\tfrac{4\pi}{1+2L}}\mathrm{Y}^L_m: \quad \int \tfrac{d^2\omega}{4\pi}\,\overline{[L']_{m'}(\varphi,\theta)}[L]_m(\varphi,\theta) = \tfrac{1}{1+2L}\delta^{LL'}\delta_{mm'}.$$

For the square integrable function classes on $[-1,1]\ni\zeta$ i.e., for the $\mathbf{SO}(2)$-functions, the *Legendre polynomials*

$$\begin{aligned}
\vec x,\vec y\ne 0,\ \zeta=\tfrac{\vec x\vec y}{|\vec x||\vec y|}=\cos\theta: &\quad \sum_{m=-L}^{L}\overline{\mathrm{Y}^L_m(\tfrac{\vec x}{|\vec x|})}\mathrm{Y}^L_m(\tfrac{\vec y}{|\vec y|}) = \sqrt{\tfrac{1+2L}{4\pi}}\mathrm{Y}^L_0(\varphi,\theta) = \mathrm{P}^L(\zeta),\\
\mathrm{P}^L(\cos\theta) &= \int_0^{2\pi}\tfrac{d\varphi}{2\pi}(\cos\theta+i\cos\varphi\sin\theta)^L,\\
\deg\mathrm{P}^L = L, &\quad \tfrac{d}{d\zeta}(\zeta^2-1)\tfrac{d}{d\zeta}\mathrm{P}^L(\zeta) = L(1+L)\mathrm{P}^L(\zeta),\\
\mathrm{P}^L(\zeta)=(-1)^L\mathrm{P}^L(-\zeta) &= \tfrac{1}{2^L L!}\tfrac{d^L}{d\zeta^L}(\zeta^2-1)^L = \tfrac{1}{2^L}\sum_{n=0}^{[\frac{L}{2}]}(-1)^n\tbinom{2L-2n}{L}\tbinom{L}{n}\zeta^{L-2n}\\
&= \left\{\begin{array}{ll}1, & L=0,\\ \zeta, & L=1,\\ \frac{3\zeta^2-1}{2}, & L=2,\end{array}\right.
\end{aligned}$$

are a Hilbert space basis

$$\begin{aligned}
&\{\mathrm{P}^L \mid L=0,1,\dots\} \text{ basis of } L^2_{d\zeta}([-1,1],\mathbb{R}) \cong L^2_{d\zeta}(\mathbf{SO}(2))\\
&\text{with}\left\{\begin{array}{rl}
\int_{-1}^{1}\frac{d\zeta}{2}\,\mathrm{P}^L(\zeta)\mathrm{P}^{L'}(\zeta) &= \frac{1}{1+2L}\delta^{LL'},\\
\sum_{L=0}^{\infty}(1+2L)\,\mathrm{P}^L(\zeta)\mathrm{P}^L(\zeta') &= \delta(\frac{\zeta-\zeta'}{2}),\\
\sum_{L=0}^{\infty}\mathrm{P}^L(\zeta)\lambda^L = \frac{1}{\sqrt{1-2\lambda\zeta+\lambda^2}} &= \frac{1}{|1-\lambda e^{i\theta}|} = \frac{|\vec x|}{|\vec x-\vec y|},\quad \lambda=|\frac{\vec y}{\vec x}|.
\end{array}\right.
\end{aligned}$$

The azimuthal part in the spherical harmonics

$$L^2_{d\varsigma}([-1,1],\mathbb{R}) \subset L^2_{d^2\omega}(\Omega^2,\mathbb{C}) \subset L^2_{d\varsigma}([-1,1],\mathbb{R}) \otimes L^2_{d\varphi}(\mathbf{U}(1),\mathbb{C})$$

uses the square integrable function classes on the group $\mathbf{U}(1) \cong [0, 2\pi]$ with the irreducible $\mathbf{U}(1)$-representations as a Hilbert space basis:

$$\mathbf{irrep}\,\mathbf{U}(1) \cong \{\varphi \longmapsto e^{im\varphi} \mid m \in \mathbb{Z}\} \text{ basis of } L^2_{d\varphi}(\mathbf{U}(1),\mathbb{C}),$$

$$\text{with} \left\{ \begin{array}{ll} \int_0^{2\pi} \frac{d\varphi}{2\pi}\, e^{im\varphi}e^{-im'\varphi} &= \delta_{mm'}, \\ \sum_{m\in\mathbb{Z}} e^{im\varphi}e^{-im\varphi'} &= \delta(\frac{\varphi-\varphi'}{2\pi}), \\ -i\frac{d}{d\varphi}e^{im\varphi} &= me^{im\varphi}. \end{array} \right.$$

8.7 The Irreducible Nonabelian Form for a Noncompact Time Representation

The irreducible nonabelian form on the Fermi quantum algebra $\mathbf{Q}_+(\mathbb{C}^4)$ can be used with definite conjugation for a Pauli representation of spin $\mathbf{SU}(2) \subset \mathbf{U}(2)$ or with indefinite $\mathbf{U}(1,1)$-conjugation for a faithful time representation (chapter "Time Representations").

The indefinite $\mathbf{U}(1,1)$-conjugation $\times$ of a 2-dimensional basic space dimension $V_{\text{doub}} \cong \mathbb{C}^2$ is appropriate for the complex quartet

$$V = \mathbb{C}\mathrm{g}, \quad V^T = \mathbb{C}\mathrm{b}^\times, \quad \overline{V} = \mathbb{C}\mathrm{g}^\times, \quad \overline{V}^T = \mathbb{C}\mathrm{b}.$$

The extension to the Fermi quantum algebra $\mathbf{Q}_+(\mathbf{V}_{\text{doub}}) \cong \mathbb{C}^{16}$

$$\begin{array}{l} \mathbf{Q}_+(\mathbb{C}^4): \ \{\mathrm{b}^\times,\mathrm{g}\} = 1 = \{\mathrm{g}^\times,\mathrm{b}\}, \ \{\mathrm{g}^\times,\mathrm{g}\} = 0 = \{\mathrm{b}^\times,\mathrm{b}\}, \\ \qquad\qquad \langle\ \rangle_{\mathrm{H}} : \mathbf{Q}_+(\mathbb{C}^4) \longrightarrow \mathbb{C}, \end{array}$$

is normalized by the value for the basic space identity

$$\begin{array}{l} \mathrm{id}_{V_{\text{doub}}} = \mathrm{g}\otimes\mathrm{b}^\times + \mathrm{b}\otimes\mathrm{g}^\times, \quad \mathrm{id}_{V^T_{\text{doub}}} = \mathrm{b}^\times\otimes\mathrm{g} + \mathrm{g}^\times\otimes\mathrm{b}, \\ \langle \mathrm{g}\otimes\mathrm{b}^\times + \mathrm{b}\otimes\mathrm{g}^\times\rangle_{\mathrm{H}} = 1 = \langle \mathrm{g}^\times\otimes\mathrm{b} + \mathrm{b}^\times\otimes\mathrm{g}\rangle_{\mathrm{H}} = \frac{1}{2}\,\mathrm{tr}\begin{pmatrix} 1 & 0 \\ 0 & 1 \end{pmatrix}. \end{array}$$

Since trivial for traceless basic endomorphisms, e.g.,

$$\langle \mathrm{g}\otimes\mathrm{b}^\times - \mathrm{b}\otimes\mathrm{g}^\times\rangle_{\mathrm{H}} = \frac{1}{2}\,\mathrm{tr}\begin{pmatrix} 1 & 0 \\ 0 & -1 \end{pmatrix} = 0,$$

one obtains for the basic endomorphism

$$\begin{array}{l} \frac{1}{2} = \langle \mathrm{g}\otimes\mathrm{b}^\times\rangle_{\mathrm{H}} = \langle \mathrm{b}\otimes\mathrm{g}^\times\rangle_{\mathrm{H}} = \langle \mathrm{g}^\times\otimes\mathrm{b}\rangle_{\mathrm{H}} = \langle \mathrm{b}^\times\otimes\mathrm{g}\rangle_{\mathrm{H}}, \\ 0 = \langle \mathrm{g}\otimes\mathrm{g}^\times\rangle_{\mathrm{H}} = \langle \mathrm{g}^\times\otimes\mathrm{g}\rangle_{\mathrm{H}} = \langle \mathrm{b}\otimes\mathrm{b}^\times\rangle_{\mathrm{H}} = \langle \mathrm{b}^\times\otimes\mathrm{b}^\times\rangle_{\mathrm{H}}, \\ \text{ghosts: } 0 = \langle \mathrm{g}|\mathrm{g}\rangle_{\mathrm{H}} = \langle \mathrm{g}^\times|\mathrm{g}^\times\rangle_{\mathrm{H}} = \langle \mathrm{b}|\mathrm{b}\rangle_{\mathrm{H}} = \langle \mathrm{b}^\times|\mathrm{b}^\times\rangle_{\mathrm{H}}, \end{array}$$

and hence the nonabelian form values of the basic (anti-)commutators

$$\begin{array}{lll} 1 & = \langle\{\mathrm{b}^\times,\mathrm{g}\}\rangle_{\mathrm{H}} & = \langle\{\mathrm{g}^\times,\mathrm{b}\}\rangle_{\mathrm{H}}, \\ 0 & = \langle[\mathrm{b}^\times,\mathrm{g}]\rangle_{\mathrm{H}} & = \langle[\mathrm{g}^\times,\mathrm{b}]\rangle_{\mathrm{H}}. \end{array}$$

The nonabelian form is indefinite on the algebra $\mathbf{Q}_+(\mathbf{V}_{\mathrm{doub}})$, e.g.,

$$\langle (\mathrm{b}^\times \pm \mathrm{g}^\times) \otimes (\mathrm{b} \pm \mathrm{g}) \rangle_{\mathrm{H}} = \langle \mathrm{b} \pm \mathrm{g} | \mathrm{b} \pm \mathrm{g} \rangle_{\mathrm{H}} = \pm 1.$$

The nondecomposable faithfully representated time is a noncompact 1-dimensional $\mathbf{U}(1,1)$-subgroup

$$\begin{aligned} &H_1 = \mu I + \nu N = \mu \tfrac{[\mathrm{g},\mathrm{b}^\times]_{-\epsilon} + [\mathrm{b},\mathrm{g}^\times]_{-\epsilon}}{2} + \nu \mathrm{g} \otimes \mathrm{g}^\times, \ \mu, \nu \in \mathbb{R}, \\ &\langle I \rangle_{\mathrm{H}} = 0, \ \ \langle N \rangle_{\mathrm{H}} = 0. \end{aligned}$$

The quantum algebra is decomposable (chapter "Quantum Algebra") into vector spaces V_k with nildimension k with respect to the adjoint action of the nil-Hamiltonian

$$\begin{array}{ll} a \text{ has nildimension } k & \iff (\,\mathrm{ad}\, N)^k(a) \neq 0, \ (\,\mathrm{ad}\, N)^{k+1}(a) = 0, \\ \mathbf{Q}_+(\mathbb{C}^4) = & V_0 \oplus V_1 \oplus V_2, \\ V_0 \cong \mathbb{C}^{10} : & \left\{ \begin{array}{l} 1, \ I - 1, \ (I-1) \otimes (I-1), \ N\, crg, \ \mathrm{g}^\times, \ \mathrm{g} \otimes I, \ I \otimes \mathrm{g}^\times, \\ \mathrm{g} \otimes \mathrm{b}, \ \mathrm{b}^\times \otimes \mathrm{g}^\times, \end{array} \right. \\ V_1 \cong \mathbb{C}^5 : & \mathrm{g} \otimes \mathrm{b}^\times - \mathrm{b} \otimes \mathrm{g}^\times, \ \mathrm{b}, \ \mathrm{b}^\times, \ \mathrm{b} \otimes I, \ I \otimes \mathrm{a}^\times, \\ V_2 \cong \mathbb{C} : & \mathrm{b} \otimes \mathrm{b}^\times. \end{array}$$

The 10-dimensional subalgebra V_0 with trivial nildimension contains all time translation eigenvectors. Thereon the nonabelian form is definite, i.e., a state

$$\langle a^\times \otimes a \rangle_{\mathrm{H}} \geq 0 \text{ for all } a \in V_0 = \mathrm{INV}_N \mathbf{Q}_+(\mathbb{C}^4) \cong \mathbb{C}^{10}.$$

As seen with anticommutators and traces, the 10 vectors above constitute an orthogonal V_0-basis. Up to the scalars, they are all ghosts, i.e., the associate $\mathbb{C}$-isomorphic Hilbert space given by the left ideal classes with $(\mathrm{g}, \mathrm{g}^\times)$,

$$\mathrm{INV}_N \mathbf{Q}_+(\mathbb{C}^4) \Big/ \mathrm{INV}_N \mathbf{Q}_+(\mathbb{C}^4)\mathrm{g} + \mathrm{INV}_N \mathbf{Q}_+(\mathbb{C}^4)\mathrm{g}^\times \cong \mathbb{C},$$

has only the state vector for the unit $1 \in \mathbf{Q}_+(\mathbb{C}^4)$ as basis.

8.8 Summary

A symmetric (Hermitian) algebra form defines an inner product. A positive algebra form (a state, if normalized) leads to a scalar product and a Hilbert space, constituted by operation classes. With a time representation, an eigenvector-spanned subalgebra with a ground state, a time invariant state with monotonic action of the Hamiltonian, allows a probability interpretation for the operation algebra with state vectors.

The trace forms of finite-dimensional vector space endomorphisms induce abelian (Fock) and nonabelian quantum algebra forms.

The irreducible Fock state of the oscillator quantum algebras $\mathbf{Q}_\epsilon(\mathbb{C}^2)$ comes with $\mathbf{U}(1)$-scalar product of the complex numbers; it is appropriate for the irreducible time representations, i.e., for time eigenvectors with nontrivial probability normalization ("objects," particles). Fock states give trivial values for traceless endomorphisms. The associate scalar product spaces are, as vector spaces, isomorphic to the Grassmann (for Fermi) and polynomial (for Bose) algebra of the basic vector space.

The normalized irreducible nonabelian form exists only for the Fermi quantum algebra $\mathbf{Q}_+(\mathbb{C}^4)$. It gives nontrivial values for traceless, not nilpotent, endomorphisms. For a positive conjugation of the basic vector space, it gives positive values for the basic space endomorphisms and their enveloping algebra elements. With indefinite $\mathbf{U}(1,1)$-inner product of the basic space, it is appropriate for noncompact time representations.

MATHEMATICAL TOOLS

8.9 Algebra Forms

A form ω of the complex algebra A, eigenform for an element a, i.e., $\omega(ab) = \omega(ba) = \omega(a)\omega(b)$ for all $b \in A$, is an algebra morphism on the polynomials $\mathbb{C}[a]_A$:

$$\omega : \mathbb{C}[a]_A \longrightarrow \mathbb{C}, \ \omega(a^n) = [\omega(a)]^n.$$

The form distinguishes a root α_{j_0} for the algebra element a with minimal polynomial $p_a(X)$ and the associated projector

$$\begin{array}{l} p_a(X) = \prod_{j=1}^{m}(X - \alpha_j)^{N_j}, \ 1_A = \sum_j \mathcal{P}_j(a), \ \ a = \sum_j [\alpha_j \mathcal{P}_j(a) + \mathcal{N}_j(a)], \\ p_a(a) = 0 \Rightarrow \omega(a) = \alpha_{j_0} \Rightarrow \left\{ \begin{array}{l} \omega(\mathcal{P}_j(a)b) = \delta_{jj_0}\omega(b) = \omega(b\mathcal{P}_j(a)), \\ \omega(\mathcal{N}_j(a)b) = 0 = \omega(b\mathcal{N}_j(a)) \text{ for all } b \in A. \end{array} \right. \end{array}$$

An invariant form of an algebra A,

$$\omega : A \longrightarrow \mathbb{K}, \ \ \omega([a,b]) = 0 \text{ for all } a, b \in A,$$

induces for an A-representation $\mathcal{D} : L \longrightarrow A$ of a Lie algebra L an invariant form ω_A of the Lie algebra

$$\omega_A : L \longrightarrow \mathbb{K}, \ \ \omega_A(l) = \omega(\mathcal{D}(l)) \Rightarrow \omega_A([l,m]) = 0.$$

With the extension of the algebra form ω on its tensor algebra $\bigotimes A$,

$$\langle a_1 \otimes \cdots \otimes a_k \rangle_\omega = \omega(a_1 \cdots a_k), \ \ a \in A,$$

one obtains an invariant form of the enveloping algebra $\mathbf{E}(L)$,

$$\begin{array}{ccc} L & \stackrel{\iota}{\longrightarrow} & \mathbf{E}(L) \\ \omega_A \downarrow & & \downarrow \langle\ \rangle_{\omega_A}, \\ \mathbb{K} & \underset{\mathrm{id}_\mathbb{K}}{\longrightarrow} & \mathbb{K} \end{array} \quad \begin{array}{l} \langle 1 \rangle_{\omega_A} = 1, \\ \langle l \rangle_{\omega_A} = \omega_A(l) = \langle \mathcal{D}(l) \rangle_\omega, \ \ l \in L, \\ \langle l_1 \otimes \cdots \otimes l_k \rangle_{\omega_A} = \langle \mathcal{D}(l_1) \circ \cdots \circ \mathcal{D}(l_k) \rangle_\omega, \\ \langle [l,a] \rangle_{\omega_A} = 0, \ \ a \in \mathbf{E}(L). \end{array}$$

8.10 Topologies

Topology works primarily with subsets $S \subseteq T$ of a set, not with the elements $x \in T$.

A nonempty subset family $\underline{T} \subseteq 2^T$ of a set T, that contains finite intersections and all unions of its members is called a *topology* with the elements $O \in \underline{T}$ called *open*. The complements $C_T O$ are *closed*. The *closure (interior)* of a set $S \subseteq T$ is the smallest closed (largest open) set with $\overline{S} \supseteq S \supseteq \overset{o}{S}$. S is *dense in* T for $\overline{S} = T$. A *basis of the topology* $\underline{T}$ is a subset family from which all open sets arise as unions. A topology with countable basis is called *separable*. A *subbasis (generating system) of the topology* $\underline{T}$ is a subset whose finite intersections yield a $\underline{T}$-basis. A set T can have more than one topology. The topologies are ordered by the inclusion

$$\text{coarsest (trivial) } \{\emptyset, T\} \subseteq \ldots \subseteq \underline{T} \subseteq \ldots \subseteq 2^T \text{ finest (discrete) topology.}$$

In general, finite sets are endowed with the discrete topology.

A mapping $f : S \longrightarrow T$ between two sets (*"element mapping"*) induces a mapping for the subsets in the reversed direction (*"subset mapping"*), stable for the empty and whole sets as for intersections and unions:

$$\begin{array}{l} f^{-1}[\ ] : 2^T \longrightarrow 2^S, \quad X \longmapsto f^{-1}[X], \\ f^{-1}[\emptyset] = \emptyset, \quad f^{-1}[T] = S, \\ f^{-1}[X \cap Y] = f^{-1}[X] \cap f^{-1}[Y], \ f^{-1}[X \cup Y] = f^{-1}[X] \cup f^{-1}[Y]. \end{array}$$

Therefore the morphisms $\mathbf{top}(S, T)$ of the *category of topological spaces* $\underline{\mathbf{top}}$ are defined by those mappings $\{f : S \longrightarrow T\}$ for which preimages of open sets are open, called *continuous mappings.* A continuous mapping $f : S \longrightarrow T$ remains continuous for a finer S-topology and a coarser T-topology.

The preimages $f : S \longrightarrow T$ of open T-sets define the *initial* topology $f^{-1}[\underline{T}]$ for a set S, the coarsest topology that makes f continuous, e.g., product topologies via continuous projections. The images $f : T \longrightarrow S$ of open T-sets define the *final* topology for S by those S-sets X for which $f^{-1}[X] \in \underline{T}$, the finest topology that makes f continuous. For a point $x \in T \in \underline{\mathbf{top}}$ all supsets of open sets $x \in O \in \underline{T}$ constitute its *neighborhoods* $\mathcal{N}(x)$. A *Hausdorff topology (space)* has disjoint neighborhoods for any pair of different points, e.g., the discrete topology. A topological space T is *quasi-compact* if every open T-covering contains a finite open T-covering (Heine-Borel). T is *compact* if Hausdorff and quasi-compact. A Hausdorff space with a compact neighborhood for each point is *locally compact.* A *filter* $\mathcal{F}$ on a set S, not necessarily a topological space, is a nonempty set of S-subsets without the empty set, stable under finite sections and containing with $F \in \mathcal{F}$ also all supersets $G \supseteq F$. A *filter basis* $\mathcal{B}$ on a set S is a nonempty set of S-subsets without the empty set that contains an element in each finite section, i.e., $B_3 \subseteq B_1 \cap B_2$ with $B_i \in \mathcal{B}$. The supersets of a filterbasis constitute a filter. The neighborhoods $\mathcal{N}(x)$ for $x \in T \in \underline{\mathbf{top}}$ are a filter with the open neighborhoods $\mathcal{O}(x)$ a filter basis.

A filter $\mathcal{F}$ on a topological space containing the neighborhood filter $\mathcal{F} \supseteq \mathcal{N}(x)$ *converges* to x, $x \in \lim \mathcal{F}$. In a Hausdorff space a filter cannot have more than one limit point (possibly none). A series $\mathbb{N} \longrightarrow S$ defines by its tails $F_k = \{x_{k+n} \mid n \in \mathbb{N}\}$, $k \in \mathbb{N}$, a filter basis that allows the convergence definition for series. Convergence persists in a coarser topology. The finer the topology, the fewer convergent filters. In the discrete topology the only convergent filters are the neighborhood filters. A filter has more limit points in a coarser topology

$$x \in T \in \underline{\mathbf{top}}: \ \underline{T}_c \subseteq \underline{T}_f \Rightarrow \mathcal{N}_c(x) \subseteq \mathcal{N}_f(x) \Rightarrow \lim_c \mathcal{F} \supseteq \lim_f \mathcal{F}$$
$$\Big(x \in \lim_f \mathcal{F} \iff \mathcal{F} \supseteq \mathcal{N}_f(x)\Big) \Rightarrow \Big(x \in \lim_c \mathcal{F} \iff \mathcal{F} \supseteq \mathcal{N}_c(x)\Big).$$

According to Hausdorff, the association of a filter $\mathcal{N}(x)$, defining neighborhoods for any point of a set $x \in S$, with $x \in N$ for all $N \in \mathcal{N}(x)$ and a neighborhood $U \in \mathcal{N}(x)$ for each $N \in \mathcal{N}(x)$ that $N \in \mathcal{N}(y)$ for all its points $y \in U$ can be used to define a topology $\underline{T}$: A set O is open if it is neighborhood for all its points $x \in O$.

8.10.1 Metric Spaces

Uniform structures are special topological structures in which the concepts uniformously continuous, Cauchy sequence, Cauchy filter, Cauchy completeness, etc. have an adequate formulation. Metric spaces constitute an important, not particularly abstract, subclass of uniform spaces. A *premetric* of a set M,

$$d : M \times M \longrightarrow \mathbb{R} \begin{cases} \textit{reflexive:} & d(x,x) = 0, \\ \textit{symmetric:} & d(x,y) = d(y,x), \\ \textit{subadditive:} & d(x,z) \leq d(x,y) + d(y,z) \\ & \textit{(triangle inequality)}, \end{cases}$$

is (with $z = x$) necessarily positive, $d(x,y) \geq 0$. For a *metric* one requires in addition

$$\textit{strictly positive:} \ d(x,y) = 0 \iff x = y.$$

A premetric d_c is coarser (and the related topology $\underline{M}_c \subseteq \underline{M}_f$ too) than a premetric d_f iff for each $\alpha > 0$ there exists $\beta > 0$ such that $d_c(x,y) \leq \alpha$ implies $d_f(x,y) \leq \beta$, e.g., if there exists a constant k with $d_c \leq k d_f$.

A premetric of a set with operator group G is invariant for

$$d(g \bullet x, g \bullet y) = d(x,y), \quad g \in G.$$

The spheres around a point $x \in M$,

$$O_{\frac{1}{n}}(x) = \{y \in M \mid d(x,y) < \tfrac{1}{n}\}, \ n \in \mathbb{N},$$

constitute a countable basis for the open x-neighborhoods defining a topology, which for a metric is Hausdorff. With a countable dense subset in M, the topology is separable, e.g., the Hausdorff topology on finite-dimensional vector spaces $V \cong \mathbb{K}^n$.

A set M with premetric defines with $x \; d \; y \iff d(x,y) = 0$ an equivalence relation. The classes M/d carry a metric with the meaningful definition $d([x],[y]) = d(x,y)$. The Hausdorff space M/d carries the final quotient topology.

A *uniformly continuous* mapping for two metric spaces $f : M \longrightarrow N$, i.e., for all $\epsilon > 0$ there exists $\delta > 0$ such that $d(x,y) < \delta$ implies $d(f(x), f(y)) < \epsilon$, independent of $x, y \in M$, is a morphism for the *category of metric spaces* **metr**. An *isometry* satisfies

$$d(f(x), f(y)) = d(x,y) \text{ for all } x, y \in M.$$

It is injective. A surjective isometry is a metric isomorphism. A continuous mapping does not have to be uniformly continuous.

If each Cauchy series in a metric space $\hat{M}$ has a limit, $\hat{M}$ is called *(Cauchy) complete*. A space M with metric (more generally, a uniform space) can be completed by the set of its Cauchy series $C(M) = \{(x_n)_{n\in\mathbb{N}} \mid \text{Cauchy series in } M\}$. With the induced premetric

$$\tilde{d} : C(M) \times C(M) \longrightarrow \mathbb{R}, \quad \tilde{d}\Big((x_n)_{n\in\mathbb{N}}, (y_m)_{m\in\mathbb{N}}\Big) = \lim_{n\in\mathbb{N}} d(x_n, y_n),$$

the equivalence classes $\mathbf{C}(M) = C(M)/\tilde{d}$ constitute the *(Cauchy) completion of M in the metric d.*

An original $x \in M$ can be taken as representative $(x_n = x)_{n\in\mathbb{N}}$ for the class of all in M to x converging Cauchy series, this embeds M in its completion $\mathbf{C}(M)$. There exists an isometric injection of M into $\mathbf{C}(M)$ with dense image

$$\iota : M \longrightarrow \mathbf{C}(M), \quad \overline{\iota[M]} = \mathbf{C}(M).$$

The completion $\mathbf{C}(M)$ is unique up to isometry. $(\mathbf{C}(M), \iota)$ is the solution for the universal mapping problem of the metric in complete metric spaces: Each uniformly continuous mapping F from the metric space M to a complete one $\hat{S}$ is factorizable in the isometric embedding ι and a unique uniformly continuous mapping $\tilde{F}$:

$$\begin{array}{ccc} M & \xrightarrow{\iota} & \mathbf{C}(M) \\ {\scriptstyle F}\downarrow & & \downarrow{\scriptstyle \tilde{F}}, \\ \hat{S} & \xrightarrow[\mathrm{id}_{\hat{S}}]{} & \hat{S} \end{array} \qquad \begin{array}{l} M \text{ metric space}, \\ \mathbf{C}(M), \hat{S} \text{ complete metric spaces}, \\ \iota, F, \tilde{F} \text{ uniformly continuous.} \end{array}$$

Hence uniformly continuous mappings can be uniquely extended to their completions with the covariant idempotent *completion functor* to complete metric spaces:

$$\underline{\mathbf{metr}} \longrightarrow \underline{\hat{\mathbf{metr}}}, \qquad \begin{array}{ccc} M & & \mathbf{C}(M) \\ {\scriptstyle f}\downarrow & \longmapsto & \downarrow{\scriptstyle \mathbf{C}(f)}. \\ N & & \mathbf{C}(N) \end{array}$$

$$\mathbf{C} \circ \mathbf{C} = \mathbf{C},$$

A Cauchy filter (sequence) in a uniform topology remains one in a coarser uniform topology:

$$\underline{M}_c \subseteq \underline{M}_f \Rightarrow \mathcal{F} \text{ Cauchy for } \underline{M}_f \text{ is Cauchy for } \underline{M}_c.$$

8.10.2 Topological Vector Spaces

The topology of a vector space $V \in \underline{\mathbf{vec}}_{\mathbb{K}}$ has to be compatible with addition and scalar multiplication. It is defined by the neighborhood filter of the neutral element $\mathcal{N}(v) = v + \mathcal{N}(0)$. Topological vector spaces have a uniform topology. Continuous linear mappings are uniformly continuous; they are the morphisms for the *category of topological vector spaces* $\underline{\mathbf{tvec}}_{\mathbb{K}}$.

A linearly independent subset $B \subseteq V$ whose finite linear combinations are dense in V is called a *basis for the topological vector space*

$$V = \overline{\bigoplus_{e \in B} \mathbb{K}e}.$$

A finite-dimensional vector space $V \cong \mathbb{K}^n$ carries a unique Hausdorff topology, definable as final topology for the n projections on the components for any basis $V \ni v \longmapsto v_i = \langle \check{e}_i, v \rangle \in \mathbb{K}$.

The quotient $V/\overline{\{0\}}$ of a topological vector space with the closure of the neutral element is Hausdorff in the final topology.

The *topological dual* with the continuous forms ($\mathbb{K}$-carries the natural topology)

$$V' = \{\omega \in V^T \mid \omega : V \longrightarrow \mathbb{K} \text{ continuous}\}$$

coincides with the algebraic one in the finite-dimensional Hausdorff case. A priori, the topological dual has no topology.

The transposition of a continuous linear mapping $f \in \mathbf{tvec}_{\mathbb{K}}(V, W)$ can be restricted to the topological duals $f^T \in \mathbf{vec}_{\mathbb{K}}(W', V')$, e.g., for a closed subspace

$$f : V \longrightarrow W \Rightarrow f^T : W' \longrightarrow V', \quad V_1 \subseteq V_2 \Rightarrow V_2' \subseteq V_1'.$$

The image of f is dense iff the transpose is injective:

$$\overline{f[V]} = W \iff \text{kern} f^T = \{0\}.$$

For example, the injection of a metric space M into its Cauchy completion involves the injection of the topological dual of the completion into the topological dual M'

$$M \hookrightarrow \overline{M} = \mathbf{C}(M), \quad \mathbf{C}(M)' \hookrightarrow M'.$$

The natural $\mathbb{K}$-topology can be used to define the *weak topology for a vector space* W *in duality with a topological vector space* V via a bilinear form

$$\gamma(\ , \ ) : W \times V \longrightarrow \mathbb{K}$$

by the initial (coarsest) topology that makes all induced linear mappings continuous:

$$v \in V, \quad \gamma(\ ,v) : W \longrightarrow \mathbb{K}.$$

This defines the weak topology on the algebraic and topological dual using the dual product.

8.11 Ordered Vector Spaces

In a preordered vector space $V \in \underline{\mathbf{vec}}_{\mathbb{K}}$ with forward cone V_+ containing the positive vectors

$$V_+ = \{v \succeq 0\}, \quad \begin{cases} v \succeq 0, \quad \alpha \geq 0 \Rightarrow \alpha v \succeq 0, \\ v, w \succeq 0 \Rightarrow v + w \succeq 0, \end{cases}$$

the order trivial elements constitute a vector subspace

$$V_0 = V_+ \cap V_- = \{n \in V \mid n \succeq 0 \text{ and } n \preceq 0\} \in \underline{\mathbf{vec}}_{\mathbb{K}}.$$

The quotient vector space V/V_0 carries an order.

With respect to a basis $\{e^j\}_{j \in I}$ an order can be defined for the vector space $V \in \underline{\mathbf{vec}}_{\mathbb{K}}$ by positivity of the coefficients:

$$v = \alpha_k e^k \succeq 0 \iff \alpha_k \geq 0 \text{ for all } \alpha_k \in \mathbb{K}.$$

With respect to a generating system (not necessarily linearly independent) this definition leads, in general, only to a preorder

$$v \succeq 0 \iff \text{There exists a combination } v = \alpha_k e^k \text{ with } \alpha_k \geq 0.$$

A *form* $\omega \in V^T$ *of a preordered vector space* V is called positive for

$$\omega(v) \geq 0 \text{ for all } v \succeq 0 \iff \omega \succeq 0.$$

A preorder of a vector space V defined by a generating system gives rise to an order of the dual space V^T.

A *(pre)ordered algebra* has to be (pre)ordered as a vector space.

In a complex algebra A with unit 1_A and conjugation the *symmetric domains* a^*a for $a \in A$ define a generating system for the isomorphic real vector subspaces $A_\pm$:

$$\begin{aligned}
&A = A_+ \oplus A_-, \quad A_+ = iA_-, \\
&A_+ = \{a = a^*\} = \{\sum_{k=1}^n \epsilon_k c_k^* c_k \mid \epsilon_k = \pm 1, \; c_k \in A\} \in \underline{\mathbf{vec}}_{\mathbb{R}}, \\
&\text{since } a = a^* \Rightarrow 4a = (a + 1_A)(a^* + 1_A) - (a - 1_A)(a^* - 1_A), \\
&A_- = \{a = -a^*\} = \{i \sum_{k=1}^n \epsilon_k c_k^* c_k \mid \epsilon_k = \pm 1, \; c_k \in A\} \in \underline{\mathbf{vec}}_{\mathbb{R}}, \\
&\text{since } a = -a^* \Rightarrow 4a = i(a + i)(a^* - i) - i(a - i)(a^* + i).
\end{aligned}$$

Hence all algebra elements have a not necessarily unique domain representation that defines the *natural preorder* of the algebra $A \in *\underline{\mathbf{aag}}_{\mathbb{C}}$:

$$a \succeq 0 \iff \text{There exists a domain combination } a = \sum_{k=1}^{n} c_k^* c_k \in A_+, \ c_k \in A.$$

The dual space A^T contains the real vector subspace with the reflection-compatible linear forms

$$\{\omega \in A^T \mid \omega(a) = \overline{\omega(a^*)} = \omega^*(a) \text{ for all } a \in A\} \in \underline{\mathbf{vec}}_{\mathbb{R}},$$

and the forward cone of the *positive (monotonic) linear forms*

$$\omega \succeq 0 \iff \omega(a^* a) \geq 0 \text{ for all } a \in A.$$

In contrast to the algebra A, the linear forms A^T always carry an order via the positive linear forms.

A *convex set* contains with any two vectors their segment

$$\{\alpha v + (1-\alpha) w \mid 0 \leq \alpha \leq 1\} \subseteq C \subseteq V \in \underline{\mathbf{vec}}_{\mathbb{K}},$$

e.g., the closed balls $\mathbf{O}^{1+s}$ in Euclidean $\mathbf{SO}(1+s) \vec{\times} \mathbb{R}^{1+s}$ or the filled-up closed future hyperboloids $\mathbf{Y}^{1+s}$ in Minkowski $\mathbf{SO}_0(1,s) \vec{\times} \mathbb{R}^{1+s}$. A vector in a convex set is *extremal* if it does not lie inside a segment, it is not combinable with $0 < \alpha < 1$. The sphere $\Omega^s = \partial \mathbf{O}^{1+s}$ and the hyperboloid $\mathcal{Y}^s = \partial \mathbf{Y}^{1+s}$ are the extremal points of $\mathbf{O}^{1+s}$ and $\mathbf{Y}^{1+s}$.

8.11.1 Order Topologies

A preordered abelian group V, especially a vector space, is called *directed (filtered)* if any two elements have a larger one:

$$v, w \in V \Rightarrow \text{There exists } a \in V \text{ with } v, w \preceq a.$$

In this case V is also directed in the opposite direction. For a nontrivial directed group $V \neq \{0\}$, there exists a strictly positive and negative element $a \succ 0, -a \prec 0$.

V is called *topologically filtered* if any two strictly positive elements have a strictly positive smaller one:

$$v, w \succ 0 \Rightarrow \text{There exists } c \succ 0 \text{ with } c \preceq v, w.$$

A topologically filtered group $V \neq \{0\}$ carries the following *order topology*: The strictly positive elements define a filter basis $\mathcal{O}(0)$ for the *open neighborhoods of* 0:

$$\begin{aligned} &a \succ 0, \quad O_a(0) = \{v \in V \mid -a \prec v \prec a\} \neq \emptyset, \text{ since } 0 \in O_a(0), \\ &\mathcal{O}(0) = \{O_a(0) \mid a \succ 0\} \neq \emptyset, \\ &O_a(0), O_b(0) \in \mathcal{O}(0) \Rightarrow \left\{ \begin{array}{l} \text{There exist } 0 \preceq c \preceq a, b \\ \text{and } O_c(0) \subseteq O_a(0) \cap O_b(0). \end{array} \right. \end{aligned}$$

On a preordered abelian group V a *premodulus (prenorm)* is defined by a mapping on the positive elements that satifies the triangle (Minkowski) inequality:

$$| \; | : V \longrightarrow V, \quad \begin{cases} \text{for a group:} & |v| = |-v| \succeq 0, \; |0| = 0, \\ \text{for a vector space:} & \\ \textit{abolute homogeneous:} & |\alpha v| = |\alpha||v|, \\ \text{for both:} & \\ \textit{subadditive:} & |v+w| \preceq |v| + |w|. \end{cases}$$

In the case of a vector space V positivity is a consequence of subadditivity. For strict positivity, $|v| \succ 0 \iff v \neq 0$, one calls the mapping a *modulus (norm)*. With the subgroup $V_0 = \{n \in V \bigm| |n| = 0\}$ the classes V/V_0 have the modulus $|[v]| = |v|$. For a modulus (norm) the topology is Hausdorff.

8.12 Normed Vector Spaces

Metric and vector space structures join in normed vector spaces. A vector space $V \in \underline{\mathbf{vec}}_{\mathbb{K}}$ carries a *prenorm* with a real-valued mapping

$$\| \; \| : V \longrightarrow \mathbb{R} \quad \begin{cases} \textit{(absolute) homogeneous:} & \| \alpha v \| = |\alpha| \, \| v \|, \\ \textit{subadditive:} & \| v + w \| \leq \| v \| + \| w \| \,. \end{cases}$$

Scalar multiplication and vector addition are continuous in the prenorm topology. A prenorm has as "its" premetric, from which it may inherit structures, e.g., positivity:

$$d : V \times V \longrightarrow \mathbb{R}, \quad d(v,u) = \| v - u \| \,.$$

If a premetric of a vector space V is invariant under addition and absolutely invariant under scalar multiplication, it defines a prenorm

$$\left. \begin{array}{l} d(v+w, u+w) = d(v,u), \\ d(\alpha v, \alpha u) = |\alpha| d(v,u) \end{array} \right\} \Rightarrow \| v \| = d(v, 0).$$

A strictly positive prenorm

$$\| v \| > 0 \iff v \neq 0$$

is called a *norm*. A vector space V with prenorm defines a norm on the quotient V/V_0 with the vector subspace V_0 of the norm-trivial elements

$$V_0 = \{n \in V | \; \| n \| = 0\} \in \underline{\mathbf{vec}}_{\mathbb{K}}, \quad [v] \in V/V_0, \quad \| [v] \| = \| v \| \,.$$

For a closed vector subspace $U \subseteq V$ the classes V/U are normed with the infimum norm (minimal distance):

$$\| \; \| : V/U \longrightarrow \mathbb{R}, \quad \| [v] \| = \inf_{u \in U} \| v + u \| \,.$$

The order of the prenorm-defined topologies is given above, e.g., for two prenorms with $\| x \|_c \leq k \| x \|_f$ the norm $\| \;\|_c$ defines the coarser topology. Two norms defining the same topology are *equivalent*. That is the case iff there exist two positive numbers $k, l > 0$ with

$$k \| x \|_1 \leq \| x \|_2 \leq l \| x \|_1 \text{ for all } x \in V.$$

For the scalars $\mathbb{K}$, the norms $k|\alpha|$ with $k > 0$ are equivalent.

The morphisms for the *category of normed spaces* $\underline{\mathbf{nvec}}_{\mathbb{K}} \subset \underline{\mathbf{tvec}}_{\mathbb{K}} \cap \underline{\mathbf{metr}}$ are the continuous linear mappings. They are are continuous or, equivalently, norm bounded:

$$\begin{array}{ccc} V & \xrightarrow{f} & W \\ \| \; \| \downarrow & & \downarrow \| \; \|, \\ \mathbb{R} & \xrightarrow[\leq k]{} & \mathbb{R} \end{array} \qquad \begin{array}{l} \text{linear } f \text{ continuous} \\ \iff \text{There exists } k \in \mathbb{R} \\ \text{with } \| f(v) \| \leq k \| v \| \\ \text{for all } v \in V. \end{array}$$

The smallest bound k defines a *norm of the morphism* f:

$$\begin{aligned} f \in \mathbf{nvec}_{\mathbb{K}}(V, W) \Rightarrow \| f \| &= \inf\{k \in \mathbb{R} \mid \| f(v)) \| \leq k \| v \| \text{ for all } v \in V\} \\ &= \sup\{\| f(v) \| \mid \| v \| \leq 1\}. \end{aligned}$$

Hence the category of normed vector spaces is morphism stable, i.e., $\mathbf{nvec}_{\mathbb{K}}(V, W) \in \underline{\mathbf{nvec}}_{\mathbb{K}}$. The *norm topology of* $\mathbf{nvec}_{\mathbb{K}}(V, W)$ *(topology of uniform convergence)* uses as 0-neighborhood basis the open sets

$$O_\epsilon(0) = \{f \in \mathbf{nvec}_{\mathbb{K}}(V, W) \mid \| f \| \leq \epsilon\}, \quad \epsilon > 0.$$

In addition to the norm topology on the endomorphisms the *strong operator topology* is defined as initial (coarsest) topology for all norms defined with a vector

$$\begin{array}{l} \mathbf{nvec}_{\mathbb{K}}(V, W) \times V \longrightarrow W, \quad (f, v) \longmapsto f(v), \\ v \in V, \quad \mathbf{nvec}_{\mathbb{K}}(V, W) \longrightarrow \mathbb{R}, \quad f \longmapsto \| f(v) \| . \end{array}$$

8.12.1 Scalar Product Vector Spaces

A symmetric bilinear form for $\mathbb{K} = \mathbb{R}$ and sesquilinear form for $\mathbb{K} = \mathbb{C}$ on $V \in \underline{\mathbf{vec}}_{\mathbb{K}}$ is called a *square*:

$$\langle \; | \; \rangle : V \times V \longrightarrow \mathbb{K}, \quad \left\{ \begin{array}{ll} \text{sesquilinear:} & \left\{ \begin{array}{l} \langle v | \alpha w + \beta u \rangle = \alpha \langle v | w \rangle + \beta \langle v | u \rangle, \\ \langle \alpha w + \beta u | v \rangle = \overline{\alpha} \langle w | v \rangle + \overline{\beta} \langle u | v \rangle, \end{array} \right. \\ \text{symmetric:} & \langle v | w \rangle = \overline{\langle w | v \rangle}. \end{array} \right.$$

If the square is

$$\text{positive: } \langle v | v \rangle \geq 0,$$

then it is called a *prescalar product.* It fulfills in this case the *Cauchy-Schwarz inequality* (submultiplicativity). The symmetry is a consequence of positivity;

$$\begin{matrix}\langle v+\alpha w|v+\alpha w\rangle \geq 0 \\ \text{for all } \alpha \in \mathbb{K}\end{matrix} \Rightarrow \left\{\begin{array}{rl} \langle v|w\rangle & = \overline{\langle w|v\rangle}, \\ \langle v|v\rangle\langle w|w\rangle & \geq \left(\frac{e^{i\beta}\langle v|w\rangle + e^{-i\beta}\langle w|v\rangle}{2}\right)^2 \\ & = (\operatorname{Re} e^{i\beta}\langle v|w\rangle)^2 \\ & \text{for all } e^{i\beta} \in \mathbf{U}(1) \cap \mathbb{K}, \\ \text{e.g., } \langle v|v\rangle\langle w|w\rangle & \geq |\langle v|w\rangle|^2 \\ & \text{with } e^{i\beta}\langle v|w\rangle = |\langle v|w\rangle|. \end{array}\right.$$

A prescalar product space (also called *pre-Hilbert space*) inherits with its prenorm the norm structures

$$\| v \| = \sqrt{\langle v|v\rangle} \geq 0, \quad \left\{\begin{array}{l} |\langle v|w\rangle| \leq \| v \| \| w \|, \\ \langle v|w\rangle = \langle v|w\rangle = \cos(v,w) \| v \| \| w \| \quad \text{for } \mathbb{K} = \mathbb{R}. \end{array}\right.$$

This definition satifies the triangle inequality as a consequence of the Cauchy-Schwarz inequality:

$$\| v+w \|^2 = \| v \|^2 + \| w \|^2 + 2\operatorname{Re}\langle v|w\rangle \leq \| v \|^2 + \| w \|^2 + 2|\langle v|w\rangle|.$$

In the opposite direction: On a space with (pre)norm a (pre)scalar product is induced if and only if the *parallelogram equation* holds:

$$\| v+w \|^2 + \| v-w \|^2 = 2 \| v \|^2 + 2 \| w \|^2$$

Hence the (pre)scalar product is given by

$$4\langle v|w\rangle = \left\{\begin{array}{ll} \sum_{\iota=\pm 1} \iota \| v+\iota w \|^2 & \text{for } \mathbb{K} = \mathbb{R}, \quad \iota \in \mathbb{I}(2), \\ \sum_{\iota=\pm 1, \pm i} \iota \| v+\iota w \|^2 & \text{for } \mathbb{K} = \mathbb{C}, \quad \iota \in \mathbb{I}(4). \end{array}\right.$$

A prescalar product is invariant under Lie algebra action for

$$\langle l \bullet u|v\rangle + \langle u|l \bullet v\rangle = 0, \quad l \in L.$$

If the product is

$$\text{strictly positive: } \langle v|v\rangle = 0 \iff v = 0,$$

V is a *scalar product space.* The invariance group in the V-automorphisms is the orthogonal group in the real case $\mathbf{O}(V) \cong \mathbf{O}(\dim_{\mathbb{R}} V)$ and the unitary group in the complex case $\mathbf{U}(V) \cong \mathbf{U}(\dim_{\mathbb{C}} V)$.

For V with prescalar product, V/V_0 with $V_0 = \{v \in V \mid \langle v|v\rangle = 0\}$ is a scalar product space. The morphisms for the *category of scalar product spaces* $\underline{\mathbf{svec}}_{\mathbb{K}} \subset \underline{\mathbf{nvec}}_{\mathbb{K}}$ are still the continuous, i.e., norm-bounded, linear mappings, the isomorphisms satisfy $\langle f(v)|f(u)\rangle = \langle v|u\rangle$.

An *orthonormal system* $O \subseteq V \in \underline{\mathbf{svec}}_{\mathbb{K}}$ consists of normalized orthogonal vectors:

$$e, f \in O \quad \Rightarrow \begin{cases} \langle e|e\rangle = 1 = \langle f|f\rangle, \\ e \neq f, \quad \langle e|f\rangle = 0. \end{cases}$$

With a countable orthonormal system O one has *Bessel's inequality*:

$$\text{orthonormal system: } O = \{e^i | i \in \mathbb{N}\}, \quad \sum_{i=1}^{\infty} |\langle v|e^i\rangle|^2 \leq \| v \|^2 \text{ for all } v \in V.$$

Each series of linearly independent vectors can be orthonormalized.

There is the *weak operator topology* on the endomorphisms of a scalar product space as initial (coarsest) topology for all prenorms defined with two vectors,

$$\begin{array}{l} \mathbf{svec}_{\mathbb{K}}(V,V) \times V \times V \longrightarrow \mathbb{K}, \quad (f,v,w) \longmapsto \langle w|f(v)\rangle, \\ v, w \in V, \quad \mathbf{svec}_{\mathbb{K}}(V,V) \longrightarrow \mathbb{R}, \quad f \longmapsto |\langle w|f(v)\rangle|, \end{array}$$

in addition to the norm topology and the strong operator topology

$$f \in \mathbf{nvec}_{\mathbb{K}}(V,V), \begin{cases} \qquad\qquad \text{norm}: & \| f \|, \\ \qquad\qquad \text{strong}: & \{\| f \|_v = \| f(v) \| \mid v \in V\}, \\ f \in \mathbf{svec}_{\mathbb{K}}(V,V), \ \text{weak}: & \{|f|_{wv} = |\langle w|f(v)\rangle| \mid v, w \in V\}. \end{cases}$$

The following inclusions hold for the topological spaces above:

$$\begin{array}{ccccccc} \underline{\mathbf{top}} & \supset & \underline{\mathbf{uniform}} & \supset & \underline{\mathbf{metr}} & & \\ & & \cup & & \cup & & \\ \underline{\mathbf{vec}}_{\mathbb{K}} & \supset & \underline{\mathbf{tvec}}_{\mathbb{K}} & \supset & \underline{\mathbf{nvec}}_{\mathbb{K}} & \supset & \underline{\mathbf{svec}}_{\mathbb{K}} \end{array}$$

8.12.2 Banach and Hilbert Spaces

A complete normed space is called a *Banach space*, e.g., $V \cong \mathbb{K}^n$ with the natural topology. The completion $\mathbf{C}(V)$ of a normed space V is normed with

$$\| (v_n)_{n \in \mathbb{N}} \| = \lim_{n \in \mathbb{N}} \| v_n \| .$$

Continuous linear mappings are uniquely extended.

If $\hat{W}$ is a Banach space, e.g., $\mathbb{K}$, the morphisms $\mathbf{nvec}_{\mathbb{K}}(V, \hat{W})$ are complete in the norm topology, especially the topological dual space

$$V \in \underline{\mathbf{nvec}}_{\mathbb{K}} \Rightarrow V' = \mathbf{nvec}_{\mathbb{K}}(V, \mathbb{K}) \in \underline{\mathbf{n\hat{vec}}}_{\mathbb{K}}, \quad V' \subseteq V^T.$$

If $V \cong \mathbb{K}^n$ is Hausdorff, then a linear f is always bounded and the topologial dual is the full algebraic one $V' = V^T$.

A complete scalar product space is called a *Hilbert space*:

$$\underline{\hat{\mathbf{metr}}} \supset \underset{\text{(Banach)}}{\underline{\mathbf{nv\hat{e}c}_{\mathbb{K}}}} \supset \underset{\text{(Hilbert)}}{\underline{\mathbf{sv\hat{e}c}_{\mathbb{K}}}}.$$

The completion $\mathbf{C}(V)$ of a scalar product space V carries a scalar product

$$\langle (v_n)_{n\in\mathbb{N}} | (u_n)_{n\in\mathbb{N}} \rangle = \lim_{n\in\mathbb{N}} \langle v_n | u_n \rangle.$$

Continuous linear mappings are uniquely extended. A scalar product space may arise from a prescalar product space which justifies its name "pre-Hilbert space."

Finite-dimensional vector spaces with a *positive* nondegenerate bilinear (for $\mathbb{R}$) and sesquilinear (for $\mathbb{C}$) form, invariant under $\mathbf{O}(n)$ and $\mathbf{U}(n)$ respectively, are Hilbert spaces.

The complexification of a real Hilbert space is a complex Hilbert space.

The (conjugate) linear mapping of a Hilbert space H in its topological dual space with the continuous, i.e., norm-bounded, linear forms

$$H \ni |v\rangle \longmapsto \langle v| \in H' = \mathbf{nvec}_{\mathbb{K}}(H, \mathbb{K}),$$
$$\langle v| : \; H \longrightarrow \mathbb{K}, \; |w\rangle \longmapsto \langle v|w\rangle,$$

with Dirac's *bra* $\langle v|$ and *ket* $|v\rangle$ notation, is bijective and isometric in the norm topology of H', therefore a (conjugate) linear dual isomorphism

$$H \in \underline{\mathbf{sv\hat{e}c}_{\mathbb{K}}} \Rightarrow H \cong H' = \mathbf{nvec}_{\mathbb{K}}(H, \mathbb{K}) \in \underline{\mathbf{sv\hat{e}c}_{\mathbb{K}}}.$$

Hilbert spaces generalize the algebraic self-duality with a scalar product for $V \cong V^T \cong \mathbb{K}^n$ to a topological self-duality $V \cong V'$.

An orthonormal system $B \subseteq H$ in a Hilbert space H is a *Hilbert space basis* if the closure of its span gives H:

$$H = \overline{\underset{|e\rangle\in B}{\perp} \mathbb{K}|e\rangle}.$$

Then B is also called a *complete orthonormal system.* This property of B can be characterized equivalently as follows:

(1) $|v\rangle = \sum_{|e\rangle\in B} \langle e|v\rangle |e\rangle$ for all $|v\rangle \in H$;

(2) $\langle v|w\rangle = \sum_{|e\rangle\in B} \langle v|e\rangle\langle e|w\rangle$ for all $|v\rangle, |w\rangle \in H$;

(3) $\| v \|^2 = \sum_{|e\rangle\in B} |\langle v|e\rangle|^2$ for all $|v\rangle \in H$ (*Parseval equation*);

(4) $|v\rangle = 0 \iff \langle e|v\rangle = 0$ for all $|e\rangle \in B$;

(5) B is a maximal orthonormal system with respect to inclusion,

or in analogy to finite-dimensional spaces,

$$B \text{ is Hilbert space } H \text{ basis} \iff \begin{cases} \langle e|e'\rangle = \delta_{ee'} & \text{(orthonormality)}, \\ \sum_{|e\rangle\in B} |e\rangle\langle e| \sim \mathrm{id}_H & \text{(completeness)}, \\ & \text{(sesquilinear for } \mathbb{C}). \end{cases}$$

Each Hilbert space has a Hilbert space basis; all Hilbert space bases of a Hilbert space have equal cardinality, the *dimension of the Hilbert space.* Two bases are related by an automorphism. H is separable iff its dimension is countable. Any basis can be orthonormalized (Gram-Schmidt orthonormalization).

8.13 Banach Algebras

The *resolvent* for an element a in a unital algebra $A \in \underline{\mathbf{aag}}_{\mathbb{K}}$ is a mapping from the scalars in the algebra

$$\mathbb{K} \ni \alpha \longmapsto \mathrm{Res}(a, \alpha) = \tfrac{1}{\alpha 1 - a} \in A$$

defining as domain the *resolvent set*, where the resolvent is defined, i.e., $\alpha 1 - a$ has an inverse. The complement is the *spectrum*; there the resolvent is not defined, i.e., $\alpha 1 - a$ does not have an inverse

$$1 \in A \in \underline{\mathbf{aag}}_{\mathbb{K}} : \quad \begin{array}{l} \mathrm{spec}\, a = \{\alpha \in \mathbb{K} \,\big|\, \alpha 1 - a \text{ not invertible}\} \\ \mathrm{Res}(a, \ ) : \mathbb{K} \setminus \mathrm{spec}\, a \longrightarrow A, \quad \alpha \longmapsto \frac{1}{\alpha 1 - a} \end{array}$$

One has for a unital complex algebra with conjugation

$$*\underline{\mathbf{aag}}_{\mathbb{C}} \Rightarrow \begin{cases} \mathrm{spec}\,(\alpha 1 - a) = \alpha - \mathrm{spec}\, a, \quad \mathrm{spec}\, a^{\star} = \overline{\mathrm{spec}\, a}, \\ a \in A^{\diamond} \Rightarrow \mathrm{spec}\, a^{-1} = (\,\mathrm{spec}\, a)^{-1}, \\ a, b \in A \Rightarrow \mathrm{spec}\, ab \cup \{0\} = \mathrm{spec}\, ba \cup \{0\}. \end{cases}$$

An associative *normed algebra* is a normed vector space that comes with a norm-continuous multiplication

$$\underline{\mathbf{naag}}_{\mathbb{K}} : \quad \| ab \| \leq \| a \| \| b \| .$$

A complete normed complex algebra is called a *Banach algebra.*

The *spectral radius* of an element a of a normed algebra is given by the limit (it exists)

$$a \in A \in \underline{\mathbf{naag}}_{\mathbb{C}} \Rightarrow \rho(a) \ = \lim_{n \to \infty} \| a^n \|^{\frac{1}{n}} = \inf_{n \to \infty} \| a^n \|^{\frac{1}{n}} \leq \| a \| .$$

The name is justified in a unital complex Banach algebra B:

$$a \in B \in \underline{\mathbf{n\hat{a}ag}}_{\mathbb{C}} \Rightarrow \rho(a) \ = \sup\{|\alpha| \,\big|\, \alpha \in \mathrm{spec}\, a\}.$$

There, $\mathrm{spec}\, a$ is nonempty and compact. The resolvent is holomorphic on the resolvent set and trivial at infinity:

$$\mathrm{Res}(a, \ ) : \mathbb{C} \setminus \mathrm{spec}\, a \longrightarrow B, \ \ \tfrac{d^k}{d\alpha^k} \mathrm{Res}(a, \alpha) = (-1)^k k! \mathrm{Res}(a, \alpha)^{k+1}.$$

For an associative *involutive normed complex algebra* the norm has to be, in addition, conjugation compatible, i.e.,

$$\star\underline{\mathbf{naag}}_{\mathbb{C}} : \quad \| a \| = \| a^{\star} \| .$$

8.13.1 Stellar or C*-Algebras

A *stellar algebra (C*-algebra)* is a Banach algebra with conjugation and the norm property

$$\underline{\star\mathbf{sa\hat{a}g}}_{\mathbb{C}} \ni S \ni a: \quad \| a \|^2 = \| a^\star a \|,$$

whence $\| a \| = \| a^\star \|$.

What Banach algebras are for normed vector spaces ("Banach" stands for "complete with norm"), stellar algebras are for scalar product vector spaces, they could be also called *Hilbert algebras*: The stellar algebra defining property relates the norm to something like a scalar product (also next section):

$$\begin{array}{ccccccc} \underline{\star\mathbf{aag}}_{\mathbb{C}} & \supset & \underline{\star\mathbf{naag}}_{\mathbb{C}} & \supset & \underline{\star\mathbf{na\hat{a}g}}_{\mathbb{C}} & \supset & \underline{\star\mathbf{sa\hat{a}g}}_{\mathbb{C}}. \\ & & & & \text{(Banach)} & & \text{(C*, stellar, Hilbert)} \end{array}$$

The stellar norm is closely related to the spectral radius, for a normal element given even by the spectral radius:

$$\begin{aligned} a \in S \Rightarrow & \quad \| a \| = \sqrt{\| a^\star a \|} = \sqrt{\rho(a^\star a)}, \\ aa^\star = a^\star a \Rightarrow & \quad \| a \| = \rho(a). \end{aligned}$$

The full matrix algebra $\mathbf{AL}(\mathbb{C}^n)$ with $\mathbf{U}(n)$-conjugation has as stellar algebra norm the maximal absolute square root eigenvalue of its square:

$$\underline{\star\mathbf{sa\hat{a}g}}_{\mathbb{C}} \ni \mathbf{AL}(\mathbb{C}^n) \ni f, \quad \| f \| = \max \sqrt{\text{spec}\, f^* \circ f}.$$

For $n \geq 2$, the stellar norm $\| f \|$ is different from the *Hilbert-Schmidt norm* $\| f \|_2$ induced by the double trace scalar product and given by the square root of the sum of all absolute squared matrix elements

$$\begin{aligned} f, g \in \mathbf{AL}(\mathbb{C}^n): \quad & \langle f | g \rangle = \text{tr}\, f^\star \circ g, \\ & \| f \|_2 = \sqrt{\text{tr}\, f^\star \circ f} = \sqrt{\langle f | f \rangle} = \sqrt{f_j^k \overline{f_j^k}} \geq \| f \|. \end{aligned}$$

.

In a stellar algebra a normal element generates a commutative stellar subalgebra

$$\begin{aligned} a \in S \in \underline{\star\mathbf{sa\hat{a}g}}_{\mathbb{C}}, \quad & aa^\star = a^\star a \Rightarrow \overline{\mathbb{C}[a, a^\star]} \in \underline{\star\mathbf{sa\hat{a}g}}_{\mathbb{C}}, \\ a = \tfrac{a+a^\star}{2} + \tfrac{a-a^\star}{2} & = a_R + i a_I, \quad [a_R, a_I] = 0. \end{aligned}$$

A stellar norm $\| a \|$ is smaller than any other norm $\| a \|_{\text{other}}$, and therefore the stellar norm is a minimal norm and unique as a stellar norm:

$$\text{stellar algebra } ||a|| \leq ||a||_{\text{other}} \text{ normed algebra.}$$

It gives the coarsest norm topology.

A *stellar prenorm* $A \in \underline{\star\mathbf{aag}}_{\mathbb{C}}$ is a prenorm $p: A \longrightarrow \mathbb{C}$ with

$$p(ab) \leq p(a)p(b), \; p(a^\star) = p(a), \quad p(a^2) = p(a^\star a).$$

The set P of all stellar prenorms of an involutive Banach algebra (including the trivial one $p(a) = 0$) defines the largest stellar prenorm

$$\star\underline{\mathbf{na\hat{a}g}}_{\mathbb{C}} \ni B \ni a \longmapsto \| a \|_\star = \sup_{p \in P} p(a) \leq \| a \| \, .$$

The completion of the ideal classes $B/\{a \bigm| \| a \|_\star = 0\}$ with respect to the induced norm is the universal *enveloping stellar algebra* leading to the corresponding covariant functor with the unique factorization property:

$$\mathbf{C}^\star : \star\underline{\mathbf{na\hat{a}g}}_{\mathbb{C}} \longrightarrow \star\underline{\mathbf{sa\hat{a}g}}_{\mathbb{C}}, \quad B \longmapsto \mathbf{C}^\star(B),$$

$$\begin{array}{ccc} B & \stackrel{\pi}{\longrightarrow} & \mathbf{C}^\star(B) \\ f\big\downarrow & & \big\downarrow \\ S & \underset{\mathrm{id}_S}{\longrightarrow} & S \end{array} \quad \tilde{f}\,, \quad S \in \star\underline{\mathbf{sa\hat{a}g}}_{\mathbb{C}}.$$

If B is commutative or unital, $\mathbf{C}^\star(B)$ is commutative or unital too.

In a unital stellar algebra S one has the properties

$$\begin{array}{llll} a \textit{ partially isometric} & \iff aa^\star a = a & \Rightarrow \| a \| = 1, \\ a \text{ unitary} & \iff a^\star = a^{-1} & \Rightarrow \operatorname{spec} a \subseteq \mathbf{U}(1), \\ a \text{ Hermitian} & \iff a = a^\star & \Rightarrow - \| a \| \leq \operatorname{spec} a \leq \| a \|, \\ \text{polynomial } P(a) & \Rightarrow \operatorname{spec} P(a) = & P(\operatorname{spec} a). \end{array}$$

A stellar algebra carries an *order* by the spectrum

$$\begin{array}{rl} p \succeq 0 & \iff p = p^\star \text{ and } \operatorname{spec} p \subset \mathbb{R}_+ \\ & \iff \text{There exists } h \in S \text{ with } p = h^2 \\ & \iff \text{There exists } a \in S \text{ with } p = a^\star a \end{array}$$

with the property

$$a \in S \Rightarrow a^\star a \succeq 0.$$

The open double cones for positive vectors define the topology

$$p \succ 0, \quad U_p(0) = \{a \in A \bigm| -p \prec a \prec p\}.$$

Any invertible element of a unital stellar algebra allows a unique *polar decomposition* in *unitary phase and strictly positive absolute value*

$$\begin{array}{rl} a \in S^\diamond \iff & a = u(a)|a| \text{ with } S \ni |a| \succ 0, \quad u(a)^\star = u(a)^{-1} \\ & |a|^2 = a^\star a, \quad |a| = \sqrt{a^\star a} \end{array}$$

with $\| a \| = \| \, |a| \, \|$. Unitary elements have as absolute value the algebra unit $|u| = 1_S$.

For a normal element $a \in S$ in a unital stellar algebra, continuous complex functions can be defined via the values on the spectrum

$$a \in S,\ \ aa^\star = a^\star a: \quad \begin{array}{rcl} f: \operatorname{spec} a & \longrightarrow & \mathbb{C}, \\ a & \longmapsto & f(a) = \{f(\alpha) \mid \alpha \in \operatorname{spec} a\}. \end{array}$$

All continuous linear mappings are isomorphic to the unital commutative stellar subalgebra $\overline{\mathbb{C}[a, a^\star]}$. Therefore a unique positive element defines the *positive real powers* of a positive element:

$$S \ni p \succeq 0,\ \beta > 0 \Rightarrow \begin{array}{l} \text{There exists a unique } r \succeq 0 \text{ with } p = r^\beta, \\ r = \{|\alpha^{\frac{1}{\beta}}| \mid \alpha \in \operatorname{spec} p \geq 0\}. \end{array}$$

One has the hierarchy

$$\begin{array}{ccccc} \boxed{\begin{array}{c} a \succeq 0 \\ \text{positive} \end{array}} & \Rightarrow & \boxed{\begin{array}{c} a = a^\star \\ \text{Hermitian} \end{array}} & \Rightarrow & \boxed{\begin{array}{c} aa^\star = a^\star a \\ \text{normal} \end{array}} \\ & & & & \Uparrow \\ & & & & \boxed{\begin{array}{c} a^{-1} = a^\star \\ \text{unitary} \end{array}} \end{array}$$

8.13.2 Morphisms of Hilbert Spaces

A continuous linear Hilbert space mapping has an *adjoint* as expressed by an involutive contravariant functor for Hilbert spaces

$$\star: \underline{\mathbf{sv\hat{e}c}}_{\mathbb{K}} \longrightarrow \underline{\mathbf{sv\hat{e}c}}_{\mathbb{K}}, \qquad \begin{array}{ccc} H & & H \cong H' \\ f \big\downarrow & \longmapsto & \big\uparrow f^\star \\ K & & K \cong K' \end{array},$$
$$\text{with } \langle f(h)|k\rangle = \langle h|f^*(k)\rangle,$$
$$(f \circ g)^\star = g^\star \circ f^\star,\ \ \mathrm{id}_H^\star = \mathrm{id}_H,\ \ f^{\star\star} = f,$$
$$(f + g)^\star = f^\star + g^\star,\ \ (\alpha f)^\star = \overline{\alpha} f^\star,$$

with the norm property

$$\| f^\star \circ f \| = \| f \circ f^\star \| = \| f \|^2 = \| f^\star \|^2 .$$

The Hilbert space isomorphisms are the unitary morphisms, $u^\star = u^{-1}$. The full matrix algebra $\mathbf{AL}(\mathbb{C}^n)$ with the Hilbert space $H \cong \mathbb{C}^n$ is an example.

With the norm topology the continuous endomorphisms of a normed (Banach) vector space and of a complex Hilbert space (bounded) constitute a normed (Banach) unital algebra and a unital stellar algebra with the adjoint defining the algebra reflection, which is denoted by $\mathbf{AL}(V)$, etc.

$$\begin{array}{lll} V \in \underline{\mathbf{nvec}}_{\mathbb{C}} & \Rightarrow \mathbf{AL}(V) = \mathbf{nvec}_{\mathbb{C}}(V, V) \in \underline{\mathbf{naag}}_{\mathbb{C}}, \\ \hat{V} \in \underline{\mathbf{nv\hat{e}c}}_{\mathbb{C}} & \Rightarrow \mathbf{AL}(\hat{V}) = \mathbf{nvec}_{\mathbb{C}}(\hat{V}, \hat{V}) \in \underline{\mathbf{na\hat{a}g}}_{\mathbb{C}}, \\ H \in \underline{\mathbf{sv\hat{e}c}}_{\mathbb{C}} & \Rightarrow \mathbf{AL}(H) = \mathbf{nvec}_{\mathbb{C}}(H, H) \in \underline{\star\mathbf{sa\hat{a}g}}_{\mathbb{C}}. \end{array}$$

And conversely, any stellar algebra is isomorphic to a norm-closed self-adjoint subalgebra of a continuous linear endomorphism algebra of a Hilbert space:

$$S \in \star\underline{\mathbf{sa\hat{a}g}}_{\mathbb{C}} \Rightarrow S \cong A = A^\star = \overline{A} \subseteq \mathbf{AL}(H), \quad H \in \underline{\mathbf{sv\hat{e}c}}_{\mathbb{C}}.$$

Stellar algebra element properties may be characterized for continuous linear endomorphisms $f : H \longrightarrow H$ of a complex Hilbert space via mapping properties:

$$\left.\begin{array}{r} f \text{ partially isometric,} \\ f \circ f^\star \circ f = f \end{array}\right\} \iff \| f|v\rangle \| = \| v \| \text{ for all } |v\rangle \text{ with } f|v\rangle \neq 0,$$

$$\left.\begin{array}{r} f \text{ normal,} \\ f \circ f^\star = f^\star \circ f \end{array}\right\} \iff \| f|v\rangle \| = \| f^\star|v\rangle \| \text{ for all } |v\rangle \in H,$$

$$\left.\begin{array}{r} f \text{ self-adjoint,} \\ f = f^\star \end{array}\right\} \Rightarrow \| f \| = \sup \ \{|\langle v|f(v)\rangle| \mid \| v \| \leq 1\},$$

$$f \text{ essentially self-adjoint} \iff f^\star \text{ selfadjoint, i.e., } f^\star = f^{\star\star},$$

$$\left.\begin{array}{r} f \text{ positive,} \\ f \succeq 0 \end{array}\right\} \iff \left\{\begin{array}{l} \langle v|f|v\rangle \geq 0 \text{ for all } |v\rangle \in H \iff \\ \text{There exists a continuous} \\ \text{morphism to a Hilbert space } K, \\ g : H \longrightarrow K \text{ with } g^\star \circ g = f. \end{array}\right.$$

The orthogonal of the kernel and the closed image of a Hilbert space mapping define the *initial* and the *final subspace* respectively (both Hilbert subspaces):

$$f : H \longrightarrow K, \quad H = \operatorname{kern} f \perp \operatorname{in} f, \quad \operatorname{out} f = \overline{f[H]} \subseteq K,$$

$$\begin{array}{lccccl} \text{initial} & H & \stackrel{f}{\longrightarrow} & K & & \text{final} \\ \text{orthoprojector } \mathcal{P}_f^H & \downarrow & & \downarrow & \mathcal{P}_f^K & \text{orthoprojector} \\ \text{of } f & \operatorname{in} f & \underset{\mathcal{P}(f)}{\longrightarrow} & \operatorname{out} f & & \text{of } f \end{array} .$$

The arising Hilbert space isomorphism $\mathcal{P}(f)$ is unitary. A continuous linear mapping is partially isometric iff $f^\star \circ f$ and $f \circ f^\star$ are the orthoprojectors to the initial and final subspaces.

Any continuous linear mapping of two Hilbert spaces has a unique polar decomposition with phase mapping $u(f)$ and absolute value endomorphism

$|f|$:

$$f: H \longrightarrow K, \quad f = u(f) \circ |f|, \quad \begin{cases} |f| : H \longrightarrow H, & \begin{cases} |f| = \sqrt{f^\star \circ f}, \\ \| \, |f| \, \| = \| f \|, \end{cases} \\ u(f) : H \longrightarrow K, & \text{partially isometric,} \end{cases}$$
$$H \supseteq \text{kern}\, f = \text{kern}\, |f| = \text{kern}\, u(f),$$
$$H \supseteq \text{in}\, f = \text{in}\, |f| = \text{out}\, |f| = \text{in}\, u(f), \quad \text{out}\, u(f) = \text{out}\, f \subseteq K.$$

The tensor product $H \otimes H'$ is injected into the continuous linear mappings

$$H \otimes H' \ni |w\rangle\langle v| : H \longrightarrow H, \quad |u\rangle \longmapsto \langle v|u\rangle |w\rangle.$$

For a Hilbert space basis $\{e^\iota\}_{\iota \in I}$ one has the projectors on subset-spanned subspaces

$$J \subseteq I: \quad \mathcal{P}_{H_J} = \sum_{\iota \in J} |e^\iota\rangle\langle e^\iota| \quad \mathcal{P}_{H_J}[H] = H_J = \bigoplus_{\iota \in J} \mathbb{C} e^\iota \subseteq H.$$

The continuous Hilbert space endomorphism with finite-dimensional image constitute the algebra with the *finite rank operators*

$$\mathbf{AL}^{\mathbb{N}}(H) = \{ f \in \mathbf{AL}(H) \; \big| \; \dim_{\mathbb{C}} f[H] \text{ finite } \} \in \star\underline{\mathbf{aag}}_{\mathbb{C}}.$$

Positive finite rank operators have a finite trace as defined by the scalar product:

$$H \otimes H' \ni |w\rangle\langle v| \longmapsto \text{tr}\, |w\rangle\langle v| = \langle v|w\rangle = \sum_{\iota \in I} \langle v|e^\iota\rangle\langle e^\iota|w\rangle \in \mathbb{C},$$
$$\mathbf{AL}^{\mathbb{N}}(H) \ni p \succeq 0: \quad \text{tr}\, p = \sum_{\iota \in I} \langle e^\iota|p|e^\iota\rangle \in \mathbb{R}_+,$$
$$\text{tr}\,(p + p') = \text{tr}\, p + \text{tr}\, p', \quad \text{tr}\, \alpha p = \alpha\, \text{tr}\, p.$$

The completion of the finite-rank operators with trace, double trace, stellar norm, and strong topology,

$$\begin{matrix} \text{completion of} \\ \mathbf{AL}^{\mathbb{N}}(H) \text{ with} \end{matrix} \begin{cases} \| f \|_1 = \text{tr}\, \sqrt{f^\star \circ f} & \textit{trace class operators,} \\ \| f \|_2 = \sqrt{\text{tr}\, f^\star \circ f} & \textit{Hilbert-Schmidt operators,} \\ \| f \| = \max \sqrt{\text{spec}\, f^\star \circ f} & \textit{compact operators,} \\ \{ \| f \|_v = \| f(v) \| \; \big| \; v \in H \} & \textit{bounded operators,} \end{cases}$$

defines algebras that together with the finite-rank operators are all ideals:

$$\begin{matrix} \| f \|_1 \geq \| f \|_2 \geq \| f \|, \\ \underbrace{\mathbf{AL}^{\mathbb{N}}(H) \subseteq \mathbf{AL}_1(H) \subseteq \mathbf{AL}_2(H) \subseteq \mathbf{AL}_c(H)}_{\text{ideals in } \mathbf{AL}(\text{H})} \subseteq \mathbf{AL}(H). \end{matrix}$$

A positive finite trace mapping has a unique positive square root r that is Hilbert-Schmidt:

$$p \succeq 0, \quad \text{tr}\, p < \infty \Rightarrow p = r \circ r \text{ with Hilbert-Schmidt } r \succeq 0.$$

With two Hilbert spaces $f \in \mathbf{nvec}_{\mathbb{C}}(H, K)$ one defines the vector spaces with the trace class and Hilbert-Schmidt operators by finite $\| f \|_{1,2}$ for the positive operator $f^\star \circ f \in \mathbf{AL}(H)$.

8.13.3 Eigenvalues of Hilbert Space Endomorphisms

The eigenvalues of a symmetric Hilbert space endomorphism have to be real. Eigenvectors for different eigenvalues are orthogonal:

$$\begin{aligned}\mathbf{AL}(H) \ni f = f^\star, \quad & f|v\rangle = \alpha|v\rangle, \ f|w\rangle = \beta|w\rangle, \ |v\rangle, |w\rangle \neq 0 \\ \Rightarrow \quad & \langle v|f|v\rangle = \alpha\langle v|v\rangle = \overline{\alpha}\langle v|v\rangle \Rightarrow \alpha \in \mathbb{R}, \\ & \langle w|f|v\rangle = \alpha\langle w|v\rangle = \beta\langle w|v\rangle \Rightarrow \langle w|v\rangle = 0 \text{ for } \alpha \neq \beta.\end{aligned}$$

For finite dimension $H \cong \mathbb{C}^n$ any endomorphism f can be triagonalized, a normal f can be unitarily diagonalized.

For a Hilbert-Schmidt mapping $f : H \longrightarrow K$, there exists a Hilbert space basis $\{e^\iota\}_{\iota\in I}$ in H where $\{f(e^\iota)\}_{\iota\in I}$ is an orthogonal family in K.

The principal spaces and the eigenspaces for an endomorphism

$$\begin{aligned} f : H \longrightarrow H, \ \alpha \in \mathbb{C}: \quad & H^\alpha(f) = \{|v\rangle \in H \,\big|\, (f - \alpha\, \mathrm{id}_H)^N|v\rangle = 0 \text{ for } N \geq 0\}, \\ & H_\alpha(f) = \{|v\rangle \in H \,\big|\, f|v\rangle = \alpha|v\rangle\} \subseteq H^\alpha(f), \end{aligned}$$

are identical for a normal Hilbert space endomorphism,

$$f \circ f^\star = f^\star \circ f \Rightarrow \quad H_\alpha(f) = H_{\overline{\alpha}}(f^\star) = H^\alpha(f) = H^{\overline{\alpha}}(f^\star).$$

For a positive, therefore normal, Hilbert space endomorphism p with finite trace there exists a Hilbert space basis $\{e^\iota\}_{\iota\in I}$ of eigenvectors with the trace being the positive eigenvalue sum,

$$p \succeq 0, \quad \mathrm{tr}\, p < \infty \Rightarrow p|e^\iota\rangle = \alpha^\iota|e^\iota\rangle \text{ (no sum)}, \ \alpha^\iota \geq 0, \ \ \mathrm{tr}\, p = \sum_{\iota\in I} \alpha^\iota,$$

and a *spectral projector decomposition* with orthogonal eigenspaces for the different eigenvalues ($J \subseteq I$):

$$H = \bigoplus_{j\in J} H_{\alpha^j}, \quad \langle H_{\alpha^j}|H_{\alpha^k}\rangle = \{0\} \text{ for } j \neq k, \quad p = \sum_{j\in J} \alpha^j \mathcal{P}_{\alpha^j}.$$

Bibliography

[1] N. Bourbaki, *Topological Vector Spaces, Chapter 1-5* (1987), Springer, Berlin, Heidelberg, New York, London, Paris, Tokyo.

[2] N. Bourbaki, *Théories Spectrales* (1967), Hermann, Paris.

[3] O Bratelli, D.W. Robinson, *Operator Algebras and Quantum Statistical Mechanics 1* (1979) Springer-Verlag, Berlin etc.

[4] C. Rickart, *General Theory of Banach Algebras* (1960), van Nostrand, New York.

[5] F. Treves, *Topological Vector Spaces, Distributions and Kernels* (1967), Academic Press, New York, London.

9

THE KEPLER FACTOR

With the operator prejudice for physical theories, both objects and interactions should have an operational origin. The basic interactions, we have so far, involve in their nonrelativistic approximation the *Kepler factor* $\frac{1}{r}$. A position radial dependence $\frac{1}{r}$ arises at characteristic points: A $\frac{1}{r}$-proportional potential describes the most important nonrelativistic interactions: gravitation and electrostatics. Historically, it led to Newton's understanding of Kepler's laws by the gravitation of mass points, which was the first step to Einstein's general relativity. Furthermore, as Coulomb potential, it played a decisive role in the discovery of Maxwell's electrodynamics with its field structure, introduced by Faraday, and its extension to quantum electrodynamics and to the standard model of the electroweak gauge interactions. Its quantum-mechanical application led to the understanding of the atoms and the periodic system. In addition to nonrelativistic potentials the $\frac{1}{r}$-dependence also characterizes spherical waves $\frac{e^{iqr}}{r}$ in electrodynamics (optics).

Already for Newton the Kepler factor $\frac{1}{r}$ was geometrically motivated. The exact power $\gamma = 1$ in $\frac{1}{r^\gamma}$, not, e.g., $\frac{1}{r^{1.007}}$, is no accident. The square of the Kepler factor is, up to a constant, the inverse of the 2-sphere area $|\Omega^2(r)| = 4\pi r^2$. $\frac{e^{\kappa r}}{r}$ will be interpreted as the 2-sphere spread for a coefficient $e^{\kappa r}$ of a representation of position translations where κ can be imaginary, $\kappa = \pm i|\vec{q}|$, for scattering waves or negative $\kappa = -|Q|$ for bound waves or trivial in $\frac{1}{r}$.

The small-distance singularity of the Kepler factor for $r = 0$ is known to prevent an interpretation of the electron mass as of electromagnetic origin. It is the nonrelativistic precursor of the divergences arising, e.g., in Feynman integrals in quantum theory for particle fields (chapter "Spectrum of Spacetime").

A Kepler potential is the infinite range $\ell \to \infty$ limit of a *Yukawa potential* with $\frac{e^{-\frac{r}{\ell}}}{r}$, which, with respect to second order differential equations with $\vec{\partial} = \frac{\partial}{\partial \vec{x}}$, is the special relativistic position supplement of an irreducible causal time representation $d_t = \frac{d}{dt}$ with frequency μ:

$$\frac{e^{-\frac{r}{\ell}}}{r} \leftrightarrow \sin|t\mu| \text{ since } \left\{ \begin{array}{lll} (d_t^2 + \mu^2) & \frac{\sin|t|\mu}{\mu} & = 2\delta(t), \\ (-\vec{\partial}^2 + \frac{1}{\ell^2}) & \frac{e^{-\frac{r}{\ell}}}{2\pi r} & = 2\delta(\vec{x}). \end{array} \right.$$

Also, such a space-time parallelism determines the power in the Kepler potential $\frac{1}{r^\gamma}$ to be exactly $\gamma = 1$. If the frequency and the potential range combine the speed of light $\mu\ell = c$, the equations above arise as time and position projections of a Lorentz-invariant inhomogeneous Klein-Gordon equation with the infinite range potential related to a mass-zero structure.

The irreducible unitary representations of time $\mathbb{R} \ni t \longmapsto e^{iEt} \in \mathbf{U}(1)$ in quantum mechanics that decompose the action of the Hamiltonian with energy E eigenvalues induce a Hilbert space structure. The inclusion of position representations spreads the state vectors to position orbits (Schrödinger wave functions) $\mathbb{R}^3 \ni \vec{x} \longmapsto \psi_E(\vec{x}) \in \mathbb{C}$. The radial translations $r = |\vec{x}|$ as rotation invariant part constitute the positive cone $\mathbb{R}_+$ of a 1-dimensional noncompact group $\mathbb{R}$ which, for a potential $V(r)$, comes in representations with the eigenvalues determined by the difference $E - V$. There arise scattering waves for kinetic energy $E - V > 0$ (compact position representations) and imaginary radial translation eigenvalue $\pm i|\vec{q}| = \pm i\sqrt{2(E-V)}$ (real momentum) and bound waves for $E - V < 0$ (noncompact position representations) with strictly negative eigenvalue $-|Q| = -\sqrt{-2(E-V)}$ (imaginary "momentum") and binding energy $|E - V|$.

9.1 Center of Mass Transformation

A Lagrangian determines the time development of two classical mass points in Euclidean position space with dual position-momentum pairs $(\vec{x}_i, \vec{p}_i)_{i=1,2}$ by a Hamiltonian H

$$L(1,2) = \vec{p}_1 d_t\vec{x}_1 + \vec{p}_2 d_t\vec{x}_2 - H(1,2), \quad H(1,2) = \tfrac{\vec{p}_1^2}{2m_1} + \tfrac{\vec{p}_2^2}{2m_2} + V(\vec{x}_1, \vec{x}_2).$$

The time action is induced by the gravitational interaction with Newton's potential and by the electrostatic interaction with Coulomb's potential:

$$V(\vec{x}_1, \vec{x}_2) = \tfrac{g_0}{|\vec{x}_1 - \vec{x}_2|}, \quad g_0 = \left\{ \begin{array}{ll} -G_N m_1 m_2, & \text{Newton's constant } G_N \\ & \text{with masses } m_{1,2}, \\ \frac{1}{4\pi\epsilon_0} Q_1 Q_2, & \text{vacuum dielectricity constant } \epsilon_0 \\ & \text{and charges } Q_{1,2}. \end{array} \right.$$

If the potential depends of the mass point distance only $V(\vec{x}_1, \vec{x}_2) = V(\vec{x}_1 - \vec{x}_2)$, the center of mass dynamics can be separated by an orthogonal transformation of the two momenta, leaving invariant the kinetic energy, and the contragredient transformation of the positions as dual variables:

$$\left. \begin{array}{rl} \frac{\vec{p}_1^2}{m_1} + \frac{\vec{p}_2^2}{m_2} & = \frac{\vec{P}^2}{m} + \frac{\vec{p}^2}{M} \\ \vec{p}_1 d_t\vec{x}_1 + \vec{p}_2 d_t\vec{x}_2 & = \vec{P} d_t\vec{X} + \vec{p} d_t\vec{x} \end{array} \right\} \iff \left\{ \begin{array}{rl} O(\beta)\begin{pmatrix} \frac{\vec{p}_1}{\sqrt{m_1}} \\ \frac{\vec{p}_2}{\sqrt{m_2}} \end{pmatrix} & = \begin{pmatrix} \frac{\vec{p}}{\sqrt{m}} \\ \frac{\vec{P}}{\sqrt{M}} \end{pmatrix}, \\ \check{O}(\beta)\begin{pmatrix} \sqrt{m_1}\vec{x}_1 \\ \sqrt{m_2}\vec{x}_2 \end{pmatrix} & = \begin{pmatrix} \sqrt{m}\vec{x} \\ \sqrt{M}\vec{X} \end{pmatrix}, \end{array} \right.$$

$$\text{with } O(\beta) = \begin{pmatrix} \cos\beta & -\sin\beta \\ \sin\beta & \cos\beta \end{pmatrix} = O^{-1T}(\beta) = \check{O}(\beta) \in \mathbf{SO}(2);$$

$\sqrt{\mu}$ and $\frac{1}{\sqrt{\mu}}$ for the masses $\mu \in \{m_1, m_2, m, M\}$ arise as normalizations, inverse to each other, for the corresponding dual position-momentum pairs.

The distance dependence determines the rotation angle

$$\vec{x} \sim \vec{x}_1 - \vec{x}_2 \Rightarrow \tan^2\beta = \tfrac{m_1}{m_2}.$$

Its normalization

$$\vec{x} = \vec{x}_1 - \vec{x}_2 \Rightarrow \sin^2\beta = \frac{m_1}{m_1+m_2}, \quad \vec{X} = \sin^2\beta\, \vec{x}_1 + \cos^2\beta\, \vec{x}_2$$

determines the normalizations of the collective variables, i.e., the value of the reduced mass m and the center mass (sum of the masses) M whose square roots are the height and hypotenuse respectively in the *center of mass triangle* $(\sqrt{m_1}, \sqrt{m_2}|\sqrt{M}, \sqrt{m})$ with the square roots of the individual masses as orthogonal sides:

$$\tan\beta = \sqrt{\frac{m_1}{m_2}}, \quad m_1 m_2 = Mm, \quad \begin{cases} \frac{1}{m} &= \frac{1}{m_1} + \frac{1}{m_2}, \\ M &= m_1 + m_2. \end{cases}$$

The Lagrangian is the sum for free center of mass motion and the reduced dynamics

$$\begin{aligned} L(1,2) &= \vec{P} d_t \vec{X} - \frac{\vec{P}^2}{2M} + L(\vec{x}, \vec{p}), \\ L(\vec{x}, \vec{p}) &= \vec{p} d_t \vec{x} - H, \quad H = \frac{\vec{p}^2}{2m} + V(\vec{x}). \end{aligned}$$

The contragredient transformations lead to the decomposition of the full angular momentum into center of mass and intrinsic angular momentum:

$$\vec{\mathcal{L}}_1 + \vec{\mathcal{L}}_2 = \vec{x}_1 \times \vec{p}_1 + \vec{x}_2 \times \vec{p}_2 = \vec{X} \times \vec{P} + \vec{x} \times \vec{p} = \vec{\mathcal{L}}_c + \vec{\mathcal{L}}.$$

The angular momenta implement the infinitesimal rotations, classically via the Poisson Lie bracket $[f,g]_P = \frac{\partial f}{\partial \vec{p}}\frac{\partial g}{\partial \vec{x}} - \frac{\partial f}{\partial \vec{x}}\frac{\partial g}{\partial \vec{p}}$ with $[p^a, x^b]_P = \delta^{ab}$. The Euclidean action groups for the two mass points are rearranged:

$$\begin{aligned} \{\vec{\mathcal{L}}_1, \vec{p}_1\} \cup \{\vec{\mathcal{L}}_2, \vec{p}_2\} &\leftrightarrow \{\vec{\mathcal{L}}_c, \vec{P}\} \cup \{\vec{\mathcal{L}}, \vec{p}\}, \\ [\mathbf{SO}(3)\ \vec{\times}\ \mathbb{R}^3]_1 \times [\mathbf{SO}(3)\ \vec{\times}\ \mathbb{R}^3]_2 &\leftrightarrow [\mathbf{SO}(3)\ \vec{\times}\ \mathbb{R}^3]_c \times [\mathbf{SO}(3)\ \vec{\times}\ \mathbb{R}^3], \end{aligned}$$

with $[\mathbf{SO}(3)\ \vec{\times}\ \mathbb{R}^3]_c$ the invariance group for the center of mass motion.

9.2 Intrinsic and ad hoc Units

All that we measure is ultimately qualifiable by apparently three basic units. A dynamics comes with its own *intrinsic units*, in general different from human order-of-magnitude related units for length, mass, and time, e.g., meter (m), kilogram (kg), and second (s).[1] This will be exemplified with the Kepler potential.

For Newton's potential in the reduced form two intrinsic units are given by the reduced mass and the product of Newton's constant with the center mass, e.g., for our planetary system with the masses $m_{1,2}$ of Sun and Earth,

$$m_\bullet \sim 6 \times 10^{24}\ \text{kg}, \quad G_N M_\circ \sim 1.3 \times 10^{20}\ \frac{\text{m}^3}{\text{s}^2}.$$

[1] Ultimately, also the human order-of- magnitude units are intrinsic: Meter and kilogram quantify the size and mass of things, e.g., babies and stones, human beings work with. Their dynamics, affected by Earth's surface gravity with acceleration $g_\bullet \sim 10\frac{\text{m}}{\text{s}^2}$, has the second as appropriate time unit. With Eratostenes' Earth circumference $2\pi R_\bullet \sim 4 \times 10^7$m and its average density (e.g., iron) $\rho_\bullet \sim 5.5 \times 10^3 \frac{\text{kg}}{\text{m}^3}$, one comes close to Newton's constant $G_N = \frac{3}{2}\frac{g_\bullet}{2\pi R_\bullet \rho_\bullet} \sim 6.7 \times 10^{-8} \frac{\text{m}^3}{\text{s}^2\text{kg}}$.

In contrast to relativity with the speed of light and the Schwarzschild radius, e.g., $\frac{G_N M_\odot}{c^2} \sim 1.5 \times 10^3$ m, there does not exist a third intrinsic unit in the nonrelativistic classical formulation. To determine all order of magnitudes one ad hoc unit has to be added, e.g., via the boundary conditions or a known unit, e.g., a time $T_\bullet$ (Earth year with $\frac{T_\bullet}{2\pi} \sim 5 \times 10^6$ s) or a distance $\ell_\bullet$ (Sun-Earth distance) obtained from measurements. Then all units are determined.[2]

For the electrostatic Coulomb interaction in the quantum case a basis for all units is given with the three intrinsic units $\{\hbar, \alpha c, m\}$ involving Planck's constant $\hbar$, the speed of light c multiplied with Sommerfeld's constant α, then only integer charge numbers are left,

$$\begin{aligned} g_0 &= \frac{1}{4\pi\epsilon_0} Q_1 Q_2 = \hbar\alpha c z_1 z_2, \\ \alpha c &= \frac{e^2}{4\pi\epsilon_0\hbar} \sim \frac{1}{137} c \sim 2.2 \times 10^6 \ \tfrac{\text{m}}{\text{s}}, \quad z_{1,2} = \frac{Q_{1,2}}{e} \in \mathbb{Z}, \end{aligned}$$

and, in the case of the atoms, the reduced mass for the electron-proton system:

$$\frac{m_e}{m_p} \sim \frac{1}{1837}, \quad m \sim m_e \sim 9.1 \times 10^{-31} \text{ kg}.$$

Hence characteristic atomic units for length and energy are, with 1 eV $\sim 1.6 \times 10^{-19} \frac{\text{kg m}^2}{\text{s}^2}$,

$$\begin{array}{l} \text{atomic units} \\ g_0 = \hbar\alpha c z_1 z_2 \end{array} \left\{ \begin{array}{lll} \hbar & \sim 1.05 \times 10^{-34} \frac{\text{kg m}^2}{\text{s}}, & \\ \ell & = \frac{\hbar}{m\alpha c |z_1 z_2|}, & \frac{\hbar}{m\alpha c} \sim 0.5 \times 10^{-10} \text{ m (Bohr length)}, \\ \mathcal{E} & = m(\alpha c z_1 z_2)^2, & \frac{m}{2}\alpha^2 c^2 \sim 13.6 \text{ eV (Rydberg energy)}. \end{array} \right.$$

With the two intrinsic units $\{m, G_N M\}$ and one ad hoc unit for Newton's gravitation, e.g., a length ℓ, and the three intrinsic units $\{m, \alpha c, \hbar\}$ for electrostatics in the quantum case the Kepler Hamiltonian is written in dimenionless variables:

$$H = \frac{\vec{p}^2}{2} + V(r), \quad V(r) = \frac{\delta}{r} \text{ with } \left\{ \begin{array}{ll} \delta = -1, & \text{Newton potential} \\ & \text{(attractive)}, \\ \delta = \mp 1, & \text{Coulomb potential} \\ & \text{(attractive, repulsive)}. \end{array} \right.$$

The variables with nontrivial intrinsic units arise from the number-valued ones by appropriate multiplication.

9.3 Symmetries of the Kepler Dynamics

A Hamiltonian $H = \frac{\vec{p}^2}{2} + V(\vec{x})$ implements a time development:

$$\begin{aligned} d_t \vec{x} &= [H, \vec{x}]_P = \vec{p}, \\ d_t \vec{p} &= [H, \vec{p}]_P = -\frac{\partial V}{\partial \vec{x}}. \end{aligned}$$

[2]Equalizing energy units yields, without Kepler's third law, orders of magnitude $\frac{\mathcal{E}}{m_\bullet} = \frac{G_N M_\odot}{\ell_\bullet} = \frac{\ell_\bullet^2}{2T_\bullet^2}$ and $\ell_\bullet \sim 1.5 \times 10^{11}$ m.

The rotation invariance of a dynamics with $V(|\vec{x}_1 - \vec{x}_2|)$ is, as seen in the center of mass transformation above, an $\mathbf{SO}(3) \times \mathbf{SO}(3)$-invariance for center of mass and intrinsic rotation for the center of mass system.

Rotation invariance is equivalent to angular momentum conservation:

$$H = \frac{\vec{p}^2}{2} + V(r),\ \vec{\mathcal{L}} = \vec{x} \times \vec{p} \Rightarrow \begin{cases} [\mathcal{L}^a, \mathcal{L}^b]_P &= -\epsilon^{abc}\mathcal{L}^c, \\ [H, \vec{\mathcal{L}}]_P &= 0. \end{cases}$$

Therefore, in classical physics, all orbits with conserved angular momentum are planar, $\vec{\mathcal{L}}\vec{x} = 0 \Rightarrow \vec{\mathcal{L}}\vec{p} = 0$.

Polar coordinates with classical radial momentum p_r have the momentum square

$$r = |\vec{x}|,\quad p_r = \vec{p}\frac{\vec{x}}{r},\quad \vec{p}^2 = p_r^2 + \frac{\vec{\mathcal{L}}^2}{r^2}.$$

The planar orbits with time translation and rotation invariants, $H = E$ and $\mathcal{L}^2 = L^2$, are obtained by one integration:

$$H = \frac{p_r^2}{2} + \frac{\vec{\mathcal{L}}^2}{2r^2} + V(r),\quad \vec{\mathcal{L}}^2 = L^2 \Rightarrow \begin{cases} d_t r &= p_r = \sqrt{2(E - V(r)) - \frac{L^2}{r^2}}, \\ d_t p_r &= \frac{L^2}{r^3} - \frac{dV(r)}{dr}, \end{cases}$$

$$\text{orbit: } t - t_0 = \int_{r(t_0)}^{r(t)} \frac{rdr}{\sqrt{2Er^2 - 2V(r)r^2 - L^2}} \Rightarrow r(t).$$

For the Kepler dynamics with quadratic polynomial $2Er^2 - 2\delta r - L^2$ under the root, attractive with $\delta = -1$, the phase space curves $E = H(r, p_r)$ are finite (compact) for $E < 0$ and infinite (noncompact) for $E \geq 0$, which is always the case for the repulsive potential $\delta = +1$.

The planetary orbits in the solar system are not only planar, they are not rosettes: In general, the Kepler Hamiltonian

$$H = \frac{\vec{p}^2}{2} + \frac{\delta}{r},\quad \delta = \pm 1 \text{ (repulsion, attraction)}$$

has, in addition to rotation invariance with conserved angular momentum $\vec{\mathcal{L}}$, the *Lenz-Runge invariance* with conserved *Lenz-Runge vector* $\vec{\mathcal{P}}$, perihelion vector in the solar system, given with the Poisson bracket

$$\vec{\mathcal{L}} = \vec{x} \times \vec{p},\quad \vec{\mathcal{P}} = \vec{p} \times \vec{\mathcal{L}} + \delta\frac{\vec{x}}{r} \Rightarrow \begin{cases} [H, \vec{\mathcal{P}}]_P &= 0, \\ [\mathcal{L}^a, \mathcal{P}^b]_P &= -\epsilon^{abc}\mathcal{P}^c, \\ [\mathcal{P}^a, \mathcal{P}^b]_P &= 2H\epsilon^{abc}\mathcal{L}^c. \end{cases}$$

The Kepler Hamiltonian should be compared in the following with the free Hamiltonian (constant potential)

$$H_0 = \frac{\vec{p}^2}{2} + V_0 \Rightarrow \begin{cases} [H_0, \vec{p}]_P &= 0, \\ [\mathcal{L}^a, p^b]_P &= -\epsilon^{abc}p^c, \\ [p^a, p^b]_P &= 0. \end{cases}$$

As seen below, the Lenz-Runge vector $\vec{\mathcal{P}}$ expands the translations $\vec{p}$ from a flat Euclidean to a hyperbolic position structure.

With the Hamiltonian in the Lenz-Runge vector brackets, three different types for the energy values from the spectrum of the Kepler Hamiltonian have to be distinguished to classify its possible symmetries. In all three cases, $E = 0$, $E > 0$, and $E < 0$, the dynamics is characterized by a real 6-dimensional invariance Lie algebra of rank 2, i.e., with two independent invariants.

For trivial energy, the symmetry is, as for the free Hamiltonian, the semi-direct Euclidean structure with rotations and translations in three dimensions:

$$\operatorname{spec} H \ni E = 0: \quad \begin{cases} [l^a, l^b] = -\epsilon^{abc} l^c, \ [l^a, p^b] = -\epsilon^{abc} p^c, \ [p^a, p^b] = 0, \\ \text{invariants: } \vec{p}^2, \ \vec{l}\vec{p}, \\ \text{representation: } l^a \longmapsto \mathcal{L}^a, \ p^a \longmapsto \mathcal{P}^a, \end{cases}$$
$$\text{Lie algebra:} \quad A_1^c \ \vec{\oplus} \ \mathbb{R}^3 = \log[\mathbf{SO}(3) \ \vec{\times} \ \mathbb{R}^3] \cong \mathbb{R}^6.$$

The angular momentum invariant $\vec{l}^2$ is not translation invariant.

A nontrivial energy can be used to renormalize the Lenz-Runge vector:

$$\operatorname{spec} H \ni E \neq 0, \ \vec{\mathcal{B}} = \frac{\vec{\mathcal{P}}}{\sqrt{2|H|}}: \quad \epsilon(E) = \frac{E}{|E|} = \frac{H}{|H|} \quad \begin{cases} [\mathcal{L}^a, \mathcal{L}^b]_P &= -\epsilon^{abc} \mathcal{L}^c, \\ [\mathcal{L}^a, \mathcal{B}^b]_P &= -\epsilon^{abc} \mathcal{B}^c, \\ [\mathcal{B}^a, \mathcal{B}^b]_P &= \epsilon(E) \epsilon^{abc} \mathcal{L}^c. \end{cases}$$

Positive energies lead to scattering orbits (classical and quantal); there arises the Lie algebra of the noncompact Lorentz group with the Lenz-Runge vector defining the "boosts" (thus called by analogy, not special relativistic transformations)

$$E > 0: \quad \begin{cases} [l^a, l^b] = -\epsilon^{abc} l^c = -[b^a, b^b], \ [l^a, b^b] = -\epsilon^{abc} b^c, \\ \text{invariants: } \vec{l}^2 - \vec{b}^2, \ \vec{l}\vec{b}, \\ \text{representation: } l^a \longmapsto \mathcal{L}^a, \ b^a \longmapsto \mathcal{B}^a, \end{cases}$$
$$\text{Lie algebra:} \quad A_1^c \ \oplus \ iA_1^c \cong \log \mathbf{SO}(1,3) \cong \mathbb{R}^6.$$

For negative energies, leading to bound orbits (classical and quantal), the symmetries constitute the compact Lie algebra of $\mathbf{SO}(4)$, locally isomorphic to $\mathbf{SO}(3) \times \mathbf{SO}(3)$:

$$E < 0: \quad \begin{cases} [l^a, l^b] = -\epsilon^{abc} l^c = [m^a, m^b], \ [l^a, m^b] = -\epsilon^{abc} m^c, \\ \vec{l}_\pm = \frac{\vec{l} \pm \vec{m}}{2} \Rightarrow [l^a_\pm, l^b_\pm] = -\epsilon^{abc} l^c_\pm, \ [l^a_+, l^b_-] = 0, \\ \text{invariants: } \vec{l}^2 + \vec{m}^2 = 2(\vec{l}_+^2 + \vec{l}_-^2), \ \vec{l}\vec{m} = \vec{l}_+^2 - \vec{l}_-^2, \\ \text{representation: } l^a \longmapsto \mathcal{L}^a, \ m^a \longmapsto \mathcal{B}^a, \end{cases}$$
$$\text{Lie algebra:} \quad A_1^c \ \oplus \ A_1^c \cong \log[\mathbf{SO}(3) \times \mathbf{SO}(3)] \cong \mathbb{R}^6.$$

The transition from positive to negative energies can be formulated as a transition from real momenta to imaginary "momenta":

$$\sqrt{2E} = \begin{cases} |\vec{q}|, & E > 0, \ \text{scattering orbits, real momenta,} \\ i|Q|, & E < 0, \ \text{bound orbits, imaginary "momenta."} \end{cases}$$

The Kepler Lie groups can be decomposed with rotation group classes

$$\begin{aligned}\mathbf{SO}(4) &\cong \mathbf{SO}(3) \times \mathbf{SO}(4)/\mathbf{SO}(3),\\ \mathbf{SO}_0(1,3) &\cong \mathbf{SO}(3) \times \mathbf{SO}_0(1,3)/\mathbf{SO}(3).\end{aligned}$$

The rotation group itself is a product of axial rotations $\mathbf{SO}(2)$ with the 2-sphere $\Omega^2 \cong \mathbf{SO}(3)/\mathbf{SO}(2)$ for the angular momentum direction, i.e., Ω^2 is the orientation manifold of the axial rotations. The Lenz-Runge operations, compact as a 3-sphere $\Omega^3 \cong \mathbf{SO}(4)/\mathbf{SO}(3)$ and noncompact as a 3-hyperboloid $\mathcal{Y}^3 \cong \mathbf{SO}_0(1,3)/\mathbf{SO}(3)$, are the orientation manifolds, or equivalence classes, of the rotation group. Also, these symmetric spaces contain a characteristic abelian subgroup and a 2-sphere for the Lenz-Runge vector direction in the Cartan factorization (chapter "Spacetime Translations")

$$\begin{pmatrix}\text{spherical}\\ \text{flat, Euclidean}\\ \text{hyperbolic}\end{pmatrix} : \begin{pmatrix}\mathbf{SO}(4)\\ \mathbf{SO}(3)\,\vec{\times}\,\mathbb{R}^3\\ \mathbf{SO}_0(1,3)\end{pmatrix} \cong \mathbf{SO}(2) \circ \Omega^2 \circ \begin{pmatrix}\mathbf{SO}(2)\\ \mathbb{R}_+\\ \mathbf{SO}_0(1,1)\end{pmatrix} \circ \Omega^2.$$

The 2-dimensional abelian subgroups reflect the rank 2 with the two independent invariants. The rotation group $\mathbf{SO}(3)$ in all groups determines the angular momentum. The representation of the second abelian factor decides on the spherical, parabolic (or flat), and hyperbolic orbits.

9.4 Classical Time Orbits

In classical theories one is primarily interested in the time orbits of the mass points in position space $\mathbb{R} \ni t \longmapsto \vec{x}(t) \in \mathbb{R}^3$, mathematically in the irreducible realizations of the time translation group $\mathbb{R}$. The characterizing eigenvalues, i.e., the invariant energies, are classically imposed by boundary or initial conditions.

9.4.1 Time Orbits in Position Space

The functions of position and momentum $(\vec{x}, \vec{p})$ build an associative unital algebra, for a classical framework commutative with the pointwise product. The problems with the Kepler potential $\frac{1}{r}$ at the origin $\vec{x} = 0$ are neglected, they deserve a more careful discussion. The commutative algebra has a noncommutative Lie algebra structure with the Poisson bracket. Therefore the three Kepler Lie algebras $\log \mathbf{SO}(4)$, $\log \mathbf{SO}_0(1,3)$ and $\log[\mathbf{SO}(3)\,\vec{\times}\,\mathbb{R}^3]$ act adjointly on this algebra.

The products of angular momentum with position, momentum, and Lenz-Runge vector vanish (orthogonality)

$$\vec{\mathcal{L}}\vec{x} = 0, \quad \vec{\mathcal{L}}\vec{p} = 0, \quad \vec{\mathcal{L}}\vec{\mathcal{P}} = 0,$$

i.e., position $\vec{x}$, momentum $\vec{p}$, and Lenz-Runge vector $\vec{\mathcal{P}}$ are in the position orbit plane. For gravity in the solar system, the orbit planarity following from rotation invariance constitutes Kepler's first law.

The invariant squares of angular momentum and perihelion vector combine the Hamiltonian and determine the energy E by angular momentum value L and perihelion value P:

$$\begin{aligned} \vec{\mathcal{P}}^2 &= 1 + 2H\vec{\mathcal{L}}^2 \Rightarrow E = \frac{P^2-1}{2L^2} \text{ with } |\vec{\mathcal{P}}| = P,\ |\vec{\mathcal{L}}| = L, \\ -\frac{1}{2H} &= \vec{\mathcal{L}}^2 - \epsilon(E)\vec{\mathcal{B}}^2. \end{aligned}$$

The time orbits in position space are conic sections, described by polar equations with one focus as origin (second Kepler law):

$$\begin{aligned} \vec{\mathcal{P}}\vec{x} = Pr\cos\varphi &= (\vec{p} \times \vec{\mathcal{L}})\vec{x} + \delta r = L^2 + \delta r \\ \Rightarrow r(\varphi) &= \frac{L^2}{P\cos\varphi - \delta} \text{ with } \delta = \pm 1, \end{aligned}$$

$\vec{x}$ directs to the peri- and aphelion for $\varphi = 0$ and $\varphi = \pi$ respectively. The invariants can be expressed by perihelion distance r_0 and momentum p_0 as possible initial conditions:

$$\text{for } \varphi = 0: \quad \vec{x}_0\vec{p}_0 = 0 \Rightarrow \begin{cases} L &= r_0 p_0, \\ P &= r_0 p_0^2 + \delta, \\ E &= \frac{p_0^2}{2} + \frac{\delta}{r_0}. \end{cases}$$

The connection between Cartesian $\vec{x} = (x, y, 0)$ and polar equations is given in the following table:

energy $E = \frac{P^2-1}{2L^2}$	$E < 0$	$E > 0$	$E = 0$
group orbit	$\mathbf{SO}(2)$, ellipse	$\mathbb{I}(2) \times \mathbf{SO}_0(1,1)$, hyperbola	$\mathbb{R}$, parabola
Cartesian equation	$\frac{x^2}{a^2} + \frac{y^2}{b^2} = 1$ $a \geq b$	$\frac{x^2}{a^2} - \frac{y^2}{b^2} = 1$ (right and left branch)	$y^2 = -2dx$ $d > 0$
foci distance $2c$	$c^2 = a^2 - b^2$	$c^2 = a^2 + b^2$	(one focus)
polar equation (with pole in the right focus)	$r(\varphi) = \frac{L^2}{P\cos\varphi+1}$ $0 < P < 1$	$r_{R,L}(\varphi) = \frac{L^2}{P\cos\varphi\pm 1}$ $P > 1$	$r(\varphi) = \frac{L^2}{\cos\varphi+1}$ $P = 1$
distance of pole to peri- and aphelion	$(a-c, a+c)$ $= (r(0), r(\pi))$ $= (\frac{L^2}{1+P}, \frac{L^2}{1-P})$	$(c-a, c+a)$ $= (r_R(0), r_L(0))$ $= (\frac{L^2}{P+1}, \frac{L^2}{P-1})$	$\frac{d}{2}$ $= r(0)$ $= \frac{L^2}{2}$
$(a, c, b) =$	$(-\frac{1}{2E}, -\frac{P}{2E}, \frac{L}{\sqrt{-2E}})$	$(\frac{1}{2E}, \frac{P}{2E}, \frac{L}{\sqrt{2E}})$	
$(L^2, P) =$	$(\frac{b^2}{a}, \frac{c}{a})$	$(\frac{b^2}{a}, \frac{c}{a})$	$(d, 1)$

Cartesian and polar equations for conic sections

The squared length of the perihelion vector is the discriminant of the second order polynomial that occurs in the equation of motion:

$$\begin{aligned} \frac{dr}{dt} = [H, r]_P = p_r,\ p_r^2 &= \frac{2Er^2 - 2\delta r - L^2}{r^2}, \\ -\det\begin{pmatrix} 2E & \delta \\ \delta & -L^2 \end{pmatrix} &= 1 + 2EL^2 = P^2 \text{ for } \delta^2 = 1. \end{aligned}$$

Attraction $\delta = -1$, e.g., in gravity or with charge numbers of opposite sign $z_1 z_2 < 0$ in electrostatics, can come with negative and positive energies for compact and noncompact orbits respectively:

$$\delta = -1: \ P^2 = 1 + 2EL^2 \begin{cases} < 1 \iff & -\frac{1}{2L^2} \le E < 0 \\ & \text{(ellipse)}, \\ = 1 \iff & E = 0 \\ & \text{(parabola)}, \\ > 1 \iff & E > 0 \\ & \text{(hyperbola branch around pole)}. \end{cases}$$

Repulsion $\delta = 1$, e.g., for charge numbers of equal sign $z_1 z_2 > 0$, has positive energies only (noncompact orbits):

$$\delta = 1: \ P^2 = 1 + 2EL^2 > 1 \iff \quad E > 0 \\ \text{(hyperbola branch, not around pole)}.$$

Since the angular momentum is twice the time change of the orbit area

$$\vec{\mathcal{L}} = \vec{x} \times d_t\vec{x} = 2d_t\vec{\mathcal{A}}$$

the orbit area $A = |\vec{\mathcal{A}}|$ and the orbit time T for ellipses are related to the conserved angular momentum, which relates the major axis of the ellipse to the orbit time (Kepler's third law):

$$L = 2\frac{A}{T} = 2\frac{\pi ab}{T} \Rightarrow a^3 = (\frac{T}{2\pi})^2 = -\frac{1}{8E^3}.$$

For the free theory the orbits are lines:

$$\delta = 0: \quad E = \frac{\vec{p}^2}{2}, \ r(\varphi) = \frac{r(0)}{\cos\varphi}, \ \begin{pmatrix} x \\ y \end{pmatrix} = \begin{pmatrix} 1 & 0 \\ \xi & 1 \end{pmatrix} \begin{pmatrix} 1 \\ 0 \end{pmatrix}.$$

9.4.2 Orbits as Time Classes

A dynamics leads to realizations of the time translation group. The solutions in classical physics are irreducible time orbits in position space. They have to be isomorphic to quotient groups of time $\mathbb{R}$ with the kernel K of the time realization,

$$\mathbb{R} \ni t \longmapsto \vec{x}(t) \in \mathbb{R}^3, \ \ \vec{x}[\mathbb{R}] \simeq \mathbb{R}/K \subset \mathbb{R}^3.$$

The $\mathbb{R}$-subgroups that can arise as kernels for Lie quotient groups are the full group $\mathbb{R}$, the trivial group $\{0\}$, and, up to isomorphism, the discrete group $\mathbb{Z}$.

All possible quotient groups $\mathbb{R}/K$ (time equivalence classes) for time realizations are seen in the sky as solutions of the Kepler dynamics. The trivial time representation is seen in our Sun, assumed infinitely heavy, with trivial point orbit $\mathbb{R}/\mathbb{R} \cong \{1\}$ and energy $E = -\infty$. The compact unfaithful time realizations with kernel $\mathbb{Z}$ give bound orbits (planets on ellipses), isomorphic to the circle $\mathbf{U}(1)$. For the Earth, $\mathbb{Z}$ counts the years, e.g., in Charlemagne's crowning date with cyclic units for days and months,

$$\underbrace{\text{25 december}}_{\text{for } \mathbb{R}/\mathbb{Z}\cong\mathbf{U}(1)} \quad \underbrace{+800.}_{\mathbb{Z}}$$

The noncompact faithful realizations, isomorphic to all of $\mathbb{R}$, give scattering orbits (never-returning comets on the branch of an hyperbola around the Sun or on a parabola).

The conic sections (ellipses and hyperbolas) for the planar orbits have the metric tensors

$$\frac{x^2}{a^2} \pm \frac{y^2}{b^2} = 1, \ \begin{pmatrix} \frac{1}{a^2} & 0 \\ 0 & \pm\frac{1}{b^2} \end{pmatrix} = \begin{pmatrix} 4E^2 & 0 \\ 0 & -\frac{2E}{L^2} \end{pmatrix} = \frac{4E^2}{L\sqrt{2|E|}} \begin{pmatrix} \eta & 0 \\ 0 & \pm\frac{1}{\eta} \end{pmatrix}.$$

The ratio of the units for the two directions $\eta = \frac{b}{a} = L\sqrt{2|E|}$ is the product of angular momentum with energy. The orbits in position space can be parametrized as follows:

$$\begin{array}{lll} \text{ellipses:} & \mathbf{SO}(2) \cong \mathbb{R}/\mathbb{Z}: & \begin{pmatrix} x \\ y \end{pmatrix} = \begin{pmatrix} \cos\theta & -\eta\sin\theta \\ \frac{1}{\eta}\sin\theta & \cos\theta \end{pmatrix}\begin{pmatrix} 1 \\ 0 \end{pmatrix}, \\ \text{hyperbolas:} & \mathbf{SO}_0(1,1) \cong \mathbb{R}: & \begin{pmatrix} x \\ y \end{pmatrix} = \begin{pmatrix} \cosh\psi & \eta\sinh\psi \\ \frac{1}{\eta}\sinh\psi & \cosh\psi \end{pmatrix}\begin{pmatrix} \pm 1 \\ 0 \end{pmatrix}. \end{array}$$

The time parametrization comes via time-dependent group parameters, $t \longmapsto \theta(t), \psi(t)$.

How are the time orbits, i.e., the quotient groups $\mathbb{R}/K$, related to the full Kepler invariance groups $\mathbf{SO}(4)$, $\mathbf{SO}_0(1,3)$, and $\mathbf{SO}(3) \vec{\times} \mathbb{R}^3$ acting on the algebra with the position-momentum functions? An individual solution of a dynamics (time realizations), here one orbit $\vec{x}$ in position space, need not have all the invariances G of the Hamiltonian. Equivalent solutions are on the orbit $\{g \bullet \vec{x} \mid g \in G\} = G \bullet \vec{x} \cong G/H$ of the invariance group G in the solution space (not in position space) with a remaining subgroup H-symmetry. For example, by rotating the initial conditions of one solution for a rotation invariant Hamiltonian one obtains an equivalent solution. The fixgroup H of a solution is the G-subgroup that leaves the orbit in position space invariant, as a whole, not its individual points. The 6-parametric invariance of the Kepler dynamics is, up to the trivial solution with position and momentum $(\vec{x}, \vec{p}) = (0,0)$, broken to a 1-parametric fixgroup symmetry for a solution since the choice of angular momentum and Lenz-Runge vectors $(\vec{\mathcal{L}}, \vec{\mathcal{P}})$ (six parameters) with $\vec{\mathcal{L}}\vec{\mathcal{P}} = 0$ (one condition) to determine an orbit comes from a 5-parametric manifold:

$$\begin{array}{rlll} H & = \mathbf{SO}(2), & \mathbf{SO}_0(1,1), & \mathbb{R}, \\ G/H & = \mathbf{SO}(4)/\mathbf{SO}(2), & \mathbf{SO}_0(1,3)/\mathbf{SO}_0(1,1), & [\mathbf{SO}(3) \vec{\times} \mathbb{R}^3]/\mathbb{R}. \end{array}$$

The nontrivial orbits $t \longmapsto \vec{x}(t)$ in position as quotient groups $\mathbb{R}/K$ of time are isomorphic to the characterizing fixgroups H in the Kepler groups. In all three cases, the solution degeneracy is $G/H \cong \mathbf{SO}(3) \times \Omega^2$ with all three compact rotation parameters for angular momentum $\vec{\mathcal{L}}$ and two compact parameters for the direction of the Lenz-Runge vector $\vec{\mathcal{P}}$.

9.4.3 Two-Sided Contraction to the Free Theory

The complex 6-dimensional group $\mathbf{SO}(\mathbb{C}^4)$, e.g., in the defining 4-dimensional representation for its Lie algebra,

$$l(\vec{\varphi},\vec{\psi}) = \vec{\varphi}\vec{L} + \vec{\psi}\vec{B} = \left(\begin{array}{c|ccc} 0 & \psi_1 & \psi_2 & \psi_3 \\ \psi_1 & 0 & \varphi_3 & -\varphi_2 \\ \psi_2 & -\varphi_3 & 0 & \varphi_1 \\ \psi_3 & \varphi_2 & -\varphi_1 & 0 \end{array}\right) \in \log \mathbf{SO}(\mathbb{C}^4),$$

with complex rank 2 and the corresponding two invariant bilinear forms $\kappa_{1,2}$ from the coefficients of the characteristic polynomial

$$\begin{aligned} \det[l(\vec{\varphi},\vec{\psi}) - \lambda\mathbf{1}_4] &= \lambda^4 + \lambda^2(\vec{\varphi}^2 - \vec{\psi}^2) - (\vec{\varphi}\vec{\psi})^2, \\ \Rightarrow &\left\{\begin{array}{ll} \kappa_1(l,l) & = -\operatorname{tr} l \circ l = \vec{\varphi}^2 - \vec{\psi}^2, \\ \kappa_2(l,l) & = \sqrt{-\det l} = \vec{\varphi}\vec{\psi}, \end{array}\right. \end{aligned}$$

has as real forms the two Kepler groups

$$\begin{array}{ll} \mathbf{SO}_0(1,3) & \text{with } \varphi_a \in \mathbb{R},\ \psi_a \in \mathbb{R}, \\ \mathbf{SO}(4) & \text{with } \varphi_a \in \mathbb{R},\ \psi_a = i\chi_a \in i\mathbb{R}. \end{array}$$

Both groups are expansions of the nonsemisimple Euclidean group $\mathbf{SO}(3)\ \vec{\times}\,\mathbb{R}^3$,

$$\left(\begin{array}{c|ccc} 0 & 0 & 0 & 0 \\ \xi_1 & 0 & \varphi_3 & -\varphi_2 \\ \xi_2 & -\varphi_3 & 0 & \varphi_1 \\ \xi_3 & \varphi_2 & -\varphi_1 & 0 \end{array}\right) \in \log[\mathbf{SO}(3)\ \vec{\times}\,\mathbb{R}^3],$$

with the "boosts" $\vec{\psi}$ or the additional "internal rotations" $\vec{\chi}$ as "unflattened, expanded translations".

Concversely, the translations $\mathbb{R}^3$ arise by Inönü-Wigner contraction to the Galileo group (chapter "Spacetime Translations") as tangent space both of the compact 3-sphere $\mathbf{SO}(4)/\mathbf{SO}(3)$, related to bound structures, and of the noncompact 3-hyberboloid $\mathbf{SO}_0(1,3)/\mathbf{SO}(3)$, related to scattering structures, to the free theory with nonsemisimple symmetry

$$\begin{array}{ccccc} \text{spherical} & & \text{flat} & & \text{hyperbolic} \\ \begin{pmatrix} \mathbf{SO}(4) \\ \cup \\ \mathbf{SO}(4)/\mathbf{SO}(3) \\ \cup \\ \mathbf{SO}(2) \end{pmatrix} & \stackrel{\eta\to 0}{\longrightarrow} & \begin{pmatrix} \mathbf{SO}(3)\ \vec{\times}\,\mathbb{R}^3 \\ \cup \\ \mathbb{R}^3 \\ \cup \\ \mathbb{R} \end{pmatrix} & \stackrel{\eta\to 0}{\longleftarrow} & \begin{pmatrix} \mathbf{SO}_0(1,3) \\ \cup \\ \mathbf{SO}_0(1,3)/\mathbf{SO}(3) \\ \cup \\ \mathbf{SO}_0(1,1) \end{pmatrix}. \end{array}$$

The contraction procedure will be given explicitly for the decisive abelian subgroups (last line): The relevant contraction parameter $\eta^2 = \frac{b^2}{a^2} = 2|E|L^2$ with energy and angular momentum is the ratio of the units for the two directions in the conic sections. It is the analogue to the ratio of a time unit to a position unit $\frac{1}{c^2} = \frac{\tau^2}{\ell^2}$ in the archetypical Inönü-Wigner contraction. For the Coulomb interaction $V(r) = \frac{1}{4\pi\epsilon_0}\frac{z_1 z_2 e^2}{r}$ with $\delta = \epsilon(z_1 z_2) = \pm 1$ and the unit for the product of energy and angular momentum $[EL^2] = \frac{me^2}{\epsilon_0}$ the contraction limit is realizable by switching off the interaction $\frac{e^2}{\epsilon_0} \to 0$. For the planetary system with gravitational interaction $-\frac{G_N Mm}{r}$ and $[EL^2] = m(G_N Mm)^2$ the contraction limit is realized by $G_N \to 0$.

The ratio of the units is used for a renormalization of the Lie parameters $(\chi, \psi) \overset{\eta}{\longmapsto} \xi$, the analogue for the reparametrization from rapidity to velocity $\tanh\psi = \frac{v}{c}$. The contraction of the length ratio $\eta = \frac{b}{a} = L\sqrt{2|E|} \to 0$ is the analogue to the contraction to an infinite velocity $\frac{1}{c} = \frac{\tau}{\ell} \to 0$:

$$\left.\begin{array}{r} E < 0 \\ \tan\chi = \eta\xi \end{array}\right\}: \quad \mathbf{SO}(2) \ni \begin{pmatrix} \cos\chi & \eta\, i\sin\chi \\ \frac{1}{\eta}\, i\sin\chi & \cos\chi \end{pmatrix}$$

$$= \frac{1}{\sqrt{1+\eta^2\xi^2}}\begin{pmatrix} 1 & \eta^2 i\xi \\ i\xi & 1 \end{pmatrix} \to \begin{pmatrix} 1 & 0 \\ i\xi & 1 \end{pmatrix} \text{ for } \eta \to 0,$$

$$\left.\begin{array}{r} E > 0 \\ \tanh\psi = \eta\xi \end{array}\right\}: \mathbf{SO}_0(1,1) \ni \begin{pmatrix} \cosh\psi & \eta\, \sinh\psi \\ \frac{1}{\eta}\, \sinh\psi & \cosh\psi \end{pmatrix}$$

$$= \frac{1}{\sqrt{1-\eta^2\xi^2}}\begin{pmatrix} 1 & \eta^2\xi \\ \xi & 1 \end{pmatrix} \to \begin{pmatrix} 1 & 0 \\ \xi & 1 \end{pmatrix} \text{ for } \eta \to 0.$$

The contracted additive group $\mathbb{R}$ comes in a multiplicative representation with nilpotent operations (chapter "Time Representations"), typical for the non-semisimplicity of semidirect groups

$$\mathbb{R} \ni \xi \longmapsto \begin{pmatrix} 1 & 0 \\ \xi & 1 \end{pmatrix} = \exp\begin{pmatrix} 0 & 0 \\ \xi & 0 \end{pmatrix} \text{ with } \begin{pmatrix} 0 & 0 \\ \xi & 0 \end{pmatrix}^2 = 0.$$

The contraction limit describes the line orbits of the free theory, not the parabolas of the Kepler potential.

9.5 Kepler Bound State Vectors

In the quantum case, the Kepler Hamiltonian with angular momentum and Lenz-Runge vectors

$$H = \frac{\vec{p}^2}{2} + \frac{\delta}{r}, \quad \vec{\mathcal{L}} = \vec{x} \times \vec{p}, \quad \vec{\mathcal{P}} = \frac{\vec{p}\times\vec{\mathcal{L}} - \vec{\mathcal{L}}\times\vec{p}}{2} + \delta\frac{\vec{x}}{r}$$

build the same three Lie algebra structures as in the classical case:

$$[H, \vec{\mathcal{L}}] = 0, \quad [H, \vec{\mathcal{P}}] = 0, \quad \left\{\begin{array}{ll} [i\mathcal{L}^a, i\mathcal{L}^b] & = -\epsilon^{abc} i\mathcal{L}^c, \\ [i\mathcal{L}^a, i\mathcal{P}^b] & = -\epsilon^{abc} i\mathcal{P}^c, \\ [i\mathcal{P}^a, i\mathcal{P}^b] & = 2H\epsilon^{abc} i\mathcal{L}^c. \end{array}\right.$$

The additional i-factor is related to the different dual normalization in the Poisson bracket $[p, x]_P = 1$ and the quantum commutator $i[p, x] = 1$.

Again, the squares of angular momentum and Lenz-Runge vector determine the Hamiltonian

$$\vec{\mathcal{P}}^2 = 1 + 2H(\vec{\mathcal{L}}^2 + 1) \Rightarrow -\frac{1}{2H} = 1 + \mathcal{L}^2 - \frac{\mathcal{P}^2}{2H}$$

with an additional constant compared to the classical case.

For $\delta = -1$ and negative energies one has representations of the compact symmetry Lie algebra

$$\text{spec}\, H \ni E < 0: \left\{\begin{array}{l} \vec{\mathcal{B}} = \frac{\vec{\mathcal{P}}}{\sqrt{-2H}}, \quad \vec{\mathcal{J}}_\pm = \frac{\vec{\mathcal{L}} \pm \vec{\mathcal{B}}}{2}, \\ [i\mathcal{J}^a_\pm, i\mathcal{J}^b_\pm] = -\epsilon^{abc} i\mathcal{J}^c_\pm, \quad [\mathcal{J}^a_+, \mathcal{J}^b_-] = 0, \\ \text{invariants: } \vec{\mathcal{L}}^2 + \vec{\mathcal{B}}^2 = 2(\vec{\mathcal{J}}^2_+ + \vec{\mathcal{J}}^2_-), \quad \vec{\mathcal{L}}\vec{\mathcal{B}} = \vec{\mathcal{J}}^2_+ - \vec{\mathcal{J}}^2_-, \end{array}\right.$$

$$\log[\mathbf{SO}(3) \times \mathbf{SO}(3)] \cong A_1^c \oplus A_1^c.$$

The quantum algebra $\mathbf{Q}_-(\mathbb{C}^8) = \mathbf{Q}_-(\mathbb{C}^4) \otimes \mathbf{Q}_-(\mathbb{C}^4)$ for the representation space of the defining 4-dimensional representation of $A_1^c \oplus A_1^c$ is generated by two pairs of Pauli spinors with Bose statistics and Euclidean conjugation

$$\begin{array}{l} \mathbf{Q}_-(\mathbf{V} \oplus \mathbf{U}) \cong \mathbf{Q}_-(\mathbf{V}) \otimes \mathbf{Q}_-(\mathbf{U}), \quad V \cong \mathbb{C}^2 \cong U, \\ \text{nontrivial: } [\mathrm{u}_A^\star, \mathrm{u}^B] = \delta_A^B, \quad [\mathrm{a}_A^\star, \mathrm{a}^B] = \delta_A^B. \end{array}$$

The double "spin" Lie algebra is implemented by the six basic vectors

$$\begin{array}{rll} \log[\mathbf{SU}(2) \times \mathbf{SU}(2)] : & i\vec{\mathcal{J}}_+ = i\mathrm{u}\frac{\vec{\sigma}}{2}\mathrm{u}^\star, & i\vec{\mathcal{J}}_- = i\mathrm{a}\frac{\vec{\sigma}}{2}\mathrm{a}^\star, \\ \log \mathbf{SO}(4) : & \vec{\mathcal{L}} = \vec{\mathcal{J}}_+ + \vec{\mathcal{J}}_-, & \vec{\mathcal{B}} = \vec{\mathcal{J}}_+ - \vec{\mathcal{J}}_-. \end{array}$$

The product Fock state leads to a scalar product space isomorphic to the product of the two polynomial algebras in the basic vectors (four creation operators)

$$\begin{array}{rl} \mathrm{FOCK}_-(V \oplus U) & \cong \mathrm{FOCK}_-(V) \otimes \mathrm{FOCK}_-(U) \\ & \cong \bigvee V \otimes \bigvee U \cong \mathbb{C}[\mathrm{u}^A] \otimes \mathbb{C}[\mathrm{a}^A], \\ \multicolumn{2}{l}{\text{basis } \{(\mathrm{u}^1)^{n_1}(\mathrm{u}^2)^{n_2}(\mathrm{a}^1)^{m_1}(\mathrm{a}^2)^{m_2}|0\rangle \bigm| n_{1,2}, m_{1,2} = 0, 1, \dots\},} \\ \multicolumn{2}{l}{\text{with } \mathrm{u}_A^\star|0\rangle = 0 = \mathrm{a}_A^\star|0\rangle, \quad \langle 0|\mathrm{u}_A^\star \mathrm{u}^B|0\rangle = \delta_A^B = \langle 0|\mathrm{a}_A^\star \mathrm{a}^B|0\rangle.} \end{array}$$

The eigenvalue of a time-conserved operator Q, i.e., $[H, Q] = 0$, for a simultaneous eigenvector of the Hamiltonian H and Q is written in the following as $\langle Q \rangle = \langle E|Q|E \rangle$.

The weight diagrams of the irreducible $\mathbf{SU}(2) \times \mathbf{SU}(2)$-representations

$$\begin{array}{rl} \mathbf{irrep}\,[\mathbf{SU}(2) \times \mathbf{SU}(2)] & = \mathbf{irrep}\,\mathbf{SU}(2) \times \mathbf{irrep}\,\mathbf{SU}(2) \\ & = \{(2J_1, 2J_2) \bigm| J_{1,2} = 0, \frac{1}{2}, 1, \dots\} \end{array}$$

occupy $(1+2J_1)(1+2J_2)$ points of a rectangular grid. The two invariants determine the occurring representations. The triviality of the invariant $\vec{\mathcal{L}}\vec{\mathcal{P}} = 0$ (classical orthogonality of angular momentum and Lenz-Runge vector) "synchronizes" the centers $\mathbb{I}_2 = \{\pm 1\}$ of both $\mathbf{SU}(2)$'s (central correlation; two cycles give one bicycle) and leads to the relevant group $\mathbf{SO}(4)$ (chapter "Rational Quantum Numbers"):

$$\frac{\mathbf{SU}(2)\times\mathbf{SU}(2)}{\mathbb{I}(2)} \cong \mathbf{SO}(4) \quad \text{with } \mathbb{I}(2) = \{(1,1), (-1,-1)\} \subset \mathbf{SU}(2) \times \mathbf{SU}(2).$$

It enforces even the equality of both $\mathbf{SU}(2)$-invariants $J_+ = J_- = J$:

$$0 = \vec{\mathcal{L}}\vec{\mathcal{B}} = \vec{\mathcal{J}}_+^2 - \vec{\mathcal{J}}_-^2 \Rightarrow \langle \vec{\mathcal{J}}_+^2 \rangle = \langle \vec{\mathcal{J}}_-^2 \rangle = J(1+J), \quad J = 0, \tfrac{1}{2}, 1, \tfrac{3}{2}, \dots.$$

Therefore the energy-degenerated representations are of the type $(2J, 2J)$; the multiplets of both A_1^c-representations have equal dimension $1+2J$. The $\mathbf{SU}(2)$-multiplet dimension is the principal quantum number $k = 1+2J$. The weight diagrams occupy $(1+2J)^2$ points of a square grid:

$$\mathbf{irrep}\,\mathbf{SO}(4)|_{\mathrm{Kepler}} = \{(2J, 2J) \bigm| J = 0, \tfrac{1}{2}, 1, \dots\}, \quad (2J, 2J) = \bigvee^{2J}(1,1).$$

The Kepler representations are the totally symmetrized products of the defining 4-dimensional $\mathbf{SO}(4)$-representation $(1,1)$.

Similar to the 1-dimensional harmonic oscillator with the 1-quantum state vector $|1\rangle$ as defining $\mathbf{U}(1)$-orbit, there is the state vector with the defining $\mathbf{SO}(4)$-representation $(1,1)$ for the atomic bound state vectors. The highest-weight vector in an irreducible representation space comes with highest "spins" $j_\pm = J$:

$$(\mathrm{u}^1\mathrm{a}^1)^{2J}|0\rangle = |J;J\rangle|J;J\rangle.$$

Its extremality involves the triviality for the action of the two raising operators

$$\begin{array}{llll}\text{raising: } \mathcal{J}_\pm^+ & = (\mathrm{u}^1\mathrm{u}_2^\star, \mathrm{a}^1\mathrm{a}_2^\star), & (\mathcal{L}^+, \mathcal{P}^+) & = \mathrm{u}^1\mathrm{u}_2^\star \pm \mathrm{a}^1\mathrm{a}_2^\star, \\ \text{lowering: } \mathcal{J}_\pm^- & = (\mathrm{u}^2\mathrm{u}_1^\star, \mathrm{a}^2\mathrm{a}_1^\star), & (\mathcal{L}^-, \mathcal{P}^-) & = \mathrm{u}^2\mathrm{u}_1^\star \pm \mathrm{a}^2\mathrm{a}_1^\star.\end{array}$$

By the two lowering operators $(\mathcal{J}_\pm^-)^{J-j_\pm}|J;J\rangle = |J;j_\pm\rangle$ (in the weight diagram: horizontal to the left and vertical downwards) one reaches all eigenvectors of a square grid:

$$\text{basis of } \bigvee^{2J} V \otimes \bigvee^{2J} U \cong \mathbb{C}^{(1+2J)^2}: \quad \{|J;j_+\rangle|J;j_-\rangle \mid j_\pm = -J,\dots,J\}.$$

The energy eigenvalues are given with the value of the Casimir operator:

$$\begin{array}{ll}-\frac{1}{2\langle H\rangle} & = 1 + 2\langle \vec{\mathcal{J}}_+^2 + \vec{\mathcal{J}}_-^2\rangle = 1 + 4J(1+J), \quad J = 0, \frac{1}{2}, 1, \frac{3}{2}, \dots, \\ E_k & = -\frac{1}{2k^2}, \quad \text{multiplicity: } k^2 = (1+2J)^2 = 1, 4, 9, 16, \dots.\end{array}$$

As seen in experiments, there is an additional twofold degeneracy in the atoms. It originates from an additional "internal" spin $\mathbf{SU}(2)$-property of the electron not contained in the nonrelativistic scheme above. It can be added by an ad hoc doubling leading to doubled multiplicities $2, 8, 18, \dots$.

For a nucleus with positive charge number z and the electron with charge number -1 and the reduced mass $m = \frac{m_e m_N}{m_e + m_N}$ the energy eigenvalues are, in atomic units,

$$\underline{E}_{1+2J} = -\frac{1}{(1+2J)^2}\frac{\mathcal{E}}{2}, \quad \frac{\mathcal{E}}{2} = z^2\frac{m}{2}(\alpha c)^2 \sim z^2 \times 13.6 \text{ eV}.$$

The energy sum of all $k^2 = (1+2J)^2$ energy-degenerated eigenvectors is always one-half of the intrinsic energy unit: Induced by the Kepler potential $\frac{1}{r}$, the spherical spread of the energy goes with the radial scaling $\frac{r}{k} = r\sqrt{-2E}$ for the weight "area" k^2 ($\mathbf{SO}(4)$-multiplicity):

$$\sum_{j_\pm=-J}^{J} \underline{E}_{1+2J} = -\frac{\mathcal{E}}{2} \quad \text{for all} \quad J = 0, \tfrac{1}{2}, 1, \dots.$$

The $\mathbf{SO}(4)$-representations are decomposable with respect to the position rotation $\mathbf{SO}(3)$-properties into irreducible representations of dimension $(1{+}2L)$ with integer $L = 0, 1, \ldots$ for angular momentum $\vec{\mathcal{L}} = \vec{\mathcal{J}}_+ + \vec{\mathcal{J}}_-$:

$$\begin{aligned}
(2J, 2J) &\overset{\mathbf{SO}(3)}{\cong} \bigoplus_{L=0}^{2J} [2L],\\
2J = L + N &\Rightarrow (L, N) = (2J, 0), (2J-1, 1), \ldots, (0, 2J).
\end{aligned}$$

The Lenz-Runge invariance-related difference $2J - L = N$ characterizing the classes $\mathbf{SO}(4)/\mathbf{SO}(3) \cong \Omega^3$ is the radial quantum number or knot number. For a graphical decomposition one has to project the $(1{+}2J)\times(1{+}2J)$ points of the $\mathbf{SO}(4)$-weight square on the square diagonal $\mathbf{SO}(3)$-angular momentum axis, $\mathcal{L}^3 = \mathcal{J}_+^3 + \mathcal{J}_-^3$, and to collect the – up the highest eigenvalues $m = \pm(1{+}2J)$ – degenerated diagonal points into $(1 + 2L)$-multiplets starting with the largest angular momentum value $L = 2J$:

$$\sum_{L=0}^{2J} (1 + 2L) = (1 + 2J)^2.$$

The degeneracy for $m = j_+ + j_-$ is $1 + 2J - |m|$.

There is an orthogonal basis transformation from eigenvectors of double "spin" $\{\mathcal{J}_\pm^3\}$ to eigenvectors of angular momentum $\{\vec{\mathcal{L}}^2, \mathcal{L}^3\}$:

$$\begin{aligned}
k = 1 + 2J: \quad |J; j_+\rangle|J; j_-\rangle &\sim |k; L, m\rangle, \quad m = j_+ + j_-,\\
\vec{\mathcal{L}}^2|k; L, m\rangle &= L(1 + L)|k; L, m\rangle, \ L = 0, \ldots, 2J,\\
\mathcal{L}^3|k; L, m\rangle &= m|k; L, m\rangle, \qquad m = -L, \ldots, +L.
\end{aligned}$$

All vectors of this basis are obtained from the highest vector $|J; J\rangle|J; J\rangle = |k; 2J, 2J\rangle$ with the angular momentum and Lenz-Runge lowering operators (in the weight diagram: diagonal and skew-diagonal downwards respectively):

$$k = 1 + 2J: \quad \begin{aligned} |k; L, L\rangle &= (\mathcal{P}^-)^{2J-L}|k; 2J, 2J\rangle,\\ |k; L, m\rangle &= (\mathcal{L}^-)^{L-m}|k; L, L\rangle, \end{aligned} \quad \text{with} \left\{ \begin{array}{l} L = 0, \ldots, k-1,\\ m = -L, \ldots, +L. \end{array}\right.$$

9.5.1 Hydrogen Atom à la Pauli

The Lenz-Runge operators act on the highest weight vectors $|k; 2J, 2J\rangle$ in the position representation with a radial derivative $d_r = \frac{d}{dr}$ as follows:

$$\begin{aligned}
\mathcal{P}^\pm &= i\tfrac{p^\mp \mathcal{L}^2 - \mathcal{L}^2 p^\mp}{2} - \tfrac{x^\mp}{r},\\
\mathcal{P}^\pm|k; k-1, k-1\rangle &= [\pm k d_r + \tfrac{k(k-1)}{r} - 1]|k; k-1, k-1\rangle.
\end{aligned}$$

Raising and lowering operators $\mathrm{P}_{k-1}^\pm$ for each principal number $k = 1 + 2J$ were introduced by Pauli in analogy to the algebraic treatment of the harmonic oscillator with creation and annihilation operator $[\mathrm{u}^\star, \mathrm{u}] = 1$ with $\frac{\mathrm{u}+\mathrm{u}^\star}{\sqrt{2}} \cong x$ and $\frac{\mathrm{u}-\mathrm{u}^\star}{\sqrt{2}} \cong d_x$ and $\{\mathrm{u}, \mathrm{u}^\star\} \cong -d_x^2 + x^2$:

$$\begin{array}{rlll} \mathrm{P}^{\pm}_{k-1} & = \pm k d_r - \frac{k(k-1)}{r} + 1 & = \pm d_R - \frac{k-1}{R} + 1 \text{ with } R = \frac{r}{k}, \\ \frac{\mathrm{P}^{+}_{k-1}-\mathrm{P}^{-}_{k-1}}{2} & = d_R, \quad \frac{\mathrm{P}^{+}_{k-1}+\mathrm{P}^{-}_{k-1}}{2} & = -\frac{k-1}{R} + 1, \\ \frac{[\mathrm{P}^{+}_{k-1},\mathrm{P}^{-}_{k-1}]}{2k^2} & = \frac{k-1}{r^2}, \quad \frac{\{\mathrm{P}^{+}_{k-1},\mathrm{P}^{-}_{k-1}\}}{2k^2} & = -d_r^2 + \frac{(k-1)^2}{r^2} - \frac{2(k-1)}{kr} + \frac{1}{k^2}. \end{array}$$

The multiplet-dependent renormalization of the radial position $R^2 = \frac{r^2}{k^2}$ with the energy $-2E = -\frac{1}{k^2}$ corresponds to the renormalization of the Lenz-Runge vector $\vec{\mathcal{P}} = \vec{\mathcal{B}}\sqrt{-2H}$. The representation with the noncompact position representation coefficient $R^{k-1}e^{-R}$ (below) gives immediately the position function representation of the highest-weight vector as the solution of $\mathrm{P}^{+}_{k-1}\psi(R) = 0$:

$$\begin{array}{l} \mathrm{P}^{+}_{k-1}|k; k-1, k-1\rangle = 0 \Rightarrow |k; k-1, k-1\rangle \cong R^{k-1}e^{-R} \\ \Rightarrow |k; k-1, k-1\rangle \cong e^{-\frac{r}{k}}(\frac{r}{k})^{k-1}\mathrm{Y}^{k-1}_{k-1}(\varphi, \theta). \end{array}$$

The Rodriguez form of Lenz-Runge raising and lowering operator

$$\mathrm{P}^{+}_{k-1} = R^{k-1}e^{-R}d_R R^{-(k-1)}e^{R}, \quad \mathrm{P}^{-}_{k-1} = -R^{-(k-1)}e^{R}d_R R^{k-1}e^{-R}$$

leads to the *Laguerre polynomials*:

$$\begin{array}{rl} \mathrm{L}^{N}_{1+2L}(\rho) = (\mathrm{P}^{-}_{1+2L})^{N}\frac{(-\rho)^N}{N!} & = \frac{1}{N!}\rho^{-(1+2L)}e^{\rho}\frac{d^N}{d\rho^N}\rho^{1+2L+N}e^{-\rho}, \\ & = \sum_{n=0}^{N}\binom{1+2L+N}{1+2L+n}\frac{(-\rho)^n}{n!}, \\ \deg \mathrm{L}^{N}_{1+2L}(\rho) & = N = 0, 1, \ldots \\ \int_0^\infty \rho^{1+2L}e^{-\rho}\, d\rho\, \mathrm{L}^{N}_{1+2L}(\rho)\, \mathrm{L}^{N'}_{1+2L}(\rho) & = \frac{(1+2L+N)!}{N!}\delta_{NN'}. \end{array}$$

The full wave functions, orthonormalized with $\int d^3x$, are obtained by Lenz-Runge lowering with $(\mathcal{P}^-)^{k-1-L} = (\mathcal{P}^-)^N$,

$$\begin{array}{rl} k = 1 + L + N: \quad |k; L, m\rangle & \cong \psi^{2J}_{Lm}(\vec{x}) \\ & = \frac{2}{k^2}\sqrt{\frac{N!}{(1+2L+N)!}}\,(\frac{2r}{k})^L\mathrm{Y}^L_m(\varphi, \theta)\; \mathrm{L}^{N}_{1+2L}(\frac{2r}{k})e^{-\frac{r}{k}}, \end{array}$$

with the explicit examples where the skew-diagonals $L + N = k - 1 = 2J$, e.g., • for $k = 3$, are used for bound-state vectors with equal energy:

	$L = 0$	$L = 1$	$L = 2$
$N = 0$	$\mathrm{L}^0_1(\rho) = 1$	$\mathrm{L}^0_3(\rho) = 1$	$\mathrm{L}^0_5(\rho) = 1$ •
$N = 1$	$\mathrm{L}^1_1(\rho) = 2 - \rho$	$\mathrm{L}^1_3(\rho) = 4 - \rho$ •	$\mathrm{L}^1_5(\rho) = 6 - \rho$
$N = 2$	$\mathrm{L}^2_1(\rho) = 3 - 3\rho + \frac{\rho^2}{2}$ •	$\mathrm{L}^2_3(\rho) = 10 - 5\rho + \frac{\rho^2}{2}$	$\mathrm{L}^2_5(\rho) = 21 - 7\rho + \frac{\rho^2}{2}$
$N = 3$	$\mathrm{L}^3_1(\rho) = 4 - 6\rho + 2\rho^2 - \frac{\rho^3}{6}$	$\mathrm{L}^3_3(\rho) = 20 - 15\rho + 3\rho^2 - \frac{\rho^3}{6}$	$\mathrm{L}^3_5(\rho) = 56 - 28\rho + 4\rho^2 - \frac{\rho^3}{6}$

Laguerre polynomials L^{N}_{1+2L}

There occurs the doubled position $\rho = 2R = \frac{2r}{k}$ in the Laguerre polynomials compared with the exponential; more below.

Summarizing: The *double spin* $\mathbf{SU}(2) \times \mathbf{SU}(2)$ vectors $|J; j_+\rangle|J; j_-\rangle$ can be transformed into vectors $|k; L, m\rangle$ for *Lenz-Runge classes with angular momentum* $\mathbf{SO}(3) \circ \Omega^3$. Their wave functions $(\frac{2r}{k})^L \mathrm{Y}^L_m(\varphi, \theta)\ \mathrm{L}^N_{1+2L}(\frac{2r}{k}) e^{-\frac{r}{k}}$ involve, up to the exponential, a product of two polynomials (harmonic and Laguerre). The degree of the spherical harmonic is the angular momentum L. The radial quantum number N as the degree of the Laguerre polynomial L^N_{1+2L} gives the number of zeros (knot number) in the radial wave functions:

$$\begin{array}{ll} \begin{matrix} k=1 \\ \text{(singlet)} \end{matrix} & \text{ground state } |0\rangle = |0;0\rangle|0;0\rangle = |1;0,0\rangle \sim e^{-r}, \end{array}$$

$$\begin{matrix} k=2 \\ \text{(quartet)} \end{matrix} \left\{ \begin{array}{lll} \frac{|\frac{1}{2};\frac{1}{2}\rangle|\frac{1}{2};-\frac{1}{2}\rangle - |\frac{1}{2};-\frac{1}{2}\rangle|\frac{1}{2};\frac{1}{2}\rangle}{\sqrt{2}} & = |2;0,0\rangle & \sim (2-r)e^{-\frac{r}{2}}, \\ \begin{pmatrix} |\frac{1}{2};\frac{1}{2}\rangle|\frac{1}{2};\frac{1}{2}\rangle \\ \frac{|\frac{1}{2};\frac{1}{2}\rangle|\frac{1}{2};-\frac{1}{2}\rangle + |\frac{1}{2};-\frac{1}{2}\rangle|\frac{1}{2};\frac{1}{2}\rangle}{\sqrt{2}} \\ |\frac{1}{2};-\frac{1}{2}\rangle|\frac{1}{2};-\frac{1}{2}\rangle \end{pmatrix} & = \begin{pmatrix} |2;1,1\rangle \\ |2;1,0\rangle \\ |2;1,-1\rangle \end{pmatrix} & \sim r\mathrm{Y}^1_m(\varphi,\theta)e^{-\frac{r}{2}}, \end{array} \right.$$

$$\begin{matrix} k=3 \\ \text{(nonet)} \end{matrix} \left\{ |1;j_+\rangle|1;j_-\rangle = \left\{ \begin{array}{l} |3;0,0\rangle \\ |3;1,m\rangle \\ |3;2,m\rangle \end{array} \right. \sim \begin{array}{r} \mathrm{L}^2_1(\frac{2r}{3})e^{-\frac{r}{3}}, \\ \frac{2r}{3}\ \mathrm{Y}^1_m(\varphi,\theta)\ \mathrm{L}^1_3(\frac{2r}{3})e^{-\frac{r}{3}}, \\ (\frac{2r}{3})^2\mathrm{Y}^1_m(\varphi,\theta)\ \mathrm{L}^0_5(\frac{2r}{3})e^{-\frac{r}{3}}. \end{array} \right.$$

9.6 Position Representations

In contrast to the classical description, the time orbits in quantum theory are not valued in position space. They are complex-valued in a Hilbert space with the scalar product for "probability amplitudes". In a Schrödinger picture the time orbits are spread to "information valued" position orbits:

$$\begin{aligned} \mathbb{R} \ni t &\longmapsto e^{iEt}|E\rangle \cong e^{iEt}\psi_E, \quad e^{iEt} \in \mathbf{U}(1), \\ \mathbb{R} \oplus \mathbb{R}^s \ni (t, \vec{x}) &\longmapsto e^{iEt}\psi_E(\vec{x}) \in \mathbb{C}. \end{aligned}$$

Schrödinger functions ψ_E are position representation coefficients with a Schrödinger equation $H\psi_E = E\psi_E$ for a time translation generator iH with the energy as H-eigenvalue.

The representation invariants for time (energy) and position are related to each other: The difference $\frac{p^2}{2m} = E - V$ with a potential V connects the eigenvalues iE for the time-translation representation with the energy E and the eigenvalues ip for the position translation representation with the momentum p. It is the nonrelativistic precursor of the relativistic energy-momentum relation $\vec{p}^2 = p_0^2 - m^2$ as used for quantum fields. The compact position representations or scattering waves come for positive kinetic energies, i.e., for $E > V$ ($p_0^2 > m^2$) with real momentum invariant $\vec{p}^2 > 0$ ("on shell" real particles). The noncompact position representations for bound waves come for $E < V$ ($p_0^2 < m^2$) with imaginary "momentum" invariant $(iQ)^2 < 0$ ("off shell" virtual particles).

The operator for time representations is the time translation implementing Hamiltonian. In contrast to the position-momentum pairs with $[ip, x] = \hbar$

there is no dual operator pair (iH, t). The so-called *"time-energy uncertainty relation"* is a reformulation of the position-momentum uncertainty relation involving the velocity $v = \frac{p}{M}$:

$$\begin{aligned} E - V_0 = \tfrac{p^2}{2M} \Rightarrow \Delta E &\stackrel{\text{def}}{=} \tfrac{p}{M}\Delta p = v\Delta p, \\ \tfrac{\hbar}{2} \le \Delta x \Delta p = \Delta t \Delta E \text{ with } \Delta t &\stackrel{\text{def}}{=} \tfrac{M}{p}\Delta x = \tfrac{\Delta x}{v}. \end{aligned}$$

9.6.1 Hilbert Spaces for Heisenberg Groups

The representations of the noncompact Heisenberg group $\mathbf{H}(s)$ with $s = 1, 2, 3, \ldots$ position-momentum pairs and characteristic Lie brackets $[\mathbf{x}^a, \mathbf{p}^b] = \delta^{ab}\mathbf{I}$ determine the quantum-mechanical Hilbert spaces (chapter "Harmonic Analysis"). For each nontrivial value $0 \neq \hbar \in \mathbb{R}$ of the invariant central operator there is an infinite-dimensional irreducible faithful representation with $\mathbf{I} \longmapsto i\hbar\mathbf{1}$. The representations are inequivalent for $\hbar \neq \hbar'$. This has to be seen in analogy to the irreducible $\mathbf{SU}(2)$-representations $2J \in \mathbb{N}$, inequivalent for different spin $J \neq J'$. In contrast to the occurrence of rotation group representations for different invariants $J \neq J'$, quantum mechanics is formulated with only one Heisenberg group representation and only one invariant $\hbar$ (Planck's constant). An $\mathbf{H}(s)$-representation space for a fixed $\hbar \neq 0$ (as intrinsic unit) can be built with the square integrable functions[3] of the position translation eigenvalues (momenta). $L^2(\mathbb{R}^s)$ is an orthogonal direct integral with Lebesgue measure over 1-dimensional Hilbert spaces $\mathbb{C}|\vec{p}\rangle \cong \mathbb{C}$ for each momentum:

$$L^2(\mathbb{R}^s) = \{f \mid \int \tfrac{d^s p}{(2\pi)^s} |f(\vec{p})|^2 < \infty\} = {}^{\perp}\!\!\int \tfrac{d^s p}{(2\pi)^s} \mathbb{C}|\vec{p}\rangle.$$

Integrals without boundary are understood to go over the full integration space, here $\int d^s p = \int_{\mathbb{R}^s} d^s p$. A distributive basis $\{|\vec{p}\rangle \bigm| \vec{p} \in \mathbb{R}^s\}$ of $L^2(\mathbb{R}^s)$ involves the irreducible translation representations ($\mathbf{U}(1)$-characters) $|\vec{p}\rangle = \{\vec{x} \longmapsto e^{i\vec{p}\vec{x}}\} \in L^\infty(\mathbb{R}^s)$ (no Hilbert space vectors) with eigenvalues $i\vec{p}$, e.g., the plane waves for $s = 3$, with distributive completeness and orthogonality:

$$\begin{aligned} \int d^s x \, e^{i\vec{p}'\vec{x}} e^{-i\vec{p}\vec{x}} &= \langle \vec{p}' | \vec{p} \rangle = \delta(\tfrac{\vec{p}-\vec{p}'}{2\pi}) = (2\pi)^s \delta(\vec{p} - \vec{p}'), \\ \int \tfrac{d^s p}{(2\pi)^s} e^{i\vec{p}\vec{x}} e^{-i\vec{p}\vec{x}'} &= \delta(\vec{x} - \vec{x}'), \\ \mathrm{id}_{L^2(\mathbb{R}^s)} &= {}^{\perp}\!\!\int \tfrac{d^s p}{2\pi} |\vec{p}\rangle\langle\vec{p}|, \\ f &= {}^{\perp}\!\!\int \tfrac{d^s p}{(2\pi)^s} f(\vec{p})|\vec{p}\rangle, \quad f(\vec{p}) = \langle \vec{p} | f \rangle. \end{aligned}$$

The Hilbert spaces for the position and momentum functions are Fourier isomorphic

$$L^2_{d^s p}(\mathbb{R}^s) \cong L^2_{d^s x}(\mathbb{R}^s) \text{ with } \left\{ \begin{aligned} \int \tfrac{d^s p}{(2\pi)^s} e^{i\vec{p}\vec{x}} f(\vec{p}) &= \psi(\vec{x}), \\ \langle f | f' \rangle &= \int d^s x \, \overline{\psi}(\vec{x}) \psi'(\vec{x}) \\ = \int \tfrac{d^s p}{(2\pi)^s} \overline{f(\vec{p})} f'(\vec{p}) &= \langle \psi | \psi' \rangle, \end{aligned} \right.$$
$$(\mathbf{x}^a, \mathbf{p}^a, \mathbf{I}) \longmapsto (x^a, -i\partial^a, i\mathbf{1}).$$

[3] The function space $L^p_{d\mu}(S, \mathbb{C})$ will be denoted by $L^p(S)$ if the positive S-measure $d\mu$ is unique up to a factor, e.g., for a locally compact group with Haar measure.

With the maximal compact homogeneous group in the affine Heisenberg group $\mathbf{GL}(\mathbb{R}^s) \vec{\times} \mathbf{H}(s)$ (chapter "Simple Lie Operations"), the Hilbert spaces $L^2(\mathbb{R}^s)$ are acted on by representations of the real $\binom{s+2}{2}$-dimensional *orthogonal Heisenberg group*

$$s = 1, 2, \cdots : \quad \mathbf{SO}(s) \vec{\times} \mathbf{H}(s) = \mathbf{SO}(s) \vec{\times} [\mathbb{R}^s \vec{\times} (\mathbb{R}^s \oplus \mathbb{R})].$$

The Hilbert space $L^2(\mathbb{R}^s)$ for the Heisenberg group is reducible with respect to irreducible subgroup representations, e.g., with 1-dimensional spaces for the position translations $\mathbb{R}^s$, with finite-dimensional spaces for the nontrivial rotation groups $\mathbf{SO}(s)$, $s \geq 2$, and with infinite-dimensional spaces for the faithful representations of the Euclidean group $\mathbf{SO}(s) \vec{\times} \mathbb{R}^s$, $s \geq 2$.

The orthogonal Heisenberg group is an Inönü-Wigner contraction of orthogonal groups that expand the momentum operations:

$$\left(\begin{array}{c||c} \mathbf{SO}(1+s) & \mathbb{R}^{1+s} \\ \hline\hline 0 & 1 \end{array}\right) \longrightarrow \left(\begin{array}{c|c||c} 1 & \mathbb{R}^s & \mathbb{R} \\ \hline 0 & \mathbf{SO}(s) & \mathbb{R}^s \\ \hline\hline 0 & 0 & 1 \end{array}\right) \longleftarrow \left(\begin{array}{c||c} \mathbf{SO}_0(1,s) & \mathbb{R}^{1+s} \\ \hline\hline 0 & 1 \end{array}\right).$$

$L^2(\mathbb{R}^s)$ is an action space also for the expanded groups.

The action on $L^2(\mathbb{R}^s)$ involves spherical, flat Euclidean and hyperbolic operations $\mathbf{SO}(1+s)$, $\mathbf{SO}(s) \vec{\times} \mathbb{R}^s$, and $\mathbf{SO}_0(1,s)$. In the following, the abelian case $s = 1$ is considered first where flat and hyperbolic are isomorphic, and then the nonabelian one $s = 3$, where the three structures are different.

9.7 Orbits of 1-Dimensional Position

In the energy eigenvalue problem of a Hamiltonian $H = \frac{p^2}{2} + V$, parametrized by 1-dimensional position translations

$$ip \cong \tfrac{d}{dx} = d_x : \quad [-\tfrac{1}{2}d_x^2 + V(x)]\psi(x) = E\psi(x),$$

the real energies for the time translation eigenvalues are given in terms of real or imaginary "momenta" for the position translation eigenvalues.

The differential equation for constant potential with position translation invariance of the Hamiltonian

$$H_0 = \tfrac{p^2}{2} + V_0, \quad [H_0, p] = 0, \quad [d_x^2 + 2(E - V_0)]\psi_0(x) = 0$$

gives rise to two types of representation coefficients of the noncompact position group $\mathbb{R}$, either with imaginary eigenvalues for spherical orbits or with real eigenvalues for hyperbolic orbits:

$$\mathbb{R} \ni x \longmapsto \left\{ \begin{array}{rcl} \begin{pmatrix} \cos Px & i\sin Px \\ i\sin Px & \cos Px \end{pmatrix} & \cong & \begin{pmatrix} e^{iPx} & 0 \\ 0 & e^{-iPx} \end{pmatrix} \in \mathbf{SO}(2) \subset \mathbf{SU}(2), \\ E - V_0 & = & \frac{P^2}{2} > 0 \text{ (free scattering waves)}, \\ \begin{pmatrix} \cosh Qx & \sinh Qx \\ \sinh Qx & \cosh Qx \end{pmatrix} & \cong & \begin{pmatrix} e^{Qx} & 0 \\ 0 & e^{-Qx} \end{pmatrix} \in \mathbf{SO}_0(1,1) \subset \mathbf{SU}(1,1), \\ E - V_0 & = & -\frac{Q^2}{2} < 0 \text{ (bound waves)}. \end{array} \right.$$

Coeffients of reducible nondecomposable representations come with strictly positive position translation powers (nildimensions) $x^N e^{\pm iPx}$ and $x^N e^{\pm Qx}$, $N = 1, 2, \dots$. These representations are indefinite unitary. The order structure of the reals, i.e., the bicone property $\mathbb{R} = \mathbb{R}_+ \uplus \mathbb{R}_-$, can be represented with additional factors $\vartheta(\pm x)$ and $\epsilon(x)$, e.g., in $|x| = \epsilon(x)x$.

Both the free scattering and the bound waves can be Fourier expanded with irreducible representations of the position translations

$$\begin{array}{ll}\cos Px &= \int dp\ |P|\delta(p^2 - P^2)e^{-ipx},\\ e^{-|Qx|} &= \int \frac{dp}{\pi}\ \frac{|Q|}{p^2+Q^2}e^{-ipx}.\end{array}$$

The functions come as residues of dual real momentum poles $p = \pm P$ for the compact representations and imaginary dual "momentum" poles $p = \pm i|Q|$ for the noncompact ones.

The groups realized by time and position orbits are the product of $\mathbf{U}(1)$ for time with the corresponding real 1-dimensional groups for position:

$$\begin{array}{ll}H_0 = \frac{p^2}{2} + V_0: & \mathbb{R}\times\mathbb{R} \longrightarrow \mathbf{U}(1)\times\begin{pmatrix}\mathbf{SO}(2)\\ \mathbf{SO}_0(1,1)\end{pmatrix},\\ \begin{pmatrix}\text{free waves}\\ \text{bound waves}\end{pmatrix}: & (t,x) \longmapsto e^{iEt}\begin{pmatrix}\cos Px\\ e^{-|Qx|}\end{pmatrix},\quad E - V_0 = \begin{pmatrix}\frac{P^2}{2}\\ -\frac{Q^2}{2}\end{pmatrix}.\end{array}$$

9.7.1 Scattering Orbits

Free scattering waves as compact spherical position orbits are essentially bounded, but not square integrable functions:

$$\{x \longmapsto e^{-ipx}\} \in L^\infty(\mathbb{R}).$$

Hilbert space vectors for scattering need square integrable momentum wave packets f in the Fourier isomorphism

$$L^2_{dp}(\mathbb{R}) \ni f \leftrightarrow \psi \in L^2_{dx}(\mathbb{R}),\quad \psi(x) = \int \frac{dp}{2\pi}\ f(p)e^{-ipx}.$$

9.7.2 Bound Orbits

Bound waves are square integrable functions $L^2(\mathbb{R})$ of noncompact position. The matrix elements of finite-dimensional indefinite unitary representations, here for translations $\mathbb{R} \ni x \longmapsto \begin{pmatrix}e^{Qx} & \\ 0 & e^{-Qx}\end{pmatrix} \in \mathbf{SU}(1,1)$, can be written as definite unitary representations of the position translations, necessarily on infinite-dimensional Hilbert spaces (chapter "Harmonic Analysis"). The ordered sum of the homogeneous solutions $(d_x^2 - Q^2)e^{\pm|Q|x} = 0$ obeys an inhomogeneous equation

$$\begin{array}{ll}e^{-|Qx|} = \int \frac{dp}{\pi}\ \frac{|Q|}{p^2+Q^2}e^{-ipx} &= \vartheta(x)e^{-|Q|x} + \vartheta(-x)e^{|Q|x},\\ (d_x^2 - Q^2)e^{-|Qx|} &= -2\pi|Q|\delta(x).\end{array}$$

Especially in nonrelativistic quantum mechanics where time and position translations are not connected with each other in Lorentz group actions, the bound waves may come with a Lie parameter ξ for the position variable

$$\mathbb{R} \ni x \longmapsto \xi(x) \in \mathbb{R}.$$

If, for the variable ξ, the bound wave solution of the Schrödinger equation is required to be a nondecomposable $\mathbb{R}$-representation coefficient as the product of a polynomial p^N (combining powers ξ^n for nontrivial nildimensions) and an irreducible exponential (hyperbolic coefficients)

$$\mathbb{R} \ni \xi \longmapsto \psi^N(x) = \mathrm{p}^N(\xi)\, e^{-\xi},$$

there remains from the Schrödinger equation in the variable x the equation for the polynomial

$$(d_x\xi)^2 d_\xi^2 \mathrm{p}^N + [d_x^2\xi - 2(d_x\xi)^2] d_\xi \mathrm{p}^N + [(d_x\xi)^2 - d_x^2\xi + 2(E - V)]\mathrm{p}^N = 0.$$

An irreducible solution with constant polynomial determines the reparametrization $x \longmapsto \xi(x)$ by the potential

$$\psi_0(x) \sim e^{-\xi(x)} \iff d_x^2\xi - (d_x\xi)^2 = 2[E_0 - V(x)].$$

If for a power reparametrization, with momentum unit Q,

$$\xi(x) = \tfrac{(Qx)^n}{n} \Rightarrow (n-1)Q^n x^{n-2} - Q^{2n} x^{2(n-1)} = 2[E_0 - V(x)],$$

the two terms on the left hand side are equalized with the constant energy and an x-dependent potential on the right hand side, two solutions are possible: The solution $n = 1$ with constant potential $V(x) = V_0$ and $\psi_0(x) = e^{-x}$ is not square integrable. The solution $n = 2$ has quadratic x^2-dependence

$$n = 2 \Rightarrow \xi(x) = \tfrac{(Qx)^2}{2}, \quad \begin{cases} V(x) &= \tfrac{(Qx)^2}{2} Q^2, \quad E_0 = \tfrac{Q^2}{2}, \\ \psi_0(x) &= e^{-\frac{(Qx)^2}{2}} = \int \tfrac{dp}{\pi}\, \tfrac{|Q|}{p^2+Q^2} e^{-ip|Q|\frac{x^2}{2}} \\ & = \int \tfrac{dp}{\sqrt{\pi^3}}\, \tfrac{e^{-\frac{p^2}{2Q^2}}}{|Q|} e^{-ipx}. \end{cases}$$

The momentum unit can be chosen conveniently, e.g., $Q = 1$. The square integrable ground state wave function of the harmonic oscillator can be written also as Fourier transformed Gauss measure $dp\ e^{-\frac{p^2}{2}}$. The equation for the harmonic oscillator polynomials in the wave functions ψ^N with $N = E_N - \frac{1}{2} = 0, 1, \dots,$

$$\left.\begin{array}{ll} V(x) & = \frac{x^2}{2} = \xi, \\ \psi^N(x) & \sim \mathrm{p}^N(\xi) e^{-\xi}, \end{array}\right\} \Rightarrow [2\xi d_\xi^2 + (1 - 4\xi) d_\xi - 1 + 2E_N]\mathrm{p}^N(\xi) = 0,$$

is equivalent to the equation for the Hermite polynomials with degree $k = E_k - \frac{1}{2} = 0, 1, \dots,$

$$\psi^k(x) \sim \mathrm{H}^k(x) e^{-\frac{x^2}{2}} \Rightarrow [d_x^2 - 2x d_x - 1 + 2E_k]\mathrm{H}^k(x) = 0.$$

The Hermite polynomials represent the bicone structure of the position translations

$$\mathbb{R} \ni x = \epsilon(x)|x| \ni \mathbb{I}(2) \times \mathbb{R}_+$$

as seen in their factorization with an x^2-dependent Laguerre polynomial

$$\left.\begin{array}{lcr} \mathrm{H}^{2n}(x) & \sim & \mathrm{L}^n_{-\frac{1}{2}}(x^2) \\ \mathrm{H}^{1+2n}(x) & \sim & \epsilon(x)|x| \; \mathrm{L}^n_{+\frac{1}{2}}(x^2) \end{array}\right\}, \quad \deg \mathrm{L}^n_{\pm\frac{1}{2}}(x^2) = n.$$

9.8 Scattering Orbits of 3-Dimensional Position

The 3-dimensional position translations with rotation group $\mathbf{O}(3)$ action are, in polar coordinates, the product of the totally ordered cone $\mathbb{R}_+$ with the radial translations and the compact 2-sphere. Both factors will be presented by corresponding orbits. The 1-dimensional case is embedded as abelian substructure

$$s \geq 1: \quad \mathbb{R}^s \cong \mathbb{R}_+ \times \Omega^{s-1}.$$

The 0-sphere consists of two points $\Omega^0 = \{\pm 1\} \cong \mathbb{I}(2) \cong \mathbf{O}(1)$, forward and backward. In a Schrödinger equation the position wave functions can be decomposed with respect to its radial and rotation representation part.

The rotation-invariant momentum square $\vec{p}^2$ contains the Hermitian radial momentum p_r that involves a 2-sphere spread factor $\frac{1}{r}$:

$$\mathbb{R}^s: \left\{\begin{array}{ll} \pi_r^\star & = \left(\frac{\vec{x}}{r}\vec{p}\right)^\star = \vec{p}\frac{\vec{x}}{r} = \frac{\vec{x}}{r}\vec{p} - \frac{(s-1)i}{r}, \\ p_r & = p_r^\star = \frac{\pi_r + \pi_r^\star}{2} = \frac{1}{2}\{\vec{p}, \frac{\vec{x}}{r}\} \Rightarrow [ip_r, r] = 1, \\ ip_r & \cong \frac{1}{r^{s-2}} d_r r^{s-2} = \frac{s-2}{r} + d_r, \quad s \geq 2. \end{array}\right.$$

In a derivative representation the radial and angular momentum squares act on differentiable complex functions as follows:

$$\begin{array}{lll} \mathbb{R}^3: & \vec{p}^2 & = p_r^2 + \frac{\vec{\mathcal{L}}^2}{r^2}, \quad [\vec{p}^2, \vec{\mathcal{L}}] = 0, \\ & \vec{\mathcal{L}}^2 & \cong \frac{1}{\sin^2\theta}[(\sin\theta \frac{\partial}{\partial\theta})^2 + (\frac{\partial}{\partial\varphi})^2] = \frac{\partial^2}{\partial\theta^2} + \frac{1}{\tan\theta}\frac{\partial}{\partial\theta} + (\frac{1}{\sin\theta}\frac{\partial}{\partial\varphi})^2. \end{array}$$

Essentially bounded radial functions are scattering representations, bound waves are square integrable. Their Fourier transforms involve a Dirac measure supported by real momentum poles $|\vec{p}| = \pm P > 0$ for free scattering (spherical) and a *dipole at dual imaginary "momenta"* $|\vec{p}| = \pm i|Q|$ for bound state vectors (hyperbolic):

$$\begin{array}{lll} L^\infty(\mathbb{R}^3): & \frac{\sin Pr}{Pr} & = \int \frac{d^3p}{2\pi P} \, \delta(\vec{p}^2 - P^2) e^{-i\vec{p}\vec{x}}, \\ L^2(\mathbb{R}^3): & e^{-|Q|r} & = \int \frac{d^3p}{\pi^2} \, \frac{|Q|}{(\vec{p}^2+Q^2)^2} e^{-i\vec{p}\vec{x}}. \end{array}$$

9.8.1 Euclidean Group Coefficients

Nonrelativistic scattering theory is formulated with the faithful representations of the Euclidean group $\mathbf{SO}(3) \vec{\times} \mathbb{R}^3$ (more mathematical details in the chapter "Harmonic Analysis").

The solutions of

$$(\vec{\partial}^2 + P^2)D(\vec{x}) = 0$$

for trivial invariant $P = 0$ are $\mathbf{SO}(3)$-representations with harmonic polynomials (next section). For nontrivial momentum invariant $P > 0$, the solutions are coefficients of the irreducible Hilbert space representations of the Euclidean group, induced with fixgroup $\mathbf{SO}(2)$-representations

$$P > 0, \quad L = 0, 1, 2, \cdots: \quad \{|\vec{p}, \pm L\rangle \,\big|\, \vec{p} \in \mathbb{R}^3, \ \vec{p}^2 = P^2\}.$$

There is no rotation-invariant $\vec{l}^2$, only dual $\mathbf{SO}(2)$-eigenvalues $\pm L \in \mathbb{Z}$ (*helicity*) for axial rotations around the momentum direction $\frac{\vec{p}}{P}$.

The orthonormalized *spherical harmonics* $\{\mathrm{Y}^L_m\}$ are a Hilbert basis for functions on the 2-sphere $\Omega^2 \cong \mathbf{SO}(3)/\mathbf{SO}(2)$ as fixgroup manifold, decomposable into Hilbert spaces with irreducible action of the rotations

$$L^2(\Omega^2) = \bigoplus_{L=0}^{\infty} \mathbb{C}^{1+2L}(\Omega^2), \quad \vec{\mathcal{L}}^2 \mathrm{Y}^L_m(\varphi, \theta) = L(1+L)\mathrm{Y}^L_m(\varphi, \theta).$$

With the wave function decomposition with the spherical harmonics

$$\psi(\vec{x}) = \sum_{L=0}^{\infty} \sum_{m=-L}^{L} \mathrm{Y}^L_m(\varphi, \theta)\psi^m_L(r),$$

a rotation-invariant Hamiltonian is decomposable into the generators H_L for each angular momentum:

$$II - \bigoplus_{L=0}^{\infty} H_L \text{ with } \begin{cases} H &= \frac{\vec{p}^2}{2} + V(r), \quad [H, \vec{\mathcal{L}}] = 0, \\ H_L &= \frac{p_r^2}{2} + \frac{L(1+L)}{2r^2} + V(r) \\ &= r^n[-\frac{1}{2}d_r^2 - \frac{1+n}{r}d_r + \frac{L(1+L)-n(1+n)}{2r^2} + V(r)]\frac{1}{r^n}, \\ n &= 0, \pm 1, \pm 2, \ldots. \end{cases}$$

Attention has to be paid to the small distance $r \to 0$ behavior, prepared with the powers r^n. Compact radial representations are spread to the 2-sphere with a factor $\frac{1}{r}$:

$$\psi(\vec{x}) = \sum_{L=0}^{\infty} \sum_{m=-L}^{L} (\tfrac{\vec{x}}{r})^L_m \frac{D_L(r)}{r} \Rightarrow [d_r^2 - \frac{L(1+L)}{r^2} + 2(E - V(r))]D_L(r) = 0,$$
$$D_L(r) = r\psi_L(r).$$

An irreducible compact radial position representation requires a constant potential

$$D_0(r) = e^{\pm iPr} \Rightarrow H = \frac{\vec{p}^2}{2} + V_0, \quad E - V_0 = \frac{P^2}{2}.$$

For large radial translations $r \to \infty$ the *centrifugal potential* $\frac{L(1+L)}{r^2}$, and hence the angular momentum dependence, vanishes. For a potential leveling off stronger than $\frac{1}{r}$ the product $D_L(r) = r\psi_L(r)$ of the translation parameter r with the wave function is a coefficient for the representation of the radial translations. The scattering eigenfunctions for trivial angular momentum and, for large distances $r \to \infty$, for any L,

$$\psi_0(r) = \frac{D_0(r)}{r} = \alpha\frac{\sin Pr}{r} + \beta\frac{\cos Pr}{r},$$

involve representation coefficients of the position cone $\mathbb{R}_+ \longrightarrow \mathbf{SO}(2)$. In analogy to the 1-dimensional waves as orbits with $e^{iPx} \in \mathbf{U}(1)$ and momentum $P \in \mathbb{R}$ one has for the cone $\mathbb{R}_+$,

$$\int_0^\infty dr\; e^{iPr} = -\frac{i}{P-io} = \pi\delta(P) - \frac{i}{P_{\rm P}}$$
$$\Rightarrow \left\{ \begin{array}{l} \int_0^\infty dr\;\; \sin P'r \sin Pr = \pi\frac{\delta(P-P')-\delta(P+P')}{2}, \\ \int_0^\infty dr\;\; \cos P'r \cos Pr = \pi\frac{\delta(P-P')+\delta(P+P')}{2}, \\ \int_0^\infty dr\;\; \sin P'r \cos Pr = \frac{1}{2}[\frac{1}{(P-P')_{\rm P}} + \frac{1}{(P+P')_{\rm P}}]. \end{array} \right.$$

The $r = 0$ regular radial translation eigenfunctions with momenta either both positive or both negative are orthogonal

$$PP' \geq 0: \quad \int_0^\infty r^2 dr\; \frac{\sin P'r}{r}\; \frac{\sin Pr}{r} = \frac{\pi}{2}\delta(P-P').$$

The radial coefficients of plane waves define the *spherical Bessel functions* $j_L \in L^\infty(\mathbb{R}_+)$ in the expansion of the irreducible translation $\mathbb{R}^3$-representation coefficient with repect to representations of the Euclidean group $\mathbf{SO}(3) \vec{\times} \mathbb{R}^3$:

$$e^{i\vec{p}\vec{x}} = e^{iPr\cos\theta} = e^{iR\zeta} = \sum_{L=0}^{\infty}(1+2L)\mathrm{P}^L(\zeta)\; i^L j_L(Pr)$$
$$\Rightarrow\; j_L(R) = i^{-L}\int_{-1}^{1}\frac{d\zeta}{2}\;\mathrm{P}^L(\zeta)e^{iR\zeta} = i^{-L}\mathrm{P}^L(\frac{d}{diR})\int_{-1}^{1}\frac{d\zeta}{2}\; e^{iR\zeta} = R^L(-\frac{1}{R}\frac{d}{dR})^L\frac{\sin R}{R}.$$

The Legendre polynomials $\mathrm{P}^L(\cos\theta) = \sqrt{\frac{4\pi}{1+2L}}\mathrm{Y}_0^L(\varphi,\theta)$ are the spherical harmonics in the momentum direction.

The essentially bounded representation coefficients of the Euclidean group in $L^\infty(\mathbb{R}^3)$ are products $\mathrm{Y}_m^L(\varphi,\theta)j_L(Pr)$ of matching spherical Bessel functions with spherical harmonics

$$P > 0: \quad \mathbb{R}^3 \ni \vec{x} \longmapsto \int\frac{d^3p}{2\pi P}\begin{pmatrix} 1 \\ \frac{\vec{p}}{|\vec{p}|}\end{pmatrix}\delta(\vec{p}^2 - P^2)e^{-i\vec{p}\vec{x}}$$
$$= \begin{pmatrix} j_0(Pr) \\ i\frac{\vec{x}}{r}j_1(Pr)\end{pmatrix} = \begin{pmatrix} \frac{\sin Pr}{Pr} \\ i\frac{\vec{x}}{r}\frac{\sin Pr - Pr\cos Pr}{P^2r^2}\end{pmatrix}.$$

The direct integral Hilbert space

$$L^2(\mathbb{R}_+) = \{f \mid \int_0^\infty \frac{2P^2dP}{\pi}\,|f(P)|^2 < \infty\} = {}^\oplus\!\!\int_0^\infty \frac{P^2dP}{\pi}\;\mathbb{C}|P\rangle$$

has the spherical Bessel functions for each angular momentum as distributive basis (not Hilbert space vectors) with distributive orthogonality and completeness (mathematically the same relations):

$$L=0,1,2,\cdots:\ L^2(\mathbb{R}_+)\ \text{ has distributive basis } \{r\longmapsto j_L(Pr)\,\big|\, P>0\}$$
$$\text{with } \left\{\begin{array}{ll} \int_0^\infty r^2dr\ j_L(P'r)\ j_L(Pr) & =\frac{\pi}{2P^2}\delta(P-P'),\\ \int_0^\infty \frac{2P^2dP}{\pi}\ j_L(Pr)\ j_L(Pr') & =\frac{1}{r^2}\delta(r-r'). \end{array}\right.$$

Via momentum measures (wave packets) one obtains Hilbert space vectors in the *Fourier-Bessel transformation*

$$L^2_{dP}(\mathbb{R}_+)\cong L^2_{dr}(\mathbb{R}_+) \text{ with } \left\{\begin{array}{rl} \psi_L(r)= & \int_0^\infty \frac{2P^2dP}{\pi}\ j_L(Pr)f_L(P),\\ \int_0^\infty r^2dr\ \overline{\psi_L(r)}\psi_L'(r)= & \int_0^\infty \frac{2P^2dP}{\pi}\ \overline{f_L(P)}f_L'(P). \end{array}\right.$$

9.8.2 Spherical Bessel and Hyperbolic Macdonald Functions

The Schrödinger equation for constant potential with the sign $\epsilon=\epsilon(E-V_0)$,

$$[d_R^2+\tfrac{2}{R}d_R-\tfrac{L(1+L)}{R^2}+\epsilon]\psi_L(R)=\tfrac{1}{R}[d_R^2-\tfrac{L(1+L)}{R^2}+\epsilon]D_L(R)=0$$
$$\Rightarrow \psi_L(R)=\frac{D_L(R)}{R}=\left\{\begin{array}{lll} \alpha j_L(R) & +\beta n_L(R), & \epsilon=+1,\\ \alpha R^L & +\beta R^{-(1+L)}, & \epsilon=0,\\ \alpha k_L(R) & +\beta k_L(-R), & \epsilon=-1, \end{array}\right.$$

is solved by the *hyperbolic Macdonald functions* and the *spherical Hankel, Neumann, and Bessel functions* for integer $L=0,1,\ldots$:

$$\begin{array}{rcl} k_L(R)=(\frac{R}{2})^L\left(-\frac{d}{d\frac{R^2}{4}}\right)^L\frac{e^{-R}}{R} & = & \frac{e^{-R}}{R}\sum_{n=0}^{L}\frac{(2L-n)!}{(L-n)!n!}(2R)^{n-L}\\ & = & \frac{e^{-R}}{R},\ \frac{1+R}{R^2}e^{-R},\ldots,\\ h_L^{1,2}(R)=j_L(R)\pm i n_L(R) & = & \frac{e^{\mp iR}}{\mp iR},\ -\frac{1\mp iR}{R^2}e^{\mp iR},\ldots,\\ n_L(R) & = & \frac{\cos R}{R},\ \frac{\cos R+R\sin R}{R^2},\ldots,\\ j_L(R) & = & \frac{\sin R}{R},\ \frac{\sin R-R\cos R}{R^2},\ldots. \end{array}$$

They arise by derivation from the scalar functions and are called

$$\begin{array}{ll} \textit{scattering waves} & \left\{\begin{array}{rl} j_0(Pr) & =\frac{\sin Pr}{Pr}=\int\frac{d^3p}{\pi 2P}\delta(\vec p^2-P^2)e^{i\vec p\vec x},\\ & \quad \textit{standing},\\ n_0(Pr) & =\frac{\cos Pr}{Pr}=\int\frac{d^3p}{\pi^2 2P}\frac{1}{\vec p_{\rm P}^2-P^2}e^{i\vec p\vec x},\\ & \quad r=0\text{-singular, only for } r\to\infty,\\ h_0^{1,2}(Pr) & =\frac{e^{\mp iPr}}{\mp iPr}=\pm i\int\frac{d^3p}{\pi^2 2P}\frac{1}{\vec p^2-P^2\pm io}e^{i\vec p\vec x},\\ & \quad \textit{in- and outgoing}, \text{ only for } r\to\infty, \end{array}\right. \\ \frac{P^2}{2}=E-V_0>0 & \\ \textit{Yukawa and Coulomb potential} & k_0(|Q|r)=\frac{e^{-|Q|r}}{|Q|r}=\int\frac{d^3p}{\pi^2 2|Q|}\frac{1}{\vec p^2+Q^2}e^{i\vec p\vec x}.\\ -\frac{Q^2}{2}=E-V_0<0 & \end{array}$$

The Kepler factor for the 2-sphere spread in position, i.e., for the transition $\psi(r) \longmapsto \frac{\psi(r)}{r}$, can be related to the *2-sphere momentum integration* in the rotation-invariant integrals

$$\begin{array}{rll} \int \frac{d^3p}{4\pi} f(\vec{p}^2) e^{i\vec{p}\vec{x}} & = \frac{i}{2r} \int dp\ p f(p^2) e^{irp} & = -\frac{1}{2r} \frac{d}{dr} \int dp\ f(p^2) e^{irp} \\ & & = -\frac{d}{dr^2} \int dp\ f(p^2) e^{irp}, \\ (k_0, j_0, n_0)(r) & = \frac{(e^{-r}, \sin r, \cos r)}{r} & = -2\frac{d}{dr^2}(e^{-r}, \cos r, -\sin r). \end{array}$$

The *large distance* behavior gives, up to the 2-sphere spread by the Kepler factor, position representation coefficients with an angular momentum dependent phase shift, i.e., alternating between sine and cosine for Bessel and Neumann functions

$$R \to \infty: \quad \left\{ \begin{array}{lll} k_L(R) & \to \frac{e^{-R}}{R}, & h_L^{1,2}(R) \to (\pm i)^L \frac{e^{\mp iR}}{\mp iR}, \\ n_L(R) & \to \frac{\cos(R - \frac{L\pi}{2})}{R} & = \frac{\cos R,\ \sin R,\ -\cos R,\ -\sin R,\ \dots}{R}, \\ j_L(R) & \to \frac{\sin(R - \frac{L\pi}{2})}{R} & = \frac{\sin R,\ -\cos R,\ -\sin R,\ \cos R,\ \dots}{R}. \end{array} \right.$$

The *small distance behavior* depends on the angular momentum value

$$R \to 0: \quad \left\{ \begin{array}{rlll} j_L(R) & \to \frac{2^L L!}{(1+2L)!} & R^L & [1 + \mathcal{O}(R^2)], \\ n_L(R),\ k_L(R) & \to \frac{(2L)!}{2^L L!} & R^{-(1+L)} & [1 + \mathcal{O}(R^2)]. \end{array} \right.$$

Only the spherical Bessel functions are from $L^\infty(\mathbb{R}_+)$ and regular at the origin $r = 0$. They compensate in the Euclidean group representation coefficients the $r = 0$-ambiguity in the spherical harmonics

$$\mathrm{Y}_m^L(\theta, \varphi) j_L(r) \sim (\tfrac{\vec{x}}{r})^L r^L + \dots \sim (\vec{x})^L + \dots$$

9.9 Bound Orbits of 3-Dimensional Position

In contrast to the spherical harmonics with the $r \to 0$ ambiguity in $\frac{\vec{x}}{r}$, the *harmonic polynomials* (chapter "Spin, Rotations, and Position") as product with the corresponding radial power are defined also for $\vec{x} \to 0$. They are eigenfunctions for a trivial translation invariant P^2

$$(\vec{x})_m^L = r^L \mathrm{Y}_m^L(\varphi, \theta): \quad \left. \begin{array}{rl} \vec{\mathcal{L}}^2 \mathrm{Y}_m^L(\varphi, \theta) & = L(1+L) \mathrm{Y}_m^L(\varphi, \theta) \\ p_r^2 r^L & = -L(1+L) r^{L-2} \end{array} \right\} \Rightarrow \vec{\partial}^2 (\vec{x})_m^L = 0.$$

The $\vec{x}$-homogeneous harmonic polynomials span the irreducible **SO**(3)-representation spaces $\mathbb{C}^{1+2L}$. Position polynomials as direct sum of homogeneous polynomials can be decomposed into harmonic polynomials and r^2-parametrized invariant coefficients.

Noncompact radial representations in bound waves come after the separation of the harmonic polynomials for the irreducible **SO**(3)-representations.

This leads to the Schrödinger equations for the radial position representation coefficients

$$\psi(\vec{x}) = \sum_{L=0}^{\infty}\sum_{m=-L}^{L} (\vec{x})^L_m d_L(r) \quad \Rightarrow [d_r^2 + \tfrac{2(1+L)}{r}d_r + 2(E - V(r))]d_L(r) = 0,$$
$$d_L(r) = \tfrac{\psi_L(r)}{r^L}, \quad r = |\vec{x}|.$$

As above for 1-dimensional position, the noncompact representations by bound waves can involve a reparametrization with a monotonic function ρ

$$\mathbb{R}_+ \ni r \longmapsto \rho(r) \in \mathbb{R}_+.$$

The $\mathbb{R}_+$-representation coefficient is a product of a polynomial and an irreducible exponential[4]

$$\mathbb{R}_+ \ni \tfrac{\rho}{2} \longmapsto \tfrac{\psi_L^N(r)}{r^L} = d_L(r) = \mathrm{p}_L^N(\rho)\, e^{-\frac{\rho}{2}}$$

with the Schrödinger equation for the polynomial

$$\begin{aligned}(d_r\rho)^2 d_\rho^2 \mathrm{p}_L^N \;& + [d_r^2\rho - (d_r\rho)^2 + \tfrac{2(1+L)}{r}d_r\rho]d_\rho \mathrm{p}_L^N \\ & + [(d_r\tfrac{\rho}{2})^2 - d_r^2\tfrac{\rho}{2} + 2(E - V - \tfrac{1+L}{r}d_r\tfrac{\rho}{2})]\mathrm{p}_L^N = 0.\end{aligned}$$

A representation with constant polynomial relates the reparametrization to the potential

$$\tfrac{\psi_L^0(r)}{r^L} \sim e^{-\frac{\rho}{2}} \Rightarrow d_r^2\tfrac{\rho}{2} - (d_r\tfrac{\rho}{2})^2 = 2[E_{L0} - V(r) - \tfrac{1+L}{r}d_r\tfrac{\rho}{2}].$$

For a power reparametrization with possibly angular momentum dependent momentum unit Q_L

$$\tfrac{\rho(r)}{2} = \tfrac{(Q_L r)^n}{n} \Rightarrow (1 + 2L + n)Q_L^n r^{n-2} - Q_L^{2n} r^{2(n-1)} = 2[E_{L0} - V(r)],$$

with also the l.h.s. the sum of a constant for the energy and an r-dependent term for the potential, two solutions are possible. The solution with linear radial dependence $n = 1$ requires the Kepler potential which - by the Kepler factor - spreads the 1-dimensional free theory on the 2-sphere for each angular momentum $L = 0, 1, \ldots$. For quadratic radial dependence, $n = 2$, there arises the harmonic oscillator

$$n = 1 \;\Rightarrow \tfrac{\rho(r)}{2} = Q_L r, \quad \begin{cases} V(r) & = -\tfrac{(1+L)Q_L}{r}, \quad E_{L0} = \tfrac{Q_L^2}{2}, \\ \tfrac{\psi_L^0(r)}{r^L} & \sim e^{-Q_L r} = \int \tfrac{d^3p}{\pi^2}\tfrac{Q_L}{(\vec{p}^2+Q_L^2)^2}e^{-i\vec{p}\vec{x}}, \end{cases}$$

$$n = 2 \;\Rightarrow \rho(r) = (Q_L r)^2, \quad \begin{cases} V(r) & = \tfrac{(Q_L r)^2}{2}Q_L^2, \quad E_{L0} = (\tfrac{3}{2} + L)Q_L^2, \\ \tfrac{\psi_L^0(r)}{r^L} & \sim e^{-\frac{(Q_L r)^2}{2}}. \end{cases}$$

The positive "ground" state functions for each L are the Fourier transformed 3-sphere measure ($n = 1$) and Gauss measure ($n = 2$) on the momenta.

[4]The normalization $\frac{\rho}{2}$ is chosen with respect to the Laguerre polynomials below.

For the polynomials, both for harmonic oscillator and Kepler potential, there remains a Laplace differential equation

$$[\rho d_\rho^2 + (1+\lambda-\rho)d_\rho + \nu]\mathrm{p}_\lambda^\nu(\rho) = 0,$$
$$(\lambda,\nu) = \begin{cases} (1+2L, \ \frac{1}{|Q|} - 1 - L), & \text{Kepler } (n=1), \\ (\frac{1+2L}{2}, \ \frac{E-\frac{3}{2}-L}{2}), & \text{oscillator } (n=2). \end{cases}$$

9.9.1 Laguerre Polynomials

The Laplace equation for integer $\nu = N$

$$[\rho\frac{d^2}{d\rho^2} + (1+\lambda-\rho)\frac{d}{d\rho} + N]\mathrm{L}_\lambda^N(\rho) = 0, \quad N = 0,1,\ldots$$

is solved by the Laguerre plolynomials of degree N (nildimension, radial quantum number, knot number) and order λ. The Rodrigues formula contains as factor for Lebesgue measure $d\rho$ the positive function $\rho^\lambda e^{-\rho}$:

$$\begin{aligned} \mathbb{R} \ni \lambda \neq -1,-2,\cdots: \quad \mathrm{L}_\lambda^N(\rho) &= (\rho^{-\lambda}e^\rho \tfrac{d}{d\rho}\rho^\lambda e^{-\rho})^N \tfrac{\rho^N}{N!} = \tfrac{1}{N!}\rho^{-\lambda}e^\rho \tfrac{d^N}{d\rho^N}\rho^{\lambda+N}e^{-\rho} \\ &= \sum_{n=0}^N \binom{\lambda+N}{\lambda+n}\frac{(-\rho)^n}{n!}, \\ \deg \mathrm{L}_\lambda^N = N, \quad \frac{d\mathrm{L}_\lambda^N(\rho)}{d\rho} &= -\mathrm{L}_{\lambda+1}^{N-1}(\rho). \end{aligned}$$

The bound waves involve the noncompact representations coefficients

$$\mathbb{R}_+ \ni \tfrac{\rho}{2} \longmapsto \mathrm{L}_\lambda^N(\rho)\, e^{-\frac{\rho}{2}},$$

which give a basis for a Hilbert space for each $\lambda \notin -\mathbb{N}$

$$\lambda \notin -\mathbb{N}: \quad \mathbb{L}^2(\mathbb{R}_+,\mathbb{R}) \text{ has basis } \{\rho \longmapsto \rho^{\frac{\lambda}{2}}\mathrm{L}_\lambda^N(\rho)e^{-\frac{\rho}{2}} \mid N = 0,1,\ldots\}$$

$$\text{with } \begin{cases} \int_0^\infty \rho^\lambda e^{-\rho}d\rho\ \mathrm{L}_\lambda^N(\rho)\ \mathrm{L}_\lambda^{N'}(\rho) &= \frac{\Gamma(1+\lambda+N)}{N!}\delta_{NN'}, \\ \sum_{N=0}^\infty \frac{N!}{\Gamma(1+\lambda+N)}\mathrm{L}_\lambda^N(\rho)\ \mathrm{L}_\lambda^N(\rho') &= \rho^{-\lambda}e^\rho\delta(\rho-\rho'), \\ \zeta^2 < 1: \ \sum_{N=0}^\infty \mathrm{L}_\lambda^N(\rho)\zeta^N &= \frac{e^{\rho\frac{\zeta}{1-\zeta}}}{(1-\zeta)^{1+\lambda}}. \end{cases}$$

The λ-dependence is used for the angular momentum L-dependence.

9.9.2 Multipoles for Kepler Bound State Vectors

For an irreducible noncompact radial representation $r \longmapsto e^{-r}$ as bound solution of 3-position, an attractive Kepler potential is necessary. It defines an associated angular-momentum-dependent momentum unit

$$d_L(r) = e^{-|Q|r} \Rightarrow V(r) = -\tfrac{1}{r} \Rightarrow (1+L)Q_L = (1+L)\sqrt{-2E_{L0}} = 1.$$

There arises the quantum analogue to the classical parameter $L\sqrt{-2E}$. For the attractive Kepler potential and negative energy the imaginary radial "momentum" $\sqrt{2E}$ is "quantized" (integer wave numbers k):

$$\begin{array}{r} [d_r^2 + \frac{2(1+L)}{r} d_r + 2(E + \frac{1}{r})] d_L(r) = 0, \\ \frac{\rho(r)}{2} = |Q| r, \\ E = -\frac{Q^2}{2} < 0, \end{array} \Rightarrow \left\{ \begin{array}{rl} d_L(r) & = \mathrm{L}_{1+2L}^N(\rho) e^{-\frac{\rho}{2}}, \\ \frac{1}{Q_k} = k & = 1 + 2J = 1 + L + N, \\ & L, N = 0, 1, \ldots, \\ \psi_{Lm}^{2J}(\vec{x}) & \sim (\frac{\vec{x}}{k})_m^L \, \mathrm{L}_{1+2L}^N(\frac{2r}{k}) \, e^{-\frac{r}{k}}. \end{array} \right.$$

The $\mathbf{SO}(4)$ multiplets comprise all wave functions ψ^{2J} with equal sum $L + N = 2J$ for the principal quantum number $k = 1 + 2J$ with angular momentum $(1 + 2L)$-multiplets for $\mathbf{SO}(3)$ and radial quantum numbers N for Lenz-Runge classes, parametrizable by the 3-sphere $\Omega^3 \cong \mathbf{SO}(4)/\mathbf{SO}(3)$. The 3-sphere measure can be parametrized (chapter "Propagators") by 3-momenta $\vec{p} \in \mathbb{R}^3$ for the Lenz-Runge vector eigenvalues with dipoles

$$\int d^3\omega = \int d^3p \frac{2}{(1+\vec{p}^2)^2} = 2\pi^2 = |\Omega^3|.$$

The singularities are on a 2-sphere $\Omega^2 \cong \{\vec{p} \in \mathbb{R}^3 \,\big|\, \vec{p}^2 = -1\}$ with negative invariant (imaginary "momentum" eigenvalues).

The positive ground state function for the Kepler poptential is the Fourier transformed 3-sphere measure

$$e^{-r} = \int \frac{d^3p}{\pi^2} \frac{1}{(1+\vec{p}^2)^2} e^{-i\vec{p}\vec{x}}.$$

The rotation dependence $\vec{x}$ is effected by momentum derivation of the measure

$$\vec{x} e^{-r} = \int \frac{d^3p}{\pi^2} \frac{4i\vec{p}}{(1+\vec{p}^2)^3} e^{-i\vec{p}\vec{x}} \text{ with } \frac{4i\vec{p}}{(1+\vec{p}^2)^3} = -i \frac{\partial}{\partial \vec{p}} \frac{1}{(1+\vec{p}^2)^2}.$$

The 3-vector factor $\frac{2i\vec{p}}{1+\vec{p}^2}$ is uniquely supplemented to a normalized 4-vector on the 3-sphere:

$$\frac{1}{1+\vec{p}^2} \begin{pmatrix} \vec{p}^2 - 1 \\ 2i\vec{p} \end{pmatrix} = \begin{pmatrix} q_0 \\ i\vec{q} \end{pmatrix} = \begin{pmatrix} \cos\chi \\ i\sin\chi \frac{\vec{p}}{|\vec{p}|} \end{pmatrix} \in \Omega^3, \quad q \in \mathbb{R}^4 \text{ with } q^2 = q_0^2 + \vec{q}^2 = 1.$$

The 4-vector direction $\mathrm{Y}^{(1,1)}(q) \sim \begin{pmatrix} q_0 \\ i\vec{q} \end{pmatrix} \in \Omega^3$ is the analogue to the 3-vector direction $\mathrm{Y}^1(\frac{\vec{p}}{|\vec{p}|}) \sim \frac{\vec{p}}{|\vec{p}|} \in \Omega^2$ used for the buildup of the 2-sphere harmonics $\mathrm{Y}^L(\frac{\vec{p}}{|\vec{p}|}) \sim (\frac{\vec{p}}{|\vec{p}|})^L$. Analoguously, the higher order Ω^3-harmonics arise from the totally symmetric traceless products $\mathrm{Y}^{(2J,2J)}(q) \sim (q)^{2J}$, e.g., the nine independent components in the (4×4) matrix

$$\begin{aligned} \mathrm{Y}^{(2,2)}(q) \sim (q)_{jk}^2 & = q_j q_k - \frac{\delta_{jk}}{4} \cong \left(\begin{array}{c|c} \frac{3q_0^2 - \vec{q}^2}{4} & i q_0 q_a \\ \hline i q_0 q_b & q_a q_b - \frac{\delta_{ab}}{4} \end{array} \right), \\ \text{with } q_a q_b - \frac{\delta_{ab}}{4} & = q_a q_b - \frac{\delta_{ab}}{3} \vec{q}^2 - \frac{\delta_{ab}}{3} \frac{3q_0^2 - \vec{q}^2}{4} \text{ for } q^2 = 1. \end{aligned}$$

The Kepler bound state vectors are representation coefficients of position as noncompact hyperboloid $\mathbf{SO}_0(1,3)/\mathbf{SO}(3) \cong \mathcal{Y}^3$ in $L^2(\mathcal{Y}^3)$ in the form of

Fourier transformed Ω^3-measures (chapter "Residual Spacetime Representations"). Hyperbolic position is as manifold, not as $\mathbf{SO}(4)$-symmetric space, isomorphic to the translations $\mathcal{Y}^3 \cong \mathbb{R}^3$. The bound waves come with $2J$-dependent multipoles $\frac{1}{(1+\vec{p}^2)^{2+2J}}$:

$$\text{for } \mathcal{Y}^3: \ \vec{x} \longmapsto \psi_L^{2J}(\vec{x}) \sim \int \frac{d^3p}{\pi^2} \frac{1}{(1+\vec{p}^2)^2} (q)^{2J} e^{-i\vec{p}Q\vec{x}} \text{ with } \left\{ \begin{array}{ll} q & = \frac{1}{1+\vec{p}^2} \begin{pmatrix} \vec{p}^2 - 1 \\ 2i\vec{p} \end{pmatrix}, \\ Q & = \frac{1}{1+2J}, \\ J & = L + N. \end{array} \right.$$

The Fourier transformations with the 3-sphere measure

$$\mu(\vec{x}) e^{-r} = \int \frac{d^3p}{\pi^2} \frac{1}{(1+\vec{p}^2)^2} \tilde{\mu}(\vec{p}) e^{-i\vec{p}\vec{x}} :$$

$\mu(\vec{x})$	$\tilde{\mu}(\vec{p})$
1	1
r	$\frac{3-\vec{p}^2}{1+\vec{p}^2}$
$\frac{\vec{x}}{2}$	$\frac{2i\vec{p}}{1+\vec{p}^2}$
$\frac{r^2}{3}$	$\frac{4(1-\vec{p}^2)}{(1+\vec{p}^2)^2}$
$r\vec{x}$	$\frac{2i\vec{p}\,(5-\vec{p}^2)}{(1+\vec{p}^2)^2}$
$\vec{x} \otimes \vec{x} - \mathbf{1}_3 \frac{r^2}{3}$	$-\frac{6(\vec{p}\otimes\vec{p} - \mathbf{1}_3 \frac{\vec{p}^2}{3})}{(1+\vec{p}^2)^2}$

are used for the ground state (dipole scalar measure):

$$Q = 1: \quad \psi_0^0(\vec{x}) \sim e^{-Qr} = \int \frac{d^3p}{Q^3\pi^2} \frac{Q^2}{(Q^2+\vec{p}^2)^2} e^{-i\vec{p}\vec{x}} = \int \frac{d^3p}{\pi^2} \frac{1}{(1+\vec{p}^2)^2} e^{-i\vec{p}Q\vec{x}}.$$

The $k = 2$ bound state vector quartet has tripole vector measure:

$$\begin{aligned} Q = \tfrac{1}{2}: \quad \begin{pmatrix} \psi_0^1 \\ \psi_1^1 \end{pmatrix}(\vec{x}) \sim \begin{pmatrix} \frac{1}{4}\mathrm{L}_1^1(2Qr) \\ \frac{Q\vec{x}}{2}\,\mathrm{L}_2^0(2Qr) \end{pmatrix} & = \begin{pmatrix} \frac{1-Qr}{2} \\ \frac{Q\vec{x}}{2} \end{pmatrix} e^{-Qr} \\ & = \int \frac{d^3p}{\pi^2} \frac{1}{(1+\vec{p}^2)^3} \begin{pmatrix} \vec{p}^2 - 1 \\ 2i\vec{p} \end{pmatrix} e^{-i\vec{p}Q\vec{x}}. \end{aligned}$$

The $k = 3$ bound state vector nonet comes with quadrupole tensor measure:

$$\begin{aligned} Q = \tfrac{1}{3}: \quad \begin{pmatrix} \psi_0^2 \\ \psi_1^2 \\ \psi_2^2 \end{pmatrix}(\vec{x}) & \sim \begin{pmatrix} \frac{1}{3}\mathrm{L}_1^2(2Qr) \\ \frac{Q\vec{x}}{2}\,\frac{1}{6}\mathrm{L}_3^1(2Qr) \\ \frac{Q^2}{2}(\mathbf{1}_3 \frac{r^2}{3} - \vec{x}\otimes\vec{x})\mathrm{L}_5^0(2Qr) \end{pmatrix} = \begin{pmatrix} 1 - 2Qr + \frac{2Q^2r^2}{3} \\ \frac{2-Qr}{3}\frac{Q\vec{x}}{2} \\ \frac{Q^2}{2}(\mathbf{1}_3 \frac{r^2}{3} - \vec{x}\otimes\vec{x}) \end{pmatrix} e^{-Qr} \\ & = \int \frac{d^3p}{\pi^2} \frac{4}{(1+\vec{p}^2)^4} \begin{pmatrix} 3(\frac{\vec{p}^2-1}{2})^2 - \vec{p}^2 \\ i\vec{p}\,\frac{\vec{p}^2-1}{2} \\ 3\vec{p}\otimes\vec{p} - \mathbf{1}_3\vec{p}^2 \end{pmatrix} e^{-i\vec{p}Q\vec{x}} \\ & = \int \frac{d^3p}{\pi^2} \frac{1}{(1+\vec{p}^2)^2} \begin{pmatrix} 3q_0^2 - \vec{q}^2 \\ iq_0\vec{q} \\ 3\vec{q}\otimes\vec{q} - \mathbf{1}_3\vec{q}^2 \end{pmatrix} e^{-i\vec{p}Q\vec{x}}. \end{aligned}$$

The Kepler dynamics also has scattering solutions with imaginary translation eigenvalues, i.e., real momenta, and a corresponding normalization of the radial position in the radial Laplace differential equations:

$$\begin{aligned} & [\rho d_\rho^2 + (2(1+L) - \rho)d_\rho - (1 + L + \tfrac{1}{\sqrt{-2E}})]D_L(\rho) = 0, \\ & 2E = \left\{ \begin{array}{ll} -Q^2 < 0 & \text{bound waves,} \\ \vec{q}^2 > 0 & \text{scattering waves} \end{array} \right. \text{ and } -\tfrac{\rho}{2} = r\sqrt{-2E} = \left\{ \begin{array}{l} -|Q|r, \\ i|\vec{q}|r, \end{array} \right. \\ & E = 0: \quad [d_r^2 + \tfrac{2(1+L)}{r}d_r + \tfrac{2}{r}]d_L(r) = 0. \end{aligned}$$

The two scattering solutions for $E > 0$, $RJ_L(R)$ and $RN_L(R)$, $R = qr$ have the same leading radial behavior for small distances as spherical Bessel and Neumann functions, but a logarithmically modified large-distance behavior

$$\begin{pmatrix} R^L \\ R^{-1-L} \end{pmatrix} \xleftarrow{R\to 0} \begin{pmatrix} J_L(R) \\ N_L(R) \end{pmatrix} \xrightarrow{R\to\infty} \frac{1}{R}\begin{pmatrix} \sin(R - q\log 2R - \frac{L\pi}{2} + \varphi_L) \\ \cos(R - q\log 2R - \frac{L\pi}{2} + \varphi_L) \end{pmatrix}$$

with additional changes in the phase φ_L and the normalization c_L. More details in the literature.

9.9.3 Nonrelativistic Color Symmetry

Separating in the position wave functions the harmonic polynomials from a function with squared radial dependence, there arise the radial equations

$$\psi(\vec{x}) = \sum_{L=0}^{\infty}\sum_{m=-L}^{L} (\vec{x})^L_m \Delta_L(\rho) \quad \Rightarrow [\rho d_\rho^2 + (\tfrac{3}{2} + L - \rho)d_\rho + \tfrac{E-V(r)}{2}]\Delta_L(\rho) = 0,$$
$$\rho = r^2 = \vec{x}^2.$$

An irreducible exponential with squared radial dependence as solution determines the harmonic oscillator potential, normalized[5] with a momentum unit $|Q|$:

$$\Delta_L(r^2) = e^{-\frac{Q^2 r^2}{2}} \Rightarrow V(r) = \tfrac{(Q_L r)^2}{2}Q_L^2, \quad E = (\tfrac{3}{2} + L)Q_L^2.$$

The momentum unit can be chosen L-independent, e.g., $Q_L = 1$. The general solution is the product of a Laguerre polynomial of degree N and, in contrast to the Kepler bound waves with principal-quantum-number-dependent exponentials $e^{-\frac{r}{k}}$, one exponential only:

$$[d_r^2 + \tfrac{2(1+L)}{r}d_r + 2(E - \tfrac{r^2}{2})]\Delta_L(r^2) = 0 \Rightarrow \begin{cases} \Delta_L(r^2) &= L^N_{\frac{1+2L}{2}}(r^2)e^{-\frac{r^2}{2}}, \\ E_k &= \frac{3}{2} + k = \frac{3}{2} + L + 2N, \\ & \quad L, N = 0, 1, \dots, \\ \psi^k_{Lm}(\vec{x}) &\sim (\vec{x})^L_m \, \mathrm{L}^N_{\frac{1+2L}{2}}(r^2)\, e^{-\frac{r^2}{2}}. \end{cases}$$

The harmonic oscillator solutions for each angular momentum $L = 0, 1, 2, \dots$ constitute a Hilbert space basis:

$$\begin{aligned} L^2(\mathbb{R}_+, \mathbb{R})\text{-basis}: \quad & \{\rho \longmapsto \rho^{\frac{1+2L}{4}}\mathrm{L}^N_{\frac{1+2L}{2}}(\rho)e^{-\frac{\rho}{2}} \mid N = 0, 1, \dots\}, \\ \int_0^\infty dr\, \mathrm{L}^N_{\frac{1+2L}{2}}(r^2)\, e^{-r^2}r^{1+2L}\mathrm{L}^{N'}_{\frac{1+2L}{2}}(r^2) &= \tfrac{1}{2}\int_0^\infty \rho^{\frac{1+2L}{2}} e^{-\rho}d\rho\, \mathrm{L}^N_{\frac{1+2L}{2}}(\rho)\, \mathrm{L}^{N'}_{\frac{1+2L}{2}}(\rho) \\ &= \frac{\Gamma(N+\frac{3}{2}+L)}{2N!}\delta_{NN'}. \end{aligned}$$

This Hilbert space is the direct integral of $\mathbb{R}$-isomorphic Hilbert spaces for each radial translation r with L-dependent ground-state-induced measure μ_L:

$$L^2(\mathbb{R}_+, \mathbb{R}) \cong {}^{\oplus}\!\!\int_0^\infty \mu_L(r)\, dr\, \mathbb{R}(r), \quad \mu_L(r) = \frac{2}{\Gamma(\frac{3}{2}+L)}r^{2+2L}e^{-r^2}, \quad \mathbb{R}(r) \cong \mathbb{R}.$$

[5]Usually, the additive term $V_0 = \frac{3}{2}Q^2$ is introduced as ground state energy.

There are only bound waves, no scattering solutions.

The harmonic oscillator Hamiltonian

$$H = \frac{\vec{p}^2+\vec{x}^2}{2} = \frac{\{\mathrm{u}^a, \mathrm{u}^\star_a\}}{2} \text{ with creation operators } \mathrm{u}^a = \frac{x^a - ip^a}{\sqrt{2}}, \quad [\mathrm{u}^\star_b, \mathrm{u}^a] = \delta^b_a$$

generates time orbits of the Hilbert vectors with k quanta. They have a Schrödinger representation as position orbits

$$\begin{aligned} |a_1, \dots a_k\rangle &= \frac{\mathrm{u}^{a_1} \cdots \mathrm{u}^{a_k}}{\sqrt{k!}}|0\rangle \cong \{\vec{x} \longmapsto x^{a_1} \cdots x^{a_k} e^{-\frac{r^2}{2}}\} \in L^2(\mathbb{R}^3), \\ \mathrm{u}(t)^a &= e^{it}\mathrm{u}^a, \quad |a_1, \dots a_k\rangle(t) = e^{ikt}|a_1, \dots, a_k\rangle. \end{aligned}$$

The complex representation of the three position translations $\mathbb{R}^3 \hookrightarrow \mathbb{C}^3$ leads to a color $\mathbf{SU}(3)$-invariance (chapter "Quantum Algebras"), with Gell-Mann matrices

$$\chi_A \lambda^A = \begin{pmatrix} \chi_3 + \frac{\chi_8}{\sqrt{3}} & \chi_1 - i\chi_2 & \chi_4 - i\chi_5 \\ \chi_1 + i\chi_2 & -\chi_3 + \frac{\chi_8}{\sqrt{3}} & \chi_6 - i\chi_7 \\ \chi_4 + i\chi_5 & \chi_6 + i\chi_7 & -2\frac{\chi_8}{\sqrt{3}} \end{pmatrix}, \quad \mathcal{C} = \frac{i}{2}\mathrm{u}^a \lambda^b_a \mathrm{u}^\star_b, \quad [\mathcal{C}, H] = 0.$$

The $\mathbf{SU}(3)$-representations $[2C_1, 2C_2]$ are characterized by two integers; they have the dimension

$$\dim_{\mathbb{C}}[2C_1, 2C_2] = (1 + 2C_1)(1 + 2C_2)(1 + C_1 + C_2).$$

The harmonic oscillator representations $[k, 0]$ (singlet, triplet, sextet, etc.) are the totally symmetric products of $\mathbf{SU}(3)$-triplets $[1, 0]$

$$\bigvee^k \mathbb{C}^3 \cong \mathbb{C}^{\binom{2+k}{2}}, \quad \dim_{\mathbb{C}}[k, 0] = \binom{2+k}{k} = 1, 3, 6, \dots, \quad k = 0, 1, 2, \dots.$$

The rotation group, generated by the transposition antisymmetric Lie subalgebra

$$\mathbf{SO}(3) \hookrightarrow \mathbf{SU}(3) \text{ with } \begin{cases} \chi_A i \frac{\lambda^A - (\lambda^A)^T}{2} = \begin{pmatrix} 0 & \chi_2 & \chi_5 \\ -\chi_2 & 0 & \chi_7 \\ -\chi_5 & -\chi_7 & 0 \end{pmatrix} = \begin{pmatrix} 0 & \varphi_3 & -\varphi_2 \\ -\varphi_3 & 0 & \varphi_1 \\ \varphi_2 & -\varphi_1 & 0 \end{pmatrix}, \\ \mathcal{L}^a = \epsilon^{abc}\mathrm{u}^b \mathrm{u}^\star_c, \quad (\mathcal{L}^1, \mathcal{L}^2, \mathcal{L}^3) = (\mathcal{C}^7, -\mathcal{C}^5, \mathcal{C}^2), \end{cases}$$

comes with the real 5-dimensional orientation manifold, i.e., the rotation group orbits $\mathbf{SU}(3)/\mathbf{SO}(3)$ in the color group. which describes the $\binom{4}{2} - 1$ relative phases of the three position directions in complex quantum structures. The principal quantum number $k = L + 2N$ is the sum of the angular momentum quantum number L for $\mathbf{SO}(3)$ and the radial quantum number (knot number) N for the rotation group classes in $\mathbf{SU}(3) \cong \mathbf{SO}(3) \times \mathbf{SU}(3)/\mathbf{SO}(3)$. One has with the angular momentum degeneracy $1 + 2L$ the energy degeneracy given by the dimensions of the $\mathbf{SU}(3)$-representations:

$$\text{multiplicity } \binom{k+2}{2}, \quad [k, 0] \overset{\mathbf{SO}(3)}{\cong} \begin{cases} \bigoplus_{L=0}^{k} [2L], & k = 0, 2, \dots \text{ (even)}, \\ \bigoplus_{L=1}^{k} [2L], & k = 1, 3, \dots \text{ (odd)}, \end{cases}$$

$$E_{LN} - \frac{3}{2} = L + 2N = k \Rightarrow (L, N) = \begin{cases} (k, 0), (k-2, 1), \dots, (0, \frac{k}{2}), \\ (k, 0), (k-2, 1), \dots, (1, \frac{k-1}{2}). \end{cases}$$

9.10 Scattering

Incoming free particles ("flying in") undergoing an interaction and "flying out" again as, possibly different, free particles are described by scattering structures. Scattering is formalized in a Hilbert space V by transition amplitudes from state vectors $|\alpha_-\rangle$ in the distant past $t_- \to -\infty$ to state vectors $|\alpha_+\rangle$ in the distant future $t_+ \to +\infty$ with the absolute squares as the associated transition probabilities

$$p_{|\alpha_-\rangle\to|\beta_+\rangle} = \frac{\langle\beta_+|\alpha_-\rangle\langle\alpha_-|\beta_+\rangle}{\langle\beta_+|\beta_+\rangle\langle\alpha_-|\alpha_-\rangle} = |\langle\alpha_-^0|\beta_+^0\rangle|^2 \text{ with normalized } |\alpha_-^0\rangle, |\beta_+^0\rangle.$$

9.10.1 Evolution and Scattering Operators

Scattering in the interaction picture considers the matrix elements of a Hamiltonian H with interaction with eigenvectors of a free Hamiltonian H_0:

$$\mathbb{R} \ni t \longmapsto D(t) = e^{iHt} \text{ and } D_0(t) = e^{iH_0t}.$$

The relation of the finite-time developments (if defined) is expressed with the associated *evolution operator* $\Omega(t)$ for a Hilbert space V:

$$\begin{aligned} \Omega(t) &= D(t)D_0(-t) = e^{iHt}e^{-iH_0t}, \quad V \xrightarrow{e^{-iH_0t}} V \xrightarrow{e^{iHt}} V \\ |\alpha(t)\rangle &= D(t)|\alpha\rangle = \Omega(t)|\alpha(t)\rangle_0, \quad |\alpha(t)\rangle_0 = D_0(t)|\alpha\rangle, \end{aligned}$$

where $\Omega(t)$ is the identity for the free theory $H = H_0$.

The strong limit (norm topology), if it exists, defines the *Moeller operators* for future and past:

$$\Omega_\pm = \lim_{t\to\pm\infty} \Omega(t) = \lim_{t\to\pm\infty} e^{iHt}e^{-iH_0t}.$$

Their product connects "in" ($t \to -\infty$) with "out" ($t \to \infty$). The double limit defines the *scattering operator* S

$$S(t_+, t_-) = \Omega(t_+)^\star\Omega(t_-), \quad S = \lim_{t_\pm\to\pm\infty} S(t_+, t_-) = \Omega_+^\star\Omega_-.$$

The nontrivial part is called *transition operator* T

$$S = \mathbf{1} - 2i\pi T.$$

The sesquilinear S*-matrix* with corresponding T*-matrix*

$$S_{\beta\alpha} = \langle\beta|S|\alpha\rangle = \langle\Omega_+\beta|\Omega_-\alpha\rangle = \langle\beta_+|\alpha_-\rangle = \langle\beta|\alpha\rangle - 2i\pi T_{\beta\alpha}$$

contains scalar products as matrix elements. It involves a Hilbert space product, linear in $|\alpha\rangle$ and antilinear in $\langle\beta|$. It can be identified, up to conjugation properties, with a transformation matrix $S_{\beta\alpha} \cong S_\alpha^\beta$ only for an orthonormal Hilbert basis. Unstable states may lead to nonorthogonal bases and higher-dimensional collectives for energy eigenstates, e.g., for the collective with the two unstable neutral kaons.

To prove, for a given Hamiltonian, the existence and unitarity of the double limit S-operator may be a difficult task where the necessary hermiticity of the Hamiltonians may not suffice. If one assumes the unitarity property

$$H_0 = H_0^\star,\ H = H^\star \Rightarrow \Omega_\pm(t), S(t_+, t_-) \in \mathbf{U}(V)$$

to survive the limits, then the Moeller operators and hence the S-operator are in the unitary group of the Hilbert space:

$$\Omega_\pm \in \mathbf{U}(V) \Rightarrow S \in \mathbf{U}(V).$$

The S-unitarity, considered for the T-operator,

$$\text{if } SS^\star = \mathbf{1} = S^\star S \Rightarrow \tfrac{i}{2\pi}(T - T^\star) = TT^\star = T^\star T,$$

involves the optical theorem.

With respect to energy conservation $[S, H] = 0$ and possibly other invariances, e.g., rotation invariance, an S-operator and the Hilbert space (taking into account the symmetry properties of the ground state) is decomposable into direct integrals, e.g., for energy or momenta, and direct sums, e.g., for rotation properties:

$$V = \bigoplus_k {}^\perp\!\!\int dE\ V_k(E), \quad \begin{cases} D(t) &= \bigoplus_k {}^\perp\!\!\int dE\ e^{iEt} f_k(E), \\ S &= \bigoplus_k {}^\perp\!\!\int dE\ S_k(E). \end{cases}$$

9.10.2 In- and Out-Vectors

Starting from a (distributive) basis for a Hilbert space V the Moeller automorphisms for future and past $\Omega_\pm \in \mathbf{GL}(V)$, if they exist, define two other (distributive) bases, called *out- and in-vectors*

$$\begin{aligned} S = \Omega_+^\star \Omega_- \in \mathbf{GL}(V),\ & \text{Hilbert bases: } \{|\alpha\rangle\} \text{ and } \{|\alpha_\pm\rangle = \Omega_\pm|\alpha\rangle\} \\ & S_{\beta\alpha} = \langle\beta|S|\alpha\rangle = \langle\beta_+|\alpha_-\rangle. \end{aligned}$$

With a distributive Hilbert basis of eigenvectors for the free time development, indexed by the energy and additional degrees of freedom

$$H_0|\alpha\rangle = E_\alpha|\alpha\rangle \text{ with } |\alpha\rangle = |E_\alpha, k_\alpha\rangle,$$

the distributive out- and in-vectors are assumed to be eigenvectors for the full time development with *equal energies*. This is the case iff free and interaction Hamiltonian are related to each other by inner Moeller automorphisms:

$$H|\alpha_\pm\rangle = E_\alpha|\alpha_\pm\rangle \iff H = \Omega_\pm H_0 \Omega_\pm^{-1}.$$

For an orthonormal distributive basis and unitary Moeller operators the orthogonal direct integral decomposition of the unit operator in the Hilbert space V is given with distributive in- and out-vectors:

$$\left.\begin{array}{l}\text{for } \langle\beta|\alpha\rangle = \delta(\alpha-\beta),\\ \mathrm{id}_V \cong {}^{\perp}\!\!\int d\alpha\ |\alpha\rangle\langle\alpha|,\\ \text{and } \Omega_\pm \in \mathbf{U}(V),\end{array}\right\} \Rightarrow \quad \mathrm{id}_V \cong {}^{\perp}\!\!\int d\alpha\ |\alpha_\pm\rangle\langle\alpha_\pm| \begin{array}{l} = \Omega_\pm {}^{\perp}\!\!\int d\alpha\ |\alpha\rangle\langle\alpha|\ \Omega_\pm^\star \\ = \Omega_\pm\Omega_\pm^\star .\end{array}$$

The (direct) summation ${}^{\perp}\!\!\int d\alpha$ includes an energy integration ${}^{\perp}\!\!\int dE_\alpha$.

For this case with Hermitian Hamiltonians, i.e., real energies $E_\alpha \in \mathbb{R}$, the Lippmann-Schwinger equations relate to each other the different distributive bases:

$$\begin{array}{rll} & H|\alpha_\pm\rangle = E_\alpha|\alpha_\pm\rangle & \\ \Rightarrow & |\alpha_\pm\rangle = |\alpha\rangle - \frac{1}{H_0 - E_\alpha \mp io} H_I |\alpha_\pm\rangle & \text{with } H_I = H - H_0, \\ & = |\alpha\rangle - {}^{\perp}\!\!\int d\beta\ \frac{\langle\beta|H_I|\alpha_\pm\rangle}{H_0 - E_\alpha \mp io}|\beta\rangle & \\ & = |\alpha\rangle - {}^{\perp}\!\!\int d\beta\ \frac{T^\pm_{\beta\alpha}}{E_\beta - E_\alpha \mp io}|\beta\rangle & \text{with } T^\pm_{\beta\alpha} = \langle\beta|H_I|\alpha_\pm\rangle. \end{array}$$

The distributional prescription $\pm io$ allows the operational formulation of the distributive out- and in-vectors $|\alpha_\pm\rangle$ as time-independent distributive Hilbert vectors: Their time orbits

$${}^{\perp}\!\!\int d\alpha\ \mu(\alpha)e^{iE_\alpha t}|\alpha_\pm\rangle = {}^{\perp}\!\!\int d\alpha\ \mu(\alpha)e^{iE_\alpha t}|\alpha\rangle - \int d\alpha\ \mu(\alpha)e^{iE_\alpha t}{}^{\perp}\!\!\int d\beta\ \frac{T^\pm_{\beta\alpha}}{E_\beta - E_\alpha \mp io}|\beta\rangle$$

have the desired behavior if one assumes that the relation for the energy integration

$$\pm\int \frac{dE_\alpha}{2i\pi} e^{iE_\alpha t}\frac{1}{E_\beta - E_\alpha \mp io} = \vartheta(\mp t)e^{itE_\beta} = \vartheta(\mp t)\int dE_\alpha e^{iE_\alpha t}\delta(E_\beta - E_\alpha)$$

can be used for the limiting time behavior of these integrals,

$$\begin{array}{l} t \to \pm\infty : {}^{\perp}\!\!\int d\alpha\ \mu(\alpha)e^{iE_\alpha t}|\alpha_\pm\rangle \to {}^{\perp}\!\!\int d\alpha\ \mu(\alpha)e^{iE_\alpha t}|\alpha\rangle \\ \qquad\qquad \mp\vartheta(\mp t)2i\pi\int d\alpha\ \mu(\alpha)e^{iE_\alpha t}{}^{\perp}\!\!\int d\beta\ T^\pm_{\beta\alpha}\delta(E_\beta - E_\alpha)|\beta\rangle, \end{array}$$

where the interaction-related term disappears in the corresponding limit.

Expanding a distributive in-vector with the S-matrix in out-vectors,

$$\begin{array}{rl} {}^{\perp}\!\!\int d\alpha\ \mu(\alpha)e^{iE_\alpha t}|\alpha_-\rangle & = {}^{\perp}\!\!\int d\alpha\ \mu(\alpha)e^{iE_\alpha t}{}^{\perp}\!\!\int d\beta\ |\beta_+\rangle S_{\beta\alpha} \\ & \to \int d\alpha\ \mu(\alpha)e^{iE_\alpha t}{}^{\perp}\!\!\int d\beta\ |\beta\rangle S_{\beta\alpha} \text{ for } t \to +\infty, \end{array}$$

one has by comparison

$$S_{\beta\alpha} = \delta(\alpha - \beta) - 2i\pi\delta(E_\alpha - E_\beta)T^+_{\beta\alpha},$$

where $\delta(\alpha - \beta)$ involves a Kronecker symbol for the discrete indices.

9.10.3 Potential Scattering

The plane waves are translation eigenvectors (no Hilbert space vectors) of the free Hamiltonian $H_0 = \frac{\vec{p}^2}{2} + V_0$ with constant potential

$$(\vec{\partial}^2 + P^2)e^{i\vec{p}\vec{x}} = 0, \quad P^2 = 2(E - V_0) > 0.$$

A plane wave is decomposed into irreducible $\mathbf{SO}(3) \vec{\times} \mathbb{R}^3$-representation coefficients, characterized by continuous momentum-invariant $P^2 > 0$ and discrete polarization L

$$e^{i\vec{p}\vec{x}} = e^{iPr\cos\theta} = e^{iPz} = \sum_{L=0}^{\infty} p^L(\theta)\, j_L(Pr), \quad p^L(\theta) = i^L(1+2L)\mathrm{P}^L(\cos\theta).$$

It contains standing spherical waves for each $\mathbf{SO}(2)$-polarization L:

$$P = |\vec{p}| > 0, \quad L = 0, 1, 2, \cdots : \quad \begin{array}{l} |L, P\rangle \cong \mathrm{Y}_0^L(\varphi, \theta) j_L(Pr), \\ \langle L', P'|L, P\rangle = \delta_{LL'} \frac{1}{4P^2} \delta(\frac{P-P'}{2\pi}). \end{array}$$

The infinite-dimensional Hilbert space for each $|L, P\rangle$ is constituted by functions $L^2(\Omega^2)$, square integrable on the momentum directions $\vec{\omega} = \frac{\vec{p}}{P} \in \Omega^2$ (chapter "Harmonic Analysis").

The position translations in the Euclidean group come in self-dual representations, e.g., $j_0(Pr) = \frac{\sin Pr}{Pr} = \frac{e^{iPr} + e^{-iPr}}{2iPr}$. This leads to the decomposition of the real standing spherical waves (Bessel functions) into Hankel functions $h_L^{1,2}(Pr)$, conjugated to each other, for in- and out-vectors:

$$e^{i\vec{p}\vec{x}} = \sum_{L=0}^{\infty} p^L(\theta)\, \frac{h_L^2(Pr) + h_L^1(Pr)}{2} \;\overset{r\to\infty}{\longrightarrow}\; \sum_{L=0}^{\infty} p^L(\theta)\, \frac{e^{i(Pr - \frac{L\pi}{2})} - e^{-i(Pr - \frac{L\pi}{2})}}{2iPr}.$$

A unitary scattering operator is decomposable with respect to the two scattering invariants momentum modulus and polarization

$$S = \bigoplus_{L=0}^{\infty} \oplus\!\!\int_0^\infty \frac{2P^2 dP}{\pi}\, S_L(P), \quad \langle L', P'|S|L, P\rangle = \delta_{LL'} \frac{1}{4P^2} \delta(\frac{P-P'}{2\pi}) S_L(P).$$

The irreducible components define the *scattering phases* $\delta_L(P)$ in the the product of the unitary Moeller transformations for the two position hemispheres

$$S_L(P) = \Omega_L^+(P)^\star \Omega_L^-(P) = e^{2i\delta_L(P)} \in \mathbf{U}(1).$$

For scattering solutions of a rotation invariant dynamics $H = \frac{\vec{p}^2}{2} + V(r)$ and Schrödinger equation

$$\psi(\vec{x}) = \sum_{L=0}^{\infty} p^L(\theta)\, \psi_L(r) : \quad [d_r^2 + \tfrac{2}{r} d_r - \tfrac{L(1+L)}{r^2} + 2(E - V(r))]\psi_L(r) = 0$$

the potential is assumed to level off for the asymptotic region:

$$\lim_{r\to\infty} V(r) = V_0 \in \mathbb{R}, \quad \lim_{r\to\infty} r[V(r) - V_0] = 0.$$

In the simplest case, one has $V(r) = V_0$ outside the interaction $r > a$. Coulomb scattering with $V(r) \sim \frac{1}{r}$ requires a more complicated treatment. The scattering solutions ψ go over into a free solution $\psi^>$ outside the interaction, i.e., for constant potential

$$\begin{aligned} &[d_r^2 + \tfrac{2}{r}d_r - \tfrac{L(1+L)}{r^2} + P^2]\psi_L^>(r) = 0 \text{ for } V(r) = V_0, \\ &r \to \infty: \quad \psi(\vec{x}) \to \psi^>(\vec{x}) = \sum_{L=0}^{\infty} p^L(\theta)\ \frac{\Omega_L^+(P)h_L^2(Pr) + \Omega_L^-(P)h_L^1(Pr)}{2}. \end{aligned}$$

Any solution for constant potential can be written as a linear combination of in- and outgoing spherical Hankel waves. The expansion involves the unitary Moeller transformations $\Omega_L^\pm(P) = e^{\pm i\delta_L(P)} \in \mathbf{U}(1)$. The scattering phases depend on the potential $V(r)$ and are determined by the full solution ψ_L.

The scattering solution can be experimentally interpreted with boundary conditions for $r \to \infty$ as a linear superposition of an in- and throughgoing plane wave (elastic scattering with the free Hamiltonian H_0) and an outgoing nontrivially scattered spherical wave with *scattering amplitude* $f(\theta, P)$:

$$\begin{aligned} r \to \infty: \quad \psi(\vec{x}) &\to \psi^>(\vec{x}) \to e^{i\vec{p}\vec{x}} + f(\theta, P)\frac{e^{iPr}}{r} \\ &= \sum_{L=0}^{\infty} p^L(\theta)\ [j_L(Pr) + f_L(P)\frac{e^{i(Pr - L\pi)}}{r}] \\ &\to \sum_{L=0}^{\infty} p^L(\theta)\ \frac{(1 + 2iPf_L(P))e^{i(Pr - L\pi)} - e^{-iPr}}{2iPr} \\ \text{with } f(\theta, P) &= \sum_{L=0}^{\infty} (1 + 2L)\mathrm{P}^L(\cos\theta)\ f_L(P). \end{aligned}$$

The factor $\frac{1}{r}$ for the spherical wave compared with the plane wave leads to a length dimension for the complex scattering amplitude. Therefore, its absolute square gives a squared length for the cross section $\sigma(\theta, P)$:

$$\sigma(\theta, P) = |f(\theta, P)|^2.$$

The (L, P)-decomposition of scattering and transition operator reads

$$\begin{aligned} S_L(P) = 1 - 2i\pi T_L(P) = &\ 1 + 2iPf_L(P) = e^{2i\delta_L(P)}, \\ -\pi T_L(P) = &\ \quad Pf_L(P) = \frac{e^{2i\delta_L(P)} - 1}{2i} = \sin\delta_L(P) e^{i\delta_L(P)}. \end{aligned}$$

The unitarity of the S-operator is equivalent to a momentum P-parametrized circle in the complex plane for the scattering amplitude:

$$|S_L(P)|^2 = 1 \iff [\mathrm{Re}\ Pf_L(P)]^2 + [\mathrm{Im}\ Pf_L(P) - \tfrac{1}{2}]^2 = (\tfrac{1}{2})^2.$$

9.10.4 Scattering by Perturbation

A unitary evolution operator

$$S(t,t_0)=e^{iH_0t}e^{-iH(t-t_0)}e^{-iH_0t_0},\quad H_I=H-H_0,$$

can be given a perturbative Dyson expansion with free time development by an iterative solution of its differential equation

$$\frac{\partial}{\partial t}S(t,t_0)=-iH_I(t)S(t,t_0)\text{ with}\left\{\begin{array}{ll}H_I(t)&=e^{iH_0t}H_Ie^{-iH_0t},\\ \frac{\partial}{\partial t}H_I(t)&=[iH_0,H_I(t)].\end{array}\right.$$

It sums up the "H_I-pushes"

$$S(t,t_0)=\mathbf{1}-i\int_{t_0}^t dt_1H_I(t_1)S(t_1,t_0)=\mathbf{1}+\sum_{n=1}^{\infty}(-i)^nI^n(t,t_0).$$

The expansion terms can be written with *time-ordered products* of the interaction

$$\begin{array}{rl}I^n(t,t_0)=&\int_{t_0}^t dt_1H_I(t_1)\int_{t_0}^{t_1}dt_2H_I(t_2)\quad\cdots\int_{t_0}^{t_{n-1}}dt_nH_I(t_n)\\ =&\int_{t_0}^t dt_1H_I(t_1)\int_{t_0}^{t}dt_2\vartheta(t_1-t_2)H_I(t_2)\quad\cdots\int_{t_0}^{t}dt_n\vartheta(t_{n-1}-t_n)H_I(t_n)\\ =&\frac{1}{n!}\int_{t_0}^t dt_1\cdots\int_{t_0}^t dt_n\mathbf{T}[H_I(t_1)\cdots H_I(t_n)],\end{array}$$

where the average over all permutations is used:

$$\begin{array}{l}\mathbf{T}[H_I(t_1)\cdots H_I(t_n)]\\ \quad=\sum_{\text{perm}}H_I(t_{\pi(1)})\cdots H_I(t_{\pi(n)})\vartheta(t_{\pi(1)}-t_{\pi(2)})\cdots\vartheta(t_{\pi(n-1)}-t_{\pi(n)})\\ \quad=H_I(t_i)\cdots H_I(t_j)\cdots H_I(t_k)\text{ for }t_i\geq\cdots\geq t_j\geq\cdots\geq t_k.\end{array}$$

Thus one has a formal expression for the evolution and scattering operator:

$$\begin{array}{rl}S(t,t_0)&=\mathbf{T}e^{-i\int_{t_0}^t d\tau H_I(\tau)},\\ S&=\mathbf{T}e^{-i\int_{-\infty}^{\infty}d\tau H_I(\tau)}=\mathbf{1}-i\int_{-\infty}^{\infty}d\tau H_I(\tau)+\cdots.\end{array}$$

9.11 Summary

Schrödinger equations formulate the diagonalization problem of a Hamiltonian (energy eigenvalues for time translations) in the form of position representations with momentum eigenvalues, real momenta for scattering representations, and imaginary "momenta" for bound representations. Schrödinger wave functions are position orbits in a Hilbert space with probability amplitudes.

The infinite-dimensional Hilbert space $L^2(\mathbb{R}^s)$, determined by the faithful representations of the group $\mathbf{H}(s)\cong\mathbb{R}^s\ \vec{\times}\ \mathbb{R}^{1+s}$ for the Heisenberg Lie algebra $[\mathbf{x}^a,\mathbf{p}^b]=\delta^{ab}\mathbf{I}$, can also be used for representations of the Euclidean group $\mathbf{SO}(s)\ \vec{\times}\ \mathbb{R}^s$ and their expansions, the simple groups $\mathbf{SO}(1+s)$ and $\mathbf{SO}_0(1,s)$.

For 3-dimensional position, the Kepler factor $\frac{1}{r}$ describes a 2-sphere spread of the 1-dimensional position structure as used in spherical waves $-\frac{\partial}{\partial\frac{r^2}{4\pi}}\cos Pr = 2\pi P^2\frac{\sin Pr}{Pr}$ and in Yukawa potentials $-\frac{\partial}{\partial\frac{r^2}{4\pi}}e^{-|Q|r} = 2\pi Q^2\frac{e^{-|Q|r}}{|Q|r}$.

Three-dimensional scattering waves are coefficients of Hilbert space representations of the Euclidean group $\mathbf{SO}(3)\ \vec{\times}\ \mathbb{R}^3$. The full wave function can be written as the product of a spherical harmonic and a spherical Bessel function. The Bessel functions have large-distance behavior with Kepler factor $j_L(R) \sim \frac{\sin(R-\frac{L\pi}{2})}{R}$ and polarization L matching regular small distance behavior $j_L(R) \sim R^L$. Spherical Bessel functions are the solutions for constant potential $E-V_0 = \frac{P^2}{2}$; they are essentially bounded, but not square integrable. They display, for each polarization L, distributive orthogonality and completeneness.

Three-dimensional bound waves are square integrable coefficients of position representations, where position is modeled with nonabelian groups (not position translations). The nonrelativistic Kepler potential implements position as the orthogonal group classes in the 3-hyperboloid $\mathcal{Y}^3 \cong \mathbf{SO}_0(1,3)/\mathbf{SO}(3) \cong \mathbf{SO}_0(1,1)\circ\Omega^2$ which has the 3-sphere as compact partner $\Omega^3 \cong \mathbf{SO}(4)/\mathbf{SO}(3) \cong \mathbf{SO}(2)\circ\Omega^2$. The wave function is the product $d_L(r)(\vec{x})^L_m$ of the radial representation coefficient and the corresponding harmonic polynomial $(\vec{x})^L_m \sim r^L\mathrm{Y}^L_m(\varphi,\theta)$ as irreducible $\mathbf{SO}(3)$-representation.

MATHEMATICAL TOOLS

9.12 Lattices and Logics

A set with two associative and commutative inner compositions (*join* $\sqcup$ and *meet* $\sqcap$) is a *lattice* if both compositions have an *absorptive* relationship to each other:

$$\begin{aligned}(L,\sqcap,\sqcup)\in \underline{\mathbf{latt}}: &\quad (a\sqcap b)\sqcup a = a = (a\sqcup b)\sqcap a \text{ (absorptive)}\\ \Rightarrow &\quad a\sqcap a = a = a\sqcup a \text{ for all } a\in L \text{ (idempotent)}.\end{aligned}$$

Lattice morphisms are compatible with both compositions. Each lattice carries its *natural order*

$$a\sqsubseteq b \iff a\sqcap b = a.$$

Join and meet are least upper bound $a\sqcup b = \sup(a,b)$ and greatest lower bound $a\sqcap b = \inf(a,b)$.

A lattice with an *origin*, it is unique;

$$\square \sqsubseteq a, \text{ i.e., } \square = \square\sqcap a \text{ for all } a\in L)$$

allows the definition of

$$\begin{aligned}\textit{a and b disjoint elements:} &\quad a\sqcap b = \square,\\ \textit{a elementary:} &\quad a\neq\square,\quad b\sqsubseteq a\Rightarrow b=\square.\end{aligned}$$

A *complementary* lattice has an involutive contramorphism relating meet and join with the origin as meet for each lattice element and its *complement*:

$$L \longrightarrow L, \quad a \longmapsto a^c, \quad a^{cc} = a, \quad \begin{cases} (a \sqcup b)^c &= a^c \sqcap b^c, \\ a \sqcap a^c &= \square \text{ for all } a \in L. \end{cases}$$

The complement of the origin is the unique *end*:

$$\square^c = \blacksquare \sqsupseteq a, \quad a \sqcup a^c = \blacksquare \text{ for all } a \in L.$$

With an appropriate language for the logical concepts a complementary lattice is called a *logic*:

$$(L, \sqcup, \sqcap, \square, c) \in \underline{\mathbf{logic}} : \begin{cases} a \in L: & \textit{proposition,} \\ \sqcap: & \textit{conjunction (and, et),} \\ \sqcup: & \textit{adjunction (or, aut),} \\ \sqsubseteq: & \textit{implication (then, ergo), also} \Rightarrow, \\ \square: & \textit{absurd proposition (falsehood, falsum),} \\ a^c \in L: & \textit{negation (not, non),} \\ \blacksquare: & \textit{self-evident proposition (truth, verum).} \end{cases}$$

An *elementary proposition* is nontrivial minimal. One obtains for the *disjunction (exclusive or, vel)* $\vee$

$$\begin{aligned} a \text{ and not } b\text{: } a \setminus b &= a \sqcap b^c, \\ \text{either } a \text{ or } b\text{: } a \vee b &= (a \sqcap b^c) \sqcup (b \sqcap a^c). \end{aligned}$$

A lattice is *distributive* (both conditions are equivalent in a logic):

$$\begin{aligned} a \sqcup (b \sqcap c) &= (a \sqcup b) \sqcap (a \sqcup c), \\ a \sqcap (b \sqcup c) &= (a \sqcap b) \sqcup (a \sqcap c). \end{aligned}$$

In a distributive logic, called *Boolean*, one has for the disjunction

$$L \in \underline{\mathbf{logic}}(\text{Boole}) \Rightarrow a \vee b = (a \sqcup b) \sqcap (a \sqcap b)^c.$$

Weaker than distributivity is *modularity* for a lattice, a partial associativity for meet and join:

$$a \sqsubseteq c \Rightarrow a \sqcup (b \sqcap c) = (a \sqcup b) \sqcap c.$$

9.13 Measure Rings and Borel Spaces

The power set $2^S = \{X \mid X \subseteq S\}$ of any set is its finest Boolean logic with the contravariant *subset functor*

$$\underline{\mathbf{set}} \longrightarrow \underline{\mathbf{logic}}(\text{Boole}), \quad f \Big\downarrow \begin{matrix} S \\ \\ T \end{matrix} \longmapsto \begin{matrix} 2^S \\ \Big\uparrow f^{-1}[\ ] \\ 2^T \end{matrix},$$

$$\text{with } (\sqcap, \sqcup, \sqsubseteq, \square, c) \sim (\cap, \cup, \subseteq, \emptyset, C).$$

The subset mapping $f^{-1}[\ ]$ is a morphism

$$\begin{array}{l} f^{-1}[\emptyset] = \emptyset, \quad f^{-1}[C_T X] = C_S f^{-1}[X], \quad f^{-1}[T] = S, \\ f^{-1}[X \cap Y] = f^{-1}[X] \cap f^{-1}[Y] \Rightarrow f^{-1}[X \setminus Y] = f^{-1}[X] \setminus f^{-1}[Y], \\ f^{-1}[X \cup Y] = f^{-1}[X] \cup f^{-1}[Y] \Rightarrow f^{-1}[X \vee Y] = f^{-1}[X] \vee f^{-1}[Y]. \end{array}$$

As for topological structures, also measure structures deal with subsets of a set, not primarily with individual elements $x \in S$ (possibly with $\{x\} \in 2^S$). The set difference is $X \setminus Y = X \cap C_S Y$ with the Y-complement $C_S Y$ in the full set S.

A subset family $\mathcal{S} \subseteq 2^S$ having the structure of a lattice with origin is called a *measure ring* $\mathcal{S} \in \underline{\textbf{rng}}$, its elements *measurable sets*, the pair $(S, \mathcal{S})$ a *measurable space*. For the measure ring property of a subset family it is enough that it includes the empty set and is stable with respect to differences and unions;

$$\mathcal{S} \in \underline{\textbf{rng}} \iff \left\{ \begin{array}{l} \emptyset \in \mathcal{S}, \\ X, Y \in \mathcal{S} \Rightarrow \left\{ \begin{array}{l} C_X Y = X \setminus Y \in \mathcal{S}, \\ X \cup Y \in \mathcal{S}, \end{array} \right. \end{array} \right.$$

which entails the measurability of the intersections

$$X \cap Y = X \setminus (X \setminus Y) \in \mathcal{S},$$

i.e., $\mathcal{S}$ is a lattice with origin. If the full set S is measurable, one has a *unital ring* (Boolean logic). It contains all complements $X, S \in \mathcal{S} \Rightarrow S \setminus X = C_S X \in \mathcal{S}$ and can be dually characterized:

$$\begin{array}{rcl} \mathcal{S} \in \underline{\textbf{rng}}(\text{unital}) & \iff & \emptyset,\ C_S X,\ X \cap Y \in \mathcal{S} \text{ for } X, Y \in \mathcal{S} \\ & \iff & S,\ C_S X,\ X \cup Y \in \mathcal{S} \text{ for } X, Y \in \mathcal{S}. \end{array}$$

The coarsest (smallest with respect to the inclusion) measure ring is $\{\emptyset\}$, coarsest unital $\{\emptyset, S\}$, the finest one is 2^S. Any measure ring contains a unital ring for each measurable set:

$$\mathcal{S} \in \underline{\textbf{rng}} \Rightarrow \mathcal{S} \cap X \in \underline{\textbf{rng}}(\text{unital}) \text{ for all } X \in \mathcal{S}.$$

For an $\aleph_0$-*measure ring* $\mathcal{S}$ (also called σ-measure ring) the union of countably many measurable subsets has to be measurable $\bigcup_{i=1}^{\infty} X_i \in \mathcal{S}$ too. Most of the following definitions and structures can be extended on $\aleph_0$-measure rings.

Each subset family $\mathcal{X}$ in $S \in \underline{\textbf{set}}$ has a unique embracing coarsest (unital) measure ring and coarsest (unital) $\aleph_0$-measure ring on S.

If a mapping f between two measurable spaces $(S, \mathcal{S})$ and $(T, \mathcal{T})$ gives measurable inverse images of measurable sets, i.e., $f^{-1}[\mathcal{T}] \subseteq \mathcal{S}$, then it is called a *measure morphism or measurable*, $f \in \textbf{mes}(S, T)$. In this case $f^{-1}[\ ]$ is a ring morphism

$$\underline{\textbf{mes}} \longrightarrow \underline{\textbf{rng}}, \quad f \left\downarrow \begin{array}{c} S \\ \\ T \end{array} \right. \longmapsto \left\uparrow \begin{array}{c} \mathcal{S} \\ \\ \mathcal{T} \end{array} \right. f^{-1}[\ ] \ .$$

In analogy to topological spaces $\underline{\mathbf{top}}$, the condition of measurability of mappings defines *initial (coarsest) and final (finest) measure rings*, e.g., the measure ring $\mathcal{U}$ of a subset by measurability of the injection $U \hookrightarrow S$ or the finite *product of measure rings* $\mathcal{S}_1 \otimes \mathcal{S}_2$ by the measurability of all projections $S_1 \times S_2 \longrightarrow S_{1,2}$. In general, the measure ring $\mathcal{S}_1 \otimes \mathcal{S}_2$ contains not only the "rectangle products" $X_1 \times X_2$.

For a topological space $S \in \underline{\mathbf{top}}$ with the open sets $\underline{S}$ the *Borel ring* with the *Borel sets* is the coarsest $\underline{S}$ embracing $\aleph_0$-measure ring $\mathcal{S} \supseteq \underline{S}$, or, equivalently, embracing all closed sets. Hence also the point sets $\{x\}$ for $x \in S$ are Borel sets. If not stated otherwise, a topological space as measurable space carries always its unital Borel measure ring (*Borel space*), hence $\underline{\mathbf{top}} \subset \underline{\mathbf{mes}}$. *Stone's theorem* implies that each Boolean logic is isomorphic to a lattice with the simultaneously open and closed sets of a topological space.

A continuous mapping $f \in \mathbf{top}(S,T)$, $f^{-1}[\underline{T}] \subseteq \underline{S}$, is Borel measurable, $\mathbf{top}(S,T) \subseteq \mathbf{mes}(S,T)$ (*Borel mappings*), leading to the covariant functor $\underline{\mathbf{top}} \longrightarrow \underline{\mathbf{mes}}$.

The natural unique Hausdorff topology of $\mathbb{R}$, hence of $\mathbb{C} \cong \mathbb{R}^2$ and of the products $\mathbb{K}^d$ and all finite-dimensional $\mathbb{K}$-vector spaces, leads to the *Borel rings* on them, denoted by $\mathcal{K}^d$, $\mathcal{R}^d$, $\mathcal{C}^d$. They are stable with respect to the affine group $\mathbf{GL}(\mathbb{K}^d) \vec{\times} \mathbb{K}^d$-action.

9.14 Disjoint-Additive Mappings (Measures)

A *disjoint-additive mapping (M-valued measure)* $\Phi \in \mathbf{meas}(S, M)$ from the measurable space $(S, \mathcal{S}) \in \underline{\mathbf{mes}}$ into an additive monoid $M \in \underline{\mathbf{mon}}$ relates subset operations to algebraic ones:

$$\Phi : \mathcal{S} \longrightarrow M, \quad \begin{cases} \Phi(\emptyset) &= 0, \\ \Phi(X \cup Y) &= \Phi(X) + \Phi(Y) \text{ for } X \cap Y = \emptyset \\ & \quad \text{(disjoint additivity).} \end{cases}$$

With

$$\begin{aligned} \Phi(X) &= \Phi(X \cap Y) + \Phi(X \setminus Y), \\ \Phi(X \cup Y) &= \Phi(X \setminus Y) + \Phi(Y), \end{aligned}$$

one obtains for a monoid with cancellation rule $(\alpha + \beta = \gamma + \beta \Rightarrow \alpha = \gamma)$

$$\Phi(X) + \Phi(Y) = \Phi(X \cup Y) + \Phi(X \cap Y).$$

Two disjoint-additive mappings are *equivalent* if they vanish on the same measurable sets:

$$\Phi_1 \sim \Phi_2 \iff \{X \in \mathcal{S} \mid \Phi_1(X) = 0\} = \{X \in \mathcal{S} \mid \Phi_2(X) = 0\}.$$

$\aleph_0$-additivity (σ-additivity) in the case of an $\aleph_0$-measure ring requires additivity for countably many pairwise disjoint measurable sets.

One can define a disjoint-additive mapping with restricted validity, e.g., a *compact disjoint-additive* mapping.

Measures of S are numerically valued in the additive monoid $\mathbb{C}_\infty = \mathbb{C} \cup \{\infty\}$ with the rules $\alpha + \infty = \infty$ for $\alpha \in \mathbb{C}_\infty$ and $\overline{\infty} = \infty$:

$$\mu : \mathcal{S} \longrightarrow \mathbb{C}_\infty.$$

With the injection topology $\mathbb{C} \hookrightarrow \mathbb{C}_\infty$ the monoid $\mathbb{C}_\infty$ is a topological space with Borel ring. A measure of a topological space may be finite on all compact Borel sets. $\mathbb{C}_\infty$ has many submonoids, especially $\mathbb{C}$ (finite monoid), $\mathbb{R}_\infty$ (real), $\mathbb{R}$ (real and finite), $\mathbb{R}_\infty^+$ (positive), and $\mathbb{R}_+$ (positive and finite), giving the name for measures valued in such a submonoid. The cardinality is a measure, for finite subsets valued in the monoid $\mathbb{N}_0$ (*counting measure*):

$$\text{card} : \mathcal{S} \longrightarrow \mathbb{N}_0.$$

For the finest measure ring $\mathbf{2}^S$ a *discrete measure* is given by a complex number for each element

$$\mu : \mathbf{2}^S \longrightarrow \mathbb{C}_\infty, \quad \{x\} \longmapsto \mu(x).$$

A *probability measure* is positive and normalized (measure ring has to be unital),

$$S \in \underline{\mathbf{logic}} \text{ (Boolean)} : \quad \mu : \mathcal{S} \longrightarrow \mathbb{R}_+, \quad \mu(S) = 1$$

e.g., the *Dirac probability measure* for any element $a \in S$:

$$\delta_a : \mathcal{S} \longrightarrow \mathbb{R}_+, \quad \delta_a(X) = \begin{cases} 1, & a \in X, \\ 0, & a \notin X. \end{cases}$$

A probability measure for the coarsest ring has to be the *yes-no probability measure*

$$\mathcal{S} = \{\emptyset, S\} \longrightarrow \{0, 1\}.$$

A property for a space with a disjoint-additive mapping (S, Φ), especially for measures μ, is valid Φ-*almost everywhere* in the following situation: There exists a set with trivial Φ-value, $\Phi(N) = 0$, where the property in question is valid on the complementary set $C_S N$ (it may be valid even in N). For example, one works with μ-almost everywhere defined measurable mappings $f : S \longrightarrow T$.

Disjoint-additive mappings into $M \in \underline{\mathbf{mon}}$ inherit M-properties by pointwise definition, e.g., reflection (conjugation), order, vector spaces for fields $M = \mathbb{R}, \mathbb{C}$, modules for algebras $M \in \underline{\mathbf{aag}}_{\mathbb{K}}$, etc.

With each measurable mapping $f \in \mathbf{mes}(S, T)$ and a fixed monoid M one obtains a monoid morphism of the disjoint-additive mappings and the following functor from the measurable spaces over the measure rings to the monoids of the disjoint-additive mappings:

$$\begin{array}{llcllcll} \mathbf{meas}(\ , M) : & \underline{\mathbf{mes}} & \longrightarrow & \underline{\mathbf{rng}} & \longrightarrow & \underline{\mathbf{mod}}_M & \\ & S & & \mathcal{S} & & \mathbf{meas}(S, M) & \\ & f \downarrow & \longmapsto & \uparrow f^{-1}[\ ] & \longmapsto & \downarrow & \circ f^{-1}[\ ], \\ & T & & \mathcal{T} & & \mathbf{meas}(T, M) & \end{array}$$

$$f \bullet \Phi(X) = \Phi\big(f^{-1}[X]\big).$$

For example, a measurable mapping $f \in \mathbf{mes}(S,T)$ defines an *image measure* $f \bullet \mu$:

$$(\mu : \mathcal{S} \longrightarrow \mathbb{C}_\infty) \longmapsto \quad (f \bullet \mu : \mathcal{T} \longrightarrow \mathbb{C}_\infty),$$
$$f \bullet \mu(X) = \mu(f^{-1}[X]).$$

9.14.1 Integration (Expectation Values)

For a space with positive measure (S, μ) the *integral* of a positive μ-almost everywhere defined measurable mapping f^+ on a measurable set $X \in \mathcal{S}$ is defined by the supremum of the summed up mapping values, which are obtained with an X-partition (decomposition $X = \biguplus_{k=1}^{n} X_k$ into pairwise disjoint sets $X_k \in \mathcal{S}$). For a given partition one takes the infimum on the individual subsets X_k, multiplied by the measure $\mu(X_k)$:

$$\begin{array}{rl} f^+ : & S \longrightarrow \mathbb{R}_+, \qquad x \longrightarrow f^+(x), \\ f^+[\ ] : & \mathcal{S} \longrightarrow \mathbb{R}_+, \qquad X \longmapsto f^+[X] = \inf_{x\in X} f^+(x), \\ X \in \mathcal{S} : & \int_X d\mu(x) f^+(x) \ = \sup_{\{X_1,\dots,X_n\}} \sum_{k=1}^{n} \mu(X_k) f^+[X_k]. \end{array}$$

Only the subset mapping $f^+[\ ]$ is used, which has not to be defined for sets with measure 0.

A real mapping is decomposable into two positive mappings $f = f^+ - (-f^-)$, its integral is the difference of both integrals. A complex mapping is decomposed into real and imaginary parts to define its integral. A measurable mapping $f : S \longrightarrow \mathbb{K}$ is called *μ-integrable* if its μ-integral is finite for all measurable sets

$$f \in \int_\mu (S, \mathbb{K}) \iff \int_X d\mu(x) f(x) \in \mathbb{K} \text{ for all } X \in \mathcal{S}.$$

One obtains disjoint additivity:

$$\begin{aligned} \int_\emptyset d\mu &= 0, \\ \int_{X\cup Y} d\mu &= \int_X d\mu + \int_Y d\mu \text{ for } X \cap Y = \emptyset. \end{aligned}$$

All μ-integrable mappings constitute a vector space, $\int_\mu (S, \mathbb{K}) \in \underline{\mathbf{vec}}_{\mathbb{K}}$, with conjugation for $\mathbb{K} = \mathbb{C}$. Again, the measurable sets with finite integral may be restricted, e.g., for topological spaces to the compact Borel sets. For a probability measure $\mu(S) = 1$ the integral is the *expectation value of f in X*:

$$\mathcal{E}_X^\mu(f) = \int_X d\mu\ f.$$

The integration set $X \in \mathcal{S}$ can be extended to the full measure space using the *characteristic function* χ_X for X:

$$\begin{aligned} \chi_X : S \longrightarrow \{0,1\}, \quad \chi_X(x) &= \begin{cases} 1, & x \in X, \\ 0, & x \notin X, \end{cases} \\ \mu : \mathcal{S} \longrightarrow \mathbb{R}_\infty^+, \quad \mu(X) &= \int_S d\mu(x) \chi_X(x), \\ \int_X d\mu(x) f(x) &= \int_S d\mu(x) \chi_X(x) f(x). \end{aligned}$$

For $\int_S$ the space S is omitted:

$$\int_S d\mu(x) f(x) = \int d\mu(x) f(x).$$

The integral defines a linear form of the μ-integrable functions:

$$\mu : \int_\mu(S, \mathbb{K}) \longrightarrow \mathbb{K}, \quad \langle \mu, f \rangle = \int d\mu(x) f(x).$$

Each μ-integrable mapping g defines a *measure* $\int d\mu\ g$ *on the basis* μ. Hence each measure μ explains a linear mapping from the corresponding μ-integrable mappings into the *measures on the basis* μ:

$$\begin{array}{ll} \int_\mu(S, \mathbb{K}) \hookrightarrow \mathbf{meas}(S, \mathbb{K}), & g \longmapsto \mu_g, \\ & \langle \mu_g, f \rangle = \int d\mu\ gf. \end{array}$$

9.14.2 Haar and Lebesgue Measure

A measure on a set with group G action is called invariant for

$$\begin{array}{ccc} \mathcal{S} & \stackrel{g\bullet}{\longrightarrow} & \mathcal{S} \\ {\scriptstyle\mu}\downarrow & & \downarrow{\scriptstyle\mu} \\ \mathbb{R}_+ & \xrightarrow[\mathrm{id}_{\mathbb{R}_+}]{} & \mathbb{R}_+ \end{array}, \quad \mu(X) = \mu(g \bullet X).$$

For a *locally compact group* G there exists a *definite* measure, invariant with respect to all left translations, $\mu^G(X) = \mu^G(gX)$, its – up to a nontrivial multiplicative constant unique – *Haar measure* $\mu^G \in \mathbf{meas}(G, \mathbb{R})$. It is finite on all compact sets. The analogous statement holds with respect to right translations. With a left Haar measure μ, also μ_g with $\mu_g(X) = \mu(Xg^{-1}) = \Delta(g)\mu(X)$ is a left Haar measure, defining the *modular function* as group realization

$$\Delta : G \longrightarrow \mathbb{R}_+, \quad \Delta(g_1 g_2) = \Delta(g_1)\Delta(g_2).$$

If a group is unimodular for $\Delta(g) = 1$ for all g, then left Haar measures are right Haar measures. That is the case for the finite, discrete, abelian, and semisimple groups. For a compact group G, the positive normalized Haar measure, $\mu^G(G) = 1$, is a probability measure.

A Haar measure for a finite group G with discrete topology uses the counting measure

$$\begin{array}{l} \mu^G : 2^G \longrightarrow \mathbb{N}_0, \quad \mu^G(X) = \mathrm{card}\, X, \\ f : G \longrightarrow \mathbb{K}, \quad \int_G d^G g\ f(g) = \sum_{g \in G} f(g). \end{array}$$

Generalizing to Lie groups, a Haar measure will be written with an integral (volume form)

$$\mu^G : \mathcal{G} \longrightarrow \mathbb{R}_+, \quad \mu^G(X) = \int_X d^G g.$$

Haar integrable functions on the group $f : G \longrightarrow \mathbb{K}$ define measures $d^G g f(g)$ with a Haar measure as basis. For a Dirac measure one uses as notation

$$h \in G : \delta_h \cong d^G g \; \delta(gh^{-1}).$$

A Haar measure on the additive Lie group $\mathbb{R}^d$ is called a *Lebesgue measure.* It is given by the volume elements

$$\begin{array}{l} \lambda^d : \mathcal{R}^d \longrightarrow \mathbb{R}_+, \quad X \longmapsto \lambda^d(X) = \int_X d^d q, \\ \text{Dirac measure: } p \in \mathbb{R}^d : \; \delta_p \cong d^d q \; \delta(q-p). \end{array}$$

A Lebesgue measure is $\mathbf{SL}(\mathbb{R}^d) \times \mathbb{R}^d$ invariant, $q \longmapsto s.q + a$, $dq \longmapsto s.dq$. Integrable functions give measures $d^d q f(q)$.

Starting from Lebesgue measures one can construct measures on differentiable manifolds $M \in \underline{\mathbf{dif}}_{\mathbb{R}}$ (chapter "Spin, Rotations, and Position"), especially measures on manifolds invariant with respect to a Lie group action (Klein spaces).

The volume element on the cotangent space $d^d q \in \bigwedge^d \mathbf{T}_q^T(M)$ with a chart $q = (q_j) \in O \subseteq \mathbb{R}^d$ transforms under chart change $q_j \longmapsto \varphi_j(q)$ with the inverse Jacobi determinant and, for equal orientation, with its absolute value

$$d^d q \longmapsto \frac{1}{|\det \partial^k \varphi_j(q)|} d^d q.$$

To have a measure, not only $\mathbf{SL}(\mathbb{R}^d)$-invariant, but invariant even under the oriented linear group $\mathbf{GL}_+(\mathbb{R}^d) = \mathbf{GL}(\mathbb{R}^d)/\mathbb{I}(2)$ with the additional dilatation group $\mathbf{D}(1)$ at each point $q \in O$, a compensation normalization factor $c|\det \partial^k \varphi_j(q)|$ with any nontrivial constant c has to be provided for. There are different possibilities:

If the tangent space comes with a nondegenerate multilinear form, e.g., linear or bilinear,

$$\begin{array}{rlrl} \mathbf{T}_q(M) & \longrightarrow \mathbb{R}, & \lambda(\partial_q^j) & = \lambda^j(q), \\ \mathbf{T}_q(M) \times \mathbf{T}_q(M) & \longrightarrow \mathbb{R}, & \kappa(\partial_q^j, \partial_q^k) & = \kappa^{jk}(q), \end{array}$$

the induced form on the volumes, e.g., the product and the discriminant, provides a compensation factor

$$d\mu(q) = d^d q \; |\lambda^1(q) \cdots \lambda^d(q)| \text{ or } d\mu(q) = d^d q \; |\sqrt{\det \kappa^{jk}(q)}|.$$

For a real Lie group one can use, near the neutral element, a faithful finite-dimensional representation and a parametrization

$$\mathbb{R}^d \supseteq O \ni q \longmapsto g(q) \in G \subseteq \mathbf{GL}(\mathbb{R}^n), \; g(0) = \mathbf{1}_n,$$

which gives a Lie algebra basis at each point, related with the Lie-Jacobi transformation q_* to the basis at the neutral element:

$$\begin{aligned} \mathbb{R}^d \supseteq O \ni q \longmapsto l^j(q) &\in L = \log G \subseteq \mathbf{AL}(\mathbb{R}^n), \\ l^j(q) &= [\partial^j g(q)] \circ g(q)^{-1} = (q_*)^j_k l^k. \end{aligned}$$

Any invariant multilinear form of the Lie algebra, e.g., linear for abelian and bilinear for simple groups, e.g., the Killing form, allows the normalization of the volume element, leading to a Haar measure:

$$\begin{array}{llll} L \longrightarrow \mathbb{R}, & \lambda(l^j(q)) = \lambda^j(q), & d^G g(q) & = |\lambda^1(q) \cdots \lambda^d(q)| d^d q, \\ L \times L \longrightarrow \mathbb{R}, & \kappa(l^j(q), l^k(q)) = \kappa^{jk}(q), & d^G g(q) & = |\sqrt{\det \kappa^{jk}(q)}| d^d q. \end{array}$$

The corresponding $\mathbf{D}(1)$-normalization is also used in the q-dependent *Laplace-Beltrami operator*

$$\Delta(q) = \det \ k^{jk}(q) : \quad \frac{1}{\sqrt{\Delta(q)}} \partial^j \Delta(q) k_{jk}(q) \frac{1}{\sqrt{\Delta(q)}} \partial^k.$$

A Haar measure arises by the external product of the Lie-Jacobi form, valued in the Lie algebra:

$$\begin{aligned} q \longmapsto \omega(q) &= [\partial^j g(q)] \circ g(q)^{-1} \otimes dq_j = (q_*)^j_k l^k \otimes dq_j \in \log G \otimes \mathbf{T}^T_q(L), \\ \bigwedge^d \omega(q) &\sim \det(q_*) \, l^1 \wedge \cdots \wedge l^d \otimes d^d q \Rightarrow d^G g(q) = |\det(q_*)| d^d q, \end{aligned}$$

generalizing the abelian cases

$$d^G g = (dg) g^{-1} \cong \begin{cases} e^{-mt} de^{mt} & = m dt \quad \text{for } \mathbf{D}(1), \\ e^{-iz\alpha} de^{iz\alpha} & = iz d\alpha \quad \text{for } \mathbf{U}(1). \end{cases}$$

9.15 Generalized Mappings (Distributions)

As seen in the integration procedure, functions can define measures. As seen with the Dirac measures, measures have not to be related to functions. Measures are definable as *generalized functions (distributions)* being continuous linear forms $\mathcal{F}(S, \mathbb{K})' \subseteq \mathbf{meas}(S, \mathbb{K})$ of an appropriately defined *topological vector space of test functions*

$$\mu : \mathcal{F}(S, \mathbb{K}) \longrightarrow \mathbb{K}, \quad f \longmapsto \langle \mu, f \rangle = \int d\mu(x) \, f(x).$$

The test function spaces correspond to the measure monoids above.

For example, *Radon measures* of a locally compact space $S \in \underline{\mathbf{top}}$ are defined as linear mappings of the continuous, compactly supported, complex-valued test functions $\mathcal{C}_c(S)$

$$\mu : \mathcal{C}_c(S) \longrightarrow \mathbb{C}, \quad \langle \mu, f \rangle = \int d\mu(x) f(x).$$

The function space obtains an appropriate topology (limiting Fréchet topology) with respect to which the Radon measures are required to be continuous:

$$\mathcal{M}(S) = \mathcal{C}_c(S)' \subseteq \mathbf{meas}(S, \mathbb{C}) \in \underline{\mathbf{vec}}_{\mathbb{C}}.$$

Positivity (μ is positive if $f \geq 0 \Rightarrow \langle \mu, f \rangle \geq 0$), conjugation, derivatives, Fourier transforms, etc. are definable via the test function spaces, e.g., via transposition. A Radon measure can be multiplied by the corresponding functions and is g-invariant iff $\langle \mu, f \rangle = \langle \mu, f \circ g \rangle$ for all f.

It is a suggestive, but sometimes dangerous, notation to write a disjoint-additive mapping $\Phi \in \mathbf{meas}(S, A)$ into an algebra $A \in \underline{\mathbf{aag}}_{\mathbb{K}}$ (subset mapping) as an "element mapping" (*generalized mapping*) using an S-measure μ as basis:

$$\Phi : \mathcal{S} \longrightarrow A, \quad \Phi(X) = \int_X d\mu(x) \boldsymbol{\Phi}(x).$$

The *measure associated Dirac distribution* is defined by

$$\begin{aligned} \delta_a : \mathcal{S} \longrightarrow \mathbb{C}, \quad \delta_a(X) &= \int_X d\mu(x) \delta(\mu(a, x)), \\ f : S \longrightarrow \mathbb{C}, \quad \langle \delta_a, f \rangle &= \int_X d\mu(x) \delta(\mu(a, x)) f(x) = f(a). \end{aligned}$$

In general, generalized mappings are not defined for elements. With regard to their origin and the algebra properties, they can be added and multiplied by a scalar, i.e., they form a vector space:

$$\boldsymbol{\Phi} : S \longrightarrow A, \quad x \longmapsto \boldsymbol{\Phi}(x), \quad \begin{cases} \boldsymbol{\Phi}_1(x) + \boldsymbol{\Phi}_2(x), \\ \alpha \boldsymbol{\Phi}(x). \end{cases}$$

In general, however, the pointwise product of generalized mappings, e.g., $\boldsymbol{\Phi}_1(x)\boldsymbol{\Phi}_2(x)$ for $\boldsymbol{\Phi}_i : S \longrightarrow A$, is not defined. The algebra product can be used to define a Cartesian product for measures μ_i of S_i, $i = 1, 2$, with generalized mappings $\boldsymbol{\Phi}_i$

$$\begin{aligned} \Phi_i &\cong d\mu_i(x_i) \, \boldsymbol{\Phi}_i(x_i) \text{ with } \boldsymbol{\Phi}_i : S_i \longrightarrow A, \\ \Phi_1 \diamond \Phi_2 &\cong d\mu_1(x_1) d\mu_2(x_2) \, \boldsymbol{\Phi}_1(x_1) \diamond \boldsymbol{\Phi}_2(x_2). \end{aligned}$$

9.15.1 Dirac Distributions on $\mathbb{R}$

The Dirac measure of the reals $\mathbb{R}$ (additive Lie group) supported at $m \in \mathbb{R}$ can be written with a generalized *Dirac function* with Lebesgue measure dq basis

$$\delta_m(X) = \int dq \, \chi_X(q) \delta(q - m), \quad \langle \delta_m, f \rangle = \int dq \, \delta(q - m) f(q) = f(m).$$

If defined, one has for a real function f

$$\begin{aligned} \delta\big(f(q)\big) &= \sum_{f(q)-\text{zeros } m} \tfrac{1}{|f'(m)|} \delta(q - m), \\ \text{e.g., } \delta(\alpha q) &= \tfrac{1}{|\alpha|} \delta(q), \quad \delta\Big((q - m_1)(q - m_2)\Big) = \tfrac{\delta(q-m_1)+\delta(q-m_2)}{|m_1 - m_2|}. \end{aligned}$$

For a Hermitian matrix one defines the support of the Dirac function at the zeros of the determinant (eigenvalues):

$$A = A^\star \in \mathbf{AL}(\mathbb{C}^d): \quad \delta(A) = \tfrac{\det A}{A}\delta(\det A).$$

Derivations of the Dirac function are shifted to the integrated functions

$$\begin{aligned}\int dq\ \delta^{(N)}(q-m)f(q) &= (-\tfrac{d}{dm})^N \int dq\ \delta(q-m)f(q)\\ &= \int dq\ \delta(q-m)(-\tfrac{d}{dq})^N f(q),\\ \delta^{(N)}(-q) &= (-1)^N\delta^{(N)}(q),\quad q\in\mathbb{R},\ N=0,1,\ldots.\end{aligned}$$

The measurable *order functions (step functions)* can be written as residues of complex poles:

$$\begin{aligned}\vartheta(x) &= \int \tfrac{dq}{2i\pi}\tfrac{1}{q-io}e^{iqx} &&= \begin{cases}1, & x>0,\\ 0, & x<0,\end{cases}\\ \epsilon(x) &= \vartheta(x)-\vartheta(-x) = \tfrac{x}{|x|} = \tfrac{d|x|}{dx} &&= \begin{cases}+1, & x>0,\\ -1, & x<0,\end{cases}\\ \vartheta(x)^2 &= \vartheta(x),\ \epsilon(x)^2=1,\ \epsilon(x)\epsilon(y) &&= \epsilon(xy),\\ \delta(x) &= \tfrac{d}{dx}\vartheta(x) = \tfrac{d}{dx}\tfrac{\epsilon(x)}{2} = \int\tfrac{dq}{2\pi}e^{iqx}.\end{aligned}$$

First the integration has to be performed with real positive $o>0$; then the limit $o\longrightarrow 0$ has to be taken.

The gamma function is defined by

$$\begin{aligned}\Gamma(1+\nu) &= \int_0^\infty dx\ x^\nu e^{-x},\quad \nu\in\mathbb{R},\ \nu\neq -1,-2,\ldots\\ \Rightarrow \Gamma(1+\nu) &= \nu\Gamma(\nu),\quad \Gamma(1+N)=N!,\quad N=0,1\ldots,\\ \Gamma(\tfrac12) &= \sqrt{\pi}.\end{aligned}$$

The following complex generalized functions arise from the logarithm by derivation $\frac{d}{da}$ and using $z^\nu = e^{\nu z}$. They can be naively conjugated. The decomposition into real and imaginary part involves the principal value integration denoted by P:

$$\begin{aligned}a\in\mathbb{R}:\quad \log(a-io) &= \log|a| - i\pi\,\vartheta(-a),\\ \tfrac{d}{da}\Rightarrow \tfrac{\Gamma(1+N)}{(a-io)^{1+N}} &= \tfrac{\Gamma(1+N)}{a_{\mathrm{P}}^{1+N}} + i\pi\delta^{(N)}(-a),\quad N=0,1,\ldots,\\ (a-io)^\nu = e^{\nu[\log|a|-i\pi\vartheta(-a)]} &= |a|^\nu[\vartheta(a)+e^{-i\nu\pi}\vartheta(-a)],\\ &\text{for } \nu\in\mathbb{R},\ \nu\neq 0,-1,-2,\ldots.\end{aligned}$$

For positive definite reals there are no imaginary contributions, e.g., Dirac distributions vanish:

$$a\in\mathbb{R}:\quad (a^2-io)^\nu = a^{2\nu},\quad \nu\in\mathbb{R},\ \text{e.g., } \delta^{(N)}(a^2)=0.$$

One has the Laplace transformations

$$\begin{aligned}\tfrac{e^{-|mx|}}{|x|} &= \tfrac12\int dq\ \vartheta(q^2-m^2)e^{-|qx|},\quad x\neq 0,\\ e^{-|mx|} &= \int dq\ |q|\delta(q^2-m^2)e^{-|qx|}.\end{aligned}$$

Some $\mathbb{C}$-linear Fourier transforms for generalized functions on $\mathbb{R}$ to the dual $\check{\mathbb{R}}$ are collected in the following table: P denotes a polynomial. The formulas are valid where the Γ-functions are defined:

$\mu(q) = \int dx \tilde{\mu}(x) e^{-iqx}$	$\tilde{\mu}(x) = \int \frac{dq}{2\pi} \mu(q) e^{iqx}$
$\mu(-q)$	$\tilde{\mu}(-x)$
$\overline{\mu(q)}$	$\overline{\tilde{\mu}(-x)}$
$\mu(\alpha q),\ \alpha > 0$	$\frac{1}{\alpha} \tilde{\mu}(\frac{x}{\alpha})$
$P(iq)\mu(q)$	$P(\frac{d}{dx})\tilde{\mu}(x)$
$e^{iqy}\mu(q),\ y \in \mathbb{R}$	$\tilde{\mu}(x+y)$
1	$\delta(x)$
$\frac{\Gamma(1+\nu)}{(q-io-m)^{1+\nu}},\ \ m \in \mathbb{R}$ $\nu \in \mathbb{R},\ \ \nu \neq -1, -2, \ldots$	$i^{1+\nu}\vartheta(x) x^\nu e^{imx}$
$\frac{\Gamma'(1) - \log(-q+io+m) + i\frac{\pi}{2}}{q-io-m}$	$i\vartheta(x) \log x \ e^{imx}$
$\frac{2\ \Gamma(1+N)}{(q_{\mathrm{P}}-m)^{1+N}},\ N = 0, 1, \ldots$	$i^{1+N}\epsilon(x) x^N e^{imx}$
$\frac{2q}{q^2-io-m^2}$	$i\epsilon(x) e^{i\lvert mx \rvert}$
$\frac{2\lvert q \rvert}{q^2-io-m^2}$	$ie^{i\lvert mx \rvert}$
$\log \frac{q^2-io-m_1^2}{q^2-io-m_2^2}$	$\frac{e^{i\lvert m_1 x\rvert} - e^{i\lvert m_2 x \rvert}}{\lvert x \rvert}$
$\frac{\Gamma(\nu)}{(q^2)^\nu}$	$\frac{1}{4^\nu \sqrt{\pi}} \frac{\Gamma(\frac{1}{2}-\nu)}{(x^2)^{\frac{1}{2}-\nu}}$
$\frac{2q}{q^2+m^2}$	$i\epsilon(x) e^{-\lvert mx \rvert}$
$\frac{2\lvert q \rvert}{q^2+m^2}$	$e^{-\lvert mx \rvert}$
$\log \frac{q^2+m_1^2}{q^2+m_2^2}$	$-\frac{e^{-\lvert m_1 x \rvert} - e^{-\lvert m_2 x \rvert}}{\lvert x \rvert}$
$e^{-\frac{q^2}{2}}$	$\sqrt{2\pi} e^{-\frac{x^2}{2}}$

The involution involves a factor 2π, which can be avoided by appropriate normalization:

$$\begin{array}{llll} \tilde{\mu}(x) & = \int \frac{dq}{2\pi} \mu(q) e^{iqx}, & \mu(q) & = \int dx\ \tilde{\mu}(x) e^{-iqx}, \\ \tilde{\mu}(x) & = \int dq\ \mu(2\pi q) e^{2\pi iqx}, & \mu(2\pi q) & = \int dx\ \tilde{\mu}(x) e^{-2\pi iqx}. \end{array}$$

One obtains in Euclidean $\mathbf{SO}(3) \ \vec{\times}\ \mathbb{R}^3$ for the scalar integrals with polar coordinates after integration over the 2-sphere the characteristic derivative $\frac{d}{d\frac{r^2}{4\pi}}$:

$$\begin{array}{rl} \int \frac{d^3q}{2\pi} \mu(\vec{q}^2) e^{\alpha \vec{q}\vec{x}} & = \frac{1}{\alpha r} \int dq\ q\mu(q^2) e^{\alpha qr}, \quad \alpha \in \mathbb{C},\ \alpha\vec{x} \neq 0, \\ & = \frac{1}{\alpha^2 r} \frac{d}{dr} \int dq\ \mu(q^2) e^{\alpha qr}, \\ \int \frac{d^3q}{2\pi} \mu(\epsilon(q_3)\lvert\vec{q}\rvert) e^{\alpha \vec{q}\vec{x}} & = \frac{1}{\alpha r} \int dq\ q\mu(q)(e^{\alpha qr} - 1), \end{array}$$

with the special cases

$$\int \frac{d^3q}{(2\pi)^3} e^{i\vec{q}\vec{x}} = \delta(\vec{x}) = \delta(\vec{\omega}) \frac{1}{r^2} \delta(r),$$

$$\begin{array}{rlll} \int \frac{d^3q}{2\pi^2} \frac{1}{\vec{q}^2+m^2} e^{i\vec{q}\vec{x}} & = -\frac{1}{r}\frac{d}{dr} \int \frac{dq}{\pi} \frac{1}{q^2+m^2} e^{iqr} & = \frac{1}{r} \int \frac{dq}{i\pi} \frac{q}{q^2+m^2} e^{iqr} & = \frac{e^{-\lvert m \rvert r}}{r}, \\ \int \frac{d^3q}{2\pi^2} \frac{1}{\vec{q}^2-io-m^2} e^{i\vec{q}\vec{x}} & = -\frac{1}{r}\frac{d}{dr} \int \frac{dq}{\pi} \frac{1}{q^2-io-m^2} e^{iqr} & = \frac{1}{r} \int \frac{dq}{i\pi} \frac{q}{q^2-io-m^2} e^{iqr} & = \frac{e^{i\lvert m \rvert r}}{r}, \\ \int d^3q e^{-i\frac{\vec{q}^2}{2}t} e^{i\vec{q}\vec{x}} & = (\frac{2\pi}{it})^{\frac{3}{2}} e^{i\frac{r^2}{2t}}. & & \end{array}$$

9.16 Lebesgue Function Spaces

A measurable space $(S, \mathcal{S})$ with a positive σ-additive measure $\mu : \mathcal{S} \longrightarrow \mathbb{R}_+$, e.g., a locally compact group with Haar measure, gives rise to prenorms on the

μ-almost everywhere defined measurable $\mathbb{K}$-valued functions

$$f : (S,\mu) \longrightarrow \mathbb{K},\ 1 \le p < \infty : \ \parallel f \parallel_p = [\int d\mu(x)|f(x)|^p]^{\frac{1}{p}} \in \mathbb{R}_\infty.$$

The additive *Minkowski's inequality* for $f + g$ (if defined) holds:

$$1 \le p < \infty : \ \parallel f + g \parallel_p \le \parallel f \parallel_p + \parallel g \parallel_p . \quad \text{(subadditivity)}$$

It follows from the multiplicative *Hölder's inequality* (Cauchy-Schwarz for different indices):

$$1 \le p,q < \infty,\ \tfrac{1}{p} + \tfrac{1}{q} = 1 : \ \parallel fg \parallel_1 \le \parallel f \parallel_p \parallel g \parallel_q . \quad \text{(submultiplicity)}$$

Hence one can define *vector spaces, absolutely integrable to the power p*, i.e., with seminorm $\parallel f \parallel_p$,

$$\mathcal{L}^p_\mu(S,\mathbb{K}) = \{f : (S,\mu) \longrightarrow \mathbb{K} \mid \parallel f \parallel_p \in \mathbb{R}\} \in \underline{\mathbf{vec}}_\mathbb{K},$$

related to each other as follows:

$$1 \le p,q < \infty,\ f \in \mathcal{L}^p_\mu(S,\mathbb{K}) \iff |f|^{\frac{p}{q}-1} f \in \mathcal{L}^q_\mu(S,\mathbb{K}),$$

and the seminorm space with the *essentially bounded functions*

$$\begin{aligned} \mathcal{L}^\infty_\mu(S,\mathbb{K}) &= \{f : (S,\mu) \longrightarrow \mathbb{K} \mid |f(x)| \in \mathbb{R}\ \mu\text{-almost everywhere}\} \in \underline{\mathbf{vec}}_\mathbb{K}, \\ \parallel f \parallel_\infty &= \inf\{\alpha \mid |f(x)| \le \alpha,\ \mu\text{-almost everywhere}\}. \end{aligned}$$

For *finite measure* μ, e.g., a compact set S, there is contramonotonicity

$$1 \le p < q < \infty \Rightarrow \mathcal{L}^p_\mu(S,\mathbb{K}) \supseteq \mathcal{L}^q_\mu(S,\mathbb{K}).$$

For locally compact groups with a Haar measure one uses the measure free notation $\mathcal{L}(G,\mathbb{K})$ and $\mathcal{L}(G)$ for complex functions.

Hölder's inequality entails the *duality pairing* with *conjugated powers* (p,q) on a hyperbola $(p-1)(q-1) = 1$:

$$1 \le p,q \le \infty,\ \tfrac{1}{p} + \tfrac{1}{q} = 1 : \ \begin{aligned} &\mathcal{L}^p_\mu(S,\mathbb{K}) \times \mathcal{L}^q_\mu(S,\mathbb{K}) \longrightarrow \mathcal{L}^1_\mu(S,\mathbb{K}), \\ &(f,g) \longmapsto fg. \end{aligned}$$

The functions that vanish μ-almost everywhere constitute the subspace $\mathcal{N}_0$ with trivial prenorm for all absolutely integrable function spaces. The normed quotients, called *Lebesgue spaces*, are even Banach spaces:

$$L^p_\mu(S,\mathbb{K}) = \mathcal{L}^p_\mu(S,\mathbb{K})/\mathcal{N}_0 \in \underline{\mathbf{n\hat{v}ec}}_\mathbb{K},\ \ 1 \le p \le \infty.$$

The duality pairing leads to a bilinear product extending the product on the *compactly supported continuous functions* $\mathcal{C}_c(S,\mathbb{K})$:

$$\langle h, f\rangle = \int d\mu(s)h(s)f(s), \left\{ \begin{array}{ll} \mathcal{C}_c(S,\mathbb{K}) \times \mathcal{C}_c(S,\mathbb{K}) & \longrightarrow \mathbb{K}, \\ L^p_\mu(S,\mathbb{K}) \times L^q_\mu(S,\mathbb{K}) & \longrightarrow \mathbb{K},\ \frac{1}{p}+\frac{1}{q}=1, \end{array}\right.$$

whence the topological vector space isomorphism $L^q_\mu(S,\mathbb{K}) \cong L^p_\mu(S,\mathbb{K})'$ with the strong dual can be derived.

Of special interest is the self-dual Hilbert space L^2 and the pair (L^1, L^∞) with $(L^1)' = L^\infty$, but $(L^\infty)' \supseteq L^1$. $L^1(G)$ is a Banach algebra for a locally compact group G (chapter "Harmonic Analysis").

On a locally compact space S, the continuous functions with compact support $\mathcal{C}_c(S)$ constitute a dense vector subspace of $\mathcal{L}^p_\mu(S,\mathbb{K})$ and $L^p_\mu(S,\mathbb{K})$, $1 \le p < \infty$, for any Radon measure $\mu \in \mathcal{M}(S)$.

9.16.1 Hilbert Spaces with Square Integrable Functions

The absolutely square integrable functions $\mathcal{L}^2_\mu(S,\mathbb{K})$ constitute a pre-Hilbert space leading to the *Hilbert space of the square μ-integrable functions*

$$\begin{array}{l} S \ni s \longmapsto |f\rangle(s) = f(s) \in \mathbb{K},\quad S \ni s \longmapsto \langle f|(s) = \overline{f(s)} \in \mathbb{K}, \\ L^2_\mu(S,\mathbb{K}) \times L^2_\mu(S,\mathbb{K}) \longrightarrow \mathbb{K},\quad \langle h|f\rangle_\mu = \int d\mu(s)\overline{h(s)}f(s). \end{array}$$

It is separable (countable Hilbert space basis) for a separable Borel space S (countable basis for the topology).

For a measure invariant under a group action $\mu(g \bullet X) = \mu(X)$, for all $g \in G$ and $X \in \mathcal{S}$, the *left-regular G-representation* on the complex functions leads to a Hilbert space representation

$$\begin{array}{lll} G \times L^2_\mu(S) \longrightarrow L^2_\mu(S), & |f\rangle \longmapsto U(g)|f\rangle = |_g f\rangle, & |_g f\rangle(k) = f(g^{-1}k), \\ & \langle f| \longmapsto \langle f|U(g^{-1}) = \langle_g f|, & \langle_g f|(k) = \overline{f(g^{-1}k)}, \end{array}$$

e.g., the representation of a locally compact group G on its complex functions $L^2(G)$.

$L^2_\mu(S,\mathbb{K})$ with countable Hilbert basis $(|e^j\rangle)_{j\in\mathbb{N}}$ has as orthogonality

$$|e^j\rangle : S \longrightarrow \mathbb{K},\quad \delta^{jk} = \langle e^j|e^k\rangle_\mu = \int d\mu(s)\ \overline{e^j(s)}e^k(s).$$

The completeness

$$L^2_\mu(S,\mathbb{K}) \ni |f\rangle = \sum_j f^j|e^j\rangle,\quad f^j = \langle e^j|f\rangle_\mu = \int d\mu(s)\overline{e^j(s)}f(s)$$

can be expressed with the measure-associated Dirac distribution

$$\begin{array}{rl} f(s) & = \sum_j \int d\mu(s')\ e^j(s)\overline{e^j(s')}f(s') = \int d\mu(s')\ \delta(\mu(s,s'))f(s'), \\ \mathrm{id}_{L^2_\mu(S,\mathbb{K})} & \cong \sum\limits_j |e^j\rangle\langle e^j| \iff \delta(\mu(s,s')) = \sum_j e^j(s)\overline{e^j(s')}, \end{array}$$

e.g., with a Lebesgue-based measure involving a positive function μ:

$$\begin{array}{c} L^2_\mu(\mathbb{R}^d,\mathbb{K}) \ni |f\rangle,\quad f(x) = \int \mu(x)dx\ \delta(\mu(x,x'))f(x') \\ \text{with} \left\{ \begin{array}{rl} d\mu(x) & = \mu(x)dx, \\ \delta(\mu(x,x')) & = \frac{1}{\mu(x)}\delta(x-x') \end{array}\right. \end{array}$$

For a Borel space morphism F to lead to a Hilbert space morphism the image measure μ_F has to be used:

$$\begin{array}{ccc} (S,\mu) & & L^2_\mu(S,\mathbb{K}) \\ F\ \big\downarrow & \circ F\longmapsto & \big\uparrow \\ (T,\mu_F) & & L^2_{\mu_F}(T,\mathbb{K}) \end{array}, \quad \mu_F[X]=\mu(F^{-1}[X]),$$

$$\begin{aligned} \langle f\circ F|g\circ F\rangle_\mu &= \int d\mu(s)\ \overline{f(F(s))}g(F(s)) \\ &= \int d\mu_F(t)\ \overline{f(t)}g(t) = \langle f|g\rangle_\mu, \\ |e^\alpha\rangle \longmapsto |e^\alpha\circ F\rangle &= F^{\alpha j}|e^j\rangle,\quad F^{\alpha j}=\langle e^j|e^\alpha\circ F\rangle_\mu. \end{aligned}$$

An automorphism $F: S\longrightarrow S$ is a reparametrization. The transformation of Lebesgue-based measures involves the nonsingular Jacobi determinant of $F:\mathbb{R}^d\longrightarrow\mathbb{R}^d$:

$$\begin{aligned} d\mu(x) &= \mu(x)dx, & d\mu_F(x) &= \mu(F^{-1}(x))\ dF^{-1}(x) \\ & & &= \mu(F^{-1}(x))|\det\tfrac{\partial F^{-1}}{\partial x}|\ dx, \\ \delta(\mu(x-x')) &= \tfrac{1}{\mu(x)}\delta(x-x'), & \delta(\mu_F(x,x')) &= \tfrac{1}{\mu(F^{-1}(x))|\det\frac{\partial F^{-1}}{\partial x}|}\delta(x-x'). \end{aligned}$$

9.16.2 Hilbert Spaces for U(1) and D(1)

Definite Haar measures for the real Lie groups $\mathbf{U}(1)\cong[-\pi,\pi]$ (compact) and $\mathbf{D}(1)\cong\mathbb{R}$ (noncompact) and their $\mathbf{U}(1)$-characters (the dual groups $\mathbb{Z}$ and $i\mathbb{R}$ respectively) are given by

$$\begin{aligned} &\int_{-\pi}^{\pi}\tfrac{d\alpha}{2\pi}, & D^z:\ &\mathbf{U}(1)\ni e^{i\alpha} \longmapsto e^{i\alpha z}\in\mathbf{U}(1),\quad z\in\mathbb{Z}, \\ &\int_{\mathbb{R}}dx=\int dx, & D^{iq}:\ &\mathbf{D}(1)\ni e^{x} \longmapsto e^{iqx}\in\mathbf{U}(1),\quad iq\in i\mathbb{R}. \end{aligned}$$

In the separable Hilbert spaces

$$\begin{aligned} L^2(\mathbf{U}(1) &\text{ with } \langle f|g\rangle=\int_{-\pi}^{\pi}\tfrac{d\alpha}{2\pi}\ \overline{f(\alpha)}g(\alpha), \\ L^2(\mathbf{D}(1)) &\text{ with } \langle f|g\rangle=\int dx\ \overline{f(x)}g(x), \end{aligned}$$

the irreducible $\mathbf{U}(1)$-representations constitute an orthonormal Hilbert space basis

$$L^2(\mathbf{U}(1))\text{-basis: } \{D^z \mid z\in\mathbb{Z}\},\quad \left\{\begin{aligned} \int_{-\pi}^{\pi}\tfrac{d\alpha}{2\pi}\ e^{-i\alpha z}e^{i\alpha z'} &= \delta_{zz'}, \\ \sum_{z\in\mathbb{Z}} e^{-i\alpha z}e^{i\alpha' z} &= \delta(\tfrac{\alpha-\alpha'}{2\pi}), \end{aligned}\right.$$

whereas the irreducible $\mathbf{D}(1)$-representations in $\mathbf{U}(1)$ are not Hilbert space elements

$$D^{iq}\notin L^2(\mathbf{D}(1)):\quad \left\{\begin{aligned} \int dx\ e^{-iqx}e^{iq'x} &= \delta(\tfrac{q-q'}{2\pi}), \\ \int\tfrac{dq}{2\pi}\ e^{-iqx}e^{iqx'} &= \delta(x-x'). \end{aligned}\right.$$

The $\mathbf{U}(1)$-representations of $\mathbf{D}(1) \cong \mathbb{R}$ define the Fourier transformation and relate to each other the isomorphic Hilbert spaces with the square integrable functions on the eigenvalues $q \in \mathbf{irrep}\,\mathbf{D}(1) \cong \mathbb{R}$ on the one hand and the square integrable noncompact $\mathbf{D}(1)$-representations on the other hand:

$$L^2_{dq}(\mathbb{R}) \cong L^2_{dx}(\mathbb{R}), \quad \begin{cases} \psi(x) &= \int \frac{dq}{2\pi}\, f(q)e^{iqx}, \\ \langle\psi|\varphi\rangle = \int dx\, \overline{\psi(x)}\varphi(x) &= \int \frac{dq}{2\pi}\, \overline{f(q)}g(q) = \langle f|g\rangle. \end{cases}$$

9.17 Direct Integral Vector Spaces

A set product of vector spaces $V(R) = \prod\limits_{q\in R} V(q)$ where the index set R carries a measure $\mu : \mathcal{R} \longrightarrow \mathbb{R}$ can be given the structure of a *direct integral vector space with maesure* μ:

$$\begin{array}{l} V(q) \in \underline{\mathbf{vec}}_{\mathbb{K}}, \quad V(R) = {}^{\oplus}\!\!\int_R d\mu(q)\; V(q) \in \underline{\mathbf{vec}}_{\mathbb{K}}, \\ \text{notation: } d\mu(q) = \mu(q)dq, \end{array}$$

with the elements those $V(R)$-valued mappings

$$v: \; R \ni q \longmapsto v(q) \in V(q), \quad v = {}^{\oplus}\!\!\int_R d\mu(q)\; v(q) \in V(R)$$

that are appropriately defined as measurable.

Direct and usual integrals are the possibly continuous generalizations of finite-support direct and usual sums

$$\bigoplus_{q\in R} \mu(q) \hookrightarrow {}^{\oplus}\!\!\int_R d\mu(q), \quad \sum_{q\in R} \mu(q) \hookrightarrow \int_R d\mu(q).$$

It is easier to write down formally the direct integral structures in analogy to the finite-dimensional case than to clear up related questions of definition that are not discussed in the following (chapter "Harmonic Analysis").

In most cases considered, the individual spaces are isomorphic and finite-dimensional $V(q) = \underset{n}{V} \times \{q\} \cong \mathbb{C}^n$. A finite index set R gives a weighted (counted) direct sum $\bigoplus\limits_{q=1} \mu(q)V(q)$. A locally compact group R carries a Haar measure. Coset spaces, e.g., spheres Ω^s or hyberboloids $\mathcal{Y}^s$, have invariant positive measures, always unique up to a scalar factor.

With a measure basis, e.g., a definite Haar measure on a locally compact group R, e.g., a Lebesgue measure on $\mathbb{R}^s$, the measure is characterized by a generalized function $d\mu(q) = d^s q\; \mu(q)$, in the simplest case a number $\mu(q) = \alpha$.

The dual space will be defined by the dual spaces $V^T(q)$. The dual product involves the usual integral

$$\begin{array}{l} V^T(R) = {}^{\oplus}\!\!\int_R d\mu(q)\; V^T(q) \in \underline{\mathbf{vec}}_{\mathbb{K}}, \\ \omega: \; R \ni q \longmapsto \omega(q) \in V^T(q), \quad \omega = {}^{\oplus}\!\!\int_R d\mu(q)\; \omega(q) \in V^T(R), \\ \langle \omega, f\rangle = \int d\mu(q) \langle \omega(q), f(q)\rangle. \end{array}$$

The *measure-related Dirac distribution* projects to the subspaces:

$$\begin{aligned} \delta_q : V(R) \longrightarrow V(q), \quad & \langle \delta_q, v\rangle = {}^{\oplus}\!\!\int_R d\mu(p)\ \delta(\mu(q,p))v(p) = v(q), \\ \text{notation:} \quad & \delta(\mu(p,q)) = \tfrac{1}{\mu(q)}\delta(p,q) = \tfrac{1}{\mu(q)}\delta(q,p), \\ & {}^{\oplus}\!\!\int_R d\mu(p)\ \delta(\mu(q,p)) = {}^{\oplus}\!\!\int_R dp\ \delta(q,p), \end{aligned}$$

with the examples

$$\begin{aligned} \text{locally compact group } G:\ & d\mu(g) = d^G g, & \delta(\mu(g,h)) = \delta(hg^{-1}), \\ \mathbb{R}^s \text{ with function } \mu:\ & d\mu(q) = \mu(q)d^s q, & \delta(\mu(q,p)) = \tfrac{1}{\mu(q)}\delta(p-q). \end{aligned}$$

Dual bases for finite-dimensional spaces $V(q)$ are generalized to *measure-related dual-bases distributions*

$$\langle \check{e}_a(q), e^b(p)\rangle = \delta_a^b\ \delta(\mu(q,p)), \quad \mathrm{id}_{V(R)} = {}^{\oplus}\!\!\int_R d\mu(q)\ e^a(q) \otimes \check{e}_a(q).$$

In general, the elements of a distributive basis are not vectors $e^a(q), \check{e}_a(q) \notin V(R)$, and may be called vector distributions. Hence one obtains for vectors, forms, and a linear transformations $B : V(R) \longrightarrow V(R)$ the expansions

$$\begin{aligned} v &= {}^{\oplus}\!\!\int_R d\mu(q)\ v_a(q)e^a(q), & v_a(q) &= \langle \check{e}_a(q), v\rangle, \\ \omega &= {}^{\oplus}\!\!\int_R d\mu(q)\ \omega^a(q)\check{e}_a(q), & \omega^a(q) &= \langle \omega, e^a(q)\rangle, \\ B &= {}^{\oplus}\!\!\int_{R\times R} d\mu(q)d\mu(p)\ B_a^b(q,p)e^a(q) \otimes \check{e}_b(p), & B_a^b(q,p) &= \langle \check{e}_a(q), B\, e^b(p)\rangle, \end{aligned}$$

with the dual product and the trace

$$\langle \omega, v\rangle = \int_R d\mu(q)\ \omega^a(q)v_a(q), \quad \mathrm{tr}\, B = \int_R d\mu(q)\ B_a^a(q,q).$$

Starting from Hilbert spaces with scalar product $V(q) \times V(q) \longrightarrow \mathbb{C}$ one obtains, with a positive measure, the *orthogonal direct integral Hilbert space* $H(R)$ *with measure* μ where the individual scalar products are integrated over, for a finite index set summed with positive weight factors:

$$\begin{aligned} H(R) \times H(R) \longrightarrow \mathbb{C}, \quad \langle v_2|v_1\rangle &= \int_R d\mu(q)\langle v_2(q)|v_1(q)\rangle, \\ \text{finite } R : \langle v_2|v_1\rangle &= \sum_{q=1}^{n} \mu(q)\langle v_2(q)|v_1(q)\rangle, \quad \mu(q) \geq 0. \end{aligned}$$

The summands are orthogonal. Since only functions with finite $\langle v|v\rangle$ (square integrable) are admitted, the direct integral Hilbert space (denoted with the orthogonal sign $\perp$) is in general a subspace:

$$H(R) = {}^{\perp}\!\!\int_R d\mu(q)\ V(q) \subseteq V(R) = {}^{\oplus}\!\!\int_R d\mu(q)\ V(q)$$

with $H(R) = V(R)$ for finite R.

A *distributive Hilbert space basis* has the properties

$$\begin{aligned} \text{orthonormality:} \quad & \langle q, a|p, b\rangle = \delta^{ab}\delta(\mu(q,p)), \\ \text{completeness:} \quad & {}^{\perp}\!\!\int_R d\mu(q)|q, a\rangle\langle q, b| \cong \mathrm{id}_{H(R)}. \end{aligned}$$

In general, $|q,a\rangle \notin H(R)$. The orthonormal basis distribution leads to the expansions

$$\begin{array}{lll} v & = {}^{\perp}\!\!\int_R d\mu(q)\ |q\rangle v(q), & v(q) & = \langle q|v\rangle, \\ B & \cong {}^{\perp}\!\!\int_{R\times R} d\mu(q)d\mu(p)\ |q\rangle B(q,p)\langle p|, & B(q,p) & = \langle q|B|p\rangle. \end{array}$$

The Hilbert space with the square integrable functions on R can be considered as a direct sum of $\mathbb{C}$-isomorphic Hilbert spaces with function $\mu(q) = 1$. A positive generalized function μ may be included in the Hilbert space of the correspondingly integrable functions

$$L^2_{d\mu}(R) \cong {}^{\perp}\!\!\int_R d\mu(q)V(q), \quad V(q) \cong \mathbb{C}, \quad \langle v_2|v_1\rangle = \int_R dq\ \mu(q)\ \overline{v_2(q)}v_1(q).$$

9.18 Linear Lattices (Birkhoff-von Neumann Logics)

Vector subspaces define *linear lattices* with the intersection for the meet (the logical "et"), the span for the join (the logical "aut") and the trivial space as origin (the logical "falsum"):

$$\begin{array}{rl} \underline{\mathbf{vec}}_{\mathbb{K}} \ni V \longmapsto & 2^V = \{W \subseteq V \mid \text{subspace}\} \quad \in \underline{\mathbf{latt}}, \\ & (\cap, +, \{0\}) \sim (\sqcap, \sqcup, \square), \\ & 2^{(V)} = \{W \subseteq V \mid \text{finite dimension}\} \in \underline{\mathbf{latt}}. \end{array}$$

For dimension $n \geq 2$ (where the vector space endomorphisms are nonabelian) these lattices are not distributive:

$$\begin{array}{l} W_i = \mathbb{K}e^i \cong \mathbb{K}, \quad i = 1,2, \quad \text{full space: } V = W_1 + W_2 \cong \mathbb{K}^2, \\ \qquad \text{diagonal space: } \Delta = \mathbb{K}(e^1 + e^2) \cong \mathbb{K}, \\ (W_1 + W_2) \cap \Delta = \Delta \neq (W_1 \cap \Delta) + (W_2 \cap \Delta) = \{0\} + \{0\} = \{0\} = W_1 \cap W_2. \end{array}$$

A lattice with finite-dimensional subspaces is modular.

Linear lattices can be related to endomorphisms: Any projector in the V-endomorphisms, for finite dimensions $\mathbf{AL}(V) \cong V \otimes V^T$, defines a subspace and, by the kernel, a direct complement

$$\begin{array}{l} \mathbf{AL}(V) \ni \mathcal{P} = \mathcal{P}^2 \longmapsto W = \mathcal{P}(V) \in 2^V, \\ V = W \oplus W' \text{ with } W' = \mathcal{P}^{-1}(0) \in 2^V. \end{array}$$

The trace is an invariant linear form which, for a projector, gives the dimension of the associate subspace

$$\begin{array}{l} \mathrm{tr}_V : \mathbf{AL}(V) \longrightarrow \mathbb{K}, \quad a \longmapsto \mathrm{tr}_V\, a, \\ \qquad \mathrm{tr}_V\, \mathcal{P} = \dim_{\mathbb{K}} \mathcal{P}(V). \end{array}$$

One subspace can be defined by different projectors and can have different complements (the logical "negation"), in the example above with two different dual bases:

$$\begin{aligned} \mathrm{id}_V &= \mathcal{P}_1 + \mathcal{P}_2 = e^1 \otimes \check{e}_1 + e^2 \otimes \check{e}_2 \\ &= \mathcal{P}'_1 + \mathcal{P}_\Delta = e^1 \otimes (\check{e}_1 - \check{e}_2) + (e^1 + e^2) \otimes \check{e}_2, \\ W_1 &= \mathcal{P}_1(V) = \mathcal{P}'_1(V), \quad V = W_1 \oplus W_2 = W_1 \oplus \Delta. \end{aligned}$$

Uniqueness is obtained with a dual isomorphism: With a nondegenerate square (inner product, symmetric bilinear, or sesquilinear form), there is a unique *orthogonal subspace.* In the case of a finite-dimensional space V orthogonality defines an involution

$$\begin{aligned} \langle\ |\ \rangle : V \times V \longrightarrow \mathbb{K}, \quad & \zeta(v, w) = \langle v|w\rangle = \overline{\langle v|w\rangle}, \\ & \langle v|w + u\rangle = \langle v|w\rangle + \langle v|u\rangle, \quad \langle v|\alpha w\rangle = \alpha\langle v|w\rangle, \\ V^\perp &= \{0\}, \\ \perp : 2^V \longrightarrow 2^V, \quad & W \longmapsto W^\perp = \{v \in V \mid \langle W|v\rangle = \{0\}, \\ W \subseteq W^{\perp\perp}, \quad & W^\perp = W^{\perp\perp\perp}, \text{ for } V \cong \mathbb{K}^n : \quad W = W^{\perp\perp}. \end{aligned}$$

With a nondegenerate square, projectors in the V-endomorphisms are bijectiveley related to subspaces

$$2^V \ni W \overset{\zeta}{\leftrightarrow} \mathcal{P}_W \in \mathbf{AL}(V).$$

The projector for a finite-dimensional subspace $W \cong \mathbb{K}^k$ can be written with a W-basis $\{e^\kappa\}_{\kappa=1}^k$ in the bra-ket notation:

$$\begin{aligned} \mathcal{P}_W &= |e^\kappa\rangle\zeta^W_{\kappa\lambda}\langle e^\lambda| \text{ with } \langle e^\lambda|e^\mu\rangle = \zeta^{\lambda\mu}_W, \quad \zeta^W_{\kappa\lambda}\zeta^{\lambda\mu}_W = \delta^\mu_\kappa, \\ \mathrm{tr}_V\, \mathcal{P}_W &= \zeta^W_{\kappa\lambda}\zeta^{\lambda\kappa}_W = \delta^\kappa_\kappa = d(W), \end{aligned}$$

especially simple for a scalar product $\zeta \succeq 0$ in Euclidean bases $\zeta^{\lambda\mu} = \delta^{\lambda\mu}$.

The involution defined by orthogonality is not complementary for an indefinite nondegenerate square. For example, in a 2-dimensional Minkowski space with Lorentz metric $\begin{pmatrix} 1 & 0 \\ 0 & -1 \end{pmatrix}$ in a basis $\{e^0, e^3\}$, time and position translations $\mathbb{T}$ and $\mathbb{S}$ respectively are orthogonal to each other, whereas the isotropic lightlike subspaces $\mathbb{L}_\pm$ are self-orthogonal:

$$\begin{aligned} \mathbb{T}^\perp &= (\mathbb{R}e^0)^\perp = \mathbb{R}e^3 = \mathbb{S}, \\ \alpha\beta &\neq 0 \Rightarrow \mathbb{R}(\alpha e^0 + \beta e^3)^\perp = \mathbb{R}(\tfrac{1}{\alpha}e^0 + \tfrac{1}{\beta}e^3) \Rightarrow \mathbb{L}_\pm = \mathbb{R}(e^0 \pm e^3) = \mathbb{L}_\pm^\perp, \\ \mathcal{P}_\mathbb{T} &= |e^0\rangle\langle e^0|, \quad \mathcal{P}_\mathbb{S} = -|e^3\rangle\langle e^3|, \quad \mathcal{P}_{\mathbb{L}_\pm} = \tfrac{1}{2}|e^0 \pm e^3\rangle\langle e^0 \mp e^3|. \end{aligned}$$

A *complementary* linear lattice has to come with a definite square, i.e., with a scalar product, which is nondegenerate for each subspace:

$$\langle v|v\rangle = 0 \iff v = 0 \Rightarrow \mathbb{K}^n \cong V = W + W^\perp = W \oplus W^\perp = W \perp W^\perp.$$

With a scalar product the vector subspaces of a finite-dimensional vector space with a nondegenerate square constitute a *Birkhoff-von Neumann logic,* (linear logic)

$$V \cong \mathbb{K}^d, \ \zeta \succeq 0 \Rightarrow (2^V, \cap, +, \{0\}, \perp) \in \underline{\mathbf{logic}}(\text{Birkhoff-von Neumann}),$$

as well as the closed subspaces of a Hilbert space with its definite scalar product,

$$\text{Hilbert space } V \Rightarrow 2^{\overline{V}} = \{W \subseteq V \mid \text{closed}\} \in \underline{\textbf{logic}}(\text{Birkhoff-von Neumann}).$$

Bibliography

[1] H. Bauer, *Wahrscheinlichkeitstheorie und Grundzüge der Maßtheorie* (1964), Sammlung Göschen, Gruyter, Berlin.

[2] G. Birkhoff, J.v. Neumann, *Annals of Mathematics* 37 (1936), 823.

[3] N. Bourbaki, *Intégration, Chapitres 1-8* (1952-1959), Hermann, Paris.

[4] I.M. Gel'fand, G.E. Shilov, *Generalized Functions I (Properties and Operations) and II (Spaces of Fundamental and Generalized Functions)* (1958, English translation 1963), Academic Press, New York and London.

[5] P.R. Halmos, *Measure Theory* (1950), Van Nostrand, Princeton, Toronto, London, Melbourne.

[6] A. Messiah, *Quantum Mechanics I,II* (1965), North Holland, Amsterdam.

[7] W. Thirring, *Lehrbuch der Mathematischen Physik* 3 (1979), Springer, Wien, New York.

[8] N.Ja. Vilenkin, A.U. Klimyk, *Representations of Lie Groups and Special Functions* (1991), Kluwer Academic Publishers, Dordrecht, Boston, London.

[9] V.S. Varadarajan, *Geometry of Quantum Theory* (1985), Springer, New York, Berlin, Heidelberg, Tokyo.

Index